Lecture Notes in Computer Science 4515

Commenced Publication in 1973
Founding and Former Series Editors:
Gerhard Goos, Juris Hartmanis, and Jan van Leeuwen

Moni Naor (Ed.)

Advances
in Cryptology -
EUROCRYPT 2007

26th Annual International Conference on the Theory
and Applications of Cryptographic Techniques
Barcelona, Spain, May 20-24, 2007
Proceedings

 Springer

Volume Editor

Moni Naor
Weizmann Institute of Science
Department of Computer Science and Applied Mathematics
Rehovot 76100 ISRAEL
E-mail: moni.naor@weizmann.ac.il

Library of Congress Control Number: 2007926705

CR Subject Classification (1998): E.3, F.2.1-2, G.2.1, D.4.6, K.6.5, C.2, J.1

LNCS Sublibrary: SL 4 – Security and Cryptology

ISSN   0302-9743
ISBN-10   3-540-72539-3 Springer Berlin Heidelberg New York
ISBN-13   978-3-540-72539-8 Springer Berlin Heidelberg New York

Springer is a part of Springer Science+Business Media

springer.com

© International Association of Cryptologic Research 2007
Printed in Germany

Typesetting: Camera-ready by author, data conversion by Scientific Publishing Services, Chennai, India
Printed on acid-free paper   SPIN: 12064380   06/3180   5 4 3 2 1 0

Preface

These are the proceedings of Eurocrypt 2007, the 26th Annual IACR Eurocrypt Conference. The conference was sponsored by the International Association for Cryptologic Research (IACR; see www.iacr.org), this year in cooperation with the Research Group on Mathematics Applied to Cryptography at UPC and the Research Group on Information Security at UMA. The Eurocrypt 2007 Program Committee (PC) consisted of 24 members whose names are listed on the next page.

The PC decided on several policies: zero PC papers - no Program Committee member could submit papers; optional anonymity - authors could choose to anonymize their papers or not. Anonymous papers were treated as usual, i.e., the author's identity was not revealed to the PC. The submission software used was "Web Submission and Review Software" written and maintained by Shai Halevi. There were 173 papers submitted to the conference and the PC chose 33 of them. Each paper was assigned to at least three PC members, who either handled it themselves or assigned it to an external referee. After the reviews were submitted, the committee deliberated both online for several weeks and finally in a face-to-face meeting held in Paris. In addition to notification of the decision of the committee, authors received reviews. Our goal was to provide meaningful comments to authors of all papers (both those selected for the program and those not selected). The default for any report given to the committee was that it should be available to the authors as well.

The committee decided to give the Best Paper Award to Shien Jin Ong and Salil Vadhan for their paper "Zero Knowledge and Soundness are Symmetric." In addition the PC chose two more notable papers for invitation to the *Journal of Cryptology*. These are "Chosen-prefix Collisions for MD5 and Colliding X.509 Certificates for Different Identities," by Marc Stevens, Arjen Lenstra and Benne de Weger, and "An $L(1/3 + \varepsilon)$ Algorithm for the Discrete Logarithm Problem for Low-Degree Curves," by Andreas Enge and Pierrick Gaudry. The conference program included two invited lectures: by Jacques Stern (IACR Distinguished Lecture) titled "Cryptography from A to Z" and by Victor Miller titled "Elliptic Curves and Cryptography: Invention and Impact."

I wish to thank all the people who made the conference possible. First and foremost the authors who submitted their papers. The hard task of reading, commenting, debating and finally selecting the papers for the conference fell on the PC members. I am indebted to the committee members' collective knowledge, wisdom and effort. I have learned a lot from the experience. The committee also used external reviewers, whose names are listed on the following pages, to extend the expertise and ease the burden. My deepest gratitude to them as well. I thank Shai Halevi for handling the submissions and reviews server and Michel Abdalla for organizing the PC Meeting in Paris. I am grateful to previous PC Chairs who have shared their experiences with me. Finally, the Eurocrypt General Chairs Javier López and Germán Sáez and the local organizing committee Monica Breitman, Paz Morillo and Jorge L. Villar deserve many thanks from all the IACR community for the organization of the conference.

March 2007   Moni Naor

Eurocrypt 2007

Barcelona, Spain, May 20–24, 2007

Sponsored by the *International Association for Cryptologic Research.*

Organized in cooperation with the
Technical University of Catalonia (UPC) and the University of Malaga (UMA).

General Chairs

Javier López and Germán Sáez

Program Chair

Moni Naor, Weizmann Institute of Science

Program Committee

Michel Abdalla	ENS and CNRS, Paris
Anne Canteaut	INRIA-Rocquencourt
Dario Catalano	University of Catania
Jung Hee Cheon	Seoul National University
Stefan Dziembowski	Warsaw University and University of Rome "La Sapienza"
Serge Fehr	CWI, Amsterdam
Marc Fischlin	TU Darmstadt
Jens Groth	UCLA
Shai Halevi	IBM T.J. Watson Research Center
Yuval Ishai	Technion
Joe Kilian	Rutgers University
Anna Lysyanskaya	Brown University
Alexander May	TU Darmstadt
Steven Myers	Indiana University
Moni Naor	Weizmann Institute of Science
Phong Nguyen	ENS and CNRS, Paris
Jesper Buus Nielsen	University of Aarhus
Giuseppe Persiano	University of Salerno
Ron Rivest	MIT
Alon Rosen	Harvard
Eran Tromer	MIT

Xiaoyun Wang   Tsinghua University
Brent Waters   SRI
Stefan Wolf   ETH, Zurich

External Reviewers

Ben Adida
Roberto Araujo
Thomas Baignères
Boaz Barak
Amos Beimel
Ian F. Blake
Carlo Blundo
Alexandra Boldyreva
Xavier Boyen
Jan Camenisch
Jean Camp
Ran Canetti
Melissa Chase
Liqun Chen
Benoît Chevallier-Mames
Joo Yeon Cho
Paul Crowley
Chris Crutchfield
Ivan Damgård
Cécile Delerablée
Alex Dent
Claus Diem
Jintai Ding
Christophe Doche
Martin Döring
Orr Dunkelman
Jean-Charles Faugère
Nelly Fazio
Matthias Fitzi
Pierre-Alain Fouque
Fabien Galand
Steven Galbraith
Clemente Galdi
Pierrick Gaudry
Rosario Gennaro
Vipul Goyal
Tim Güneysu
Robbert de Haan
Iftach Haitner

Guillaume Hanrot
Danny Harnik
Alex Healy
Martin Hirt
Dennis Hofheinz
Susan Hohenberger
Thomas Holenstein
Nick Howgrave-Graham
Vasyltsov Ihor
Stanislaw Jarecki
Antoine Joux
Pascal Junod
Jonathan Katz
Nathan Keller
Eike Kiltz
Jaeheon Kim
Matthias Kleinmann
Hugo Krawczyk
Konrad Kulikowski
Soonhak Kwon
Taekyoung Kwon
Tanja Lange
Cédric Lauradoux
Dong Hoon Lee
Hyang-Sook Lee
Anja Lehmann
Gaëtan Leurent
Pierre Loidreau
Mira Meyerovich
Marine Minier
Tal Moran
Sean Murphy
Christophe Nègre
Gregory Neven
Antonio Nicolosi
Kobbi Nissim
Shien Jin Ong
Yossi Oren
Raphael Overbeck

Adriana Palacio
Omkant Pandey
Alan Park
Rafael Pass
Michael Østergaard Pedersen
Chris Peikert
Ludovic Perret
Duong Hieu Phan
Krzysztof Pietrzak
Gilles Piret
David Pointcheval
Manoj Prabhakharan
Bartosz Przydatek
Charles W. Rackoff
Håvard Raddum
Mario Di Raimondo
Dominik Raub
Christian Rechberger
Omer Reingold
Leonid Reyzin
Thomas Ristenpart
Maike Ritzenhofen
Phil Rogaway
Amit Sahai
Louis Salvail
Berry Schoenmakers
Dominique Schröder

Gil Segev
Nicolas Sendrier
Emily Shen
Peter Shor
Alice Silverberg
William Speirs
François-Xavier Standaert
Damien Stehlé
Andreas Stein
Lakshminarayanan Subramanian
Madhu Sudan
Qiang Tang
Thomas Toft
Vinod Vaikuntanathan
Mayank Varia
Carmine Ventre
Ivan Visconti
David Wagner
Samuel Wagstaff
Shabsi Walfish
Bogdan Warinschi
Ralf-Philipp Weinmann
Enav Weinreb
Douglas Wikstrom
Jürg Wullschleger
Hyojin Yoon
Aram Yun

Table of Contents

Chosen-Prefix Collisions for MD5 and Colliding X.509 Certificates for
Different Identities 1
   Marc Stevens, Arjen Lenstra, and Benne de Weger

Non-trivial Black-Box Combiners for Collision-Resistant Hash-Functions
Don't Exist 23
   Krzysztof Pietrzak

The Collision Intractability of MDC-2 in the Ideal-Cipher Model 34
   John P. Steinberger

An Efficient Protocol for Secure Two-Party Computation in the
Presence of Malicious Adversaries 52
   Yehuda Lindell and Benny Pinkas

Revisiting the Efficiency of Malicious Two-Party Computation 79
   David P. Woodruff

Efficient Two-Party Secure Computation on Committed Inputs 97
   Stanisław Jarecki and Vitaly Shmatikov

Universally Composable Multi-party Computation Using Tamper-Proof
Hardware 115
   Jonathan Katz

Generic and Practical Resettable Zero-Knowledge in the Bare
Public-Key Model 129
   Moti Yung and Yunlei Zhao

Instance-Dependent Verifiable Random Functions and Their
Application to Simultaneous Resettability 148
   Yi Deng and Dongdai Lin

Conditional Computational Entropy, or Toward Separating
Pseudoentropy from Compressibility 169
   Chun-Yuan Hsiao, Chi-Jen Lu, and Leonid Reyzin

Zero Knowledge and Soundness Are Symmetric 187
   Shien Jin Ong and Salil Vadhan

Mesh Signatures 210
   Xavier Boyen

The Power of Proofs-of-Possession: Securing Multiparty Signatures
against Rogue-Key Attacks 228
   Thomas Ristenpart and Scott Yilek

Batch Verification of Short Signatures 246
Jan Camenisch, Susan Hohenberger, and
Michael Østergaard Pedersen

Cryptanalysis of SFLASH with Slightly Modified Parameters 264
Vivien Dubois, Pierre-Alain Fouque, and Jacques Stern

Differential Cryptanalysis of the Stream Ciphers Py, Py6 and Pypy 276
Hongjun Wu and Bart Preneel

Secure Computation from Random Error Correcting Codes 291
Hao Chen, Ronald Cramer, Shafi Goldwasser, Robbert de Haan, and
Vinod Vaikuntanathan

Round-Efficient Secure Computation in Point-to-Point Networks 311
Jonathan Katz and Chiu-Yuen Koo

Atomic Secure Multi-party Multiplication with Low Communication.... 329
Ronald Cramer, Ivan Damgård, and Robbert de Haan

Cryptanalysis of the Sidelnikov Cryptosystem 347
Lorenz Minder and Amin Shokrollahi

Toward a Rigorous Variation of Coppersmith's Algorithm on Three
Variables 361
Aurélie Bauer and Antoine Joux

An $L(1/3 + \varepsilon)$ Algorithm for the Discrete Logarithm Problem for Low
Degree Curves 379
Andreas Enge and Pierrick Gaudry

General *Ad Hoc* Encryption from Exponent Inversion IBE............. 394
Xavier Boyen

Non-interactive Proofs for Integer Multiplication 412
Ivan Damgård and Rune Thorbek

Ate Pairing on Hyperelliptic Curves 430
Robert Granger, Florian Hess, Roger Oyono, Nicolas Thériault, and
Frederik Vercauteren

Ideal Multipartite Secret Sharing Schemes.......................... 448
Oriol Farràs, Jaume Martí-Farré, and Carles Padró

Non-wafer-Scale Sieving Hardware for the NFS: Another Attempt to
Cope with 1024-Bit 466
Willi Geiselmann and Rainer Steinwandt

Divisible E-Cash Systems Can Be Truly Anonymous................... 482
Sébastien Canard and Aline Gouget

A Fast and Key-Efficient Reduction of Chosen-Ciphertext to
Known-Plaintext Security   498
   Ueli Maurer and Johan Sjödin

Range Extension for Weak PRFs; The Good, the Bad, and the Ugly....   517
   Krzysztof Pietrzak and Johan Sjödin

Feistel Networks Made Public, and Applications   534
   Yevgeniy Dodis and Prashant Puniya

Oblivious-Transfer Amplification   555
   Jürg Wullschleger

Simulatable Adaptive Oblivious Transfer   573
   Jan Camenisch, Gregory Neven, and abhi shelat

Author Index   591

Chosen-Prefix Collisions for MD5 and Colliding X.509 Certificates for Different Identities

Marc Stevens[1], Arjen Lenstra[2], and Benne de Weger[1]

[1] TU Eindhoven, Faculty of Mathematics and Computer Science
P.O. Box 513, 5600 MB Eindhoven, The Netherlands
[2] EPFL IC LACAL, Station 14, and Bell Laboratories
CH-1015 Lausanne, Switzerland

Abstract. We present a novel, automated way to find differential paths for MD5. As an application we have shown how, at an approximate expected cost of 2^{50} calls to the MD5 compression function, for any two chosen message prefixes P and P', suffixes S and S' can be constructed such that the concatenated values $P\|S$ and $P'\|S'$ collide under MD5. Although the practical attack potential of this construction of *chosen-prefix collisions* is limited, it is of greater concern than random collisions for MD5. To illustrate the practicality of our method, we constructed two MD5 based X.509 certificates with identical signatures but different public keys *and* different Distinguished Name fields, whereas our previous construction of colliding X.509 certificates required identical name fields. We speculate on other possibilities for abusing chosen-prefix collisions. More details than can be included here can be found on `www.win.tue.nl/hashclash/ChosenPrefixCollisions/`.

1 Introduction

In March 2005 we showed how Xiaoyun Wang's ability [17] to quickly construct random collisions for the MD5 hash function could be used to construct two different valid and unsuspicious X.509 certificates with identical digital signatures (see [10] and [11]). These two *colliding certificates* differed in their public key values only. In particular, their Distinguished Name fields containing the identities of the certificate owners were equal. This was the best we could achieve because

- Wang's hash collision construction requires identical Intermediate Hash Values (IHVs);
- the resulting colliding values look like random strings: in an X.509 certificate the public key field is the only suitable place where such a value can unsuspiciously be hidden.

A natural and often posed question (cf. [7], [3], [1]) is if it would be possible to allow more freedom in the other fields of the certificates, at a cost lower than 2^{64}

M. Naor (Ed.): EUROCRYPT 2007, LNCS 4515, pp. 1–22, 2007.

calls to the MD5 compression function. Specifically, it has often been suggested that it would be interesting to be able to select Distinguished Name fields that are different and, preferably, chosen at will, non-random and human readable as one would expect from these fields. This can be realized if two arbitrarily chosen messages, resulting in two different IHVs, can be extended in such a way that the extended messages collide. Such collisions will be called *chosen-prefix collisions*.

We describe how chosen-prefix collisions for MD5 can be constructed, and show that our method is practical by constructing two MD5 based X.509 certificates with different Distinguished Name fields and identical digital signatures. The full details of the chosen-prefix collision construction and the certificates can be found in [16] and [14], respectively.

Section 2 contains a bird's eye view of the chosen-prefix collision construction method and its complexity. Its potential applications are discussed in Section 3 with Section 4 containing implications and details of the application to X.509 certificates. Details of the automated differential path construction for MD5 are provided in Section 5.

2   Chosen-Prefix Collisions for MD5

The main contribution of this paper is a method to construct MD5 collisions starting from two arbitrary IHVs. Given this method one can take any two chosen message prefixes and construct bitstrings that, when appended to the prefixes, turn them into two messages that collide under MD5. We refer to such a collision as a *chosen-prefix collision*. Their possibility was mentioned already in [3, Section 4.2 case 1] and, in the context of SHA-1, in [1] and on www.iaik.tugraz.at/research/krypto/collision/.

We start with a pair of arbitrarily chosen messages, not necessarily of the same length. Padding with random bits may be applied so that the padded messages have the same bitlength which equals 416 modulo 512 (incomplete last block). Equal length is unavoidable, because Merkle-Damgård strengthening, involving the message length, is applied after the last message block has been compressed by MD5. The incomplete last block condition is a technical requirement. In our example of colliding certificates the certificate contents were constructed in such a way that padding was not necessary, to allow for shorter RSA moduli.

Given the padded message pair, we followed a suggestion by Xiaoyun Wang[1] to find a pair of 96-bit values that, when used to complete the last blocks by appending them to the messages and applying the MD5 compression function, resulted in a specific form of difference vector between the IHVs. Finding these 96-bit values was done using a birthdaying procedure.

The remaining differences between the IHVs were then removed by appending *near-collision blocks*. Per pair of blocks this was done by constructing new differential paths using an automated, improved version of Wang's original approach. This innovative differential path construction is described in detail in Section 5

---

[1] Private communication.

below. Due to the specific form of the near-collisions and the first difference vector, essentially one triple of bit differences could be removed per near-collision block, thus shortening the overall length of the colliding values. For our example 8 near-collision blocks were needed to remove all differences. Thus, a total of $96 + 8 \times 512 = 4192$ bits were appended to each of the chosen message prefixes to let them collide.

The birthdaying step can be entirely avoided, thereby making it harder to find the proper differential paths and considerably increasing the number of near-collision blocks. Or the birthdaying step could be simplified, increasing the number of near-collision blocks from 8 to about 14. Our approach was inspired by our desire to minimize the number of near-collision blocks. Using a more intricate differential path construction it should be possible to remove more than a single triple of bit differences per block, which would reduce the number of near-collision blocks. Potential enhancements and variations, and the full details of the construction as used, will be discussed in [16].

The expected complexity of the birthdaying step is estimated at 2^{49} MD5 compression function calls. Estimating the complexity of the near-collision block construction is hard, but it turned out to be a small fraction of the birthdaying complexity. Based on our observations we find it reasonable to estimate the overall expected complexity of finding a chosen-prefix collision for MD5 at about 2^{50} MD5 compression function calls. For the example we constructed, however, we had some additional requirements and also were rather unlucky in the birthdaying step, leading to about 2^{52} MD5 compression function calls. Note that, either way, this is substantially faster than the trivial birthday attack which has complexity 2^{64}.

The construction of just a single example required, apart from the development of the automated differential path construction method, substantial computational efforts. Fortunately, the work is almost fully parallelizable and suitable for grid computing. It was done in the "HashClash" project (see `www.win.tue.nl/hashclash/`) and lasted about 6 months: using BOINC software (see `boinc.berkeley.edu/`) up to 1200 machines contributed, involving a cluster of computers at TU/e and a grid of home PCs. We expect that another chosen-prefix collision can be found much faster, but that it would again require substantial effort, both human and computationally: say 2 months real time assuming comparable computational resources.

3 Applications of Chosen-Prefix Collisions

We mention some potential applications of chosen-prefix collisions.

- The example presented in the next section, namely colliding X.509 certificates with different fields before the appended bitstrings that cause the collision. Those bitstrings are 'perfectly' hidden inside the RSA moduli, where 'perfect' means that inspection of either one of the RSA moduli does not give away anything about the way it is constructed (namely, crafted such that it collides with the other one). In particular it could be of interest to be able to freely choose the Distinguished Name fields, which contain the identities of the alleged certificate owners.

– It was suggested to combine different Distinguished Names with equal public keys, to lure someone to encrypt data for one person, which can then be decrypted by another. It is unclear to us how realistic this is—or why one would need identical digital signatures. Nevertheless, if the appendages are not hidden in the public key field, some other field must be found for them, located before or after the public key field. Such a field may be specially defined for this purpose, and there is a good chance that the certificate processing software will not recognize this field and ignore it. However, as the appendages have non-negligible length, it will be hard to define a field that will not look suspicious to someone who looks at the certificate at bit level.

– A way to realize the above variant is to hide the collision-causing appendages in the public exponent. Though the public exponent is often taken from a small set (3, 17, and 65537 are common choices), a large, random looking one is in principle possible. It may even be larger than the modulus, but that may raise suspicion. In any case, the two certificates can now have identical RSA moduli, making it easy for the owner of one private key to compute the other one.

– Entirely different abuse scenarios are conceivable. In [2] (see also [4]) it was shown how to construct a pair of Postscript files that collide under MD5, and that send different messages to output media such as screen or printer. However, in those constructions both messages had to be hidden in each of the colliding files, which obviously raises suspicions upon inspection at bit level. With chosen-prefix collisions, this can be avoided. For example, two different messages can be entered into a document format that allows insertion of color images (such as Microsoft Word), with one message per document. At the last page of each document a color image will be shown—a short one pixel wide line will do, for instance hidden inside a layout element, a company logo, or in the form of a nicely colored barcode claiming to be some additional security feature, obviously offering far greater security than those old-fashioned black and white barcodes—carefully constructed such that the hashes of the documents collide when their color codes are inserted. In Figure 1 the actual 4192-bit collision-causing appendages computed for the certificates are built into bitmaps to get two different barcode examples. Each string of 4192 bits leads to one line of 175 pixels, say A and B, and the barcodes consist of the lines ABBBBB and BBBBBB respectively. Apart

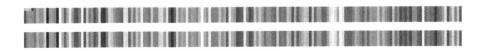

Fig. 1. A collision built into bitmap images.

from the 96 most significant bits corresponding to the 4 pixels in the upper left corner, the barcodes differ in only a few bits, which makes the resulting color differences hard to spot for the human eye. As noted above the 'obviously differing' 4 initial pixels can be avoided at the cost of more near-collision blocks (thus longer barcodes), and the barcodes can be shortened again at the cost of more elaborate differential path constructions.

- In [12] and [8] it was shown how to abuse existing MD5 collisions to mislead integrity checking software based on MD5. Similar to the colliding Postscript applications, they also used the differences in the colliding inputs to construct deviating execution flows of some programs. Here too chosen-prefix collisions allow a more elegant approach, especially since common operating systems ignore bitstrings that are appended to executables: the programs will run unaltered. Thus one can imagine two executables: a 'good' one (say Word.exe) and a bad one (the attacker's Worse.exe). A chosen-prefix collision for those executables is computed, and the collision-causing bitstrings are appended to them. The resulting altered file Word.exe, functionally equivalent to the original Word.exe, can then be offered to Microsoft's Authenticode signing program and receive an 'official' MD5 based digital signature. This signature will be equally valid for the attacker's Worse.exe, and the attacker might be able to replace Word.exe by his Worse.exe (renamed to Word.exe) on the appropriate download site. This construction affects a common functionality of MD5 hashing and may pose a practical threat, also because there is no a priori reason why the collision-causing bitstrings could not be hidden *inside* the executables.
- More ideas can be found on `www.iaik.tugraz.at/research/krypto/collision/`.

Further study is required to assess the impact of chosen-prefix collisions on applications of hash functions. Commonly used protocols and message formats such as SSL, S/MIME (CMS) and XML Signatures should be studied, with special attention to whether random looking data can be hidden in these protocols and data formats, in such a way that some or all implementations will not detect them. For instance, it was suggested by Pascal Junod to let a 'proper' certificate collide with one that contains executable code in the Distinguished Name field, thereby potentially triggering a buffer overflow, but we have not seen an actually working example of this idea yet. It also requires more study to see if there are formats that even allow the much easier random collision attacks.

4   Colliding X.509 Certificates for Different Identities

In this section we concentrate on the first application mentioned above, that of two X.509 certificates with identical digital signatures but different Distinguished Name fields, where the collisions are perfectly hidden inside the public key moduli.

4.1 Attack Scenarios

Though our current X.509 certificates construction, involving different Distinguished Names, should have more attack potential than the one with identical names fields in [11], we have not been able to find truly convincing attack scenarios yet. Ideally, a realistic attack targets the core of PKI: provide a relying party with trust, beyond reasonable cryptographic doubt, that the person indicated by the Distinguished Name field has exclusive control over the private key corresponding to the public key in the certificate. The attack should also enable the attacker to cover his trails.

Getting two certificates for the price of one could be economically advantageous in some situations, e.g. with two different owner names, or for two different validity periods. Such certificates undermine the proof of knowledge of the secret key corresponding to a certified public key. These possibilities have been noted before (cf. [10]) and do, in our opinion, not constitute attacks.

Our construction requires that the two colliding certificates are generated simultaneously. Although each resulting certificate by itself is completely unsuspicious, the fraud becomes apparent when the two certificates are put alongside, as may happen during a fraud analysis. An attacker can generate one of the certificates for a targeted person, the other one for himself, and attempt to use his own credentials to convince an external and generally trusted CA to sign the second one. If successful, the attacker can then distribute the first certificate, which will be trusted by relying parties, e.g. to encrypt messages for the targeted person. The attacker however is in control of the corresponding private key, and can thus decrypt confidential information embedded in intercepted messages meant for the targeted person. Or the attacker can masquerade as the targeted person while signing messages, which will be trusted by anyone trusting the CA. In this scenario it does not matter whether the two certificates have different public keys (as in our example) or identical ones (in which case the colliding blocks would have to be hidden somewhere else in the certificate).

A problem is, however, that the CA will register the attacker's identity. As soon as a dispute arises, the two certificates will be produced and revealed as colliding, and the attacker will be identified. Another problem is that the attacker must have sufficient control over the CA to predict all fields appearing before the public key, such as the serial number and the validity periods. It has frequently been suggested that this is an effective countermeasure against colliding certificate constructions in practice, but there is no consensus how hard it is to make accurate predictions. When this condition of sufficient control over the CA by the attacker is satisfied, colliding certificates based on chosen-prefix collisions are a bigger threat than those based on random collisions.

Obviously, the attack becomes effectively impossible if the CA adds a sufficient amount of fresh randomness to the certificate fields before the public key, such as in the serial number (as some already do, though probably for different reasons). This randomness is to be generated after the approval of the certification request. On the other hand, in general a relying party cannot verify this randomness. In our opinion, trustworthiness of certificates should not crucially depend on such secondary and circumstantial aspects. On the contrary, CAs should use a trustworthy hash function that meets the design criteria. Unfortunately, this is no longer the case for MD5, or SHA-1.

We stress that our construction (we prefer this wording to 'attack') is not a preimage attack. As far as we know, existing certificates cannot be forged by chosen-prefix collisions if they have not been especially crafted for that purpose. However, a relying party cannot distinguish any given trustworthy certificate from a certificate that has been crafted by our method to violate PKI principles. Therefore we repeat, with more urgency, our recommendation that MD5 is no longer used in new X.509 certificates. Similar work [1] is in development for the SHA-1 hash function, so we feel that a renewed assessment of the use of SHA-1 in certificate generation is also appropriate.

4.2 Certificate Construction Outline

Table 1 outlines the to-be-signed fields of the colliding certificates that were constructed.

Table 1. The to-be-signed parts of the colliding certificates

field	comments	value first certificate	value second certificate
X.509 version number	identical, standard X.509	version 3	
serial number	different, chosen by CA	0x010C0001	0x020C0001
signature algorithm identifier	identical, standard X.509	md5withRSAEncryption	
issuer distinguished name	identical, chosen by CA	CN = "Hash Collision CA" L = "Eindhoven" C = "NL"	
not valid before	identical, chosen by CA	Jan. 1, 2006, 00h00m01s GMT	
not valid after	identical, chosen by CA	Dec. 31, 2007, 23h59m59s GMT	
subject distinguished name	different, chosen by us	CN = "Arjen K. Lenstra" O = "Collisionairs" L = "Eindhoven" C = "NL"	CN = "Marc Stevens" O = "Collision Factory" L = "Eindhoven" C = "NL"
public key algorithm	identical, standard X.509	rsaEncryption	
subject public key info	different, constructed by us	modulus $S_b\|S_c\|E$ as below	modulus $S_b'\|S_c'\|E$ as below
version 3 extensions	identical, standard X.509	(irrelevant for the present description)	

Here, S_b and S_b' are 96-bit values found using birthdaying, S_c and S_c' each consist of 8 near-collision blocks found using the automated method to find differential paths, and E is a 4000-bit value such that the 8192-bit values $S_b\|S_c\|E$ and $S_b'\|S_c'\|E$ are both RSA moduli. The details of the construction are set forth below.

Before the collision search (i.e., the searches for S_b, S_b' and for S_c, S_c') is started the contents needs to be known of all to-be-signed fields of the certificate that appear before the modulus. Therefore, to be able to construct the certificates, sufficient control over the CA is necessary. This was achieved by implementing and operating this CA ourselves. In fact, we used the CA that had already been set up for [10]. It is used solely for the purposes of signing colliding certificates.

4.3   Certificate Construction Details

We provide a detailed description of our construction.

1. We construct two templates for the certificates in which all fields are filled in, with the exception of the RSA public key moduli and the signature, meeting the following three requirements:
   - The data structures must be compliant with the X.509 standard and the ASN.1 DER encoding rules (see [5], but see also the final section of [14]);
   - The byte lengths of the moduli and the public exponent (in fact, also the byte lengths of the entire to-be-signed parts of the certificates) must be fixed in advance, because these numbers have to be specified as parts of the ASN.1 structure, coming before the modulus;
   - The position where the RSA moduli start must be controlled. We chose to have this at an exact multiple of 64 bytes (512 bits) minus 96 bits, after the beginning of the to-be-signed fields. This gives convenient space for the results of the birthdaying step (described below).

   The third condition can be dealt with by adding dummy information to the subject Distinguished Name. This we did in the Organization-field (i.e., the value O in the outline above).
2. We apply MD5 to each of the first parts of the two to-be-signed fields, truncated at the last full block (thus excluding the incomplete blocks whose last 96 bits will consist of the most significant bits of the RSA moduli under construction), suppressing the padding normally used in MD5. As output we get a pair of IHVs that we use as input for the next step. These IHVs will be completely different and have no special properties built in.
3. Using the IHVs and their corresponding incomplete blocks (the ones that still fail their last 96 bits) as input, we complete these blocks by appending 96-bit values S_b and S'_b. These values are computed by birthdaying, to satisfy 96 bit conditions on the output IHV difference. For this purpose each output IHV is interpreted as 4 little endian 32-bit integers, and the difference between the output IHVs is defined as the 4-tuple of differences modulo 2^{32} between the four corresponding 32-bit integers. If we represent this IHV difference as $\delta a\|\delta b\|\delta c\|\delta d$ for 32-bit $\delta a, \delta b, \delta c, \delta d$, then the conditions are $\delta a = 0$ and $\delta b = \delta c = \delta d$, as suggested to us by Xiaoyun Wang. The reason for this choice is that it facilitates the search for near-collision blocks, as explained in Section 5.3. The resulting δb can be expressed as only 8 signed bit differences (these are not bitwise XOR but additive differences).
4. Using the techniques developed in [16] and described in Section 5, we compute two different bitstrings S_c and S'_c, of 4096 bits (8 near-collision blocks) each. Each near-collision block is used to eliminate one (triple) of the bit differences in the IHVs from the previous step, so that at the end of the 8 near-collision blocks the IHVs are equal, and a complete MD5 collision has been constructed. We now have $S = S_b\|S_c$ and $S' = S'_b\|S'_c$ that form the leading 4192 bits of the RSA moduli, such that the two to-be-signed fields up to and including S and S', respectively, collide under MD5. Therefore,

in order not to destroy the collision, everything that is to be appended from now on must be identical for the two certificates.

5. Next we used the method from [10] to craft two secure 8192-bit RSA moduli from the two bitstrings S and S' of 4192 bits each, by appending to each the same 4000-bit E. As explained in [11] this means that we could in principle construct moduli that are products of primes of sizes roughly 2000 and 6192 bits. In order to speed up the RSA modulus construction process, we aimed somewhat lower here and settled for products of 1976 and 6216-bit primes. This took about an hour on a regular laptop. The strongly unbalanced RSA moduli may be unusual, but for our parameter choices (smallest prime factor around 1976 bits for a modulus of 8192 bits) we see no reason to believe that these moduli are less secure than more balanced, regular RSA moduli of the same size, given the present state of factoring technology.

6. We insert the subject public key info into the template for the first certificate, thereby completing the to-be-signed part of the first certificate. We compute the MD5 hash of the entire to-be-signed part, and from it we compute the signature, which is added to the certificate. The first certificate is now complete. To obtain the second valid certificate, we put the proper subject public key info and the same signature at their locations in the template for the second certificate.

Finding the chosen-prefix MD5 collisions (i.e., Steps 3 and 4) is by far the computationally hardest part of the above construction, a remark that is similar to one made in [10]. However, in the meantime the methods for constructing MD5 collisions with *identical* initial IHVs have been improved considerably: such collisions can now be found within seconds, see [15] and [9]. So in the scenario of [10] the bottleneck may now have shifted from the collision search to the moduli construction.

An example pair of colliding certificates is available in full detail in [14] and on www.win.tue.nl/hashclash/ChosenPrefixCollisions/.

5   Chosen-Prefix Collision Construction

5.1   Preliminaries

MD5 operates on 32-bit words, and uses *little endian* byte ordering.

A *binary signed digit representation* (BSDR) for a 32-bit word X is defined as $(k_i)_{i=0}^{31}$, where

$$X = \sum_{i=0}^{31} 2^i k_i, \quad k_i \in \{-1, 0, +1\}.$$

Many different BSDRs may exist for any given X. The *weight* of a BSDR is the number of non-zero k_i's. A particularly useful BSDR is the Non-Adjacent Form (NAF), where no two non-zero k_i's are adjacent. The NAF is not unique since we work modulo 2^{32} (making $k_{31} = +1$ equivalent to $k_{31} = -1$), but uniqueness of the NAF can be enforced by choosing $k_{31} \in \{0, +1\}$. Among the BSDRs of an integer, the NAF has minimal weight. We use the following notation:

- Integers are denoted in hexadecimal as $\mathtt{12EF}_{16}$ and in binary as $\mathtt{000100101110}$ $\mathtt{1111}_2$;
- $X \wedge Y$ is the bitwise AND of X and Y;
- $X \vee Y$ is the bitwise OR of X and Y;
- $X \oplus Y$ is the bitwise XOR of X and Y;
- $\bar{X}$ is the bitwise complement of X;

for 32-bit integers X and Y:

- $X[i]$ is the i-th bit of the regular binary representation of X;
- $X + Y$ resp. $X - Y$ is the addition resp. subtraction modulo 2^{32};
- $RL(X, n)$ (resp. $RR(X, n)$) is the cyclic left (resp. right) rotation of X by n bit positions:

$$RL(\mathtt{10100100\ldots00000001}_2, \ 5) = \mathtt{10000000\ldots00110100}_2;$$

and for a 32-digit BSDR X:

- $X[\![i]\!]$ is the i-th signed bit of X;
- $RL(X, n)$ (resp. $RR(X, n)$) is the cyclic left (resp. right) rotation of X by n positions.

For chosen message prefixes P and P' we seek suffixes S and S' such that $P\|S$ and $P'\|S'$ collide under MD5. In this section a variable occurring during the construction of S and intermediate P-related MD5 calculations, may have a corresponding variable during the construction of S' and intermediate P'-related MD5 calculations. If the former variable is X, then the latter is denoted X'. Furthermore, $\delta X = X' - X$ for such a 'matched' 32-bit integer variable X, and $\Delta X = (X'[i] - X[i])_{i=0}^{31}$, which is a BSDR of δX. For a 'matched' variable Z that consist of tuples of 32-bit integers, say $Z = (z_1, z_2, \ldots)$, we define δZ as $(\delta z_1, \delta z_2, \ldots)$.

5.2   Description of MD5

5.2.1   MD5 Message Processing
MD5 can be split up into these parts:

1. *Padding.* Pad the message with: first the '1'-bit, next as many '0' bits until the resulting length equals 448 mod 512, and finally the bitlength of the original message as a 64-bit little-endian integer. The total bitlength of the padded message is $512N$ for a positive integer N.
2. *Partitioning.* Partition the padded message into N consecutive 512-bit blocks $M_1, M_2, \ldots, M_N$.

3. *Processing.* MD5 goes through $N + 1$ states IHV_i, for $0 \leq i \leq N$, called the *intermediate hash values.* Each intermediate hash value IHV_i consists of four 32-bit words a_i, b_i, c_i, d_i. For $i = 0$ these are initialized to fixed public values:

$$(a_0, b_0, c_0, d_0) = (67452301_{16}, \text{EFCDAB89}_{16}, 98\text{BADCFE}_{16}, 10325476_{16}),$$

and for $i = 1, 2, \ldots N$ intermediate hash value IHV_i is computed using the MD5 compression function described in detail below:

$$IHV_i = \text{MD5Compress}(IHV_{i-1}, M_i).$$

4. *Output.* The resulting hash value is the last intermediate hash value IHV_N, expressed as the concatenation of the hexadecimal byte strings of the four words a_N, b_N, c_N, d_N, converted back from their little-endian representation.

5.2.2   MD5 Compression Function

The input for the compression function $\text{MD5Compress}(IHV, B)$ is an intermediate hash value $IHV = (a, b, c, d)$ and a 512-bit message block B. There are 64 *steps* (numbered 0 up to 63), split into four consecutive *rounds* of 16 steps each. Each step uses a modular addition, a left rotation, and a non-linear function. Depending on the step t, *Addition Constants* AC_t and *Rotation Constants* RC_t are defined as follows:

$$AC_t = \lfloor 2^{32} |\sin(t + 1)| \rfloor , \quad 0 \leq t < 64,$$

$$(RC_t, RC_{t+1}, RC_{t+2}, RC_{t+3}) = \begin{cases} (7, 12, 17, 22) & \text{for } t = 0, 4, 8, 12, \\ (5, 9, 14, 20) & \text{for } t = 16, 20, 24, 28, \\ (4, 11, 16, 23) & \text{for } t = 32, 36, 40, 44, \\ (6, 10, 15, 21) & \text{for } t = 48, 52, 56, 60. \end{cases}$$

The non-linear function f_t depends on the round:

$$f_t(X, Y, Z) = \begin{cases} F(X, Y, Z) = (X \wedge Y) \oplus (\bar{X} \wedge Z) & \text{for } 0 \leq t < 16, \\ G(X, Y, Z) = (Z \wedge X) \oplus (\bar{Z} \wedge Y) & \text{for } 16 \leq t < 32, \\ H(X, Y, Z) = X \oplus Y \oplus Z & \text{for } 32 \leq t < 48, \\ I(X, Y, Z) = Y \oplus (X \vee \bar{Z}) & \text{for } 48 \leq t < 64. \end{cases}$$

The message block B is partitioned into sixteen consecutive 32-bit words $m_0, m_1, \ldots, m_{15}$ (with little endian byte ordering), and expanded to 64 words W_t, for $0 \leq t < 64$, of 32 bits each:

$$W_t = \begin{cases} m_t & \text{for } 0 \leq t < 16, \\ m_{(1+5t) \bmod 16} & \text{for } 16 \leq t < 32, \\ m_{(5+3t) \bmod 16} & \text{for } 32 \leq t < 48, \\ m_{(7t) \bmod 16} & \text{for } 48 \leq t < 64. \end{cases}$$

We follow the description of the MD5 compression function from [6] because its 'unrolling' of the cyclic state facilitates the analysis. For $t = 0, 1, \ldots, 63$, the compression function algorithm maintains a working register with 4 state words Q_t, Q_{t-1}, Q_{t-2} and Q_{t-3}. These are initialized as $(Q_0, Q_{-1}, Q_{-2}, Q_{-3}) = (b, c, d, a)$ and, for $t = 0, 1, \ldots, 63$ in succession, updated as follows:

$$F_t = f_t(Q_t,\ Q_{t-1},\ Q_{t-2}),$$
$$T_t = F_t + Q_{t-3} + AC_t + W_t,$$
$$R_t = RL(T_t, RC_t),$$
$$Q_{t+1} = Q_t + R_t.$$

After all steps are computed, the resulting state words are added to the intermediate hash value and returned as output:

$$\text{MD5Compress}(IHV, B) = (a + Q_{61},\ b + Q_{64},\ c + Q_{63},\ d + Q_{62}).$$

5.3   Outline of the Collision Construction

A chosen-prefix collision is a pair of messages M and M' that consist of arbitrarily chosen prefixes P and P' (not necessarily of the same length), together with constructed suffixes S and S', such that $M = P\|S$, $M' = P'\|S'$, and $MD5(M) = MD5(M')$. The suffixes consist of three parts: random padding bitstrings S_r, S_r', followed by 'birthday' bitstrings S_b, S_b', followed by 'near collision' blocks S_c, S_c'. The random padding bitstrings are chosen to guarantee that the bitlengths of $P\|S_r\|S_b$ and $P'\|S_r'\|S_b'$ are both equal to $512n$ for a positive integer n. (In our example of the colliding certificates we engineered the prefixes such that S_r and S_r' were both empty.) The MD5 compression function applied to $P\|S_r\|S_b$ resp. $P'\|S_r'\|S_b'$ will result in IHV_n resp. IHV_n', in the notation from Section 5.2.1. The birthday bitstrings S_b, S_b' are taken in such a way that the resulting δIHV_n has certain desirable properties, to be described below.

The idea is to eliminate the difference δIHV_n using a series of pairs of near-collision blocks that together constitute S_c, S_c'. For each near-collision we need to construct a differential path such that the NAF weight of the new δIHV_{n+j} is lower than the NAF weight of δIHV_{n+j-1}, until after r pairs of near-collision blocks we have reached $\delta IHV_{n+r} = 0$.

For the j-th pair of near-collision blocks, i.e., M_{n+j} and M_{n+j}', we fix all but one of the 32-bit words δm_i of δM_{n+j} as 0, and allow only δm_{11} to be $\pm 2^d$ with varying d, $0 \leq d < 32$. This was suggested by Xiaoyun Wang because with this type of message difference the number of bitconditions over the final two and a half rounds can be kept low. This is illustrated in Table 2, where the corresponding partial differential path is shown for the final 31 steps. For these types of message differences we try to find in an automated way a differential path with the right properties, and then try to find a pair of near-collision blocks M_{n+j}, M_{n+j}' that satisfies the differential path.

The differential paths under consideration can only add (or substract) a tuple $(0, 2^i, 2^i, 2^i)$ to δIHV_{n+j} and therefore cannot eliminate arbitrary δIHV_n. To

Table 2. Partial differential path with $\delta m_{11} = \pm 2^d$

t	δQ_t	δF_t	δW_t	δT_t	δR_t	RC_t
30	$\mp 2^d$					
31	0					
32	0					
33	0	0	$\pm 2^d$	0	0	16
$34 - 60$	0	0	0	0	0	·
61	0	0	$\pm 2^d$	$\pm 2^d$	$\pm 2^{d+10 \bmod 32}$	10
62	$\pm 2^{d+10 \bmod 32}$	0	0	0	0	15
63	$\pm 2^{d+10 \bmod 32}$	0	0	0	0	21
64	$\pm 2^{d+10 \bmod 32}$					

solve this we first use a birthday attack to find 'birthday' bitstrings S_b and S_b' such that $\delta IHV_n = (0, \delta b, \delta b, \delta b)$ for some δb. The birthday attack actually searches for a collision $(a, b - c, b - d) = (a', b' - c', b' - d')$ between $IHV_n = (a, b, c, d)$ and $IHV_n' = (a', b', c', d')$, implying indeed $\delta a = 0$ and $\delta b = \delta c = \delta d$. The search space consists of 96 bits and therefore the birthday step can be expected to require on the order of $2\sqrt{2^{96}} = 2^{49}$ calls to the MD5 compression function.

One may extend the birthdaying by searching for a δb of low NAF weight, as this weight is the number of near-collision block pairs to be found. On average one may expect to find a δb of NAF weight 11. In the case of our colliding certificates example we found a δb of NAF weight only 8, after having extended the search somewhat longer than absolutely necessary.

Let (k_i) be the NAF of δb. Then we can reduce $\delta IHV_n = (0, \delta b, \delta b, \delta b)$ to $(0, 0, 0, 0)$ by using, for each non-zero k_i, a differential path based on the partial differential path in Table 2 with $\delta m_{11} = -k_i 2^{i-10 \bmod 32}$. In other words, the signed bit difference at position i in δb can be eliminated by choosing a message difference only in δm_{11}, with just one opposite-signed bit set at position $i - 10 \bmod 32$. Let i_j for $j = 1, 2, \ldots, r$ be the indices of the non-zero k_i. Starting with n-block $M = P \| S_r \| S_b$ and $M' = P' \| S_r' \| S_b'$ and the corresponding resulting IHV_n and IHV_n' we do the following for $j = 1, 2, \ldots, r$ in succession:

1. Let $\delta m_{11} = -k_{i_j} 2^{i_j - 10 \bmod 32}$ and $\delta m_\ell = 0$ for $\ell \neq 11$ (note the slight abuse of notation, since we define just the message block differences, without specifying the message blocks themselves).
2. Starting from IHV_{n+j-1} and IHV_{n+j-1}', find a differential path.
3. Find message blocks $S_{c,j}$ and $S_{c,j}' = S_{c,j} + \delta M_{n+j}$, that satisfy the differential path. This can be done by using collision finding techniques such as Klima's tunnels, cf. [9] and [15].
4. Let $IHV_{n+j} = \text{MD5Compress}(IHV_{n+j-1}, S_{c,j})$, $IHV_{n+j}' = \text{MD5Compress}(IHV_{n+j-1}', S_{c,j}')$, and append $S_{c,j}$ to M and $S_{c,j}'$ to M'.

It remains to explain step 2 in this algorithm.

Fig. 2. δIHVs for the colliding certificates

Figure 2 visualizes the entire process. The horizontal lines represent the NAFs of δIHV_i for $i = 0, 1, \ldots, 21$. The section $P\|S_\mathrm{r}\|S_\mathrm{b}$ consists of 4 blocks (i.e., $n = 4$), so at $i = 4$ only $r = 8$ triples of bit differences are left. They are annihilated one by one by the 8 near-collision block pairs (i.e., $S_{\mathrm{c},j}$ and $S'_{\mathrm{c},j}$ for $j = 1, 2, \ldots, 8$), so that at $i = 12$ a full collision is reached. The blocks after that (which include E from Section 4.3) are identical for the two messages, so that the collision is retained.

5.4 Differential Paths and Bitconditions

Assume MD5Compress is applied to pairs of inputs for both intermediate hash value and message block, i.e., to (IHV, B) and (IHV', B'). A differential path for MD5Compress is a precise description of the propagation of differences through the 64 steps caused by δIHV and δB:

$$\delta F_t = f_t(Q'_t, Q'_{t-1}, Q'_{t-2}) - f_t(Q_t, Q_{t-1}, Q_{t-2});$$
$$\delta T_t = \delta F_t + \delta Q_{t-3} + \delta W_t;$$
$$\delta R_t = RL(T'_t, RC_t) - RL(T_t, RC_t);$$
$$\delta Q_{t+1} = \delta Q_t + \delta R_t.$$

Note that δF_t is not uniquely determined by $\delta Q_t, \delta Q_{t-1}$ and δQ_{t-2}, so it is necessary to describe the value of δF_t and how it can result from the Q_i, Q'_i in such a way that it does not conflict with other steps. Similarly δR_t is not uniquely determined by δT_t and RC_t, so also the value of δR_t has to be described.

5.4.1 Bitconditions

We use *bitconditions* on (Q_t, Q'_t) to describe differential paths, where a single bitcondition specifies directly or indirectly the values of the bits $Q_t[i]$ and $Q'_t[i]$. Thus a differential path consists of a matrix of bitconditions with 68 rows (for the possible indices $t = -3, -2, \ldots, 64$ in Q_t, Q'_t) and 32 columns (one for each bit). A *direct* bitcondition on $(Q_t[i], Q'_t[i])$ does not involve other bits $Q_j[k]$ or $Q'_j[k]$, while an *indirect* bitcondition does, and specifically one of $Q_{t-2}[i], Q_{t-1}[i], Q_{t+1}[i]$ or $Q_{t+2}[i]$. Using only bitconditions on (Q_t, Q'_t) we can specify all the values of δQ_t, δF_t and thus δT_t and $\delta R_t = \delta Q_{t+1} - \delta Q_t$ by the relations above. A bitcondition on $(Q_t[i], Q'_t[i])$ is denoted by $\mathsf{q}_t[i]$, and symbols like $0, 1, +, -, \hat{} , \ldots$ are used for $\mathsf{q}_t[i]$, as defined below. The 32 bitconditions $(\mathsf{q}_t[i])_{i=0}^{31}$ are denoted by q_t. We discern between *differential* bitconditions and *boolean*

Table 3. Differential bitconditions

$q_t[i]$	condition on $(Q_t[i], Q'_t[i])$	k_i
.	$Q_t[i] = Q'_t[i]$	0
+	$Q_t[i] = 0, \quad Q'_t[i] = 1$	$+1$
-	$Q_t[i] = 1, \quad Q'_t[i] = 0$	-1

Note: $\delta Q_t = \sum_{i=0}^{31} 2^i k_i$ and $\Delta Q_t = (k_i)$.

Table 4. Boolean function bitconditions

$q_t[i]$	condition on $(Q_t[i], Q'_t[i])$	direct/indirect	direction
0	$Q_t[i] = Q'_t[i] = 0$	direct	
1	$Q_t[i] = Q'_t[i] = 1$	direct	
^	$Q_t[i] = Q'_t[i] = Q_{t-1}[i]$	indirect	backward
v	$Q_t[i] = Q'_t[i] = Q_{t+1}[i]$	indirect	forward
!	$Q_t[i] = Q'_t[i] = \overline{Q_{t-1}[i]}$	indirect	backward
y	$Q_t[i] = Q'_t[i] = \overline{Q_{t+1}[i]}$	indirect	forward
m	$Q_t[i] = Q'_t[i] = Q_{t-2}[i]$	indirect	backward
w	$Q_t[i] = Q'_t[i] = Q_{t+2}[i]$	indirect	forward
#	$Q_t[i] = Q'_t[i] = \overline{Q_{t-2}[i]}$	indirect	backward
h	$Q_t[i] = Q'_t[i] = \overline{Q_{t+2}[i]}$	indirect	forward
?	$Q_t[i] = Q'_t[i] \wedge (Q_t[i] = 1 \vee Q_{t-2}[i] = 0)$	indirect	backward
q	$Q_t[i] = Q'_t[i] \wedge (Q_{t+2}[i] = 1 \vee Q_t[i] = 0)$	indirect	forward

function bitconditions. The former, shown in Table 3, are direct, and specify the value $k_i = Q'_t[i] - Q_t[i]$ which together specify $\delta Q_t = \sum 2^i k_i$ by how each bit changes. Note that (k_i) is also a BSDR. The boolean function bitconditions, shown in Table 4, are used to resolve any ambiguity in

$$\Delta F_t[\![i]\!] = f_t(Q'_t[i], Q'_{t-1}[i], Q'_{t-2}[i]) - f_t(Q_t[i], Q_{t-1}[i], Q_{t-2}[i]) \in \{-1, 0, +1\}$$

caused by different possible values for $Q_j[i], Q'_j[i]$ for given bitconditions. As an example, for $t = 0$ and $(q_t[i], q_{t-1}[i], q_{t-2}[i]) = (., +, -)$ there is an ambiguity:

$$\text{if } Q_t[i] = Q'_t[i] = 0 \text{ then } \Delta F_t[\![i]\!] = f_t(0, 1, 0) - f_t(0, 0, 1) = -1,$$
$$\text{but if } Q_t[i] = Q'_t[i] = 1 \text{ then } \Delta F_t[\![i]\!] = f_t(1, 1, 0) - f_t(1, 0, 1) = +1.$$

To resolve this ambiguity the bitcondition (.,+,-) can be replaced by (0,+,-) or (1,+,-).

All boolean function bitconditions include the constant bitcondition $Q_t[i] = Q'_t[i]$, so they do not affect δQ_t. Furthermore, indirect boolean function bitconditions never involve a bit with condition + or -, since then it could be replaced by one of the direct bitconditions ., 0 or 1. We distinguish in the direction of indirect bitconditions, since that makes it easier to resolve an ambiguity later on. It is quite easy to change all backward bitconditions into forward ones in a valid (partial) differential pathm, and vice versa.

When all δQ_t and δF_t are determined by bitconditions then also δT_t and δR_t can be determined, which together describe the bitwise rotation of δT_t in each step. Note that this does not describe if it is a valid rotation or with what probability the rotation from δT_t to δR_t occurs. The differential paths we constructed for our example can be found at www.win.tue.nl/hashclash/ ChosenPrefixCollisions/.

5.4.2   Differential Path Construction Overview

The basic idea in constructing a differential path is to construct a partial lower differential path over steps $t = 0, 1, \ldots, 11$ and a partial upper differential path over steps $t = 16, 17, \ldots, 63$, so that the Q_i involved in the partial paths meet but do not overlap. Then try to connect those partial paths over the remaining 4 steps into one full differential path. Constructing the partial lower path can be done by starting with bitconditions q_{-3}, q_{-2}, q_{-1}, q_0 that are equivalent to the values of IHV, IHV' and then extend this step by step. Similarly the partial upper path can be constructed by extending the partial path in Table 2 step by step. To summarize, step 2 in the algorithm of section 5.3 consist of the following substeps:

2.1 Using IHV and IHV' determine bitconditions $(q_i)_{i=-3}^0$.
2.2 Generate a partial lower differential path by extending $(q_i)_{i=-3}^0$ forward up to step $t = 11$.
2.3 Generate a partial upper differential path by extending the path in Table 2 down to $t = 16$.
2.4 Try to connect these lower and upper differential paths over $t = 12, 13, 14, 15$. If this fails generate other partial lower and upper differential paths and try again.

5.5   Extending Differential Paths

When constructing a differential path one must fix the message block differences $\delta m_0, \ldots, \delta m_{15}$. Suppose we have a partial differential path consisting of at least bitconditions q_{t-1} and q_{t-2} and that the values δQ_t and δQ_{t-3} are known. We want to extend this partial differential path forward with step t resulting in the value δQ_{t+1} and (additional) bitconditions q_t, q_{t-1}, q_{t-2}. We assume that all indirect bitconditions are forward and do not involve bits of Q_t. If we also have q_t instead of only the value δQ_t (e.g. q_0 resulting from given values IHV, IHV'), then we can skip the carry propagation and continue at Section 5.5.2.

5.5.1   Carry Propagation

First we want to use the value δQ_t to select bitconditions q_t. This can be done by choosing any BSDR of δQ_t, which directly translates into a possible choice for q_t as given in Table 3. Since we want to construct differential paths with as few bitconditions as possible, but also want to be able to randomize the process, we may choose any low weight BSDR (such as the NAF).

5.5.2   Boolean Function

For some i, let $(a, b, c) = (\mathsf{q}_t[i], \mathsf{q}_{t-1}[i], \mathsf{q}_{t-2}[i])$ be any triple of bitconditions such that all indirect bitconditions involve only $Q_t[i]$, $Q_{t-1}[i]$ or $Q_{t-2}[i]$. The triple (a, b, c) is associated with the set U_{abc} of tuples of values $(x, x', y, y', z, z') = (Q_t[i], Q'_t[i], Q_{t-1}[i], Q'_{t-1}[i], Q_{t-2}[i], Q'_{t-2}[i])$:

$$U_{abc} = \{(x, x', y, y', z, z') \in \{0, 1\}^6 \text{ satisfies bitconditions } (a, b, c)\}.$$

If $U_{abc} = \emptyset$ then (a, b, c) is said to be contradicting and cannot be part of any valid differential path. We define $\mathcal{F}_t$ as the set of all triples (a, b, c) such that all indirect bitconditions involve only $Q_t[i]$, $Q_{t-1}[i]$ or $Q_{t-2}[i]$ and $U_{abc} \neq \emptyset$.

We define V_{abc} as the set of all possible boolean function differences $f_t(x', y', z') - f_t(x, y, z)$ for given bitconditions $(a, b, c) \in \mathcal{F}_t$:

$$V_{abc} = \{f_t(x', y', z') - f_t(x, y, z) \mid (x, x', y, y', z, z') \in U_{abc}\} \subset \{-1, 0, +1\}.$$

If $|V_{abc}| = 1$ then (a, b, c) leaves no ambiguity and the triple (a, b, c) is said to be a *solution*. Let $\mathcal{S}_t$ be the set of all solutions. If $|V_{abc}| > 1$ then for each $g \in V_{abc}$ we define $W_{abc,g}$ as the set of solutions $(d, e, f) \in \mathcal{S}_t$ that are compatible with (a, b, c) and that have g as boolean function difference:

$$W_{abc,g} = \{(d, e, f) \in \mathcal{S}_t \mid U_{def} \subset U_{abc} \wedge V_{def} = \{g\}\}.$$

Note that for all $g \in V_{abc}$ there is always a triple $(d, e, f) \in W_{abc,g}$ that consists only of direct bitconditions 01+-, hence $W_{abc,g} \neq \emptyset$. The direct and forward (resp. backward) boolean function bitconditions were chosen such that for all t, i and $(a, b, c) \in \mathcal{F}_t$ and for all $g \in V_{abc}$ there exists a triple $(d, e, f) \in W_{abc,g}$ consisting of direct and forward (resp. backward) bitconditions such that

$$U_{def} \text{ is equal to } \{(x, x', y, y', z, z') \in U_{abc} \mid f_t(x', y', z') - f_t(x, y, z) = g\}.$$

In other words, these boolean function bitconditions allows one to resolve an ambiguity in an optimal way. If the triple (d, e, f) is not unique, then we prefer direct over indirect bitconditions and short indirect bitconditions (vy^!) over long indirect bitconditions (whqm#?) for simplicity reasons. For given t, bitconditions (a, b, c), and $g \in V_{abc}$ we define $FC(t, abc, g) = (d, e, f)$ and $BC(t, abc, g) = (d, e, f)$ as the preferred triple (d, e, f) consisting of direct and forward, respectively backward bitconditions. These should be precomputed for all cases.

For all $i = 0, 1, \ldots, 31$ we have by assumption valid bitconditions $(a, b, c) = (\mathsf{q}_t[i], \mathsf{q}_{t-1}[i], \mathsf{q}_{t-2}[i])$ where only c can be an indirect bitcondition. If so, it must involve $Q_{t-1}[i]$. Therefore $(a, b, c) \in \mathcal{F}_t$. If $|V_{abc}| = 1$ there is no ambiguity and we let $\{g_i\} = V_{abc}$. Otherwise, if $|V_{abc}| > 1$, then we choose any $g_i \in V_{abc}$ and we resolve the ambiguity left by bitconditions (a, b, c) by replacing them by $(d, e, f) = FC(t, abc, g_i)$, which results in boolean function difference g_i. Given all g_i, the values $\delta F_t = \sum_{i=0}^{31} 2^i g_i$ and $\delta T_t = \delta F_t + \delta Q_{t-3} + \delta W_t$ can be determined.

5.5.3 Bitwise Rotation

The integer δT_t does not uniquely determine the value of $\delta R_t = RL(T_t', n) - RL(T_t, n)$, where $n = RC_t$. Nevertheless, we simply use $\delta R_t = RL(NAF(\delta T_t), n)$ and determine $\delta Q_{t+1} = \delta Q_t + \delta R_t$ to extend our partial differential path forward with step t.

Another approach to determine δR_t uses the fact that any BSDR (k_i) of δT_t determines δR_t:

$$\delta R_t = \sum_{i=0}^{31} 2^{i+n \bmod 32}(T_t'[i] - T_t[i]) = \sum_{i=0}^{31} 2^{i+n \bmod 32} k_i$$

$$= 2^n \sum_{i=0}^{31-n} 2^i k_i + 2^{n-32} \sum_{i=32-n}^{31} 2^i k_i.$$

Different BSDRs (k_i) and (ℓ_i) of δT_t result in the same δR_t as long as

$$\sum_{i=0}^{31-n} 2^i k_i = \sum_{i=0}^{31-n} 2^i \ell_i \quad \text{and} \quad \sum_{i=32-n}^{31} 2^i k_i = \sum_{i=32-n}^{31} 2^i \ell_i.$$

In general, let $(\alpha, \beta) \in \mathbb{Z}^2$ be a partition of the integer δT_t with $\alpha + \beta = \delta T_t$ mod 2^{32}, $|\alpha| < 2^{32-n}$, $|\beta| < 2^{32}$ and $2^{32-n}|\beta$. For a BSDR (k_i) of δT_t we say that $(\alpha, \beta) \equiv (k_i)$ if $\alpha = \sum_{i=0}^{31-n} 2^i k_i$ and $\beta = \sum_{i=32-n}^{31} 2^i k_i$. The rotation of (α, β) is defined as $RL((\alpha, \beta), n) = 2^n \alpha + 2^{n-32} \beta \mod 2^{32}$.

Let $x = (\delta T_t \mod 2^{32-n})$ and $y = (\delta T_t - x \mod 2^{32})$, then $0 \le x < 2^{32-n}$ and $0 \le y < 2^{32}$. This gives rise to at most 4 partitions of δT_t:

1. $(\alpha, \beta) = (x, y)$;
2. $(\alpha, \beta) = (x, y - 2^{32})$,   if $y \ne 0$;
3. $(\alpha, \beta) = (x - 2^{32-n}, y + 2^{32-n} \mod 2^{32})$,   if $x \ne 0$;
4. $(\alpha, \beta) = (x - 2^{32-n}, (y + 2^{32-n} \mod 2^{32}) - 2^{32})$,   if $x \ne 0$ and $y + 2^{32-n} \ne 0$ mod 2^{32}.

The probability of each partition (α, β) equals

$$p_{(\alpha,\beta)} = \sum_{(k_i) \equiv (\alpha,\beta)} 2^{-\text{weight of } (k_i)}.$$

One then chooses any partition (α, β) for which $p_{(\alpha,\beta)} \ge \frac{1}{4}$ and determines δR_t as $RL((\alpha, \beta), n)$. In practice $NAF(\delta T)$ most often leads to the highest probability, which validates the simpler approach we used.

5.5.4 Extending Backward

Similar to extending forward, suppose we have a partial differential path consisting of at least bitconditions q_t and q_{t-1} and that the differences δQ_{t+1} and δQ_{t-2} are known. We want to extend this partial differential path backward with step t resulting in δQ_{t-3} and (additional) bitconditions $\mathsf{q}_t, \mathsf{q}_{t-1}, \mathsf{q}_{t-2}$. We assume that all indirect bitconditions are backward and do not involve bits of Q_{t-2}.

We choose a BSDR of δQ_{t-2} with weight at most 1 or 2 above the lowest weight, such as the NAF. We translate the chosen BSDR into bitconditions q_{t-2}.

For all $i = 0, 1, \ldots, 31$ we have by assumption valid bitconditions $(a, b, c) = (\mathsf{q}_t[i], \mathsf{q}_{t-1}[i], \mathsf{q}_{t-2}[i])$ where only b can be an indirect bitcondition. If so, it must involve $Q_{t-2}[i]$. Therefore $(a, b, c) \in \mathcal{F}_t$. If $|V_{abc}| = 1$ there is no ambiguity and we let $\{g_i\} = V_{abc}$. Otherwise, if $|V_{abc}| > 1$, then we choose any $g_i \in V_{abc}$ and we resolve the ambiguity left by bitconditions (a, b, c) by replacing them by $(d, e, f) = BC(t, abc, g_i)$, which results in boolean function difference g_i. Given all g_i, the value $\delta F_t = \sum_{i=0}^{31} 2^i g_i$ can be determined.

To rotate $\delta R_t = \delta Q_{t+1} - \delta Q_t$ over $n = 32 - RC_t$ bits, we simply use $\delta T_t = RL(NAF(\delta R_t), n)$. Or we may choose a partition (α, β) of δR_t with $p_{(\alpha, \beta)} \geq \frac{1}{4}$ and determine $\delta T_t = RL((\alpha, \beta), n)$. As in the 'forward' case, $NAF(\delta R_t)$ often leads to the highest probability. Finally, we determine $\delta Q_{t-3} = \delta T_t - \delta F_t - \delta W_t$ to extend our partial differential path backward with step t.

5.6   Constructing Full Differential Paths

Construction of a full differential path can be done as follows. Choose δQ_{-3} and bitconditions $\mathsf{q}_{-2}, \mathsf{q}_{-1}, \mathsf{q}_0$ and extend forward up to step 11. Also choose δQ_{64} and bitconditions $\mathsf{q}_{63}, \mathsf{q}_{62}, \mathsf{q}_{61}$ and extend backward down to step 16. This leads to bitconditions $\mathsf{q}_{-2}, \mathsf{q}_{-1}, \ldots, \mathsf{q}_{11}, \mathsf{q}_{14}, \mathsf{q}_{15}, \ldots, \mathsf{q}_{63}$ and differences $\delta Q_{-3}, \delta Q_{12}, \delta Q_{13}, \delta Q_{64}$. It remains to finish steps $t = 12, 13, 14, 15$. As with extending backward we can, for $t = 12, 13, 14, 15$, determine δR_t, choose the resulting δT_t after right rotation of δR_t over RC_t bits, and determine $\delta F_t = \delta T_t - \delta W_t - \delta Q_{t-3}$.

We aim to find new bitconditions $\mathsf{q}_{10}, \mathsf{q}_{11}, \ldots, \mathsf{q}_{15}$ that are compatible with the original bitconditions and that result in the required $\delta Q_{12}, \delta Q_{13}, \delta F_{12}, \delta F_{13}, \delta F_{14}, \delta F_{15}$, thereby completing the differential path. First we can test whether it is even possible to find such bitconditions.

For $i = 0, 1, \ldots, 32$, let $\mathcal{U}_i$ be a set of tuples $(q_1, q_2, f_1, f_2, f_3, f_4)$ of 32-bit integers with $q_j \equiv f_k \equiv 0 \mod 2^i$ for $j = 1, 2$ and $k = 1, 2, 3, 4$. We want to construct each $\mathcal{U}_i$ so that for each tuple $(q_1, q_2, f_1, f_2, f_3, f_4) \in \mathcal{U}_i$ there exist bitconditions $\mathsf{q}_{10}[\ell], \mathsf{q}_{11}[\ell], \ldots, \mathsf{q}_{15}[\ell]$, determining the $\Delta Q_{11+j}[\![\ell]\!]$ and $\Delta F_{11+k}[\![\ell]\!]$ below, over the bits $\ell = 0, \ldots, i - 1$, such that

$$\delta Q_{11+j} = q_j + \sum_{\ell=0}^{i-1} 2^\ell \Delta Q_{11+j}[\![\ell]\!], \quad j = 1, 2,$$

$$\delta F_{11+k} = f_k + \sum_{\ell=0}^{i-1} 2^\ell \Delta F_{11+k}[\![\ell]\!], \quad k = 1, 2, 3, 4.$$

This implies $\mathcal{U}_0 = \{(\delta Q_{12}, \delta Q_{13}, \delta F_{12}, \delta F_{13}, \delta F_{14}, \delta F_{15})\}$. The other $\mathcal{U}_i$ are constructed inductively by Algorithm 1. Furthermore, $|\mathcal{U}_i| \leq 2^6$, since for each q_j, f_k there are at most 2 possible values that can satisfy the above relations.

If we find $\mathcal{U}_{32} \neq \emptyset$ then there exists a path $u_0, u_1, \ldots, u_{32}$ with $u_i \in \mathcal{U}_i$ where each u_{i+1} is generated by u_i in Algorithm 1. Now the desired new bitconditions

---

Algorithm 1. Construction of $\mathcal{U}_{i+1}$ from $\mathcal{U}_i$.

Suppose $\mathcal{U}_i$ is constructed as desired in Section 5.6.

Let $\mathcal{U}_{i+1} = \emptyset$ and $(a, b, e, f) = (\mathsf{q}_{15}[i], \mathsf{q}_{14}[i], \mathsf{q}_{11}[i], \mathsf{q}_{10}[i])$.

For each tuple $(q_1, q_2, f_1, f_2, f_3, f_4) \in \mathcal{U}_i$ do the following:

1. For each bitcondition $d = \mathsf{q}_{12}[i] \in \begin{cases} \{.\} & \text{if } q_1[i] = 0 \\ \{\text{-,+}\} & \text{if } q_1[i] = 1 \end{cases}$ do

2.   Let $q_1' = 0, -1, +1$ for resp. $d = .,-,+$

3.   For each different $f_1' \in \{-f_1[i], +f_1[i]\} \cap V_{def}$ do

4.   Let $(d', e', f') = FC(12, def, f_1')$

5.   For each bitcondition $c = \mathsf{q}_{13}[i] \in \begin{cases} \{.\} & \text{if } q_2[i] = 0 \\ \{\text{-,+}\} & \text{if } q_2[i] = 1 \end{cases}$ do

6.   Let $q_2' = 0, -1, +1$ for resp. $c = .,-,+$

7.   For each different $f_2' \in \{-f_2[i], +f_2[i]\} \cap V_{cd'e'}$ do

8.   Let $(c', d'', e'') = FC(13, cd'e', f_2')$

9.   For each different $f_3' \in \{-f_3[i], +f_3[i]\} \cap V_{bc'd''}$ do

10.   Let $(b', c'', d''') = FC(14, bc'd'', f_3')$

11.   For each different $f_4' \in \{-f_4[i], +f_4[i]\} \cap V_{ab'c''}$ do

12.   Let $(a', b'', c''') = FC(15, ab'c'', f_4')$

13.   Insert $(q_1 - 2^i q_1', q_2 - 2^i q_2', f_1 - 2^i f_1', f_2 - 2^i f_2', f_3 - 2^i f_3', f_4 - 2^i f_4')$

into $\mathcal{U}_{i+1}$.

Keep only one of each tuple in $\mathcal{U}_{i+1}$ that occurs multiple times. By construction we find $\mathcal{U}_{i+1}$ as desired.

$(\mathsf{q}_{15}[i], \mathsf{q}_{14}[i], \ldots, \mathsf{q}_{10}[i])$ are $(a', b'', c''', d''', e'', f')$, which can be found at step 13 of Algorithm 1, where one starts with u_i and ends with u_{i+1}.

5.7 Implementation Details

Implementation of these techniques was done in C++ using the general purpose library Boost and the BOINC framework. BOINC is an open source distributed computing framework that allows volunteers on the Internet to join a project and donate cpu-time. Each project running a BOINC server automatically handles compute-client inputs and outputs specific to any number of applications, including output validation and re-assignment of jobs, if required. Volunteers, which can form teams, can monitor their own and others' progress, thus providing an inspiring competitive environment. Our BOINC project had a peak performance of approximately 400 GigaFLOPS.

To construct our chosen-prefix collision we used six applications:

1. One that generates birthday trails ending in a distinguished point [13];
2. One that collects birthday trails and computes collisions when found;
3. One that loads a set of partial lower differential paths and extends those forward with step t and saves only the paths with the fewest bitconditions;
4. One that loads a set of partial upper differential paths and extends those backward with step t and saves only the paths with the fewest bitconditions;

5. One that loads sets of lower and upper differential paths and tries to connect each combination;
6. One that searches for near-collision blocks that satisfy a given full differential path.

While extending a partial differential path we exhaustively try all BSDRs of δQ_t with weight at most 2 above the lowest weight, and all possible δF_t and all high-probability rotations. We keep only the N paths with the fewest bitconditions, for some preset value of N. Also we keep only those paths that have a preset minimum total tunnel strength over the Q_4, Q_5, Q_9, Q_{10}-tunnels [9]. With the exception of the 2nd, all applications can be fully parallelized. For the 1st and 5th application, which were by far the most cpu-time consuming, we used BOINC; the others were run on a cluster.

6   Concluding Remark

We have presented an automated way to find differential paths for MD5, have shown how to use them to construct chosen-prefix collisions, and have constructed two X.509 certificates with different name fields but idential signatures. Our construction required substantial cpu-time, but chosen-prefix collisions can be constructed much faster by using a milder birthday condition (namely, just $\delta a = 0$ and $\delta c = \delta d$) and allowing more near-collision blocks (about 14). See [16] for details.

Acknowledgements

This work benefited greatly from suggestions by Xiaoyun Wang. We are grateful for comments and assistance received from the Eurocrypt 2007 reviewers, Bart Asjes, Stuart Haber, Paul Hoffman, Pascal Junod, Vlastimil Klima, Bart Preneel, NBV, Gido Schmitz, Eric Verheul, and Yiqun Lisa Yin. Finally, we thank hundreds of BOINC enthousiasts all over the world, most unknown to us, who donated an impressive amount of cycles to the HashClash project running with BOINC software.

References

1. Christophe de Cannière and Christian Rechberger, *Finding SHA-1 Characteristics: General results and applications*, AsiaCrypt 2006, Springer LNCS 4284 (2006), 1–20.
2. M. Daum and S. Lucks, *Attacking Hash Functions by Poisoned Messages, "The Story of Alice and her Boss"*, June 2005, www.cits.rub.de/MD5Collisions/.
3. P. Gauravaram, A. McCullagh and E. Dawson, *Collision Attacks on MD5 and SHA-1: Is this the "Sword of Damocles" for Electronic Commerce?*, AusSCERT 2006 R&D Stream, May 2006, www.isi.qut.edu.au/people/subramap/AusCert-6.pdf.

4. M. Gebhardt, G. Illies and W. Schindler, *A Note on Practical Value of Single Hash Collisions for Special File Formats*, NIST First Cryptographic Hash Workshop, October/November 2005, `csrc.nist.gov/pki/HashWorkshop/2005/Oct31%5FPresentations/Illies%5FNIST%5F05.pdf` .

5. R. Housley, W. Polk, W. Ford and D. Solo, *Internet X.509 Public Key Infrastructure Certificate and Certificate Revocation List (CRL) Profile*, IETF RFC 3280, April 2002, `www.ietf.org/rfc/rfc3280.txt`.

6. Philip Hawkes, Michael Paddon and Gregory G. Rose, *Musings on the Wang et al. MD5 Collision*, Cryptology ePrint Archive, Report 2004/264, `eprint.iacr.org/2004/264`.

7. P. Hoffman and B. Schneier, *Attacks on Cryptographic Hashes in Internet Protocols*, IETF RFC 4270, November 2005, `www.ietf.org/rfc/rfc4270.txt`.

8. D. Kaminsky, *MD5 to be considered harmful someday*, December 2004, `www.doxpara.com/md5%5Fsomeday.pdf`.

9. Vlastimil Klima, *Tunnels in Hash Functions: MD5 Collisions Within a Minute*, Cryptology ePrint Archive, Report 2006/105, `eprint.iacr.org/2006/105`.

10. A.K. Lenstra, X. Wang and B.M.M. de Weger, *Colliding X.509 certificates*, Cryptology ePrint Archive, Report 2005/067, `eprint.iacr.org/2005/067`. An updated version has been published as an appendix to [11].

11. A.K. Lenstra and B.M.M. de Weger, *On the possibility of constructing meaningful hash collisions for public keys*, ACISP 2005, Springer LNCS 3574 (2005), 267–279.

12. O. Mikle, *Practical Attacks on Digital Signatures Using MD5 Message Digest*, Cryptology ePrint Archive, Report 2004/356, `eprint.iacr.org/2004/356`.

13. Paul C. van Oorschot and Michael J. Wiener, *Parallel collision search with cryptanalytic applications*, Journal of Cryptology **12**(1), 1–28, 1999.

14. Marc Stevens, Arjen Lenstra and Benne de Weger, *Target Collisions for MD5 and Colliding X.509 Certificates for Different Identities*, Cryptology ePrint Archive, Report 2006/360, `eprint.iacr.org/2006/360`.

15. Marc Stevens, *Fast Collision Attack on MD5*, Cryptology ePrint Archive, Report 2006/104, `eprint.iacr.org/2006/104`.

16. Marc Stevens, TU Eindhoven MSc thesis, in preparation. See `www.win.tue.nl/hashclash/`.

17. X. Wang and H. Yu, *How to Break MD5 and Other Hash Functions*, EuroCrypt 2005, Springer LNCS 3494 (2005), 19–35.

Non-trivial Black-Box Combiners for Collision-Resistant Hash-Functions Don't Exist

Krzysztof Pietrzak[*]

CWI Amsterdam
pietrzak@cwi.nl

Abstract. A (k, ℓ)-robust combiner for collision-resistant hash-functions is a construction which from ℓ hash-functions constructs a hash-function which is collision-resistant if at least k of the components are collision-resistant. One trivially gets a (k, ℓ)-robust combiner by concatenating the output of any $\ell - k + 1$ of the components, unfortunately this is not very practical as the length of the output of the combiner is quite large. We show that this is unavoidable as no black-box (k, ℓ)-robust combiner whose output is significantly shorter than what can be achieved by concatenation exists. This answers a question of Boneh and Boyen (Crypto'06).

1 Introduction

A function $H : \{0, 1\}^* \to \{0, 1\}^v$ is a collision-resistant hash-function (CRHF), if no efficient algorithm can find two inputs $M \neq M'$ where $H(M) = H(M')$, such a pair (M, M') is called a collision for H.[1]

In the last few years we saw several attacks on popular CRHFs previously believed to be secure [18,19]. Although provably secure[2] hash-functions exist (see e.g. [3] and references therein), they are rather inefficient and rarely used in practice. As we do not know which of the CRHFs used today will stay secure, it is natural to investigate combiners for CRHFs. In its simplest form the problem is the following: given two hash-functions

$$H_1, H_2 : \{0, 1\}^* \to \{0, 1\}^v,$$

can we construct a new hash-function which is collision-resistant if either H_1 or H_2 is? The answer is that of course we can, just concatenate the outputs:

$$H(X) = H_1(X) \| H_2(X). \tag{1}$$

---

[*] Supported by DIAMANT, the Dutch national mathematics cluster for discrete interactive and algorithmic algebra and number theory. This work was partially done while the author was a postdoc at the Ecole Normale Supérieure, Paris.

[1] This definition is very informal as there are some issues which make it hard to have a definition for collision-resistant hash-functions which is theoretically and practically satisfying, see [15] for recent discussion on that topic.

[2] Provably secure means that finding a collision can be shown to be at least as hard as solving some concrete (usually number theoretic) problem.

M. Naor (Ed.): EUROCRYPT 2007, LNCS 4515, pp. 23–33, 2007.

As any collision M, M' for H is also a collision for H_1 *and* H_2, if *either* H_1 or H_2 is collision-resistant, so is H. Unfortunately the length of the output of H is the sum of the output lengths of H_1 and H_2, this makes the combiner quite unattractive for practical purposes.

1.1   The Boneh-Boyen and Our Result

Boneh and Boyen [2] ask whether one can combine CRHFs such that the output length is (significantly) less than what can be achieved by concatenation. They prove a first negative result in this direction, namely that there is no black-box construction for combining CRHFs in such a way that the output is shorter than what can be achieved by concatenation *under the additional assumption that this combiner queries each of the components exactly once.* They ask whether a similar impossibility result can be obtained in the general case where the combiner is allowed to query the components several times. We answer this question in the affirmative: any combiner for ℓ functions with range $\{0, 1\}^v$ must have output length at least $(v - O(\log(q)))\ell$ bits[3], where q is the number of oracle calls made by the combiner. Stated in asymptotic terms, if $q \in 2^{o(v)}$ is subexponential, then the output length is in $(1 - o(1))v\ell$, and if q is constant the output length is in $v\ell - O(1)$, this must be compared to $v\ell$ which is trivially achieved by concatenation.

(k, ℓ)-ROBUST COMBINER. In this paper we will consider the more general question whether secure and non-trivial (k, ℓ)-robust combiners for collision-resistant hash-functions exist. A (k, ℓ)-robust combiner is collision-resistant, if at least k (and not just one) of the components used are secure. We trivially get a (k, ℓ)-robust combiner by concatenating any $\ell - k + 1$ of the components,[4] which gives an output length of $v(\ell - k + 1)$. We show that this cannot be significantly improved as any (k, ℓ)-robust combiner must have output length at least $(v - O(\log(q)))(\ell - k + 1) - \ell$.

The main technical contribution of this paper is Lemma 2, which generalizes (and as a special case contains the statement of) Theorem 3 from [2]. Roughly, this lemma states that there exist hash-functions and a collision for any combiner with sufficiently short output, such that this collision does not trivially lead to collisions for all[5] of the hash-functions. The proof of this lemma follows from a simple application of the probabilistic method, and in particular is much simpler than the proof of Theorem 3 in [2].

AN INFORMATION THEORETIC ARGUMENT. There is a quite intuitive information theoretic argument why (k, ℓ)-robust combiners for CHRFs $\{0, 1\}^* \to \{0, 1\}^v$ whose output is significantly shorter than $v(\ell - k + 1)$ bits can't exist. We give this argument below, it will turn out that this simple approach gives an impossibility result which is much weaker than what we prove in this paper. This argument is shown only for motivational reasons and is not relevant for the

---

[3] In this paper all logarithms are to base 2.

[4] We'll look at this construction in more detail in the next section.

[5] Or for $\ell - k + 1$ of the hash-functions if we consider (k, ℓ)-robust combiners.

rest of the paper, the reader can skip the rest of this section if this does not seem to be of interest.

Basically, the argument uses the fact that one can encode a collision for any function with output length w using roughly w bits[6] and if the function is uniformly random, then w bits are also necessary. Now if a combiner with short output is instantiated with uniformly random functions, (the encoding of) a collision for the combiner will simply be too short to encode the information necessary to find collisions for the components. A bit more precisely, for $(1,2)$-robust combiners this argument goes as follows. Assume we are given a combiner C for two functions $\{0,1\}^* \to \{0,1\}^v$ whose output length is $2v - t$ (i.e. t bits less than concatenation). Now we simply sample two uniformly random functions $H_1, H_2 : \{0,1\}^* \to \{0,1\}^v$ and output a collision M, M' for C^{H_1,H_2}, such a collision can be encoded using $2v - t$ bits. To encode a collision for a random function $\{0,1\}^* \to \{0,1\}^v$, v bits are necessary and sufficient. Thus given the collision for the combiner, we still lack about $t/2$ bits of information (i.e. we have that much min-entropy) about a collision for one of the H_i's, and would have to make about $2^{t/2}$ more queries to this H_i in order to find a collision. This argument only rules out very strong combiners, where from any collision on the combiner we expect to get a collision for both components very efficiently. For example it does not rule out the possibility of $(1,2)$-robust combiners with range $3v/2$ (which we can consider significantly less than $2v$), where each collision for the combiner gives collisions for both components if we are ready to invest an additional $O(2^{v/4})$ queries. Such a combiner would still be sufficient if we are willing to assume that at least one of the components we combine has security (slightly more than) $2^{v/4}$. This assumption is very mild, as usually v is something like 160 or 256, such that the birthday bound $2^{v/2}$ is infeasible, but if a collision can be found after $2^{v/4}$ queries, the CRHF would be considered completely broken. More generally, the above argument does not rule out $(1,2)$-robust combiners with output length $2v - t$ for a t where $2^{t/2}$ queries are considered feasible (for an attacker). In contrast, the theorem proven in this paper rules out $(1,2)$-robust combiners with output length $2v - t$, unless the combiner itself makes $2^{t/2}$ invocations to the components.

---

[6] The following is a possible encoding. To define the encoding choose values $X_1, X_2, \ldots$ in $\{0,1\}^{w+1}$ uniformly at random. Now given a function $f : \{0,1\}^* \to \{0,1\}^w$ (which can be chosen adversarialy, but independent of the X_i's) let i be minimal such that $f(X_i)$ has at least 2 preimages in $\{0,1\}^{w+1}$ and output any $X \in \{0,1\}^{w+1}$ where $X \neq X_i$ and $f(X) = f(X_i)$. The expectation of i is at most 2 (as the probability that $f(Z)$ has only one preimage in $\{0,1\}^{w+1}$ for a random Z is at most $1/2$). Thus given X, we must make an expected number of at most 3 queries to f to find a collision (i.e. first compute $f(X)$ and then try $f(X_1), f(X_2), \ldots$ until $f(X) = f(X_i)$). If we only have $w + 1 - c$ (not $w + 1$) bits for the encoding, we can simply omit the last c bits in the encoding just described, and when decoding trying all 2^c possibilities for this bits, thus we need an expected number of $2 + 2^c$ evaluations of f to find a collision given $w + 1 - c$ bits of X, which is better than no information at all if c is less than $w/2$.

1.2   Related Work

COMBINERS. The idea of combining two or more cryptographic components in order to get a system which is secure whenever at least one of the underlying primitives is secure is quite old.[7] The early results are on symmetric encryption schemes [1,6,11]. Combiners for asymmetric primitives were constructed by Dodis and Katz [5] (for CCA secure encryption schemes) and Harnik et al. [7] (for key-agreement). The general notion of a combiner was put forward by Herzberg [8] who calls them "tolerant combiners". In recent works one often calls them "robust combiners", a term introduced in [7]. Combiners have been generalized in several ways:

(k, ℓ)-ROBUST COMBINERS: [7] put forward the notion of (k, ℓ)-robust combiners as discussed in the last section. Such combiners are only guaranteed to be secure if at least k (and not just one) of the ℓ components used is secure. Interestingly, for natural primitives as statistically hiding commitments [8] and oblivious transfer [7,13] only 2-3 but no 1-2 combiners are known.

CROSS-PRIMITIVE COMBINERS: In a cross-primitive combiner the combined primitive is different from the components used, one can think of this as simultaneously being a reduction and a combiner. This notion was introduced by Meier and Przydatek [12] who construct a 1-2 private information retrieval to oblivious transfer cross-primitive combiner, which is interesting as normal 1-2 combiners for oblivious transfer might not exist [7].

EFFICIENCY AND OTHER PARAMETERS: In practice the mere existence of a combiner is not enough, as the parameters of a combiner are important. Efficiency is always of concern, fortunately for most primitives where combiners are known to exist, also efficient realizations are known [7,8], with bit-commitments being a notable exception [8] to that rule. Besides efficiency, for different primitives also other parameters are important, in particular this paper is about the output-length of combiners for CRHFs.

COLLISION RESISTANCE. collision-resistant hash-functions are very important and subtle [15] cryptographic primitives which have attracted a lot of research, even more in the recent years as widely used (presumably) collision-resistant hash-functions as MD5 or SHA-1 have been broken [18,19]. Here we only mention some of the generic results on CRHFs.

Simon shows that collision-resistant hash-functions cannot be constructed from one-way functions via a black-box reduction [17]. On the positive side, Naor and Yung [14] show that for some applications (in particular for signature schemes) collision resistance is not necessary, as universal one-way hash-functions are enough. Those can be constructed from one-way functions [10,16].

Merkle and Damgård show that by iterating a CRHF with fixed input length, one gets a CRHF for inputs of arbitrary length. Most CRHFs used today follow

---

[7] We also see many combiners in the physical world, for example one often has several *different* locks on a door. This does not to simply increase the time a burglar needs to break the k locks by a factor of k, but there's hope that some particular lock might turn out to be much harder to come by than the others.

this approach. Coron et al. [4] show that the Merkle-Damgård construction does not give a random function if instantiated with a random function (which was not the design goal of this construction), but that this can be achieved with some small modifications. Joux [9] shows that for iterated hash-functions (like the Merkle-Damgård construction) finding many values which hash to the same value is not much harder than finding an ordinary collision. As a consequence concatenating the output of such hash-functions does not increase the security: let H_1, H_2 be iterated hash-functions with v bits output, then one can find a collision for $H(X) = H_1(X) \| H_2(X)$ in time $O(v2^{v/2})$.

2   Combiners for CRHFs

Informally, a (k, ℓ)-robust combiner for CRHFs is a construction (modeled as an oracle circuit C) which, if instantiated with any ℓ hash-functions $H_1, \ldots, H_\ell :$ $\{0, 1\}^* \to \{0, 1\}^v$, is collision-resistant if at least k of the H_i's are. In order to show that a construction is a (k, ℓ)-robust combiner, one must provide an efficient procedure P which given two colliding inputs for the combiner, finds collisions for at least $\ell - k + 1$ of the underlying H_i's. In this paper we only consider black-box combiners as defined in [7], this means that C and P are only given oracle access to the H_i's.

The following definition of a (k, ℓ)-robust combiner is a generalization of the definition given in [2], where only the case $k = 1$ was considered.

Definition 1. *A combiner for ℓ collision-resistant hash-functions* $\{0, 1\}^* \to \{0, 1\}^v$ *is a pair (C, P) where C is an oracle circuit and P is an oracle probabilistic polynomial-time Turing machine (PPTM)*[8]

$$C : \{0, 1\}^m \to \{0, 1\}^n \qquad P : \{0, 1\}^{2m} \to \{0, 1\}^*.$$

There are ℓ types of oracle gates (tapes) in C (P). With $B^{H_1, \ldots, H_\ell}(X)$ (where B is C or P) we denote the output of B on input X when the ℓ types of oracle gates are instantiated with functions $H_1, \ldots, H_\ell : \{0, 1\}^* \to \{0, 1\}^v$ respectively.

We say that P k-succeeds on $M, M' \in \{0, 1\}^*$ and oracles $H_1, \ldots, H_\ell$ if its output contains collisions for all but at most $k - 1$ of the H_i's, i.e. for

$$P^{H_1, \ldots, H_\ell}(M, M') \to (U_1, \ldots, U_\ell, U'_1, \ldots, U'_\ell)$$

we have

$$\exists J \subseteq \{1, \ldots, \ell\}, |J| \geq \ell - k + 1 : (U_i, U'_i) \text{ is a collision for } H_i.$$

*Let $Adv_P^k[(H_1, \ldots, H_\ell), (M, M')]$ denote the probability (over P's coin tosses) that $P^{H_1, \ldots, H_\ell}(M, M')$ k-succeeds. Then (C, P) is an ϵ-secure (k, ℓ)-**combiner**, if for all (compatible) $H_1, \ldots, H_\ell$ and all collisions (M, M') on $C^{H_1, \ldots, H_\ell}$ we have*

$$Adv_P^k[(H_1, \ldots, H_\ell), (M, M')] > 1 - \epsilon.$$

We say that (C, P) is an (k, ℓ)-robust combiner if it is ϵ-secure for a small ϵ.[9]

---

[8] The only reason P is defined as a Turing machine and not as a circuit is that we don't want to put an a priori bound on the output length of P.

[9] Here "small" usually means negligible in some security parameter.

For example consider the following (k, ℓ)-robust combiner (C, P)

$$C^{H_1,\ldots,H_\ell}(M) \to H_1(M)\|\ldots\|H_{\ell-k+1}(M)$$

$$P^{H_1,\ldots,H_\ell}(M, M') \to (M,\ldots,M),(M',\ldots,M')$$

As any collision M, M' for $C^{H_1,\ldots,H_\ell}$ is a collision for H_i for $i = 1,\ldots,\ell-k+1$,

$$Adv_P^k[(H_1,\ldots,H_\ell),(M, M')] = 1.$$

So (C, P) can be considered a secure (k, ℓ)-robust combiner, as from any collision on $C^{H_1,\ldots,H_\ell}$ we get from P collisions for all but $k-1$ of the H_i's, thus if k of the H_i's are secure, also $C^{H_1,\ldots,H_\ell}$ must be secure. The output length of C is $n = v(\ell - t + 1)$, by the following theorem this cannot be significantly improved.

Theorem 1. *Let (C, P) be a (k, ℓ)-robust combiner, where $C : \{0,1\}^m \to \{0,1\}^n$ has q_C oracle gates and P makes at most q_P oracle calls. Suppose that*

$$n < (v - 2\log(2q_C))(\ell - k + 1) - \ell - 1 \quad \text{and} \quad m > n. \tag{2}$$

Then there exist $M, M' \in \{0,1\}^m$ and functions $\hat{H}_i : \{0,1\}^* \to \{0,1\}^v$ for $i = 1,\ldots,\ell$ relative to which

$$Adv_P^k[(\hat{H}_1,\ldots,\hat{H}_\ell),(M, M')] \leq \frac{(q_P + q_C)^2 + k}{2^v}. \tag{3}$$

For the special case where $k = 1$ and C queries each $\hat{H}_i$ exactly once (which are the constructions considered in [2]) the bound on n can be improved to

$$n < v\ell - 1 \quad \text{and} \quad m > n$$

or

$$n < v\ell \quad \text{and} \quad m - 1 > n.$$

The last statement slightly improves on the main result from [2] where a stronger $n < m - \log \ell$ bound was needed in order to get $n < v\ell$. As we can find a collision for any function with range $\{0,1\}^v$ in $2^{v/2}$ steps, in order to reason about CRHFs with range $\{0,1\}^v$ the value $2^{v/2}$ must be unfeasibly large. In particular for any reasonable combiner $q_P + q_C \ll 2^{v/2}$ and thus the advantage (3) will be very small.

Let us remark that in [2] the ranges of the H_i's were allowed to be different, for the sake of exposition we drop this generalization, but it is straight forward to adapt (the proof of) Theorem 1 to this more general case. Note that when the H_i's have different output lengths, say H_i has length v_i where $v_1 \leq v_2 \leq v_3 \leq \ldots \leq v_\ell$, then we can construct a (k, ℓ)-robust combiner by concatenating the outputs of $H_1,\ldots,H_{\ell-k+1}$ (i.e. the H_i's with the shortest outputs), which will give a combiner with output length $\sum_{i=1}^{\ell-k+1} v_i$. Again, this is basically best possible, as for this setting Theorem 1 holds by generalizing equation (2) to

$$n < \sum_{i=1}^{\ell-k+1} (v_i - 2\log(2q_C)) - \ell - 1 \quad \text{and} \quad m > n.$$

and replacing v with v_1 in (3).

Following [2], to prove Theorem 1 it is sufficient to prove that hash-functions $H_1, \ldots, H_\ell$ and a collision M, M' exists where in the computation of $C^{H_1, \ldots, H_\ell}$ on inputs M and M' at least k of the H_i's are not queried on two distinct inputs X, X' where $H_i(X) = H_i(X')$. Note that this means that one does not trivially get a collision for those H_i's when learning M, M'. Let $J \subseteq \{1, \ldots, \ell\}, |J| = k$ be the indices of these k H_i's. We prove the existence of such H_i's and M, M' in Lemma 2 below. Then, from such $H_1, \ldots, H_\ell$ and M, M' we can get the $\hat{H}_1, \ldots, \hat{H}_\ell$ as required by Theorem 1, by setting $\hat{H}_i(X) = H_i(X)$ for all inputs X which appear as input to H_i in the computation of $C^{H_1, \ldots, H_\ell}(M)$ or $C^{H_1, \ldots, H_\ell}(M')$, and $\hat{H}_i(X)$ is assigned a random value otherwise. Clearly M, M' is also a collision for $C^{\hat{H}_1, \ldots, \hat{H}_\ell}$, moreover all $\hat{H}_i$ where $i \in J$ are "very" collision-resistant, as we just randomly defined their outputs, except on a subset of inputs which itself does not contain a collision, Lemma 1 below is a formal statement of this intuitive argument.

Proof (of Theorem 1). The theorem follows from Lemmata 1 and 2.

In the lemmata below[10] let

- $\mathbf{W}_i(X)$ be the set of oracle queries to H_i made while evaluating $C^{H_1 \cdots H_\ell}(X)$.
- $\mathbf{V}_i(X) = \{H_i(W) : W \in \mathbf{W}_i(X)\}$ be the set of corresponding outputs (taken without repetition).

Lemma 1. *Let (C, P) be a (k, ℓ)-robust combiner, where C has q_C oracle gates and P makes at most q_P oracle calls. Assume there exist oracles $H_i : \{0, 1\}^* \rightarrow \{0, 1\}^v, i = 1, \ldots, \ell$ and messages M, M' such that*

- $M \neq M'$ *and* $C^{H_1, \ldots, H_\ell}(M) = C^{H_1, \ldots, H_\ell}(M')$.
- $|\mathbf{V}_j(M) \cup \mathbf{V}_j(M')| = |\mathbf{W}_j(M) \cup \mathbf{W}_j(M')|$ *for at least k different $j \in \{1, \ldots, \ell\}$.*

Then there exist deterministic $\hat{H}_i : \{0, 1\}^* \rightarrow \{0, 1\}^v, i = 1, \ldots, \ell$ relative to which

$$Adv_P^k[(\hat{H}_1, \ldots, \hat{H}_\ell), (M, M')] \leq \frac{(q_P + q_C)^2 + k}{2^v}.$$

Proof. Let $J \subseteq \{1, \ldots, \ell\}, |J| = k$ be the indices of the k hash-functions for which no collision occurs during the computation of $C^{H_1, \ldots, H_\ell}$ on input M and M', i.e.

$$\forall j \in J : |\mathbf{V}_j(M) \cup \mathbf{V}_j(M')| = |\mathbf{W}_j(M) \cup \mathbf{W}_j(M')|.$$

For $i \notin J$ we let $\hat{H}_i := H_i$, and for each $i \in J$ let $R_i : \{0, 1\}^* \rightarrow \{0, 1\}^v$ be uniformly random and

$$\hat{H}_i(W) := \begin{cases} H_i(W) \text{ if } W \in \mathbf{W}_i(M) \cup \mathbf{W}_i(M') \\ R_i(W) \text{ otherwise} \end{cases}$$

Note that $C^{\hat{H}_1, \ldots, \hat{H}_\ell}(M) = C^{\hat{H}_1, \ldots, \hat{H}_\ell}(M')$ as for each i, $H_i(W) = \hat{H}_i(W)$ for inputs $W \in \mathbf{W}_i(M) \cup \mathbf{W}_i(M')$ which come up on the computation of $C^{H_1, \ldots, H_\ell}$

---

[10] Our Lemma 1 is basically Theorem 2 from [2], the only difference is that we consider (k, ℓ)-robust combiners whereas [2] were only interested in the case $k = 1$.

on inputs M, M', let $\mathbf{Q}$ denote all those inputs together with the corresponding outputs.

$$\mathbf{Q} = \bigcup_{i=1}^{\ell} \{\mathbf{V}_i(M), \mathbf{W}_i(M), \mathbf{V}_i(M'), \mathbf{W}_i(M')\}$$

Let P' be the oracle PPTM which makes at most q_P oracle calls and maximizes the probability α defined below.

$$\alpha = \Pr_{P'^{\hat{H}_1, \ldots, \hat{H}_\ell}(\mathbf{Q}) \rightarrow \{U_1, \ldots, U_\ell, U'_1, \ldots, U'_\ell\}]} [\exists i \in J : U_i \neq U'_i \wedge \hat{H}_i(U_i) = \hat{H}_i(U'_i)] \quad (4)$$

α is an upper bound on $Adv_P^k[(\hat{H}_1, \ldots, \hat{H}_\ell), (M, M')]$, as one possible strategy for P' is to first compute M, M', which given $\mathbf{Q}$ can be done without access to the $\hat{H}_i$ oracles, and then simulate $P^{\hat{H}_1, \ldots, \hat{H}_\ell}(M, M')$ and output the output of this simulation.[11] To save on notation let P^* denote $P'^{\hat{H}_1, \ldots, \hat{H}_\ell}(\mathbf{Q})$. We say that P^* found a collision if for some[12] $\hat{H}_i, i \in J$ it makes an oracle query $\hat{H}_i(X)$ where either for a previous query $X' \neq X$ to $\hat{H}_i$ we have $\hat{H}_i(X) = \hat{H}_i(X')$ or $\hat{H}_i(X) \in \mathbf{V}_i(M) \cup \mathbf{V}_i(M')$ and $X \notin \mathbf{W}_i(M) \cup \mathbf{W}_i(M')$. For $i = 1, \ldots, q_P$ let $\mathcal{C}_i$ denote the event that P^* found a collision after the i'th oracle query is made. If the i'th oracle query is to a $\hat{H}_j$ where $j \notin J$ or a query which has already been made we cannot get a collision, so

$$\Pr[\mathcal{C}_i | \neg \mathcal{C}_{i-1}] = 0.$$

So assume that the i'th oracle query is a new query X to a $\hat{H}_j$ where $j \in J$. Then $\hat{H}_i(X) = R_i(X)$ is uniformly random and independent of any previous outputs, thus the probability that it will collide with any of the $\leq i$ previous queries to $\hat{H}_i$ or with one the $\leq 2q_C$ values in $\mathbf{V}_i(M) \cup \mathbf{V}_i(M')$ is at most $(2q_C + i)/2^v$, we get

$$\Pr[\mathcal{C}_{q_P}] = \sum_{i=1}^{q_P} \Pr[\mathcal{C}_i | \mathcal{C}_{i-1}] \leq \sum_{i=1}^{q_P} \frac{2q_C + i}{2^v} \leq \frac{q_P(2q_C + q_P)}{2^v} \leq \frac{(q_P + q_C)^2}{2^v}.$$

Even if $\neg \mathcal{C}_{q_P}$, i.e. P^* does not find a collision for some $\hat{H}_i, i \in J$, there still is a tiny chance that P^* guesses U_i, U'_i where $\hat{H}_i(U_i) = \hat{H}_i(U'_i)$ for some of the $i \in J$. The probability of this is at most $|J|/2^v \leq k/2^v$. Taking everything together:

$$Adv_P^k[(\hat{H}_1, \ldots, \hat{H}_\ell), (M, M')] \leq \alpha \leq \Pr[\mathcal{C}_{q_P}] + k/2^v \leq \frac{(q_P + q_C)^2 + k}{2^v}. \quad (5)$$

We're almost done, except that in the above inequality, the $\hat{H}_i$'s are not deterministic as required by the lemma, but randomized (as the R_i's were chosen at

---

[11] The reason we give away the full $\mathbf{Q}$ is that that M, M' will usually leak some information on $\mathbf{Q}$, and the simplest way to deal with this leakage is to simply assume that P' knows all those values.

[12] Note that we don't care about collision for $\hat{H}_i, i \notin J$ as $\mathbf{Q}$ contains collisions for those $\hat{H}_i$'s.

random). We can get fixed $\hat{H}_i$'s for which (5) holds by choosing the R_i's so they minimize the left hand side of (5). □

Lemma 2. *Let $C : \{0,1\}^m \to \{0,1\}^n$ be as in the previous lemma. Then whenever*

$$n < (v - 2\log(2q_C))(\ell - k + 1) - \ell - 1 \quad and \quad m > n$$

there exist functions $H_1, \ldots, H_\ell$ and messages M, M' such that

- $M \neq M'$ and $C^{H_1,\ldots,H_\ell}(M) = C^{H_1,\ldots,H_\ell}(M')$.
- $|\mathbf{V}_j(M) \cup \mathbf{V}_j(M')| = |\mathbf{W}_j(M) \cup \mathbf{W}_j(M')|$ for at least k different $j \in \{1, \ldots, \ell\}$.

For the special case where $k = 1$ and C queries each H_i exactly once (which are the constructions considered in [2]) the bounds on n can be improved to

$$n < v\ell - 1 \quad and \quad m > n$$

or

$$n < v\ell \quad and \quad m - 1 > n.$$

Proof. Consider the following random experiment. First we sample ℓ functions $H_i : \{0,1\}^* \to \{0,1\}^v$ uniformly at random.[13] Then $M, M' \in \{0,1\}^m$ are sampled uniformly at random. We define the events $\mathcal{E}_1$ and $\mathcal{E}_2$ as

$$\mathcal{E}_1 \iff M \neq M' \text{ and } C^{H_1,\ldots,H_\ell}(M) = C^{H_1,\ldots,H_\ell}(M')$$
$$\mathcal{E}_2 \iff \exists J \subseteq \{1, \ldots, \ell\}, |J| > \ell - k$$
$$\text{where } \forall j \in J : |\mathbf{V}_j(M) \cup \mathbf{V}_j(M')| \neq |\mathbf{W}_j(M) \cup \mathbf{W}_j(M')|$$

We will show that $\Pr[\mathcal{E}_1] > \Pr[\mathcal{E}_2]$, which then implies $\Pr[\mathcal{E}_1 \wedge \neg\mathcal{E}_2] > 0$. This will prove the lemma as it shows that random $H_1, \ldots, H_\ell$ and M, M' have the property as claimed by the lemma with non-zero probability, and thus $H_1, \ldots, H_\ell$ and M, M' with this property exist.

As $\Pr[M = M'] = 2^{-m}$, $\Pr[C^{H_1,\ldots,H_\ell}(M) = C^{H_1,\ldots,H_\ell}(M')] \geq 2^{-n}$ and $m > n$ we get

$$\Pr[\mathcal{E}_1] \geq 2^{-n} - 2^{-m} \geq 2^{-n-1}. \tag{6}$$

Let q_i denote the number of H_i oracle gates in C, note that $\sum_{i=1}^{\ell} q_i = q_C$. We can upper bound $\Pr[\mathcal{E}_2]$ by the probability that the best oracle algorithm $A^{H_1,\ldots,H_\ell}$ which can query the i'th oracle H_i at most $2q_i$ times finds a collision for at least $\ell - k + 1$ of the H_i's.[14] As the H_i's are all independent random functions, the best A can do is to query it i'th oracle on $2q_i$ distinct inputs (which ones is

---

[13] One can't simply sample a H_i as this would need infinite randomness, but one can use lazy sampling here, this means that $H_i(X)$ is only assigned a (random) value when H_i is actually invoked on input X.

[14] This is an upper bound as one possible strategy for $A^{H_1,\ldots,H_\ell}$ is to simply evaluate $C^{H_1,\ldots,H_\ell}$ on two random inputs M, M' to get success probability exactly $\Pr[\mathcal{E}_2]$.

irrelevant), by the birthday bound[15] the probability of finding a collision for any H_i is at most $2q_i(2q_i - 1)/2^{v+1}$, now

$$\Pr[\mathcal{E}_2] \leq \Pr[A^{H_1,\ldots,H_\ell} \text{ finds } \ell - k + 1 \text{ collisions }]$$

$$\leq \sum_{\substack{J \subseteq \{1,\ldots,\ell\} \\ |J| = \ell - k + 1}} \Pr[\forall i \in J : A^{H_1,\ldots,H_\ell} \text{ finds a collision for } H_i]$$

$$\leq \sum_{\substack{J \subseteq \{1,\ldots,\ell\} \\ |J| = \ell - k + 1}} \prod_{i \in J} \frac{2q_i(2q_i - 1)}{2^{v+1}}$$

$$< \sum_{\substack{J \subseteq \{1,\ldots,\ell\} \\ |J| = \ell - k + 1}} \frac{(2q_C^2)^{\ell - k + 1}}{2^{v(\ell - k + 1)}} \leq \binom{\ell - k + 1}{\ell} \frac{(2q_C^2)^{\ell - k + 1}}{2^{v(\ell - k + 1)}} < \frac{2^\ell (2q_C^2)^{\ell - k + 1}}{2^{v(\ell - k + 1)}}.$$

From the above equation, (6) and $n < (v - 2\log(2q_C))(\ell - k + 1) - \ell - 1$ we now get $\log(\Pr[\mathcal{E}_1]) > \log(\Pr[\mathcal{E}_2])$, and thus $\Pr[\mathcal{E}_1] > \Pr[\mathcal{E}_2]$, as

$$\log(\Pr[\mathcal{E}_1]) \geq \log(2^{-n-1}) = -n - 1 > -(v - 2\log(2q_C))(\ell - k + 1) + \ell$$

and

$$\log(\Pr[\mathcal{E}_2]) < \log\left(\frac{2^\ell (2q_C^2)^{\ell - k + 1}}{2^{v(\ell - k + 1)}}\right) = -(v - 2\log(2q_C))(\ell - k + 1) + \ell$$

Our estimate on $\Pr[\mathcal{E}_2]$ has some slack as to keep the expression simple. For the special case $k = 1$ and $q_i = 1, i = 1, \ldots, \ell$ which covers the constructions considered in [2] we get

$$\Pr[\mathcal{E}_2] \leq \prod_{i \in \{1,\ldots,\ell\}} \frac{2q_i(2q_i - 1)}{2^{v+1}} = 2^{-v\ell}$$

which satisfies $\Pr[\mathcal{E}_1] > \Pr[\mathcal{E}_2]$ already for $n < v\ell - 1$. If we additionally assume that $n < m - 1$ (not just $n < m$) then we can strengthen (6) to $\Pr[\mathcal{E}_1] > 2^{-n-1}$ and $\Pr[\mathcal{E}_1] > \Pr[\mathcal{E}_2]$ holds for the optimal $n < v\ell$. $\qquad\square$

References

1. C. A. Asmuth and G. R. Blakley. An efficient algorithm for constructing a cryptosystem which is harder to break than two other cryptosystems. *Computers and Mathematics with Applications*, pages 447–450, 1981.
2. Dan Boneh and Xavier Boyen. On the impossibility of efficiently combining collision resistant hash functions. In *CRYPTO*, 2006.
3. Scott Contini, Arjen K. Lenstra, and Ron Steinfeld. Vsh, an efficient and provable collision-resistant hash function. In *EUROCRYPT*, pages 165–182, 2006.

---

[15] This bound states that when randomly throwing q balls into N buckets, some bucket will contain more than one element with probability at most $q(q - 1)/2N$.

4. Jean-Sébastien Coron, Yevgeniy Dodis, Cécile Malinaud, and Prashant Puniya. Merkle-damgård revisited : How to construct a hash function. In *Advances in Cryptology — CRYPTO '05*, volume 3621 of *Lecture Notes in Computer Science*, pages 430–448, 2005.
5. Yevgeniy Dodis and Jonathan Katz. Chosen-ciphertext security of multiple encryption. In *TCC*, pages 188–209, 2005.
6. Shimon Even and Oded Goldreich. On the power of cascade ciphers. *ACM Trans. Comput. Syst.*, 3(2):108–116, 1985.
7. Danny Harnik, Joe Kilian, Moni Naor, Omer Reingold, and Alon Rosen. On robust combiners for oblivious transfer and other primitives. In *EUROCRYPT*, pages 96–113, 2005.
8. Amir Herzberg. On tolerant cryptographic constructions. In *CT-RSA*, pages 172–190, 2005.
9. Antoine Joux. Multicollisions in iterated hash functions. application to cascaded constructions. In *CRYPTO*, pages 306–316, 2004.
10. Jonathan Katz and Chiu-Yuen Koo. On constructing universal one-way hash functions from arbitrary one-way functions, 2005. Cryptology ePrint Archive: Report 2005/328.
11. Ueli M. Maurer and James L. Massey. Cascade ciphers: The importance of being first. *J. Cryptology*, 6(1):55–61, 1993.
12. Remo Meier and Bartosz Przydatek. On robust combiners for private information retrieval and other primitives. In Cynthia Dwork, editor, *CRYPTO '06*, volume 4117 of *Lecture Notes in Computer Science*, pages 555–569, 2006.
13. Remo Meier, Bartosz Przydatek, and Jürg Wullschleger. Robuster combiners for oblivious transfer. In *TCC 2007*, volume 4392 of *Lecture Notes in Computer Science*, pages 404–418, 2007.
14. Moni Naor and Moti Yung. Universal one-way hash functions and their cryptographic applications. In *STOC*, pages 33–43, 1989.
15. Phillip Rogaway. Formalizing human ignorance: Collision-resistant hashing without the keys, 2006. Cryptology ePrint Archive: Report 2006/281.
16. John Rompel. One-way functions are necessary and sufficient for secure signatures. In *STOC*, pages 387–394, 1990.
17. Daniel R. Simon. Finding collisions on a one-way street: Can secure hash functions be based on general assumptions? In *EUROCRYPT*, pages 334–345, 1998.
18. Xiaoyun Wang, Yiqun Lisa Yin, and Hongbo Yu. Finding collisions in the full sha-1. In *CRYPTO*, pages 17–36, 2005.
19. Xiaoyun Wang and Hongbo Yu. How to break md5 and other hash functions. In *EUROCRYPT*, pages 19–35, 2005.

The Collision Intractability of MDC-2 in the Ideal-Cipher Model

John P. Steinberger

Dept. of Mathematics, University of California, Davis, California 95616 USA
jpsteinb@math.ucdavis.edu

Abstract. We provide the first proof of security for MDC-2, the most well-known construction for turning an n-bit blockcipher into a $2n$-bit cryptographic hash function. Our result, which is in the ideal-cipher model, shows that MDC-2, when built from a blockcipher having block-length and keylength n, has security much better than that delivered by any hash function that has an n-bit output. When the blocklength and keylength are $n = 128$ bits, as with MDC-2 based on AES-128, an adversary that asks fewer than $2^{74.9}$ queries usually cannot find a collision.

Keywords: Collision-resistant hashing, cryptographic hash functions, ideal-cipher model, MDC-2.

1 Introduction

OVERVIEW. A double block length hash-function uses an n-bit blockcipher as the building block by which it maps (possibly long) strings to $2n$-bit ones. The classical double block length hash-function is MDC-2, illustrated in Figure 1. This nearly 20-year-old technique [5, 22] is specified in the ANSI X9.31 and ISO/IEC 10118-2 standards [1, 13], and it is implemented in popular libraries and toolkits, such as OpenSSL.

This paper gives the first proof of security for MDC-2. Our result establishes that when MDC-2 is based on an ideal blockcipher with keylength and block-length of n bits, the adversary must ask well over $2^{n/2}$ queries to find a collision. In particular, for $n = 128$, no adversry can find a collision with so much as a 50% chance if it asks fewer than $2^{74.9}$ forward-or-backward queries of a 128-bit blackbox-modeled blockcipher.

Getting a collision-resistance bound of $2^{74.9}$ queries when $n = 128$ is still far from the optimum one might hope for, which is a bound of 2^{128} queries for an output of $2n = 256$ bits (the birthday bound). But obtaining *any* bound above 2^{64} (a trivial lower bound) has proved elusive to researchers thus far, given the combinatorial complexity of the problem.

WHAT IS MDC-2? Traditionally, MDC-2 was instantiated using DES, and some people may understand MDC-2 to *mean* MDC-2 based on DES. This is not our meaning. Indeed this paper assumes a common keylength and blocklength n bits, and so our results don't directly apply to MDC-2 based on DES. (We assume that, with signficant work, one could extend our analysis to handle the

M. Naor (Ed.): EUROCRYPT 2007, LNCS 4515, pp. 34–51, 2007.

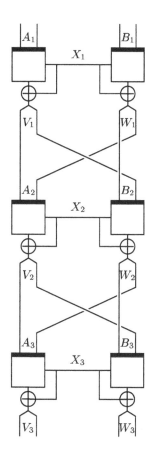

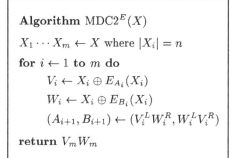

Algorithm $\mathrm{MDC2}^E(X)$

$X_1 \cdots X_m \leftarrow X$ where $|X_i| = n$

for $i \leftarrow 1$ **to** m **do**

  $V_i \leftarrow X_i \oplus E_{A_i}(X_i)$

  $W_i \leftarrow X_i \oplus E_{B_i}(X_i)$

  $(A_{i+1}, B_{i+1}) \leftarrow (V_i^L W_i^R, W_i^L V_i^R)$

return $V_m W_m$

Fig. 1. Left: Definition of the MDC-2 algorithm based on a blockcipher E with key length and block length n. The message being acted on is $X = X_1 \cdots X_m$ where $m \geq 1$ and $|X_i| = n$. Strings A_1 and B_1 are distinct n-bit constants. For an even-length sting S we let S^L and S^R be its left and right half. **Right:** Illustration of the algorithm acting on a three-block messsage $X = X_1 X_2 X_3$. The resulting hash is $H(X) = V_3 W_3$. The darkened edge of the box representing the blockcipher indicates the input that is the key.

DES parameters of 56-bit keys and 64-bit blocks, but we haven't done this.) In this paper we consider MDC-2 using a blockcipher $E: \{0,1\}^n \times \{0,1\}^n \to \{0,1\}^n$ with equal-length blocks and keys. We make this assumption for simplicity, while preserving contemporary applicability: eliminating "bit-dropping" makes the algorithm cleaner, while the usage of MDC-2 that people nowadays envisage is with the blockcipher AES-128 [30]. All future mention of MDC-2 in this paper assumes equal blocklength and keylength.

  The MDC-2 algorithm is simple and elegant: building on the usual Merkle-Damgård approach [6, 21], the compression function uses two parallel invocations of the Matyas-Meyer-Oseas compression function [20] and then swaps the right halves of the outputs. It is defined and illustrated in Figure 1. It is easy to see that the algorithm doesn't work (that is, it admits efficient attacks) if it is "over-simplified" by dropping the left/right swapping, the feed-forward XOR, or both.

The version of MDC-2 that we consider does not incorporate a "bit fixing" step like replacing the leftmost bit of each left-column blockcipher key in Figure 1 with a 0-bit and replacing the leftmost bit of each right-column blockcipher key with a 1-bit. Such bit-fixing was employed in MDC2-DES [1, 13] to overcome the key-complementation property of the primitive and also, conceivably, as a security measure.

We also comment that in the version of MDC-2 that we consider, no length-annotation or padding is used, and the domain is correspondingly restricted to $(\{0,1\}^n)^+$. It is easy and customary to use padding and length-annotation to extend MDC-2 to handle a domain of any string of less than 2^n bits. Provable-security results immediately extend: a collision-intractability result for the $(\{0,1\}^n)^+$ domain version of a hash function will always lift to give essentially the same bound for the $\{0,1\}^{<n}$ domain version one gets after padding and length annotation.

OUR RESULTS. We work in the ideal-cipher model, as in [4, 8, 15]. This is the customary model for proving the security of a blockcipher-based hash function. In the ideal-cipher model the underlying primitive, a blockcipher E, is modeled as a family of random permutations $\{E_K\}$ with a random permutation chosen independently for each key K. The adversary may make a query $E_K(X)$ to discover the corresponding value $Y = E_K(X)$, or the adversary may make a query $E_K^{-1}(Y)$ so as to learn the corresponding value $X = E_K^{-1}(Y)$ for which $E_K(X) = Y$. We are interested in the chance that an adversary can find a collision, namely a pair of distinct messages that collide under MDC2^E, by asking q queries. More formal definitions will be given below.

It is easy to show that finding a collision for MDC2 implies finding K, X, K', X' with $(K, X) \neq (K', X')$ such that $E_K(X) \oplus X = E_K(X') \oplus X'$. From this it easily follows (see [4]) that an adversary's chance of finding a collision in q queries is at most $q(q+1)/2^n \approx q^2/2^n$ where $n = |X| = |K|$ is the block size. This is a trivial upper bound, only as good as the conventional bound one expects for a hash function with n-bit output.

Ideally one would like to prove a bound of $q^2/2^{2n}$ for MDC-2, the bound corresponding to the birthday attack, since the output length of MDC-2 is $2n$. However, despite the lack of known attacks on MDC-2, no one has even been able to exhibit an improvement on the trivial bound of $q^2/2^n$. In this paper we give the first improvement by showing that an adversary has chance $O(q^5/2^{3n})$ of finding an attack and therefore needs at least $q \approx 2^{3n/5}$ queries to have an even chance of finding a collision. For example when $n = 128$ (the main case of interest) we show that an adversary needs $q = 2^{74.9}$ queries to have an even chance of obtaining a collision, which is over 2^{10} greater than the trivial bound of $2^{63.5}$. Figure 2 shows our upper bound as function of q for the case $n = 128$.

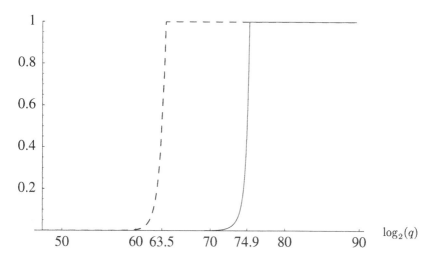

Fig. 2. Our upper bound on $\mathsf{Adv}_{128}^{\mathrm{MDC2}}(q)$ as a function of q (solid line) compared to the previous best upper bound of $q(q+1)/2^{128}$ (dotted line)

2   Preliminaries

Let $\mathsf{Bloc}(n)$ be the set of functions $E\colon \{0,1\}^n \times \{0,1\}^n \to \{0,1\}^n$ such that $E(K,\cdot) = E_K(\cdot)$ is a permutation on $\{0,1\}^n$. Given a blockcipher $E \in \mathsf{Bloc}(n)$ we define $\mathrm{MDC2}^E\colon (\{0,1\}^n)^+ \to \{0,1\}^{2n}$ by the algorithm of Fig. 1. The hash of a word X where $|X|$ is a multiple of n by $\mathrm{MDC2}^E$ is denoted by $\mathrm{MDC2}^E(X)$.

An adversary is a computationally unbounded but always-halting algorithm A with access to an oracle $E \in \mathsf{Bloc}(n)$. We can assume (by standard arguments) that A is deterministic. The adversary can make either a "forward" query $(K_i, X_i)_{\mathrm{fwd}}$ to its oracle E or a "backward" query $(K_i, Y_i)_{\mathrm{bwd}}$. The forward query is answered by $Y_i = E_{K_i}(X_i)$ and the backward query is answered by $X_i = E_{K_i}^{-1}(Y_i)$. Either way the result of the query is stored in a triple (X_i, K_i, Y_i) and the *query history* of A^E, denoted $\mathcal{Q} = \mathcal{Q}(A^E)$, is the tuple $(Q_1, \ldots, Q_q)$ where $Q_i = (X_i, K_i, Y_i)$ is the result of the i-th query made by the adversary, and where q is the total number of queries made by the adversary. If (X_i, K_i, Y_i) is an element of the query history then we refer to X_i as the "word input" of the query, to K_i as the "key" of the query, and to Y_i as the "output" of the query. The quantity $X_i \oplus Y_i$ is called the "XOR output" of the query.

The adversary's goal is to output a pair of nonempty strings X, X' such that $X \neq X'$ and $\mathrm{MDC2}^E(X) = \mathrm{MDC2}^E(X')$. Moreover we impose the condition that the adversary must have made all queries necessary to compute $\mathrm{MDC2}^E(X)$ and $\mathrm{MDC2}^E(X')$. This restriction is reasonable since otherwise the adversary can output very long words X, X' where $\mathrm{MDC2}^E(X) = \mathrm{MDC2}^E(X')$ with good probability but where computing $\mathrm{MDC2}^E(X)$, $\mathrm{MDC2}^E(X')$ is infeasible. (For example, without making any queries, the adversary could simply output

0^{Kn} and 0^{2Kn} where K is the lcm of all numbers between 1 and 2^n and have probability 1 of obtaining a collision, but this isn't a reasonable type of attack.)

Since we may tell simply from the adversary's query history $\mathcal{Q}$ whether it is possible for the adversary to output words $X \neq X'$ such that $\mathrm{MDC2}^E(X) = \mathrm{MDC2}^E(X')$ and such that $\mathcal{Q}$ contains all the queries necessary for the computation of $\mathrm{MDC2}^E(X)$, $\mathrm{MDC2}^E(X')$, we will in fact dispense the adversary from having to output X, X' and simply determine whether the adversary has been successful or not by examining its query history $\mathcal{Q}$. Formally, we say that $\mathsf{Coll}^E(\mathcal{Q})$ holds if there are two distinct nonempty words X, X' of lengths divisible by n such that $\mathrm{MDC2}^E(X) = \mathrm{MDC2}^E(X')$ and such that $\mathcal{Q}$ contains all the queries necessary to compute $\mathrm{MDC2}^E(X)$, $\mathrm{MDC2}^E(X')$ as defined by the algorithm of Fig. 1. The goal of the adversary A is thus to make some sequence of queries $\mathcal{Q} = \mathcal{Q}(A)$ such that $\mathsf{Coll}^E(\mathcal{Q})$. We define the adversary's ability to break MDC-2 by

$$\mathsf{Adv}_n^{\mathrm{MDC2}}(A) = \Pr[E \xleftarrow{\$} \mathrm{Bloc}(n); \mathcal{Q} \leftarrow A^E : \mathsf{Coll}^E(\mathcal{Q})].$$

We let $\mathsf{Adv}_n^{\mathrm{MDC2}}(q)$ be the max over all adversaries A making at most q queries of $\mathsf{Adv}_n^{\mathrm{MDC2}}(A)$. Our goal is thus to upper bound $\mathsf{Adv}_n^{\mathrm{MDC2}}(q)$. We can assume without loss of generality that A always asks exactly q queries and thus that $|\mathcal{Q}(A^E)| = q$.

Say that numbers n and q have been fixed as well as an adversary A such that $|\mathcal{Q}(A^E)| = q$ for all $E \in \mathrm{Bloc}(n)$. If P is any predicate that can be true or false for a sequence of queries $\mathcal{Q}$ (such as $\mathsf{Coll}^E(\mathcal{Q})$) then we write $\Pr[\mathsf{P}(\mathcal{Q})]$ as a shorthand for $\Pr[E \xleftarrow{\$} \mathrm{Bloc}(n); \mathcal{Q} \leftarrow A^E : \mathsf{P}(\mathcal{Q})]$. With this notation we have $\mathsf{Adv}_n^{\mathrm{MDC2}}(A) = \Pr[\mathsf{Coll}^E(\mathcal{Q})]$. We will often use this simpler notation to avoid over-complicating our formulas.

3   Our Security Bound

Our upper bound can be stated in varying degrees of generality and comprehensibility. The most general and least comprehensible statement of our upper bound is the following:

Theorem 1. *Let n, q be natural numbers with $q < 2^n$. Let $N = 2^n$, $N' = N - q$ and let m_a, m_b, m_c be any positive numbers with $eqN^{\frac{1}{2}}/N' \leq m_b \leq N^{\frac{1}{2}}$, $eq/N' \leq m_c$. Finally let $M_b = m_b N'/qN^{\frac{1}{2}}$, $M_c = m_c N'/q$ and $N'' = N'(N^{\frac{1}{2}} - m_b)/N^{\frac{1}{2}}$. Then*

$$\mathsf{Adv}_n^{\mathrm{MDC2}}(q) \leq$$

$$q^2/m_a N' + 2qN^{\frac{1}{2}} e^{qN^{\frac{1}{2}} M_b(1 - \ln(M_b))/N'} + qN e^{qM_c(1 - \ln(M_c))/N'} + \tag{1}$$

$$q(m_a^2 + m_a m_b^2 + m_b^4)/N' + \tag{2}$$

$$q(4m_a m_b)/N' + q(2m_a m_b)/N'' + \tag{3}$$

$$q(m_b^2 m_c + 5m_b^2 + m_a m_c + 6m_a)/N' + q(4m_a + 8m_b^2)/N'' + \tag{4}$$

$$q(4 + 10m_b + 2m_b m_c)/N'' + 3q/N' + 4q/N'' + q^2/N'^2 \qquad \blacksquare \tag{5}$$

q	$\mathsf{Adv}^{\mathrm{MDC2}}_{128}(q) \leq$	m_a	m_b	m_c
2^{64}	7.57×10^{-7}	2.64×10^6	44.01	3.7147
$2^{68.22}$	10^{-4}	7.01×10^6	128.09	3.9448
$2^{72.19}$	$1/100$	1.75×10^7	898.95	4.1899
$2^{74.00}$	$1/10$	2.66×10^7	2902.32	4.3082
$2^{74.72}$	$1/3$	3.14×10^7	4687.89	4.3523
$2^{74.91}$	$1/2$	3.29×10^7	5355.49	4.3640
$2^{75.21}$	1	$-$	$-$	$-$

Fig. 3. Upper bounds on $\mathsf{Adv}^{\mathrm{MDC2}}_{128}(q)$ given by Theorem 1. The right three columns specify the values m_a, m_b, and m_c used to obtain the bound of the second column.

For Theorem 1 to give a good bound one must choose suitable values for the constants m_a, m_b, m_c. Choosing large values of m_a, m_b, m_c reduces the terms of line (1) but increases the terms of lines (2)-(5). Unfortunately there is no good closed form for the optimal values of m_a, m_b, m_c (these will change with every q), hence the complex-looking form of Theorem 1. The meaning of the constants m_a, m_b, m_c is explained in the proof.

What Theorem 1 concretely means for $n = 128$ is shown in Figs. 2–3. Fig. 3 shows specific numerical upper bounds for $\mathsf{Adv}^{\mathrm{MDC2}}_{128}(q)$ for various values of q. The threshold value where Theorem 1 gives an upper bound of $1/2$ is $q = 2^{74.91}$ (to be compared with the previous best threshold of $q = 2^{63.5}$). For each value of q we also show the values of m_a, m_b, m_c which yield the stated upper bound. Fig. 2 plots our upper bounds on $\mathsf{Adv}^{\mathrm{MDC2}}_{128}(q)$ as a function of q, compared to the previous upper bound of $q(q+1)/N$. The method for optimizing m_a, m_b, m_c for given values of n, q in order to obtain the best bound on $\mathsf{Adv}^{\mathrm{MDC2}}_n(q)$ is discussed in the full version of this paper [29]. There we also show (via straightforward calculus) that Theorem 1 implies the following:

Theorem 2. *Let* $q = 2^{\frac{3}{5}n - \epsilon}$ *where* $\epsilon > 0$. *Then* $\mathsf{Adv}^{\mathrm{MDC2}}_n(q) \to 0$ *as* $n \to \infty$. ∎

Asymptotically as $n \to \infty$, thus, our bound for $\mathsf{Adv}^{\mathrm{MDC2}}_n(q)$ behaves like the function $\min(1, q^5/2^{3n})$, though the two functions still look significantly different for $n = 128$ (e.g. $q^5/2^{3n}$ has a threshold of $2^{76.6}$ for $n = 128$ whereas our bound on $\mathbf{Adv}^{\mathrm{MDC2}}_{128}(q)$ has a threshold of $2^{74.9}$). Though the two functions converge asymptotically there does not seem to be any good closed form relating our bound on $\mathsf{Adv}^{\mathrm{MDC2}}_n(q)$ to the function $q^5/2^{3n}$.

4   Analysis

OVERVIEW. Rather than analyzing the probability that the queries $\mathcal{Q}$ made by the adversary contain the means of constructing a collision we simplify the problem by analyzing the probability that the queries $\mathcal{Q}$ contain the means of constructing the last two rounds of a collision. Effectively we look to see whether there exist keys K_0, K_1, K'_0, K'_1 and n-bit words X_1, X_2, X'_1, X'_2 such that the

MDC-2 hash of X_1X_2 using the incoming keys K_0, K_1 (rather than A_1, B_1) equals the MDC-2 hash of $X_1'X_2'$ using the incoming keys K_0', K_1', and such that Q contains all the queries necessary to make both hashes. Naturally a collision does not necessarily involve two words of at least two blocks each, as either or both words may consist of a single block, and our analysis also allows for this contingency.

To upper bound the probability of the adversary obtaining queries that can be used to construct the last two rounds (or fewer) of a collision we upper bound the probability of the adversary making a query that can be used as the final query to complete such last two rounds. Namely for each i, $1 \leq i \leq q$, we upper bound the probability that the answer to the adversary's i-th query $(K_i, X_i)_{\mathsf{fwd}}$ or $(K_i, Y_i)_{\mathsf{bwd}}$ (depending) will allow the adversary to use the i-th query to complete (what looks like) the last two rounds of a collision. In the latter case we say the i-th query is "successful", and we give the attack to the adversary.

Naturally this probability will depend on the adversary's first $i - 1$ queries. In particular we need to make sure that the adversary hasn't already been too "lucky" with its first $i - 1$ queries, or else the probability of the i-th query being successful will be hard to upper bound. An example of being "lucky" would be if there exists a large subset of the first $i - 1$ queries that all have the same XOR output (there are two more ways of being lucky defined below). Our upper bound thus breaks down into two pieces: an upper bound for the probability of the adversary getting lucky in one of three specific ways defined below, and the probability of the adversary ever making a successful i-th query, conditioned on the fact that the adversary has not yet become lucky by its $(i - 1)$-th query.

DETAILS. Fix numbers n, q and an adversary A asking q queries to its oracle. We upper bound $\Pr[\mathsf{Coll}^E(Q)]$ by exhibiting predicates $\mathsf{Win0}(Q), \ldots, \mathsf{Win8}(Q)$ such that $\mathsf{Coll}^E(Q) \implies \mathsf{Win0}(Q) \vee \ldots \vee \mathsf{Win8}(Q)$ and then by upper bounding separately the probabilities $\Pr[\mathsf{Win0}(Q)], \ldots, \Pr[\mathsf{Win8}(Q)]$. Obviously $\Pr[\mathsf{Coll}^E(Q)] \leq \Pr[\mathsf{Win0}(Q)] + \cdots + \Pr[\mathsf{Win8}(Q)]$. (The event $\mathsf{Win0}(Q)$ happens if the adversary is lucky, whereas if the adversary is not lucky but makes a successful i-th query then one of the predicates $\mathsf{Win1}(Q), \ldots, \mathsf{Win8}(Q)$ will hold.)

To state the predicates $\mathsf{Win0}(Q), \ldots, \mathsf{Win8}(Q)$ we need some extra definitions. Define functions a, b, b^L, b^R and c on query sequences of length q as follows:

$$a(Q) = |\{(i,j) \in [1 \ldots q]^2 : i \neq j, X_i \oplus Y_i = X_j \oplus Y_j\}| \text{ is the number of}$$
ordered pairs of distinct queries in Q with same XOR outputs

$$b^L(Q) = \max_{Y \in \{0,1\}^{n/2}} |\{i : (X_i \oplus Y_i)^L = Y\}| \text{ is the maximum size of a set}$$
of queries in Q whose XOR outputs all have the same left $n/2$ bits

$$b^R(Q) = \max_{Y \in \{0,1\}^{n/2}} |\{i : (X_i \oplus Y_i)^R = Y\}| \text{ is the maximum size of a set}$$
of queries in Q whose XOR outputs all have the same right $n/2$ bits

$$b(Q) = \max(b^L(Q), b^R(Q))$$

$$c(Q) = \max_{Y \in \{0,1\}^n} |\{i : X_i \oplus Y_i = Y\}| \text{ is the maximum size of a set of}$$
queries in Q whose XOR outputs are all the same

The event $\mathsf{Win0}(\mathcal{Q})$ is simply defined by

$$\mathsf{Win0}(\mathcal{Q}) = (a(\mathcal{Q}) \geq m_a) \vee (b(\mathcal{Q}) \geq m_b) \vee (c(\mathcal{Q}) \geq m_c)$$

where m_a, m_b, m_c are the constants from Theorem 1. Thus as m_a, m_b, m_c are chosen larger $\Pr[\mathsf{Win0}(\mathcal{Q})]$ diminishes.

The events $\mathsf{Win1}(\mathcal{Q})$, ..., $\mathsf{Win8}(\mathcal{Q})$ are different in nature from the event $\mathsf{Win0}(\mathcal{Q})$; they concern the feasibility of fitting certain subconfigurations of MDC-2 using queries from $\mathcal{Q} = (X_1, K_1, Y_1)$, ..., (X_q, K_q, Y_q). Take for example the configuration $1a$ of Fig. 5. In this configuration, the two strings marked A are equal and the queries marked i, $!i$ are different. These are the only constraints; unmarked strings may or may not be equal, and other queries in the diagram may or may not be equal. Since the bottom left and bottom right queries are distinct fitting the diagram means using two distinct queries $Q_i = (X_i, K_i, Y_i)$ and $Q_{i'} = (X_{i'}, K_{i'}, Y_{i'})$ from $\mathcal{Q}$ for these two positions. We say that four queries $Q_i = (X_i, K_i, Y_i), Q_{i'} = (X_{i'}, Y_{i'}, Y_{i'}), Q_j = (X_j, K_j, Y_j), Q_k = (X_k, K_k, Y_k)$ in $\mathcal{Q}$ "fit" configuration $1a$ if $i \neq i'$ and if Q_i, $Q_{i'}$, Q_j, Q_k can be placed in respectively the bottom left, bottom right, top left and top right positions of configuration $1a$ such that the wiring constraints of the diagram are respected and such that the two strings marked A are equal. Formally, the four queries Q_i, $Q_{i'}$, Q_j, Q_k fit configuration $1a$ if and only if

$$(i \neq i') \wedge (X_i = X_{i'}) \wedge (X_j = X_k) \wedge (X_i \oplus Y_i = X_{i'} \oplus Y_{i'}) \wedge$$
$$((X_j \oplus Y_j)^L = K_i^L) \wedge ((X_j \oplus Y_j)^R = K_{i'}^R) \wedge$$
$$((X_k \oplus Y_k)^L = K_{i'}^L) \wedge ((X_k \oplus Y_k)^R = K_i^R).$$

Moreover we say that $\mathsf{ExistsFit}_{1a}(\mathcal{Q})$ holds if there exist $i, i', j, k \in [1..q]$ such that queries Q_i, $Q_{i'}$, Q_j, Q_k fit configuration $1a$. The predicates $\mathsf{ExistsFit}_{1b}$, $\mathsf{ExistsFit}_2$, $\mathsf{ExistsFit}_3$, $\mathsf{ExistsFit}_{4a}$, $\mathsf{ExistsFit}_{4b}$, $\mathsf{ExistsFit}_{6a}$, $\mathsf{ExistsFit}_{6b}$, $\mathsf{ExistsFit}_{6c}$, $\mathsf{ExistsFit}_{6d}$, $\mathsf{ExistsFit}_{7a}$, $\mathsf{ExistsFit}_{7b}$, whose configurations are shown in Figs. 5–6, are likewise defined. In these configurations strings marked by the same letter must be equal but strings marked with different letters may or may not be equal; likewise queries marked i, $!i$ or j, $!j$ are different but two queries marked with different letters may be the same. We also let $\mathsf{ExistsFit}_1 = \mathsf{ExistsFit}_{1a} \vee \mathsf{ExistsFit}_{1b}$, $\mathsf{ExistsFit}_4 = \mathsf{ExistsFit}_{4a} \vee \mathsf{ExistsFit}_{4b}$, and so on. Note that $\mathsf{ExistsFit}_{6a} = \mathsf{ExistsFit}_{6b}$ and that $\mathsf{ExistsFit}_{6c} = \mathsf{ExistsFit}_{6d}$, thus $\mathsf{ExistsFit}_6 = \mathsf{ExistsFit}_{6a} \vee \mathsf{ExistsFit}_{6c}$ (configurations $6b$, $6d$ are only provided to facilitate referencing).

Some additional notation is required to indicate inequality between queries in configurations 5 and 8. In these configurations, pairs of queries from the bottom row that do not both contain a '1' or both contain a '0' (namely, queries with different labels) are presumed different; there are no constraints relating top row to bottom row queries, and queries with the same label are not presumed equal (see Fig. 4 for an explanation of "top row", "bottom row"). The predicates $\mathsf{ExistsFit}_5(\mathcal{Q})$, $\mathsf{ExistsFit}_8(\mathcal{Q})$ then denote the existence of a set of queries in $\mathcal{Q}$ fitting respectively configurations 5 and 8 under these constraints.

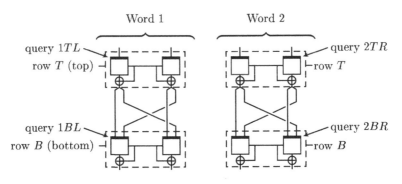

Fig. 4. The query labels

Let NotWin $j = \overline{\text{Win0}(\mathcal{Q}) \vee \cdots \vee \text{Win } j(\mathcal{Q})}$ for $1 \leq j < 8$. We now define:

$$\text{Win1}(\mathcal{Q}) = \text{NotWin0}(\mathcal{Q}) \wedge \text{ExistsFit}_1(\mathcal{Q})$$
$$\text{Win2}(\mathcal{Q}) = \text{NotWin1}(\mathcal{Q}) \wedge \text{ExistsFit}_2(\mathcal{Q})$$

$$\vdots$$

and so forth. Thus $\text{Win4}(\mathcal{Q})$, for example, is the predicate which is true if and only if $a(\mathcal{Q}) < m_a, b(\mathcal{Q}) < m_b, c(\mathcal{Q}) < m_c$ (these conditions being $\text{NotWin0}(\mathcal{Q})$) and $\mathcal{Q}$ contains queries that fit configurations $4a$ or $4b$ but $\mathcal{Q}$ does not contain queries fitting configurations $1a$, $1b$, 2 or 3.

The reader will note that all configurations in Figs. 5–6 have at most two pieces and each piece is a subportion of two rounds of MDC-2. If the configuration has two pieces (such as configurations 2, $4a$, $4b$, 5, $6a$, $6b$, $6c$, $6d$, $7a$, $7b$, 8 as opposed to configurations $1a$, $1b$, 3) then the left portion of the configuration is called "Word 1" and the right portion of the configuration is called "Word 2" (Fig. 4). Queries in the right-hand column of a two-round piece are called "right column" queries and queries in the left-hand column of a two-block portion are called "left column" queries. "Top row" and "bottom row" queries are defined the expected way. A query in the configuration is given coordinates $1TR$ for "Word 1, Top row, Right column" or $2BL$ for "Word 2, Bottom row, Left column", etc. If the configuration has only one piece then we drop the prefix "1" or "2" and simply give coordinates TL, TR, etc. for the queries. The reader should refer to Fig. 4.

We now show that $\text{Coll}^E(\mathcal{Q}) \implies \text{Win0}(\mathcal{Q}) \vee \cdots \vee \text{Win8}(\mathcal{Q})$:

Lemma 1. $\text{Coll}^E(\mathcal{Q}) \implies \text{Win0}(\mathcal{Q}) \vee \cdots \vee \text{Win8}(\mathcal{Q})$.

Proof. First note that $\text{ExistsFit}_1(\mathcal{Q}) \vee \cdots \vee \text{ExistsFit}_8(\mathcal{Q}) \implies \text{Win0}(\mathcal{Q}) \vee \cdots \vee \text{Win8}(\mathcal{Q})$, so it is sufficient to show that $\text{Coll}^E(\mathcal{Q}) \implies \text{ExistsFit}_1(\mathcal{Q}) \vee \cdots \vee \text{ExistsFit}_8(\mathcal{Q})$.

Say $\text{Coll}^E(\mathcal{Q})$. Then a collision can be constructed from the queries $\mathcal{Q}$. We can assume that the collision is earliest possible in the sense that one cannot truncate

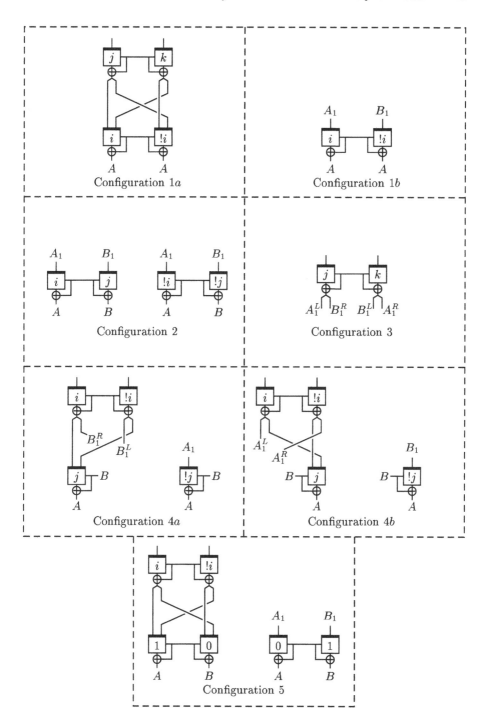

Fig. 5.

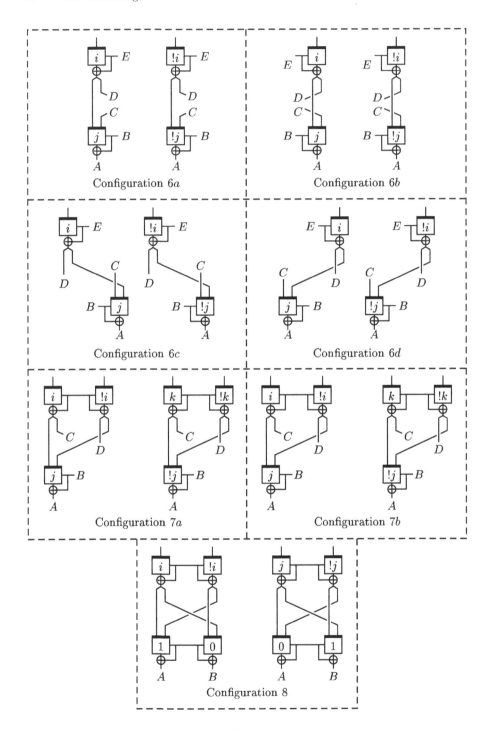

Fig. 6.

either one or both words involved to form a collision from the leftover prefixes (otherwise, take this smaller pair of words). By definition collisions involve words with at least one block, so the collision must either (i) use two words that are one block long each (ii) use one word of at least two blocks and one word of one block or (iii) use two words of at least two blocks each. If the collision uses two words that are one block long each then obviously $\mathsf{ExistsFit}_2(\mathcal{Q})$ (if query i where equal to $!i$, the two words would be the same), so we can assume either (ii) or (iii).

Say first the collision is of type (ii), namely that the collision has one word with $m \geq 2$ blocks, which is WLOG word 1, and one word of with one block, which is word 2. Note first that when word 1 is hashed via MDC-2 there can never be a round where the same query appears both on the left and right-hand sides unless $\mathsf{ExistsFit}_1(\mathcal{Q})$ holds (to see this, take the earliest such round; since the constant keys A_1, B_1 are different this is not the first round and the two queries from the round before are different but have the same XOR output, so $\mathsf{ExistsFit}_1(\mathcal{Q})$). Therefore we can assume that at every round in the hashing of word 1, different queries appear on the left and right-hand sides. Naturally the same query may appear both in the left and right columns in different rounds.

We now examine the last two rounds of the hashing of word 1. The four (not necessarily distinct) queries comprising these two rounds are labeled $1TL$, $1TR$, etc. as in Fig. 4 and as per our convention described above. The two queries making up the unique round for the hashing of word 2 are simply labeled $2L$ and $2R$, where $2L$ is the query with key input A_1 and $2R$ is the query with key input B_1. By our previous remark, queries $1TL$ and $1TR$ are distinct as well as queries $1BL$ and $1BR$. If query $1BL$ equals query $2L$ and query $1BR$ equals query $2R$ then $\mathsf{ExistsFit}_3(\mathcal{Q})$. On the other hand if query $1BL$ is not equal to query $2L$ and query $2BR$ is not equal to query $2R$ then $\mathsf{ExistsFit}_5(\mathcal{Q})$. Therefore we can assume (by symmetry) that query $1BL$ is not equal to query $2L$ but that query $1BR$ equals query $2R$. But then $\mathsf{ExistsFit}_{4a}(\mathcal{Q})$. This concludes the case when the adversary's collision is of type (ii).

We now assume that both of the words involved in the collision have at least two rounds. We examine the last two rounds of the hashing of each word; the queries for these last two rounds are labeled as in Fig. 4. By the same remark as above, the same query cannot appear in both left and right positions at the same round of the same word, so the top row constraints of configuration 8 are satisfied. If query $1BL$ equals $2BL$ and query $1BR$ equals query $2BR$ then the collision is not earliest possible, a contradiction, so we can assume (by symmetry) that query $1BL$ is not equal to query $2BL$. If queries $1BR$ and $2BR$ are equal then $\mathsf{ExistsFit}_{7a}(\mathcal{Q})$ so they too must be unequal. But then $\mathsf{ExistsFit}_8(\mathcal{Q})$ so we are done.   □

The reader may have noted that $\mathsf{ExistsFit}_6(\mathcal{Q})$ does not actually appear in the proof of Lemma 1. However $\mathsf{Win6}(\mathcal{Q})$ will be used to upper bound $\Pr[\mathsf{Win7}(\mathcal{Q})]$ (as $\Pr[\mathsf{Win7}(\mathcal{Q})] \leq \Pr[\mathsf{Win6}(\mathcal{Q})] + \Pr[\mathsf{NotWin6}(\mathcal{Q}) \wedge \mathsf{Win7}(\mathcal{Q})]$).

Let $\mathsf{WinFit}(\mathcal{Q}) = \mathsf{Win1}(\mathcal{Q}) \vee \ldots \vee \mathsf{Win8}(\mathcal{Q})$, so $\Pr[\mathsf{Coll}^E(\mathcal{Q})] \leq \Pr[\mathsf{Win0}(\mathcal{Q})] + \Pr[\mathsf{WinFit}(\mathcal{Q})]$. We show:

Lemma 2. *Let N, N', N'', m_a, m_b, M_b, m_c, M_c be as in Theorem 1. Then*
$$\Pr[\mathsf{Win0}(\mathcal{Q})] \le q^2/m_a N' + 2qN^{\frac{1}{2}} e^{qN^{\frac{1}{2}} M_b(1-\ln(M_b))/N'} + qN e^{qM_c(1-\ln(M_c))/N'}. \quad \blacksquare$$

and:

Lemma 3. *Let N, N', N'', m_a, m_b, m_c be as in Theorem 1. Then:*

$$\begin{aligned}
\Pr[\mathsf{WinFit}(\mathcal{Q})] \le\ & q(m_a^2 + m_a m_b^2 + m_b^4)/N' + \\
& q(4m_a m_b)/N' + q(2m_a m_b)/N'' + \\
& q(m_b^2 m_c + 5m_b^2 + m_a m_c + 6m_a)/N' + q(4m_a + 8m_b^2)/N'' + \\
& q(4 + 10m_b + 2m_b m_c)/N'' + 3q/N' + 4q/N'' + q^2/N'^2. \quad \blacksquare
\end{aligned}$$

Lemmas 2 and 3 imply Theorem 1 (by Lemma 1). The proof of Lemma 2 uses straightforward balls-in-bins probability and can be found in the full version of our paper [29]. The proof of Lemma 3 is more involved and in some sense constitutes the heart of our paper. Here we only give a grief glimpse of the type of analysis involved by showing how to upper bound $\Pr[\mathsf{NotWin0}(\mathcal{Q}) \wedge \mathsf{ExistsFit}_{1a}(\mathcal{Q})]$, which establishes "half" of the upper bound for $\Pr[\mathsf{Win1}(\mathcal{Q}) = \mathsf{NotWin0}(\mathcal{Q}) \wedge (\mathsf{ExistsFit}_{1a}(\mathcal{Q}) \vee \mathsf{ExistsFit}_{1b}(\mathcal{Q}))]$. (Again, the full proof of Lemma 3 is found in the full version.)

For the next proof we use the notational convention that (K_i, X_i) denotes a forward query $(K_i, X_i)_{\mathrm{fwd}}$ and that (K_i, Y_i) denotes a backward query $(K_i, Y_i)_{\mathrm{bwd}}$. The constants N, N', N'' will remain throughout as defined in Theorem 1, namely $N = 2^n$, $N' = N - q$, $N'' = N'(N^{\frac{1}{2}} - m_b)/N^{\frac{1}{2}}$.

Proposition 1. $\Pr[\mathsf{NotWin0}(\mathcal{Q}) \wedge \mathsf{ExistsFit}_{1a}(\mathcal{Q})] \le q(m_a + m_b^2)/N' + 2qm_b/N''$.

Proof. Let $\mathcal{Q}_i$ denote the first i queries made by the adversary. The term "last query" means the latest query made by the adversary (we examine the adversary's queries (K_i, X_i) or (K_i, Y_i) one at a time, in succession as they come in). The last query is always given index i. We say the last query is "successful" if the output Y_i or X_i for the last query is such that $a(\mathcal{Q}_i) < m_a$, $b(\mathcal{Q}_i) < m_b$, $c(\mathcal{Q}_i) < m_c$ and such that the adversary can use the query (X_i, K_i, Y_i) to fit configuration $1a$ using only queries in $\mathcal{Q}_i$ (in particular, the last query *must* be used in the fitting for that query to count as successful). The goal is thus to upper bound the adversary's chance of ever making a successful last query.

The strategy for upper bounding the probability of the last query being successful is to consider separately the different ways in which the last query can be used to fit the configuration and to upper bound the probability of success in each case, and finally to sum the various upper bounds. For example, the adversary may use the last query only once in the configuration or otherwise in several different positions of the configuration (such as, say, TL and BL). The basic setup for upper bounding the probability of success in a given case is to upper bound the maximum number of different outputs Y_i or X_i (depending on whether the last query is a forward or backward query) that would allow the query (X_i, K_i, Y_i) to be used to fit the configuration, and then to divide this number by $N' = N - q$ (since either Y_i or X_i, depending, is chosen randomly among a set of at least N' different values). That ratio is then multiplied by q, since the adversary makes q queries in all, each of which could become a successful last query.

Case 1: The last query is used exactly once in the configuration. We can assume WLOG that it is used in the left column.

Subcase 1.1: The last query is used in position BL. Say first that the last query is a forward query (K_i, X_i). Since the last query cannot be successful if $b(\mathcal{Q}_{i-1}) \geq m_b$ (by definition) we can assume that $b(\mathcal{Q}_{i-1}) < m_b$. Then since the left half of the XOR output of the query used in position TL must be equal to the left half of K_i there are at most m_b different queries in $\mathcal{Q}_{i-1}$ that could be used in position TL, for the given inputs (K_i, X_i) of the last query. Likewise because the right half of the XOR output of the query used in position TR must be equal to the right half of K_i there are at most m_b different queries in $\mathcal{Q}_{i-1}$ that could be used in position TR. Since X_i together with the outputs of the queries used in positions TL, TR completely determine the query used in position BR, there are therefore at most m_b^2 different queries in $\mathcal{Q}_{i-1}$ which can be used in position BR for the given inputs (K_i, X_i). Therefore there are at most m_b^2 outputs Y_i which would enable the last query be used to fit the configuration at position BL (namely which would enable the XOR output $X_i \oplus Y_i$ of query BL to be equal to the XOR output of query BR), so the chance of success of the last query if it is forward is $\leq m_b^2/N'$.

Now say the last query is a backward query (K_i, Y_i). We cannot reason like for the forward query case that there are only m_b^2 queries in $\mathcal{Q}_{i-1}$ that that can appear in position BR since we do not know the word input X_i anymore. However because the query used in position BR has same XOR output and same word input as the query in position BL it must also have the same output as the query in position BL, which means the output of the query in position BR is actually Y_i. Now because E is a blockcipher, there is exactly at most one possible query for position BR in $\mathcal{Q}_{i-1}$ for any given value of the key of the query in position BR, and since the key can take at most m_b^2 different values (as in the forward case) there are again at most m_b^2 different queries that can be used in position BR. Therefore there are at most m_b^2 different values for X_i which would make the backwards query (K_i, Y_i) successful, so the last query again has chance of success $\leq m_b^2/N'$.

Thus the last query has chance of success $\leq m_b^2/N'$ whether it is a forward or backward query. Multiplying by q, we obtain that the chance of ever making a successful last query of this type is $\leq qm_b^2/N'$. This concludes the analysis of Subcase 1.1.

Note: we will not always give as many details as in Subcase 1.1. In particular, we will not continue to remind that one can assume $a(\mathcal{Q}_{i-1}) < m_a$, $b(\mathcal{Q}_{i-1}) < m_b$, $c(\mathcal{Q})_{i-1} < m_c$ (or else the last query is by definition not successful) and we will often shorten phrases of the type "query used in position TL" to simply "query TL".

Subcase 1.2: The last query is used in position TL. Because the queries use in positions BL, BR are distinct but have the same XOR output there are at most m_a different ordered pairs of queries in $\mathcal{Q}_{i-1}$ that can be used for the pair BL, BR. But the pair of queries for BL, BR completely determines what the XOR output $X_i \oplus Y_i$ of the last query should be. Therefore the last query has chance at most m_a/N' of success and the total probability of making this type of successful last query is $\leq qm_a/N'$.

Note: Subcase 1.2 does not require a separate analysis for the forward and backward case because we can upper bound the maximum number of successful XOR outputs for the last query *without* looking at the inputs for the last query; by contrast, in Subcase 1.1 we inspected X_i in the forward case and Y_i in the backward case in order to determine the maximum possible number of successful XOR outputs. In general, whenever an upper bound on the total number of successful XOR outputs for the last query can be found without inspecting any inputs for the last query besides the key, the same analysis will work both for the forward and backward cases.

Case 2: The last query is used twice or more in the configuration. Because queries BR and BL are distinct the queries TR and TL are also distinct and so the last query must in fact appear exactly twice in the configuration. We can assume WLOG that it is used in position TL.

The type of analysis we use for this case is slightly different than the analysis for Subcases 1.1, 1.2. To estimate the probability of the last query succeeding we will first look at the left $n/2$ bits of XOR output, estimate their probability P_l of success (the left bits are "successful" if they do not preclude the last query from being successful) and then we estimate the probability of success $P_{r|l}$ of the right $n/2$ bits of XOR output being successful, conditioned on the fact that the left $n/2$ bits are successful (the right $n/2$ bits are "successful" if the last query is successful). The probability of success of the last query is then $P_l P_{r|l}$. Note that if the set of left half of XOR outputs which are successful has size T then $P_l \leq TN^{\frac{1}{2}}/N'$ since the return to any query has chance $\leq N^{\frac{1}{2}}/N'$ of having its left half of XOR output equal to any particular value (there are at most $N^{\frac{1}{2}}$ strings that have that left half, each of which is returned with chance at most $1/N'$). Then if the left half is successful and there are U different possible ways of completing the left half into a successful string, namely U different successful right halfs, the chance of the right half being successful given $\mathsf{NotWin0}(\mathcal{Q}_{i-1})$ is $\leq U/(N^{\frac{1}{2}} - m_b)$ since the XOR output could be any of at least $N^{\frac{1}{2}} - m_b$ values with equal probability (there are at most m_b values which we know will not appear because they have already appeared for this left half). So the total chance of success of the last query in this case (assuming U was independent of the left half, as it will be in our analysis) is $\leq TUN^{\frac{1}{2}}/N'(N^{\frac{1}{2}} - m_b)$ or $\leq TU/N''$.

Subcase 2.1: The last query is used in positions TL, BL. Since the last query appears in positions TL, BL the left half of the last query's XOR output must be equal to the left half of its key input, so the left half of output has chance $P_l \leq N^{\frac{1}{2}}/N'$ chances of succeeding. If it succeeds, there are at most m_b queries for BR in $\mathcal{Q}_{i-1}$ with that left half of XOR output (which must be shared with query BL), so the right half of XOR output has chance $P_{r|l} \leq m_b/(N^{\frac{1}{2}} - m_b)$ of succeeding if the the left half succeeds. Therefore the last query has chance $P_l P_{r|l} \leq m_b N^{\frac{1}{2}}/N'(N^{\frac{1}{2}} - m_b) = m_b/N''$ of succeeding and the adversary's total chance of making this kind of successful last query is $\leq qm_b/N''$.

Subcase 2.2: The last query is used in position TL and in position BR. One can apply the same type of analysis as for Subcase 2.1, showing that the total chance of a successful last query of this type is $\leq qm_b/N''$.

Subcase 2.2 concludes Case 2 and thus all possible cases of making a successful query for configuration $1a$. Summing up the probabilities we get that $\Pr[\mathsf{NotWin0}(\mathcal{Q}) \wedge \mathsf{ExistsFit}_{1a}(\mathcal{Q})] \leq q(m_a + m_b^2)/N' + 2qm_b/N''$. □

5   Conclusion

We have proved the first nontrivial security bound for MDC-2. While such a bound has been a long time coming, we expect that our result is only a first foot in the door. In particular there remains a large gap between the best-known collision-finding attack, which is the trivial attack that succeeds with chance $q^2/2^{2n}$, and the security bound of Theorem 1. Likely our security bound is far from optimal, and it remains an interesting open question to find matching upper and lower bounds.

Acknowledgments

This work was supported in part by NSF CCR-0208842 and a gift from Intel Corporation; thanks to Jesse Walker for sponsoring this research. Part of this work was carried out while the author was visiting NTT labs in Yokosuka, Japan; thanks to Tatsuaki Okamoto for his kind support. The research topic was suggested to the author by Phillip Rogaway, who also provided patient mentoring and guidance throughout the project.

References

1. ANSI X9.31. Public key cryptography using reversible algorithms for the financial services industry. American National Standards Institute, 1998.
2. B. den. Boer and A. Bosselaers. Collisions for the compression function of MD5. *Advances in Cryptology – EUROCRYPT '93*, Lecture Notes in Computer Science, vol. 765, Springer, pp. 293–304, 1993.
3. J. Black, M. Cochran, and T. Shrimpton. On the impossibility of highly efficient blockcipher-based hash functions. *Advances in Cryptology – EUROCRYPT '05*, Lecture Notes in Computer Science, vol. 3494, Springer, pp.–546-541, 2005.

4. J. Black, P. Rogaway, and T. Shrimpton. Black-box analysis of the block-cipher-based hash-function constructions from PGV. *Advances in Cryptology – CRYPTO '02*, Lecture Notes in Compuer Science, vol. 2442, Springer, pp. 320–355, 2002.

5. B. Brachtl, D. Coppersmith, M. Hyden, S. Matyas, C. Meyer, J. Oseas, S. Pilpel, and M. Schilling. Data authentication using modification detection codes based on a public one-way encryption function. US Patent #4,908,861. Awarded March 13, 1990 (filed Auguest 28, 1987).

6. I. Damgård. A design principle for hash functions. *Advances in Cryptology – CRYPTO '89*, Lecture Notes in Computer Science, vol. 435, Springer, pp. 416–427, 1990.

7. H. Dobbertin. The status of MD5 after a recent attack. *CryptoBytes* 2 (2), 1996.

8. S. Even and Y. Mansour. A construction of a cipher from a single pseudorandom permutation. *Advances in Cryptology – ASIACRYPT '91*, Lecture Notes in Computer Science, vol. 739, Springer, pp. 210–224, 1991.

9. M. Hattori, S. Hirose, and S. Yoshida. Analysis of double block lengh hash functions. *Cryptography and Coding, 9th IMA International Conference*, Lecture Notes in Computer Science, vol. 2898, Springer, pp. 290–302, 2003.

10. S. Hirose. Provably secure double-block-length hash functions in a black box model. *Information Security and Cryptology—ISISC '04*, Lecture Notes in Computer Science, vol. 3506, Springer, pp. 330-342, 2005.

11. S. Hirose. Some plausible constructions of double-block-length hash functions. *Fast Software Encryption (FSE '06)*. Lecture Notes in Computer Science, vol. 4047, Springer, pp. 210–225, 2005.

12. W. Hohl, X. Lai, T. Meier, and C. Waldvogel. Security of iterated hash functions based on block ciphers. *Advances in Cryptology – CRYPTO '93*. Lecture Notes in Computer Science, vol. 773, Springer, pp. 303–311, 1993.

13. ISO/IEC 10118-2:2000. Information technology – Security techniques – Hash functions – Hash functions using an n-bit block cipher. International Organization for Standardization, Geneva, Switzerland, 2000. First released in 1992.

14. A. Joux. Multicollisions in iterated hash functions, applications to cascaded constructions. *Advances in Cryptology – CRYPTO '04*. Lecture Notes in Computer Science, vol. 3152, Springer, pp. 306–316, 2004.

15. J. Kilian and P. Rogaway. How to Protect DES Against Exhaustive Key Search. *Journal of Cryptology*, vol. 14, no. 1, pp. 17–35, 2001.

16. L. Knudsen, X. Lai, and B. Preneel. Attacks on fast double block length hash functions. *Journal of Cryptology*, vol. 11, no. 1, pp. 59–72, 1998.

17. X. Lai and J. Massey. Hash functions based on block ciphers. *Advances in Cryptology – EUROCRYPT '92*. Lecture Notes in Computer Science, vol. 658, Springer, pp. 55–70, 1992.

18. W. Lee, M. Nandi, P. Sarkar, D. Chang, S. Lee, and K. Sakurai. PGV-style block-cipher-based hash families and black-box analysis. *IEICE Transactions 88-A(1)*, pp. 39–48, 2005.

19. S. Lucks. Design principles for iterated hash functions. Cryptology ePrint Archive, Report 2004/253, 2004.

20. S. Matyas, C. Meyer, and J. Oseas. Generating strong one-way functions with cryptographic algorithm. IBM Technical Disclosure Bulletin, 27, pp. 5658–5659, 1985.

21. R. Merkle. One way hash functions and DES. *Advances in Cryptology – CRYPTO '89*. Lecture Notes in Computer Science, vol. 435, Springer, pp. 428–446, 1990.

22. C. Meyer and S. Matyas. Secure program load with manipulation detection code. *Proceedings of the 6th Worldwide Congress on Computer and Communications Security and Protection (SECURICOM '88)*, pp. 111–130, 1988.
23. M. Nandi, W. Lee, K. Sakurai, and S. Lee. Security analysis of a 2/3-rate double length compression function in the black-box model. *Fast Software Encryption (FSE '05)*, Lecture Notes in Computer Science, vol. 3557, pp. 243–254, 2005.
24. M. Nandi. Towards optimal double-length hash functions. *Progress in Cryptography – INDOCRYPT '05*, Lecture Notes in Computer Science, vol. 3797, Springer, pp. 77–89, 2005.
25. M. Rabin. Digitalized signatures. In R. DeMillo, D. Dobkin, A. Jones, and R. Lipton, editors, *Foundations of Secure Computation*, Academic Press, pp. 155–168, 1978.
26. R. Rivest. The MD4 message digest algorithm. *Advances in Cryptology – CRYPTO '90*, Lecture Notes in Comptuer Science, vol. 537, pp. 303–311, 1991.
27. P. Rogaway and T. Shrimpton. Cryptographic hash-function basics: definitions, implications, and separations for preimage resistance, second-preimage resistance, and collision Resistance. *Fast Software Encryption (FSE '04)*, Lecture Notes in Computer Science, vol. 3017, pp. 371-388, Springer, vol. 3017, 2004.
28. T. Satoh, M. Haga, and K. Kurosawa. Towards secure and fast hash functions. *IEICE Transactions on Fundamentals of Electronics, Communications and Computer Sciences*, vol. E82–A No. 1, pp. 55–62.
29. J. Steinberger. The collision intractability of MDC-2 in the ideal-cipher model. Full version of this paper. Cryptology ePrint Archive, Report 2006/294, 2006.
30. J. Viega. The AHASH mode of operation. Manuscript, 2004. Available from www.cryptobarn.com.
31. X. Wang, X. Lai, D. Feng, H. Chen, and X. Yu. Cryptanalysis of the hash functions MD4 and RIPEMD. *Advances in Cryptology – EUROCRYPT '05*, Lecture Notes in Computer Science, vol. 3494, Springer, pp. 1–18. 2005.
32. X. Wang, Y. Yin, and H. Yu. Finding collisions in the full SHA-1. *Advances in Cryptology – CRYPTO '05*, Lecture Notes in Computer Science, vol. 3621, Springer, pp. 17–36, 2005.

An Efficient Protocol for Secure Two-Party Computation in the Presence of Malicious Adversaries

Yehuda Lindell[1,*] and Benny Pinkas[2,**]

[1] Dept. of Computer Science, Bar-Ilan University, Israel
lindell@cs.biu.ac.il
[2] Dept. of Computer Science, University of Haifa, Israel
benny@pinkas.net

Abstract. We show an efficient secure two-party protocol, based on Yao's construction, which provides security against malicious adversaries. Yao's original protocol is only secure in the presence of semi-honest adversaries. Security against malicious adversaries can be obtained by applying the compiler of Goldreich, Micali and Wigderson (the "GMW compiler"). However, this approach does not seem to be very practical as it requires using generic zero-knowledge proofs.

Our construction is based on applying cut-and-choose techniques to the original circuit and inputs. Security is proved according to the ideal/ real simulation paradigm, and the proof is in the standard model (with no random oracle model or common reference string assumptions). The resulting protocol is computationally efficient: the only usage of asymmetric cryptography is for running $O(1)$ oblivious transfers for each input bit (or for each bit of a statistical security parameter, whichever is larger). Our protocol combines techniques from folklore (like cut-and-choose) along with new techniques for efficiently proving consistency of inputs. We remark that a naive implementation of the cut-and-choose technique with Yao's protocol does *not* yield a secure protocol. This is the first paper to show how to properly implement these techniques, and to provide a full proof of security.

Our protocol can also be interpreted as a constant-round black-box reduction of secure two-party computation to oblivious transfer and perfectly-hiding commitments, or a black-box reduction of secure two-party computation to oblivious transfer alone, with a number of rounds which is linear in a statistical security parameter. These two reductions are comparable to Kilian's reduction, which uses OT alone but incurs a number of rounds which is linear in the depth of the circuit [18].

1 Introduction

Secure two-party computation. In the setting of two-party computation, two parties with respective private inputs x and y, wish to jointly compute a functionality

---

* Research supported in part by an Infrastructures grant from the Ministry of Science, Israel.
** Research supported in part by the Israel Science Foundation (grant number 860/06).

M. Naor (Ed.): EUROCRYPT 2007, LNCS 4515, pp. 52–78, 2007.

$f(x, y) = (f_1(x, y), f_2(x, y))$, such that the first party receives $f_1(x, y)$ and the second party receives $f_2(x, y)$. Loosely speaking, the security requirements are that nothing is learned from the protocol other than the output (*privacy*), and that the output is distributed according to the prescribed functionality (*correctness*). The actual definition follows the simulation paradigm and blends the above two requirements. Of course, security must be guaranteed even when one of the parties is adversarial. Such an adversary may be *semi-honest* (or passive), in which case it correctly follows the protocol specification, yet attempts to learn additional information by analyzing the transcript of messages received during the execution. In contrast, the adversary may be *malicious* (or active), in which case it can arbitrarily deviate from the protocol specification.

The first general solutions for the problem of secure computation were presented by Yao [29] for the two-party case (with security against semi-honest adversaries) and Goldreich, Micali and Wigderson [11] for the multi-party case (with security even against malicious adversaries). Thus, the results of [29] and [11] constitute important and powerful feasibility results for secure two-party and multi-party computation.

Yao's protocol. In [29], Yao presented a *constant-round* protocol for securely computing any functionality in the presence of semi-honest adversaries. Denote party P_1 and P_2's respective inputs by x and y and let f be the functionality that they wish to compute (for simplicity, assume that both parties wish to receive $f(x, y)$). Loosely speaking, Yao's protocol works by having one of the parties (say party P_1) first generate a "garbled" (or encrypted) circuit computing $f(x, \cdot)$ and then send it to P_2. The circuit is such that it reveals nothing in its encrypted form and therefore P_2 learns nothing from this stage. However, P_2 can obtain the output $f(x, y)$ by "decrypting" the circuit. In order to ensure that P_2 learns nothing more than the output itself, this decryption must be "partial" and must reveal $f(x, y)$ only. Without going into unnecessary details, this is accomplished by P_2 obtaining a series of keys corresponding to its input y, such that given these keys and the circuit, the output value $f(x, y)$, and only this value, may be obtained. Of course, P_2 must somehow receive these keys without revealing anything about y to P_1. This can be accomplished by running $|y|$ instances of a secure 1-out-of-2 Oblivious Transfer protocol [27,7]. Yao's generic protocol is highly efficient, and even practical, for functionalities that have relatively small circuits. An actual implementation of the protocol was presented in [21], with very reasonable performance.

Security against malicious behavior. Yao's protocol is only secure in the presence of relatively weak semi-honest adversaries. Thus, an important question is how to "convert" the protocol into one that is secure in the presence of malicious adversaries, while preserving the efficiency of the original protocol to the greatest extent possible. Of course, one possibility is to use the compiler of Goldreich, Micali and Wigderson [11]. This compiler converts any protocol that is secure for semi-honest adversaries into one that is secure for malicious adversaries, and as such is a powerful tool for demonstrating feasibility. However, it is based on reducing the statement that needs to be proved (in our case, the honesty of the parties' behavior) to an NP-complete problem, and using generic zero-knowledge proofs to prove this statement. The resulting secure protocol therefore runs in polynomial time but is rather inefficient. (For more details on existing methods for proving security against malicious behavior see the section on related work below.)

Malicious behavior and cut-and-choose. Consider for a moment what happens if party P_1 is malicious. In such a case, it can construct a garbled circuit that computes a function that is different to the one that P_1 and P_2 agreed to compute. A folklore solution to this problem uses the "cut-and-choose" technique. According to this technique, P_1 first constructs *many* garbled circuits and sends them to P_2. Then, P_2 asks P_1 to "open" half of them (namely, reveal the decryption keys corresponding to these circuits). P_1 opens the requested half, and P_2 checks that they were constructed correctly. If they were, then P_2 evaluates the rest of the circuits and derives the output from them. The idea behind this methodology is that if a malicious P_1 constructs the circuits incorrectly, then P_2 will detect this with high probability. Clearly, this solution solves the problem of P_1 constructing the circuit incorrectly. However, it does not suffice. First, it creates new problems within itself. Most outstandingly, once the parties now evaluate a number of circuits, some mechanism must be employed to make sure that they use the same input when evaluating each circuit (otherwise, as we show below, an adversarial party could learn more information than allowed). Second, in order to present a proof of security based on *simulation*, there are additional requirements that are not dealt with by just employing cut-and-choose (e.g., input extraction). Third, the folklore description of cut-and-choose is very vague and there are a number of details that are crucial when implementing it. For example, if P_2 evaluates many circuits, then the protocol must specify what P_2 should do if it does not receive the same output in every circuit. If the protocol requires P_2 to abort in this case (because it detected cheating from P_1), then this behavior actually yields a concrete attack in which P_1 can always learn a specified bit of P_2's input. It can be shown that P_2 must take the majority output and proceed, even if it knows that P_1 has attempted to cheat. This is just one example of a subtlety that must be dealt with. Another example relates to the fact that P_1 may be able to construct a circuit that can be opened with two different sets of keys: the first set opens the circuit correctly and the second incorrectly. In such a case, an adversarial P_1 can pass the basic cut-and-choose test by opening the circuits to be checked correctly. However, it can also supply incorrect keys to the circuits to be computed and thus cause the output of the honest party to be incorrect.

Our contributions. This paper provides several contributions:

- *Efficient protocol for malicious parties:* We present an implementation of Yao's protocol with the cut-and-choose methodology, which is secure in the presence of malicious adversaries and is computationally efficient: the protocol does not use public-key operations, except for performing oblivious transfers for every input bit of P_2. For n-bit inputs and a statistical security parameter s the protocol uses $O(\max(s, n))$ oblivious transfers. Thus, when the input is as large as the security parameter, only $O(1)$ oblivious transfers are needed per input bit.

Beyond carefully implementing the cut-and-choose technique on the circuits in order to ensure that the garbled circuits are constructed correctly, we present a new method for enforcing the parties to use the same input in every circuit. This method involves "consistency checks" that are based on cut-and-choose tests which are applied to *sets of commitments* to the garbled values associated with the input wires of the circuit, rather than to the circuits themselves.

In actuality, we combine the cut-and-choose test over the circuits together with the cut-and-choose test over the commitments in order to obtain a secure solution. The test is rather complex conceptually, but is exceedingly simple to implement. Specifically, P_1 just needs to generate a number of commitments to the garbled values associated with the input wires, and then open them based on cut-and-choose queries from P_2. (Actually, these cut-and-choose queries are chosen jointly by the parties using a simple coin-tossing protocol; this is necessary for achieving simulation.)

We note that the use of cut-and-choose inevitably incurs a higher communication overhead. We also note that in this work we emphasized providing a clear and full proof of the protocol, rather than fully optimizing its overhead at the expense of complicating the proof.

- *Simulation based proof:* We present a *rigorous* proof of the security of the protocol, based on the real/ideal-model simulation paradigm [5,9]. The proof is in the standard model, with no random oracle model or common random string assumptions. The protocol was designed to support such a proof, rather than make do with separate proofs of privacy and correctness. (It is well-known that it is strictly harder to obtain a simulation based proof rather than security under such definitions.) One important advantage of simulation based proofs is that they enable the use of the protocol as a building block in more complicated protocols, while proving the security of the latter using general composition theorems like those of [5,9]. (For example, the secure protocol of [1] for finding the kth ranked element is based on invoking several secure computations of simpler functions, and provides simulation based security against malicious adversaries if the invoked computations have a simulation based proof. However, prior to our work there was no known way, except for the GMW compiler, of efficiently implementing these computations with this level of security.) See [5,9] for more discussion on the importance of simulation-based definitions.

- *A black-box reduction:* Our protocol can be interpreted as a constant-round black-box reduction of secure two-party computation to oblivious transfer and perfectly-hiding commitments. The perfectly-hiding commitments are only used for conducting, in $O(1)$ rounds, joint coin-tossing of a string of length s, where s is a statistical security parameter. This coin-tossing can be done sequentially (bit by bit), without using perfectly-hiding commitments. We therefore also obtain an $O(s)$ round black-box reduction of secure two-party computation to oblivious transfer alone. These two reductions are comparable to Kilian's reduction, which uses OT alone but incurs a number of rounds which is linear in the depth of the circuit [18]. In addition, our reduction is much more efficient than that of [18].

Related work. As we have mentioned, this paper presents a protocol which (**1**) has a proof of security against malicious adversaries in the standard model, according to the real/ideal model simulation definition, (**2**) has essentially the same computational overhead as Yao's original protocol (which is only secure against semi-honest adversaries), and (**3**) has a somewhat larger communication overhead, which depends on a statistical security parameter s.

We compare this result to other methods for securing Yao's protocol against malicious parties. There are several possible approaches to this task:

- The parties can reduce the statement about the honesty of their behavior to a statement which has a well-known zero-knowledge proof, and then prove this statement. This is the approach taken by the GMW compiler [11]. The resulting secure protocol is not black-box, and is rather inefficient.
- Another approach is to apply a cut-and-choose modification to Yao's protocol. Mohassel and Franklin [23] show such a protocol which has about the same overhead as ours, namely a communication overhead of $O(|C|s + n^2s)$ for a circuit C with n inputs, and a statistical security parameter s. This result was improved by Woodruff [28], who describes how to reduce the communication to $O(|C|s+ns) = O(|C|s)$, using expanders. The protocol of [23] provides output to the circuit evaluator alone. It enables, however, the circuit constructor to carry out the following attack: it can corrupt, say, its OT input which corresponds to a 0 value of the first input bit of the circuit evaluator, while not corrupting the OT input for the 1 value. Other than that it follows the protocol. This behavior forces the circuit evaluator to abort if its first input bit is 0, while if its first input bit is 1 it does not learn anything at all about the attack. If the evaluator complains, then the circuit constructor can conclude that its first input bit is 0, and therefore the evaluator cannot complain if it wants to preserve its privacy. (This attack is similar to the attack we describe in Section 3.2 where we discuss the encoding of P_2's input.) The protocol therefore does not provide security according to a standard definition. (We note however that this attack can be prevented using the methods we describe in Section 3.2 for encoding P_2's input.) Another protocol which is based on cut-and-choose is described in [19]. This protocol uses committed OT to address attacks similar to the one described above. We stress that both of these papers ([23,19]) lack a full proof of security, and to our best judgment they need considerable changes in order to support security according to a simulation based definition.
- Jarecki and Shmatikov [15] designed a protocol in which the parties efficiently prove, gate by gate, that their behavior is correct. The protocol is based on

the use of a special homomorphic encryption system, which is used to encode the gates of the table (compared to the use of symmetric encryption in Yao's original protocol and in our paper). The protocol is secure in a universally composable way under the decisional composite residuosity and the strong RSA assumptions, assuming a common reference string.

In this paper, we construct an efficient protocol for *general* secure computation. Thus, we do not (and cannot) compete with protocols that are constructed for specific tasks, like voting, auctions, etcetera. We also do not discuss here the large body of work that considers the efficiency of secure *multi-party* computation.

Organization. We present standard definitions of security for secure two-party computation in Section 2.1. Then, in Section 2.2 we show that a functionality that provides outputs to both parties can be securely reduced to one which provides output for a single party, and therefore we can focus on the latter case. In Section 3 we describe our protocol, prove its security, and analyze its efficiency. The basic protocol we describe increases the number of inputs, and therefore the number of OT invocations. In Section 5.2 we show how to reduce this number of OT invocations in order to improve efficiency. We remark that a description of Yao's basic protocol for two-party computation, secure against semi-honest adversaries, is provided in [20].

2   Preliminaries

2.1   Definitions – Secure Computation

In this section we present the definition for secure two-party computation. The following description and definition is based on [9, Chapter 7], which in turn follows [12,22,4,5].

Two-party computation. A two-party protocol problem is cast by specifying a random process that maps pairs of inputs to pairs of outputs (one for each party). We refer to such a process as a functionality and denote it $f : \{0,1\}^* \times \{0,1\}^* \to \{0,1\}^* \times \{0,1\}^*$, where $f = (f_1, f_2)$. That is, for every pair of inputs (x, y), the output-pair is a random variable $(f_1(x,y), f_2(x,y))$ ranging over pairs of strings. The first party (with input x) wishes to obtain $f_1(x,y)$ and the second party (with input y) wishes to obtain $f_2(x,y)$.

Adversarial behavior. Loosely speaking, the aim of a secure two-party protocol is to protect an honest party against dishonest behavior by the other party. In this paper, we consider *malicious adversaries* who may arbitrarily deviate from the specified protocol. When considering malicious adversaries, there are certain undesirable actions that cannot be prevented. Specifically, a party may refuse to participate in the protocol, may substitute its local input (and use instead a different input) and may abort the protocol prematurely. One ramification of the adversary's ability to abort, is that it is impossible to achieve "fairness". That is, the adversary may obtain its output while the honest party does not. As is standard for two-party computation, in this work we consider a static corruption model, where one of the parties is adversarial and the other is honest.

Security of protocols (informal). The security of a protocol is analyzed by comparing what an adversary can do in the protocol to what it can do in an ideal scenario that is secure by definition. This is formalized by considering an *ideal* computation involving an incorruptible *trusted third party* to whom the parties send their inputs. The trusted party computes the functionality on the inputs and returns to each party its respective output. Loosely speaking, a protocol is secure if any adversary interacting in the real protocol (where no trusted third party exists) can do no more harm than if it was involved in the above-described ideal computation.

Execution in the ideal model. As we have mentioned, some malicious behavior cannot be prevented (for example, early aborting). This behavior is therefore incorporated into the ideal model. An ideal execution proceeds as follows:

Inputs: Each party obtains an input, denoted w ($w = x$ for P_1, and $w = y$ for P_2).

Send inputs to trusted party: An honest party always sends w to the trusted party. A malicious party may, depending on w, either abort or send some $w' \in \{0,1\}^{|w|}$ to the trusted party.

Trusted party answers first party: In case it has obtained an input pair (x, y), the trusted party first replies to the first party with $f_1(x, y)$. Otherwise (i.e., in case it receives only one valid input), the trusted party replies to both parties with a special symbol $\bot$.

Trusted party answers second party: In case the first party is malicious it may, depending on its input and the trusted party's answer, decide to *stop* the trusted party by sending it $\bot$. In this case the trusted party sends $\bot$ to the second party. Otherwise the trusted party sends $f_2(x, y)$ to the second party.

Outputs: An honest party always outputs the message it has obtained from the trusted party. A malicious party may output an arbitrary (probabilistic polynomial-time computable) function of its initial input and the message obtained from the trusted party.

Let $f : \{0,1\}^* \times \{0,1\}^* \mapsto \{0,1\}^* \times \{0,1\}^*$ be a functionality, where $f = (f_1, f_2)$, and let $\overline{M} = (M_1, M_2)$ be a pair of non-uniform probabilistic *expected* polynomial-time machines (representing parties in the ideal model). Such a pair is admissible if for at least one $i \in \{1, 2\}$ we have that M_i is honest (i.e., follows the honest party instructions in the above-described ideal execution). Then, the joint execution of f under $\overline{M}$ in the ideal model (on input pair (x, y)), denoted $\text{IDEAL}_{f, \overline{M}}(x, y)$, is defined as the output pair of M_1 and M_2 from the above ideal execution.

Execution in the real model. We next consider the real model in which a real (two-party) protocol is executed (and there exists no trusted third party). In this case, a malicious party may follow an arbitrary feasible strategy; that is, any strategy implementable by non-uniform probabilistic polynomial-time machines.

Let f be as above and let Π be a two-party protocol for computing f. Furthermore, let $\overline{M} = (M_1, M_2)$ be a pair of non-uniform probabilistic polynomial-time machines (representing parties in the real model). Such a pair is admissible if for at least one $i \in \{1, 2\}$ we have that M_i is honest (i.e., follows the strategy specified by Π). Then, the joint execution of Π under $\overline{M}$ in the real model (on input pair (x, y)), denoted $\text{REAL}_{\Pi, \overline{M}}(x, y)$, is defined as the output pair of M_1 and M_2 resulting from the protocol interaction.

Security as emulation of a real execution in the ideal model. Having defined the ideal and real models, we can now define security of protocols. Loosely speaking, the definition asserts that a secure two-party protocol emulates the ideal model (in which a trusted party exists). This is formulated by saying that admissible pairs in the ideal model are able to simulate admissible pairs in an execution of a secure real-model protocol.

Definition 1. (secure two-party computation): *Let f and Π be as above. Protocol Π is said to* securely compute f *(in the malicious model) if for every pair of admissible non-uniform probabilistic polynomial-time machines $\overline{A} = (A_1, A_2)$ for the real model, there exists a pair of admissible non-uniform probabilistic expected polynomial-time machines $\overline{B} = (B_1, B_2)$ for the ideal model, such that*

$$\left\{ \text{IDEAL}_{f, \overline{B}}(x, y) \right\}_{x,y \text{ s.t. } |x|=|y|} \overset{\text{c}}{\equiv} \left\{ \text{REAL}_{\Pi, \overline{A}}(x, y) \right\}_{x,y \text{ s.t. } |x|=|y|}$$

Namely, the two distributions are computationally indistinguishable.

We note that the above definition assumes that the parties know the input lengths (this can be seen from the requirement that $|x| = |y|$). Some restriction on the input lengths is unavoidable, see [9, Section 7.1] for discussion. We also note that we allow the ideal adversary/simulator to run in expected (rather than strict) polynomial-time. This is essential for achieving constant-round protocols; see [3].

We denote the security parameter by n and, for the sake of simplicity, unify it with the length of the inputs (thus we consider security for "all sufficiently long inputs"). Everything in the paper remains the same if a separate security parameter n is used, and we consider security for inputs of all lengths. We will also use a statistical security parameter s; see the beginning of Section 3.1 for an explanation of the use of this separate parameter.

The hybrid model. Our protocol uses a secure oblivious transfer protocol as a subprotocol. It has been shown in [5] that it suffices to analyze the security of such a protocol in a hybrid model in which the parties interact with each other *and* have access to a trusted party that computes the oblivious transfer protocol for them. We remark that the composition theorem of [5] holds for the case that the subprotocol executions are all run sequentially (and the messages of the protocol calling the subprotocol do not overlap with any execution). We also remark that if the oblivious transfer subprotocol is secure under parallel composition, then it is straightforward to extend [5] so that the subprotocols may be run in parallel (again, as long as the messages of the protocol calling the subprotocol do not overlap with any execution).

2.2   Functionalities That Provide Output to a Single Party

In the definition above, we have considered the case that both parties receive output, and these outputs may be *different*. However, the presentation of our protocol is far simpler for the case that only party P_2 receives output. We will show now that this suffices for the general case. That is, any protocol that can securely compute *any* efficient functionality $f(x, y)$ where only P_2 receives output, can be used to securely compute *any* efficient functionality $f = (f_1, f_2)$ where party P_1 receives $f_1(x, y)$ and party P_2 receives $f_2(x, y)$.

Let $f = (f_1, f_2)$ be a functionality. We wish to construct a secure protocol in which P_1 receives $f_1(x, y)$ and P_2 receives $f_2(x, y)$; as a building block we use a protocol for computing any efficient functionality with the limitation that only P_2 receives output. Let $\mathcal{F}$ be a field that contains the range of values $\{f_1(x, y)\}_{x, y \in \{0,1\}^n}$, and let p, a, b be randomly chosen elements in $\mathcal{F}$. Then, in addition to x, party P_1's input includes the elements p, a, b. Furthermore, define a functionality g (that has only a single output) as follows: $g((p, a, b, x), y) = (\alpha, \beta, f_2(x, y))$, where $\alpha = p + f_1(x, y)$, $\beta = a \cdot \alpha + b$, and the arithmetic operations are defined in $\mathcal{F}$. Note that α is a one-time pad encryption of P_1's output $f_1(x, y)$, and β is an information-theoretic message authentication tag of α (specifically, $a\alpha + b$ is a pairwise-independent hash of α). Now, the parties compute the functionality g, using a secure protocol in which only P_2 receives output. Following this, P_2 sends the pair (α, β) to P_1. Party P_1 checks that $\beta = a \cdot \alpha + b$; if yes, it outputs $\alpha - p$, and otherwise it outputs $\perp$.

It is easy to see that P_2 learns nothing about P_1's output $f_1(x, y)$, and that it cannot alter the output that P_1 will receive (beyond causing it to abort), except with probability $1/|\mathcal{F}|$. (We assume that $1/|\mathcal{F}|$ is the required probability for detecting attempts to alter the output. If it is required instead that any change by P_2 to P_1's output is detected with probability 2^{-s}, then the parameters a, b and the computation of $\beta = a \cdot \alpha + b$ can be defined in a field whose representation is s bits long.) We remark that it is also straightforward to construct a simulator for the above protocol. (Note that in order to meet Definition 1, one must actually switch the roles of P_1 and P_2 above.)

We remark that the circuit for computing g is only mildly larger than that for computing f. Thus, the construction above is also efficient and has only a mild effect on the complexity of the secure protocol.

3   The Protocol

Our protocol is based upon Yao's garbled circuit construction, which is secure in the presence of semi-honest adversaries [29]. That protocol has two parties: P_1 (who is the the *sender*, or circuit *constructor*), and P_2 (who is the *receiver*, or the circuit *evaluator*). The protocol is described and proved in [20]. Our presentation from here on assumes full familiarity with Yao's basic protocol.

There are a number of issues that must be dealt with when attempting to make Yao's protocol secure against malicious adversaries rather than just semi-honest ones (beyond the trivial observation that the oblivious transfer subprotocol must now be secure in the presence of malicious adversaries).

First and foremost, a malicious P_1 must be forced to construct the garbled circuit correctly so that it indeed computes the desired function. The method that is typically referred to for this task is called *cut-and-choose*. According to this methodology, P_1 constructs many independent copies of the garbled circuit and sends them to P_2. Party P_2 then asks P_1 to open half of them (chosen randomly). After P_1 does so, and party P_2 checks that the opened circuits are correct, P_2 is convinced that most of the remaining (unopened) garbled circuits are also constructed correctly. (If there are many incorrectly constructed circuits, then with high probability, one of those circuits will be in the set that P_2 asks to open.) The parties can then evaluate the remaining unopened garbled circuits as in the original protocol for semi-honest adversaries, and take the majority output-value.[1]

The cut-and-choose technique described above indeed solves the problem of a malicious P_1 constructing incorrect circuits. However, it also generates new problems! The primary problem that arises is that since there are now many circuits being evaluated, we must make sure that both P_1 and P_2 use the same inputs in each circuit; we call these *consistency checks*. (Consistency checks are important since if the parties were able to provide different inputs to different copies of the circuit, then they can learn information that is different from the desired output of the function. It is obvious that P_2 can do so, since it observes the outputs of all circuits, but in fact even P_1, who only gets to see the majority

---

[1] The reason for taking the majority value as the output is that the aforementioned test only reveals a single incorrectly constructed circuit with probability $1/2$. Therefore, if P_1 generates a single or constant number of "bad" circuits, there is a reasonable chance that it will not be caught. In contrast, there is only an exponentially small probability that the test reveals no corrupt circuit *and* at the same time a majority of the circuits that are not checked are incorrect. Consequently, with overwhelming probability it holds that if the test succeeds and P_2 takes the majority result of the remaining circuits, the result is correct. We remark that the alternative of aborting in case not all the outputs are the same (namely, where cheating is detected) is not secure and actually yields a concrete attack. The attack works as follows. Assume that P_1 is corrupted and that it constructs all of the circuits correctly except for one. The "incorrect circuit" is constructed so that it computes the exclusive-or of the desired function f with the first bit of P_2's input. Now, if P_2's policy is to abort as soon as two outputs are not the same then P_1 learns the first bit of P_2's input.

output, can learn additional information: information[2].) Another problem that arises when proving security is that the simulator must be able to fool P_2 and give it incorrect circuits (even though P_2 runs a cut-and-choose test). This is solved using rather standard techniques, like choosing the circuits to be opened via a coin-tossing protocol (to our knowledge, this issue has gone unnoticed in all previous applications of cut-and-choose to Yao's protocol). Yet another problem is that P_1 might provide corrupt inputs to some of P_2's possible choices in the OT protocols. P_1 might then learn P_2's input based on whether or not P_2 aborts the protocol.

We begin by presenting a high-level overview of the protocol. We then proceed to describe the consistency checks, and finally the full protocol.

3.1   High-Level Overview

We work with two security parameters. The parameter n is the security parameter for the commitment schemes, encryption, and the oblivious transfer protocol. The parameter s is a statistical security parameter which specifies how many garbled circuits are used. The difference between these parameters is due to the fact that the value of n depends on computational assumptions, whereas the value of s reflects the possible error probability that is incurred by the cut-and-choose technique and as such is a "statistical" security parameter. Although it is possible to use a single parameter n, it may be possible to take s to be much smaller than n. Recall that for simplicity, and in order to reduce the number of parameters, we denote the length of the input by n as well.

Protocol 1. (high-level overview): *Parties P_1 and P_2 have respective inputs x and y, and wish to compute the output $f(x, y)$ for P_2.*

 0. The parties decide on a circuit computing f. They then change the circuit by replacing each input wire of P_2 by a gate whose input consists of s new input wires of P_2 and whose output is the exclusive-or of these wires (such an s-bit exclusive-or gate can be implemented using $s-1$ two-bit exclusive-or gates). Consequently, the number of input wires of P_2 increases by a factor of s. (In Section 5.2, we show how to reduce the number of inputs.)
 1. P_1 commits to s different garbled circuits computing f, where s is a statistical security parameter. P_1 also generates additional commitments to the garbled values corresponding to the input wires of the circuits. These commitments are constructed in a special way in order to enable consistency checks.
 2. For every input bit of P_2, parties P_1 and P_2 run a 1-out-of-2 oblivious transfer protocol in which P_2 learns the garbled values of input wires corresponding to its input.

---

[2] Suppose, for example, that the protocol computes n invocations of a circuit computing the inner-product between n bit inputs. A malicious P_2 could provide the inputs $\langle 10 \cdots 0 \rangle$, $\langle 010 \cdots 0 \rangle$,...,$\langle 0 \cdots 01 \rangle$, and learn all of P_1's input. If, on the other hand, P_1 is malicious, it could also provide the inputs $\langle 10 \cdots 0 \rangle$, $\langle 010 \cdots 0 \rangle$,...,$\langle 0 \cdots 01 \rangle$. In this case, P_2 sends it the value which is output by the majority of the circuits, and which is equal to the majority value of P_2's input bits.

3. P_1 sends to P_2 all the commitments of Step 1.
4. P_1 and P_2 run a coin-tossing protocol in order to choose a random string that defines which commitments and which garbled circuits will be opened.
5. P_1 opens the garbled circuits and committed input values that were chosen in the previous step. P_2 verifies the correctness of the opened circuits and runs consistency checks based on the decommitted input values.
6. P_1 sends P_2 the garbled values corresponding to P_1's input wires in the unopened circuits. P_2 runs consistency checks on these values as well.
7. Assuming that all of the checks pass, P_2 evaluates the unopened circuits and takes the majority value as its output.

3.2   Checks for Correctness and Consistency

As can be seen from the above overview, P_1 and P_2 run a number of checks, with the aim of forcing a potentially malicious P_1 to construct the circuits correctly and use the same inputs in (most of) the evaluated circuits. This section describes these checks.

Encoding P_2's input: As mentioned above, a malicious P_1 may provide corrupt input to one of P_2's possible inputs in an OT protocol. If P_2 chooses to learn this input it will not be able to decode the garbled tables which use this value, and it will therefore have to abort. If P_2 chooses to learn the other input associated with this wire then it will not notice that the first input is corrupt. P_1 can therefore learn P_2's input based on whether or not P_2 aborts. (Note that checking that the circuit is well-formed will not help in thwarting this attack, since the attack is based on changing P_1's input to the OT protocol.) The attack is prevented by the parties replacing each input bit of P_2 with s new input bits whose exclusive-or is used instead of the original input (this step was described as Step 0 of Protocol 1). P_2 therefore has 2^{s-1} ways to encode a 0 input, and 2^{s-1} ways to encode a 1, and given its input it chooses the encoding to use with uniform probability. The parties then execute the protocol with the new circuit, and P_2 uses oblivious transfer to learn the garbled values of its new inputs. As is shown in the full paper, if P_1 supplies incorrect values as garbled values that are associated with P_2's input, the probability of P_2 detecting this cheating is *almost independent* (up to a bias of 2^{-s+1}) of P_2's actual input. This is not true if P_2's inputs are not "split" in the way described above. The encoding presented here increases the number of P_2's input bits and, respectively, the number of OTs, from n to ns. In Section 5.2 we show how to reduce the number of new inputs for P_2 (and thus OTs) to a total of only $O(\max(s, n))$.

An unsatisfactory method for proving consistency of P_1's input: Consider the following idea for forcing P_1 to provide the same input to all circuits. Let s be a security parameter and assume that there are s garbled copies of the circuit. Then, P_1 generates two ordered sets of commitments for every wire of the circuit. Each set contains s commitments: the "0 set" contains commitments to the garbled encodings of 0 for this wire in every circuit, and the "1 set" contains commitments to the garbled encodings of 1 for this wire in every circuit. P_2 receives these commitments from P_1 and then chooses a random subset of the circuits, which will be defined as check-circuits. These circuits will never be evaluated and are used only for checking correctness and consistency. Specifically, P_2 asks P_1 to de-garble all of the check-circuits and to open the values that correspond to the check-circuits in *both* commitment sets. (That is, if circuit i is a check-circuit, then P_1 decommits to both the 0 encoding and 1 encoding of all the input wires in circuit i.) Upon receiving the decommitments, P_2 verifies that all opened commitments from the "0 set" correspond to garbled values of 0, and that a similar property holds for commitments from the "1 set".

It now remains for P_2 to evaluate the remaining circuits. In order to do this, P_1 provides (for each of its input wires) the garbled values that are associated with the wire in all of the remaining circuits. Then, P_1 must prove that all of these values come from the same set, without revealing whether the set that they come from is the "0 set" or the "1 set" (otherwise, P_2 will know P_1's input). In this way, on the one hand, P_2 does not learn the input of P_1, and on the other hand, it is guaranteed that all of the values come from the same set, and so P_1 is forced into using the same input in all circuits. This proof can be carried out using, for example, the proofs of partial knowledge of [6]. However, this would require n proofs, each for s values, thereby incurring $O(ns)$ costly asymmetric operations which we want to avoid.

Proving consistency of P_1's input: P_1 can prove consistency of its inputs without using public-key operations. The proof is based on a cut-and-choose test for the consistency of the commitment sets, which is combined with the cut-and-choose test for the correctness of the circuits. (Note that in the previous proposal, there is only one cut-and-choose test, and it is for the correctness of the circuits.) We start by providing a high level description of the proof of consistency: The proof is based on P_1 constructing, *for each of its input wires*, s pairs of sets of commitments. One set in every pair contains commitments to the 0 values of this wire in *all* circuits, and the other set is the same with respect to 1. The protocol chooses a random subset of these pairs, and a random subset of the circuits, and checks that these sets provide consistent inputs for these circuits. Then the protocol evaluates the *remaining* circuits, and asks P_1 to open, in each of the *remaining* pairs, and only in one set in every pair, its garbled values for all evaluated circuits. (In this way, P_2 does not learn whether these garbled values correspond to a 0 or to a 1.) In order for the committed sets and circuits to pass P_2's checks, there must be large subsets C and S, of the circuits and commitment sets, respectively, such that every choice of a circuit from C and a commitment set from S results in a circuit and garbled values which compute the desired function f. P_2 accepts the verification stage only if all the circuits and sets it chooses to check are from C and S, respectively. This means that if P_2 does not abort then circuits which are not from C are likely to be a minority of the evaluated circuits, and a similar argument holds for S. Therefore the majority result of the evaluation stage is correct. The exact construction is as follows:

STAGE 1 – COMMITMENTS: P_1 generates s garbled versions of the circuit. Furthermore, it generates commitments to the garbled values of the wires corresponding to P_2's input in each circuit. These commitments are generated in ordered pairs so that the first item in a pair corresponds to the 0 value and the second to the 1 value. The procedure regarding the input bits of P_1 is more complicated (see Figure 1 for a diagram explaining this construction). P_1 generates s pairs of sets of committed values for each of its input wires. Specifically, for every input wire i of P_1, it generates s sets of the form $\{W_{i,j}, W'_{i,j}\}_{j=1}^{s}$; we call these **commitment sets**. Before describing the content of these sets, denote by $k_{i,r}^{b}$ the garbled value that is assigned to the value $b \in \{0,1\}$ in wire i of circuit r. Then, the sets $W_{i,j}$ and $W'_{i,j}$ both contain $s+1$ commitments and are defined as follows. Let $b \in_R \{0,1\}$ be a random bit, chosen independently for every $\{W_{i,j}, W'_{i,j}\}$ pair. Define $W_{i,j}$ to contain a commitment to b, as well as commitments to the garbled value corresponding to b in wire i in all of the s circuits, and define $W'_{i,j}$ similarly, but with respect to $1-b$. In other words, $W_{i,j} = \{\mathsf{com}(b), \mathsf{com}(k_{i,1}^{b}), \ldots, \mathsf{com}(k_{i,s}^{b})\}$ and $W'_{i,j} = \{\mathsf{com}(1-b), \mathsf{com}(k_{i,1}^{1-b}), \ldots, \mathsf{com}(k_{i,s}^{1-b})\}$. We stress that in each of the pairs $(W_{i,1}, W'_{i,1}), \ldots,$ $(W_{i,s}, W'_{i,s})$, the values that are committed to are the same. The only difference is that independent randomness is used in each pair for choosing b and constructing the commitments. We call the first bit committed to in a commitment set the **indicator bit**.

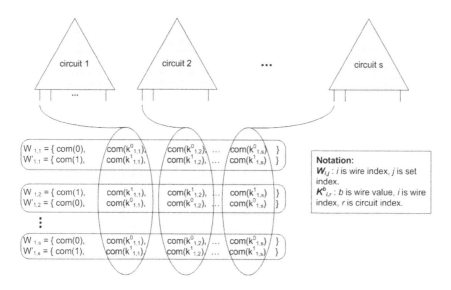

Fig. 1. The commitment sets corresponding to P_1's **first** input wire

After constructing these circuits and commitment sets, party P_1 sends to P_2 all of the s garbled circuits (i.e., the garbled gate-tables and output-tables, but *not* the garbled values corresponding to the input wires), and all the commitment sets. Note that if P_1's input is of length n, then there are sn pairs of commitment *sets*; and a total of $sn(2s + 2) = O(s^2n)$ commitments.

STAGE 2 – CHALLENGE: Two random strings $\rho, \rho' \in_R \{0,1\}^s$ are chosen and sent to P_1 (in the actual protocol, these strings are determined via a simple coin-tossing protocol). The string ρ is a challenge indicating which garbled circuits to open, and the string ρ' is a challenge indicating which commitment sets to open. We call the opened circuits check-circuits and the unopened ones evaluation-circuits. Likewise, we call the opened sets check-sets and the unopened ones evaluation-sets. A circuit (resp., commitment set) is defined to be a check-circuit (resp., check-set) if the corresponding bit in ρ (resp., ρ') equals 1; otherwise, it is defined to be an evaluation-circuit (resp., evaluation-set).

STAGE 3 – OPENING: First, party P_1 opens all the commitments corresponding to P_2's input wires in all of the check-circuits. Second, in all of the *check-sets* P_1 opens the commitments that correspond to *check-circuits*. That is, if circuit r is a check circuit, then P_1 decommits to all of the values $\mathsf{com}(k^0_{i,r}), \mathsf{com}(k^1_{i,r})$ in check-sets, where i is any of P_1's input bits. Finally, for every check-set, P_1 opens the commitment to the indicator bit, the initial value in each of the sets $W_{i,j}, W'_{i,j}$. See Figure 2 for a diagram in which the values which are opened are highlighted (the diagram refers to only one of P_1's input wires in the circuit).

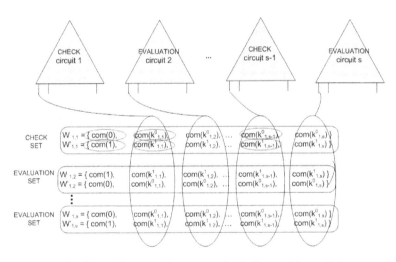

Fig. 2. In every check-set, the commitment to the indicator bit, and the commitments corresponding to check-circuits are all opened.

STAGE 4 – VERIFICATION: In this step, party P_2 verifies that all of the check-circuits were correctly constructed. In addition, it verifies that all of the opened commitments in sets whose first item is a commitment to 0 are to garbled encodings of 0; likewise for 1. These checks are carried out as follows. First, in all of the check-circuits, P_2 receives the decommitments to the garbled values corresponding to its own input, and by the order of the commitments P_2 knows which value corresponds to 0 and which value corresponds to 1. Second, for every check-circuit, P_2 receives decommitments to the garbled input values of P_1 in all the check-sets, along with a bit indicating whether these garbled values correspond to 0 or to 1. It first checks that for every wire, the garbled values of 0 (resp., of 1) are all equal. Then, the above decommitments enable the complete opening of the garbled circuits (i.e., the decryption of all of the garbled tables). Once this has been carried out, it is possible to simply check that the check-circuits are all correctly constructed. Namely, that they agree with a specific and agreed-upon circuit computing f.

STAGE 5 – EVALUATION AND VERIFICATION: Party P_1 reveals the garbled values corresponding to its input: If i is a wire that corresponds to a bit of P_1's input and r is an evaluation-circuit, then P_1 decommits to the commitments $k_{i,r}^b$ in all of the *evaluation-sets*, where b is the value of its input bit. This is depicted in Figure 3. Finally, P_2 verifies that (1) for every input wire, all of the opened commitments that were opened in evaluation-sets contain the same garbled value, and (2) for every i, j P_1 opened commitments of evaluated circuits in exactly one of $W_{i,j}$ or $W_{i,j}'$. If these checks pass, it continues to evaluate the circuit.

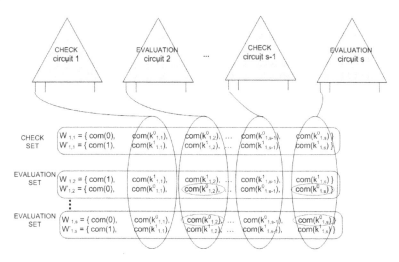

Fig. 3. P_1 opens in the evaluation-sets, the commitments that correspond to its input. In every evaluation-set these commitments come from the same item in the pair.

Intuition. Having described the mechanism for checking consistency, we now provide some intuition as to why it is correct. A simple cut-and-choose check verifies that most of the evaluated circuits are correctly constructed. The main remaining issue is ensuring that P_1's inputs to most circuits are consistent. If P_1 wants to provide different inputs to a certain wire in two circuits, then *all* the $W_{i,j}$ (or $W_{i,j}'$) sets it opens in *evaluation-sets* must contain a commitment to 0 in the first circuit and a commitment to 1 in the other circuit. However, if any of these sets is chosen to be checked, and the circuits are among the checked circuits, then P_2 aborts. This means that if P_1 attempts to provide different inputs to two circuits and they are checked, it is almost surely caught. Now, since P_2 outputs the majority output of the evaluated circuits, the result is affected by P_1 providing different inputs only if these inputs affect a constant fraction of the circuits. But since all of these circuits must not be checked, P_1's probability of success is exponentially small in s.

3.3   The Full Protocol

We now describe the full protocol in detail. We use the notation com_b to refer to a perfectly binding commitment scheme, and com_h to refer to a perfectly hiding commitment scheme (See [8] for definitions).

Protocol 2. (protocol for computing $f(x,y)$):

- **Input:** P_1 has input $x \in \{0,1\}^n$ and P_2 has input $y \in \{0,1\}^n$.

- **Auxiliary input:** a statistical security parameter s and the description of a circuit C^0 such that $C^0(x,y) = f(x,y)$.

- **Specified output:** party P_2 should receive $f(x,y)$ and party P_1 should receive no output. (Recall that this suffices for the general case where both parties receive possibly different outputs; see Section 2.2.)

- **The protocol**
  0. CIRCUIT CONSTRUCTION: *The parties replace C^0 with a circuit C which is constructed by replacing each input wire of P_2 by the result of an exclusive-or of s new input wires of P_2. (We show in Section 5.2 how the number of new input bits can be reduced.) The number of input wires of P_2 is increased from $|y| = n$ to sn. Let the bit-wise representation of P_2's original input be $y = y_1 \ldots y_n$. Denote its new input as $\hat{y} = \hat{y}_1, \ldots, \hat{y}_{ns}$. P_2 chooses its new input at random subject to the constraint $y_i = \hat{y}_{(i-1)\cdot s+1} \oplus \cdots \oplus \hat{y}_{i\cdot s}$.*

  1. COMMITMENT CONSTRUCTION: P_1 constructs the circuits and commits to them, as follows:
     (a) *P_1 constructs s independent copies of a garbled circuit of C, denoted $GC_1, \ldots, GC_s$.*

     (b) *P_1 commits to the garbled values of the wires corresponding to P_2's input to each circuit. That is, for every input wire i corresponding to an input bit of P_2, and for every circuit GC_r, P_1 computes the ordered pair $(\mathsf{com}_b(k_{i,r}^0), \mathsf{com}_b(k_{i,r}^1))$, where $k_{i,r}^b$ is the garbled value associated with b on input wire i in circuit GC_r.*

     (c) *P_1 computes commitment-sets for the garbled values that correspond to its own inputs to the circuits. That is, for every wire i that corresponds to an input bit of P_1, it generates s pairs of commitment sets $\{W_{i,j}, W'_{i,j}\}_{j=1}^s$, in the following way:*

Denote by $k_{i,r}^b$ the garbled value that was assigned by P_1 to the value $b \in \{0,1\}$ of wire i in GC_r. Then, P_1 chooses $b \in_R \{0,1\}$ and computes

$$W_{i,j} = \langle \mathsf{com}_b(b), \mathsf{com}_b(k_{i,1}^b), \ldots, \mathsf{com}_b(k_{i,s}^b) \rangle, \qquad \text{and}$$
$$W_{i,j}' = \langle \mathsf{com}_b(1-b), \mathsf{com}_b(k_{i,1}^{1-b}), \ldots, \mathsf{com}_b(k_{i,s}^{1-b}) \rangle$$

For each i, j, the sets are constructed using independent randomness, and in particular the value of b is chosen independently for every $j = 1 \ldots s$. There are a total of ns commitment-sets. We divide them into s supersets, where superset S_j is defined as $S_j = \{(W_{1,j}, W_{1,j}'), \ldots, (W_{n,j}, W_{n,j}')\}$. Namely, S_j is the set containing the jth commitment set for all wires.

2. OBLIVIOUS TRANSFERS: For every input bit of P_2, parties P_1 and P_2 run a 1-out-of-2 oblivious transfer protocol in which P_2 receives the garbled values for the wires that correspond to its input bit (in every circuit). That is, let $c_{i,r}^b$ denote the commitment to the garbled value $k_{i,r}^b$ and let $dc_{i,r}^b$ denote the decommitment value for $c_{i,r}^b$. Furthermore, let $i_1, \ldots, i_{ns}$ be the input wires that correspond to P_2's input.

Then, for every $j = 1, \ldots, ns$, parties P_1 and P_2 run a 1-out-of-2 OT protocol in which:
(a) P_1's input is the pair of vectors $([dc_{i_j,1}^0, \ldots, dc_{i_j,s}^0], [dc_{i_j,1}^1, \ldots, dc_{i_j,s}^1])$.
(b) P_2's input is its jth input bit $\hat{y}_j$ (and its output should thus be $[dc_{i_j,1}^{\hat{y}_j}, \ldots, dc_{i_j,s}^{\hat{y}_j}]$).
If the oblivious transfer protocol provides security for parallel execution, then these executions are run in parallel. Otherwise, they are run sequentially.

3. SEND CIRCUITS AND COMMITMENTS: P_1 sends to P_2 the garbled circuits (i.e., the gate and output tables), as well as all of the commitments that it prepared above.

4. PREPARE CHALLENGE STRINGS: (1) P_2 chooses a random string $\rho_2 \in_R \{0,1\}^s$ and sends $\mathsf{com}_h(\rho_2)$ to P_1. (2) P_1 chooses a random string $\rho_1 \in \{0,1\}^s$ and sends $\mathsf{com}_b(\rho_1)$ to P_2. (3) P_2 decommits, revealing ρ_2. (4) P_1 decommits, revealing ρ_1. (5) P_1 and P_2 set $\rho = \rho_1 \oplus \rho_2$. The above steps are run a second time, defining an additional string ρ'.[3]

5. DECOMMITMENT PHASE FOR CHECK-CIRCUITS: From here on, we refer to the circuits for which the corresponding bit in ρ is 1 as check-circuits,

---

[3] Recall that ρ and ρ' are used to ensure that P_1 constructs the circuits correctly and uses consistent input in each circuit. Thus, it may seem strange that they are generated via a coin-tossing protocol, and not just chosen singlehandedly by P_2. Indeed, in order to prove the security of the protocol when P_1 is corrupted, there is no need for a coin-tossing protocol here. However, having P_2 choose ρ and ρ' singlehandedly creates a problem for the simulation in the case that P_2 is corrupted. We therefore use a coin-tossing protocol instead.

and we refer to the other circuits as evaluation-circuits. *Likewise, if the jth bit of ρ' equals 1, then the commitments sets in $S_j = \{(W_{i,j}, W'_{i,j})\}_{i=1...n}$ are referred to as* check-sets; *otherwise, they are referred to as* evaluation-sets.

For every check-circuit GC_r, *party P_1 operates in the following way:*

(a) *For every input wire i corresponding to an input bit of P_2, party P_1 decommits to the pair $(\mathsf{com}(k^0_{i,r}), \mathsf{com}(k^1_{i,r}))$ (namely to both of P_2's inputs).*

(b) *For every input wire i corresponding to an input bit of P_1, party P_1 decommits to the appropriate values in the superset S_j, in the check-sets $\{W_{i,j}, W'_{i,j}\}$. Specifically, P_1 decommits to the $\mathsf{com}(k^0_{i,r})$ and $\mathsf{com}(k^1_{i,r})$ values in $(W_{i,j}, W'_{i,j})$, for every check-set S_j (see Figure 2).*

For every pair of check-sets $(W_{i,j}, W'_{i,j})$, *party P_1 decommits to the first value in each set (i.e., to the value that is supposed to be a commitment to the indicator bit, $\mathsf{com}(0)$ or $\mathsf{com}(1)$).*

6. DECOMMITMENT PHASE FOR P_1'S INPUT IN EVALUATION-CIRCUITS: *P_1 decommits to the garbled values that correspond to its inputs in evaluation-circuits. Let i be the index of an input wire that corresponds to P_1's input (the following procedure is applied to all such wires). Let b be the binary value that P_1 assigns to input wire i. In every evaluation-set $(W_{i,j}, W'_{i,j})$, P_1 chooses the set (out of $(W_{i,j}, W'_{i,j})$), which corresponds to the value b. It then opens in this set the commitments that correspond to evaluation-circuits. Namely, to the values $k^b_{i,r}$, where r is an index of an evaluation circuit (see Figure 3).*

7. CORRECTNESS AND CONSISTENCY CHECKS: *P_2 performs the following checks; if any of them fails it aborts.*

(a) Checking correctness of the check-circuits: *P_2 verifies that each check-circuit GC_i is a garbled version of C. This check is carried out by P_2 first constructing the input tables that associate every garbled value of an input wire to a binary value. The input tables for P_2's inputs are constructed by checking that the decommitments to the pairs $(\mathsf{com}(k^0_{i,r}), \mathsf{com}(k^1_{i,r}))$ (where i is a wire index and r is a circuit index) are valid, and then interpreting the first value to be associated with 0 and the second value to be associated with 1.*

Next, P_2 checks the decommitments to P_1's inputs. This check involves first checking that the decommitment values are valid. Then, P_2 verifies that in each pair of check-sets, one of $(W_{i,j}, W'_{i,j})$ begins with a commitment to 0 (henceforth the 0-tuple), and the other begins with a commitment to 1 (henceforth the 1-tuple). Then P_2 checks that for every wire, the values that are decommitted to in the 0-tuples in all check-sets are all equal, and that a similar property holds for the 1-tuples. P_2 then assigns the logical value of 0 to all of the opened *commitments in the 0-tuples, and the logical value of 1 to the opened commitments in the 1-tuples.*

Finally, given all the garbled values to the input wires and their associated binary values, P_2 decrypts the circuit and compares it with the circuit C.

(b) Verifying P_2's input in the check-circuits: P_2 verifies that P_1's decommitments to the wires corresponding to P_2's input values in the check-circuits are correct, and agree with the logical values of these wires (the indicator bits). P_2 also checks that the inputs it learned in the oblivious transfer stage for the check-circuits correspond to its actual input. Specifically, it checks that the decommitment values that it received in the oblivious transfer stage open the committed values that correspond to the garbled values of its logical input (namely, that it received the first value in the pair if the input bit is 0 and the second value if it is 1).[4]

(c) Checking P_1's input to evaluation-circuits: Finally, P_2 verifies that for every input wire i of P_1 the following two properties hold:

　i. In every evaluation-set P_1 chose one of the two sets and decommitted to all the commitments in it which correspond to evaluation-circuits.

　ii. For every evaluation-circuit, all of the commitments that P_1 opened in evaluation-sets are for the same garbled value.

8. CIRCUIT EVALUATION: *If any of the above checks fails, P_2 aborts and outputs $\perp$. Otherwise, P_2 evaluates the evaluation circuits (in the same way as for the semi-honest protocol of Yao). It might be that in certain circuits the garbled values provided for P_1's inputs, or the garbled values learned by P_2 in the OT stage, do not match the tables and so decryption of the circuit fails. In this case P_2 also aborts and outputs $\perp$. Otherwise, P_2 takes the output that appears in most circuits, and outputs it (the proof shows that this value is well defined).*

4   Proof of Security

The security of Protocol 2 is stated in the following theorem.

Theorem 2. *Let $f : \{0,1\}^* \times \{0,1\}^* \to \{0,1\}^*$ be any probabilistic polynomial-time two-party functionality and consider the instantiation of Protocol 2 for functionality f. Assume that the oblivious transfer protocol is secure, that com_b is a perfectly-binding commitment scheme, that com_h is a perfectly-hiding commitment scheme, and that the garbled circuits are constructed as in [20]. Then, Protocol 2 securely computes f.*

---

[4] This check is crucial and thus the order of first running the oblivious transfer and then sending the circuits and commitments is not at all arbitrary.

The theorem is proved in two stages: first for the case that P_1 is corrupted and next for the case that P_2 is corrupted. The proof is provided in the full version of this paper. We highlight here the basic intuition behind the proof.

Security against a Malicious P_1. The proof constructs an ideal-model adversary/simulator which has access to P_1 and to the trusted party, and can simulate the view of an actual run of the protocol. It uses the fact that the strings ρ, ρ', which choose the circuits and commitment sets that are checked, are uniformly distributed even if P_1 is malicious. The simulator runs the protocol until P_1 opens the commitments of the checked circuits and checked commitment sets, and then rewinds the execution and runs it again with new random ρ, ρ' values. We expect that about one quarter of the circuits are checked in the first execution *and* evaluated in the second execution. For these circuits, in the first execution the simulator learns the translation between the garbled values of P_1's input wires and the actual values of these wires, and in the second execution it learns the garbled values that are associated with P_1's input (this association is learned from the garbled values that P_1 sends to P_2). Combining the two, it learns P_1's input x, which can then be sent to the trusted party. The trusted party answers with $f(x, y)$, which we use to define P_2's output and complete the simulation.

Security against a Malicious P_2. Intuitively, the security in this case is derived from the fact that: (a) the oblivious transfer protocol is secure, and so P_2 only learns a single set of keys (corresponding to a single input y) for decrypting the garbled circuits, and (b) the commitment schemes are hiding and so P_2 does not know what input corresponds to the garbled values that P_1 sends it for evaluating the circuit. Of course, in order to formally prove security we construct an ideal-model simulator B_2 working with an adversary A_2 that has corrupted P_2. The simulator first extracts A_2's input bits from the oblivious transfer protocol, and then sends the input y it obtained to the trusted party and receives back $z = f(x, y)$. Given the output, the simulator constructs the garbled circuits. However, rather than constructing them all correctly, for each circuit it tosses a coin and, based on the result, either constructs the circuit correctly, or constructs it to compute the constant function outputting z (the output is received from the trusted party). In order to make sure that the simulator is not caught cheating, it biases the coin-tossing phase so that all of the correctly-constructed garbled circuits are check-circuits, and all of the other circuits are evaluation-circuits (this is why the protocol uses joint coin-tossing rather than let P_2 alone choose the circuits to be opened). A_2 then checks the correctly-constructed circuits, and is satisfied with the result as if it were interacting with a legitimate P_1. A_2 therefore continues the execution with the circuits which always output z.

5   Efficiency of the Protocol

We discuss below the efficient implementation of the different building blocks of the protocol (namely, encryption, commitment schemes, and oblivious transfer).

The overhead of the protocol depends on a statistical security parameter s. The security proof shows that the adversary's cheating probability is exponentially small in s. We note that in this paper we preferred to present a full and clear proof, rather than overly optimize the construction at the cost of complicating the proof. We have not not analyzed the exact constants affecting the dependence of the error probability on the security parameter s.

The communication overhead of the protocol is dominated by sending s copies of the garbled circuit, and $2s(s+1)$ commitments for each of the n inputs of P_1. In the protocol, the original circuit C^0 is modified by replacing each of the n original input bits of P_2 with the exclusive-or of s of the new input bits, and therefore the size of the evaluated circuit C is $|C| = |C^0| + O(ns)$ gates. The communication overhead is therefore $O(s(|C^0| + ns) + s^2n) = O(s|C^0| + s^2n)$ times the length of the secret-keys (and ciphertexts) used to construct the garbled circuit. (Note that the improved construction in Section 5.2 reduces the size of the new circuit to $|C| = |C^0| + O(\max(n, s))$ and therefore only improves the communication overhead by a constant; the significance of the improvement is with respect to computation.)

The computation overhead is dominated by the oblivious transfers. In Protocol 2 each input bit of P_2 is replaced by s new inputs and therefore $O(ns)$ OTs are required. In Section 5.2 we show how to use only $O(\max(n, s))$ new input bits, and consequently the number of OTs is reduced to $O(\max(n, s))$ (namely $O(1)$ OTs per input bit, assuming $n = \Omega(s)$).

5.1   Efficient Implementation of the Different Primitives

In this section, we describe efficient implementations of the different building blocks of the protocol.

Encryption scheme. Following [20], the construction uses a symmetric key encryption scheme that has indistinguishable encryptions for multiple messages and an elusive efficiently verifiable range. Informally, this means **(1)** that for any two (known) messages x and y, no polynomial-time adversary can distinguish between the encryptions of x and y, and **(2)** that there is a negligible probability that an encryption under one key falls into the range of encryptions under another key, and given a key k it is easy to verify whether a certain ciphertext is in the range of encryptions with k. See [20] for a detailed discussion of these properties, and for examples of easy implementations satisfying them. For example, the encryption scheme could be $E_k(x) = \langle r, f_k(r) \oplus x0^n \rangle$, where k is a pseudo-random function keyed by k, and r is a randomly chosen value.

Commitment schemes. The protocol uses both unconditionally hiding and unconditionally binding commitments. Our goal should be, of course, to use the most efficient implementations of these primitives, and we therefore concentrate on schemes with $O(1)$ communication rounds (all commitment schemes we describe here have only two rounds). Efficient unconditionally hiding commitment schemes can be based on number theoretic assumptions, and use $O(1)$ exponentiations (see, e.g., [13,26]). The most efficient implementation is probably the one due to Halevi and Micali, which uses a collision-free hashing function and no other cryptographic primitive [14]. Efficient unconditionally binding commitments can be constructed using the scheme of Naor [24], which has two rounds and is based on using a pseudo-random generator.

Oblivious transfer. The protocol needs to use an OT protocol which is secure according to the real/ideal model simulation definition. Candidate protocols can be the protocol of [7] compiled according to the GMW paradigm, or the two-round protocols of [25,2,16] with additional proofs of knowledge.

5.2   Reducing the Number of Oblivious Transfers

Protocol 2 uses a construction which replaces each input bit of P_2 with multiple input bits, providing P_2 with multiple options for encoding each of its inputs. This limits the information that P_1 can gain from corrupting OT inputs (and in particular, P_2 aborts with almost the same probability irrespective of its actual input). The construction replaces each original input wire of P_2 with s new wires, thus increasing the number of input wires of P_2 from n to ns. We show here a probabilistic construction which reduces the number of input wires of P_2 to $\max(4n, 8s)$ (we also describe how to use codes to construct an explicit construction with similar performace). The construction has a direct effect on the overhead of the protocol, since the number of OTs is equal to the number of input wires of P_2.

We denote the original input bits as $w_1, \ldots, w_n$ and the new input bits as $w'_1, \ldots, w'_m$. Our goal is to minimize m. Each w_i is defined as the exclusive-or of a subset of the new input bits. We define the indicator vector z_i as an m-bit binary string whose jth bit is 1 iff w'_j is in the subset of new input bits whose exclusive-or is w_i. The construction described in Protocol 2 corresponds to indicator vectors $z_i = (\underbrace{0\ldots0}_{(i-1)s}\underbrace{1\ldots1}_{s}\underbrace{0\ldots0}_{(n-i)s})$.

Before analyzing the constructions, let us recall how P_2 encodes its inputs: it chooses random values for the bits $w'_1, \ldots, w'_m$, subject to the constraint that the exclusive-or of any set of new bits which corresponds to an original bit w_i is equal the original value of w_i. P_2 then runs an OT for each of its new input bits. If one of the answers it receives in these OTs is corrupt, it aborts the protocol. Our goal is to make sure that the decision to abort does not reveal information about P_2's original input (this is the only place that it is used in the proof). It is clear that if P_1 corrupts the inputs of a single OT, then, since each input bit of P_2 is the exclusive-or of several new bits, the decision to abort does not reveal information about any specific input bit of P_2. This observation must be generalized for the case of P_1 corrupting more OT inputs, and hold with respect to any subset of P_2's inputs.

Warmup – reusing bits. In order to use less "new" input bits, P_2 must reuse these bits. Assume that P_2 has two input wires w_1, w_2 and that we replace them with $s+1$ new wires, $w'_1, \ldots, w'_{s+1}$. The input values are defined as $w_1 = w'_1 \oplus \cdots \oplus w'_s$, and $w_2 = w'_2 \oplus \cdots \oplus w'_{s+1}$ (namely $z_1 = 11 \cdots 10$ and $z_2 = 01 \cdots 11$). In this case, it is easy to see that any strategy used by a malicious P_1 to corrupt OT values gives it an advantage of at most 2^{-s+1} in identifying a single bit of P_2's original input (e.g., if P_1 corrupts the '1' inputs of $w'_1, \ldots, w'_s$, then if $w_1 = 1$ P_2 always aborts, whereas if $w_1 = 0$ there is a probability of 2^{-s+1} that P_2 does not abort). However, $w_1 \oplus w_2 = w'_1 \oplus w'_{s+1}$ (namely, $z_1 \oplus z_2 = 10 \ldots 01$) and therefore if P_1 corrupts the OT values of both w'_1 and w'_{s+1} it can obtain a non-negligible advantage in learning $w_1 \oplus w_2$. (For example, P_1 can corrupt the '1' inputs of w'_1 and w'_{s+1}. If P_2 does not abort P_1 can conclude that $w'_1 = w'_{s+1} = 0$ and therefore $w_1 = w_2$.)

The attack presented above can be prevented if the exclusive-or of any subset of P_2's original bits contains at least s new input bits. Namely, if, in the general case, for every non-empty subset $L \subseteq \{1, \ldots, n\}$ it holds that the Hamming weight of $\oplus_{i \in L} z_i$ is at least s. The two lemmata stated below (which are proved in the full version of the paper) show that this requirement is sufficient to prove that, up to a negligible probability, P_2's decision to abort is independent of its input values.

Lemma 1. *Suppose that for any set $L = \{i_1, \ldots, i_{|L|}\}$ (corresponding to a set $\{w_{i_1}, \ldots, w_{i_{|L|}}\}$ of original input wires), the Hamming weight of $z_{i_1} \oplus \cdots \oplus z_{i_{|L|}}$ is at least s. Fix the values of any subset of less than s new input wires arbitrarily, and choose the values of all other new input wires uniformly at random. Then for any set $L = \{i_1, \ldots, i_{|L|}\}$, the value of the vector $(w_{i_1}, \ldots, w_{i_{|L|}})$ is uniformly distributed.*

Lemma 2. *Suppose that for all sets $L = \{i_1, \ldots, i_{|L|}\}$ the Hamming weight of $z_{i_1} \oplus \cdots \oplus z_{i_{|L|}}$ is at least s. Then, for any two different inputs y and y' of P_2, the difference between the probability that P_2 aborts the protocol as a result of corrupt OT values when its input is y and when its input is y' is at most 2^{-s+1}.*

Given Lemma 2 it is possible to construct assignments of the new input values to the original input values which ensure that OT corruptions by P_1 do not reveal information about P_2's input. The constructions are based on ensuring that for any set $S = \{i_1, \ldots, i_{|L|}\}$ the Hamming weight of $z_{i_1} \oplus \cdots \oplus z_{i_{|L|}}$ is at least s. We describe below a randomized construction which achieves this property. As was pointed to us by David Woodruff, an explicit construction can be achieved using any explicit linear code from $\{0,1\}^s$ to $\{0,1\}^{O(s)}$, for which any two codewords have a distance of at least $\Omega(s)$ (Justesen codes are an example of such a code).

The randomized construction. We define $4n$ new input bits for P_2. Assume, without loss of generality, that $n > 2s$. (Otherwise add dummy input bits. Therefore the exact number of new input bits is $\max(4n, 8s)$.) The mapping between the n old input bits and the $4n$ new input bits is chosen randomly in the following way: each original input bit w_i is defined to be equal to the exclusive-or of a uniformly chosen subset of the new input bits (in other words, z_i is a uniformly distributed string of $4n$ bits).

We examine the probability that there is a subset $L \subseteq \{0,1\}^n$ for which the Hamming weight of $\oplus_{i \in L} z_i$ is less than s: Consider any subset L, then $\oplus_{i \in L} z_i$ is a uniformly distributed string with $4n > 8s$ bits, with an expected Hamming weight of $2n$. Let X_j be a random variable which is set to 1 if the jth bit in this string is 1. Note that $s/4n < 1/8$ by our assumption that $n > 2s$. We have:

$$\Pr\left[\sum_{j=1}^{4n} X_j < s\right] = \Pr\left[\frac{\sum X_j}{4n} < \frac{s}{4n}\right] < \Pr\left[\frac{\sum X_j}{4n} < \frac{1}{8}\right] \leq \Pr\left[\left|\frac{\sum X_j}{4n} - \frac{1}{2}\right| > \frac{3}{8}\right]$$

Applying the Chernoff bound, we have that $\Pr\left[\sum_{j=1}^{4n} X_j < s\right] =$ $< 2e^{-\frac{(3/8)^2}{2(1/2)(1/2)}4n} = 2e^{-9n/8}$. There are a total of 2^n subsets of the original input bits, and therefore the probability that any of them is equal to the exclusive-or of less than s new input bits is bounded by $2^n 2 e^{-9n/8} \approx 2^{(1-9/8\log(e))n} \approx 2^{-0.6n} < 2^{-1.2s}$. Lemma 2 therefore implies that with probability $1 - 2^{-1.2s}$ the construction suffices for our proof of security.

Choosing the strings z_i. In order to use the above construction, the parties must construct a circuit that has $4n$ new input bits for P_2. Furthermore, the parties must define n random strings z_i of length $4n$ and then have the circuit map P_2's i^{th} input bit according to the string z_i (as described above). This can be done in two ways. One possibility is to choose the mapping once and for all and hardwire it into the protocol specification. This is problematic because then there is a negligible probability that the protocol is not secure (in any execution). Thus, the mapping should instead be chosen as part of the protocol execution (because negligible failure in any execution is allowed). Fortunately, P_2 can singlehandedly choose the strings $z_1, \ldots, z_n$ in the first step of the protocol and send them to P_1. The reason why this is fine is because this entire issue only arises in the proof of the case that P_1 is corrupted (indeed, for the case of a corrupted P_2 there is no need to split P_2's input bits at all).

Acknowledgments

We would like to thank Yuval Ishai, Moni Naor, Adam Smith and David Woodruff for helpful discussions about this work.

References

1. G. Aggarwal, N. Mishra, and B. Pinkas. Secure Computation of the k-th Ranked Element. In *EUROCRYPT 2004*, Springer-Verlag (LNCS 3027), 40–55, 2004.
2. B. Aiello, Y. Ishai, and O. Reingold. Priced Oblivious Transfer: How to Sell Digital Goods. In *EUROCRYPT 2001*, Springer-Verlag (LNCS 2045), 119–135, 2001.
3. B. Barak and Y. Lindell. Strict Polynomial-Time in Simulation and Extraction. *SIAM Journal on Computing*, 33(4):783–818, 2004.
4. D. Beaver. Foundations of Secure Interactive Computing. In *CRYPTO'91*, Springer-Verlag (LNCS 576), pages 377–391, 1991.

5. R. Canetti. Security and Composition of Multiparty Cryptographic Protocols. *Journal of Cryptology*, 13(1):143–202, 2000.
6. R. Cramer, I. Damgard, and B. Schoenmakers. Proofs of Partial Knowledge and Simplified Design of Witness Hiding Protocols. In *CRYPTO'94*, Springer-Verlag (LNCS 839), pages 174–187, 1994.
7. S. Even, O. Goldreich and A. Lempel. A Randomized Protocol for Signing Contracts. In *Communications of the ACM*, 28(6):637–647, 1985.
8. O. Goldreich. *Foundations of Cryptography: Volume 1 – Basic Tools.* Cambridge Univ. Press, 2001.
9. O. Goldreich. *Foundations of Cryptography: Volume 2 – Basic Applications.* Cambridge Univ. Press, 2004.
10. O. Goldreich and A. Kahan. How To Construct Constant-Round Zero-Knowledge Proof Systems for NP. *Journal of Cryptology*, 9(3):167–190, 1996.
11. O. Goldreich, S. Micali and A. Wigderson. How to Play any Mental Game – A Completeness Theorem for Protocols with Honest Majority. In *19th STOC*, pages 218–229, 1987. For details see [9].
12. S. Goldwasser and L. Levin. Fair Computation of General Functions in Presence of Immoral Majority. In *CRYPTO'90,* Springer-Verlag (LNCS 537), pages 77–93, 1990.
13. S. Goldwasser, S. Micali and R.L. Rivest, A Digital Signature Scheme Secure Against Adaptive Chosen-Message Attacks. *SIAM J. Comput.* 17(2): 281-308, 1988.
14. S. Halevi and S. Micali, Practical and Provably-Secure Commitment Schemes from Collision-Free Hashing, *CRYPTO 1996*, Springer-Verlag (LNCS 1109), pages 201-215, 1996.
15. S. Jarecki and V. Shmatikov. Efficient Two-Party Secure Computation on Committed Inputs. In these proceedings (Eurocrypt '2007).
16. Y.T. Kalai. Smooth Projective Hashing and Two-Message Oblivious Transfer. In *EUROCRYPT 2005*, Springer-Verlag (LNCS 3494), pages 78–95, 2005.
17. J. Katz and Y. Lindell. Handling Expected Polynomial-Time Strategies in Simulation-Based Security Proofs. In the *2nd Theory of Cryptography Conference* (TCC), Springer-Verlag (LNCS 3378), pp. 128–149, 2005.
18. J. Kilian. Founding Cryptography on Oblivious Transfer. In *20th STOC*, pages 20–31, 1988.
19. M. Kiraz and B. Schoenmakers. A Protocol Issue for the Malicious Case of Yao's Garbled Circuit Construction. In *Proceedings of 27th Symposium on Information Theory in the Benelux*, 283–290, 2006.
20. Y. Lindell and B. Pinkas. A Proof of Yao's Protocol for Secure Two-Party Computation. To appear in the *Journal of Cryptology*. Also appeared as *Cryptology ePrint Archive,* Report 2004/175, 2004.
21. D. Malkhi, N. Nisan, B. Pinkas and Y. Sella. Fairplay – A Secure Two-Party Computation System. In the *13th USENIX Security Symposium*, pages 287–302, 2004.
22. S. Micali and P. Rogaway. Secure Computation. Unpublished manuscript, 1992. Preliminary version in *CRYPTO'91*, Springer-Verlag (LNCS 576), pages 392–404, 1991.
23. P. Mohassel and M.K. Franklin. Efficiency Tradeoffs for Malicious Two-Party Computation. In the *9th PKC conference*, Springer-Verlag (LNCS 3958), pages 458–473, 2006.
24. M. Naor. Bit Commitment Using Pseudorandomness. *Journal of Cryptology,* 4(2):151–158, 1991.

25. M. Naor and B. Pinkas. Efficient Oblivious Transfer Protocols. In the 12*th SODA*, pages 448-457, 2001.

26. T.P. Pedersen. Non-Interactive and Information-Theoretic Secure Verifiable Secret Sharing. In *CRYPTO'91*, Springer-Verlag (LNCS 576), pages 129–140, 1992.

27. M. Rabin. How to Exchange Secrets by Oblivious Transfer. Tech. Memo TR-81, Aiken Computation Laboratory, Harvard U., 1981.

28. D. Woodruff. Revisiting the Efficiency of Malicious Two-Party Computation. In these proceedings (Eurocrypt '2007).

29. A. Yao. How to Generate and Exchange Secrets. In 27*th FOCS*, pages 162–167, 1986.

Revisiting the Efficiency of Malicious Two-Party Computation

David P. Woodruff [*]

MIT Computer Science and Artificial Intelligence Laboratory
dpwood@mit.edu

Abstract. In a recent paper Mohassel and Franklin study the efficiency of secure two-party computation in the presence of malicious behavior. Their aim is to make classical solutions to this problem, such as zero-knowledge compilation, more efficient. The authors provide several schemes which are the most efficient to date. We propose a modification to their main scheme using expanders. Our modification asymptotically improves at least one measure of efficiency of all known schemes. We also point out an error, and improve the analysis of one of their schemes.

Keywords: secure function evaluation, malicious model, efficiency, expander graphs.

1 Introduction

Two parties, Alice with input x and Bob with input y, wish to evaluate a function $f(x, y)$ in such a way that neither learns more information than what can be deduced from the output $f(x, y)$. This problem, known as general two-party secure computation, generalizes many important cryptographic tasks. A celebrated result is *Yao's garbled circuit protocol* [34, 21], which provides a solution to this problem for any efficiently computable function f.

Yao's protocol provides security in the *semi-honest model*, that is, a model in which parties must follow the instructions of the protocol, though they may keep message histories in an attempt to learn more than what is prescribed. A more reaslistic security model is the *malicious model* in which parties may behave arbitrarily. The textbook solution to achieve security in the malicious model is to perform the zero-knowledge compilation of Goldreich *et al* [15, 16, 17] to Yao's protocol. This yields a protocol with communication and computation cost bounded by a polynomial in the size of a circuit for computing f. This results in optimal efficiency, up to polynomial factors, but the polynomial factors are rather large and so this approach may not be useful in practice.

This motivates alternative methods for protecting Yao's protocol against malicious behavior, as suggested in [24, 26, 28]. These techniques provide a well-defined tradeoff between security and efficiency, and are useful in practice.

---

[*] Supported by an N.D.S.E.G. fellowship.

M. Naor (Ed.): EUROCRYPT 2007, LNCS 4515, pp. 79–96, 2007.

These protocols all use the following *cut-and-choose* technique. Alice creates m independently garbled circuits $C_1, \ldots, C_m$, each computing the same function f. These garbled circuits are transmitted to Bob, along with various commitments. Bob chooses a subset $S \subset [m] = \{1, \ldots, m\}$, and asks Alice to reveal the secrets of all circuits C_i (along with their corresponding commitments) with $i \in S$. This gives Bob confidence that Alice correctly formed most of the garbled circuits and commitments. Alice then sends her garbled inputs for the circuits in $[m] \setminus S$, and Alice and Bob perform oblivious transfer for Bob to receive his garbled inputs for these circuits. Finally, Bob evaluates the garbled circuits and outputs the majority value.

There are a number of subtleties and complexities within this framework. As pointed out by Mohassel and Franklin in [26], the *Fairplay* scheme [24] designed in this framework has a subtle bug allowing one of the parties to cheat undetectably. Moreover, Kiraz and Schoenmakers [20] found an error that occurs in both Mohassel and Franklin's work and Fairplay. Recently, Lindell and Pinkas [22] have pointed out new flaws in many existing protocols. In this paper we also present an error in [26], showing a flaw in their estimated concrete costs.

This framework poses the following problems. How do we ensure Alice provides the same garbled input to most of the circuits in $[m] \setminus S$? How do we ensure Bob receives the same garbled input to most of the circuits in $[m] \setminus S$? If neither of these conditions hold, Alice can fool Bob into outputting an incorrect value or having to abort the protocol depending on his input.

Let f be a function computable by a Boolean circuit with g gates and I inputs. We want a protocol which achieves both privacy and correctness. Intuitively, the privacy aspect is that nothing is learned from the output, and the correctness aspect is that the output is distributed according to the described functionality, see [17]. Following previous work [26, 22], we will have two security parameters. The first is the input length, n, which is the security parameter for our commitment schemes, encryption, and oblivious transfer protocols. The second parameter, ϵ, is a statistical security parameter specifying the number of garbled circuits used in the cut-and-choose framework. Here, n depends on computational assumptions, whereas ϵ indicates the error probability incurred in this framework, and is therefore a "statistical" security parameter (this term was coined in [22]). Note that one can set ϵ independently of n, and this can be used to "trade" security for efficiency, as discussed below.

We note that when both parties are honest, it suffices to only have Bob output $f(x, y)$. Indeed, as shown in Section 2.2 of [22], this is without much loss of generality because given such a secure protocol, it can be efficiently transformed so that Alice also obtains $f(x, y)$, or even $f'(x, y)$ for some other function f'. We thus assume that only Bob has output in the remainder of the paper.

1.1   Our Contributions

We study the efficiency of protocols in this framework. We measure three quantities: the number of symmetric encryptions, the number of exponentiations, and the communication complexity.

We are aware of four schemes in this framework - *Fairplay* [24], *Committed-input* [26], *Equality-checker* [26], and the very recent protocol of Lindell and Pinkas [22]. These schemes differ in the way the set S is chosen, together with their methods of enforcing Alice and Bob to have consistent inputs.

The main result of this paper is a new scheme, *Expander-checker*, which asymptotically improves at least one measure of efficiency of all known schemes. It results in fewer symmetric encryptions and smaller communication complexity than *Fairplay*, *Equality-checker*, and Lindell and Pinkas' protocol, while achieving fewer exponentiations[1] than *Committed-input*. See the section on other related work below for a more detailed account of Lindell and Pinkas' new protocol. Our results are summarized by the following table.

Scheme	Symmetric Enc.	Exponentiations	Communication
Fairplay [24]	$O(\frac{1}{\epsilon}g)$	$O(I)$	$O(\frac{1}{\epsilon}g)$
Committed-input [26]	$O(\ln(\frac{1}{\epsilon})g)$	$O(\ln(\frac{1}{\epsilon})I)$	$O(\ln(\frac{1}{\epsilon})g)$
Equality-checker [26]	$O(\ln(\frac{1}{\epsilon})g + \ln(\frac{1}{\epsilon})^2 I)$	$O(I)$	$O(\ln(\frac{1}{\epsilon})g + \ln(\frac{1}{\epsilon})^2 I)$
Lindell-Pinkas [22]	$O(\ln(\frac{1}{\epsilon})g + \ln(\frac{1}{\epsilon})^2 I)$	$O(I)$	$O(\ln(\frac{1}{\epsilon})g + \ln(\frac{1}{\epsilon})^2 I)$
Expander-checker (new)	$O(\ln(\frac{1}{\epsilon})g)$	$O(I)$	$O(\ln(\frac{1}{\epsilon})g)$

Our scheme is built off of *Equality-checker*. In that scheme, S is a random subset of size $m/2$. With a suitable commitment scheme, Mohassel and Franklin ensure that Bob's garbled inputs to the different circuits correspond to the same ungarbled input in each of the oblivious transfer steps. The more interesting part is how they ensure that Alice's garbled inputs to the different circuits correspond to the same ungarbled input. Their method only assumes a generic commitment scheme and can be implemented without any exponentiations.

Alice commits to tuples $(j, j', K^j_{i,b}, K^{j'}_{i,b})$ for all distinct $j, j' \in [m]$, where $K^j_{i,b}$ refers to the key in Yao's garbled circuit protocol associated with the ith input wire of Alice with value b in circuit j. When Bob is given purported keys $K^j_{i,b}$ and $K^{j'}_{i,b'}$, which correspond to Alice's garbled ith input for circuits j and j' respectively, Bob can use the witness to verify that $b = b'$.

If Alice creates enough commitments $(j, j', K^j_{i,b}, K^{j'}_{i,b})$ with $b \neq b'$, then the set S likely contains a pair of circuits $C_j, C_{j'}$ with this property, and she will be caught when forced to reveal the circuits in S and the commitments between them. On the other hand, suppose most of the commitments $(j, j', K^j_{i,b}, K^{j'}_{i,b'})$ satisfy $b = b'$. Consider the complete graph G with vertex set $[m] \setminus S$, each vertex indexing a circuit not chosen by Bob to reveal. Since every pair of circuits C_j, C'_j with $j, j' \in [m] \setminus S$ has a commitment $(j, j', K^j_{i,b}, K^{j'}_{i,b'})$, there is a large connected component C for which for each edge $\{j, j'\} \in C$, for each i and each b, in the

---

[1] Some care needs to be taken when measuring the number of exponentiations since under certain assumptions it is possible to obtain a large number of exponentiations by only performing a small number of exponentiations and a few simpler operations, see [5, 18]. In this work we follow previous work and make the simplifying, practical assumption that there are $O(I)$ exponentiations in the oblivious transfer stage.

commitment $(j, j', K_{i,b}^j, K_{i,b'}^{j'})$, $b = b'$. By transitivity, Alice's input is the same to every circuit in C. If C is large enough, then the majority of circuits Bob evaluates (those in $[m] \setminus S$) have the same input from Alice, and the protocol will be simulatable.

The drawback of this scheme is the number of commitments computed and transmitted. This is $\Theta(mg + m^2 I)$, where I is the number of input wires owned by Alice. To achieve probability of undetected cheating at most ϵ, we need $m = \Omega(\ln(\frac{1}{\epsilon}))$, and thus $\Omega(g \ln \frac{1}{\epsilon} + I \ln^2 \frac{1}{\epsilon})$ commitments. Each commitment involves at least one symmetric encryption and one transfer from Alice to Bob, resulting in a total of $\Omega(g \ln(\frac{1}{\epsilon}) + I \ln^2(\frac{1}{\epsilon}))$.

Our idea is instead of computing commitments to all tuples $(j, j', K_{i,b}^j, K_{i,b}^{j'})$, we only commit to tuples for which $\{j, j'\}$ is an edge in an *expander graph*. Suppose G is an expander with vertex set $[m]$ and $O(m)$ edges. We commit only to pairs of circuits with a corresponding edge in G, and thus the number of symmetric encryptions and communication drop to $O(\ln(\frac{1}{\epsilon})g)$. For many circuits g is not much larger than I, and in this case we save a factor of $\ln(\frac{1}{\epsilon})$ in both efficiency measures.

Why is the new protocol secure? If Alice commits to enough $(j, j', K_{i,b}^j, K_{i,b'}^{j'})$ with $b \neq b'$, then as in *Equality-checker*, she is likely to get caught when Bob chooses a random subset of circuits and commitments to expose. On the other hand, if many of the $(j, j', K_{i,b}^j, K_{i,b'}^{j'})$ satisfy $b = b'$, then, since the corresponding graph G is an expander, it contains a large connected component of such edges. Thus, as before, the majority of circuits Bob evaluates will have the same Alice input, and the protocol will be simulatable.

Mohassel and Franklin [26] evaluated concrete costs for some practical settings of parameters. We point out an error in their analysis for Equality-checker, which is not obvious to us how to fix within their framework. We present a new graph-theoretic framework which fixes this and gives sharper bounds. We show the probability Alice can cheat is at most $2 \cdot 2^{-\frac{m}{4}}$, whereas it was previously thought this probability was at most $2 \cdot 2^{-\frac{m}{6}}$. Since the communication and number of symmetric encryptions of Equality-checker are proportional to $mg + m^2 I$, for a given security level we achieve at least a $(1/4)/(1/6) = 3/2$ factor efficiency improvement. This implies that Equality-checker is superior in practice to Fairplay for a wider range of parameters than Tables 2 and 4 of [26] suggest. To provide a good comparison with previous schemes, it is essential that we also lower bound the probability that Alice can cheat. We give a lower bound that is within a factor of 2 of our upper bound on this probability.

For Expander-checker we show the probability that Alice can cheat is at most $2^{-\frac{m}{4}+O\left(\frac{m \log d}{\sqrt{d}}\right)}$. With the present analysis, for a practical setting of parameters Equality-checker is still superior. We discuss barriers in derandomization and protocol design that need to be overcome in order to provably make Expander-checker superior in practice. We leave it as an open question to improve the analysis or provide an implementation to determine which protocol is more practical.

1.2   Other Related Work

Expanders have been used in other contexts for enforcing equality constraints. For example, see [19, 32, 10]. As far as we are aware though, this is the first time they have been used in the cut-and-choose framework.

Very recently, Lindell and Pinkas [22] have given the first rigorous proof of security of a protocol in the cut-and-choose framework that meets a simulation-based definition. Our protocol has the same security as Equality-checker [26] (with the fix pointed out by [20]), and it seems that a rigorous simulation-based security proof for Equality-checker has not appeared anywhere.

We stress that the focus of this work is a new way of proving input consistency using expander graphs, which we hope can be of use in other protocols in the cut-and-choose framework. It seems likely, though we have yet to formally verify, that one can combine our approach with Lindell and Pinkas' protocol to achieve the improved efficiency of Expander-checker while also achieving a full simulation-based proof of security. We sketch how this might be done in the proof of Theorem 3.

1.3   Organization

Section 2 reviews secure two-party computation, Yao's garbled circuit protocol, the *Equality-checker* scheme, and expander graphs. In Section 3 we present *Expander-checker*, and prove its security. In Section 4 we discuss efficiency, both in theory and in practice. This work also appears as a technical report in [33].

2   Preliminaries

2.1   Two-Party Secure Computation

For an excellent treatment of secure two-party computation, the reader is referred to [17]. Here we summarize the model. A two-party computation is a random process mapping pairs of inputs to pairs of outputs. We refer to this process as the desired *functionality*, denoted $f : \{0,1\}^* \times \{0,1\}^* \to \{0,1\}^* \times \{0,1\}^*$ where $f = (f_1, f_2)$. For any two inputs $x, y \in \{0,1\}^n$, the output $(f_1(x,y), f_2(x,y))$ is a random variable ranging over pairs of strings. The interpretation here is that the first party wants to learn $f_1(x,y)$ and the second party wants to learn $f_2(x,y)$.

In this paper we consider the *malicious* model of security. The formal definitions can be found in [17]. In this model one of the parties can behave in an arbitrary way. We will, however, assume that both parties are computationally bounded (i.e., randomized polynomial-time Turing machines). Security is achieved by comparing the adversaries in the *real model* with those in an *ideal-model* in which both parties have a *trusted party* to interact with. Informally, a two-party protocol is secure if for any admissible pair of parties (A, B) in the real-model, there is an admissible pair of parties (A', B') in the ideal model where the outputs of the two executions are indistinguishable. A pair is admissible if at least one of the parties in the pair is honest. Thus, intuitively, the protocol is secure if it provides the correct output behavior, and provides privacy to honest parties.

In our protocols we need a specific protocol called *oblivious transfer*, which has been extensively studied [11, 27, 29]. We only need *1-out-of-2* oblivious transfer. In this case, $x = (z_0, z_1)$, $y = \sigma$, $f_1(x, y) = \emptyset$, and $f_2(x, y) = z_\sigma$. Efficient oblivious transfer protocols secure in the malicious model exist [27].

2.2 Yao's Garbled Circuit Protocol

Here we review Yao's garbled circuit protocol [34]. Let f be an efficiently-computable function two parties wish to securely compute. Then f can be represented as a polynomial-size circuit. The first party computes a garbled form of this circuit as follows.

For every wire j in the circuit, she chooses two random strings $K_{j,0}$ and $K_{j,1}$. These random strings correspond to a value of 0 and a value of 1 on wire j, respectively. Next, for every gate in the circuit, she computes a garbled truth table as follows. Let E be a symmetric encryption scheme. Then she uses E together with the keys corresponding to the values on the input wires to encrypt the value of the corresponding output wire. For example, if the gate is an AND gate on two input wires j, j' with output wire ℓ, then there are four entries

$$E_{K_{j,0}}(E_{K_{j',0}}(K_{\ell,0})), E_{K_{j,0}}(E_{K_{j',1}}(K_{\ell,0})), E_{K_{j,1}}(E_{K_{j',0}}(K_{\ell,0})), E_{K_{j,1}}(E_{K_{j',1}}(K_{\ell,1})).$$

These entries should be permuted so that the second party does not learn intermediate values of the computation. Also, she creates a table which translates the garbled output values to their actual values (0 or 1). She sends the garbled circuit and her garbled inputs to the second party.

The second party learns his garbled inputs through an oblivious transfer step. This ensures that only his garbled inputs are learned, and nothing else, while the first party learns nothing about the second party's inputs. The second party then computes the garbled circuit gate by gate, obtaining his garbled output. Finally, using the translation table, he obtains the actual output of the circuit. See [21] for detail, and a proof of security in the semi-honest model.

It is well-known that Yao's garbled protocol is not secure in the malicious model. The standard way of fixing this is to apply the zero-knowledge compiler of [15, 16]. The first party needs to supply a zero-knowledge proof that her circuit was constructed correctly and computes the desired functionality. The second party needs to supply zero-knowledge proofs that show he correctly evaluated the circuit. These zero-knowledge proofs, though theoretically feasible, are very inefficient and motivate the search for practical solutions.

2.3 Equality-Checker Scheme

The following is the *Equality-checker* scheme of [26]. Here, $z_{j,j',i,b}$ is Alice's commitment to the tuple $(j, j', i, K_{i,b}^j, K_{i,b}^{j'})$ for every distinct pair of circuits $j, j' \in [m]$, every input wire i of Alice, and every input value $b \in \{0, 1\}$. $w_{j,j',i,b}$ is the corresponding witness for decommittal. $z_{j,i,b}$ is the commitment of the tuple $(j, i, b, K_{i,b}^j)$ for every circuit $j \in [m]$, every input wire i of Bob, and every input value $b \in \{0, 1\}$. $w_{j,i,b}$ is the corresponding witness for decommittal.

In the original paper [26], a generic oblivious transfer scheme was chosen in step 6, and this was shown to be insecure [20]. One fix is to use a committed oblivious transfer scheme (as stated below), or a *committing* scheme. See [20] for the details. Yet another approach will be discussed in the proof of Theorem 3.

We note that this does not affect our asymptotic analysis, and only marginally affects our concrete costs.

Equality-checker:

1. Alice creates m garbled circuits $C_1, \ldots, C_m$. She sends the C_j, $(j, j', i, z_{j,j',i,b})$, and $(j, i, b, z_{j,i,b})$ to Bob. The $(j, j', i, z_{j,j',i,b})$ should be sent in a random order so that Alice cannot distinguish $z_{j,j',i,0}$ from $z_{j,j',i,1}$.
2. Bob chooses a random subset $S \subset [m]$ with $|S| = m/2$ and sends S to Alice.
3. Alice exposes the secrets of the C_i for every $i \in S$. She also sends witnesses $w_{j,j',i,b}$ and $w_{j,i,b}$ for all i, b and all $j, j' \in S$. Bob verifies the garbled circuits and commitments are correct.
4. Renumber the remaining garbled circuits $C_1, \ldots, C_{m/2}$. Alice sends the keys K_{i,b_i}^j and the witnesses w_{j,j',i,b_i} for every distinct $j, j' \in [m/2]$ and each of her input wires i, where b_i is her input for wire i.
5. Bob uses the witnesses w_{j,j',i,b_i} to verify that Alice's input to all the circuits is the same.
6. Alice and Bob engage in committed oblivious transfers in order for Bob to receive his garbled input bits. For every input wire i of Bob, Alice uses a single oblivious transfer to give Bob one of two tuples: $(K_{i,0}^1, w_{1,i,0}, K_{i,0}^2, w_{2,i,0}, \ldots, K_{i,0}^{m/2}, w_{m/2,i,0})$ or $(K_{i,1}^1, w_{1,i,1}, K_{i,1}^2, w_{2,i,1}, \ldots, K_{i,1}^{m/2}, w_{m/2,i,1})$, depending on Bob's value for input i.
7. Bob evaluates the $m/2$ garbled circuits and prints the majority output.

We assume, as in [26], that computing the commitments $z_{j,j',i,b}$ and $z_{j,i,b}$ does not require exponentiation, but rather, just a symmetric encryption. We also assume a single oblivious transfer requires $O(1)$ exponentiations.

Theorem 1. *([26]) Equality-checker is secure in the malicious model with inverse exponential (in m) probability of undetected cheating. The number of symmetric encryptions and the communication complexity are $O(mg + m^2 I)$, and the number of exponentiations is $O(I)$, where g and I are the number of gates and inputs of the circuit to be computed, respectively.*

2.4 Expander Properties

Let $G = (V, E)$ be a d-regular graph on n vertices. Let $A = (a_{uv})$, $u, v \in V$, be its adjacency-matrix, that is, $a_{uv} = 1$ if $(u, v) \in E$ and $a_{uv} = 0$ otherwise. Since G is

d-regular, the largest eigenvalue of A is d, corresponding to the all 1s eigenvector. Let $\lambda = \lambda(G)$ denote the second largest *absolute value* of an eigenvalue of G. We need the following discrepancy theorem, known as the *expander-mixing lemma* (see, e.g, [1, 7], for the proof).

Theorem 2. *For any subsets* $X, Y \subseteq V$,

$$|e(X,Y) - \frac{d}{n}|X||Y|| \leq \frac{\lambda}{n}\sqrt{|X|(n-|X|)|Y|(n-|Y|)},$$

where $e(X,Y)$ *is the number of edges with one endpoint in* X *and one endpoint in* Y.

In our asymptotic analysis, we use explicit expander graphs known as *Ramanujan graphs*. The construction we use is essentially due to Lubotzky, Phillips, and Sarnak [23], and independently discovered by Margulis [25]. However, the form of these graphs [23, 25] is not so convenient to work with. We use a slight variant of these graphs described in section II of [2].

Fact 1. *[2] Let* p, q *be any distinct primes congruent to* 1 *modulo* 4, *with* p *a quadratic residue modulo* q, *and* $q \geq 2\sqrt{p}$. *Let* $d = p+1$. *Then for every positive integer* ℓ, *there is an explicit* $(p+1)$-*regular graph on* $\frac{1}{2}(q^{3\ell} - q^{3\ell-2})$ *vertices such that* $\lambda \leq 2\sqrt{p}$.

For fixed p, q as we vary ℓ we get an infinite family of graphs, and there is a positive constant α such that for any integer m, there is a graph in the family with m' vertices, where $m' \leq m \leq \alpha m'$. For a description of how to efficiently compute these graphs, see section II of [2].

We note that one can also obtain Ramanujan graphs by random sampling, and testing with Gaussian elimination. See [13] for how to sample such graphs.

2.5   Combinatorial Identities

Fact 2. *(see [12, 30]) For integers* $n > 0$, $\sqrt{2\pi} n^{n+\frac{1}{2}} e^{-n} \leq n! \leq \sqrt{2\pi} n^{n+\frac{1}{2}} e^{-n} e^{\frac{1}{12}}$.

3   Expander-Checker

Alice associates her m garbled circuits with the vertices of a d-regular Ramanujan graph $G = (V, E)$ on m vertices. The difference between our protocol and Equality-checker is that instead of committing to every pair of circuits $\{j, j'\}$, Alice only commits to the edges of G. Equality-checker is a special case of our protocol, which corresponds to setting $d = m - 1$. Since G has $dm/2$ edges, Alice performs $dm/2$ commitments.

We borrow some notation from Equality-checker, as described in Section 2.3. Let $z_{j,i,b}, w_{j,i,b}, z_{j,j',i,b}$, and $w_{j,j',i,b}$ be the commitments and witnesses as defined in that section. Alice only computes $z_{j,j',i,b}$ and $w_{j,j',i,b}$ for those $\{j, j'\}$ for which $\{j, j'\}$ is an edge of G.

For a subset S of the vertices V, let $G(S)$ denote the induced subgraph of G on vertex set S.

Expander-checker:

1. Alice creates m garbled circuits $C_1, \ldots, C_m$. For edges $\{j, j'\}$ in G, she sends the C_j, $(j, j', i, z_{j,j',i,b})$, and $(j, i, b, z_{j,i,b})$ to Bob. The $(j, j', i, z_{j,j',i,b})$ should be sent in a random order so that Alice cannot distinguish $z_{j,j',i,0}$ from $z_{j,j',i,1}$.

2. Bob chooses a (uniformly) random subset $S \subseteq [m]$ of size $m/2$. Bob sends S to Alice.

3. Alice exposes the secrets of the C_i for every $i \in S$. She also sends witnesses $w_{j,j',i,b}$ and $w_{j,i,b}$ for all i, b, all $j \in S$, and all $\{j, j'\} \in G(S)$. Bob verifies the garbled circuits and commitments are correct.

4. Renumber the remaining garbled circuits $C_1, \ldots, C_{m/2}$. Alice sends the keys K^j_{i,b_i} and the witnesses w_{j,j',i,b_i} for every $j \in V \setminus S$, every edge $\{j, j'\} \in G(V \setminus S)$, and each of her input wires i, where b_i is her input for wire i.

5. Bob uses the witnesses w_{j,j',i,b_i} to verify that Alice's input to all the circuits is the same.

6. Alice and Bob engage in committed oblivious transfers in order for Bob to receive his garbled input bits. For every input wire i of Bob, Alice uses a single oblivious transfer to give Bob one of the two tuples $(K^1_{i,0}, w_{1,i,0}, K^2_{i,0}, w_{2,i,0}, \ldots, K^{m/2}_{i,0}, w_{m/2,i,0})$ or $(K^1_{i,1}, w_{1,i,1}, K^2_{i,1}, w_{2,i,1}, \ldots, K^{m/2}_{i,1}, w_{m/2,i,1})$, depending on Bob's value for input i.

7. Bob evaluates the $m/2$ garbled circuits and prints the majority output.

If both parties are honest, the above protocol is correct, so we turn to security. We first develop a framework for proving the security of Equality-checker that is more powerful than that given in [26] (leading to better bounds, see Section 1), and which generalizes to Expander-checker.

3.1   Security Analysis for Equality-checker

We will show that in order for a malicious Alice to cheat with non-negligible probability, the following must be true: *Alice does not provide the same input for more than $\frac{m}{4}$ of the correctly-garbled circuits that Bob will evaluate.* If this is not true then Bob will respond with the output corresponding to the majority input of Alice, in which case the protocol will be simulatable in the ideal model by sending the majority input to the trusted third party.

Let $\mathcal{F}$ be a family of complete graphs where each $G \in \mathcal{F}$ has some of its edges labeled bad, and some of its vertices labeled incorrect. We will use the observation above to construct a family $\mathcal{F}$ containing all of the (labeled) complete graphs G for which a malicious Alice can cheat with non-negligible probability.

If Alice can cheat by sending a graph G with exactly ϵm incorrect cicuits, then there must be some subset S of $\frac{m}{2}$ vertices of G which Bob can sample, so that if we remove S from G, Alice can assign her inputs to the remaining vertices so that no more than $\frac{m}{4}$ of the remaining vertices are assigned the same input. Partition the set of remaining vertices into groups $B, C_1, C_2, \ldots, C_r$, where B denotes the set of incorrect circuits (here, $|B| = \epsilon m$), and for each C_i, all vertices in C_i are assigned the same input. Then, all of the edges connecting C_i to C_j, for any $i \neq j$, must be bad edges, as otherwise Alice will get caught. Moreover, by the observation above, $|C_i| \leq \frac{m}{4}$ for all i. For a given G, there may be more than one choice of S, each giving rise to different sets $B, C_1, C_2, \ldots, C_r$ with the above properties. For our purposes, what matters is that there is at least one such $S, B, C_1, C_2, \ldots, C_r$ for the graph G. Let $\mathcal{F}$ be the family of all such graphs G.

Lemma 1. *If Alice chooses any graph $G \in \mathcal{F}$, she will get caught when Bob samples $\frac{m}{2}$ vertices of G with probability at least $1 - 2\binom{\frac{3m}{4}}{\frac{m}{2}}/\binom{m}{\frac{m}{2}}$.*

Proof. Fix any $G \in \mathcal{F}$, and let $S, B, C_1, C_2, \ldots, C_r$ be a partition of the vertices of G with the properties described above. We compute the probability that Alice does not get caught. Note that $|S| = \frac{m}{2}$ and $|B| = \epsilon m$. For all i, let $c_i = |C_i| \leq \frac{m}{4}$. As observed above, all of the edges between C_i and C_j for $i \neq j$ are bad, and therefore in order for Alice not to get caught, Bob can sample vertices from at most one C_i. Since B contains only incorrect circuits, Bob's samples must all be drawn from S and at most one C_i. Define an *elusive* set E to be a set of vertices of G not containing any incorrect vertices and such that no two endpoints of a bad edge lie in E. For Alice not to get caught, Bob must sample an elusive set. The number of elusive sets is at most $\sum_{j=0}^{\frac{m}{2}} \binom{\frac{m}{2}}{j} \sum_{i=1}^{r} \binom{c_i}{\frac{m}{2}-j}$.

We claim this expression is maximized when $r = 2$, $c_1 = \frac{m}{4}$, and $c_2 = \frac{m}{4} - \epsilon m$ (recall that $\sum_{i=1}^{r} c_i = \left(\frac{1}{2} - \epsilon\right) m$). First, if $r = 0$, the number of elusive sets is 1, namely, the set S. Second, if $r = 1$, then since $c_1 \leq \frac{m}{4}$, the expression evaluates to at most $\binom{\frac{3m}{4}}{\frac{m}{2}}$. This follows from the identity: $\sum_{j=0}^{\ell} \binom{n_1}{j}\binom{n_2}{\ell-j} = \binom{n_1+n_2}{\ell}$. For the remainder of the proof, assume $r \geq 2$.

We now use the identity for $a \geq b$: $\binom{a}{x} + \binom{b}{x} \leq \binom{a+1}{x} + \binom{b-1}{x}$. Since $c_i \leq \frac{m}{4}$ for all i, we may inductively apply the identity so that $r = 2$, $c_1 = \frac{m}{4}$, and $c_2 = \frac{m}{4} - \epsilon m$. It follows that the number of elusive sets is at most

$$\sum_{j=0}^{\frac{m}{2}} \binom{\frac{m}{2}}{j}\left(\binom{\frac{m}{4}}{\frac{m}{2}-j} + \binom{\frac{m}{4}-\epsilon m}{\frac{m}{2}-j}\right) = \binom{\frac{3m}{4}}{\frac{m}{2}} + \binom{\frac{3m}{4}-\epsilon m}{\frac{m}{2}} \leq 2\binom{\frac{3m}{4}}{\frac{m}{2}}.$$

It follows that the probability that Alice does not get caught is at most $2\binom{\frac{3m}{4}}{\frac{m}{2}}/\binom{m}{\frac{m}{2}}$.

Corollary 1. *With probability at least $1 - 2\binom{\frac{3m}{4}}{\frac{m}{2}}/\binom{m}{\frac{m}{2}}$, there are more than $\frac{m}{4}$ correctly-garbled circuits that Bob evaluates for which Alice will provide the same input, or Alice will get caught.*

Proof. If Alice does not use the same input for more than $\frac{m}{4}$ of the correctly-garbled circuits that Bob will evaluate, she will be caught unless she sends some graph $G \in \mathcal{F}$. But then, by the previous lemma, she will get caught with probability at least $1 - 2\binom{\frac{3m}{4}}{\frac{m}{2}}/\binom{m}{\frac{m}{2}}$, as needed.

Theorem 3. Equality-checker *is secure when Alice is malicious with probability of undetected cheating by Alice at most* $2\left(\frac{3m}{4}\right)/\left(\frac{m}{2}\right) \leq 2 \cdot 2^{-\frac{m}{4}}$.

Proof. By the previous corollary, with probability at least $1 - 2\left(\frac{3m}{4}\right)/\left(\frac{m}{2}\right) \geq 1 - 2 \cdot 2^{-\frac{m}{4}}$, the majority of inputs to the correctly-garbled circuits that Bob evaluates have the same input, or Alice will get caught, and thus Bob will output the value outputted by the circuits on this input.

The security, at this point, reduces to the original argument for Equality-checker given in Claim 3 of [26]. As the proof in [26] is incomplete, we refer the reader to [22]. To make the protocol simulatable, one needs to change step 2 of the protocol so that Alice and Bob run a standard coin-tossing protocol to generate the subset of circuits to evaluate. This ensures that if Bob is malicious, the circuits evaluated are still uniformly chosen (this sub-protocol is very efficient, and doesn't affect the overall efficiency). Also, instead of using committed oblivious transfer, another approach (analyzed in [22]) is for Bob to receive his inputs before Alice sends the garbled circuits. This amounts to removing step 6, and inserting it after step 1 in the protocol. Since the circuits to be evaluated have not yet been chosen, Bob should simply receive his inputs for every circuit.

Theorem 4. Equality-checker *is secure when Bob is malicious.*

Proof. The security reduces to the original argument for Equality-checker given in Claim 4 of [26]. For a formal proof, we refer the reader to [22].

Theorem 5. *In* Equality-checker, *Alice can cheat with probability at least* $\left(\frac{3m}{4}\right)/\left(\frac{m}{2}\right)$.

Proof. Alice will send the following labeled graph $G \in \mathcal{F}$ to Bob. She will not create any incorrect circuits. She will partition the vertices into two groups V_1, V_2, with $|V_1| = \frac{3m}{4}$ and $|V_2| = \frac{m}{4}$ (assume m is a multiple of 4). An edge is labeled bad if and only if it connects V_1 to V_2. Consider the following event $\mathcal{E}$: Bob samples all $\frac{m}{2}$ of his circuits from V_1. This occurs with probability $\left(\frac{3m}{4}\right)/\left(\frac{m}{2}\right)$.

Assume the circuit being evaluated has only one bit of input from Alice. Suppose $\mathcal{E}$ occurs. Alice may then assign all remaining vertices in V_1 the input 0 and all vertices in V_2 the input 1. If the function being evaluated differs on its output (for a given Bob input) when Alice's input is a 0 or a 1, then there is no majority output of Bob's evaluations (there are two outputs, and each one occurs for exactly half of the circuits). Thus, Bob will have to abort (and this behavior cannot be hidden from Alice), and this may reveal information to Alice about Bob's input. For instance, there may be another possible input of Bob which is insensitive to the input of Alice, in which case all circuits will have the same output, and Bob will not abort.

In Appendix 4.1, we present a counterexample to Lemma 3 in [26], from which their Table 4, which analyzes the performance of Equality-checker for different security levels, is constructed.

3.2   Security Analysis for Expander-checker

We generalize the analysis of the previous section. The difficulty is that now the family $\mathcal{F}$ of graphs for which Alice can cheat with non-negligible probability is more complex. The graphs are no longer labeled complete graphs, but rather labeled expander graphs. We bound the new probability that Alice gets caught if she chooses a graph $G \in \mathcal{F}$ to send to Bob.

As before, for Alice to cheat, she cannot provide the same input for more than $\frac{m}{4}$ of the correctly-garbled cicuits that Bob will evaluate. Corollary 1, Theorem 3, and Theorem 4 are unchanged, except for the probability that Alice does not get caught, which will increase. We prove the new version of Lemma 1 in Theorem 6 below.

In Expander-checker, if Alice can cheat by sending a graph G, then as before, we can find a vertex partition $S, B, C_1, C_2, \ldots, C_r$ with $|S| = \frac{m}{2}$, $|B| = \epsilon m$ for some ϵ where B denotes the set of incorrect circuits, all edges in the expander connecting C_i to C_j for $i \neq j$ are bad, and $|C_i| \leq \frac{m}{4}$ for all i. Let $\mathcal{F}$ be the family of all such labeled graphs G.

We assume the expander graph satisfies $\lambda \leq 2\sqrt{d}$.

Theorem 6. *Let G be a d-regular Ramanujan graph for a sufficiently large constant d. If Alice chooses any graph $G \in \mathcal{F}$, she will get caught when Bob samples $\frac{m}{2}$ vertices of G with probability at least*

$$1 - 3\left(\frac{m}{4} + 1\right)\sqrt{\frac{\pi m e^{1/3}}{2}} \cdot 2^{-\frac{m}{4} + 2m\sqrt{\frac{2}{d}}\log\left(\frac{e}{4}\sqrt{\frac{d}{2}}\right)}.$$

Remark 1. Recall that our bound on the probability of undetected cheating by Alice for Equality-checker was $2 \cdot 2^{-\frac{m}{4}}$. Comparing this to our bound for Expander-checker, we see that when the degree $d = \omega(1)$, our new bound has the form $2^{-\frac{m}{4} + o(m)}$, close to that of Equality-checker.

Proof. Fix a graph $G \in \mathcal{F}$ with corresponding $S, B, C_1, C_2, \ldots, C_r$, where $|S| = \frac{m}{2}$, $|B| = \epsilon m$, and $c_i \overset{\text{def}}{=} |C_i| \leq \frac{m}{4}$ for all i. The difference between this proof and the previous is that now Bob can actually sample vertices from more than one C_i without Alice getting caught. This is because the graph G is not complete, so there may not be any edges connecting Bob's samples in the different C_i. However, using the expander-mixing lemma, we will show that if Bob samples too many vertices from different C_i, there will be bad edges connecting some of them, and Alice will get caught.

Define an elusive set as in the proof of Lemma 1. In order for Alice not to get caught, Bob must sample an elusive set, i.e., his vertices must come from $S \cup C_1 \cup C_2 \cup \cdots \cup C_r$ and there must be no edge between any of his samples lying in different C_i. We seek an upper bound on the number of elusive sets in G.

If $r = 0$, the number of elusive sets of G is 1. If $r = 1$, since $c_1 \leq \frac{m}{4}$, as in the proof of Lemma 1 for $r = 1$, the number of elusive sets is at most $\binom{\frac{3m}{4}}{\frac{m}{2}}$. For the remainder of the proof, $r \geq 2$.

We consider a labeled graph G' which has at least as many elusive sets as G. It will be easier to upper bound the number of elusive sets of G'. We want G' to have the property that its vertices can be partitioned into sets S, B, D_0, D_1, D_2 or sets S, B, D_0, D_1 such that $|S| = \frac{m}{2}$, $|B| = \epsilon m$, all edges between D_i and D_j with $i \neq j$ are bad, and $d_i = |D_i| \leq \frac{m}{4}$ for all i.

If $r = 2$ or $r = 3$, then put $G' = G$. Otherwise, $r \geq 4$. By averaging, there exist distinct C_i and C_j in G with $c_i + c_j \leq \frac{m}{4}$. Suppose we create G' from G by removing all bad edges between C_i and C_j, and by grouping vertices in C_i and C_j into a single set D of size $d = c_i + c_j \leq \frac{m}{4}$. It follows that r has decreased by 1. If r is still more than 3, repeat this process on G'. We eventually end up with the desired labeled graph G'. We will assume that $r = 3$. If actually $r = 2$, we may just set $D_2 = \emptyset$. We introduce some notation.

Definition 1. *We say that three integers i_0, i_1, i_2, where $i_0 \leq d_0$, $i_1 \leq d_1$, and $i_2 \leq d_2$, are* **harmonious** *if there exist sets $S_0 \subseteq D_0$, $S_1 \subseteq D_1$, and $S_2 \subseteq D_2$, where $|S_j| = i_j$ for $j = 0, 1, 2$, such that $e(S_0, S_1) = e(S_0, S_2) = e(S_1, S_2) = 0$. That is, there are no edges in G' between them.*

The number of elusive sets in G', and thus in G, is at most

$$\sum_{j=0}^{\frac{m}{2}} \binom{\frac{m}{2}}{\frac{m}{2} - j} \sum_{\substack{i_0+i_1+i_2=j \\ \text{harmonious } i_0,i_1,i_2}} \binom{d_0}{i_0}\binom{d_1}{i_1}\binom{d_2}{i_2} \leq$$

$$\sum_{j=0}^{\frac{m}{2}} \binom{\frac{m}{2}}{j} \sum_{r=0}^{2} \sum_{\substack{i_0+i_1+i_2=j \\ i_r=\max(i_0,i_1,i_2) \\ \text{harmonious } i_0,i_1,i_2}} \binom{d_0}{i_0}\binom{d_1}{i_1}\binom{d_2}{i_2}$$

We will choose d_0, d_1, d_2 to maximize this expression, subject to $\sum_i d_i = \frac{m}{4} - \epsilon m$ and $d_i \leq \frac{m}{4}$ for all i. As before, it is clear that the expression is maximized when $\epsilon = 0$. We start by bounding the following expression.

$$\sum_{j=0}^{\frac{m}{2}} \binom{\frac{m}{2}}{j} \sum_{\substack{i_0+i_1+i_2=j \\ i_0 \geq i_1,i_2 \\ \text{harmonious } i_0,i_1,i_2}} \binom{d_0}{i_0}\binom{d_1}{i_1}\binom{d_2}{i_2}. \tag{1}$$

The following is the only place where we use the fact that G is an expander.

Claim. For fixed harmonious i_0, i_1, i_2 with $i_0 + i_1 + i_2 = j$ and $i_0 \geq i_1, i_2$, we have,

$$\binom{d_1}{i_1}\binom{d_2}{i_2} \leq \binom{\frac{m}{2}}{2m\sqrt{2/d}}.$$

Proof. Suppose first that $i_0 \leq m\sqrt{2/d}$. Then since $i_0 \geq i_1, i_2$, we have $i_1 + i_2 \leq 2i_0 \leq 2m\sqrt{2/d}$. We arrive at

$$\binom{d_1}{i_1}\binom{d_2}{i_2} \leq \binom{d_1 + d_2}{i_1 + i_2} \leq \binom{\frac{m}{2}}{2m\sqrt{2/d}},$$

where we have used that $i_1 + i_2 \leq 2m\sqrt{2/d} \leq \frac{m}{4}$ since d is sufficiently large. Now suppose that $i_0 \geq m\sqrt{2/d}$. This is where we use the fact that G is an expander. Suppose T is a subset of $D_0 \cup D_1 \cup D_2$, and set $X = T \cap D_0$ and $Y = T \cap (D_1 \cup D_2)$. Suppose $|X| = i_0$ and $|Y| = i_1 + i_2$. We first note that the edgeset in G' connecting X to Y is identical to that in G. By the expander-mixing lemma, there is at least one edge from X to Y provided[2] that

$$\frac{d}{m}|X||Y| > \frac{\lambda}{m}\sqrt{|X|(m - |X|)|Y|(m - |Y|)}.$$

This is equivalent to the condition $|X||Y| > \left(\frac{\lambda}{d}\right)^2 (m - |X|)(m - |Y|)$. As we will choose λ so that $\lambda \leq 2\sqrt{d}$, this is in turn implied by the simpler $|X||Y| > \frac{4m^2}{d}$. This is just $i_0(i_1 + i_2) > \frac{4m^2}{d}$. Since $i_0 \geq m\sqrt{2/d}$, this holds if $i_1 + i_2 > 2m\sqrt{2/d}$. Thus, i_0, i_1, and i_2 are not harmonious if $i_1 + i_2 > 2m\sqrt{2/d}$, and so we again have $\binom{d_1}{i_1}\binom{d_2}{i_2} \leq \binom{d_1+d_2}{i_1+i_2} \leq \binom{\frac{m}{2}}{2m\sqrt{2/d}}$.

By the previous claim, expression 1 simplifies to

$$\sum_{j=0}^{\frac{m}{2}} \binom{\frac{m}{2}}{j} \sum_{\substack{i_0+i_1+i_2=j \\ i_0 \geq i_1,i_2 \\ harmonious\ i_0,i_1,i_2}} \binom{d_0}{i_0}\binom{\frac{m}{2}}{2m\sqrt{2/d}} =$$

$$\binom{\frac{m}{2}}{2m\sqrt{2/d}} \sum_{j=0}^{\frac{m}{2}} \binom{\frac{m}{2}}{j} \sum_{\substack{i_0+i_1+i_2=j \\ i_0 \geq i_1,i_2 \\ harmonious\ i_0,i_1,i_2}} \binom{d_0}{i_0}$$

In expression 1, we took $i_0 \geq i_1, i_2$, but we could've equally well taken $i_1 \geq i_0, i_2$ or $i_2 \geq i_0, i_1$. It follows that the number of elusive sets in G is at most

$$\binom{\frac{m}{2}}{2m\sqrt{2/d}} \sum_{j=0}^{\frac{m}{2}} \binom{\frac{m}{2}}{j} \sum_{r=0}^{2} \sum_{\substack{i_0+i_1+i_2=j \\ i_r \geq i_{r+1},i_{r+2} \\ harmonious\ i_0,i_1,i_2}} \binom{d_r}{i_r}, \tag{2}$$

where the subscripts should be understood modulo 3. At this point, our task is to maximize expression 2 subject to $\sum_i d_i = \frac{m}{2}$ and $d_i \leq \frac{m}{4}$ for all i.

By switching the order of summations, we have shown that the number of elusive sets is at most

$$\binom{\frac{m}{2}}{2m\sqrt{2/d}} \sum_{r=0}^{2} \sum_{j=0}^{\frac{m}{2}} \sum_{\substack{i_0+i_1+i_2=j \\ i_r \geq i_{r+1},i_{r+2} \\ harmonious\ i_0,i_1,i_2}} \binom{\frac{m}{2}}{j}\binom{d_r}{i_r}. \tag{3}$$

---

[2] One can do slightly better than the expander-mixing lemma by using Tanner's inequality [31]. This does not affect our bound much, so we omit this improvement.

Then, since there are at most $j + 1$ pairs (i_{r+1}, i_{r+2}) for a given i_r for which $i_r + i_{r+1} + i_{r+2} = j$, we can bound the inner sum by $\binom{m}{2}(j+1)2^{d_r}$. We may then pull out the 2^{d_r} term and, ignoring the terms that we have pulled out, we are left with $\sum_{j=0}^{\frac{m}{2}} \binom{\frac{m}{2}}{j}(j+1)$. We recall the identity: $\sum_{i=0}^{n} i\binom{n}{i} = n2^{n-1}$. This implies

$$\sum_{j=0}^{\frac{m}{2}}(j+1)\binom{\frac{m}{2}}{j} = \frac{m}{2} \cdot 2^{\frac{m}{2}-1} + 2^{\frac{m}{2}} = \left(\frac{m}{4}+1\right)2^{\frac{m}{2}}.$$

We can now simplify expression 3 to the following,

$$\binom{\frac{m}{2}}{2m\sqrt{2/d}}\left(\frac{m}{4}+1\right)2^{\frac{m}{2}}\left(2^{d_0}+2^{d_1}+2^{d_2}\right).$$

This expression is clearly maximized when $d_r = d_{r+1} = \frac{m}{4}$ and $d_{r+2} = 0$ for some value of r. Since $2^{\frac{m}{4}} \geq 1$ for any $m \geq 0$, this expression is at most

$$3\binom{\frac{m}{2}}{2m\sqrt{2/d}}\left(\frac{m}{4}+1\right)2^{\frac{3m}{4}}.$$

Using the identity $\binom{a}{b} \leq \left(\frac{ae}{b}\right)^b$, we further upper bound this expression as

$$3\left(\frac{m}{4}+1\right)2^{\frac{3m}{4}+2m\sqrt{\frac{2}{d}}\log\left(\frac{e}{4}\sqrt{\frac{d}{2}}\right)},$$

which upper bounds the total number of elusive sets. Thus, the probability that Alice does not get caught is at most this quantity divided by $\binom{m}{\frac{m}{2}}$. Using Fact 2, after some algebraic manipulation, $\binom{m}{m/2} = \frac{m!}{(m/2)!^2} \geq 2^m \left(\frac{2}{\pi m e^{1/3}}\right)^{1/2}$. We conclude that the probability that Alice does not get caught is at most

$$3\left(\frac{m}{4}+1\right)\sqrt{\frac{\pi m e^{1/3}}{2}} \cdot 2^{-\frac{m}{4}+2m\sqrt{\frac{2}{d}}\log\left(\frac{e}{4}\sqrt{\frac{d}{2}}\right)}.$$

and the proof of the theorem is complete.

4   Efficiency

To compare Expander-checker with Equality-checker, we would like to achieve inverse exponential (in m) probability of undetected cheating, where m is an input parameter we use to measure our protocol's efficiency. m corresponds to the number of garbled circuits in the above.

The probability Alice can cheat in Expander-checker is at most

$$3\left(\frac{m}{4}+1\right)\sqrt{\frac{\pi m e^{1/3}}{2}} \cdot 2^{-\frac{m}{4}+2m\sqrt{\frac{2}{d}}\log\left(\frac{e}{4}\sqrt{\frac{d}{2}}\right)}.$$

One can write a short computer program to find a constant $d = p + 1$ with p a prime congruent to 1 mod 4, for which we can instantiate the graphs G in the previous section with those of Fact 1 on $\Theta(m)$ vertices, so that this probability is at most $2^{-\Omega(m)}$. One can also find such a graph by random sampling [13].

To achieve error probability ϵ, we may set $m = O(\ln \frac{1}{\epsilon})$. Recall that g and I denote the number of gates and inputs to the circuit to be computed, respectively.

Step 1 requires $O(mg) = O(\ln(\frac{1}{\epsilon})g)$ symmetric encryptions and communication for the garbled circuits. The commitments require $O(dmI + 2mI) = O(mI) = O(mg)$ symmetric encryptions and communication. Step 2 requires communication $O(m)$. Similar to step 1, step 3 requires $O(mg)$ communication. Step 4 requires $O(mI)$ communication. Step 6 requires $O(I)$ exponentiations.

Theorem 7. Expander-checker *is secure in the malicious model with inverse exponential (in m) probability of undetected cheating. The number of symmetric encryptions and communication complexity are $O(mg)$, and the number of exponentiations is $O(I)$.*

Recall that Equality-checker achieves $2^{-\Omega(m)}$ probability of undetected cheating with $O(mg + m^2 I)$ communication and number of symmetric encryptions, while the number of exponentiations is $O(I)$ (see Theorem 1). Suppose we want error probability ϵ. Let m be such that we achieve error probablity ϵ in Equality-checker. Then in Expander-checker we achieve error probability ϵ for $m' = O(m)$. Moreover, our communication and number of symmetric encryptions is $O(m'g) = O(mg)$, which improves the $\Omega(mg + m^2 I)$ of Equality-checker for sufficiently large m and I.

4.1   Practical Issues and Open Questions

For a practical setting of parameters our bounds on the probability that Alice can cheat in Expander-checker may not be good enough to make Expander-checker favorable to Equality-checker. This is due in part to a certain suboptimality of our Ramanujan graphs. In Claim 3.2 we argued that any two disjoint sets of vertices in a Ramanujan graph on m vertices, one of size at least $m\sqrt{2/d}$ and one of size at least $2m\sqrt{2/d}$, have an edge between them. However, a counting argument shows there exist graphs on m vertices for which there is an edge between any two disjoint sets of vertices of size at least $2m \ln d/d$. Such an explicit graph would significantly reduce the $2^{2m\sqrt{2/d}\log(\frac{e}{4}\sqrt{\frac{d}{2}})}$ factor in our probability bound.

We cannot even rule out that there exist graphs on m vertices for which there is an edge between any two disjoint subsets of $\Theta(m/d)$ vertices. As far as we are aware, the best explicit construction of such graphs can be obtained from [6], and show there exist graphs on m vertices for which any two disjoint sets of vertices of size $\Omega(m \cdot \text{polylog}(d)/d)$ have an edge between them. We leave it as an open problem to see if the work of [6] can be of practical use in this context.

Besides directly trying to construct such graphs, it may be possible to slightly change the protocol. The natural thing to do would be to have Bob sample a d-regular graph on m vertices at random, and send it to Alice to use instead of our explicit Ramanujan graph. Then with high probability it is such that any two disjoint subsets of vertices of size $2m \ln d/d$ have an edge between them. The problem with this approach is that the probability of sampling such a graph is only $1 - 2^{-\Theta(m/d)}$, which is smaller than the $1 - 2^{-\Theta(m)}$ we are looking for. We leave it as an open problem to see if a probabilistic approach can be effective here.

Acknowledgments. We thank Payman Mohassel, Benny Pinkas, and the anonymous referees for many helpful comments.

References

[1] N. Alon, *Eigenvalues and expanders,* Combinatorica **6**, 1986, pp. 86–96.

[2] N. Alon, J. Bruck, J. Naor, M. Naor, and R. Roth, *Construction of asymptotically good, low-rate error-correcting codes through pseudo-random graphs,* IEEE Transactions on Information Theory **38** (192), pp. 509-516.

[3] N. Alon and V. D. Milman. *Eigenvalues, expanders, and superconcentrators,* FOCS, 1984.

[4] N. Alon and J. Spencer. *The Probabilistic Method,* 2000.

[5] D. Beaver. *Correlated pseudorandomness and the complexity of private computations,* STOC, 1996.

[6] M. Capalbo, O. Reingold, S. Vadhan, and A. Wigderson. *Randomness conductors and constant-degree lossless expanders,* STOC, 2002.

[7] F. Chung, *Spectral Graph Theory,* CBMS Lecture Notes, AMS Publications, 1997.

[8] F. Chung and L. Lu. *Concentration inequalities and martingale inequalities - a survey,* Internet Mathematics, to appear.

[9] R. Diestel. *Graph Theory,* Springer-Verlag, 2005.

[10] I. Dinur. *The PCP Theorem by Gap Amplification,* STOC, 2006.

[11] S. Even, O. Goldreich and A. Lempel. *A randomized protocol for signing contracts,* Communications of the ACM, 1985.

[12] W. Feller, *Stirling's Formula,* Section 2.9 in An Introduction to Probability Theory and its Applications **1**, 3rd edition, New York: Wiley, pp. 50 -53, 1968.

[13] J. Friedman, *A Proof of Alon's Second Eigenvalue Conjecture,* STOC, 2003.

[14] O. Gabber and Z. Galil. *Explicit constructions of linear-sized superconcentrators,* JCSS, **22(3)**:407-420, 1981.

[15] O. Goldreich, S. Micali, and C. Rackoff. *Proofs that yield nothing but their validity or all languages in NP have zero-knowledge proofs,* FOCS, 1986.

[16] O. Goldreich, S. Micali, and A. Wigderson. *How to play any mental game or a completeness theorem for protocols with honest majority,* STOC, 1987.

[17] O. Goldreich. *Foundations of cryptography* - volume 2, ch. 7, 2004.

[18] Y. Ishai, J. Kilian, K. Nissim, and E. Petrank. *Extending oblivious transfers efficiently,* Crypto, 2003.

[19] J. Kilian and E. Petrank. *An efficient noniteractive zero-knowledge proof system for NP with general assumptions,* Journal of Cryptology, **11**:1–27, 1998.

[20] M. Kiraz and B. Schoenmakers, *A protocol issue for the malicious case of Yao's garbled circuit construction,* in the 27th Symposium on information theory in the BENELUX (WIC), 2006.

[21] Y. Lindell, and B. Pinkas. *A proof of Yao's protocol for secure two-party computation,* Cryptology ePrint Archive, Report 2004/175, 2004.

[22] Y. Lindell, and B. Pinkas. *An efficient protocol for secure two-party computation in the presence of malicious adversaries,* to appear in these proceedings, Eurocrypt, 2007.

[23] A. Lubotzky, R. Phillips, and P. Sarnak. *Explicit expanders and the Ramanujan conjectures*, STOC, 1986. See also: A. Lubotzky, R. Phillips, and P. Sarnak, Ramanujan graphs, Combinatorica 8, 1988, pp. 261-277.

[24] D. Malkhi, N. Nisan, B. Pinkas, and Y. Sella. *Fairplay - a secure two-party computation system*, Usenix, 2004.

[25] G. A. Margulis. *Explicit group-theoretical constructions of combinatorial schemes and their application to the design of expanders and superconcentrators*, Problemy Peredachi Informatsii **24**: 51-60 (Russian). English translation in Problems of Information Transmission **24**, 1988, 39-46.

[26] P. Mohassel and M. Franklin. *Efficiency Tradeoffs for Malicious Two-Party Computation*, PKC, 2006.

[27] M. Naor and B. Pinkas. *Efficient oblivious transfer.* SODA, 2001.

[28] B. Pinkas. *Fair secure two-party computation*, Eurocrypt, 2003.

[29] M. Rabin. *How to exchange secrets by oblivious transfer*, Technical Report Tech., Memo. TR-81, Aiken Computation Laboratory, Harvard University, 1981.

[30] H. Robbins. *A remark of Stirling's Formula.*, Amer. Math Monthly **62**, pp. 26-29, 1955.

[31] R. M. Tanner. *Explicit Construction of Concentrators from Generalized N-Gons*, SIAM J. Alg. Discr. Math **5**, 1984, pp. 287-293.

[32] L. Trevisan. *Inapproximability of Combinatorial Optimization Problems*, Optimisation Combinatiore 2.

[33] D. Woodruff *Revisiting the efficiency of malicious two-party computation*, Cryptology ePrint Archive, Report 2006/397, 2006.

[34] A. C. Yao. *How to generate and exchange secrets*, FOCS, 1986.

Appendix: A Counterexample

We've restated the lemma of [26] in our language (in this paper we have swapped the roles of Alice and Bob):

Lemma 3 of [26]: *With probability* $\geq 1 - 2^{-\frac{m}{6}}$, *at least* $\frac{5}{6}$ *of Alice's* $\frac{m}{2}$ *inputs are the same, or Alice will get caught.*

Consider the following behavior of a malicious Alice. Label the garbled circuits $C_1, ..., C_m$. Suppose m is a multiple of 8. For the first $\frac{7m}{8}$ circuits $C_1, ..., C_{7m/8}$, Alice will use the input 0 (assume Alice has only one input to the circuits), and for every other circuit, Alice will use the input 1. Thus, the bad edges are exactly those between one of the first $\frac{7m}{8}$ circuits and one of the last $\frac{m}{8}$ circuits.

Since all the circuits are correctly garbled, Alice only gets caught if a bad commitments is exposed in step 3. Consider the following event $\mathcal{E}$: Bob samples all $\frac{m}{2}$ of his circuits from the first $\frac{7m}{8}$ garbled circuits. Observe that if $\mathcal{E}$ occurs, no bad commitment is exposed in step 3, and therefore Alice does not get caught. Moreover, if $\mathcal{E}$ occurs, Bob will use $\frac{7m}{8} - \frac{m}{2} = \frac{3m}{8}$ 0 inputs when he performs verification, and $\frac{m}{8}$ 1 inputs. Thus, at most $\frac{3}{4}$ of Alice's $\frac{m}{2}$ inputs are the same, contrary to the $\frac{5}{6}$ claimed by Lemma 3.

For the counterexample to go through, it remains to show $\Pr[\mathcal{E}] > 2^{-\frac{m}{6}}$. But $\Pr[\mathcal{E}]$ is just $\binom{\frac{7m}{8}}{\frac{m}{2}} / \binom{m}{\frac{m}{2}}$. It is then straightforward to show $\binom{\frac{7m}{8}}{\frac{m}{2}} / \binom{m}{\frac{m}{2}} > 2^{-\frac{m}{6}}$, as needed.

The above presentation was done for simplicity. One can replace $\frac{7m}{8}$ by any value less than $\frac{11m}{12}$ in the above to get a "stronger" counterexample.

Efficient Two-Party Secure Computation
on Committed Inputs

Stanisław Jarecki and Vitaly Shmatikov

University of California, Irvine
The University of Texas at Austin

Abstract. We present an efficient construction of Yao's "garbled circuits" proto-col for securely computing any two-party circuit on committed inputs. The pro-tocol is secure in a universally composable way in the presence of *malicious* adversaries under the decisional composite residuosity (DCR) and strong RSA assumptions, in the common reference string model. The protocol requires a con-stant number of rounds (four-five in the standard model, two-three in the ran-dom oracle model, depending on whether both parties receive the output), $O(|C|)$ modular exponentiations per player, and a bandwidth of $O(|C|)$ group elements, where $|C|$ is the size of the computed circuit.

Our technical tools are of independent interest. We propose a homomorphic, semantically secure variant of the Camenisch-Shoup verifiable cryptosystem, which uses shorter keys, is *unambiguous* (it is infeasible to generate two keys which successfully decrypt the same ciphertext), and allows efficient proofs that a committed plaintext is encrypted under a *committed key*.

Our second tool is a practical four-round (two-round in ROM) protocol for *committed* oblivious transfer on *strings* (string-COT) secure against malicious participants. The string-COT protocol takes a few exponentiations per player, and is UC-secure under the DCR assumption in the common reference string model. Previous protocols of comparable efficiency achieved either committed OT on *bits*, or standard (non-committed) OT on strings.

1 Introduction

Informally, a two-party protocol for computing a circuit is *secure* if participants do not learn anything from the protocol execution beyond what is revealed by the output of the circuit. In a seminal paper, Andrew Yao showed a "garbled circuit" protocol [Yao86] for secure two-party computation (2PC) of any circuit in the *semi-honest* model, *i.e.*, assuming that participants faithfully follow the protocol specification. Yao's protocol requires $O(|C|)$ symmetric-key operations, and its bandwidth is $O(|C|)$ symmetric-key ciphertexts, in addition to the cost of n instances of an oblivious transfer (OT) protocol, where n is the size of the circuit's inputs. Using a 2-round OT protocol, Yao's protocol takes only two communication rounds (assuming only one player receives the output).

The main contribution of this paper is a new variant of Yao's protocol, which replaces $O(|C|)$ symmetric-key operations with $O(|C|)$ public-key operations, and at this cost achieves security against *malicious* participants in the common reference string (CRS) model. Specifically, our protocol operates on a multiplicative group $\mathbb{Z}_{n^2}^*$ where n is

M. Naor (Ed.): EUROCRYPT 2007, LNCS 4515, pp. 97–114, 2007.

a safe RSA modulus which satisfies DCR and strong RSA assumptions. The protocol requires $O(|C|)$ modular exponentiations, its bandwidth is $O(|C|)$ elements in $\mathbb{Z}_{n^2}^*$, and it takes four rounds in the standard model and two in ROM. Moreover, our protocol is universally composable, and securely computes any circuits on *committed* inputs.

A fundamental primitive in Yao's protocol is *oblivious transfer* (OT). Informally, OT is a two-party protocol in which the receiver (a.k.a. the "chooser") receives a value of his choice from among several values sent by the sender, while learning nothing about the other values. The sender does not learn anything from the protocol, and in particular he does not learn which of the values he sent was received by the chooser. *Committed oblivious transfer* (COT) is a variant of oblivious transfer, introduced by Crépeau [Cré89] as a "verifiable OT," in which both the sender and the chooser are committed to their inputs, and the oblivious transfer proceeds on the committed values. The second contribution of our paper is a new protocol for committed oblivious transfer on *strings* ("string-COT"). The protocol requires $O(1)$ exponentiations and has the bandwidth of $O(1)$ elements in $\mathbb{Z}_{n^2}^*$, which is comparable to the cost of previous protocols for standard (non-committed) OT on strings or previous COT protocols that operated only on bits. This new string-COT protocol is also universally composable in the CRS model.

A *committed* OT protocol secure against malicious players is a much more useful tool in a security protocol than a standard OT. For example, unless the OT protocol runs on committed inputs, it is fundamentally non-robust against network failures because re-running the protocol after a failure allows the cheating receiver to learn both of the sender's values. Similarly, secure *committed 2PC* protocol is a much more useful tool than a standard 2PC protocol. In general, universally composable string-OT and general 2PC *on committed data* makes it easy to ensure that multiple instances of these protocols are executed on consistent inputs, for example as prescribed by some larger protocol.

Technical roadmap. Both protocols we present in this paper, the protocol for secure two-party computation on committed inputs ("committed 2PC") and the string-COT protocol, rely on a modification of the verifiable encryption given by Camenisch and Shoup [CS03]. The efficiency of these two protocols is essentially due to the very strong properties that this encryption offers. We will refer to the original scheme of [CS03] as *CS encryption*, and we call our modification *sCS encryption*, where "s" stands for both "short" and "simplified," because the modification consists of (1) stripping off the chosen-ciphertext security check in the CS encryption, and (2) using significantly shorter private keys. Below we explain how several interesting properties of this encryption enable the efficient string-COT and committed 2PC protocols.

The sCS encryption scheme is *additively homomorphic*, *i.e.*, given ciphertexts of two values, one can obtain a ciphertext of their sum without decrypting the ciphertexts, and it is *verifiable*, *i.e.*, there is a very efficient ZK proof system due to [CS03] for showing that the encrypted message corresponds to a previously committed one. These two features together enable an efficient string-COT protocol. First, we use additive homomorphism of the sCS encryption to build an efficient protocol for OT on strings in a way that is similar to how Aiello *et al.* [AIR01] build a standard (*i.e.*, non-committed) OT on strings from the multiplicatively homomorphic ElGamal encryption. Then, by adapting the ZK proof systems given for the CS encryption in [CS03], we add efficient

ZK proofs for showing that the parties run this string-OT protocol on the previously committed inputs.

The sCS encryption has further useful properties which allow us to extend the string-COT protocol to an efficient committed 2PC protocol. First, it is *unambiguous*, in the sense that it is committing not only to the plaintext, but also to the encryption key: it is infeasible to produce a ciphertext that can be successfully decrypted, even to the same plaintext, under two different decryption keys. This property is crucial in the maliciously secure version of Yao's protocol. Otherwise, the player who creates the garbled circuit could embed all sorts of faults into the circuit. If the circuit evaluator encounters a fault which causes him to stop, the malicious player will learn information about the evaluator's inputs that he is not supposed to learn.

Second, we extend the Camenisch-Shoup ZK proof system to an efficient ZK proof that *a ciphertext encrypts a committed plaintext under a committed key*. (Technically, this proof system is defined for a symmetric-key version of the sCS encryption, where the key is both an encryption and a decryption key.) This proof system is a crucial component of proving that Yao's "garbled circuit" is formed correctly. Yao's construction of the garbled circuit involves encrypting, for every circuit gate, the keys corresponding to the output wires under the keys corresponding to the input wires. In our version of Yao's protocol, the sender commits to the keys he created for every circuit wire. For the wires corresponding to the receiver's inputs, the sender sends to the receiver the appropriate key values using our efficient string-COT protocol operating on these commitments. Furthermore, the sender must prove, for each gate, that the ciphertexts that are supposed to encrypt the appropriate output-wire keys under the appropriate input-wire keys are formed correctly. This is accomplished precisely by the above proof system, because the input-wire keys appear as *keys* in these ciphertexts, while the output-wires keys appear as *plaintexts*.

Giving an efficient ZK proof system for this statement for some version of the CS encryption scheme is an interesting technical challenge, because in the CS cryptosystem plaintexts and keys "live" in different groups (and are acted upon by different moduli). It is not immediately obvious how to encrypt one CS encryption key under another CS encryption key and have an efficient proof of correctness for this encryption, because the efficient proof systems given for the CS encryption require that the plaintext be significantly smaller than the encryption key. One solution is to extend these proof systems to handle larger plaintexts (namely, plaintexts of the same size as the key), using proofs of equality of elements of two different groups represented as integers (*e.g.*, [Bou00]). We propose a simpler solution based on the observation that, from the results of Håstad, Schrift and Shamir [HSS93] on simultaneous bit security of exponentiation in groups of unknown order, it follows that one can shorten the private keys used in the CS encryption to $\frac{|n|}{2}$ bits. This significantly speeds up the CS encryption, but, more importantly, this modification allows for a very efficient ZK proof that a ciphertext encrypts a *committed plaintext* under a *committed key*.

Organization of the paper. In Section 2 we discuss related work. In Section 3, we describe our cryptographic toolkit. In Section 4, we present the string-COT protocol, and in Section 5, the protocol for general two-party secure computation on committed inputs. All proofs have been delegated to the full version of the paper.

2   Related Work on Constant-Round 2PC and Committed OT

2PC protocols. The first constructions for secure two-party computation are Yao's "garbled circuits" protocol [Yao86] and the protocol of [GMW87]. Of the two, only Yao's protocol is constant-round, but secure only in the semi-honest model. Most subsequent constant-round protocols for secure computation in the malicious model, such as [Kil88, Lin03, KO04], employ generic zero knowledge proofs (*i.e.*, proofs for any NP statement). The overhead of this approach is likely to remain prohibitive for practical applications.

There are secure 2PC protocols that avoid generic zero-knowledge proofs (*e.g.*, see [JJ00, GMY04] and references therein), but the round complexity of these protocols is linear in the (boolean or arithmetic) circuit depth. On the other hand, Damgård and Ishai [DI05] showed the first constant-round *multi-party* protocol with $O(|C|n^2k)$ bandwidth and computation (here n is the number of parties, k is the security parameter), assuming a trusted preprocessing stage, but this protocol is secure only with an honest majority, and its techniques (*e.g.*, verifiable secret sharing) do not seem applicable to two-party computation.

2PC using verifiable encryption. Like our protocol, the constant-round 2PC protocol of Cachin and Camenisch [CC00] uses a verifiable public-key encryption scheme, but unlike in our scheme, their zero-knowledge proofs require s cut-and-choose repetitions where s is the statistical security parameter. Hence their 2PC protocol requires $O(s|C|)$ group elements in bandwidth and the same number of exponentiations (vs. $O(|C|)$ in our construction). It is worth mentioning, however, that our ciphertexts are elements of $\mathbb{Z}_{n^2}^*$, for n satisfying the DCR and strong RSA assumptions, while [CC00] can use any group where the Diffie-Hellman assumption holds.

2PC using cut-and-choose approach. A recent series of works on efficient constant-round 2PC protocols [Pin03, MF06, LP07, Woo07] shows that security in the malicious model can be achieved by cut-and-choose verification of the entire garbled circuit, at the cost of $O(s|C| + s^2n)$ [LP07] or $O(s|C|)$ [Woo07] symmetric-key operations, where s is the statistical security parameter of cut-and-choose and n is the input size. These cut-and-choose constructions probably require less computation than our protocol to achieve similar levels of security based on common assumptions, but our protocol may require less bandwidth, especially for small circuits whose size is comparable to the input size. Also, our protocol can be made non-interactive in the random oracle model at no extra cost, while the security parameter s in the cut-and-choose solutions increases if they are made non-interactive using the Fiat-Shamir heuristic.

COT. Committed OT (COT) was introduced by Crépeau [Cré89], where it was used to construct a general 2PC protocol (but not constant-round one) following the approach of [GMW87]. Crépeau constructed COT using black-box invocations of $\Omega(n^3)$ OTs. This was improved by [CvdGT95] to $O(n)$ OT's and $O(n^2)$ bit commitments. Both COT protocols, however, operate on bits rather than strings. Based on the concrete assumptions of Computational or Decisional Diffie-Hellman, Cramer and Damgård [CD97] and then Garay *et al.* [GMY04] give COT protocols which require $O(1)$ exponentiations but still operate only on bits, while Camenisch and Cachin [CC00] give a string-COT protocol, but it requires $O(k)$ modular exponentiations where k is the security parameter.

Lipmaa [Lip03] proposed to extend the (non-committed) string-OT protocol of Aiello *et al.* [AIR01] to a committed OT protocol on strings at the cost of $O(1)$ exponentiations. While this protocol does ensure that the *received* string is consistent with the sender's commitment, the sender can successfully cheat on the string that has *not* been transferred during the OT. This can be used to break chooser's privacy in any application (such as 2PC) where the sender can observe whether the chooser succesfully completed the protocol. Stronger verifiability can potentially be achieved by extending this protocol with zero-knowledge proofs, but the resulting protocol would not beat the $O(k)$ modular exponentiations bound because the commitment schemes (*e.g.*, [CGHGN01]) suggested in [Lip03] seem to have only cut-and-choose ZK proofs.

3   Cryptographic Tools

3.1   Camenisch-Shoup (CS) Encryption Scheme [CS03]

Common reference string. A trusted third party generates a safe RSA modulus $n = pq$, where $p = 2p' + 1, q = 2q' + 1, |p| = |q|, p \neq q$, and p, q, p', q' are all primes, a random element g' in $\mathbb{Z}_{n^2}^*$ and an element $g = (g')^{2n}$. The common reference string is (n, g), which also implicitly defines element $\alpha = 1 + n$. For standalone applications of CS encryption, pair (n, g) can be thought of as part of the public key. However, placing (n, g) in the CRS enables soundness of some very useful proof systems associated with this encryption scheme, *e.g.*, those used in our COT and 2PC protocols.

The group $\mathbb{Z}_{n^2}^*$ defined by the safe RSA modulus n can be decomposed into a cross-product of four subgroups: $\mathbb{Z}_{n^2}^* = G_n \times G_{n'} \times G_2 \times T$, where group G_n, generated by $\alpha = n + 1$, has order n, group $G_{n'}$ has order $n' = p'q'$, and G_2 and T are subgroups of order 2. As one consequence of this structure of $\mathbb{Z}_{n^2}^*$, the above procedure of picking g as a $2n$-power of a random element implies that, with an overwhelming probability, g is a generator of subgroup $G_{n'}$. In the following we treat all multiplications and exponentiations as operations in $\mathbb{Z}_{n^2}^*$, unless stated otherwise.

Key generation. The private key is a random triple x_1, x_2, x_3 chosen in $[0, \frac{n^2}{4}]$. The public key is $PK = (n, g, \mathfrak{g}, \mathfrak{h}, \mathfrak{f}, \mathsf{hk})$ where $\mathfrak{g} = g^{x_1}, \mathfrak{h} = g^{x_2}, \mathfrak{f} = g^{x_3}$, and hk is a key of a collision-resistant keyed hash function $\mathcal{H}$.

Encryption. Consider plaintext m as an integer in $[-\frac{n}{2}, \frac{n}{2}]$. (Note that one can encode elements m' in $\mathbb{Z}_n$ in this range as $m = m'$ rem n, *i.e.*, $m = m'$ if $m' \leq \frac{n}{2}$ and $m = m' - n$ if $m' > \frac{n}{2}$. Observe that $m = m' \mod n$.) A CS encryption of m under key PK with label L, denoted $\mathsf{CSenc}_{PK}^L(m)$, is a tuple (u, e, v) where $u = g^r$, $e = \alpha^m \mathfrak{g}^r$, and $v = \mathrm{abs}((\mathfrak{h}\mathfrak{f}^{\mathcal{H}_{\mathsf{hk}}(u,e,L)})^r)$, for a randomly chosen $r \in [0, \frac{n}{4}]$. Operation $abs(a)$ returns a for $a < \frac{n}{2}$ and $n - a$ for $a \geq \frac{n}{2}$.

Decryption. Given a ciphertext (u, e, v), check $abs(v) = v$ and $u^{2(x_2 + \mathcal{H}_{\mathsf{hk}}(u,e,L)x_3)} = v^2$. If this holds, compute $\hat{m} = (e/u^{x_1})^2$. Note that $e/u^{x_1} = \alpha^m$ for correctly formed ciphertexts. If $\hat{m} \notin \langle \alpha \rangle$, *i.e.*, if n does not divide $\hat{m} - 1$, reject. Otherwise, set $\hat{m}' = \frac{\hat{m}-1}{n}$ (over the integers), $m' = \hat{m}'/2 \mod n$, and $m = m'$ rem n.

This encryption is CCA secure under the DCR assumption on safe RSA moduli [CS03]:

Assumption 1. (DCR) *[Pai99]: Given RSA modulus n, random elements of $\mathbb{Z}_{n^2}^*$ are computationally indistinguishable from elements of a subgroup formed by n-th powers of elements in $\mathbb{Z}_{n^2}^*$.*[1]

3.2 Simplified Camenisch-Shoup (sCS) Encryption Scheme

The group setting (n, g) is the same. Denote $k'' = \frac{|n|}{2}$, and let k, k' be parameters that control the quality of soundness and zero-knowledge of proof systems associated with the sCS encryption. We require that $2k + k' < k''$ and $k < p', q'$. For 80-bit security, one can take $k'' = 512$ and $k = k' = 80$.

Key generation. The private key is $x \in [0, 2^{k''}]$. The public key is $y = g^x$.

Encryption. The sCS encryption under key y of m, an integer in $[-\frac{n}{2}, \frac{n}{2}]$, denoted $\mathsf{sCSenc}_y(m)$, is (u, e) s.t. $e = \alpha^m y^r \bmod n^2$ and $u = g^r$ for a random r in $[0, \frac{n}{4}]$.

Decryption. Proceeds exactly like CS decryption, but omitting the CCA checks on v (since there's no v here), and using x instead of x_1 in decrypting (u, e).

Apart from stripping the CCA check, the only difference between CS and sCS encryption is the shortened private key. The fact that the scheme remains semantically secure with such modification follows from adapting the results of [HSS93] on simultaneous bit security of exponentiation modulo a Blum integer (and a safe RSA modulus is Blum integer) to exponentiation in $\mathbb{Z}_{n^2}^*$.[2] It follows that under the factoring assumption, the entire upper half of the bits of exponent x is simultaneously hidden under the exponentiation function $y = g^x \bmod n^2$, and therefore key $y = g^x$ for x random in $\mathbb{Z}_{n'}$ is indistinguishable from $y = g^x$ for x random in $[0, \frac{|n|}{2}]$.[3]

Theorem 1. *sCS encryption is semantically secure under DCR assumption on safe RSA moduli.*

Symmetric-key version of sCS encryption scheme. The sCS cryptosystem can also be used as a symmetric encryption scheme if the private key $x \in [0, 2^{k''}]$ is treated as a symmetric key. Encryption of m under key x is a pair (e, u), where $e = \alpha^m u^x \bmod n^2$, $u = g^r$ for random $r \in [0, \frac{n}{4}]$. The decryption procedure does not change, nor does the security of the encryption scheme.

Unambiguity of sCS encryption. We introduce a very strong notion of *unambiguous encryption*, which applies to both public-key and symmetric schemes. It says that a ciphertext that passes a certain proof system, denoted ZKUnEnc, cannot decrypt to two different plaintexts under two different private keys. Moreover, no two distinct decryption keys can decrypt a ciphertext even to the same plaintext. Therefore, in an unambiguous encryption scheme, the ciphertext is committing not only to the plaintext, but also to the decryption key. This notion of encryption unambiguity is essential for our

---

[1] For the safe RSA moduli n, the subgroup of n-th residues in $\mathbb{Z}_{n^2}^*$ is the subgroup $G_{n'} \times G_2 \times T$.

[2] Cf. similar observation in [CGHG01] for Paillier encryption, on which CS encryption is based.

[3] Note that in this way one can also shorten keys x_2, x_3 in CS encryption and the randomness r.

version of Yao's 2PC protocol, because otherwise a malicious creator of the garbled circuit could introduce errors in this circuit, and then learn something extra about the receiver's inputs by observing whether the receiver successfully completes his computation on this circuit.

Definition 1. *An encryption scheme is* unambiguous *if there exists a zero-knowledge proof system* ZKUnEnc *s.t. for every efficient probabilistic algorithm* A, *the following event has only negligible probability: (1)* A *outputs tuple* (c, x_1, x_2) *s.t.* $x_1 \neq x_2$, *(2)* A *passes the* ZKUnEnc *proof system on ciphertext* c, *(3)* x_1, x_2 *are valid private keys, i.e., they are accepted by the decryption procedure, and (4) both* $Dec_{x_1}(c)$ *and* $Dec_{x_2}(c)$ *output a valid message (or messages). In the CRS model, the probability is also taken over the randomness of the common reference string generation.*

Theorem 2. *sCS encryption is unambiguous under the factoring assumption on safe RSA moduli, in the CRS model.*

The ZK proof system ZKUnEnc for the sCS encryption is the proof that u^2 belongs to the group generated by g, *i.e.*, ZKUnEnc(u, e) = ZKDL(g, u). (See section 3.4.)

3.3 CS Commitments and sCS Commitments

Our COT and 2PC protocols could be adapted to work with standard Pedersen-like commitment schemes of [Ped91, FO97, DF02] at the cost of additional mappings, via range proofs [CM99, Bou00, DF02], between commitments with different ranges of plaintexts. Instead, we use the full (*i.e.*, adaptive chosen-ciphertext secure) CS encryption as a commitment scheme, because it operates on the same group as the encryption we use, and hence is well-suited for both the COT and 2PC protocols of Sections 4 and 5.[4] Moreover, using a CCA-secure encryption as a commitment helps in showing that the COT and 2PC schemes are secure in the strong sense of universal composability.

An instance of a CS commitment scheme is a CS encryption public key $PK =$ $(n, g, \mathfrak{g}, \mathfrak{h}, \mathfrak{f}, hk)$. The public key is chosen by a trusted third party, and security of this commitment scheme requires the CRS model. The CS commitment on message m, an integer in range $[-\frac{n}{2}, \frac{n}{2}]$ (with an obvious mapping to $\mathbb{Z}_n$), with label L, is the ciphertext Com = CSenc$_{PK}^L(m)$. For notational convenience of the COT and 2PC protocols, we denote the tuple forming commitment Com as (u, C, v), *i.e.*, $u = g^r$, $C = \alpha^m \mathfrak{g}^r$, and $v = \text{abs}((\mathfrak{h}\mathfrak{f}^{\mathcal{H}_{hk}(u,C,L)})^r)$. The decommitment is the (r, m, L) tuple. In the COT and 2PC protocols, we often treat value C in the CS commitment as a commitment to m by itself. This shortened commitment is used very heavily in the 2PC protocol, thus we refer to value $C = \alpha^m \mathfrak{g}^r$ by itself as an *sCS commitment*. The corresponding decommitment is (m, r).

3.4 Efficient Concurrently Secure ZK Proof Systems in the CRS Model

All proof systems used in our COT and Committed 2PC protocols are concurrently secure ZK proofs in the CRS model. Specifically, each proof system is computationally

---

[4] Note that instances of other commitment schemes can be mapped to this one using the verifiable encryption proof system that accompanies the Camenisch-Shoup encryption [CS03].

sound and statistical zero-knowledge with a straight-line simulator. The latter is important for showing that the protocols are universally composable. Each of these proof systems is built from efficient HVZK proof systems for the languages listed below by a series of compilations which preserve the efficiency of the underlying HVZK protocols.

The compilations start from 3-round HVZK proof systems with the properties of *special honest-verifier zero-knowledge* and *(weak) special soundess* (we discuss these below). First, with the techniques of Cramer *et al.* [CDS94], HVZK systems of this class can be combined, at no extra cost, into HVZK proof systems of the same class for any (monotonic) disjunctive and/or conjuctive formula over statements proved in the component proof systems. Then, using Damgård [Dam02], the resulting HVZK proof system can be compiled into a three-round concurrently secure ZK proof systems with statistical zero-knowledge, computational soundness, and a straight-line simulator in the CRS model. This latter technique requires statistically hiding trapdoor commitments, and using Pedersen's commitment scheme it incurs a computational overhead of just one extra exponentiation per player. The computational soundness of the resulting ZK proof system is subject to the same assumption as the computational binding of the commitment scheme, which can be Strong RSA if Pedersen's trapdoor commitment is adapted to the $\mathbb{Z}_{n^2}^*$ setting, *e.g.*, as in Damgård-Fujisaki commitments [DF02]. Note that in ROM, using the Fiat-Shamir heuristic, the HVZK proof systems of this class can be converted at no extra cost to *non-interactive* ZKs with the same properties of computational soundness and statistical zero-knowledge with straight-line simulation.

We denote the statements being proved as X, Y, Z, and the corresponding "atomic" HVZK proof systems as HVZKX, HVZKY, HVZKZ. We use a notation derived from boolean formulas for the ZK proof systems resulting from this series of compilations. For example, the resulting ZK proof system for language $X \wedge (Y \vee Z)$ will be denoted ZKX∧(ZKY∨ZKZ). We catalog the proof systems used in the COT and 2PC protocols by the statements they prove, namely, membership in the languages DL, DLEQ, NotEq, Cot, Com, and PlainEq. Each of these is parameterized by tuple $(n, g, \mathfrak{g}, \mathfrak{h}, \mathfrak{f}, \mathsf{hk})$, which forms an instance of the CS commitment scheme. Triple $(n, g, \mathfrak{g})$ also defines an instance of the sCS commitment. Parameters k, k', k'' are as in Section 3.2.

DL $= \{(g, X) \mid$ there exists x s.t. $X^2 = g^{2x}\}$.

DLEQ $= \{(g, X, \tilde{g}, \tilde{X}) \mid$ there exists x s.t. $X^2 = g^{2x}, \tilde{X}^2 = \tilde{g}^{2x}\}$.

NotEq $= \{(C_a, C_b) \mid$ there exist a, b, r_a, r_b s.t. $a \neq b \bmod n$, $C_a = \alpha^a \mathfrak{g}^{r_a}$, and $C_b = \alpha^b \mathfrak{g}^{r_b}\}$. In other words, C_a and C_b are sCS commitments to two different values.

Cot $= \{(i, e', u', e, u, y, C) \mid$ there exist m, w, s, r s.t. $C^2 = \alpha^{2m}\mathfrak{g}^{2w}$, $e'^2 = e^{2s}\alpha^{2m-i*2s}y^{2r}$, and $u'^2 = u^{2s}g^{2r}\}$. In other words, m rem n is committed in sCS commitment C, and (u', e') is a correct "re-encryption" of m performed by the sender in the COT protocol, given the (y, u, e) tuple sent by the receiver.

Com $= \{(\mathsf{Com}, \mathsf{ids}) \mid$ there exist m, r s.t. $\mathsf{Com} = (u, C, v)$ where $u = g^r$, $C = \alpha^m \mathfrak{g}^r$, and $v = \mathrm{abs}((\mathfrak{h}\mathfrak{f}^{\mathcal{H}_{\mathsf{hk}}(u,C,\mathsf{ids})})^r)\}$. In other words, Com is a properly formed CS commitment to some message m with label ids.

PlainEq $= \{((e, u), C_x, C_m) \mid$ there exist x, m, r_x, r_m s.t. $e = \alpha^m u^x$, $C_x = \alpha^x \mathfrak{g}^{r_x}$, and $C_m = \alpha^m \mathfrak{g}^{r_m}\}$. In other words, (e, u) is an sCS encryption of the plaintext m committed in (sCS commitment) C_m under the key x committed in C_x.

All of the above languages have efficient 3-round HVZK proof systems HVZKDL, HVZKDLEQ, *etc.*, which unconditionally satisfy the two properties we need: (1) special HVZK, and (2) weak special soundness. The only exception is HVZKPlainEq, for which we show that weak special soundness holds under the strong RSA assumption. All systems are efficient: the players make only a few exponentiations (between one and four) modulo n^2, and communication complexity ranges from $3|n|$ in HVZKDL to at most $20|n|$ bits in HVZKPlainEq. We show the HVZKPlainEq proof system in Appendix A, because it has the most novelty. We delegate the other proof systems to the full version of the paper, but most of them are either standard, or simple modifications of the proofs that appear in [CS03]. The HVZKPlainEq proof system shown in Appendix A gives a good idea of how all of these HVZKs work.

Special HVZK and (weak) special soundness. Let (P_1, P_2, V) be a specification of a 3-round public coin proof system for language L. The prover's message in the first round on instance x, witness w for $x \in L$, and randomness r is computed as $a = P_1(x, w, r)$, its response in the third round is computed as $z = P_2(x, w, r, e)$ where e is the verifier's challenge, and the verifier accepts if and only if $V(x, a, e, z) = 1$. We call this proof system *special (statistical) HVZK* if there exists a simulator S s.t. for every challenge e and every witness (x, w) for $x \in L$, the tuple (a, z) output by $S(z, e)$ is distributed statistically close to tuple (a, z) where $a = P_1(x, w, r)$ and $z = P_2(x, w, r, e)$. The probability is over the coins of S and over r. We say that this proof system has *(weak) special soundness* if for every $x \notin L$, and for every PPT algorithm $\hat{P}$, the probability that $\hat{P}(x)$ outputs (a, e, z, e', z') s.t. $e \neq e'$ and $V(x, a, e, z) = V(x, a, e', z') = 1$, is negligible. Since the HVZK proof systems we use are parametrized by a reference string, the adversary $\hat{P}$ takes the CRS as an input and the probability is taken over the choice of the CRS and the adversary's coins. This notion of *(weak) special soundness* is weaker than the *special soundness* assumed by the compilers of [CDS94, Dam02], but it's easy to see that the same compilers still apply to this weaker class of HVZKs.

4 UC-Secure Committed Oblivious Transfer on Strings

Our protocol $\mathcal{P}_{cot}$ for 1-out-of-2 committed oblivious transfer (COT) on strings is similar to the 1-out-of-2 non-committed string-OT protocol of Aiello *et al.* [AIR01], but instead of multiplicatively homomorphic ElGamal encryption, $\mathcal{P}_{cot}$ uses *additively* homomorphic and *verifiable* sCS encryption, which enables succinct (constant number of exponentiations) proofs that receiver's and sender's inputs into OT match their previous commitments. Moreover, $\mathcal{P}_{cot}$ is universally composable in the CRS model.

We define the ideal functionality $\mathcal{F}_{COT}$ for a COT scheme, and show that $\mathcal{P}_{cot}$ securely realizes it. In contrast to the ideal COT functionality proposed by Garay *et al.* [GMY04], our functionality $\mathcal{F}_{COT}$ runs on *strings* rather than bits. However, $\mathcal{F}_{COT}$ is more restricted than the functionality of [GMY04] in that (1) the obliviously

---

Ideal functionality $\mathcal{F}_{\mathrm{COT}}$ for committed oblivious transfer on strings (COT)

Commit: Upon receiving a $\langle \mathsf{ComMsg}, (P_i, cid), m \rangle$ message from P_i, $\mathcal{F}_{\mathrm{COT}}$ records the $((P_i, cid), m)$ pair and broadcasts $\langle \mathsf{Committed}, (P_i, cid) \rangle$. Here m can be either a message in the prescribed message space or a special symbol $\bot$.

StartCOT: Upon receiving $msg = \langle \mathsf{StartCOT}, (P_S, P_R, sid, cid_R, cid_{S,0}, cid_{S,1}) \rangle$ from P_R, $\mathcal{F}_{\mathrm{COT}}$ verifies that it has records $((P_R, cid_R), m_R)$, $((P_S, cid_{S,0}), m_{S,0})$, and $((P_S, cid_{S,1}), m_{S,1})$, and that $m_R \neq \bot$. If this fails, $\mathcal{F}_{\mathrm{COT}}$ ignores this message; otherwise, $\mathcal{F}_{\mathrm{COT}}$ records msg and forwards it to P_S.

CompleteCOT: Upon receiving $\langle \mathsf{CompleteCOT}, (P_S, P_R, sid, cid_R, cid_{S,0}, cid_{S,1}) \rangle$ from P_S, $\mathcal{F}_{\mathrm{COT}}$ verifies that it has a record $\langle \mathsf{StartCOT}, ids \rangle$, where $ids = (P_S, P_R, sid, cid_R, cid_{S,0}, cid_{S,1})$. $\mathcal{F}_{\mathrm{COT}}$ looks up records $((P_S, cid_{S,0}), m_{S,0})$ and $((P_S, cid_{S,1}), m_{S,1})$, and checks if $m_{S,0} \neq \bot$ and $m_{S,1} \neq \bot$. If anything fails, $\mathcal{F}_{\mathrm{COT}}$ ignores this message.

Otherwise $\mathcal{F}_{\mathrm{COT}}$ looks up the record $((P_R, cid_R), m_R)$ (observe that such a record must exist). If $m_R \notin \{0,1\}$, $\mathcal{F}_{\mathrm{COT}}$ sends a special message $\langle \mathsf{COTFailed}, P_S, P_R, sid \rangle$ to P_R. Otherwise $\mathcal{F}_{\mathrm{COT}}$ sends $\langle \mathsf{CompleteCOT}, ids, (m_{S,b}, b) \rangle$ to P_R for $b = m_R$.

Note: Additionally, $\mathcal{F}_{\mathrm{COT}}$ screens outs duplicates in commitment identifiers cid for every P_i, and in COT instance identifiers sid for every (P_S, P_R) pair.

Fig. 1. $\mathcal{F}_{\mathrm{COT}}$ ideal functionality

transferred values are the plaintexts of commitments, not full decommitments; and (2) $\mathcal{F}_{\mathrm{COT}}$ does not support opening of the committed values. Nevertheless, $\mathcal{F}_{\mathrm{COT}}$ can ensure that any combination of COT instances is executed on same committed inputs, and thus it can ensure that whenever COT is used as part of *any* security protocol, the parties' inputs into COT are consistent across multiple COT instances.

The COT protocol $\mathcal{P}_{\mathrm{cot}}$ is given in fig. 2. It assumes a common reference string picked by the trusted third party, which defines an instance PK of the CS commitment scheme. The message space for this COT scheme is $[-\frac{n}{2}, \frac{n}{2}]$, the message space of the CS commitment scheme. The commitment, identified as cid, of player P_i on message m is a CS commitment $\mathsf{Com} = \mathsf{CSenc}_{PK}^{\mathrm{ids}}(m)$ with label $\mathrm{ids} = (P_i, cid)$. As we will argue, $\mathcal{P}_{\mathrm{cot}}$ is a secure realization of $\mathcal{F}_{\mathrm{COT}}$; in particular, the receiver either outputs message m_σ committed in $\mathsf{Com}_{S,\sigma}$, or rejects.

The two proof systems used in $\mathcal{P}_{\mathrm{cot}}$ involve conjunctions of Com, DLEQ, and Cot statements. As explained in Section 3.4, such proofs are computationally sound ZK proofs which are concurrently secure in the CRS model. Each takes only a few exponentiations and three communication rounds. Moreover, the messages in both proofs (P_R to P_S and P_S to P_R) can be piggy-backed, with the statements proved by the two players delayed to the last messages, which results in a 4-round protocol. In the random oracle model these proofs are non-interactive and the protocol takes only 2 rounds.

Theorem 3. *Under the DCR assumption, protocol $\mathcal{P}_{\mathrm{cot}}$ is a UC-secure realization of the Committed-OT functionality $\mathcal{F}_{\mathrm{COT}}$ in the CRS model, if the proof systems involved*

Protocol $\mathcal{P}_{\text{cot}}$ for committed oblivious transfer on strings

Common Reference String: CS commitment instance $PK = (n, g, \mathfrak{g}, \mathfrak{h}, \mathfrak{f}, \mathsf{hk})$.

Commit: For player P_i, on commitment instance cid and message m: Player P_i sets $\mathsf{ids} = (P_i, cid)$, $\mathsf{Com} = \mathsf{CSenc}^{\mathsf{ids}}_{PK}(m)$, and broadcasts $\langle \mathsf{ComMsg}, \mathsf{ids}, \mathsf{Com} \rangle$.

Receiver P_R executes a COT instance sid with sender P_S. P_R's bit σ is committed in Com_R, P_S's messages m_0, m_1 are committed in $\mathsf{Com}_{S,0}, \mathsf{Com}_{S,1}$. Let $cid_R, cid_{S,0}, cid_{S,1}$ be the identifiers for these commitments.

COT Step 1: P_R sets $\mathsf{ids} = (P_S, P_R, sid, cid_R, cid_{S,0}, cid_{S,1})$, retrieves $\mathsf{Com}_R = (\tilde{u}, C, \tilde{v})$ and its decommitment $r \in [0, \frac{n}{4}]$. Note that $C = \alpha^\sigma \mathfrak{g}^r$. P_R picks $x \in [0, \frac{n}{4}]$, and computes

$$y = g^x, \quad u = g^r, \quad e = \alpha^\sigma y^r$$

P_R sends $\langle \mathsf{COTMsg1}, \mathsf{ids}, (u, e, y) \rangle$ to P_S, and performs as the prover in the proof system $\mathsf{ZKDLEQ}(g, u, \mathfrak{g}/y, C/e) \wedge \mathsf{ZKCom}(PK, \mathsf{Com}_R, (P_R, cid_R))$ with P_S.

COT Step 2: Upon receiving $\langle \mathsf{COTMsg1}, \mathsf{ids}, (u, e, y) \rangle$ from P_R, P_S retrieves messages m_0, m_1 committed in $\mathsf{Com}_{S_0} = (\tilde{u}_0, C_0, \tilde{v}_0)$ and $\mathsf{Com}_{S_1} = (\tilde{u}_1, C_1, \tilde{v}_1)$. Note that $C_i = \alpha^{m_i} \mathfrak{g}^{r_{m_i}}$ for some r_{m_i}. P_S creates two "COT-encryptions" for $i = 0, 1$:

$$e_i = e^{s_i} \alpha^{m_i - i*s_i} y^{r_i} \quad \text{and} \quad u_i = u^{s_i} g^{r_i}$$

for random *even* values $s_i \in [0, 2n]$ and $r_i \in [0, \frac{n}{2}]$. If P_R passed its proof in Step 1, P_S sends message $\langle \mathsf{COTMsg2}, \mathsf{ids}, (u_0, e_0, u_1, e_1) \rangle$ to P_S, and performs with P_R as the verifier a proof system $\mathsf{ZKCot}(0, e_0, u_0, e, u, y, C_0) \wedge \mathsf{ZKCot}(1, e_1, u_1, e, u, y, C_1) \wedge \mathsf{ZKCom}(\mathsf{Com}_{S,0}, (P_S, cid_{S_0})) \wedge \mathsf{ZKCom}(\mathsf{Com}_{S,1}, (P_S, cid_{S_1}))$.

COT Step 3: P_R decrypts the sCS ciphertext (u_σ, e_σ) and obtains m_σ. If P_S passed its proof in step 2, then P_R outputs m_σ; otherwise P_R rejects.

<u>Note</u>: Either player rejects if the values he receives are *visibly* not in $\mathbb{Z}^*_{n^2}$, i.e., they are outside the $[1, n^2]$ range or are divisible by n.

Fig. 2. Protocol $\mathcal{P}_{\text{cot}}$ for committed OT on strings

---

are computationally sound and statistically zero-knowledge with straight-line simulators in the CRS model.

Due to lack of space, we present only the crucial aspects of the proof.

Verifiability of inputs. By computational soundness of the proof systems, the players cannot, except with negligible probability, enter different values σ, m_0, m_1 into the OT protocol than those they previously committed. This is easy to see for the cheating receiver P_R. For the cheating sender P_S, by soundness of ZKCot, if P_R accepts, then, with overwhelming probability, for each i there exists a tuple (m_i, r_{m_i}, s_i, r_i) s.t. $(C_i)^2 = \alpha^{2m_i} \mathfrak{g}^{2r_{m_i}}$, $e_i^2 = e^{2s_i} \alpha^{2m_i - i*2s_i} y^{2r_i}$, and $u_i^2 = u^{2s_i} g^{2r_i}$, where $\mathsf{Com}_i = (\tilde{u}_i, C_i, \tilde{v}_i)$ is P_S's commitment whose id is $cid_{S,i}$. In particular, m_i is the message committed in Com_i. Since for honest P_R, $e = \alpha^\sigma y^r$ and $u = g^r$, it follows that for

$i = \sigma$ we have $e_\sigma^2 = \alpha^{2m_\sigma} y^{2r''}$ and $u_\sigma^2 = g^{2r''}$ where $r'' = s_\sigma r + r_\sigma$. Therefore, message m_σ decrypted by P_R from the ciphertext (u_σ, e_σ) is the message committed in Com_σ.

Receiver's and sender's privacy. Receiver's privacy follows from semantic security of CS encryption, while the sender's privacy relies on the fact that if P_R's commitment $\mathsf{Com}_R = (\tilde{u}, C, \tilde{v})$ and the tuple (u, e, y) in P_R's COT message are correctly formed (and they are, except for negligible probability, if P_S accepts P_R's ZKCom and ZKDLEQ proofs, and if the factoring assumption holds), and if σ is a value that satisfies $e^2 = \alpha^{2\sigma} g^{2r}$ for some r (there exists such σ for every $e \in \mathbb{Z}_{n^2}^*$), then the pairs (e_0, u_0) and (e_1, u_1) sent by P_S reveal m_σ, but information-theoretically hide m_i for $i \neq \sigma$. Observe first that if tuples $(\tilde{u}, C, \tilde{v})$ and (u, e, y) are accepted by the verifier (*i.e.*, each element is in $\mathbb{Z}_{n^2}$, but is not a multiple of n), then under the factoring assumption, which is implied by the DCR assumption, all these elements are also in $\mathbb{Z}_{n^2}^*$, except for negligible probability. Second, if P_R passes the ZKCom proof on Com_R and the ZKDLEQ proof on (u, e, y), then except for negligible probability we have $e = \omega_0 \alpha^\sigma g^r$, $u = \omega_1 g^r$, and $y = \omega_2 g^x$ for some (σ, r, x) and some elements $\omega_0, \omega_1, \omega_2$ of order 2 in $\mathbb{Z}_{n^2}^*$. Therefore, values (u_i, e_i) sent by P_S are equal to $e_i = \alpha^{m_i + s_i(\sigma - i)} y^{s_i r + r_i}$ and $u_i = g^{s_i r + r_i}$, because s_i is even. Note that for any σ, $\gcd(\sigma - i, n) = 1$ for either $i = 0$ or $i = 1$ (or for both). Since the order of α is n, and $(s_i \bmod n)$ is distributed uniformly in $\mathbb{Z}_n$, value $\alpha^{m_i + s_i(\sigma - i)}$ is distributed uniformly in the subgroup generated by α in $\mathbb{Z}_{n^2}^*$. Because (1) the orders of g and y are both divisors of $2n'$, (2) $s_i r + r_i$ is even, and (3) $(r_i \bmod n')$ is distributed statistically close to uniform over $\mathbb{Z}_{n'}$, it follows that pair $(g^{s_i r + r_i}, y^{s_i r + r_i})$ is distributed statistically close to $(g^{2r'}, y^{2r'})$ for r' uniform in $\mathbb{Z}_{n'}$. Taken together, it follows that pair (e_i, u_i), for $i \neq \sigma$, is distributed statistically close to $(\alpha^{m'} y^{2r'}, g^{2r'})$ for random $(m', r') \in (\mathbb{Z}_n \times \mathbb{Z}_{n'})$, and thus it is statistically independent of m_i.

Construction of the straight-line simulator. The proof that protocol $\mathcal{P}_{\mathrm{cot}}$ UC-realizes the COT functionality $\mathcal{F}_{\mathrm{COT}}$ involves construction of a straight-line simulator, which pretends to follow the protocol on behalf of the uncorrupted parties by executing it on some fixed values unrelated to the real inputs of these parties, and simulates their proof systems using their straight-line simulators. Moreover, the simulator straight-line extracts the effective inputs contributed by the corrupted players by choosing the Camenisch-Shoup public key PK embedded in the CRS and decrypting these players' inputs from their commitments. The simulator submits these extracted inputs to the ideal functionality if the corrupted players pass the associated ZK proofs. CCA security of Camenisch-Shoup encryption implies that the ciphertexts contained in the commitments and COT messages created by the simulator remain indistinguishable from the corresponding ciphertexts created in the real protocol, even if the simulator accesses the decryption oracle (to extract the values committed by the corrupt players). Finally, the proof systems performed by the corrupted players are sound even if the simulator picks the CRS because as long as the adversary passes its proofs only on correct statements, the simulation is distributed statistically close to the real execution. Hence, by the standard soundness of the proof systems involved, the adversary has only negligible probability of passing some proof on an incorrect statement in the simulation.

5   UC-Secure Two-Party Computation on Committed Inputs

We present an efficient version of Yao's "garbled circuits" protocol for secure two-party computation (2PC). The protocol operates on committed inputs and is universally composable (in the CRS model). In addition to any two-party secure computation in the malicious model, our protocol can be used, for example, to ensure that multiple instances of secure computation are executed on consistent inputs.

The ideal functionality $\mathcal{F}_{2PC}$ for secure two-party computation on committed inputs in shown in fig. 3. Abstracting from the bookkeeping details, $\mathcal{F}_{2PC}$ is a simple generalization of the standard secure computation functionality where two players send their respective inputs x and y to the trusted third party $\mathcal{F}$, who returns the result of evaluating some circuit $C(x,y)$ to one or both players.

The *committed 2PC* functionality $\mathcal{F}_{2PC}$ accepts any number of commitments from parties $P_1, \ldots, P_n$, which are intended to represent the commitments to the bits encoding these parties' inputs into some two-party computation protocols. For every commitment, $\mathcal{F}_{2PC}$ records the committed bit. If some party P_R requests secure computation of some circuit C with another party P_S, the request specifies C and a vector of commitments to P_R's and P_S's inputs into this circuit. If party P_S accedes to this request, $\mathcal{F}_{2PC}$ sends to P_R the output of circuit C computed on the inputs committed in the specified commitments. Note that our $\mathcal{F}_{2PC}$ sends the output only to P_R, but since this is a *committed* 2PC functionality, the players can simply reverse the roles and request that the same C be computed on the same vector of commitments, in order to enable P_S to receive the output. (Our actual 2PC protocol allows P_S to receive the output with no computational overhead and one extra communication round.)

We assume that the circuit $\mathcal{C}$ consists of binary two-input gates $G = \{g_1, \ldots, g_c\}$ with unbounded fan-out but no cycles, connected by wires $W = \{w_1, \ldots, w_m\}$. Some subset W_S of n_s input wires are designated as P_S's inputs, and n_r input wires form the set W_R of P_R's inputs. Some subset W_O of the output wires is designed as outputs for P_R. (Optionally, some output wires can also be designated as outputs for P_S.)

The Committed 2PC protocol is in fig. 4. It is similar to the COT protocol of Section 4, and uses the same commitments and same message pattern, requiring 4 rounds in CRS and 2 rounds in ROM. In the first message, the receiver uses the proof systems of the $\mathcal{P}_{\text{cot}}$ protocol and an additional proof system $\mathsf{ZKBit}(C) = (\mathsf{ZKDL}(\mathfrak{g}, C) \vee \mathsf{ZKDL}(\mathfrak{g}, C/\alpha))$ for proving that the CS commitment $\mathsf{Com} = (u, C, v)$ or the sCS commitment C are commitments to a bit. In the second message, the sender creates the garbled circuit and uses the $\mathsf{CorrectYao}$ proof system to prove that it has been formed correctly. This step encompasses the entire Yao's construction and is discussed below. In the following, we denote sender P_S as S and receiver P_R as R.

Wire keys and commitments: S picks two random (symmetric) sCS private keys x_0^w, x_1^w for every wire $w \in W$, and for each x_i^w computes an sCS commitment C_i^w to x_i^w. Also, S makes a set of wire keys corresponding to his inputs, $\{x_{b_w}^w\}_{w \in W_S}$, where b_w is S's input bit on $w \in W_S$.

COTs on receiver's wire keys: S completes n_r instances of the COT protocol on the wire keys corresponding to receiver's wires: for each $i = 1, .., n_r$, S enters keys $(x_0^{w_i}, x_1^{w_i})$ as a sender in the COT protocol, where w_i designates the receiver's i^{th} input

---

Ideal functionality $\mathcal{F}_{2\mathrm{PC}}$ for two-party secure computation on committed inputs

Commit: Upon receiving a $\langle \mathsf{ComMsg}, (P_i, cid), m \rangle$ message from P_i, $\mathcal{F}_{2\mathrm{PC}}$ verifies that this cid has not been used by P_i before, records the $((P_i, cid), m)$ pair and broadcasts a $\langle \mathsf{Committed}, (P_i, cid) \rangle$ message. Message m is either a message in the prescribed message space, or a special symbol $\perp$.

Start2PC: Upon receiving

$$msg = \langle \mathsf{Start2PC}, (P_S, P_R, sid, cid_{S1}, \ldots, cid_{Sn_s}, cid_{R1}, \ldots, cid_{Rn_r}, C) \rangle$$

from P_R, $\mathcal{F}_{2\mathrm{PC}}$ verifies that (i) this sid has not been used by P_S and P_R before; (ii) for every index k such that $1 \leq k \leq n_s$, $\mathcal{F}_{2\mathrm{PC}}$ has a unique record $((P_S, cid_{Sk}), m_{Sk})$ (these commitments correspond to P_S's inputs into the protocol); (iii) for every index l such that $1 \leq l \leq n_r$, $\mathcal{F}_{2\mathrm{PC}}$ has a unique record $((P_R, cid_{Rl}), m_{Rl})$ and that $m_{Rl} \in \{0, 1\}$ (these commitments correspond to P_R's inputs into the protocol), and (iv) C is a description of a circuit that takes $n_s + n_r$ bits as inputs. If this fails, $\mathcal{F}_{2\mathrm{PC}}$ ignores this message; otherwise, it records msg and forwards it to P_S.

Complete2PC: Upon receiving

$$msg = \langle \mathsf{Complete2PC}, (P_S, P_R, sid, cid_{S1}, \ldots, cid_{Sn_s}, cid_{R1}, \ldots, cid_{Rn_r}, C) \rangle$$

from P_S, $\mathcal{F}_{2\mathrm{PC}}$ verifies that it has a record $\langle \mathsf{Start2PC}, \mathsf{ids} \rangle$, where ids $=$ $(P_S, P_R, sid, cid_{S1}, \ldots, cid_{R1}, \ldots, C)$. If not, $\mathcal{F}_{2\mathrm{PC}}$ ignores this message. $\mathcal{F}_{2\mathrm{PC}}$ looks up the records $((P_S, cid_{S1}), m_{S1}), \ldots, ((P_S, cid_{Sn_s}), m_{Sn_s})$ and $((P_R, cid_{R1}), m_{R1}), \ldots, ((P_R, cid_{Rn_r}), m_{Rn_r})$. If $m_{Sk} \notin \{0, 1\}$ for some index k, $\mathcal{F}_{2\mathrm{PC}}$ ignores this instance of the 2PC protocol. Otherwise, $\mathcal{F}_{2\mathrm{PC}}$ evaluates circuit C on inputs $m_{S1}, \ldots, m_{Sn_s}, m_{R1}, \ldots, m_{rn_r}$. $\mathcal{F}_{2\mathrm{PC}}$ sends $\langle \mathsf{Complete2PC}, \mathsf{ids}, b \rangle$ to P_R, where b is the output of the circuit.

<u>Note</u>: This is a functionality for *one-directional* two-party computation, where only the receiver P_R learns the output. Because both parties are committed to their inputs, they can run another instance of the same protocol with the roles of P_S and P_R reversed.

Fig. 3. $\mathcal{F}_{2\mathrm{PC}}$ ideal functionality

wire. This way, for every $w \in W_r$, the receiver obtains the wire key $x_{b_w}^w$ where b_w is his input bit on wire w. Technically, S computes tuple $(u_0^{(w_i)}, e_0^{(w_i)}, u_1^{(w_i)}, e_1^{(w_i)})$ by following the sender's algorithm in Step 2 of $\mathcal{P}_{\mathrm{cot}}$ on tuple $(u^{(i)}, e^{(i)}, y^{(i)})$ and a pair of messages $(x_0^{w_i}, x_1^{w_i})$, and their corresponding sCS commitments $(C_0^{w_i}, C_1^{w_i})$.

Receiver's output wires: For every receiver's output wire $w \in W_0$, S creates a pair of ciphertexts E_0^w, E_1^w that enables R to interpret the corresponding wire keys. Namely, $E_0^w = \mathsf{sCSenc}_{x_0^w}(0)$ and $E_1^w = \mathsf{sCSenc}_{x_1^w}(1)$.

Forming the garbled truth tables: The following process is repeated for every gate $g \in G$. Let A and B be the input wires of g, and C the output wire. Let $C_{0,1}^A, C_{0,1}^B, C_{0,1}^C$ be the six sCS commitments to the respective wire keys (two per wire). These commitments form the truth table for the gate g in which the input bits b_A, b_B and the output bit $b_C = g(b_A, b_B)$ are replaced by commitments to the corresponding wire keys. As in the

---

Committed 2PC Protocol

Common Reference String: CS commitment instance $PK = (n, g, \mathfrak{g}, \mathfrak{h}, \mathfrak{f}, \mathsf{hk})$.

Commit: As in $\mathcal{P}_{\mathrm{cot}}$ of fig. 2, player P_i on commitment instance cid and message m broadcasts $\langle \mathsf{ComMsg}, \mathsf{ids}, \mathsf{Com} \rangle$ where $\mathsf{Com} = \mathsf{CSenc}_{PK}^{\mathsf{ids}}(m)$ for $\mathsf{ids} = (P_i, cid)$.

2PC Step 1: To trigger instance sid of the protocol in order to compute circuit C on commitment instances $cid_{S1}, \ldots, cid_{Sn_s}$ made by P_S and commitments $cid_{R1}, \ldots, cid_{Rn_r}$ made by P_R, the *receiver* P_R prepares n_r messages, each computed as in Step 1 of $\mathcal{P}_{\mathrm{cot}}$ (fig. 2): for each $i = 1, .., n_r$, P_S computes a tuple $(y^{(i)}, u^{(i)}, e^{(i)})$ on bit σ_i committed in $\mathsf{Com}_{cid_{Ri}} = (\tilde{u}^{(i)}, C^{(i)}, \tilde{v}^{(i)})$ and its decommitment $r^{(i)}$. P_S sends to P_R message

$$\langle \mathsf{Start2PC}, \mathsf{ids}, C, \{y^{(i)}, u^{(i)}, e^{(i)}\}_{i=1..n_r} \rangle$$

where ids is the above vector of commitment ids. P_R then performs the ZK proof system $\mathsf{ZKR_{2PC}}$, which is a conjunction of n_r instances of the $\mathsf{ZKDLEQ}(\ldots) \wedge \mathsf{ZKCom}(\ldots)$ proof system used in Step 1 of $\mathcal{P}_{\mathrm{cot}}$, one per each tuple $(r^i, C^{(i)}, y^{(i)}, u^{(i)}, e^{(i)})$, and n_r instances of the $\mathsf{ZKBit}(C^{(i)})$ proof.

2PC Step 2: On receiving the $\langle \mathsf{Start2PC}, \mathsf{ids}, C, \ldots \rangle$ message and verifying the ZK proofs, P_S retrieves its commitments $\mathsf{Com}_{cid_{S1}}, \ldots, \mathsf{Com}_{cid_{Sn_s}}$ specified in the ids string, and sends to P_R a garbled version of circuit C computed on these inputs:

$$\mathsf{Complete2PC}\langle \; \mathsf{ids}, \{C_b^w\}_{b \in \{0,1\}, \, w \in W}, \{E_{\alpha\beta}^g\}_{\alpha\beta \in \{00,01,10,11\}, \, g \in G},$$

$$\{x_{b_w}^w\}_{w \in W_S}, \{E_0^w, E_1^w\}_{w \in W_0}, \{u_0^{(w)}, e_0^{(w)}, u_1^{(w)}, e_1^{(w)}\}_{w \in W_R} \; \rangle$$

These values are defined in Section 5. P_S also performs the ZK proof $\mathsf{CorrectYao}$.

2PC Step 3: P_R verifies the ZK proof $\mathsf{CorrectYao}$, evaluates the garbled circuit and outputs its result. (Optionally, P_R can send back to P_S the wire keys corresponding to P_S's output wires.)

Fig. 4. Committed 2PC Protocol

---

original Yao's protocol, S creates a ciphertext for each row of the truth table, encrypting the output-wire key corresponding to this row's output bit under the two input-wire keys corresponding to this row's input bits. The ciphertexts must be randomly shuffled to prevent R from learning which row $(b_A, b_B, g(b_A, b_B))$ of the truth table he succeeds in decrypting. S picks two random bits, σ_A and σ_B, which determine, intuitively, if the values corresponding to the A and B wires are "switched" or not. (If w is S's input wire, than σ_w is equal to S's input bit on that wire.) If the rows are denoted in binary as $00, 01, 10, 11$, then the first ciphertext received by R corresponds to row $\sigma_A \sigma_B$, the second to row $\bar{\sigma}_A \sigma_B$, the third to row $\sigma_A \bar{\sigma}_B$, and the fourth to row $\bar{\sigma}_A \bar{\sigma}_B$.

S creates the ciphertext list $(E_{00}, E_{01}, E_{10}, \text{ and } E_{11})$ using a two-key encryption scheme $E_{\alpha\beta} = \mathsf{2KEnc}_{x_1, x_2}(x)$, where for each α, β, $x_1 = x_{\alpha \oplus \sigma_A}^A$, $x_2 = x_{\beta \oplus \sigma_B}^B$, and $x = x_{g(\alpha \oplus \sigma_A, \beta \oplus \sigma_B)}^C$. For example, if $\sigma_A = \sigma_B = 0$, then each $E_{\alpha\beta}$ is a two-key encryption under keys x_α^A and x_β^B of the output-wire key $x_{g(\alpha,\beta)}^C$. If $\sigma_A = 1, \sigma_B = 0$, then each $E_{\alpha\beta}$ is a two-key encryption under keys $x_{\bar{\alpha}}^A$ and x_β^B of key $x_{g(\bar{\alpha},\beta)}^C$, and so on. Note that tuple $(\sigma_A, \sigma_B, \alpha, \beta)$ uniquely defines the commitments C_1, C_2, C that correspond to the above keys x_1, x_2, x: $C_1 = C^A_{\alpha \oplus \sigma_A}$, $C_2 = C^B_{\beta \oplus \sigma_B}$, and $C = C^C_{g(\alpha \oplus \sigma_A, \beta \oplus \sigma_B)}$.

The two-key encryption $2\mathsf{KEnc}_{x_1, x_2}(x)$ is created as follows. The key $x \in [0, 2^{k''}]$ is split in two parts, x'_1 and x'_2, by choosing x'_1 at random in $[-2^{k''+k}, 2^{k''+k}]$ (recall that k'', k are security parameters, where $k'' = \frac{|n|}{2}$ and k can be 80), and setting $x'_2 = x - x'_1$ (over integers). S also computes an sCS commitment D to x'_1. Observe that if C is an sCS commitment to x, then C/D is an sCS commitment to x'_2. The ciphertext E is a triple $\langle D, F^{(1)}, F^{(2)} \rangle$, where $F^{(i)} = \mathsf{sCSenc}_{x_i}(x'_i)$. Let $E_{\alpha\beta}$ denote $\langle D_{\alpha\beta}, F^{(1)}_{\alpha\beta}, F^{(2)}_{\alpha\beta} \rangle$.

Proving circuit correctness: CorrectYao is a (concurrent ZK, with straight-line simulator) proof system formed by *conjunction* of the following proof systems:

$$\bigwedge_{g \in G} \mathsf{CorrectGarble}_g \ \wedge \ \bigwedge_{w \in W} \mathsf{GoodKeys}_w \ \wedge \ \bigwedge_{w \in W_S} \mathsf{CorrectInput}_w$$
$$\wedge \ \bigwedge_{w \in W_R} \mathsf{ZKS}_w \qquad \wedge \ \bigwedge_{w \in W_O} \mathsf{CorrectOutput}_w$$

where

$$\begin{aligned}
\mathsf{GoodKeys}_w &= \mathsf{ZKNotEq}(C^w_0, C^w_1) \\
\mathsf{CorrectInput}_w &= (\mathsf{ZKDL}(\mathfrak{g}, C^w_0/\alpha^{x^w_{bw}}) \wedge \mathsf{ZKDL}(\mathfrak{g}, C_b)) \vee \\
&\quad (\mathsf{ZKDL}(\mathfrak{g}, C^w_1/\alpha^{x^w_{bw}}) \wedge \mathsf{ZKDL}(\mathfrak{g}, C_b/\alpha)), \text{ where } C_b \text{ is the} \\
&\quad \text{sCS commitment inside } \mathsf{Com}_{cid_{Si}} \text{ if } w \text{ is the } i^{th} \text{ input wire of } S \\
\mathsf{CorrectOutput}_w &= \mathsf{ZKPlainEq2}(E^w_0, C^w_0, 0) \wedge \mathsf{ZKPlainEq2}(E^w_1, C^w_1, 1)
\end{aligned}$$

Here ZKS_w refers to the proof performed by the sender in the instance of the COT protocol that corresponds to receiver's wire $w \in W_R$. $\mathsf{ZKPlainEq2}(E, C_k, m)$ is the proof system for showing that E is an sCS encryption of plaintext m under key k committed in C_k, and is a trivial simplification of the $\mathsf{ZKPlainEq}(E, C_k, C_m)$ proof system for proving the same about commitment C_m to m. Finally, $\mathsf{CorrectGarble}_g$ proves that the ciphertext table $E_{00}, E_{01}, E_{10}, E_{11}$ corresponding to garbled gate g is formed correctly, where $E_{\alpha\beta} = (D_{\alpha\beta}, F^{(1)}_{\alpha\beta}, F^{(2)}_{\alpha\beta})$:

$$\begin{aligned}
\mathsf{CorrectGarble}_g = \ &\mathsf{CorrectShuffle}(0,0) \vee \mathsf{CorrectShuffle}(0,1) \vee \\
&\mathsf{CorrectShuffle}(1,0) \vee \mathsf{CorrectShuffle}(1,1)
\end{aligned}$$

$$\begin{aligned}
\mathsf{CorrectShuffle}(\alpha, \beta) = \ &\mathsf{CorrectCipher}(0,0,\alpha,\beta) \wedge \mathsf{CorrectCipher}(0,1,\alpha,\beta) \wedge \\
&\mathsf{CorrectCipher}(1,0,\alpha,\beta) \wedge \mathsf{CorrectCipher}(1,1,\alpha,\beta)
\end{aligned}$$

$$\begin{aligned}
\mathsf{CorrectCipher}(\sigma_A, \sigma_B, \alpha, \beta) = \ &\mathsf{ZKPlainEq}(F^{(1)}_{\alpha\beta}, C^A_{\alpha \oplus \sigma_A}, D_{\alpha\beta}) \wedge \\
&\mathsf{ZKPlainEq}(F^{(2)}_{\alpha\beta}, C^B_{\beta \oplus \sigma_B}, (C^C_{g(\alpha \oplus \sigma_A, \beta \oplus \sigma_B)}/D_{\alpha\beta}))
\end{aligned}$$

Circuit evaluation: R obtains his input-wire keys via COT and evaluates the entire circuit gate by gate. Unambiguity of sCS encryption and soundness of the proof systems ensures that for each gate, R decrypts exactly one of the four ciphertexts forming that gate's garbled truth table and obtains the key corresponding to the gate's output wire.

Theorem 4. *Under the strong RSA and DCR assumptions, the 2PC protocol of fig. 4 is a UC-secure realization of the Committed 2PC functionality $\mathcal{F}_{2\mathrm{PC}}$ in the CRS model.*

References

[AIR01]   W. Aiello, Y. Ishai, and O. Reingold. Priced oblivious transfer: How to sell digital goods. In *Proc. EUROCRYPT*, pages 119–135, 2001.

[Bou00]   F. Boudot. Efficient proofs that a committed number lies in an interval. In *Proc. EUROCRYPT*, pages 431–444, 2000.

[CC00]   J. Camenisch and C. Cachin. Optimistic fair secure computation. In *Proc. CRYPTO*, pages 93–111, 2000.

[CD97]   R. Cramer and I. Damgård. Linear zero-knowledge – a note on efficient zero-knowledge proofs and arguments. In *Proc. STOC*, pages 436–445, 1997.

[CDS94]   R. Cramer, I. Damgård, and B. Schoenmakers. Proofs of partial knowledge and simplified design of witness hiding protocols. In *Proc. CRYPTO*, pages 174–187, 1994.

[CGHG01]   D. Catalano, R. Gennaro, and N. Howgrave-Graham. The bit security of Paillier's encryption scheme and its applications. In *Proc. EUROCRYPT*, pages 229–243, 2001.

[CGHGN01]   D. Catalano, R. Gennaro, N. Howgrave-Graham, and P. Nguyen. Paillier's cryptosystem revisited. In *Proc. CCS*, pages 206–214, 2001.

[CM99]   J. Camenisch and M. Michels. Proving in zero-knowledge that a number is a product of two safe primes. In *Proc. EUROCRYPT*, pages 107–122, 1999.

[Cré89]   C. Crépeau. Verifiable disclosure of secrets and applications. In *Proc. EUROCRYPT*, pages 181–191, 1989.

[CS03]   J. Camenisch and V. Shoup. Practical verifiable encryption and decryption of discrete logarithms. In *Proc. CRYPTO*, pages 126–144, 2003.

[CvdGT95]   C. Crépeau, J. van de Graaf, and A. Tapp. Committed oblivious transfer and private multiparty computation. In *Proc. CRYPTO*, pages 110–123, 1995.

[Dam02]   I. Damgård. Efficient concurrent zero-knowledge in the auxiliary string model. In *Proc. EUROCRYPT*, pages 418–430, 2002.

[DF02]   I. Damgård and E. Fujisaki. A statistically hiding integer commitment scheme based on groups with hidden order. In *Proc. ASIACRYPT*, pages 125–142, 2002.

[DI05]   I. Damgård and Y. Ishai. Constant-round multiparty computation using a black-box pseudorandom generator. In *Proc. CRYPTO*, pages 378–394, 2005.

[FO97]   E. Fujisaki and T. Okamoto. Statistical zero knowledge protocols to prove modular polynomial relations. In *Proc. CRYPTO*, pages 16–30, 1997.

[GMW87]   O. Goldreich, S. Micali, and A. Wigderson. How to play any mental game. In *Proc. STOC*, pages 218–229. ACM, 1987.

[GMY04]   J. Garay, P. MacKenzie, and K. Yang. Efficient and universally composable oblivious transfer and applications. In *Proc. TCC*, pages 297–316, 2004.

[HSS93]   J. Håstad, A. Schrift, and A. Shamir. The discrete logarithm modulo a composite hides $o(n)$ bits. *J. Comput. Syst. Sci.*, 47:850–864, 1993.

[JJ00]   M. Jakobsson and A. Juels. Mix and match: Secure function evaluation via ciphertexts. In *Proc. ASIACRYPT*, pages 162–177, 2000.

[Kil88]   J. Kilian. Founding cryptography on oblivious transfer. In *Proc. STOC*, pages 20–31, 1988.

[KO04]   J. Katz and R. Ostrovsky. Round-optimal secure two-party computation. In *Proc. CRYPTO*, pages 335–354, 2004.

[Lin03]   Y. Lindell. Parallel coin-tossing and constant-round secure two-party computation. *J. Cryptology*, 16(3):143–184, 2003.

[Lip03]   H. Lipmaa. Verifiable homomorphic oblivious transfer and private equality test. In *Proc. ASIACRYPT*, pages 416–433, 2003.

[LP07]   Y. Lindell and B. Pinkas. An efficient protocol for secure two-party computation in the presence of malicious adversaries. In *Proc. EUROCRYPT*, 2007.

[MF06]   P. Mohassel and M. Franklin. Efficiency tradeoffs for malicious two-party computation. In *Proc. PKC*, pages 458–473, 2006.

[Pai99]   P. Paillier. Public-key cryptosystems based on composite degree residuosity classes. In *Proc. EUROCRYPT*, pages 223–238, 1999.

[Ped91]   T. P. Pedersen. Non-interactive and information-theoretic secure verifiable secret sharing. In *Proc. CRYPTO*, pages 129–140, 1991.

[Pin03]   B. Pinkas. Fair secure two-party computation. In *Proc. EUROCRYPT*, pages 87–105, 2003.

[Woo07]   D. Woodruff. Revisiting the efficiency of malicious two-party computation. In *Proc. EUROCRYPT*, 2007.

[Yao86]   A. Yao. How to generate and exchange secrets. In *Proc. FOCS*, pages 162–167, 1986.

A   HVZK Proof System for Statement PlainEq

This is an HVZK proof system for language $\mathsf{PlainEq} = \{((e, u), C_x, C_m) \mid$ there exist x, m, r_x, r_m s.t. $e = \alpha^m u^x$, $C_x = \alpha^x \mathfrak{g}^{r_x}$, and $C_m = \alpha^m \mathfrak{g}^{r_m}\}$, *i.e.*, for the language of tuples $((e, u), C_x, C_m)$ s.t. (e, u) is an sCS encryption of the plaintext m committed in sCS commitment C_m under the key x committed in sCS commitment C_x. It is special HVZK with weak special soundness under the strong RSA assumption. All the parameters are as in section 3.4, except for two additional elements G, H which are assumed to be random in $\mathbb{Z}_{n^2}^*$ and can be included in the CRS.

1. The private inputs of the prover are

$$m \in [-2^{k''+k}, 2^{k''+k}], \quad x \in [0, 2^{k''}], \quad r_m, r_x \in [0, \frac{n}{4}]$$

2. The prover picks $t_x \in [0, \frac{n}{4}]$ and sends $T_x = G^x H^{t_x}$ to the verifier. He also picks

$$m', r'_m, x', r'_x, t'_x \in [0, 2^{k+k'+2k''}]$$

and sends the following commitments to the verifier:

$$e' = \alpha^{2m'} u^{2x'}, \quad C'_x = \alpha^{2x'} \mathfrak{g}^{2r'_x}, \quad C'_m = \alpha^{2m'} \mathfrak{g}^{2r'_m}, \quad T'_x = G^{x'} H^{t'_x}$$

3. Verifier responds with a random challenge $c \in \{0, 1\}^k$
4. Prover sends the following responses, all computed over integers:

$$\tilde{m} = m' - cm, \quad \tilde{r}_m = r'_m - cr_m, \quad \tilde{x} = x' - cx, \quad \tilde{r}_x = r'_x - cr_x, \quad \tilde{t}_x = t'_x - ct_x$$

5. Verifies accepts if $\tilde{x} \in [-\frac{n}{4}, \frac{n}{4}]$ and if the following equations hold:

$$e' = e^{2c} \alpha^{2\tilde{m}} u^{2\tilde{x}},$$
$$C'_m = (C_m)^{2c} \alpha^{2\tilde{m}} \mathfrak{g}^{2\tilde{r}_m}, \quad C'_x = (C_x)^{2c} \alpha^{2\tilde{x}} \mathfrak{g}^{2\tilde{r}_x},$$
$$T'_x = (T_x)^c G^{\tilde{x}} H^{\tilde{t}_x}$$

Universally Composable Multi-party Computation Using Tamper-Proof Hardware

Jonathan Katz*

Dept. of Computer Science, University of Maryland
jkatz@cs.umd.edu

Abstract. Protocols proven secure within the *universal composability (UC) framework* satisfy strong and desirable security properties. Unfortunately, it is known that within the "plain" model, secure computation of general functionalities without an honest majority is impossible. This has prompted researchers to propose various "setup assumptions" with which to augment the bare UC framework in order to bypass this severe negative result. Existing setup assumptions seem to inherently require *some* trusted party (or parties) to initialize the setup in the real world.

We propose a new setup assumption — more along the lines of a *physical* assumption regarding the existence of tamper-proof hardware — which also suffices to circumvent the impossibility result mentioned above. We suggest this assumption as potentially leading to an approach that might alleviate the need for trusted parties, and compare our assumption to those proposed previously.

1   Motivation

For many years, researchers considered the security of protocols in a stand-alone setting where a single protocol execution was considered in isolation. Unfortunately, a proof of stand-alone security for a protocol does not, in general, provide any guarantees when the protocol is executed multiple times in a concurrent fashion (possibly by different sets of parties), or in a network where other protocol executions are taking place. This realization has motivated a significant amount of work aimed at providing models and security definitions that explicitly address such concerns.

The universal composability (UC) framework, introduced by Canetti [6], gives strong security guarantees in exactly such a setting. (Other frameworks with similar guarantees also exist [20], but we adopt the UC model in this work.) We refer the reader to Canetti's paper for a full discussion of the advantages of working within this framework, and focus instead on the question of *feasibility*. Canetti's initial work already demonstrates broad feasibility results for realizing any (polynomial-time computable) multi-party functionality in the presence of

---

* This research was supported by NSF Trusted Computing grant #0310751, NSF CAREER award #0447075, and US-Israel Binational Science Foundation grant #2004240. Portions of this work were done while visiting IPAM.

M. Naor (Ed.): EUROCRYPT 2007, LNCS 4515, pp. 115–128, 2007.

a strict majority of honest players. Unfortunately, this was soon followed by results of Canetti and Fischlin [8] showing that the setting without an honest majority is substantially different: even for the case of two parties (one of whom may be malicious) there exist natural functionalities that cannot be securely computed within the UC framework. Subsequent work of Canetti, et al. [9], further characterizing those two-party functionalities which cannot be securely realized in the UC framework, rules out essentially all non-trivial functions.

The impossibility results mentioned above hold for the so-called "plain model" where there is no additional infrastructure beyond the communication channels available to the parties. (The term "plain model" is actually a bit misleading, since even the plain model usually incorporates quite strong — though standard — assumptions about the communication channels, such as the existence of authenticated channels between all pairs of parties as well as a broadcast channel/"bulletin board" [6, Sect. 6.2]. We stress that the impossibility results hold even in this case.) In contrast, the impossibility results can be bypassed if one is willing to assume some stronger form of "setup" in the network. This idea was first proposed in the UC framework by Canetti and Fischlin [8], who suggest using a common reference string (CRS) in order to circumvent the impossibility results shown in their paper. (The use of a CRS in other contexts has a long history going back to [4].) In fact, a CRS turns out to suffice for universally composable multi-party computation of any (well-formed) functionality, for any number of corrupted parties [10].

If universally composable protocols are ever to be used in practice, one important research direction is to further explore setup assumptions that suffice to obtain feasibility results in the UC framework similar to those of [10]. Having multiple setup assumptions available would offer options to protocol designers; furthermore, some assumptions may be more attractive than others depending on the scenario in which protocols are to be run. Indeed, a variety of setup assumptions have been investigated recently including variations of trusted "public-key registration" services [2,7] (see also [6, Sect. 6.6]), or the use of government-issued "signature cards" [15]; these are discussed further in the following section.

From a high-level perspective, one very important research direction is to determine whether (or to what extent) *trusted parties* are needed for obtaining broad feasibility results in the UC framework. It appears in particular that all existing solutions require some trusted party to initialize the setup in the real world. (See the discussion in the following section.) It might be possible, however, to replace this trust with some *physical* assumption about the environment in which the protocol is run. Using physical assumptions to circumvent impossibility results is not without precedent in cryptography (though it has not been considered previously in the context of the UC framework); examples analogous to what we have in mind include the assumption of a physical broadcast channel (or even "multicast channels" [12]) to circumvent impossibility results [18] regarding the fraction of malicious players that can be tolerated; or the assumption of noisy channels [21,14,16] or the laws of quantum mechanics [3] to achieve information-theoretically secure key agreement over public channels.

We present in this paper what is intended to be a partial step toward this goal. Specifically, we introduce an assumption that has the flavor of a physical assumption regarding the possibility of tamper-proof hardware, and show that UC multi-party computation is realizable with respect to this assumption. A difficulty, of course, is that although there may be some intuitive idea of the properties possessed by tamper-proof hardware in the real world, it is not at all clear what is the most appropriate way to mathematically *model* tamper-proof hardware in the UC framework. We do not claim to have found the "right" formalization. Instead, we intend this work only to serve as an indication of what might be possible, and as inspiration for subsequent work in this direction.

1.1   A Brief Review of Existing Solutions

As mentioned earlier, a variety of setup assumptions have been explored in an attempt to circumvent the impossibility results of [8]. We briefly discuss these now.

Common reference string (CRS). The use of a CRS was suggested by [8] (in the UC setting) and has been used in much subsequent work. It is fair to say that this is the setup assumption that has so far received the most attention. In the CRS model, a string is generated according to some prescribed distribution by a trusted party and given to the parties running an execution of a protocol. We remark that only the parties running the protocol are supposed to have access to the string, and so this is not quite a "common" reference string as in the original work of [4] (see [7] for a discussion of this point).

If the party publishing the CRS is malicious, this party can potentially set things up so that it can learn all private data in the network or cheat undetectably in protocol executions in which it is involved. These problems can be mitigated to some extent by having the CRS generated in a threshold manner so that, say, security holds as long as a majority of the parties involved in generation of the CRS are honest. Nevertheless, this still requires all parties in the network to jointly agree to place their trust in a small set of parties, and also assumes the availability of some set of parties willing to take responsibility for generating a CRS.

For some protocols, the CRS is simply a uniformly-random string (this is often called the common *random* string model) and here one might hope that the string could be generated based on naturally-occurring (random) events and without relying on a trusted party. The main drawback of this approach is that although certain natural events can be viewed as producing bit-sources with high min-entropy, the resulting bit-sources may not be uniformly random (and, furthermore, may not allow for deterministic extraction).

Public-key registration services. Existing proposals for public-key registration services [2,7] that can be used to circumvent the impossibility results in the UC framework go beyond the "traditional" model in which parties simply publish their public keys. (The latter corresponds to the "basic" registration functionality described in [6, Sect. 6.2], for which the impossibility results regarding secure computation still hold.) The functionality in [2], for example, essentially prevents an adversary from registering any public key that is not "well-formed" and for which the adversary does not know the corresponding secret key. It is unclear how this would be implemented in practice without a significant assumption of trust on the part of existing certification authorities.

Signature cards. An interesting idea pursued by [15] is to use government-issued signature cards as a form of global setup. Roughly speaking, such cards hold an honestly-generated public-/secret- key pair for a secure signature scheme; sign any message given to them; and never reveal the secret key to anyone (including the legitimate owner of the card) under any circumstances. Cards with this functionality are apparently being issued by some European governments [15] indicating that such cards may, in fact, represent a realistic assumption. The main drawback of these signature cards is that the producer and/or issuer of these cards must be completely trusted.

1.2   Relying on Tamper-Proof Hardware

As discussed above, all existing setup assumptions that imply general feasibility results in the UC framework seem to inherently require a great deal of *trust* in at least some parties in the system. It is natural to wonder whether such trust is essential, or whether other setup assumptions — perhaps of a slightly different character — might also suffice.

We suggest that it might be possible to eliminate the need for any trusted parties if one is willing instead to rely on a *physical* assumption regarding the existence (and practicality) of tamper-proof hardware. We hasten to add that a complete elimination of all trusted parties using our approach may not be practical, possible, or desirable in realistic scenarios; nevertheless, our proposals indicates that at least in theory this might be achievable. Alternately, one can view the approach explored here as allowing a reduced level of trust that we might be comfortable with; after all, we generally trust that our packets will be routed correctly over the Internet, but may not be willing to trust a corporation to generate a CRS.

Our assumption is that tamper-proof hardware exists, in the sense that (1) an honest user can construct a hardware token T_F implementing any desired (poly-time) functionality F but (2) an adversary given T_F can do no more than observe the input/output characteristics of this token. An honest player given a token $T'_{F'}$ by an adversary has no guarantee whatsoever regarding the function F' that this token implements (other than what the honest user can deduce from the input/output of this device). We show how this, seemingly-basic primitive can be used along with standard cryptographic assumptions to realize the commitment functionality (and hence general secure computation [10]) in the UC framework.

The above is a rather informal summary of the properties we assume; a more formal discussion of how to model tamper-proof hardware (as well as a concrete ideal functionality meant to capture that requirements evidenced in that discussion) is given in Section 2.

The idea of using secure hardware to achieve stronger security properties is not entirely new; it was directly inspired by work of Chaum, Pedersen, Brands, and Cramer [11,5,13] who propose the use of *observers* in the context of e-cash. Roughly speaking, it was suggested in that line of work that a bank could issue each user an "observer" (i.e., a smartcard) T_F implementing some functionality F, and a user would interact with both the bank and T_F whenever it executed an instance of a blind signature protocol to withdraw an e-coin (the bank and T_F could not communicate directly). The observer, by monitoring the actions of the user, could enforce some sort of honest behavior on the part of the user in the protocol execution. On the other hand, the user was guaranteed that even if the bank were malicious (and, e.g., sent a smartcard that was programmed in some arbitrary manner), anonymity of the user could not be violated. In some sense our work can be seen as formalizing this earlier work on observers, and extending its applicability from the domain of e-cash to the case of secure computation of arbitrary functionalities.

1.3   Have We Gained Anything?

Our assumption regarding tamper-proof hardware does not seem to trivially imply any of the setup assumptions discussed in Section 1.1. For example, two parties A and B cannot generate a CRS by simply having A send to B a token implementing a coin-tossing protocol: if A is malicious, a simulator will indeed be able to "rewind" the hardware token provided by A to B, and thus be able to "force" the value of the CRS output in this case to any desired value. On the other hand, if B is malicious then the simulator must (informally speaking) send some token to A, but then cannot "rewind" A to force the value of the CRS. We also do not see any way to trivially implement key-registration using our approach: for example, if each party sends to the other a token that checks public keys for validity (and then, say, outputs a signed receipt) then even the honest party will have to produce a secret key corresponding to its public key, which is not the case in the key-registration functionality of [2] (indeed, the security proofs in that work break down if this is the case). Another problem, unrelated to this, is that the signed receipt output by the device might be used as a "covert channel" to leak information about honest users' private keys.

As for whether we fundamentally gain anything by introducing our new assumption, this is (of course) subject to debate though we hope to convince the reader that the answer is "yes." In what follows we will simply assume that tamper-proof hardware is (or will someday be) available; clearly, if this assumption is false (and it may well be) then the entire discussion is meaningless. Under this assumption, we summarize our arguments in favor of relying on tamper-proof hardware as follows:

Possible elimination of trust. An advantage of our approach is that it seems to potentially allow for the elimination of trust in anyone but oneself. This is because, in theory, each user could construct the hardware token itself (or, more likely, buy a "blank" token and program it itself) without having to rely on anyone else. This distinguishes our approach from the "signature card" approach described earlier, where it is essential that a specific third party produce the cards (and a user cannot produce cards by itself).

An objection here is that we still assume secure channels and also secure distribution of tokens, and so trust has not been completely eliminated. We first emphasize that existing impossibility results hold even if secure channels are available, and so in that sense being able to eliminate the additional need for a trusted CRS represents progress in the right direction. If secure channels do not exist (and if secure distribution of tokens is not possible) we would seem to degenerate to a security model like that of [1] which still guarantees a non-trivial level of security. Finally, secure distribution of tokens is possible if a physical meeting of parties can be arranged; given this, a key can be stored on the token at the same time so as to bootstrap secure channels.

We do not mean to minimize the above concerns, only to suggest how they might be overcome. Developing a complete solution eliminating all trust in a practical manner remains an interesting direction for future work.

Possible reduction of trust. The above is a bit of an extreme scenario. But it is indicative of the fact that our approach may allow for *more relaxed requirements on trust*. In particular, under our approach each party could choose to buy pre-programmed tokens from any vendor of their choice; other parties executing the protocol do not need to approve of this choice, and can in turn buy from any vendors of their choice. This is not the case for any of the other setup assumptions mentioned in Section 1.1: parties must agree on which CRS to use; must approve of the registration authorities used by other parties; or must be sure that other parties use signature cards produced by a trusted entity.

Accountability. A final important point is the *accountability* present in our approach, which does not seem to be present when parties use a CRS or a key-registration authority. (It does seem to be available in the signature card scenario.) In the case of a CRS, for example, it seems impossible to prove that a CRS generated by some party is "bad" — in particular, the CRS might come from the exactly correct distribution except that the party has "neglected" to erase the trapdoor information associated with this CRS. Similarly in the case of a registration authority: how would one prove that an adversary's key is *not* well-formed?

On the other hand, one could imagine independent labs demonstrating that hardware sold by some vendor is *not* tamper-proof, or that supposedly blank tokens contained some (hidden) embedded code. This is in some sense reminiscent of the distinction suggested by Naor [17] between "falsifiable" assumptions and "unfalsifiable" ones.

2   Modeling Tamper-Proof Hardware

In this section, we suggest an ideal functionality that is intended to model tamper-proof hardware. More accurately, we define a "wrapper" functionality

---

Functionality $\mathcal{F}_{\mathsf{wrap}}$

$\mathcal{F}_{\mathsf{wrap}}$ is parameterized by a polynomial p and an implicit security parameter k

"Creation" Upon receiving (create, sid, P, P', M) from P, where P' is another user in the system and M is an interactive Turing machine, do:
1. Send (create, sid, P, P') to P'.
2. If there is no tuple of the form $(P, P', \star, \star, \star)$ stored, then store $(P, P', M, 0, \emptyset)$.

"Execution" Upon receiving (run, sid, P, msg) from P', find the unique stored tuple $(P, P', M, i, \mathsf{state})$ (if no such tuple exists, then do nothing). Then do:
  Case 1 ($i = 0$): Choose random $\omega \leftarrow \{0, 1\}^{p(k)}$. Run $M(\mathsf{msg}; \omega)$ for at most $p(k)$ steps, and let out be the response (set out $= \perp$ if M does not respond in the allotted time). Send (sid, P, out) to P'. Store $(P, P', M, 1, (\mathsf{msg}, \omega))$ and erase $(P, P', M, i, \mathsf{state})$.
  Case 2 ($i = 1$): Parse state as (msg_1, ω). Run $M(\mathsf{msg}_1 \| \mathsf{msg}; \omega)$ for at most $p(k)$ steps, and let out be the response (set out $= \perp$ if M does not respond in the allotted time). Send (sid, P, out) to P'. Store $(P, P', M, 0, \emptyset)$ and erase $(P, P', M, i, \mathsf{state})$.

Fig. 1. The $\mathcal{F}_{\mathsf{wrap}}$ functionality, specialized for the case when M is a 2-round (i.e., 4-message) protocol

which is intended to model the following sequence of events in the real world: (1) a party takes some software and "seals" it inside a tamper-proof hardware token; (2) this party gives the token to another party, who can then access the embedded software in a black-box manner. We will sometimes refer to the first party as the *creator* of the token, and the other party as the token's *user*.

The wrapper functionality is presented in Figure 1. The formalism in the description obscures to some extent what is going on, so we give a high-level description here. The functionality accepts two types of messages: the first type is used by a party P to create a hardware token (encapsulating an interactive protocol M) and to "give" this token to another party P'. The functionality enforces that P can send at most one token to P' which is used for all their protocol interactions throughout their lifetimes (and not just for the interaction labeled by the sid used when the token is created); since this suffices for honest parties we write the functionality this way in an effort to simplify things.

Once the token is "created" and "given" to P', this party can interact with the token in an arbitrary black-box manner. This is formalized by allowing P' to send messages of its choice to M via the wrapper functionality $\mathcal{F}_{\mathsf{wrap}}$. Note that each time a new copy of M is invoked, a fresh random tape is chosen for M.

To simplify the description of the functionality, we have assumed that M represents a 2-round (4-message) protocol since our eventual construction of commitment will use an M of this form. It should be clear how the functionality can be extended for the more general case.

A real-world action that is not modeled here is the possible (physical) *transference* of a token from one party to another. An honest party is never supposed to transfer a token; furthermore, in our eventual construction, tokens created by honest parties allow easy identification of their creator. Thus, transference does not represent a viable adversarial action, and so for simplicity we have not modeled such an action within $\mathcal{F}_{\text{wrap}}$.

The following real-world assumptions underly the existence of $\mathcal{F}_{\text{wrap}}$:

- We assume that the party creating a hardware token "knows" the code corresponding to the actions the token will take. This is evidenced by the fact that the creator P must explicitly provide $\mathcal{F}_{\text{wrap}}$ with a description of M. Looking ahead, this property will allow the simulator to "extract" the code within any adversarially-created token.

- The hardware token must be completely tamper-proof, so that the user P' cannot learn anything about M that it could not learn given black-box access. Furthermore, P' cannot cause M to use a "bad" (i.e., non-uniform) random tape, or to use the same random tape more than once. We are thus also assuming that the token has access to a built-in source of randomness. This latter requirement is not needed if we are willing to assume that the token can maintain state — in that case, we can use a hard-coded key for a pseudorandom function to generate the random tape as needed. Unfortunately we do not know how to prove security of this approach (for our particular protocol) without relying on complexity leveraging.

- We also assume that the creator of a token cannot send messages to the token once it is given to another party. (On the other hand, the token can send messages to its creator, either directly or via a covert channel.)

Our results are meaningful only to the extent that one is prepared to accept these assumptions as reasonable, or at least more reasonable than the existence of a common reference string or the other setup assumptions discussed earlier.

3   Using Tamper-Proof Hardware for Secure Computation

We now show how to securely realize the multiple commitment functionality $\mathcal{F}_{\text{mcom}}$ (see [8]) in the $\mathcal{F}_{\text{wrap}}$-hybrid model, for static adversaries. By the results of [8,10], this implies the feasibility of computing any (well-formed) two-party functionality, again fir static adversaries. It is also not hard to see that the techniques used in [10] can be used to show that our results imply the feasibility of computing any (well-formed) multi-party functionality as well. We omit further details from the present abstract.

For convenience, the multiple commitment functionality is given in Figure 2. Although we could optimize our construction to allow commitment to strings, for simplicity we focus on commitment to a single bit.

Before describing our protocol we introduce some notation. A tuple $(p, g, h, \hat{g}, \hat{h})$ is called a *Diffie-Hellman tuple* if (1) p and $q \overset{\text{def}}{=} \frac{p-1}{2}$ are prime; (2) $g, h, \hat{g}, \hat{h}$

Functionality $\mathcal{F}_{\mathsf{mcom}}$

Commit phase Upon receiving (commit, sid, cid, P, P', b) from P, where $b \in \{0, 1\}$, record (cid, P, P', b) and send (receipt, sid, cid, P, P') to P' and the adversary. Ignore subsequent values (commit, sid, cid, P, P', $\star$) from P.

Decommitment phase Upon receiving (open, sid, cid, P, P') from P, if the tuple (cid, P, P', b) is recorded then send (open, sid, cid, P, P', b) to P' and the adversary. Otherwise do nothing.

Fig. 2. The $\mathcal{F}_{\mathsf{mcom}}$ functionality

are in the order-q subgroup $\mathbb{G} \subset \mathbb{Z}_p^*$, with g, h generators; and (3) $\log_g \hat{g} = \log_h \hat{h}$. If the first two conditions hold but $\log_g \hat{g} \neq \log_h \hat{h}$, then we refer to the tuple as a *random tuple*. Given $\mathsf{tuple} = (p, g, h, \hat{g}, \hat{h})$ with q as defined above, we let $\mathsf{Com}_{\mathsf{tuple}}(b)$ denote the commitment defined by the two group elements $g^{r_1} h^{r_2}$, $\hat{g}^{r_1} \hat{h}^{r_2} g^b$, for randomly-chosen $r_1, r_2 \in \mathbb{Z}_q$. It is well-known (and easy to check) that if tuple is a random tuple then this commitment scheme is perfectly hiding; on the other hand if tuple is a Diffie-Hellman tuple and $r = \log_g \hat{g} = \log_h \hat{h}$ is known, then b can be efficiently recovered from the commitment.

We now describe a complete protocol for realizing $\mathcal{F}_{\mathsf{mcom}}$ for a sender P and a receiver P'. The security parameter is denoted by k.

Commitment phase. The parties perform the following steps:

1. P generates a public-key/secret-key pair (PK, SK) for a secure digital signature scheme, and constructs and sends a token to P' encapsulating the following functionality M:
   (a) Wait for a message (p, g, h). Check that p and $\frac{p-1}{2} = q$ are prime, that p has length k, and that g, h are generators of the order-q subgroup $\mathbb{G} \subset \mathbb{Z}_p^*$, and aborts if these do not hold.
   (b) Choose random elements $g_1, h_1 \in \mathbb{G}$. Using the Pedersen (perfectly-hiding) commitment scheme [19] and the generators received in the previous step, commit to g_1, h_1.
   (c) Wait for a message (g_2, h_2) where $g_2, h_2 \in \mathbb{G}$. (Abort if an invalid message is received.)
   (d) Set $\hat{g} = g_1 g_2$ and $\hat{h} = h_1 h_2$. Define $\mathsf{tuple}_{P \to P'} \stackrel{\text{def}}{=} (p, g, h, \hat{g}, \hat{h})$, and compute $\sigma_{P \to P'} = \mathsf{Sign}_{SK}(P, P', \mathsf{tuple}_{P \to P'})$. As the final message, send $\sigma_{P \to P'}$ as well as decommitment information for the commitments sent in the previous round.
   P' symmetrically constructs and sends a token to P.

2. P interacts with the token sent to it by P' and in this way obtains $\mathsf{tuple}_{P' \to P}$ and $\sigma_{P' \to P}$. (If cheating on the part of the token is detected, then P aborts

the entire protocol.) Party P' acts symmetrically. From now on, the parties communicate directly with each other and no longer need to access their tokens.

3. P sends $\mathsf{tuple}_{P'\to P}$ and $\sigma_{P'\to P}$ to P', and P' acts symmetrically. Then P checks that $\mathsf{Vrfy}_{PK}(\mathsf{tuple}_{P\to P'}, \sigma_{P\to P'}) = 1$ and, if not, it aborts the protocol. Party P' acts symmetrically.

   At the end of this step each party holds $\mathsf{tuple}_{P'\to P}$ and $\mathsf{tuple}_{P\to P'}$.

4. This step is the first that depends on the input bit b to be committed. P first commits to b using any statistically-binding commitment scheme; let C denote the resulting commitment. P also chooses random r_1, r_2 and computes $\mathsf{com} = \mathsf{Com}_{\mathsf{tuple}_{P\to P'}}(b)$. It sends C and com to P', and then gives an (interactive) witness indistinguishable proof that either (1) both C and com are commitments to the same bit b, or (2) $\mathsf{tuple}_{P'\to P}$ is a Diffie-Hellman tuple.

5. Upon successful completion of the previous step, party P' outputs (receipt, sid, cid, P, P').

We remark that steps 1–3 need only be carried out once by parties P and P', after which the values $\mathsf{tuple}_{P\to P'}$ and $\mathsf{tuple}_{P'\to P}$ can be used by these same parties to commit to each other (with either party acting as the sender) arbitrarily-many times.

Decommitment phase. P sends b to P' and gives a witness indistinguishable proof that (1) C is a commitment to b, or (2) $\mathsf{tuple}_{P'\to P}$ is a Diffie-Hellman tuple. Upon successful completion of this step, P' outputs (open, sid, cid, P, P', b).

3.1   Proof Intuition

The intuition underlying the security of the scheme is as follows. We need to argue that for any real-world adversary $\mathcal{A}$ (interacting with parties running the above protocol in the $\mathcal{F}_{\mathsf{wrap}}$-hybrid model), there exists an ideal-model simulator S (running in the $\mathcal{F}_{\mathsf{mcom}}$-hybrid model), such that no PPT $\mathcal{Z}$ can distinguish whether it is interacting with $\mathcal{A}$ or with S. When party P is honest and party P' is malicious, the simulator S will be unable to "rewind" P' (specifically, in the interaction of P' with the token that S must provides on behalf of P), and so the simulator cannot "force" the value of $\mathsf{tuple}_{P\to P'}$ to some desired value. On the other hand, an information-theoretic argument shows that, with all but negligible probability, a value $\mathsf{tuple}_{P\to P'}$ obtained by an interaction of P' with P's token is always a *random tuple* regardless of the behavior of P' (note that P' might interact with the token provided by P polynomially-many times, even though it is supposed to interact with it only once). Security of the signature scheme used (within the token) on behalf of the honest party P implies that P' can only send a value $\mathsf{tuple}_{P\to P'}$ that was output by the token. The upshot is that, with all but negligible probability, a value $\mathsf{tuple}_{P\to P'}$ used by P in step 4 of the protocol will be a random tuple.

On the other hand, continuing to assume that P is honest and P' is malicious, S *can* force the value of $\text{tuple}_{P' \to P}$ to any desired value in the following way. By simulating $\mathcal{A}$'s access to the $\mathcal{F}_{\text{wrap}}$ functionality, S obtains from $\mathcal{A}$ the code $M_{P'}$ that is "placed" in the token that $\mathcal{A}$ provides (on behalf of the malicious party P') to the honest party P. By rewinding $M_{P'}$, it is possible for S to "force" the output $\text{tuple}_{P' \to P}$ to, in particular, a (random) *Diffie-Hellman tuple* (for which it knows the necessary discrete logarithms evidencing this fact). Under the assumption that Diffie-Hellman tuples and random tuples are indistinguishable, this difference will not be detectable to $\mathcal{A}$ or $\mathcal{Z}$. The upshot is that S can set $\text{tuple}_{P' \to P}$ to be a Diffie-Hellman tuple in an undetectable manner.

Given the above, simulation follows in a fairly straightforward manner. Say the honest party P is committing to some value. The simulator, who does not yet know the value being committed to, will simply set C to be a commitment to "garbage" while choosing the elements of com uniformly at random. Note that Com here is perfectly hiding since $\text{tuple}_{P \to P'}$ is a random tuple, so this aspect of the simulation is fine. Furthermore, S can give a successful witness indistinguishable proof that it prepared the commitments correctly since $\text{tuple}_{P' \to P}$ is a Diffie-Hellman tuple (and S knows an appropriate witness to this fact).

In the decommitment phase, when S learns the committed value, it can simply send this value and again give a successful witness indistinguishable proof that it acted correctly (using again the fact that $\text{tuple}_{P' \to P}$ is a Diffie-Hellman tuple with discrete logarithm known to S).

The second case to consider is when the honest party P is the receiver. Say the malicious sender P' sends values C and com in step 4, and also gives a successful witness indistinguishable proof in that round. Since $\text{tuple}_{P \to P'}$ is a random tuple, this means that (with all but negligible probability) C and com are indeed commitments to the same value. Furthermore, since $\text{tuple}_{P' \to P}$ is a Diffie-Hellman tuple (with the appropriate discrete logarithm known to S), it is possible for S to extract the committed value b from com. Arguing similarly shows that in the decommitment phase P' will only be able to successfully decommit to the value b thus extracted.

Further details are provided in the following section.

3.2   Proof of Security

In this section, we sketch the proof that the protocol given earlier securely realizes the $\mathcal{F}_{\text{mcom}}$ functionality. Let $\mathcal{A}$ be a static adversary interacting with parties running the above protocol in the $\mathcal{F}_{\text{wrap}}$-hybrid model. We describe an ideal-model simulator S running in the $\mathcal{F}_{\text{mcom}}$-hybrid model, such that no PPT environment $\mathcal{Z}$ can distinguish whether it is interacting with $\mathcal{A}$ or with S.

S runs an internal copy of $\mathcal{A}$, forwarding all messages from $\mathcal{Z}$ to $\mathcal{A}$ and vice versa. We now specify the actions of S in response to messages received from $\mathcal{F}_{\text{mcom}}$:

Initialization. When a commitment is about to be carried out between parties P, P' for the first time, the simulator S does the following: say P is honest and

P' is corrupted (situations when both parties are corrupted or both parties are honest are easy to simulate).

1. Adversary $\mathcal{A}$ submits a message of the form (create, sid, P', P, M) to the (simulated copy of the) $\mathcal{F}_{\text{wrap}}$ functionality on behalf of P', and this message is intercepted by S. Simulator S chooses coins for M at random and runs an honest execution (on behalf of P) with M. If this leads to an abort on the part of P, then no further action is needed. Otherwise, in the standard way, S "rewinds" M and tries to generate an execution in which the output tuple$_{P' \to P}$ is a (randomly-chosen) Diffie-Hellman tuple, with discrete logarithm known to S. Using standard techniques and assuming the hardness of the decisional Diffie-Hellman problem, this can be done with all but negligible probability in expected polynomial time.

2. S, simulating the $\mathcal{F}_{\text{wrap}}$ functionality, sends the message (create, sid, P) to P'. It then runs an honest execution of the token functionality with $\mathcal{A}$ (who is acting on behalf of P'). We stress that S does no rewinding here — indeed, it cannot since it is not given the ability to rewind $\mathcal{A}$ (or, equivalently, $\mathcal{Z}$). S continues to simulate the actions of an honestly-generated token as many times as $\mathcal{A}$ chooses (note that $\mathcal{A}$ may even request further interactions at some later point in time).

Commitment when the sender is corrupted. Say S receives a message (receipt, sid, cid, P', P), the initialization as described above has already been carried out, and the sender P' is corrupted but the receiver P is honest. S begins by sending the value tuple$_{P' \to P}$ and the corresponding signature (generated as discussed above) to P'. Then S receives values tuple$_{P \to P'}$ and $\sigma_{P \to P'}$ from P' (this corresponds to step 3 of the commitment phase). If the signature does not verify, then P can abort and no further action is needed. If the signature verifies but tuple$_{P \to P'}$ was not generated in one of the executions of P' in the initialization phase described above, then S aborts. (This does not correspond to a legal action in the real world, but occurs with only negligible probability by security of the signature scheme.) Otherwise, in the following round S will receive values C and com from P'. It then acts as an honest receiver in the witness indistinguishable proof given by P'. If the proof fails, P will again abort as in the real world. Otherwise, S extracts the committed bit b from com (this is possible since tuple$_{P' \to P}$ is a Diffie-Hellman tuple) and sends (commit, sid, cid, P', P, b) to $\mathcal{F}_{\text{mcom}}$ (on behalf of corrupted party P').

In the decommitment phase, S again acts an an honest verifier. If the proof fails, then no further action is required. Otherwise, assuming the bit b sent by P' in this phase matches the bit extracted by S in the commitment phase, S simply sends (open, sid, cid, P, P') to $\mathcal{F}_{\text{mcom}}$. The other possibility is that the proof succeeds but the bit b is *different*; however, as argued informally in the previous section, this will occur with only negligible probability.

Commitment when the receiver is corrupted. Say S receives a notification (commit, sid, cid, P, P') from $\mathcal{F}_{\text{mcom}}$ that the honest party P has committed to a bit, and the initialization described earlier has already been carried out. S begins by sending the value $\mathsf{tuple}_{P' \to P}$ and the corresponding signature to P'. Then S receives values $\mathsf{tuple}_{P \to P'}$ and $\sigma_{P \to P'}$ from P' (this corresponds to step 3 of the commitment phase). If the signature does not verify, then P can abort and no further action is needed. If the signature verifies but $\mathsf{tuple}_{P \to P'}$ was not generated in one of the executions of P' in the initialization phase described above, then S aborts. (This does not correspond to a legal action in the real world, but occurs with only negligible probability by security of the signature scheme.) Otherwise, S proceeds as follows: it computes C as a commitment to the all-0 string, and sets the two components of commitment com to random elements of the appropriate group. It sends these values to P' and then acts as the honest prover in the witness indistinguishable proof, but using the witness (that it knows) for the fact that $\mathsf{tuple}_{P' \to P}$ is a Diffie-Hellman tuple.

When S later receives notification $(\mathsf{open}, \mathsf{sid}, \mathsf{cid}, P, P', b)$ that P has opened to the bit b, the simulator simply sends this value b and then, again, acts as the honest prover in the witness indistinguishable proof, but using the witness (that it knows) for the fact that $\mathsf{tuple}_{P' \to P}$ is a Diffie-Hellman tuple.

We defer to the full version of this paper the details of the proof that S as described above provides a good simulation of $\mathcal{A}$.

4   Conclusions and Future Directions

UC multi-party computation is impossible without some extension to the so-called "plain model." We now know of a variety of extensions, or "setup assumptions," that enable this impossibility result to be circumvented. An important direction of research is to find *realistic* setup assumptions that could feasibly be implemented and used. The suggestion made in this paper is to consider physical assumptions instead of (or possibly in addition to) trust-based assumptions. A particular example based on tamper-proof hardware was proposed, and shown to be sufficient for realizing UC multi-party computation.

Some intriguing questions are left open by this work. Of course, alternate (weaker?) models of tamper-proof hardware could be explored in an effort to obtain easier-to-realize conditions under which UC multi-party computation exists. One interesting possibility here is to use tamper-*evident* tokens (that could be returned to their creator at some intermediate point of the protocol) in place of tamper-*resistant* ones. (This idea was suggested by an anonymous referee.) Coming back to the model proposed here, it would be nice to show a protocol secure against adaptive adversaries, and it would be especially gratifying to construct a protocol based on general assumptions. (It is not hard to see that the protocol can be based on a variety of standard number-theoretic assumptions other than the DDH assumption.)

Acknowledgments

I am very grateful to the anonymous referees for their supportive comments and helpful suggestions.

References

1. B. Barak, R. Canetti, Y. Lindell, R. Pass, and Tal Rabin. Secure Computation Without Authentication. Crypto 2005.
2. B. Barak, R. Canetti, J.B. Nielsen, and R. Pass. Universally Composable Protocols with Relaxed Set-Up Assumptions. FOCS 2004.
3. C. Bennett and G. Brassard. Quantum Cryptography: Public Key Distribution and Coin Tossing. Intl. Conf. on Computers, Systems, and Signal Processing, 1984.
4. M. Blum, P. Feldman, and S. Micali. Non-Interactive Zero-Knowledge and its Applications. STOC '88.
5. S. Brands. Untraceable Off-line Cash in Wallets with Observers. Crypto '93.
6. R. Canetti. Universally Composable Security: A New Paradigm for Cryptographic Protocols. FOCS 2001. Full version available at http://eprint.iacr.org/2000/067.
7. R. Canetti, Y. Dodis, R. Pass, and S. Walfish. Universally Composable Security with Global Setup. TCC 2007.
8. R. Canetti and M. Fischlin. Universally Composable Commitments. Crypto 2001.
9. R. Canetti, E. Kushilevitz, and Y. Lindell. On the Limitations of Universally Composable Two-Party Computation Without Set-Up Assumptions. *J. Cryptology* 19(2): 135–167, 2006.
10. R. Canetti, Y. Lindell, R. Ostrovsky and A. Sahai. Universally Composable Two-Party and Multi-Party Secure Computation. STOC 2002. Full version available at http://eprint.iacr.org/2002/140.
11. D. Chaum and T. Pedersen. Wallet Databases with Observers. Crypto '92.
12. J. Considine, M. Fitzi, M. Franklin, L.A. Levin, U. Maurer, and D. Metcalf. Byzantine Agreement Given Partial Broadcast. *J. Cryptology* 18(3): 191–217, 2005.
13. R. Cramer and T. Pedersen. Improved Privacy in Wallets with Observers. Eurocrypt '93.
14. I. Csiszár and J. Körner. Broadcast Channels with Confidential Messages. *IEEE Trans. Info. Theory* 24(3): 339–348, 1978.
15. D. Hofheinz, J. Müller-Quade, and D. Unruh. Universally Composable Zero-Knowledge Arguments and Commitments from Signature Cards. 5th Central European Conference on Cryptology, 2005. A version is available at http://homepages.cwi.nl/~hofheinz/card.pdf.
16. U. Maurer Secret Key Agreement by Public Discussion from Common Information. *IEEE Trans. Info. Theory* 39(3): 733–742, 1993.
17. M. Naor. On Cryptographic Assumptions and Challenges. Crypto 2003.
18. M. Pease, R. Shostak, and L. Lamport. Reaching Agreement in the Presence of Faults. *J. ACM* 27(2): 228–234, 1980.
19. T. P. Pedersen. Non-Interactive and Information-Theoretic Secure Verifiable Secret Sharing. Crypto '91.
20. B. Pfitzmann and M. Waidner. Composition and Integrity Preservation of Secure Reactive Systems. ACM CCCS 2000.
21. A.D. Wyner. The Wire-Tap Channel. *Bell System Technical Journal* 54(8): 1355–1387, 1975.

Generic and Practical Resettable Zero-Knowledge in the Bare Public-Key Model*

Moti Yung[1] and Yunlei Zhao[2],**

[1] RSA Laboratories and Department of Computer Science, Columbia University,
New York, NY, USA
`moti@cs.columbia.edu`
[2] Software School, Fudan University, Shanghai 200433, China
`ylzhao@fudan.edu.cn`

Abstract. We present a generic construction for constant-round concurrently sound resettable zero-knowledge (rZK-CS) arguments for $\mathcal{NP}$ in the bare public-key (BPK) model under any (sub-exponentially strong) one-way function (OWF), which is a traditional assumption in this area. The generic construction in turn allows round-optimal implementation for $\mathcal{NP}$ still under general assumptions, and can be converted into a highly practical instantiation (under specific number-theoretic assumptions) for any language admitting Σ-protocols. Further, the rZK-CS arguments developed in this work also satisfy a weak (black-box) concurrent knowledge-extractability property as proofs of knowledge, in which case some super-polynomial-time assumption is intrinsic.

1 Introduction

Resettable zero-knowledge (rZK) is the strongest version of the remarkable notion of zero-knowledge (ZK) [13] to date. It was put forth by Canetti, Goldreich, Goldwasser and Micali [5], motivated by implementing zero-knowledge provers using smart-cards or other devices that may be (maliciously) reset to their initial conditions and/or cannot afford to generate fresh randomness for each new invocation. rZK also preserves the prover's security when the protocol is executed concurrently in an asynchronous network like the Internet. In fact, rZK is a generalization and strengthening of the notion of concurrent zero-knowledge (cZK) introduced by Dwork, Naor and Sahai [10].

A major measure of efficiency for interactive protocols is the round-complexity. Unfortunately, there are no constant-round rZK in the standard model, at least for the black-box case, as implied by the works of Canetti, Kilian, Petrank and Rosen [6]. To get constant-round rZK protocols, [5] introduced a simple model with very appealing trust requirement, the *bare public-key* (BPK) model.

A protocol in the BPK model simply assumes that all verifiers have deposited a public key in a public file before any interaction takes place among the users.

---

* The second author is supported by 973 project No. 2007CB807900.
** Corresponding author.

M. Naor (Ed.): EUROCRYPT 2007, LNCS 4515, pp. 129–147, 2007.

(Actually, the BPK model also allows dynamic key registration with a reasonable amount time between key registration and key usage [5].) But, no assumption is made on whether the public-keys deposited are unique or valid. That is: no trusted third party is assumed, preprocessing is reduced to users noninteractively posting public-keys in a public file, and the underlying communication network is assumed to be adversarially asynchronous. In many cryptographic settings, availability of a public key infrastructure (PKI) is assumed or required, in which case the BPK model that is weaker than PKI is natural.

Soundness in public-key models, when verifiers register public-keys, turns out to be more complicated and subtle than in other models as was shown by Micali and Reyzin. They showed that under standard intractability assumptions there are four distinct meaningful notions of soundness, i.e., from weaker to stronger: one-time, sequential, concurrent and resettable soundness [16]. In this work, we focus on concurrent soundness, which roughly means that a malicious prover P^* cannot convince the honest verifier V of a *false* statement even when P^* is allowed multiple interleaving interactions with V. They also showed that any (resettable or not) black-box ZK protocols with concurrent soundness in the BPK model (for non-trivial languages outside $\mathcal{BPP}$) must run at least four rounds [16]. The recent work of [21] formulates a new concurrent verifier security in the public-key model, named concurrent knowledge-extraction (CKE), and shows that CKE is strictly stronger than concurrent soundness in the public-key model when proofs of knowledge are considered.

A direct application of rZK is to achieve (smartcard based) identification schemes secure against resetting attacks [5,2]. Despite its significant importance to practice, especially to smartcard based e-commerce over the Internet, most existing rZK systems are only theoretical feasible solutions which cannot be directly employed in practice and are not implementable by smartcards. That is, there is a gap between the significant importance and motivation for rZK as a mode suitable for practice and the present theoretical constructions of rZK systems. (Note that it is natural to investigate general feasibility prior to practical solutions.) Given the state of protocols, it is an important issue to develop highly practical rZK systems (say, with only a very small constant number of exponentiations, which are within reach for coming smartcard environments) for languages widely used in cryptography.

1.1   Our Contributions

The main result of this work is a generic construction for constant-round concurrently sound rZK (rZK-CS) arguments for $\mathcal{NP}$ in the BPK model under any generic sub-exponentially strong OWF (sub-exponential assumptions in order to enable ZK protocols in this highly constrained resettable setting have been employed from the introduction of the model). The structure and techniques of the generic rZK-CS construction, in turn, allow round-optimal (still under general assumptions) and highly practical instantiation (under specific number-theoretic assumptions) implementations. Further, the rZK-CS arguments developed in this work also satisfy a weak (black-box) concurrent knowledge-extractability

(CKE) property in the public-key model. (Roughly, a malicious prover not only cannot convince of a *false* statement by concurrent interactions as required by concurrent soundness, but also cannot convince of a true statement in its concurrent interactions without knowing a witness if the underlying language is sub-exponentially hard.) This answers several open problems left over in the field of round-efficient rZK in the BPK model [16,22,9].

Specifically, the generic construction allows the following round-optimal or highly practical implementations, which involve novel uses of a number of cryptographic tools:

- Round-optimal (i.e., 4-round) rZK-CS arguments for $\mathcal{NP}$ in the BPK model under any sub-exponentially strong one-way permutation (OWP) and any (standard polynomially secure) preimage-verifiable OWF. Note that preimage-verifiable OWF is a generic and actually quite weak hardness assumption that includes, in particular, any certified one-way permutation and any 1-1 length-preserving one-way function. This implies, in particular, that round-optimal rZK-CKE arguments for $\mathcal{NP}$ in the BPK model can be based on any certified one-way permutation.
- A *generic practical* transformation achieving 5-round rZK-CS arguments in the BPK model. By "generic" we mean applicability to any language that admits Σ-protocols. By "practical", we mean that the transformation does not go through general $\mathcal{NP}$-reductions, and if the starting Σ-protocol and the underlying pseudorandom function (PRF) are practical then the transformed rZK-CS arguments are also practical. For example, when instantiated with DL or RSA functions, together with the Naor-Reingold practical PRFs [18], the transformed rZK-CS arguments (for the languages of DL or RSA respectively) employ a very small constant number of exponentiations.

Discussions on related works are deferred to the full version.

2   Preliminaries

We briefly recall some basic definitions and tools, with detailed presentations deferred to the full version.

Preimage-verifiable one-way functions. A OWF f is called preimage-verifiable if there exists a polynomial-time computable predicate $D_f : \{0,1\}^* \longrightarrow \{0,1\}$ such that for any string y, $D_f(y) = 1$ if and only if there exists an x such that $y = f(x)$.

Statistically-binding commitment schemes. We employ both the OWP based one-round perfectly-binding commitment scheme [12], and Naor's OWF-based 2-round scheme [17]. Note that the first-round message of Naor's commitment scheme can be fixed once and for all and, in particular, can be posted as part of a public-key in the public-key setting. *We remark that if the underlying OWP or OWF are secure against 2^{n^c}-time adversaries for some constant* *c*, $0 < c < 1$, *on a security parameter n, then the hiding property of the corresponding commitment schemes above also holds against* 2^{n^c}-*time adversaries.*

Public-coin witness indistinguishability (WI) proof of knowledge (POK) systems for $\mathcal{NP}$. One is Blum's protocol for directed Hamiltonian cycle DHC [3], and another is the Lapidot-Shamir protocol for DHC [15]. The salient feature of the Lapidot-Shamir protocol is that the prover sends the first-round message without knowing the statement to be proved other than its size. We remark that the WI property of Blum's protocol or the Lapidot-Shamir protocol for HC relies on the hiding property of the underlying statistically-binding commitment scheme (used in its first-round). If the hiding property of the underlying statistically-binding commitment scheme is secure against 2^{n^c}-time adversaries for some constant c, $0 < c < 1$, on a security parameter n, then the WI property also holds against 2^{n^c}-time adversaries.

Trapdoor commitment schemes. Normal trapdoor commitment schemes run in two rounds, in which the commitment receiver generates and sends the trapdoor commitment public key ($TCPK$) in the first-round (while keeping the trapdoor secret key $TCSK$ in private). For the Feige-Shamir trapdoor commitment scheme (FSTC) [11], $TCPK$ consists of ($y = f(x), G$) (for OWF-based solution, the $TCPK$ also includes a random string R serving as the first-round message of Naor's OWF-based statistically-binding commitment scheme), where f is a OWF and G is a graph that is reduced from y by the Cook-Levin $\mathcal{NP}$-reduction. The corresponding trapdoor is x (or equivalently, a Hamiltonian cycle in G). Note that the first-round message, i.e., $TCPK$, can be fixed once and for all. The commitment sender forms the second-round message by using (either OWP-based one-round or Naor's OWF-based two-round) statistically-binding commitment scheme. Again, if the hiding property of the underlying statistically-binding commitment scheme is secure against sub-exponential-time adversaries, then both the hiding property and the trapdoorness property of the FSTC scheme hold also against sub-exponential-time adversaries.

Σ-protocols and Σ_{OR}-protocols. Informally, a Σ-protocol is itself a 3-round public-coin *special* honest verifier zero-knowledge (SHVZK) protocol with special soundness in the knowledge-extraction sense. A very large number of Σ-protocols have been developed in the literature. One basic construction with Σ-protocols is the OR of a real and simulated transcript, called Σ_{OR}, that allows a prover to show that given two inputs x_0, x_1, it knows a w such that either $(x_0, w) \in R_0$ or $(x_1, w) \in R_1$, but without revealing which is the case [7] (i.e., witness indistinguishable WI). For a good survey of Σ-protocols and their applications, the reader is referred to [8].

The malicious resetting verifier and rZK in the BPK model. A malicious s-resetting malicious verifier V^* in the BPK model, where s is a positive polynomial, is a PPT Turing machine working in two stages so that on input 1^n,

Stage-1. V^* receives $s(n)$ *distinct* strings $\bar{\mathbf{x}} = \{x_1, \cdots, x_{s(n)}\}$ of equal length $poly(n)$ each, and outputs an arbitrary public-file F and a list of (without loss of generality) $s(n)$ identities $id_1, \cdots, id_{s(n)}$.

Stage-2. Starting from the final configuration of Stage-1, $s(n)$ random tapes, $\gamma_1, \cdots, \gamma_{s(n)}$, are randomly selected and then fixed for P, resulting in $s(n)^3$ deterministic prover strategies $P(x_i, id_j, \gamma_k)$, $1 \leq i, j, k \leq s(n)$. V^* is given oracle access to these $s(n)^3$ provers, and finally outputs its "view" of the interactions (i.e., its random tapes and messages received from all its oracles).

Definition 1 (black-box resettable zero-knowledge [5]). *A protocol $\langle P, V \rangle$ is black-box resettable zero-knowledge for a language $L \in \mathcal{NP}$ if there exists a PPT black-box simulator S such that for every s-resetting verifier V^*, the following two probability distributions are indistinguishable. Let each distribution be indexed by a sequence of* **distinct** *common inputs $\bar{\mathbf{x}} = \{x_1, \cdots, x_{s(n)}\}$, $x_i \in L \cap \{0,1\}^{poly(n)}$ for $1 \leq i \leq s(n)$, and their corresponding NP-witnesses $aux(\bar{\mathbf{x}}) = \{w_1, \cdots, w_{s(n)}\}$:*

Distribution 1. *The output of V^* obtained from the experiment of choosing $\gamma_1, \cdots, \gamma_{s(n)}$ uniformly at random, running the first stage of V^* to obtain F, and then letting V^* interact in its second stage with the following $s(n)^3$ instances of P: $P(x_i, w_i, F, id_j, \gamma_k)$ for $1 \leq i, j, k \leq s(n)$. Note that V^* can oracle access to these $s(n)^3$ instances of P.*

Distribution 2. *The output of $S(\bar{\mathbf{x}})$.*

Remark. In Distribution 1 above, since V^* oracle accesses to $s(n)^3$ instances of P: $P(x_i, w_i, F, id_j, \gamma_k)$, $1 \leq i, j, k \leq s(n)$, it means that V^* may invoke and interact with the same $P(x_i, w_i, F, id_j, \gamma_k)$ in multiple protocols (sessions). We remark that, as clarified in [5], in the resettable setting interleaving interactions do not help the malicious resetting verifier get more advantages on learning "knowledge" from its oracles than it can do by sequential interactions. Without loss of generality, in the rest of this paper we assume the resetting malicious verifier V^* works in the sequential version.

3   The Generic rZK-CS Construction

The high-level overview of the protocol. We first convey basic ideas and a high-level overview of the protocol. Let f_V be any (*sub-exponentially strong*) OWF, each (honest) verifier V randomly selects an element x_V from the domain of f_V, and publishes $y_V = f_V(x_V)$ as its public-key with x_V as its secret-key. Let L be an $\mathcal{NP}$-language and $x \in L$ be the common input, the main-body of the protocol goes as follows: The honest prover P first generates and sends a hard-instance using a standard *polynomially-secure* OWF f_P. The hard-instance is then fixed once and for all. Then, P proves to V the existence of the preimage of the hard-instance, by executing a OWF-based resettable witness-hiding (rWH) protocol. After that, V proves to P that it knows either the preimage of y_V

(i.e., its secret-key x_V) or the preimage of the hard-instance generated by P, by executing a OWF-based constant-round WIPOK protocol for $\mathcal{NP}$. Finally, P proves to V that it knows either a witness for $x \in L$ or the preimage of y_V (i.e., V's secret-key), by executing another OWF-based constant-round rWI argument for $\mathcal{NP}$. The detailed protocol description is depicted in Figure 1 (page 135).

The underlying complexity-leveraging. For provable security and for the weak CKE security, we employ the complexity-leveraging technique (originally introduced in [5]). Specifically, the verifier V uses a security parameter N (in generating messages from it) that is also the system security parameter. But, the prover P uses a relatively smaller security parameter n (still polynomially related to N). The justification and discussions of the complexity-leveraging technique are given in [5]. *Here, we additionally remark that, pragmatically speaking, letting the verifier and the prover use different security parameters is quite reasonable in the resettable setting, in which the prover is implemented by smart-cards or clients that have relatively limited computational resources and power and the verifier is normally implemented by servers that have much more computational resources and power.*

Specifically, the security parameters are set as follows. On the system parameter N, suppose f_V is secure against $2^{N^{c_V}}$-time adversaries for some constant c_V, $0 < c_V < 1$. This implies that the hiding property of the underlying statistically-binding commitment scheme used by the verifier holds also against any $2^{N^{c_V}}$-time adversary, which in turn guarantees that the WI property of the underlying WI protocol for $\mathcal{NP}$ executed in Stage-2 of Phase-1 and Phase-3, and the hiding and trapdoorness properties of the underlying trapdoor commitment scheme all hold against any $2^{N^{c_V}}$-time adversary. The prover uses a relatively smaller security parameter n and uses a standard polynomially-secure OWF f_P that can be broken (brute-force wise) in time $2^{n^{c_P}}$ for some constant c_P, $c_P \geq 1$. Specifically, c_P is the constant that: for all sufficiently large n's, the size of G_P (reduced from $(y_P^{(0)}, y_P^{(1)})$ at Stage-1 of Phase-1) is bounded by n^{c_P}, which in turn implies that the statistically-binding commitment scheme used by the prover (that is run on the security parameter n) can be brute-force decommitted in time $poly(n) \cdot 2^{n^{c_P}}$. Let c_L, $0 < c_L \leq 1$, be a constant specific to the underlying language L (the use of c_L is specified in Section 3.1 for the weak CKE property). Let c be any constant such that $0 < c < \min\{c_V, c_L\}$, in other words, $\min\{c_V, c_L\} = c + c'$ for another constant c', $0 < c' < 1$. Let ε be any constant such that $\varepsilon > \frac{c_P}{c}$, then we set $N = n^\varepsilon$. Note that N and n are polynomially related. That is, any quantity that is a polynomial of N is also another polynomial of n. This complexity leveraging guarantees that although any $poly(n) \cdot 2^{n^{c_P}}$-time adversary can break f_P on a security parameter n, it is still infeasible to break the one-wayness of f_V, because $poly(n) \cdot 2^{n^{c_P}} \ll 2^{N^c} \ll 2^{N^{c_V}}$ (also note that $poly(n) \cdot 2^{n^{c_P}} \ll 2^{N^{c_L}}$).

The OWF-based protocol depicted in Figure 1 (page 135) runs in 7 rounds after some round combinations. In particular, the first two rounds of Phase-4 can be combined into previous phases. Actually, the round-complexity can be further reduced to 6 but under any (sub-exponentially strong) OWP.

Key generation. On the system security parameter N, each honest verifier V randomly selects an element x_V of length N, computes $y_V = f_V(x_V)$, publishes y_V as its public-key PK while keeping x_V as its secret-key SK. If P uses Naor's OWF-based statistically-binding commitment scheme in Phase-2 or Phase-4 (that is run on security parameter n), V also deposits a random string R_V of length $3n$.

Common input. An element $x \in L \cap \{0,1\}^{poly(N)}$, the public-file F and an index j that specifies the j-th entry of F, i.e., $PK_j = (y_V^{(j)}, R_V^{(j)})$.

P **private input.** An $\mathcal{NP}$-witness w for $x \in L$, a pair of random strings (γ_1, γ_2), where γ_1 is a $poly(n)$-bit string and γ_2 is the n-bit randomness seed of a PRF.

V **private input.** SK_j. For presentation simplicity, we denote $PK_j = f_V(SK_j)$.

Phase-1. Phase-1 consists of two stages:

    Stage-1. Let f_P be any polynomially-secure OWF. On security parameter n, P randomly selects two elements $x_P^{(0)}$ and $x_P^{(1)}$ of length n each in the domain of f_P, computes $y_P^{(b)} = f_P(x_P^{(b)})$ for $b \in \{0,1\}$, reduces $(y_P^{(0)}, y_P^{(1)})$ to a directed graph G_P by Cook-Levin $\mathcal{NP}$-reduction such that finding a Hamiltonian cycle in G_P is equivalent to finding the preimage of either $y_P^{(0)}$ or $y_P^{(1)}$. For OWF-based solution, P also randomly selects a string R_P of length $3N$ serving as the first-round message of Naor's OWF-based statistically-binding commitment scheme. Finally, P sends $(y_P^{(0)}, y_P^{(1)}, G_P, R_P)$ to V. The randomness used by P in this process is γ_1, *which means* $(y_P^{(0)}, y_P^{(1)}, G_P, R_P)$ *is fixed once and for all.*

    Stage-2. V first checks whether or not G_P is reduced from $(y_P^{(0)}, y_P^{(1)})$ and R_P is of length $3N$. If the checking is successful, V randomly chooses two random strings $e_V^{(0)}$ and $e_V^{(1)}$ from $\{0,1\}^n$, computes $c_V^{(0)} = Com(1^N, R_P, e_V^{(0)})$ *by using the underlying statistically-binding commitment scheme* Com, and $c_V^{(1)} = TCCom(1^N, (G_P, R_P), e_V^{(1)})$ by using the underlying FSTC trapdoor commitment scheme. Then, on common input $((y_P^{(0)}, y_P^{(1)}, G_P, R_P), PK_j)$ V computes the first-round message, denoted a_V, of (n-parallel repetitions of) Blum's WIPOK for $\mathcal{NP}$ for showing the knowledge of either SK_j or a Hamiltonian cycle in G_P (equivalently, the preimage of either $y_P^{(0)}$ or $y_P^{(1)}$). Finally, V sends $(c_V^{(0)}, c_V^{(1)}, a_V)$ to P. *From then on, all randomness used by P in the remaining computation is got by applying* $PRF(\gamma_2, \cdot)$ *on the "determining" message* $D = (x, F, (j, PK_j), (y_P^{(0)}, y_P^{(1)}, G_P, R_P), (c_V^{(0)}, c_V^{(1)}, a_V))$.

Phase-2. P proves to V the existence of a Hamiltonian cycle in G_P by executing the (n-parallel repetitions of) Blum's WI protocol for $\mathcal{NP}$ on $(y_P^{(0)}, y_P^{(1)}, G_P, R_V^{(j)})$, in which V sends the assumed random challenge by just revealing $e_V^{(0)}$ committed to $c_V^{(0)}$. Note that the first-round message of Phase-2 (from P to V) consists of n committed adjacency matrices committed by running the underlying statistically-binding commitment scheme *on security parameter* n. If P successfully finishes this phase and V accepts, then goto Phase-3. Otherwise, V aborts.

Phase-3. V and P continue the WIPOK protocol for $\mathcal{NP}$ suspended at Stage-2 of Phase-1. If V successfully convinces P of the knowledge of either SK_j or a Hamiltonian cycle in G_P, then goto Phase-4. Otherwise, P aborts. We denote by e_V, z_V, the first-round message and the second-round message of Phase-3 respectively.

Phase-4. P proves that it "knows" either the witness w for $x \in L$ or the secret-key SK_j, by executing Blum's WI protocol for $\mathcal{NP}$ on common input (x, PK_j), in which V sends the assumed random challenge by just revealing $e_V^{(1)}$ committed to $c_V^{(1)}$.

Fig. 1. The generic rZK-CS argument $\langle P, V \rangle$ for $\mathcal{NP}$

Theorem 1. *Assuming the OWF f_P (used by the prover) is secure against standard polynomial-time adversaries, and the OWF f_V (used by the verifier) is secure against sub-exponential-time adversaries, the protocol depicted in Figure 1 is a constant-round rZK-CS argument for $\mathcal{NP}$ in the BPK model.*

Proof (sketch)

Black-box resettable zero-knowledge

For any s-resetting adversary V^* who receives $s(N)$ *distinct* strings $\bar{\mathbf{x}} = \{x_1, \cdots, x_{s(N)}\}$, $x_i \in L \cap \{0,1\}^{poly(N)}$ for each i ($1 \le i \le s(N)$), and outputs an arbitrary public-file F containing $s(N)$ entries $PK_1, \cdots, PK_{s(N)}$ in its first stage, we say a public-key PK_j in F, $1 \le j \le s(N)$, is "covered" if the rZK simulator S has already learned (extracted) the corresponding secret-key SK_j (if such exists). In its second stage, V^* is given oracle access to $(s(N))^3$ prover instances $P(x_i, PK_j, \gamma_k)$, $1 \le i, j, k \le s(N)$. We denote by $D_t = (x_i, F, (j, PK_j), (y_P^{(0)}, y_P^{(1)}, G_P, R_P)^k, (c_{V^*}^{(0)}, c_{V^*}^{(1)}, a_{V^*})^t)$ the "determining" message of the t-th session with respect to common input x_i and public-key PK_j and the honest prover instance $P(\cdot, \cdot, \gamma_k)$, $1 \le i, j, k \le s(N)$ and $1 \le t \le (s(N))^3$. As discussed in [5], w.l.o.g., we use the convention that V^* works in the sequential version in its second stage, and the rZK simulator utilizes a truly random function rather than a pseudorandom one.

The rZK simulation procedure is similar to, but more complicated than, that of [5]. Specifically, the rZK simulator S runs V^* as a subroutine, and works in at most $s(N) + 1$ phases such that in each phase it either successfully finishes its simulation or "covers" a new public-key in F. In each phase, S makes a simulation attempt from scratch with a new truly random function that is to be defined adaptively, and works session by session sequentially in at most $(s(N))^3$ sessions. The difficulties lie in that for such rZK simulation to be successful, the rZK simulator S needs to have the ability to cover new uncovered public-keys within time inversely proportional to the probability that it encounters a success of Phase-3 relative to a yet uncovered public-key in its simulation. Pending on S's such ability, the rZK property follows from the pseudorandomness of PRF and the rWI property of Phase-4 combined with Phase-1 (according to the CGGM general paradigm for achieving rWI [5]).

Specifically, we want to argue that the underlying Blum's WIPOK protocol on $((y_P^{(0)}, y_P^{(1)}, G_P, R_P)^k, PK_j)$ (executed in Stage-2 of Phase-1 and Phase-3) is actually an *argument of knowledge* of the preimage of PK_j (i.e., the secret-key SK_j). But, the subtle and complicated situation here is that before V^* finishes Phase-3, S has already proved the knowledge of the Hamiltonian cycle of $(y_P^{(0)}, y_P^{(1)}, G_P, R_P)^k$ in Phase-2. Note that the $(y_P^{(0)}, y_P^{(1)}, G_P, R_P)^k$ is fixed once and for all (*that can be viewed as the public-key of the honest prover instance $P(\cdot, \cdot, \gamma_k)$*), and furthermore V^* is resettingly (more than concurrently) interacting with the honest prover instances. As demonstrated in [21], normal argument of knowledge and even concurrent soundness do not guarantee *correct* knowledge-extractability in such setting. In particular, one may argue that, by rewinding the honest prover instances arbitrarily, V^* may potentially malleate the interactions on $(y_P^{(0)}, y_P^{(1)}, G_P, R_P)^k$ provided by the honest prover in Phase-2 of one session into successful but "false" interactions on $((y_P^{(0)}, y_P^{(1)}, G_P, R_P)^k, PK_j)$ in Stage-2 of Phase-1 and Phase-3 of another session with respect to public-key PK_j, in the sense that although the interactions are valid but V^* actually does not know the corresponding secret-key SK_j. This means that, in such a case the interactions on $((y_P^{(0)}, y_P^{(1)}, G_P, R_P)^k, PK_j)$ executed in Phase-3 together with Stage-2 of Phase-1 are no longer *arguments of knowledge* of the preimage of PK_j, although it is always a system for proof of knowledge of either SK_j or a Hamiltonian cycle of $(y_P^{(0)}, y_P^{(1)}, G_P, R_P)^k$. What save us here is the (concurrent) WI property of the Blum's protocol for HC.

Below, we construct an algorithm $\hat{S}$ that emulates the real rZK simulator while *concurrently* (not resettingly) running the Blum's protocol for HC. That is, on common inputs $\{(y_P^{(0)}, y_P^{(1)}, G_P, R_P)^1, \cdots, (y_P^{(0)}, y_P^{(1)}, G_P, R_P)^{s(N)}\}$ $\hat{S}$ concurrently interacts with $s(N)$ instances of the knowledge prover, denoted $\hat{P}$, of Blum's protocol for HC by playing the role of knowledge verifier. We denote each of the $s(N)$ instances of $\hat{P}$ by $\hat{P}((y_P^{(0)}, y_P^{(1)}, G_P, R_P)^k)$, $1 \le k \le s(N)$; At the same time, $\hat{S}$ runs the s-resetting malicious V^* as a subroutine by playing the role of the honest prover, and sends $(y_P^{(0)}, y_P^{(1)}, G_P, R_P)^k$ as the Stage-1 message of Phase-1 whenever V^* initiates a session with the honest prover instance $P(\cdot, \cdot, \gamma_k)$. $\hat{S}$ emulates the rZK simulator S but with the following modification: whenever $\hat{S}$ needs to send a "fresh" first-round message of Blum's protocol for HC on $(y_P^{(0)}, y_P^{(1)}, G_P, R_P)^k$ in Phase-2 with respect to a "determining" message, it initiates a new session with $\hat{P}((y_P^{(0)}, y_P^{(1)}, G_P, R_P)^k)$, and forwards the first-round message received from $\hat{P}((y_P^{(0)}, y_P^{(1)}, G_P, R_P)^k)$ to V^*. This "fresh" message happens due to either V^* sends a distinct "determining" message in one session or $\hat{S}$ needs rewinding V^* and redefining the underlying random function f to extract knowledge used by V^* in a successful execution of Stage-2 of Phase-1 and Phase-3 with respect to an uncovered public-key. Then, $\hat{S}$ runs V^* further, and in case V^* successfully reveals the assumed challenge (that is *statistically-bindingly* committed to the underlying "determining" message in question) then $\hat{S}$ returns back the revealed challenge to $\hat{P}$ as its own challenge in the corresponding simultaneous session of Blum's protocol for HC, and returns back the third-round message received from $\hat{P}((y_P^{(0)}, y_P^{(1)}, G_P, R_P)^k)$ to V^*. For a session with a "determining" message that is identical to that of some previous sessions, $\hat{S}$ just copies what was sent in the previous sessions. Note that in this case, $\hat{S}$ may still possibly need to interact with $\hat{P}$ in some *existing* concurrent session to get some third-round message (in case V^* did not reveal or invalidly revealed the random challenge statistically-bindingly committed to the underlying "determining" message in all previous sessions, but correctly reveals it in the current session). However, the key point here is that in this case S does not need to initiate a new concurrent session with $\hat{P}$.

Note that from the viewpoint of V^*, the behavior of $\hat{S}$ is identical to the behavior of the real rZK simulator, where the real rZK simulator S generates

$(y_P^{(0)}, y_P^{(1)}, G_P, R_P)^k$'s and provides the corresponding Phase-2 messages by itself (rather than get them by externally interacting with the knowledge prover instances $\hat{P}((y_P^{(0)}, y_P^{(1)}, G_P, R_P)^k)$'s). The key observation here is that although V^* is actually resettingly interacting with $\hat{S}$, $\hat{S}$ only concurrently interacts with the instances of $\hat{P}$ and never rewinds $\hat{P}$. *The underlying reason is just that in any session, Phase-2 interactions take place only after V^* sent the "determining" message at Stage-2 of Phase-1 that determines the subsequent behaviors of V^* in that session.* Note that in this case, the (concurrent) WI property of the Blum's protocol for HC on common input $(y_P^{(0)}, y_P^{(1)}, G_P, R_P)^k$ actually implies witness hiding (WH), which means that no PPT algorithm can output a Hamiltonian cycle in $(y_P^{(0)}, y_P^{(1)}, G_P, R_P)^k$ even by concurrently interacting with $\hat{P}((y_P^{(0)}, y_P^{(1)}, G_P, R_P)^k)$'s. Also note that on common input $((y_P^{(0)}, y_P^{(1)}, G_P, R_P)^k, PK_j)$, Phase-3 together with Stage-2 of Phase-1 is always a system for proving the knowledge of either a Hamiltonian cycle in $(y_P^{(0)}, y_P^{(1)}, G_P, R_P)^k$ or the preimage of PK_j (i.e., SK_j), which means that with overwhelming probability $\hat{S}$ (or the real rZK simulator S) can always extract either a Hamiltonian cycle in $(y_P^{(0)}, y_P^{(1)}, G_P, R_P)^k$ or the corresponding secret-key SK_j within time inversely proportional to the probability that V^* successfully finishes Phase-3 (by rewinding V^* and redefining the underlying random function as is done in [5]). But, the WH property of Blum's protocol for HC shows that with overwhelming probability, $\hat{S}$ (or the real rZK simulator S) never outputs a Hamiltonian cycle in $(y_P^{(0)}, y_P^{(1)}, G_P, R_P)^k$ in its simulation that is done in expected polynomial-time. Here, a subtle point needs to be further addressed. Specifically, the normal WH property is defined with respect to probabilistic (strict) polynomial-time algorithms, but here $\hat{S}$ works in expected polynomial-time. But, by Markov inequality, it is easy to see that if the WH property of a protocol holds with respect to any strict polynomial-time algorithms, then it also holds with respect to any expected polynomial-time algorithms.

Concurrent soundness

We show that for any (whether true or not) common input $x \in \{0, 1\}^{poly(N)}$, if a PPT s-concurrent malicious P^*, on a public-key PK, can convince an honest verifier V (with public-key PK and secret-key SK) of the statement "$x \in L$" with non-negligible probability p_x in one of the $s(N)$ concurrent interactions, then there exists an algorithm E that, on the *same* public-key PK with oracle accessing P^*, works in $poly(n) \cdot 2^{n^{c_P}}$-time and outputs either a witness for $x \in L$ or the preimage of y_V also with non-negligible probability. Note that according to the underlying complexity leveraging on the security parameters N and n, no $poly(n) \cdot 2^{n^{c_P}}$-time algorithm can break the one-wayness of f_V used by V in forming its public-key on security parameter N (because $poly(n) \cdot 2^{n^{c_P}} \ll 2^{N^{c_V}}$). This implies that $x \in L$.

On the same public-key PK, E runs P^* as a subroutine by playing the role of the honest verifier with public-key PK. *Note that E does not know the corresponding secret-key SK.* In each session t, $1 \le t \le s(N)$, after receiving the Stage-1 message of Phase-1, denoted $((y_{P*}^{(0)}, y_{P*}^{(1)})^t, G_{P*}^t, R_{P*}^t)$, E first checks whether or not G_{P*}^t is $\mathcal{NP}$-reduced from $(y_{P*}^{(0)}, y_{P*}^{(1)})^t$ and R_{P*}^t is of length $3N$. If the checking is successful, then E tries to find a Hamiltonian cycle in G_{P*}^t by brute-force searching in $2^{n^{c_P}}$-time.

- If E finds a Hamiltonian cycle in G_{P*}^t, then E sets the Stage-2 message of Phase-1 of the t-th session, denoted $((c_V^{(0)})^t, (c_V^{(1)})^t, a_V^t)$, as follows: it randomly chooses one random string $(e_V^{(0)})^t$ from $\{0,1\}^n$, computes $(c_V^{(0)})^t = Com(1^N, R_{P*}^t, (e_V^{(0)})^t)$ by using the underlying Naor's statistically-binding commitment scheme Com, and computes $(c_V^{(1)})^t = TCCom(1^N, (G_{P*}^t, R_{P*}^t), 0^n)$ by using the underlying Feige-Shamir trapdoor commitment scheme (note that, $(c_V^{(1)})^t$ commits to 0^n rather than a random string in $\{0,1\}^n$ as the honest verifier does). Then, on common input $(((y_{P*}^{(0)}, y_{P*}^{(1)})^t, G_{P*}^t, R_{P*}^t), PK)$ E computes the first-round message, denoted a_V^t, of (n-parallel repetitions of) Blum's WIPOK for $\mathcal{NP}$ for showing the knowledge of either SK or a Hamiltonian cycle in G_{P*}^t. *Note that the first-round message of Blum's WIPOK for $\mathcal{NP}$ is computed without using any witness knowledge (i.e., either SK or a Hamiltonian cycle in G_{P*}^t)*; In case P^* successfully finishes Phase-2 of the t-th session, E moves into Phase-3. After receiving the first-round message of Phase-3 of the t-th session, denoted e_V^t, E computes the second-round message of Phase-3, denoted z_V^t (i.e., the third-round message of Blum's WIPOK for showing the knowledge of either SK or a Hamiltonian cycle in G_{P*}^t), *by using the extracted Hamiltonian cycle in G_{P*}^t as its witness*; Finally, in Phase-4 of the t-th session, E decommits $(c_V^{(1)})^t$ to a random string $(e_V^{(1)})^t$ of length n, *by using the extracted Hamiltonian cycle in G_{P*}^t as the trapdoor*.
- If there exists *no* Hamiltonian cycle in G_{P*}^t, then E sets and sends the Stage-2 message of Phase-1 of the t-th session, i.e., $((c_V^{(0)})^t, (c_V^{(1)})^t, a_V^t)$, just as above. But, whenever P^* successfully finishes Phase-2 of the t-th session and sends to E the first-round message of Phase-3 of the t-th session (i.e., e_V^t), E aborts with an error message (as it has no witness for generating the next message).

Whenever P^* stops, E also stops and outputs the simulated transcript str (i.e., the view of P^* interacting with E). Denote by $view_{P*}^{E(PK)}(1^n, PK)$ the view of P^* (i.e., str) in the above run of $E(1^n, PK)$. We first establish that the simulated transcript is indistinguishable from the view of P^* in real execution with honest verifier instances. The purpose of E is to extract witnesses to all accepting sessions in str, which will be demonstrated later.

Lemma 1. *For any sufficiently large n, and for all (except for a negligible fraction of) (PK, SK) outputted by the key-generation stage of the honest verifier, the view of P^* in the run of $E(1^n, PK)$ (i.e., $view_{P*}^{E(PK)}(1^n, PK)$) is indistinguishable from the view of P^* in real execution with honest verifier instances.*

Proof (of Lemma 1). This is done by establishing a series of hybrid experiments.

We first consider a mental experiment in which P^* concurrently interacts with an imaginary verifier $\widehat{V}$ with the same public-key PK and secret-key SK. $\widehat{V}$ mimics the real honest verifier V with public-key PK and secret-key SK but with the following modifications: For any session t, $1 \leq t \leq s(N)$, in case P^* successfully finishes Phase-2 and sends to $\widehat{V}$ the first-round message of Phase-3, $\widehat{V}$ enumerates all possible Hamiltonian cycles of $G_{P^*}^t$ by brute-force searching in $2^{n^{c_P}}$-time, where $((y_{P^*}^{(0)}, y_{P^*}^{(1)})^t, G_{P^*}^t, R_{P^*}^t)$ is the Stage-1 message of Phase-1 of the t-th session. If there exists *no* Hamiltonian cycle in $G_{P^*}^t$, $\widehat{V}$ aborts with an error message, *although it can continue the execution with SK as its witness!*

Note that the only difference between the interactions between P^* and $\widehat{V}$ and the interactions between P^* and the real honest verifier V is that: for any session $t, 1 \leq t \leq s(N)$, the real honest verifier always continues the execution of Phase-3 by using SK as its witness in forming the second-round message of Phase-3, in case P^* successfully finished Phase-2 and sent the first-round message of Phase-3; but $\widehat{V}$ may abort in this case if it finds that $G_{P^*}^t$ is "false" (i.e. there exists no Hamiltonian cycle in $G_{P^*}^t$) by brute-force searching in $2^{n^{c_P}}$-time. That the view of P^* interacting with $\widehat{V}$ is indistinguishable from its view in real execution with honest verifier instances is from the following lemma.

Lemma 2. *For all positive polynomials $s(\cdot)$ and all s-concurrent malicious P^*, the probability that there exists a t, $1 \leq t \leq s(N)$, such that P^* can successfully finish Phase-2 with respect to a false $G_{P^*}^t$ (i.e., $G_{P^*}^t$ contains no Hamiltonian cycle) in the t-th session of the $s(N)$ concurrent sessions (against the real honest verifier V with public-key PK) is negligible in n.*

Proof (of Lemma 2). We show that if a PPT s-concurrent adversary P^* can convince V (with public-key PK) of a false $G_{P^*}^t$ with non-negligible probability $p'(n)$ in Phase-2 of one of the $s(N)$ concurrent sessions, then this will violate the hiding property of the underlying *statistically-binding* commitment scheme, denoted Com, used by V in Phase-1 that is run on security parameter N. Note that according to the hiding property of the underlying statistically-binding commitment scheme Com, given two strings $\hat{e}_0$ and $\hat{e}_1$ that are taken uniformly at random from $\{0,1\}^n$ and $C = Com(1^N, R_{P^*}^t, \hat{e}_b)$ for a randomly chosen bit $b \in \{0,1\}$, no $2^{N^{c_V}}$-time (non-uniform) algorithm can distinguish whether C commits to $\hat{e}_0$ or to $\hat{e}_1$ (i.e., guess the bit b correctly) with non-negligible advantage over $1/2$, even with $\hat{e}_0$, $\hat{e}_1$ and the secret-key of V (i.e., SK) as its non-uniform inputs.

We construct a (non-uniform) algorithm A that takes $(1^n, (\hat{e}_0, \hat{e}_1, SK), C)$ as input and attempts to guess b with a non-negligible advantage over $1/2$, where $\hat{e}_0$ and $\hat{e}_1$ are taken uniformly at random from $\{0,1\}^n$ and $C = Com(1^N, R_{P^*}, \hat{e}_b)$ for a randomly chosen bit $b \in \{0,1\}$. E randomly selects j from $\{1, \cdots, s(N)\}$, runs P^* as a subroutine by playing the role of the honest verifier V with secret-key SK in any session other than the j-th session. In the j-th session, after receiving $G_{P^*}^j$ from P^* at Stage-1 of Phase-1, E first checks whether there exists a Hamiltonian cycle in G_{P*}^j or not by brute-force searching in time $2^{n^{c_P}}$. If E finds a Hamiltonian cycle in G_{P*}^j, then E randomly guesses the bit b and stops. Otherwise (i.e., there exists no Hamiltonian cycle in G_{P*}^j), E runs P^* further and continues the interactions of the j-th session as follows: E gives C to P^* as the assumed commitment to $(e_V^{(0)})^j$ at Stage-2 of Phase-1. After receiving the first-round message of Phase-2 (i.e., the first-round of Blum's protocol for proving the existence of a Hamiltonian cycle in G_{P*}^j) that contains n committed adjacency matrices, E first opens all the committed adjacency matrices by brute-force in $poly(n) \cdot 2^{n^{c_P}}$-time (note that E can do this since the underlying statistically-binding commitment scheme used by the prover in forming these n committed adjacency matrices is run on security parameter n). For each revealed graph G_k^j $(1 \leq k \leq n)$ (described by the corresponding opened adjacency matrix entries) we say that G_k^j is a 0-valid graph if it is isomorphic to G_{P*}^j, or a 1-valid graph if it contains a Hamiltonian cycle of the same size of G_{P*}^j. We say that the set of revealed graphs $\{G_1^j, \cdots, G_n^j\}$ is $\hat{e}_b$-valid ($b \in \{0,1\}$) if for all k, $1 \leq k \leq n$, G_k^j is a $\hat{e}_b^{(k)}$-valid graph, where $\hat{e}_b^{(k)}$ denotes the k-th bit of $\hat{e}_b$. Note that for the set of revealed graphs $\{G_1^j, \cdots, G_n^j\}$, E can determine whether it is $\hat{e}_0$-valid or $\hat{e}_1$-valid in time $poly(n) \cdot 2^{n^{c_P}}$. Then, E outputs 0 if the set $\{G_1^j, \cdots, G_n^j\}$ is $\hat{e}_0$-valid but not $\hat{e}_1$-valid. Similarly, E outputs 1 if the set $\{G_1^j, \cdots, G_n^j\}$ is $\hat{e}_1$-valid but not $\hat{e}_0$-valid. In other cases, E just randomly guesses the bit b.

The key observation here is that if G_{P*}^j is false (i.e., containing no Hamiltonian cycle), then for each revealed graph it cannot be both a 0-valid graph and a 1-valid graph. Similarly, for false G_{P*}^j, the set of revealed graphs $\{G_1^j, \cdots, G_n^j\}$ cannot be both $\hat{e}_0$-valid and $\hat{e}_1$-valid for different $\hat{e}_0 \neq \hat{e}_1$. Furthermore, suppose C commits to $\hat{e}_b$ ($b \in \{0,1\}$), then for false G_{P*}^j with probability $1 - 2^{-n}$ the set of revealed graphs $\{G_1^j, \cdots, G_n^j\}$ is not $\hat{e}_{1-b}$-valid (since $\hat{e}_{1-b}$ is taken uniformly at random from $\{0,1\}^n$). Since the value j is randomly chosen from $\{1, \cdots, s(N)\}$, we conclude that E can successfully guess the bit b with probability at least $(1 - 2^{-n}) \cdot \frac{p'(n)}{s(N)} + \frac{1}{2}(1 - \frac{p'(n)}{s(N)}) = \frac{1}{2} + \frac{1}{2} \cdot \frac{p'(n)}{s(N)} - 2^{-n} \cdot \frac{p'(n)}{s(N)}$ in time $poly(n) \cdot 2^{n^{c_P}}$. That is, E successfully guesses the bit b with non-negligible advantage over $1/2$ in time $poly(n) \cdot 2^{n^{c_P}} \ll 2^{N^{c_V}}$, which violates the hiding property of the underlying statistically-binding commitment scheme Com used by V that is run on the security parameter N. This finishes the proof of Lemma 2.

Now, we want to show that the view of P^* with $\widehat{V}$ is indistinguishable the view of P^* with E. This is established by conducting another hybrid experiment.

Specifically, we consider the following hybrid experiment. An algorithm $\widehat{E}$ takes (PK, SK) as its input (that is, $\widehat{E}$ takes both the verifier's public-key and the corresponding secret-key as its input), and runs P^* as a subroutine by mimicking the knowledge-extractor E (who only takes PK as input) but with the following modification: For any session t, $1 \leq t \leq s(N)$, in case P^* successfully finishes Phase-2 and sends to $\widehat{E}$ the first-round message of Phase-3, $\widehat{E}$ enumerates all possible Hamiltonian cycles of G_{P*}^t by brute-force searching in $2^{n^{c_P}}$-time, where $((y_{P*}^{(0)}, y_{P*}^{(1)})^t, G_{P*}^t, R_{P*}^t)$ is the Stage-1 message of Phase-1 of the t-th session. If there *exists* a Hamiltonian cycle in G_{P*}^t, then $\widehat{E}$ continues the execution by forming the second-round message of Phase-3 of the t-th session (for showing the knowledge of either SK or a Hamiltonian cycle of G_{P*}^t) *but using SK as its witness just as the real honest verifier does* (note that in this case E continues the execution with the extracted Hamiltonian cycle of G_{P*}^t as the corresponding witness). If there exists *no* Hamiltonian cycle in G_{P*}^t, then $\widehat{E}$ aborts with an error message just as E (or $\widehat{V}$) does (*although in this case $\widehat{E}$ can continue the execution with SK as its witness*).

Note that the difference between the interactions between P^* and the imaginary verifier $\widehat{V}$ in the first hybrid experiment and the interactions between P^* and $\widehat{E}$ is that: in any session t, $1 \leq t \leq s(N)$, of the interactions between P^* and $\widehat{V}$, $\widehat{V}$ always commits (and accordingly decommits to) a random string of length n (i.e., $(e_V^{(1)})^t$) by using the underlying FSTC scheme (just as the honest verifier V does), but in the interactions between P^* and $\widehat{E}$, $\widehat{E}$ always commits 0^n and then decommits to a random string of length n by using the brute-force extracted Hamiltonian cycle of G_{P*}^t as the trapdoor (just as E does). We show the view of P^* with $\widehat{E}$ is indistinguishable from the view of P^* with $\widehat{V}$. Otherwise, by hybrid arguments, we can construct a $poly(n) \cdot 2^{n^{cP}} \ll 2^{N^{cV}}$-time algorithm that breaks the hiding and trapdoorness properties of FSTC.

The difference between the interactions between P^* and $\widehat{E}$ and the interactions between P^* and E is that: E always uses the brute-force extracted Hamiltonian cycle of G_{P*}^t as its witness in Phase-3 of any session t, $1 \leq t \leq s(N)$, but $\widehat{E}$ always uses the verifier's secret-key SK as its witness (just as the honest verifier does). Similarly, the view of P^* with $\widehat{E}$ is indistinguishable from the view of P^* with E is indistinguishable. Otherwise, by hybrid arguments, we can break the WI property of Blum's protocol for $\mathcal{NP}$ in time $poly(n) \cdot 2^{n^{cP}} \ll 2^{N^{cV}}$. This finishes the proof of Lemma 1.

Now, E wants to extract the corresponding witness to each accepting session in the simulated transcript str. For any t, $1 \leq t \leq s(N)$, suppose the t-session is accepting in str, we define an experiment E_t that emulates E with the fixed random coins of E, but with the following exception: the random n-bit string $(e_V^{(1)})^t$ (i.e., the decommitted value to $(c_V^{(1)})^t$) is no longer emulated internally, but received externally. Note that the experiment E_t actually amounts to the *stand-alone* execution of the Blums's WIPOK of Phase-4 on common input x_t between a (stand-alone) sub-exponential-time prover (combining all internal emulation of E with running P^* as the subroutine, except for $(e_V^{(1)})^t$ to be received externally) and a public-coin honest verifier that sends $(e_V^{(1)})^t$. By applying the stand-alone knowledge-extractor on E_t, except for the probability 2^{-n} we can get one of the following within time $poly(n) \cdot 2^n \ll 2^{n^{cP}}$ (actually, within expected polynomial-time): a witness w_t for $x_t \in L$ or the corresponding secret-key SK such that $PK = f_V(SK)$. As E_t runs in $poly(n) \cdot 2^{n^{cP}}$-time, we conclude that E can extract either w_t or SK within time $poly(n) \cdot 2^{n^{cP}}$ in total. As $poly(n) \cdot 2^{n^{cP}} \ll 2^{N^{cV}}$ and f_V is secure against any $2^{N^{cV}}$-time adversary, we know with overwhelming probability (except for a negligible fraction of (PK, SK)'s output by the key-generation stage of V) the extracted witness must be w_t. This means within time $poly(n) \cdot 2^{n^{c_P}} \ll 2^{N^{c_V}}$ E will output all witnesses to common inputs of accepting sessions in str with overwhelming probability.

As the simulated transcript str is indistinguishable from the view of P^* in real execution with honest verifier instances, this implies that for any x and for all (except for a negligible fraction of) (PK, SK) outputted by the key-generation stage of V, if P^* can convince the honest verifier $V(SK)$ of "$x \in L$" in one of the $s(N)$ sessions with non-negligible probability p_x, then P^* will also convince $E(PK)$ of this statement with probability negligibly close to p_x. According to the knowledge-extraction ability of E, E will output a witness to $x \in L$ with probability negligibly close to p_x. Then, the concurrent soundness of the protocol depicted in Figure 1 follows. This finishes the proof of Theorem 1.

3.1   Discussion: On the Weak Concurrent Knowledge-Extractability

We remark that the above proof for concurrent soundness actually establishes a (*black-box*) weak CKE property, roughly as follows: there exists a *sub-exponential-time* (specifically, $poly(n) \cdot 2^{n^{c_P}}$-time) *black-box* simulator/extractor E such that for any concurrent malicious PPT prover P^* against verifier instances with public-key PK, on the *same* public-key PK E outputs a simulated indistinguishable transcript, together with all witnesses to common inputs of accepting sessions in str. Note that, according to the parameter specifications in Section 3, $poly(n) \cdot 2^{n^{c_P}} \ll 2^{N^c} \ll 2^{N^{c_L}}$. Suppose the underlying language L is $2^{N^{c_L}}$-hard for some constant c_L, $0 < c_L < 1$, such weak CKE property essentially says that P^* "knows" witnesses to common inputs whose validations are successfully conveyed by concurrent interactions, rather than only convincing the verity (i.e., membership) of common inputs. Formal formulation of the weak CKE property and detailed discussions are deferred to the full version (in particular, the weak CKE property is strictly stronger than concurrent soundness in the public-key model under any sub-exponentially strong OWF).

We remark that super-polynomial-time is intrinsic to *black-box* knowledge-extraction for rZK arguments, as rZK (black-box) arguments of knowledge exist only for $\mathcal{BPP}$ languages [1]. Also, we believe that the weak CKE property is still very useful in practice. In particular, it allows highly practical rZK implementations for specific languages, e.g., DLP and RSA, that are widely assumed to be sub-exponentially hard.

4   Simplified, Practical, Round-Optimal Implementations

4.1   Simplified Implementation

We further investigate the interactions combining Phase-1 and Phase-2 of the OWF-based rZK-CS protocol (depicted in Figure 1) when the messages $c_V^{(1)}$ and a_V are removed from Stage-2 of Phase-1 (i.e., V only sends $c_V^{(0)}$ at Stage-2 of Phase-1). The key observation here is that if the OWF f_P used by the prover is *preimage-verifiable*, then such interactions can be replaced by only letting P

send, at the start, the initialization messages (y_P, G_P, R_P): a *unique* value $y_P = f_P(x_P)$ (rather than a pair of values $(y_P^{(0)}, y_P^{(1)})$), the graph G_P (reduced from y_P by $\mathcal{NP}$-reduction) and the random string R_P. Note that the initialization messages (y_P, G_P, R_P) is fixed once and for all. Thereby, we obtain a much more simplified 5-round implementation. In this case, the proof of Theorem 1 remains essentially unchanged (other than being simplified). *We remark that the preimage-verifiability property plays a critical role in the proof of concurrent soundness, as otherwise the malicious P^* can distinguish whether it is interacting with honest verifier instances (who always continue the interactions w.r.t. a false G_{P^*} in which no Hamiltonian cycle exists) or with the knowledge extractor (who always stops by brute-force checking the validity of G_{P^*}).*

4.2 Generic Yet Practical Transformation

We first recall some key tools used in the generic practical transformation: We assume the OWF f_V used in key-generation admits Σ-protocols. Note that the set of OWFs admitting Σ-protocols is large, which in particular includes the popular DLP and RSA functions [20,14]. The PRF used by the prover is the Naor-Reingold PRFs that can be based on the factoring (Blum integers) or the decisional Diffie-Hellman hardness assumptions [18]. The computational complexity of computing the value of the Naor-Reingold functions at a given point is about two modular exponentiations and can be further reduced to only two multiple products modulo a prime (without any exponentiations!) with natural preprocessing, which is great for practices involving PRFs.

Verifiable and Σ-provable trapdoor commitments (VPTC). For our purpose, we need TC schemes satisfying the following additional requirements:

- Public-key verifiability. The validity of $TCPK$ (even generated by a malicious commitment receiver) can be efficiently verified. In particular, given any $TCPK$, one can efficiently verify whether or not $TCSK$ exists. *Actually, in the generic practical transformation the public-key verifiability property just serves the role of preimage-verifiable OWF in the above preimage-verifiable OWF-based simplified implementation.*
- Public-key Σ-provability. On common input $TCPK$ and private input $TCSK$, one can prove, by Σ-protocols, the knowledge of $TCSK$.

The first round of a VPTC scheme is denoted by $VPTCPK$ and the corresponding trapdoor is denoted by $VPTCSK$. We note both the DLP-based [4] and the RSA-based [19] *perfectly-hiding* trapdoor commitment schemes are VPTC.

The generic practical transformation from any Σ-protocol. We highlight the modifications, in comparison with the preimage-verifiable OWF-based simplified implementation. The generic practical implementation is for any language L that admits Σ-protocols. The RRF is replaced by Naor-Reingold PRF; The OWF f_V used in key-generation stage is replaced by any OWF admitting Σ-protocols; The trapdoor commitment scheme is replaced by the VPTC scheme, and the sending of the y_P using the preimage-verifiable OWF f_P is just replaced by the sending of $VPTCPK$ on the top (note that we no longer need to reduce $VPTCPK$ to a Hamiltonian Graph by $\mathcal{NP}$-reductions); All WI protocols are replaced by Σ_{OR}-protocols (without $\mathcal{NP}$-reductions).

4.3   Round-Optimal Implementation

For the above 5-round preimage-verifiable OWF simplified implementation, to further reduce the round-complexity, we want to fold the prover's initialization message, i.e., (y_P, G_P, R_P), into the third-round of the 5-round protocols (that is from the prover to the verifier). This would render us 4-round (that is optimal) rZK-CS arguments for $\mathcal{NP}$ in the BPK model. To this end, we let the verifier use OWP-based one-round perfectly-binding commitment scheme at Stage-2 of Phase-1 (thus waiving the value R_P), and replace the Blum's WIPOK protocol (executed on common input (y_P, G_P, y_V) with the verifier playing the role of knowledge prover) by the Lapidot-Shamir WIPOK protocol (as in this case the verifier sends the first-round message without knowing the statement, i.e, (y_P, G_P, y_V), to be proved). But, the challenge here is that, for our purpose, we need the following cryptographic tool (to replace the two-round FSTC scheme): A *one-round* OWP-based trapdoor commitment scheme based on DHC, in which the committer sends the one-round commitments without knowing the graph G_P (serving as $TCPK$) other than the lower and upper bounds of its size (guaranteed by the underlying $\mathcal{NP}$-reduction from y_P to G_P), and G_P is only sent in the decommitment stage after the commitment stage is finished. We develop a trapdoor commitment scheme of this type in this work, described below:

One-round commitment stage. To commit a bit 0, the committer sends a q-by-q adjacency matrix of commitments with each entry of the adjacency matrix committing to 0. To commit a bit 1, the committer sends a q-by-q adjacency matrix of commitments such that the entries committing to 1 constitute a randomly-labeled cycle C. We remark that the underlying commitment scheme used in this stage is the one-round OWP-based perfectly-binding commitment scheme.

Two-round decommitment stage. The commitment receiver sends a Hamiltonian graph $G = (V, E)$ with size $q = |V|$ to the committer. Then, to decommit to 0, the committer sends a random permutation π, and for each non-edge of G $(i, j) \notin E$, the committer reveals the value (that is 0) that is committed to the $(\pi(i), \pi(j))$ entry of the adjacency matrix sent in the commitment stage (and the receiver checks all revealed values are 0 and the unrevealed positions in the adjacency matrix constitute a graph that is isomorphic to G via the permutation π). To decommit to 1, the committer only reveals the committed cycle (and the receiver checks that all revealed values are 1 and the revealed entries constitute a q-cycle).

Acknowledgments. We thank Di Crescenzo, Persiano and Visconti for helpful discussions.

References

1. B. Barak, O. Goldreich, S. Goldwasser and Y. Lindell. Resettably-Sound Zero-Knowledge and Its Applications. In *IEEE Symposium on Foundations of Computer Science*, pages 116-125, 2001.
2. M. Bellare, M. Fischlin, S. Goldwasser and S. Micali. Identification protocols secure against reset attacks. In *B. Pfitzmann (Ed.): Advances in Cryptology-Proceedings of EUROCRYPT 2001, LNCS 2045*, pages 495–511. Springer-Verlag, 2001.
3. M. Blum. How to Prove a Theorem so No One Else can Claim It. In Proceedings of the International Congress of Mathematicians, pages 1444-1451, 1986.
4. Brassard, D. Chaum and C. Crepeau. Minimum Disclosure Proofs of Knowledge. *Journal of Computer Systems and Science*, 37(2): 156-189, 1988.
5. R. Canetti, O. Goldreich, S. Goldwasser and S. Micali. Resettable Zero-Knowledge. In *ACM Symposium on Theory of Computing*, pages 235-244, 2000.
6. R. Canetti, J. Kilian, E. Petrank and A. Rosen. Black-Box Concurrent Zero-Knowledge Requires (Almost) Logarithmically Many Rounds. In *SIAM Journal on Computing*, 32(1): 1-47, 2002.
7. R. Cramer, I. Damgard and B. Schoenmakers. Proofs of Partial Knowledge and Simplified Design of Witness Hiding Protocols. In *Y. Desmedt (Ed.): Advances in Cryptology-Proceedings of CRYPTO 1994, LNCS 839*, pages 174-187, 1994.
8. I. Damgard. Lecture Notes on Cryptographic Protocol Theory, Aarhus University.
9. G. Di Crescenzo, G. Persiano and I. Visconti. Constant-Round Resettable Zero-Knowledge with Concurrent Soundness in the Bare Public-Key Model. In *M. Franklin (Ed.): Advances in Cryptology-Proceedings of CRYPTO 2004, LNCS 3152*, pages 237-253, Springer-Verlag, 2004.
10. C. Dwork, M. Naor and A. Sahai. Concurrent Zero-Knowledge. In *ACM Symposium on Theory of Computing*, pages 409-418, 1998.
11. U. Feige and Shamir. Zero-Knowledge Proofs of Knowledge in Two Rounds. In *G. Brassard (Ed.): Advances in Cryptology-Proceedings of CRYPTO 1989, LNCS 435*, pages 526-544, Springer-Verlag, 1989.
12. O. Goldreich. *Foundation of Cryptography-Basic Tools*. Cambridge University Press, 2001.
13. S. Goldwasser, S. Micali and C. Rackoff. The Knowledge Complexity of Interactive Proof-Systems In *ACM Symposium on Theory of Computing*, pages 291-304, 1985.
14. L. Guillou and J. J. Quisquater. A Practical Zero-Knowledge Protocol Fitted to Security Microprocessor Minimizing both Transmission and Memory. In *C. G. Gnther (Ed.): Advances in Cryptology-Proceedings of EUROCRYPT 1988, LNCS 330* , pages 123-128, Springer-Verlag, 1988.
15. D. Lapidot and A. Shamir. Publicly-Verifiable Non-Interactive Zero-Knowledge Proofs. In *A.J. Menezes and S. A. Vanstone (Ed.): Advances in Cryptology-Proceedings of CRYPTO 1990, LNCS 537*, pages 353-365. Springer-Verlag, 1990.
16. S. Micali and L. Reyzin. Soundness in the Public-Key Model. In *J. Kilian (Ed.): Advances in Cryptology-Proceedings of CRYPTO 2001, LNCS 2139*, pages 542–565, Springer-Verlag, 2001.
17. M. Naor. Bit Commitment Using Pseudorandomness. *Journal of Cryptology*, 4(2): 151-158, 1991.
18. M. Naor and O. Reingold. Number-Theoretic Constructions of Efficient Pseudo-Random Functions. *Journal of the ACM*, 1(2): 231-262 (2004).
19. T. Okamoto. Provable Secure and Practical Identification Schemes and Corresponding Signature Schemes In *E. F. Brickell (Ed.): Advances in Cryptology-Proceedings of CRYPTO 1992, LNCS 740*, pages 31-53, Springer-Verlag, 1992.

20. C. Schnorr. Efficient Signature Generation by Smart Cards. *Journal of Cryptology*, 4(3): 24, 1991.
21. A. C. C. Yao, M. Yung and Y. Zhao. Concurrent Knowledge-Extraction in the Public-Key Model. Manuscript, 2007.
22. Y. Zhao, X. Deng, C. H. Lee and H. Zhu. Resettable Zero-Knowledge in the Weak Public-Key Model. In *E. Biham (Ed.): Advances in Cryptology-Proceedings of EUROCRYPT 2003, LNCS 2656* , pages 123-140, Springer-Verlag, 2003.

Instance-Dependent Verifiable Random Functions and Their Application to Simultaneous Resettability*

Yi Deng and Dongdai Lin

The state key laboratory of information security,Institute of software,
Chinese Academy of sciences, Beijing, 100080, China
{ydeng,ddlin}@is.iscas.ac.cn

Abstract. We introduce a notion of *instance-dependent verifiable random functions* (InstD-VRFs for short). Informally, an InstD-VRF is, in some sense, a verifiable random function [23] with a special public key, which is generated via a (possibly)*interactive* protocol and contains an instance $y \in L \cap \{0,1\}^*$ for a specific NP language L, but the security requirements on such a function are relaxed: we only require the *pseudorandomness* property when $y \in L$ and only require the *uniqueness* property when $y \notin L$, instead of requiring both pseudorandomness and uniqueness to hold *simultaneously*. We show that this notion can be realized under standard assumption.

Our motivation is the conjecture posed by Barak et al.[2], which states there exist resettably-sound resettable zero knowledge arguments for NP. The instance-dependent verifiable random functions is a powerful tool to tackle this problem. We first use them to obtain two interesting instance-dependent argument systems from the Barak's public-coin bounded concurrent zero knowledge argument [1], and then, we

1. Construct the *first* (constant round) zero knowledge arguments for NP enjoying a *certain* simultaneous resettability under standard hardness assumptions in the plain model, which we call bounded-class resettable ZK arguments with weak resettable-soundness Though the malicious party (prover or verifier) in such system is limited to a kind of bounded resetting attack, We put NO restrictions on the number of the total resets made by malicious party.
2. show that, under standard assumptions, if there exist public-coin concurrent zero knowledge arguments for NP, there exist the resettably-sound resetable zero knowledge arguments for NP.

Keywords: instance-dependent verifiable random functions, simultaneous resettability, zero knowledge.

1 Introduction

Pseudorandom functions, introduced by Goldreich, Goldwasser and Micali [14], are basic cryptographic primitives and have been used in a wide range of

---

* This work is supported by the National Natural Science Foundation of China under Grant No. 60673069.

M. Naor (Ed.): EUROCRYPT 2007, LNCS 4515, pp. 148–168, 2007.

cryptographic applications. Loosely speaking, pseudorandom functions are efficient functions that cannot be tell apart from truly random functions by any polynomial-time observer that given a black-box access to those functions.

In some applications, the seed (the description of a specific pseudorandom function) owner needs to convince the observer (querier) that his reply to the the observer's query is correctly computed in order to protect the observer. To serve this need, Micali et al. put forward the concept of verifiable random functions [23]. Informally, a verifiable random function is described by a public/secret key pair, and its output consists of two part, a pseudorandom value and a proof that proving this value is correct. The security requirements for such a function are: 1)uniqueness. Except with a exponentially small probability, there is only one value for a fixed query can be proved correct with respect to the public key; 2)pseudorandomness. After several queries, a polynomial-time observer can not distinguish between a value that is computed by evaluating the function on his new query and a value picked at random *without the help of proof of correctness*. These distinguishing features make it useful in protocol design, as demonstrated in [21, 22].

Zero knowledge (ZK for short) proof [16, 15], a proof that reveals nothing but the validity of the assertion, is another fundamental tool in design of cryptographic protocols. In recent years, several notable notions have emerged to capture some new security concerns that arise in modern maliciously asynchronous communication environment, such as concurrent ZK[11]and universal composable ZK[7]. An other notable concept is resettable ZK (rZK) introduced by Canetti et al.[8]. The rZK formalizes security in a scenario in which the verifier is allowed to reset the prover in the middle of proof to any previous stage. From the randomness point of view, this notion is a strongest security measure. Obviously the notion of rZK is stronger than that of concurrent ZK and therefore we can not construct a constant round black-box rZK protocol in the plain model for non-trivial languages[9].

Following the above work, Barak et al. [2] initiated the study of soundness in a setting where the prover can reset the honest verifier, and showed that the public-coin constant round ZK argument of knowledge [1, 3] can be easily transformed into constant-round resettably-sound ZK argument of knowledge. Barak et al. also made a fascinating conjecture in [2]: there exist resettably-sound resettable ZK arguments for NP. Unfortunately, no progress on this conjecture has been made so far. The known results either achieved only resettable zero knowledge, such as the (non-constant round) resettable ZK proof system of [8] and the constant round public-coin bounded-resettable ZK argument system (we call it BLV's protocol) of [5], or achieved only resettable-soundness, such as resettably-sound ZK argument system of [2].

It is shown that psedurandom functions are crucial ingredients in the constructions of all known rZK or resettably-sound ZK protocols. However, the pseudorandom functions, even the stronger primitive of verifiable random functions, seem not powerful enough to tackle the simultaneous resettability problem, and this lead us to develop the new primitive—instance-dependent verifiable random functions (InstD-VRFs for short).

Motivation behind InstD-VRFs. Let's return to the resettably-sound ZK argument in [2]. We first note that if we modify this protocol in such a way that the verifier's messages (except for the first one) satisfy some kind of binding property, for example, the verifier's responses are determined by the first verifier's message and the history messages so far, then the resetting attack from the malicious verifier can be trivialized if the verifier do not reset the prover past its first message, therefore it seems the resulting protocol achieves some certain simultaneous resettability. Apparently, This can be done by plugging a verifiable random function in Barak's public-coin bounded concurrent ZK protocol: the first verifier's message includes a public key of verifiable random function along with the description of a hash, and all subsequent verifier's messages are computed by applying the verifiable random function to the history messages in a session.

At a first look, the above resulting protocol enjoys certain desirable simultaneous resettability: besides achieving a stronger ZK property than bounded resettable ZK that is achieved in the BLV protocol [5], it seems to be resettably-sound due to the pseudorandomness of the verifiable random function. Indeed, we can prove this protocol is ZK against somewhat restricted resetting verifier. However, we do not know how to prove the soundness in standard way (i.e., proving by reduction), let alone the resettably-soundness, because for the analysis of soundness to go through we typically need some freedom in verifier's responses to the same history of messages (i.e., verifier can choose different messages to reply the same history of messages)in the WI universal argument of Barak's protocol, and unfortunately, the uniqueness of the verifiable random functions precludes this possibility.

Inspired by the previous interesting instance-dependent commitment scheme [19, 20], We introduce the notion of instance-dependent verifiable random functions (InstD-VRFs), to achieve a certain simultaneous resettability. Informally, an InstD-VRF is, in some sense, a verifiable random function [23] with a special public key, which is generated via a (possibly)*interactive* protocol and contains an instance $y \in L \cap \{0,1\}^*$ for a specific NP language L, but the security requirements on such a function are relaxed: we only require the *pseudorandomness* property when $y \in L$ and only require the *uniqueness* property when $y \notin L$, instead of requiring both pseudorandomness and uniqueness to hold *simultaneously*. The reason why such functions are useful is that we use only the uniqueness to justify resettable ZK property and use only the pseudorandomness to justify resettable-soundness.

Our contributions. In this paper We introduce a notion of *instance-dependent verifiable random functions* and realize them under standard assumption. These functions yields two interesting instance-dependent argument systems, which we call key_instance-dependent resettably-sound bounded-class resettable ZK argument and resettable witness indistinguishable argument with instance-dependent weak resettably-soundness. These results extend the study of instance-dependent primitives (protocols).

The instance-dependent verifiable random functions, together with the above instance-dependent protocols, are powerful tools to tackle the simultaneous resettability conjecture. By using them, we construct the first (constant round) zero knowledge arguments for NP enjoying *certain* simultaneous resettability in the plain model. In our argument system, both the prover and the verifier are protected from some kinds of *restricted* resetting attacks: For the malicious resetting prover, we put a priori bound on the number of the first messages sent by it to each incarnation of the honest verifier, and for the malicious resetting verifier, in addition to putting the aforementioned restriction on it, We further put a priori bound on the number of incarnations of prover with which it is allowed to interact. We call this protocol the bounded-class resettable ZK argument with weak resettable-soundness. We stress our arguments assume standard (polynomial time) hardness assumptions and their resettable security for the prover (the verifier) does not rely on any restriction on the number of the total resets made by malicious verifiers (provers). This is in contrast to the BLV protocol [5] (that achieves only standard soundness), which relies on a priori bound of the total resets made by malicious verifiers and exponential hardness assumptions.

We also show that, if there exist public-coin concurrent ZK arguments for NP, the idea behind the above construction can be applied to the *unbounded* simultaneously resettable setting, and this leads to resettably-sound resetable ZK arguments for NP.

The resettable witness indistinguishable argument with instance-dependent weak resettably-soundness is a crucial and delicate component in our main constructions (in section 5), in which both the prover and the verifier use a instance-dependent verifiable random function. For the prover, the instance to be proven serves as the key_instance for its InstD-VRF directly. The most interesting but difficult task is to produce a key_instance for the verifier's InstD-VRF. Our solution to this problem is to have the prover generate a NO key_instance with respect to a hard-to-decide language for the verifier and prove to the verifier that the statement to be proven is true or this key_instance is an YES instance via a resettable witness indistinguishable argument with instance-dependent weak resettably-soundness. A glaring property of this argument is that it is *argument of knowledge* when the statement to be proven (in the global system) is false, and this is the crux in the analysis of the soundness of our main argument systems presented in section 5.

Subsequent work. The instance-dependent verifiable random functions seem to have potential beyond what we demonstrate in this paper. Very recently, by using these functions in a novel way, we construct resettable witness indistinguishable argument with instance-dependent *unbounded* (in contrast to "*weak*") resettably-soundness, and this immediately yields a (unbounded) resettably-sound bounded-class resettable ZK argument, which gets close to the simultaneous resettability conjecture.

Outline. The definition of bounded-class resettable ZK and weak resettable-soundness are presented in section 2. In section 3, we introduce the notion of instance-dependent verifiable random functions and show a construction of this primitive under standard assumption. The application of the new primitive are described in section 4, 5. In section 4 we present two interesting instance-dependent protocols that are crucial building block for our construction of bounded-class resettable ZK argument with weak resettable-soundness, and the latter argument, along with a sufficient condition for the simultaneous resettability conjecture, are presented in section 5.

2   Definitions

In this section we mainly define bounded-class resettable ZK and weak resettable-soundness, which can be viewed as intermediate notions between their full resettable analogues and bounded resettable analogues. Due to lack of space, we refer readers to [13] for some basic concepts, such as computational indistinguishability, (statistically-biding) commitment scheme, hybrid argument, and so on.

In the following We denote by $\delta \leftarrow_R \Delta$ the process of picking a random element δ from Δ, and abbreviate probabilistic polynomial time as PPT. A function $f(n)$ is said to be negligible if for every polynomial $q(n)$ there exists an N such that for all $n \geq N$, $f(n) \leq 1/q(n)$.

We follows the standard definition of zero knowledge argument in [13]. Note that for such a protocol the soundness is required to hold only against PPT adversaries.

Resettable prover and bounded-class resettable ZK. Resettable ZK was introduced in [8]. In essence, it guarantees the security of a prover with fixed random tape in a scenario the verifier is allowed to run polynomial number sessions with this fixed prover.

We introduce the notion of *bounded-class* resettable ZK. We call a fixed prover strategy $P^{(i,j)} = P_{x_i,w_i,r_j}$ an incarnation. We categorize all sessions between a verifier and a fixed incarnation of prover into a *class* if they share the *same* verifier's first message msg. We denote a class associated with the incarnation $P^{(i,j)}$ and the verifier's first message msg with $\text{Class}_{P^{(i,j)},\text{msg}}$. *Note that it is possible that a class contains (unbounded) any polynomial number sessions because the verifier is allowed to reset the prover.*

Definition 1. *(Bounded-class resettable ZK argument) Let t be a polynomial. An interactive argument (P, V) for a language L is said to be t^3-bounded-class resettable ZK if for every every PPT adversary V^* there exists a PPT M so that the following two distributions are computational indistinguishable, where each distribution is indexed by a sequence of distinct common inputs $\overline{x} = x_1, \cdots, x_t \in L \cap \{0,1\}^n$ and a corresponding sequence of prover's auxiliary inputs $\overline{w} = w_1, \cdots, w_t,$*

Distribution 1. *is defined by the following random process depending on P and V.*

1. *Randomly pick and fix t random tapes, $r_1, \cdots, r_t$, resulting in t^2 deterministic incarnations $P^{(i,j)} = P_{x_i, w_i, r_j}$ defined by $P_{x_i, w_i, r_j}(\alpha) = P(x_i, w_i, r_j, \alpha)$, for $(i, j) \in \{1, \cdots, t\} \times \{1, \cdots, t\}$.*

2. *The adversary V^* is allowed to run polynomial many sessions with the $P^{(i,j)}$'s, but for each $P^{(i,j)}$, the verifier can not reset $P^{(i,j)}$ past its first message more than $t - 1$ times, that is, the number of different V^*'s first message to each incarnation $P^{(i,j)}$ is a priori bounded by t. Under this restriction, the verifier is allowed to schedule all sessions in interleaving way: V^* can send arbitrary messages to each of the $P^{i,j}$, and obtain the responses of $P^{(i,j)}$ to such messages immediately.*
   This results in at most t^3 classes in the whole interaction.

3. *Once V^* decides it is done interacting with the $P^{(i,j)}$'s, it produces an output based on its view of the whole interaction. We denote this output by $(P(\overline{w}), V^*)(\overline{x})$.*

Distribution 2. *is the output of $M(\overline{x})$.*

The resetting attack performed by the restricted malicious verifier in the above definition is called *bounded-class resetting attack*. We stress there is essential difference between bounded-resettable ZK [5] and bounded-class resettable ZK: We impose no restriction on the number of the total resets (sessions) that malicious verifiers can make.

Resettable verifier and weak resettable-soundness. Following the definitions of resettablly-sound arguments in [2], we consider a *weak resetting attack*, in which a malicious prover is not allowed to reset an incarnation of the verifier past its first message more than $t - 1$ times, but still can interact with *arbitrary polynomial number* of verifier's incarnations. This kind of attack corresponds to the following notion of soundness.

Definition 2. *(Weak resettably-sound argument of knowledge.) for some a-priori fixed polynomial t, a weak resetting attack of a malicious prover P^* on a resettable verifier V is defined by the following random process, indexed by a security parameter n.*

1. *Uniformly picks and fix poly(n) random-tapes, denoted $r_1, \cdots, r_{poly(n)}$, for V, resulting in deterministic strategies $V^{(j)}(x) = V_{x, r_j}$, $x \in \{0, 1\}^n$ and $j \in \{1, \cdots, poly(n)\}$, defined by $V_{x, r_j}(\alpha) = V(x, r_j, \alpha)$. We call each $V^{(j)}(x)$ an incarnation of V.*

2. *Taking as input 1^n, P^* is allowed to initiate any polynomial number sessions with the $V^{(j)}(x)$'s, but the number of different P^*'s first message to each incarnation $V^{(j)}(x)$ is a priori bounded by t. Under this restriction, the prover P^* is allowed to schedule all sessions in interleaving way as usual: P^* can send arbitrary messages to each of the $V^{(j)}(x)$, and obtain the responses of $V^{(j)}(x)$ to such messages immediately.*

We say an argument system (P, V) is a weak resettably-sound argument of knowledge system if it satisfies:

1. *Resttable-completeness: Considering an arbitrary resetting attack of a PPT P^*. If P^* follows the strategy of P in some sessions after selecting an incarnation $V^{(j)}(x)$ and $x \in L$, then $V^{(j)}(x)$ rejects with negligible probability.*
2. *weak Resettably-soundness: For every weak resetting attack of a PPT P^*, the probability that in some sessions the corresponding $V^{(j)}(x)$ has accepted an false statement $(x \notin L)$ is negligible.*
3. *Argument of knowledge: For every PPT P^*, there exists a PPT machine E such that for every weak resetting attack of P^*, the probability that E, upon input the description of P^*, outputs a witness for the statement in a session is negligibly close to the probability that P^* convinces V in a session.*

We remark that in a weak resetting attack the malicious prover is entitled to interacts with *unbounded* number of incarnations of the verifier. So we have four types of resetting attack, full resetting attack, weak resetting attack, bounded-class resetting attack and bounded resetting attack [5], each of which is more powerful than the previous one.

Resettably-sound resettable WI. Roughly speaking, witness indistinguishability arguments [12] are arguments with property that nobody can tell which witness was used to prove a statement in an interaction. Analogously, we define Resettably-sound resettable WI (cf. [2]), and its variants according to our restrictions on the number of class and/or the number of a malicious party's first messages to each incarnation of the honest party. Due to space limitations, we omit it here.

A note on terminology. Let A be a security property or a type of attack. In the rest of the paper, the notion "unbounded A" means an unrestricted version of A.

3   Instance-Dependent Verifiable Random Functions

In this section we will present the formal definition of instance-dependent verifiable random functions and show how to implement it.

3.1   InstD-VRFs: Definition

As is hinted by its name, the public key for a instance-dependent verifiable function contains a instance $y \in L \cap \{0, 1\}^*$ for a NP language L, and unlike the verifiable random functions, whose *pseudorandomness* and *uniqueness* are required to hold at the same time, we require an instance-dependent verifiable random function satisfies only the *pseudorandomness* when $y \in L$, and satisfies only the *uniqueness* when $y \notin L$.

Let $d, l : N \rightarrow N$ be two polynomial. Formally, an instance-dependent verifiable random function with respect to an NP language L associates with the following (interactive) algorithms:

- KGProt, the key generation protocol between two parties, the querier A and the owner of the function B, each party taking security parameter n and an random string as input, produces a public/secret key pair (PK, SK), and PK is of form $(y, \cdot)$, where $y \in L \cap \{0,1\}^n$ is called key_instance.
- $F = (f, prov)$, the function evaluator, the first component is a deterministic algorithm while the second component $prov$ is a $probabilistic$ algorithm. Given a (PK, SK), on input an element $a \in \{0,1\}^{d(n)}$ (the domain of f_{SK}), it outputs a function value $b \in \{0,1\}^{l(n)}$ (the range of f_{SK}) and a proof π. That is, $F_{(PK,SK)}(a) = (f_{SK}(a), prov(a, f_{SK}(a), PK, SK)) = (b, \pi)$, where $F_{(PK,SK)}(\cdot) = F(PK, SK, \cdot)$.
- Ver, the verification (deterministic) algorithm, on input a, b, PK and a proof π, Ver outputs 1 or 0.
- FakeF, the $fake$ function evaluator. Assume y is in L. Given PK and a witness w_y for $y \in L$, for every $a \in \{0,1\}^{d(n)}$, $\mathsf{FakeF}_{(PK,w_y)}$ can validate an arbitrary false function value, i.e., for an arbitrary $b \in \{0,1\}^{l(n)}$, taking a as input, $\mathsf{FakeF}_{(PK,w_y)}$ can output $(b, prov(a, b, PK, w_y)) = (b, \pi)$ such that $\mathsf{Ver}(a, b, PK, \pi) = 1$.

The property of the fake function evaluator guarantees for a function $h : \{0,1\}^{d(n)} \rightarrow \{0,1\}^{l(n)}$ that deviates arbitrarily from the function f_{SK} specified by the secret key, We can run the algorithm $prov$ using w_y to produce a valid proof of correctness for the function value $h(a)$. We define the following two useful fake functions:

- $\mathsf{FakeF}_{(PK,w_y,s)}(a) \triangleq (f_s(a), prov(a, f_s(a), PK, w_y))$, where $f_s : \{0,1\}^{d(n)} \rightarrow \{0,1\}^{l(n)}$ is an arbitrarily (independent of f_{SK}) pseudorandom function.
- $\mathsf{FakeF}_{(PK,w_y,h)}(a) \triangleq (h(a), prov(a, h(a), PK, w_y))$, where $h : \{0,1\}^{d(n)} \rightarrow \{0,1\}^{l(n)}$ is an arbitrarily (truly) random function.

Note that the algorithm $prov$ in $\mathsf{FakeF}_{(PK,w_y)}$ produces a valid proof without knowledge of the secret key SK, or the seed s (the description) of the function f_s (h) plugged in.

We say $F_{(PK,SK)}(\cdot)$ is a $instance\text{-}dependent\ verifiable\ function$ if it satisfies the following conditions:

1. $Provability.$ If $(b, \pi) = F_{(PK,SK)}(a)$, then $\mathsf{Ver}(a, b, PK, \pi) = 1$
2. $Uniqueness\ on\ NO\ key\_instance.$ If $y \notin L$, then except with a negligible probability, there exist no values $(a, b_1, b_2, PK, \pi_1, \pi_2)$ such that $\mathsf{Ver}(a, b_1, PK, \pi_1) = \mathsf{Ver}(a, b_2, PK, \pi_2) = 1$
3. $Pseudorandomness\ on\ YES\ key\_instance.$ If $y \in L$ and w_y is a witness for $y \in L$, then for every PPT oracle machine M, every polynomial p, and all sufficient large $n's$,

$$|[Pr[M^{F_{(PK,SK)}}(1^n) = 1] - Pr[M^{\mathsf{FakeF}_{(PK,w_y,h)}}(1^n) = 1 : h \leftarrow_R H_n]| < 1/p(n)$$

where H_n is the ensemble of all functions mapping $d(n)$-bit-long strings to $l(n)$-bit-long strings.

Remarks

On key generation protocol. In contrast to the verifiable random functions [23] whose keys are generated by the function owner alone, our instance-dependent verifiable random functions allows interaction between the querier and the owner of the function in the key generation process. Note also that in above definition we do not make any requirement on the key generation protocol. This will allow us to design different key generation protocols for different purposes, see details in section 4.2 and section 5.

On pseudorandomness when $y \in L$. We remark that in the testing experiment, M obtains the function value *along with its proof of correctness* for every string that the oracle machine M queried. This is different from the testing experiment used to demonstrate the pseudorandomness of verifiable random functions, in which providing the proof of correctness along with the function value of the last query for judgement to the testing machine will trivialize this test because of the uniqueness of the verifiable random functions.

3.2 InstD-VRFs: Constructions

We begin with an informal description of our construction. The querier A and the function owner B execute a protocol and produce an key_instance y, then B selects a pseudorandom function f at random and commits to the description of this function. On input a string a in the domain of f, B returns $f(a)$ and a witness indistinguishable proof in which he proves that the function value is computed correctly or $y \in L$ using the knowledge of description of f he committed to. The public key of this instance-dependent verifiable random function consists of the instance y, the commitment and the setup information for the WI proof, the secret key is the decommitment.

For our applications, we require that the proof in use satisfies both resettable-soundness and resettable WI. To this end, the 2-round ZAPs introduced in [10] and the non-interactive ZAPs suggested in [17] are good candidates. Here we adopt 2-round ZAPs just for the purpose of basing our results on more general assumptions.

In the key generation protocol KGProt, the way to produce an key_instance for the owner of the function is a subtle problem and may vary depending on specific applications. In our applications there are two approaches to do this: for the function owned by the prover, the key_instance in its public key is the instance to be proven (therefore the honest prover supposedly knows the corresponding witness w_y); for the function owned by the verifier, the key_instance will be generated by the prover and the prover gives a special WI argument in which it proves the statement to be proven is true or the key_instance generated by itself is an YES instance(therefore the instance generated by an honest prover is a NO instance, and this guarantees the uniqueness property of this function). See details in section 5.

Bearing the above in mind, we omit the formal description of KGProt here for greater flexibility. Now, we simply assume that the key_instance y has been generated already. The rest components of the key pair are created in following way: the function owner B picks a pseudorandom functions f_{s_0} from the ensemble $\{f_s : \{0,1\}^{d(n)} \rightarrow \{0,1\}^{l(n)}\}_{s \in \{0,1\}^n}$, and commits to the seed s_0 using a statistically binding commitment scheme Com, let $c = Com(s_0, r)$ (r is the randomness required by the commitment scheme). The querier A selects a random string ρ as the first round message of ZAP and send it to B. On received ρ, the function owner B publish the public key $PK = (y, c, \rho)$ and keep $SK = (s_0, r)$ as the secret key.

Given (PK, SK), on input $a \in \{0,1\}^{d(n)}$, $F_{(PK,SK)}$ returns a function value b and the second round message of ZAP π (the proof) in which it proves that there exist (s_0, r) such that $f_{s_0}(a) = b$ and $c = Com(s_0, r)$ or $y \in L$. i.e., $F_{(PK,SK)}(a) = (f_{s_0}(a), prov(a, f_{s_0}(a), PK, SK)) = (b, \pi)$. Here we can view the probabilistic algorithm $prov$ as the prover in a ZAP system.

In the fake function evaluator $\mathsf{FakeF}_{(PK,w_y)}$, by using the witness w_y to the YES instance y, the algorithm $prov$ can always generate the valid proof of correctness regardless of whether the function value is correct or not.

Theorem 1. If there exist trapdoor permutations, there exist instance-dependent verifiable random functions.

Proof. We prove that The function evaluator $F_{(PK,SK)}$ described above is an instance-dependent verifiable random function. Note that the statistically-binding commitment scheme and pseudorandom functions can be constructed based on one-way fucntions [24, 18], and ZAPs assumes only trapdoor permutations.

The Provability is straightforward. *Uniqueness on NO key_instance* follows immediately from the statistically-binding property of Com and the soundness of ZAPs.

We prove the *Pseudorandomness on YES key_instance* using hybrid arguments. For every PPT oracle machine M, we consider the following sequences of hybrids, in each hybrid M makes a polynomial number of queries to a function that is slightly different from the one in previous hybrid. We complete the proof by showing M distinguishes each hybrid from its neighbor with at most a negligible probability.

Hybrid 0 M queries $F_{(PK,SK)}$.

Hybrid 1 M queries $F'_{(PK,SK)}$, where $F'_{(PK,SK)}(\cdot) = (f_{s_0}(\cdot), prov(\cdot, f_{s_0}(\cdot), PK, w_y))$. That is, $F'_{(PK,SK)}$ behaviors as the same as $F_{(PK,SK)}$ except it produces the proof of correctness using the witness w_y to $y \in L$. The fact that $F_{(PK,SK)}$ and $F'_{(PK,SK)}$ are indistinguishable follows immediately from the witness indistinguishability of ZAPs.

Hybrid 2 M queries $\mathsf{FakeF}_{(PK,w_y,s)}$, where the *pseudorandom* function seed s is selected at random. We claim that M cannot distinguish $F'_{(PK,SK)}$ from $\mathsf{FakeF}_{(PK,w_y,s)}$ with non-negligible probability. Assume otherwise, we construct a non-uniform algorithm D to break the hiding property of the statistically binding commitment Com. We give the detailed proof later.

Hybrid 3 M queries $\mathsf{FakeF}_{(PK,w_y,h)}$, where h is a *truly random* function. Note that the proof of correctness has nothing to do with the seed s or the description of h, so if M distinguishes $\mathsf{FakeF}_{(PK, w_y, s)}$ from $\mathsf{FakeF}_{(PK, w_y, h)}$, it distinguishes a pseudorandom function form a truly random function.

Now we give the description of algorithm D to prove the claim in Hybrid 2. D runs as follows. It takes s_0, s_1 and $w_y{}^1$ as input. On received the target commitment c' that is the commitment to s_0 or s_1, D uses $PK = (y, c', \rho)$ as public key, and for any query a made by M, it returns $f_{s_0}(a)$ and $prov(a, f_{s_0}(a), PK, w_y))$ (Note that D always uses f_{s_0} to compute the function value, then when $c' = Com(s_0)$, D performs as $\mathsf{F}'_{(PK, SK)}$; when $c' = Com(s_1)$, D performs as $\mathsf{FakeF}_{(PK, w_y, s_0)}$). At end, if M outputs $b \in \{0, 1\}$, D output $1 - b$. We show D breaks the hiding property of Com. We assume for some polynomial p,

$$|[Pr[M^{\mathsf{F}'_{(PK, SK)}}(1^n) = 1] - Pr[M^{\mathsf{FakeF}_{(PK, w_y, s)}}(1^n) = 1 : s \leftarrow_R \{0, 1\}^n]| > 1/p(n)$$

Then we have

$$
\begin{aligned}
&|[Pr[D \text{ outputs } 1 | c' = Com(s_1)] - Pr[D \text{ outputs } 1 | c' = Com(s_0)] \\
=&|[Pr[M^{\mathsf{FakeF}_{(PK, w_y, s_0)}}(1^n) = 0] - Pr[M^{\mathsf{F}'_{(PK, SK)}}(1^n) = 0]| \\
=&|[Pr[M^{\mathsf{F}'_{(PK, SK)}}(1^n) = 1] - Pr[M^{\mathsf{FakeF}_{(PK, w_y, s_0)}}(1^n) = 1]| \\
>& 1/p(n)
\end{aligned}
$$

A note on input length. We remark that an InstD-VRF $\mathsf{F}_{(PK, SK)}$ with domain $\{0, 1\}^{d(n)}$ can also be applied to inputs of length shorter than $d(n)$ by simply encoding the shorter inputs into the ones of desired length (cf. [13]) and using a prefix of ρ with suitable length as the first round message of a ZAP for the proof of correctness.

4   Two Instance-Dependent Protocols

With the instance-dependent verifiable random functions we developed, We are ready to construct two interesting instance-dependent protocols, which we call key-instance-dependent resettabley-sound bounded-class resettable ZK argument (KInstD rs-rZK argument) and resettable WI argument with instance-dependent weak resettable-soundness (InstD rs-rWI argument). Though these protocols do not even satisfy ZK (WI) and (knowledge) soundness at the same time, they are crucial tools for our main constructions presented in next sections.

---

[1] In case y is generated by the querier in the key generation protocol, it seems that y must be generated before the commitment c is seen by the querier, otherwise, our non-uniform algorithm D does not work because it needs to take a fixed advice (i.e., the witness to y) in advance (before seeing the target commitment) in breaking the hiding property of the commitment. However, in our applications, we do not need comply with this order. To enable the above analysis, we require the querier give a special argument of knowledge of the witness to y which is generated by itself.

4.1   Key_Instance-Dependent Resettabley-Sound Bounded-Class Resettable ZK Arguments for NP

We first show how to transform a public-coin bounded concurrent ZK argument into a key_instance-dependent resettabley-sound bounded-class resettable ZK argument by equipping the verifier in the former system with an instance-dependent verifiable random function. Similar to any other instance-dependent primitive, the key_instance-dependent resettabley-sound bounded-class resettable ZK argument satisfies only the resettable-soundness when the key_instance is a YES instance, and satisfies only the bounded-class resettable ZK when the key_instance is a NO instance.

As showed in [2], we can transform a constant-round public-coin bounded concurrent ZK argument (P_B, V_B) into a constant round resettably-sound bounded concurrent ZK argument (P_R, V_R) by simply equipping V_R with a pseudorandom function and letting V_R emulate V_B except that it generate the current round message by applying a pseudorandom function to the transcript so far. With the argument (P_R, V_R), We construct a key_instance-dependent resettably-sound bounded-class resettable ZK argument (KInstD rs-rZK argument, for short) (P, V) as follows. The prover P and the verifier V first execute a key generation protocol KGProt aimed at setting up a key pair (PK, SK) of an instance-dependent verifiable random function $F_{(PK,SK)} = (f_{s_0}, prov)$ with respect to a hard language L' (the choice of language L' see section 5) for V, and then they execute the protocol (P_R, V_R) in following way: 1) In each P's step, P checks whether the message sent by V is computed correctly, if so, it replies according to the instruction of P_R; 2) V feeds V_R with the randomness s_0, and in each V's step, V generates its message by running V_R and using the algorithm $prov$ in $F_{(PK,SK)} = (f_{s_0}, prov)$ to give a proof of correctness for the output by V_R. Note that V_R always generates its message by applying f_{s_0} to the history produced by P_R and V_R, so each V's message can be viewed as the output by $F_{(PK,SK)} = (f_{s_0}, prov)$ on the *the transcript of the underlying protocol* (P_R, V_R) so far. See Fig.1 for the formal description.

We assume the transcript size of an execution of the resettably-sound t^3-bounded concurrent ZK argument (P_R, V_R) is bounded by a polynomial d, and assume the longest message sent by V_R is $t^3 n^3$. Without of loss generality, We assume all verifier's messages are of equal length.

Theorem 2. The KInstD rs-rZK argument (P, V) depicted in Fig.1 satisfies following conditions:

1. Unbounded resettable-soundness when $y \in L'$: for any $x \notin L$, if all key_instance y's generated by an incarnation $V^{(j)}(x) = V_{x, r_{v_j}}$ are in L, then for any PPT P^* mounting unbounded resetting attack, the probability that $V^{(j)}(x)$ accept in some session is negligible.
2. t^3-Bounded-class resettable ZK when $y \notin L'$: For all PPT V^* mounting bounded-class resetting attack, if $y \notin L'$ holds for all sessions, then there exists a PPT M satisfying the requirement of Definition 1.

KInstD rs-rZK Argument (P, V)

Common input: $x \in L$ ($|x| = n$)
The Prover's private input: a witness w for $x \in L$.
Prover's randomness: r_p, a seed of a pseudorandom function f_{r_p}
Verifier's randomness: r_v, a seed of a pseudorandom function f_{r_v}

Phase 1: the key generation protocol KGProt
$V \to P$ V sets $(r_v^1, r_v^2) = f_{r_v}(x)$, selects $f_{s_0} \leftarrow_R \{f_s : \{0,1\}^{\le d(n)} \to \{0,1\}^{t^3 n^3}\}_{s \in \{0,1\}^n}$ and $r_0 \leftarrow_R \in \{0,1\}^n$ using randomness r_v^1, computes $c_0 = Com(s_0, r_0)$ using the statistically-binding commitment scheme Com and generates an instance $y \in L' \cap \{0,1\}^n$, stores $SK = (s_0, r_0)$. Sends c_0, y;

$P \to V$ P sets $(r_p^1, r_p^2) = f_{r_p}(x, c_0, y)$, selects the first message ρ for a ZAP using randomness r_p^1. at random. At the end of this step, the InstD-VRF's key pair $(PK, SK) = ((y, c_0, \rho), (s_0, r_0))$ is set up for V. Sends ρ;

Phase 2: the *Modified* resettably-sound t^3-bounded concurrent ZK argument
$V \Leftrightarrow P$ Let (P_R, V_R) be the resettably-sound t^3-bounded concurrent ZK argument. P writes r_p^2 on P_R's random tape. V writes s_0 on V_R's random tape (the description of f_{s_0}). P and V perform the following strategy.

> **V's Strategy** In each V's step, V runs $V_R(s_0, \cdot)$ on input $hist$, where $hist$ is the history (including the common input) produced by P_R and V_R (*not the history produced by P and V*) so far. V obtains $m_v = V_R(s_0, hist)$, and computes $\pi = prov(hist, m_v, PK, SK)$ using randomness $f_{r_v^2}(hist)$, then V sends (m_v, π) to P.
> Note that $(m_v, \pi) = (f_{s_0}(hist), prov(hist, m_v, PK, SK)) = F_{(PK,SK)}(hist)$ (V_R always generates m_v by applying f_{s_0} to the history so far).
>
> **P's Strategy** In each P's step, P checks whether the message (m_v, π) sent by V is correct by using the algorithm Ver associated with $F_{(PK,SK)}$, if not, aborts; if so, runs $P_R(r_p^2, \cdot)$ on input $(hist, m_v)$, here $hist$ is the history produced by P_R and V_R before the message m_v was sent by V. P sends $m_p = P_R(r_p^2, hist, m_v)$ to V.
>
> **V's Decision** V accepts if only if V_R accepts the transcript that generated by P_R and V_R.

Fig. 1. The key-instance-dependent resettably-sound bounded-class resettable ZK argument for a NP language L

Intuitively, the property of resettable-soundness when $y \in L'$ follows from the fact that once the verifier has the witness, it can send arbitrary messages to a same history of a session without being detected by the prover, so we can reduce the soundness of the KInstD rs-rZK argument (P, V) to the underlying protocol (P_R, V_R). On the other hand, if all y's are NO instance, then the first verifier's message essentially determines the verifier's behavior, this will make our protocol enjoy bounded-class resettable ZK. The actual proof is omitted here, and will be found in the full version of this paper.

4.2 The Resettable WI Arguments with Instance-Dependent Weak Resettable-Soundness

In this subsection, we construct resettable WI arguments with instance-dependent weak resettable-soundness (InstD rs-rWI argument, for short), in which the prover proves that one of the two instances x_0 and x_1 is in the language L. Though there are resetably-sound resettable WI arguments for NP such as ZAPs, which achieves more stronger security than our InstD rs-rWI arguments, however, the InstD rs-rWI arguments have a distinguishing property that ZAPs do not enjoy: they are *arguments of knowledge* on some special instances. This property is crucial for the analysis of soundness of our main construction presented in next section.

We assume that there are 3 round public-coin WI arguments of knowledge for all NP languages. Let (a, e, z) is the three messages exchanged in a session. Furthermore, we assume these arguments have the following property: it is easy to extract the witness for the statement from two different transcripts (a, e, z) and (a, e', z') when $e \neq e'$. To this end, the parallelized version of Blum's proof of knowledge for Hamiltonian Cycle is a good candidate, which assumes one-way permutations exist.

Our construction is inspired by the protocol 6.2 in [2], in which the verifier first commits to a seed of a pseudorandom function and generates the query e by applying this function to the first round message a. The important deviation is that the verifier in our system uses a KInstD rs-rZK argument to prove the query e is computed correctly. In the KInstD rs-rZK argument, one instance, say x_0, serves as the key_instance for a InstD-VRF used by the verifier (the prover in the global system). The formal description appears in Fig.2.

Theorem 3. The InstD rs-rWI argument (P_W, V_W) depicted in Fig.2 satisfies following properties:

1. *Unbounded* Resettable witness indistinguishability: For any PPT V_W^* mounting *unbounded* resetting attack, the distribution $(P_W(\overline{w_0}), V_W^*)(\overline{x})$ is computationally indistinguishable from $(P_W(\overline{w_1}), V_W^*)(\overline{x})$, where $\overline{x} = x^1, \cdots, x^{poly(n)}$, $x^i = (x_0^i, x_1^i)$, $\overline{w_b} = w_b^1, \cdots, w_b^{poly(n)}$ such that $(x_b^i, w_b^i) \in R_L$, $i = 1, \cdots, poly(n)$, $b = 0, 1$.
2. Weak resettably-sound argument of knowledge property when $x_0 \notin L$: For every PPT P_W^* mounting *weak* resetting attack, if P_W^* convinces V_W on statement (x_0, x_1) such that $x_0 \notin L$ with probability p in a session, then there exists a PPT machine E, upon input the description of P_W^*, outputs a witness for the instance x_1 with probability negligibly close to p.

The *Unbounded* Resettable witness indistinguishability follows from the fact that the underlying KInstD rs-rZK argument satisfies resettably-soundness when $x_0 \in$

L. For the Weak resettably-sound argument of knowledge property when $x_0 \notin L$, We can construct a extractor E to justify it. Assume P_{W}^* convinces an incarnation $V_{\mathsf{W}}^j(x)$ on statement (x_0, x_1) such as $x_0 \notin L$ with high probability in a session. The extractor E first plays the role of $V_{\mathsf{W}}^j(x)$ and get an accepting transcript

---

InstD rs-rWI Argument $(P_{\mathsf{W}}, V_{\mathsf{W}})$

Common input: two instance $x_0, x_1 \in L$, a security parameter n.
The Prover's private input: the witness w such that $(x_0, w) \in R_L$ or $(x_1, w) \in R_L$.
Prover's randomness: r_p, a seed of a pseudorandom function f_{r_p}
Verifier's randomness: r_v, a seed of a pseudorandom function f_{r_v}

$V_{\mathsf{W}} \rightarrow P_{\mathsf{W}}$ V_{W} sets $(r_v^1, r_v^2) = f_{r_v}(x_0, x_1)$. Using the randomness r_v^1, V_{W} selects a pseudorandom function $f_s : \{0,1\}^{\leq poly(n)} \rightarrow \{0,1\}^{|e|}$ and r, computes $c = Com(s, r)$ using a statistically-binding commitment scheme Com. Sends c;

$P_{\mathsf{W}} \rightarrow V_{\mathsf{W}}$ P_{W} sets $(r_p^1, r_p^2) = f_{r_p}(x_0, x_1, c)$.
Using the randomness r_p^1 , P_{W} invokes the 3 round WI argument in which it proves $x_0 \in L$ or $x_1 \in L$, produces the first message a of this protocol .
using the randomness r_p^2, P_{W} invokes a KInstD rs-rZK argument (in which P_{W} plays the role of verifier), produces the first message c_0 (i.e., the commitment to the description of pseudorandom function) and uses x_0 as the key_instance.
Send a, c_0;

$V_{\mathsf{W}} \rightarrow P_{\mathsf{W}}$ V_{W} computes $e = f_s(x_0, x_1, c, a, c_0)$. Using the randomness r_v^2, V_{W} and selects the first message ρ for a ZAP according to the KInstD rs-rZK argument.
Sends e, ρ;

$P_{\mathsf{W}} \Leftrightarrow V_{\mathsf{W}}$ P_{W} and V_{W} continue to run the KInstD rs-rZK argument in which V_{W} proves there exist s, r such that $e = f_s(x_0, x_1, c, a, c_0)$ and $c = Com(s, r)$. The public key for P_{W}'s InstD-VRF is $PK = (x_0, c_0, \rho)$ and the corresponding secret key is the decommitment to c_0.

$P_{\mathsf{W}} \rightarrow V_{\mathsf{W}}$ Sends the answer z to the query e according to the 3 round WI argument if the above transcript is accepting.

V_{W}'s **Decision** V_{W} accepts if only if the transcript (a, e, z) is accepting.

Fig. 2. The resettable WI arguments with instance-dependent weak resettable-soundness

(a, e, z) of the underlying 3-round WI argument, and then rewinds P_{W}^* to the point that the message that contains a was first sent by P_{W}^*, and then sends another challenge e' and runs the simulator associated with the KInstD rs-rZK argument to prove that e' is correctly computed. At the end E will receive another accepting transcript (a, e', z') with probability close to the probability that P_{W}^* convinces $V_{\mathsf{W}}^j(x)$, and this allows E to compute a witness for x_1 (x_0 is assumed to be NO instance). Due to space limitations, the formal analysis of this extractor are omitted here and will appear the full version of this paper.

We stress that the InstD rs-rWI argument achieves *weak* resettably-sound argument of knowledge property when $x_0 \notin L$, rather than *bounded-class* resettably-sound argument of knowledge property when $x_0 \notin L$. There are two reasons for this: 1) it is sufficient to consider only one incarnation of the verifier in the analysis of soundness; 2) the simulation performed by E will be run successful (due to the bounded resettable zero knowledge property when $x_0 \notin L$ of the underlying KInstD rs-rZK argument, note x_0 is the key-instance for P_W^*'s InstD-VRF).

However, we failed to achieve *unbounded* resettable-soundness. This is because, for justifying the extraction, we need to simulate all proofs given by $V_W^j(x)$, not just those proofs given by $V_W^j(x)$ with the first P_W^*'s (P_W^* plays the role of the verifier in the underlying KInstD rs-rZK argument) first message (a, c_0) (therefore, we need to put a priori bound on the number of the P_W^*'s first messages). Recently, we overcome this obstacle by using the the instance-dependent verifiable random functions in a novel way.

5   Transforming Public-Coin (Bounded) Concurrent ZK Arguments to (Bounded-Class) Resettable ZK Arguments with (Weak) Resettable-Soundness

We now show how to transform a public-coin bounded concurrent ZK argument into a bounded-class resettable ZK argument with weak resettable-soundness, using the two instance-dependent protocols developed in last section. Furthermore, if public-coin (unbounded) concurrent ZK argument exists, the same transformation yields a (unbounded) resettably-sound resettable ZK argument immediately.

We obtain a bounded-class resettable ZK argument with weak resettable-soundness from a public-coin bounded concurrent ZK argument in the following way: we first transform the public–coin bounded concurrent ZK argument into a key-instance-dependent resettably-sound bounded-class resettable ZK argument, then we modify the resulting protocol in such a way taht, in the key generation protocol, instead of having the verifier generates a key-instance itself, we have the prover generates a NO instance with respect to some hard-to-decide language L' as the key-instance for the verifier's InstD-VRF and gives a proof that the statement $x \in L$ to be proven is true or this key-instance is a YES instance via a resettable WI argument with instance-dependent weak resettable-soundness in which the prover uses x as the key-instance for its own InstD-VRF, The second phase of the key-instance-dependent resettably-sound bounded-class resettable ZK argument remains unchanged. For the language L', we choose the one defined by a pseudorandom generator [6, 25]: $L' = \{y | \exists \delta, y = G(\delta), |y| = 2n, |\delta| = n\}$, where $G : \{0, 1\}^n \rightarrow \{0, 1\}^{2n}$ is a pseudorandom generator. This protocol is formally depicted in Fig.3.

The key idea behind this protocol is that for an honest prover, it always generates NO instance as key-instance of the verifier's InstD-VRF, and for a false statement $x \notin L$, the cheating prover must generate YES instance due to the weak resettably-sound argument of knowledge property when $x \notin L$ of the InstD rs-rWI argument used in the key generation protocol. Note that the common input x itself serves as the key_instance for the prover's InstD-VRF in the InstD rs-rWI argument.

Theorem 4. If there exist trapdoor permutations and collision-resistant hash functions, there exist bounded-class resettable ZK arguments with weak resettable-soundness for all NP languages.

We prove this theorem by showing the wrs-brZK argument described in Fig.3 is a bounded-class resettable ZK argument with weak resettable-soundness for any NP language L. For the complexity assumption, we note that one-way functions are sufficient for the pseudorandom generator, so, our protocol assumes trapdoor permutations and collision-resistant hash functions (for Barak's protocol). *Proof.* **Completeness** is straightforward.

Weak resettable-soundness. Assume a PPT $\mathbf{P}^*$ mounting a *weak* resetting attack, convinces an incarnation $\mathbf{V}^j(x)$ on a false statement $x \notin L$ with non-negligible probability p. By the weak resettably-sound argument of knowledge property when $x \notin L$ (note that x serves as the key_instance of $\mathbf{P}^*$'s InstD-VRF) of the InstD rs-rWI argument in the key generation protocol, we have with probability essentially close to p there exists $\exists \delta, |\delta| = n$ such that $y = G(\delta)$ (assume y is the instance generated by $\mathbf{P}^*$), and use the extractor associated with the InstD rs-rWI argument, we will extract the witness δ. Furthermore, note that y serves as the key_instance of the verifier's InstD-VRF in phase 2, then using this witness and the strategy $\mathbf{P}^*$, we can break the resettable-soundness of the underlying resettably-sound bounded-concurrent ZK argument in phase 2 in a way similar to the analysis of soundness for the KInstD rs-rZK argument, which leads to a contradiction.

Bounded-class resettable ZK. This property follows from the next lemma.

Lemma 1. Let (P_R, V_R) be a resettably-sound t^3-bounded concurrent ZK argument system, and $(\mathbf{P}, \mathbf{V})$ be the wrs-brZK argument transformed from (P_R, V_R). Then for every PPT $\mathbf{V}^*$ bounded-class resettable model, there exists a PPT V_R^* in the bounded concurrent model such that $(P_R(\overline{w}), V_R^*)(\overline{x})$ is computationally indistinguishable from $(\mathbf{P}(\overline{w}), \mathbf{V}^*)(\overline{x})$, where $\overline{x} = x_1, \cdots, x_t \in L$, $\overline{w} = w_1, \cdots, w_t$ such that $(x_i, w_i) \in R_L$, $i = 1, \cdots, t$.

Proof. We construct V_R^* in bounded concurrent model using the following strategy to handle $\mathbf{V}^*$'s message.

1. $\mathbf{V}^*$ sends a new first message msg to $\mathbf{P}^{(i,j)}$: Assume this is the kth new first message to $\mathbf{P}^{(i,j)}$ ($1 \leq i, j, k \leq t$). V_R^* chooses δ randomly, generates $y = G(\delta)$ itself, and stores (y, δ). It acts as honest prover but uses δ as the witness to execute the InstD rs-rWI argument, and forwards a message to $\mathbf{V}^*$. Furthermore, V_R^* maintains a table, in which the row with index (i, j, k) contains those messages belonging to the class $\text{Class}_{\mathbf{P}^{(i,j)}, \text{msg}_k}$.

2. $\mathbf{V}^*$ repeats a first message msg to $\mathbf{P}^{(i,j)}$. Assume msg equals msg_k that V_R^* received before. V_R^* retrieves its response to this message from $\mathrm{row}_{(i,j,k)}$ in its table and forwards it to $\mathbf{V}^*$.

---

wrs-brZK Argument ($\mathbf{P}, \mathbf{V}$)

Common input: $x \in L$ ($|x| = n$).
The Prover's private input: the witness w such that $(x, w) \in R_L$.
Prover's randomness: r_p, a seed of a pseudorandom function f_{r_p}
Verifier's randomness: r_v, a seed of a pseudorandom function f_{r_v}

Phase 1: the key generation protocol KGProt
$\mathbf{V} \to \mathbf{P}$ $\mathbf{V}$ sets $(r_v^1, r_v^2, r_v^3) = f_{r_v}(x)$, selects f_{s_0} and $r_0 \in \{0,1\}^n$ using randomness r_v^1, computes $c_0 = Com(s_0, r_0)$ using the statistically-binding commitment scheme Com, and stores $SK = (s_0, r_0)$.
Sends c_0;
$\mathbf{P} \Leftrightarrow \mathbf{V}$ $\mathbf{P}$ sets $(r_p^1, r_p^2, r_p^3, r_p^4) = f_{r_p}(x, c_0)$ and generates a random string y ($|y| = 2n$) using randomness r_p^1.
$\mathbf{P}$ and $\mathbf{V}$ run a InstD rs-rWI argument in which $\mathbf{P}$ proves that $x \in L$ or $\exists \delta, |\delta| = n$ such that $y = G(\delta)$. In the underlying the KInstD rs-rZK argument used in this InstD rs-rWI argument, x serves as the InstD-VRF's (owned by the prover in global system) key_instance, the randomness used by $\mathbf{P}$ is r_p^2 and the randomness used by $\mathbf{V}$ is r_v^2.
$\mathbf{P} \to \mathbf{V}$ $\mathbf{P}$ selects the first message ρ for a ZAP using randomness r_p^3. At the end of this step, the InstD-VRF's key pair $(PK, SK) = ((y, c_0, \rho), (s_0, r_0))$ is set up for $\mathbf{V}$.
Sends ρ;

Phase 2: the *Modified* resettably-sound bounded concurrent ZK argument
$\mathbf{V} \Leftrightarrow \mathbf{P}$ $\mathbf{P}$ and $\mathbf{V}$ follows the same strategy as described in the phase 2 in the KInstD rs-rZK argument (see Fig.1), in which the verifier uses an InstD-VRF described by $(PK, SK) = ((y, c_0, \rho), (s_0, r_0))$. In this phase, $\mathbf{P}$ uses randomness r_p^4 and $\mathbf{V}$ uses randomness r_v^3.

Fig. 3. The bounded-class resettable ZK argument with weak resettable-soundness for a NP language L

3. $\mathbf{V}^*$ sends a *valid* non-first message belonging to the key generation protocol KGProt to $\mathbf{P}^{(i,j)}$. V_R^* produces the response to this message[2] according to the key generation protocol KGProt as the honest prover but in the execution of InstD rs-rWI argument it uses the δ as the witness. V_R^* stores this response, and forwards it to $\mathbf{V}^*$.

Note that $\mathbf{V}^*$ is free in the key generation protocol, Thus a class recorded by V_R^* may contain many different sessions due to $\mathbf{V}^*$'s resetting in phase 1.

---

[2] Without loss of generality, we assume each message sent by V^* is prepended with a session ID.

4. $\mathbf{V}^*$ sends a *invalid* message belonging to the the key generation protocol KGProt to $\mathbf{P}^{(i,j)}$. V_R^* sends an abort message to $\mathbf{V}^*$ to end this session.

5. $\mathbf{V}^*$ sends a *valid* message belonging to the *Modified* resettably-sound bounded concurrent zero knowledge argument (i.e., a message sent in phase 2) to $\mathbf{P}^{(i,j)}$. Assume that this message is the lth round message belonging to $\mathrm{Class}_{\mathbf{P}^{(i,j)},\mathsf{msg}_k}$. Parse this message into (m_l, π_l). We distinguish three cases:

   Case 1. The lth round message was sent by $\mathbf{V}^*$ before in some previous session in this class and the current message m_l does not equal the lth round message m_l' recorded in the row row$_{(i,j,k)}$. In this case, V_R^* **terminates**.

   Case 2. The lth round message was never sent by $\mathbf{V}^*$ before in this class. In this case, V_R^* forwards m_l to the incarnation P_R^{i,j_k}, stores m_l and P_R^{i,j_k}'s response in row$_{(i,j,k)}$ and forwards it to $\mathbf{V}^*$.

   Case 3. The lth round message was sent by $\mathbf{V}^*$ before in some previous session in this class and the current message m_l equals the lth round message m_l' recorded in the row row$_{(i,j,k)}$. In this case, V_R^* retrieves the P_R^{i,j_k}'s response to this message in the row$_{(i,j,k)}$ and forwards it to $\mathbf{V}^*$. We stress that in this case V_R^* does not interact with any incarnation of P_R.

   Observe that for two different first verifier $\mathbf{V}^*$'s messages $\mathsf{msg}_m \neq \mathsf{msg}_n$ to the same incarnation $\mathbf{P}^{(i,j)}$, V_R^* initiates two independent incarnations P_R^{i,j_m} and P_R^{i,j_n} to generate the $\mathbf{V}^*$'s view, and this strategy makes V_R^* look like the real incarnation $\mathbf{P}^{(i,j)}$.

6. $\mathbf{V}^*$ sends a *invalid* message belonging to the *Modified* resettably-sound bounded concurrent zero knowledge argument to $\mathbf{P}^{(i,j)}$. V_R^* sends an abort message to $\mathbf{V}^*$ to end this session.

7. $\mathbf{V}^*$ terminates. Without loss of generality, $\mathbf{V}^*$ outputs its view in the whole interaction. V_R^* outputs what $\mathbf{V}^*$ outputs.

It is easy to see that the strategy V_R^* works in bounded concurrent models if $\mathbf{V}^*$ works in bounded-class resettable model. Since V_R^* runs only one session with each P_R^{i,j_k}'s, we can identify each class $\mathrm{Class}_{P^{(i,j)},\mathsf{msg}_k}$ with the single session of V_R^* with P_R^{i,j_k} (though a class contains many different sessions, but all those sessions have the same tail, i.e., the transcript produced by P_R^{i,j_k} and V_R^* in phase 2).

Note that in sessions having different first verifier's message or sessions between $\mathbf{V}^*$ and different honest incarnations of $\mathbf{P}$, $\mathbf{P}$ will generate (almost) independent random tapes to emulate the action of P_R in the second phase of the wrs-brZK argument in different sessions, and V_R^*, incorporating with P_R^{i,j_k}'s, uses the same strategy as $\mathbf{P}$ in the second phase of this argument. So, if the case 1 in item 5 does not occur, the only difference between $\mathbf{V}^*$'s view during the interaction with V_R^* and its view during the real interaction with many incarnations of $\mathbf{P}$ is that in the former interaction V_R^* uses pseudorandom generator to produce YES instances y and uses the corresponding witness to execute the InstD rs-rWI argument. We can use a standard hybrid algorithm[3] to show $\mathbf{V}^*$'s view in these two scenario are indistinguishable, furthermore, note that V_R^*'s view is just the copy of $\mathbf{V}^*$'s view, so $(P_R(\overline{w}), V_R^*)(\overline{x})$ is computationally indistinguishable from $(\mathbf{P}(\overline{w}), \mathbf{V}^*)(\overline{x})$.

Note that the case 1 in item 5 occurs with only negligible probability when y is NO instance due to the uniqueness property of the InstD-VRF on NO key_instance. Therefore, if the case 1 occurs with non-negligible property in our setting that all y's are YES instances, then we construct an algorithm to break the pseudorandomness of the generator. This algorithm takes all witness for these statements $\overline{x}$ as inputs and uses them to execute the underlying InstD rs-rWI argument, if case 1 occurs, outputs "yes". It is not hard to see that this algorithm works.

Public-Coin Concurrent ZK Implies Simultaneously Resettable Secure ZK. If public-coin concurrent ZK argument exists, then we can construct key_instance-dependent resettably-sound *unbounded* resettable ZK argument and resettable WI argument with instance-dependent *unbounded* resettable soundness, therefore, using the above transformation, we get a resettably-sound resettable ZK argument. So, we have

Theorem 5. If there exist public-coin concurrent zero knowledge argument for a NP language L, and if trapdoor permutations exist, there exist resettably-sound resettable ZK argument for L.

We don't know if the public-coin concurrent ZK argument for non-trivial language exists, and if Barak's technique can be extended to (unbounded) concurrent setting. Theorem 5 shows this question deserves our attention: if such protocols (regardless of the number of rounds) for NP exist, the simultaneous resettability conjecture is true.

Acknowledgements. Yi Deng is grateful to Boaz Barak and Yehuda Lindell for helpful conversations on concurrent and resettable zero knowledge. We thank anonymous reviewers for their encouragement and valuable comments.

References

[1] B. Barak. How to go beyond the black-box simulation barrier. In Proc. of IEEE FOCS 2001, pp.106-115.

[2] B. Barak, O. Goldreich, S. Goldwasser, Y. Lindell. Resettably sound Zero Knowledge and its Applications. In Proc. of IEEE FOCS 2001, pp. 116-125.

---

[3] Consider an algorithm H takes the witnesses for all statement $\overline{x}$ as input, generates YES instances using a pseudorandom generator, and interacts with $\mathbf{V}^*$ using the witnesses for $\overline{x}$ as witnesses to execute the underlying InstD rs-rWI argument. Observe that $\mathbf{V}^*$'s view in the interaction with H is distinguishable form both its view in the interaction with V_R^* (due to resettable WI property of the underlying InstD rs-rWI argument) and its view in the interaction with $\mathbf{P}$ (due to pseudorandomness of the generator), thus we conclude the latter two are indistinguishable.

[3] B. Barak, O. Goldreich. Universal Arguments and Their Applications. In Proc. of IEEE CCC 2002, pp. 194-203.

[4] M. Blum. How to Prove a Theorem so No One Else can Claim It. In Proc. of ICM'86, pp. 1444-1451, 1986.

[5] B. Barak, Y. Lindell, S. Vadhan. Lower Bounds for Non-Black-Box Zero Knowledge. In Proc. of IEEE FOCS 2003, pp.384-393

[6] M. Blum, S. Micali. How to Generate Cryptographically Strong Sequences of Pseudo Random Bits. In Proc. of IEEE FOCS 1982, pp. 112-117

[7] R. Canetti. Universally Composable Security: A New Paradigm for Cryptographic Protocols. In Proc. of IEEE FOCS 2001, pp.136-145

[8] R. Canetti, O. Goldreich, S. Goldwasser, S. Micali. Resettable Zero Knowledge. In Proc. of ACM STOC 2000, pp.235-244

[9] R. Canetti, J. Kilian, E. Petrank and A. Rosen. Concurrent Zero-Knowledge requires $\Omega(log n)$ rounds. In Proc. of ACM STOC 2001, pp.570-579.

[10] C. Dwork, M. Naor. Zaps and Their Applications. In Proc. of IEEE FOCS 2000, pp.283-293

[11] C. Dwork, M. Naor and A. Sahai. Concurrent Zero-Knowledge. In Proc. of ACM STOC 1998, pp.409-418.

[12] U.Feige and A. Shamir. Witness Indistinguishability and Witness Hiding Protocols. In Proc. of ACM STOC 1990, pp.416-426.

[13] O. Goldreich. Foundation of Cryptography-Basic Tools. Cambridge University Press, 2001.

[14] O. Goldreich, S. Goldwasser, S. Micali. How to construct random functions. J. ACM 33(4), pp.792-807

[15] O. Goldreich, S. Micali and A. Wigderson. Proofs that yield nothing but their validity or All languages in NP have zero-knowledge proof systems. J. ACM, 38(3), pp.691-729, 1991.

[16] S. Goldwasser, S. Micali, and C. Rackoff. The knowledge complexity of interactive proof systems. SIAM. J. Computing, 18(1):186-208, February 1989.

[17] J. Groth, R. Ostrovsky and A. Sahai. Non-interactive Zaps and New Techniques for NIZK. In Advances in Cryptology-Crypto'o6, LNCS 4117, pp.97-111.

[18] J. Hastad, R. Impagliazzo, L. A. Levin, M. Luby. A Pseudorandom Generator from Any One-Way Functions. SIAM Journal on Computing 28(4):1364-1396, 1999.

[19] T. Itoh, Y. Ohta. A language-dependent cryptographic primitive. Journal of Cryptology 10(1) pp.37-49, 1997

[20] Daniele Micciancio, Shien Jin Ong, Amit Sahai, Salil P. Vadhan. Concurrent Zero Knowledge Without Complexity Assumptions. TCC 2006, LNCS3876, pp.1-20

[21] S. Micali, L. Reyzin. Soundness in the public-key model. In Advances in Cryptology-Crypto'o2, LNCS2139, pp.542C565, 2001.

[22] S. Micali, R. Rivest. Micropayments revisited. In CT-RSA, pp.149C163, 2002.

[23] S. Micali, M. Rabin, and S. Vadhan. Verifiable random functions. In Proc. of IEEE FOCS, pp. 120C130, 1999.

[24] M. Naor. Bit Commitment using Pseudorandomness. Journal of Cryptology 4(2): 151-158, 1991.

[25] A. Yao. Theory and Applications of Trapdoor Functions. In Proc. of IEEE FOCS 1982, pp.80-91

Conditional Computational Entropy, or Toward Separating Pseudoentropy from Compressibility

Chun-Yuan Hsiao[1], Chi-Jen Lu[2], and Leonid Reyzin[1]

[1] Boston University, Boston, MA 02215, USA
{cyhsiao,reyzin}@cs.bu.edu
Work performed in part while visiting the
Institute for Pure and Applied Mathematics at UCLA
[2] Academia Sinica, 128 Academia Road, Section 2, Nankang, Taipei 115, Taiwan
cjlu@iis.sinica.edu.tw

Abstract. We study conditional computational entropy: the amount of randomness a distribution appears to have to a computationally bounded observer who is given some correlated information. By considering conditional versions of HILL entropy (based on indistinguishability from truly random distributions) and Yao entropy (based on incompressibility), we obtain:

- a separation between conditional HILL and Yao entropies (which can be viewed as a separation between the traditional HILL and Yao entropies in the shared random string model, improving on Wee's 2004 separation in the random oracle model);
- the first demonstration of a distribution from which extraction techniques based on Yao entropy produce more pseudorandom bits than appears possible by the traditional HILL-entropy-based techniques;
- a new, natural notion of unpredictability entropy, which implies conditional Yao entropy and thus allows for known extraction and hard-core bit results to be stated and used more generally.

1 Introduction

The various information-theoretic definitions of entropy measure the amount of randomness a probability distribution has. As cryptography is able to produce distributions that appear, for computationally bounded observers, to have more randomness than they really do, various notions of *computational* entropy attempt to quantify this *appearance* of entropy. The commonly used HILL entropy (so named after [HILL99]) says that a distribution has computational entropy k if it is indistinguishable (in polynomial time) from a distribution that has information-theoretic entropy k.[1] The so-called Yao entropy [Yao82, BSW03], says that a distribution has computational entropy k if it cannot be efficiently compressed to below k bits and then efficiently decompressed. Other computational notions of entropy have been considered as well [BSW03, HILL99].

---

[1] The specific notion of information-theoretic entropy depends on the desired application; for the purposes of this paper, we will use min-entropy, defined in Section 2.

M. Naor (Ed.): EUROCRYPT 2007, LNCS 4515, pp. 169–186, 2007.
© International Association for Cryptologic Research 2007

Computational notions of entropy are useful, in particular, for extracting strings that are pseudorandom (i.e., look uniform to computationally bounded observers) from distributions that appear to have entropy. Indeed, generation of pseudorandom bits is the very purpose of computational entropy defined in [HILL99], and its variant considered in [GKR04]. Pseudorandom bits have many uses, for example, as keys in cryptographic applications.

The adversary in cryptographic applications (or, more generally, an observer) often possesses information related to the distribution whose entropy is being measured. For example, in the case of Diffie-Hellman key agreement [DH76] the adversary has g^x and g^y, and the interesting question is the amount of computational entropy of g^{xy}. Thus, the entropy of a distribution for a particular observer (and thus the pseudorandomness of the extracted strings) depends on what other information the observer possesses. Because notions of computational entropy necessarily refer to computationally-bounded machines (e.g., the distinguisher for the HILL entropy or the compressor and decompressor for the Yao entropy), they must also consider the information available to these machines. This has sometimes been done implicitly (e.g., in [GKR04]); however, most commonly used definitions do not do so explicitly.

In this work, we explicitly put forward notions of *conditional* computational entropy. This allows us to:

1. Separate conditional Yao entropy from conditional HILL entropy by demonstrating a joint distribution (X, Z) such that X has high Yao entropy but low HILL entropy when conditioned on Z.
2. Demonstrate (to the best of our knowledge, first) application of Yao entropy by extracting more pseudorandom bits from a distribution using Yao-entropy-based techniques than seems possible from HILL-entropy-based techniques.
3. Define a new, natural notion of unpredictability entropy, which can be used, in particular, to talk about the entropy of a value that is unique, such as g^{xy} where g^x and g^y are known to the observer, and possibly even verifiable, such as the preimage x of a one-way permutation f, where $y = f(x)$ is known to the observer.

HILL-Yao Separation. The first contribution (Section 3) can be seen as making progress toward the open question of whether Yao entropy implies HILL entropy, attributed in [TVZ05] to Impagliazzo [Imp99] (the converse is known to be true: HILL entropy implies Yao entropy, because compressibility implies distinguishability). Wee [Wee04] showed that Yao entropy does not imply HILL entropy in the presence of a random oracle and a membership testing oracle. Our separation of conditional Yao entropy from conditional HILL entropy can be seen as an improvement of the result of [Wee04]: it shows that Yao entropy does not imply HILL entropy in the presence of a (short) random string, because the distribution Z on which X is conditioned is simply the uniform distribution on strings of polynomial length. The separation holds under the quadratic residuosity assumption.

Randomness Extraction. Usually, pseudorandomness extraction is analyzed via HILL entropy, because distributions with HILL entropy are indistinguishable from distributions with the same statistical entropy, and we have tools (namely, randomness extractors [NZ96]) to obtain uniform strings from the latter. Tools are also available to extract from Yao entropy: namely, extractors with a special *reconstruction* property [BSW03]. Our second contribution (Section 4) is to show that considering the Yao entropy and applying a reconstructive extractor can yield many more pseudorandom bits than the traditional analysis, because, according to our first result, Yao entropy can be much higher than HILL entropy. This appears to be the first application of Yao entropy, and also demonstrates the special power of reconstructive extractors.

It is worth mentioning that while our separation of entropies is conditional, the extraction result holds even for the traditional (unconditional) notion of pseudorandomness. The analysis of pseudorandomness of the resulting string, however, relies on the notion of conditional entropy, thus demonstrating that it can be a useful tool even in the analysis of pseudorandomness of unconditional distributions.

Unpredictability Entropy. Unpredictability entropy is a natural formalization of a previously nameless notion that was implicitly used in multiple works.. Our definition essentially says that if some value cannot be predicted from other information with probability higher than 2^{-k}, then it has entropy k when conditioned on that information. For example, when a one-way permutation f is hard to invert with probability higher than 2^{-k}, then conditioned on $f(x)$, the value x has entropy k. The use of *conditional* entropy is what makes this definition meaningful for cryptographic applications.

We demonstrate that almost k pseudorandom bits can be extracted from distributions with unpredictability entropy k, by showing that unpredictability entropy implies conditional Yao entropy, to which reconstruction extractors can be applied. Thus, unpredictability entropy provides a simple language that allows, in particular, known results on hardcore bits of one-way functions to be stated more generally.

We also prove other (fairly straightforward) relations between unpredictability entropy and HILL and Yao conditional entropies.

2   Definitions and Notation

In this section we recall the HILL and Yao definitions of computational entropy (or pseudoentropy) and provide the new, conditional definitions.

Notation. We will use n for the length parameter; our distributions will be on strings of length polynomial in n. We will use s as the circuit size parameter (or running time bound when dealing with Turing machines instead of circuits). To denote a value x sampled from a distribution X, we write $x \leftarrow X$. We denote by $M(X)$ the probability distribution on the outputs of a Turing machine M, taken over the coin tosses (if any) of M and the random choice of the input x according to the distribution X. We use U_n to denote the uniform distribution on $\{0,1\}^n$. For a joint distribution (X, Z), we write X_z to denote the conditional distribution of X when $Z = z$; conversely, given a collection of distributions X_z and a distribution Z, we use (X, Z) to denote the joint distribution given by $\mathbf{Pr}[(X, Z) = (x, z)] = \mathbf{Pr}[Z = z]\mathbf{Pr}[X_z = x]$.

We may describe more complicated distributions by describing the sampling process and then the sampled outcome. For example, $\{a \leftarrow X; b \leftarrow X : (a, b)\}$ denotes two independent samples from X, while $\{a \leftarrow X : (a, M(a, Y))\}$ denotes the distribution obtained by sampling X to get a, sampling Y to get b, running $M(a, b)$ to get c, and outputting (a, c).

The statistical distance between two distributions X and Y, denoted by $\mathtt{dist}(X, Y)$, is defined as $\max_T |\mathbf{Pr}[T(X) = 1] - \mathbf{Pr}[T(Y) = 1]|$ where T is any test (function). (This is equivalent to the commonly seen $\mathtt{dist}(X, Y) = \frac{1}{2}\sum_a |\mathbf{Pr}[X = a] - \mathbf{Pr}[Y = a]|$.) The computational distance with respect to size s circuits, denoted by $\mathtt{cdist}_s(X, Y)$, limits T to be any circuit of size s.

Unconditional Computational Entropy. The min-entropy of a distribution X, denoted by $\mathbf{H}_\infty(X)$, is defined as $-\log(\max_x \mathbf{Pr}[X = x])$. Although min-entropy provides a rather pessimistic view of a distribution (looking only at its worst-case element), this notion is useful in cryptography, because even a computationally unbounded predictor can guess the value of a sample from X with probability at most $2^{-\mathbf{H}_\infty(X)}$. Most results on randomness extractors are formulated in terms of min-entropy of the source distribution.

The first definition says that a distribution has high computational min-entropy if it is *indistinguishable* from some distribution with high statistical min-entropy. It can thus be seen as generalization of the notion of pseudorandomness of [Yao82], which is defined as indistinguishability from uniform.

Definition 1 ([HILL99, BSW03]). *A distribution X has* **HILL entropy** *at least k, denoted by $\mathbf{H}_{\epsilon,s}^{\mathsf{HILL}}(X) \geq k$, if there exists a distribution Y such that $\mathbf{H}_\infty(Y) \geq k$ and $\mathtt{cdist}_s(X, Y) \leq \epsilon$.*

(In [HILL99] Y needs to be efficiently samplable; however, for our application, as well as for [BSW03], samplability is not required.)

Another definition of computational entropy considers compression length. Shannon's theorem [Sha48] says that the minimum compression length of a distribution, by all possible compression and decompression functions, is equal to its average entropy (up to small additive terms). Yao [Yao82] proposed to measure computational entropy by imposing computational constraints on the compression and decompression algorithms.[2] In order to convert this into a worst-case (rather than average-case) metric similar to min-entropy, Barak et al. [BSW03] require that any subset in the support of X (instead of only the entire X) be hard to compress.

---

[2] Yao called it "effective" entropy.

Definition 2 ([Yao82, BSW03]). *A distribution X has* **Yao entropy** *at least k, denoted by $\mathbf{H}^{\mathsf{Yao}}_{\epsilon,s}(X) \geq k$, if for every pair of circuits c, d (called "compressor" and "decompressor") of total size s with the outputs of c having length ℓ,*

$$\Pr_{x \leftarrow X}[d(c(x)) = x] \leq 2^{\ell-k} + \epsilon.$$

Note that just like HILL entropy, for $\epsilon = 0$ this becomes equivalent to min-entropy (this can be seen by considering the singleton set of the most likely element).

Conditional Computational Entropy. Before we provide the new conditional definitions of computational entropy, we need to consider the information-theoretic notion of conditional min-entropy.

Let (Y, Z) be a distribution. If we take the straightforward average of the min-entropies $\mathbb{E}_{z \leftarrow Z}[\mathbf{H}_\infty(Y_z)]$ to be the conditional min-entropy, we will lose the relation between min-entropy and prediction probability, which is important for many applications (see e.g. Lemma 4 and Lemma 7). For instance, if for half of Z, $\mathbf{H}_\infty(Y_z) = 0$ and the other half $\mathbf{H}_\infty(Y_z) = 100$, then, given a random z, Y can be predicted with probability over $1/2$, much more than 2^{-50} the average would suggest. A conservative approach, taken in [RW05], would be to take the minimum (over z) of $\mathbf{H}_\infty(Y_z)$. However, this definition may kill "good" distributions like $Y_z = U_n$ for all $z \neq 0^n$ and $Y_z = 0^n$ for $z = 0^n$; although this problem can be overcome by defining a so-called "smooth" version [RW05, RW04], we follow a different approach.

For the purposes of randomness extraction, Dodis et al. [DORS06] observed that because Z is not under adversarial control, it suffices that the *average*, over Z, of the maximum probability is low. They define average min-entropy: $\tilde{\mathbf{H}}_\infty(Y|Z) \stackrel{\text{def}}{=} -\log(\mathbb{E}_{z \leftarrow Z}[2^{-\mathbf{H}_\infty(Y|Z=z)}]) = -\log(\mathbb{E}_{z \leftarrow Z}[\max_y \Pr[Y_z = y]])$. This definition averages prediction probabilities before taking the logarithm and ensures that for any predictor P, $\Pr_{(y,z) \leftarrow (Y,Z)}[P(z) = y] \leq 2^{-\tilde{\mathbf{H}}_\infty(Y|Z)}$. It also ensures that randomness extraction works almost as well as it does for unconditional distributions; see Section 4.

Using this definition of conditional min-entropy, defining conditional HILL-entropy is straightforward.

Definition 3 (Conditional HILL entropy). *For a distribution (X, Z), we say X has HILL entropy at least k conditioned on Z, denoted by $\mathbf{H}^{\mathsf{HILL}}_{\epsilon,s}(X|Z) \geq k$, if there exists a collection of distributions Y_z (giving rise to a joint distribution (Y, Z)) such that $\tilde{\mathbf{H}}_\infty(Y|Z) \geq k$ and $\mathtt{cdist}_s((X, Z), (Y, Z)) \leq \epsilon$.*

For conditional Yao entropy, we simply let the compressor and decompressor have z as input.

Definition 4 (Conditional Yao entropy). *For a distribution (X, Z), we say X has Yao entropy at least k conditioned on Z, denoted by $\mathbf{H}^{\mathsf{Yao}}_{\epsilon,s}(X|Z) \geq k$, if for every pair of circuits c, d of total size s with the outputs of c having length ℓ,*

$$\Pr_{(x,z) \leftarrow (X,Z)}[d(c(x, z), z) = x] \leq 2^{\ell-k} + \epsilon.$$

We postpone the discussion of unpredictability entropy until Section 5.

Asymptotic Definitions. All above definitions are with respect to a single distribution and fixed-size circuits. We are also interested in their asymptotic behaviors, so we consider *distribution ensembles*. In this case, everything is parameterized by n: $X^{(n)}$, $s(n)$, and $\epsilon(n)$. In such a case, whether circuits in our definitions are determined after n is chosen (the nonuniform setting), or whether an algorithm of running time $s(n)$ is chosen independent of n (the uniform setting) makes a difference. We consider the nonuniform setting.

We omit the subscripts $s(n)$ and $\epsilon(n)$ when they "denote" any polynomial and negligible functions, respectively ($\epsilon(n)$ is negligible if $\epsilon(n) \in n^{-\omega(1)}$). More precisely, we write $\mathbf{H}^{\mathsf{HILL}}(X^{(n)}) \geq k(n)$, if there is a distribution ensemble $Y^{(n)}$ such that $\mathbf{H}_\infty(Y^{(n)}) \geq k(n)$ for all n, and for every polynomial $s(n)$, there exists a negligible $\epsilon_s(n)$ such that $\mathtt{cdist}_{s(n)}(X^{(n)}, Y^{(n)}) \leq \epsilon_s(n)$. Similarly for the other definitions.

3   Separating HILL Entropy from Yao Entropy

In this section we construct a joint distribution (X, Z),[3] such that given Z, the distribution X has high Yao but low HILL entropy; namely, $\mathbf{H}^{\mathsf{Yao}}(X|Z) \gg \mathbf{H}^{\mathsf{HILL}}(X|Z)$. This is a separation of conditional HILL and Yao entropies. Since Z will be simply a polynomially long random string, this result can also be viewed as a separation of Yao entropy and HILL entropy in the Common Reference String (CRS) model. (In this model one assumes that a uniformly-distributed string of length $q(n)$, for some fixed polynomial q, is accessible to everyone.)

Our construction uses a non-interactive zero knowledge proof system, so we describe it briefly in the following subsection.

3.1   Non-interactive Zero Knowledge (NIZK)

NIZK was introduced by Blum et al. [BFM88, BDMP91]. For our purposes, a single-theorem variant suffices. Let λ be a positive polynomial and $L \in \mathbf{NP}$ be a language that has witnesses of length n for theorems of lengths $(\lambda(n-1), \lambda(n)]$. (It is easier for us to measure everything in terms of witness length rather than the more traditional theorem length, but they are anyway polynomially related for the languages we are interested in.) NIZK works in the CRS model. Let q be a positive polynomial, and let the CRS be $r \leftarrow U_{q(n)}$ when witnesses are of length n. A NIZK proof system for L is a pair of polynomial-time Turing machines (P, V), called the *prover* and the *verifier* (as well as the polynomial q) such that the following three conditions hold.

1. Completeness: $\forall \phi \in L$ with NP witness w, if $\pi = \mathsf{P}(\phi, w, r)$ is the proof generated by P, then $\mathbf{Pr}_{r \leftarrow U_{q(n)}}[\mathsf{V}(\phi, \pi, r) = 1] = 1$.[4]

---

[3] Actually, (X, Z) should be defined as a distribution ensemble $(X^{(n)}, Z^{(n)})$, but we'll omit the superscript for ease of notation.

[4] If P is probabilistic, the probability is taken over the choice r and random choices made by P.

2. Soundness: Call r *bad* if $\exists \phi \notin L$, $\exists \pi'$, such that $V(\phi, \pi', r) = 1$ (and *good* otherwise). Then $\mathbf{Pr}_{r \leftarrow U_{q(n)}}[r$ is bad$]$ is negligible in n.
3. Zero-knowledgeness: There is a probabilistic polynomial time Turing machine SIM called the simulator, such that $\forall \phi \in L$ and every witness w for ϕ, $\mathsf{SIM}(\phi) = (\phi, \Pi_{\mathsf{SIM}}, R_{\mathsf{SIM}})$ is computationally indistinguishable from $(\phi, \Pi, R) = \{r \leftarrow U_{q(n)} ; \pi \leftarrow P(\phi, w, r) : (\phi, \pi, r)\}$.

For our analysis, we need two additional properties. First, we need the proofs π not to add too much entropy. For this, we use ideas on unique NIZK by Lepinski, Micali and shelat [LMS05]. We do not need the full-fledged uniZK system; rather, the single-theorem system described as the first part of the proof of [LMS05, Theorem 1] suffices (it is based on taking away most of the prover freedom for the single-theorem system of [BDMP91]). The protocol of [LMS05] is presented in the public-key model, in which the prover generates the public key (x, y) consisting of an n-bit modulus x and n-bit value $y \in \mathbb{Z}_x^*$. To make it work for our setting, we simply have the prover generate the public key during the proof and put it into π. Once the public key is fixed, the prover has no further choices in generating π, except choosing a witness w for $\phi \in L$ (note that this actually requires a slight modification to the proof of [LMS05], which we describe in Appendix A).

The second property we need is that the simulated shared randomness R_{SIM} is independent of the simulator input ϕ. It is satisfied by the [LMS05] proof system (as well as by the [BDMP91] system on which it is based).

The zero-knowledge property of the [LMS05] proof system is based on the following assumption (the other properties are unconditional).

Assumption 1 (Quadratic Residuousity [GM84] for Blum Integers). *For all probabilistic polynomial time algorithms P, if p_1 and p_2 are random $n/2$-bit primes congruent to 3 modulo 4, y is a random integer between 1 and $p_1 p_2$ with Jacobi symbol $\left(\frac{y}{p_1 p_2}\right) = 1$, and $b = 1$ if y is a quadratic residue modulo $p_1 p_2$ and 0 otherwise, then $|1/2 - \mathbf{Pr}[P(y, p_1 p_2) = b]|$ is negligible in n.*

The formal statement of the properties we need from [LMS05] follows.

Lemma 1 ([LMS05]+Appendix A). *If the above assumption holds, then there exists an NIZK proof system for any language $L \in \mathbf{NP}$ with the following additional properties: (1) if r is good and ϕ has t distinct witnesses w, then the number of proofs π for ϕ that are accepted by V is at most $t 2^{2n}$, and (2) the string R_{SIM} output by the simulator is independent of the simulator input ϕ.*

3.2   The Construction

Our intuition is based on the separation by Wee [Wee04], who demonstrated an oracle relative to which there is a random variable that has high Yao and low HILL entropy. His oracle consists of a random length-increasing function and an oracle for testing membership in the sparse range of this function. The random variable is simply the range of the function. The ability to test membership in the range helps distinguish it from uniform, hence HILL entropy is low. On the other hand, knowing that a random variable is in the range of a random function does not help to compress it, hence Yao entropy is high.

We follow this intuition, but replace the length-increasing random function and the membership oracle with a pseudorandom generator and an NIZK proof of membership, respectively. Our distribution X consists of two parts: 1) output of a pseudorandom generator and, 2) an NIZK proof that the first part is as alleged. However, an NIZK proof requires a polynomially long random string (shared, but not controlled, by the prover and the verifier). So we consider the computational entropy of X, *conditioned* on a polynomially long random string r chosen from the uniform distribution $Z = U_{q(n)}$.

Let $G : \{0,1\}^n \to \{0,1\}^{\lambda(n)}$, for some polynomial λ, be a pseudorandom generator (in order to avoid adding assumptions, we can build based on Assumption 1), and let $((\mathsf{P}, \mathsf{V}), q)$ be the NIZK proof system guaranteed by Lemma 1 for the **NP** language $L = \{\phi \mid \exists \alpha \text{ such that } \phi = G(\alpha)\}$. Let $Z = R = U_{q(n)}$. Our random variable X consists of two parts $(G(U_n), \pi)$, where π is the proof, generated by P, that the first part is an output of G. More precisely, the joint distribution (X, Z) is defined as $\{\alpha \leftarrow U_n \; ; \; r \leftarrow U_{q(n)} \; ; \; \pi \leftarrow \mathsf{P}(G(\alpha), \alpha, r) : ((G(\alpha), \pi), r)\}$. Note that because X contains a proof relative to the random string r, it is defined only after the value r of Z is fixed.

Lemma 2 (Low HILL entropy). $\mathbf{H}^{\mathsf{HILL}}(X|Z) < 3n + 1$.

Proof. Suppose there is some collection $\{Y_r\}_{r \in Z}$ for which $\tilde{\mathbf{H}}_\infty(Y|Z) \geq 3n + 1$. We will show that there is a distinguisher that distinguishes (X, Z) from (Y, Z). In fact, we will use the verifier V of the NIZK proof system as a universal distinguisher, which works for every such Y.

Let $p(r) \overset{\text{def}}{=} \max_y \mathbf{Pr}[Y_r = y]$ be the probability of most likely value of the random variable Y_r.

When r is good, the number of (ϕ, π) pairs for which $\mathsf{V}(\phi, \pi, r) = 1$ is at most 2^{3n}: the total number 2^n of witnesses times the number of proofs 2^{2n} for each witness. Now, parse y as a theorem-proof pair. The number of y such that $\mathsf{V}(y, r) = 1$ is at most 2^{3n}, and each of these y happens with probability at most $p(r)$. Therefore, when r is good, $\mathbf{Pr}_{y \leftarrow Y_r}[\mathsf{V}(y, r) = 1] \leq 2^{3n} p(r)$, by the union bound. Hence, for any r, $\mathbf{Pr}_{y \leftarrow Y_r}[\mathsf{V}(y, r) = 1 \ \wedge \ r \text{ is good}] \leq 2^{3n} p(r)$ (for good r this is the same as above, and for bad r this probability is trivially 0, because of the conjunction).

Now consider running V on a sample from (Y, Z).

$$
\begin{aligned}
\mathbf{Pr}_{(y,r) \leftarrow (Y,Z)}[\mathsf{V}(y,r) = 1] &\leq \mathbf{Pr}_{r \leftarrow Z}[r \text{ is bad}] + \mathbf{Pr}_{(y,r) \leftarrow (Y,Z)}[\mathsf{V}(y,r) = 1 \ \wedge \ r \text{ is good}] \\
&\leq \mathrm{negl}(n) + \mathbb{E}_{r \leftarrow Z}[\mathbf{Pr}_{y \leftarrow Y_r}[\mathsf{V}(y,r) = 1 \ \wedge \ r \text{ is good}]] \\
&\leq \mathrm{negl}(n) + \mathbb{E}_{r \leftarrow Z}[2^{3n} p(r)] \\
&\leq \mathrm{negl}(n) + \frac{1}{2}
\end{aligned}
$$

(the last inequality follows from the definition of $\tilde{\mathbf{H}}_\infty$: $2^{-\tilde{\mathbf{H}}_\infty(Y|Z)} = \mathbb{E}_{r \leftarrow Z}[p(r)]$
$\leq 2^{-(3n+1)}$).

Since $\mathbf{Pr}_{(x,r) \leftarrow (X,Z)}[\mathsf{V}(x,r) = 1] = 1$, V distinguishes (X, Z) from (Y, Z) with advantage close to $1/2$. $\qquad \square$

Lemma 3 (High Yao entropy). *If Assumption 1 holds, then* $\mathbf{H}^{\mathsf{Yao}}(X|Z) \geq \lambda(n)$.

Proof. Let $s(n)$ be a polynomial. The following two statements imply that under Assumption 1, $\epsilon_s(n) \overset{\text{def}}{=} \mathrm{cdist}_{s(n)}((X, Z), \mathsf{SIM}(U_{\lambda(n)}))$ is negligible, by the triangle inequality.

1. $\mathrm{cdist}_{s(n)}((X, Z), \mathsf{SIM}(G(U_n)))$ is negligible. Indeed, fix a seed $\alpha \in \{0,1\}^n$ for G, and let $(X_\alpha, Z) = \{r \leftarrow U_{q(n)}; \pi \leftarrow \mathsf{P}(G(\alpha), \alpha, r) : ((G(\alpha), \pi), r)\}$. By the zero-knowledge property, we know that $\mathrm{cdist}_{s(n)}((X_\alpha, Z), \mathsf{SIM}(G(\alpha)))$ is negligible. Since it holds for every $\alpha \in \{0,1\}^n$, it also holds for a random α; we conclude that $\mathrm{cdist}_{s(n)}((X, Z), \mathsf{SIM}(G(U_n)))$ is negligible.
2. $\mathrm{cdist}_{s(n)}(\mathsf{SIM}(U_{\lambda(n)}), \mathsf{SIM}(G(U_n)))$ is negligible, because G is a pseudorandom generator.

By definition of $\epsilon_s(n)$, if the compressor and decompressor c and d have total size t, then

$$\left| \Pr_{(x,z) \leftarrow (X,Z)}[d(c(x, z), z) = x] \;-\; \Pr_{(x,z) \leftarrow \mathsf{SIM}(U_{\lambda(n)})}[d(c(x, z), z) = x] \right| \leq \epsilon_s(n),$$

where $s = t + $ (size of circuit to check equality of strings of length $|x|$), because we can use $d(c(\cdot, \cdot), \cdot)$ together with the equality operator as a distinguisher.

Let the output length of c be ℓ. Then $\mathbf{Pr}_{(x,z) \leftarrow \mathsf{SIM}(U_{\lambda(n)})}[d(c(x, z), z) = x] \leq 2^{\ell - \lambda(n)}$, because for every fixed z, x contains $\phi \in U_{\lambda(n)}$ (because by Lemma 1, z is independent of ϕ in the NIZK system we use). Hence $\mathbf{Pr}_{(x,z) \leftarrow (X,Z)}[d(c(x, z), z) = x] \leq 2^{\ell - \lambda(n)} + \epsilon_s(n)$, and $\mathbf{H}^{\mathsf{Yao}}_{\epsilon_s(n), t(n)}(X|Z) \geq \lambda(n)$. For every polynomial $t(n)$, the value $s(n)$ is polynomially bounded, and therefore $\epsilon_s(n)$ is negligible, so $\mathbf{H}^{\mathsf{Yao}}(X|Z) \geq \lambda(n)$. $\qquad \square$

Remark 1. In the previous paragraph, we could consider also the simulated proof π (recall $x = (\phi, \pi)$) when calculating $\mathbf{Pr}_{(x,z) \leftarrow \mathsf{SIM}(U_{\lambda(n)})}[d(c(x, z), z) = x]$ for even higher Yao entropy. A simulated proof π contains many random choices made by the simulator. Although the simulator algorithm for [LMS05] is not precisely specified, but rather inferred from the simulator in [BDMP91], it is quite clear that the simulator will get to flip at least three random coins per clause in the 3-CNF formula produced out of ϕ in the reduction to 3-SAT (these three coins are needed in order to simulate the location of the $(0,0,0)$ triple [LMS05, proof of Theorem 1, step 9] among the eight triples). This more careful calculation of $\mathbf{Pr}_{(x,z) \leftarrow \mathsf{SIM}(U_{\lambda(n)})}[d(c(x, z), z) = x]$ will yield the slightly stronger statement $\mathbf{H}^{\mathsf{Yao}}(X|Z) \geq \lambda(n) + 3\gamma(n)$, where $\gamma(n)$ is the number of clauses in the 3-CNF formula. This more careful analysis is not needed here, but will be used in Section 4.3.

Since for any polynomial $\lambda(n)$, we have pseudorandom generators of stretch λ, Lemma 2 and Lemma 3 yield the following theorem.

Theorem 1 (Separation). *Under the Quadratic Residuosity Assumption, for every polynomial λ, there exists a joint distribution ensemble $(X^{(n)}, Z^{(n)})$ such that $\mathbf{H}^{\mathsf{Yao}}(X^{(n)} \mid Z^{(n)}) \geq \lambda(n)$ and $\mathbf{H}^{\mathsf{HILL}}(X^{(n)} \mid Z^{(n)}) \leq 3n+1$. Moreover, $Z^{(n)} = U_{q(n)}$ for some polynomial $q(n)$.*

4   Randomness Extraction

As mentioned in the introduction, one of the main applications of computational entropy is the extraction of pseudorandom bits. Based on Theorem 1, in this section we show that the analysis based on Yao entropy can yield many more pseudorandom bits than the traditional analysis based on HILL entropy. Although Theorem 1 is for the conditional setting, we will see an example of extraction that benefits from the conditional-Yao-entropy analysis for the unconditional setting as well.

Before talking about extracting pseudorandom bits from computational entropy, let us look at a tool for analogous task in the information-theoretic setting: an *extractor* takes a distribution Y of min-entropy k, and with the help of a uniform string called the seed, "extracts" the randomness contained in Y and outputs a string of length m that is *almost uniform* even given the seed.

Definition 5 ([NZ96]). *A polynomial-time computable function $E : \{0,1\}^n \times \{0,1\}^d \to \{0,1\}^m \times \{0,1\}^d$ is a strong (k, ϵ)-extractor if the last d outputs of bits of E are equal to the last d input bits (these bits are called seed), and $\mathtt{dist}((E(X, U_d), U_m \times U_d) \leq \epsilon$ for every distribution X on $\{0,1\}^n$ with $\mathbf{H}_\infty(X) \geq k$. The number of extracted bits is m, and the entropy loss is $k - m$.*

There is a long line of research on optimizing the parameters of extractors: minimizing seed length, minimizing ϵ, and maximizing m. For applications of primary interest here—using extracted randomness for cryptography—seed length is less important, because strong extractors can use non-secret random seeds, which are usually much easier to create than the secret from which the pseudorandom bits are being extracted. It is more important to maximize m (as close to k as possible), while keeping ϵ negligible.[5]

4.1   Extracting from Conditional HILL Entropy

It is not hard to see that applying an extractor on distributions with HILL entropy yields pseudorandom bits; because otherwise the extractor together with the distinguisher violate the definition of HILL entropy. We show the same for the case of conditional HILL entropy. We reiterate that in the conditional case,

---

[5] This is in contrast to the derandomization literature, where a small constant ϵ suffices, and one is more interested in (simultaneously) maximizing m and minimizing d.

the variable Z is given to the distinguisher who is trying to tell the output of the extractor from random.

Lemma 4. *If* $\mathbf{H}^{\mathsf{HILL}}_{\epsilon_1,s}(X|Z) \geq k$, *then for any* $(k - \log \frac{1}{\delta}, \epsilon_2)$*-extractor* $E : \{0,1\}^n \times \{0,1\}^d \to \{0,1\}^m$,

$$\operatorname*{cdist}_{s'}\left(\{(x,z) \leftarrow (X,Z) : (E(x,U_d),z)\}, U_m \times U_d \times Z\right) \leq \epsilon_1 + \epsilon_2 + \delta,$$

where $s' = s - size(E)$.

Proof. $\mathbf{H}^{\mathsf{HILL}}_{\epsilon_1,s}(X|Z) \geq k$ means that there exists a collection of $\{Y_z\}_{z \in Z}$ such that $\operatorname{cdist}_s((X,Z)(Y,Z)) \leq \epsilon_1$, and $\tilde{\mathbf{H}}_\infty(Y|Z) \geq k$. By Markov's inequality, $\mathbf{Pr}_{z \in Z}[\mathbf{H}_\infty(Y_z) \leq k - \log \frac{1}{\delta}] \leq \delta$. Hence, the extractor works as expected in all but δ fraction of the cases; that is, for all but δ fraction of z values, $\operatorname{dist}(E(Y_z,U_d),U_m \times U_d) \leq \epsilon_2$. Taking expectation over $z \in Z$, we get

$$\operatorname{dist}\left(\{(y,z) \leftarrow (Y,Z) : (E(y,U_d),z)\}, U_m \times U_d \times Z\right) \leq \epsilon_2 + \delta,$$

because dist is bounded by 1. The desired result follows by triangle inequality. $\square$

Remark 2. The entropy loss of E is at least $2\log\frac{1}{\epsilon_2} - O(1)$, by a fundamental constraint on extractors [RT00], giving us a total entropy loss of at least $\log\frac{1}{\delta} + 2\log\frac{1}{\epsilon_2} - O(1)$. The loss of $\log\frac{1}{\delta}$ can be avoided for some specific E, such as pairwise-independent (a.k.a. strongly universal) hashing [CW79], as shown in [DORS06, Lemma 4.2]; because pairwise-independent hashing has optimal entropy loss of $2\log\frac{1}{\epsilon_2} - 2$, this gives us the maximum possible number of extracted bits. The loss of $\log\frac{1}{\delta}$ can be also avoided when $\min_{z \in Z} \mathbf{H}_\infty(Y_z) \geq k$ (as is the case in, e.g., [GKR04]).

Using an extractor on distributions with HILL entropy (the method that we just showed extends to conditional HILL entropy) is a common method for extracting pseudorandom bits. HILL entropy is used, in particular, because it is easier to analyze than Yao entropy. In fact, in the unconditional setting, the only way we know how to show that a distribution has high Yao entropy (incompressibility) is by arguing that it has high HILL entropy (indistinguishability). Nevertheless, Barak et al. [BSW03] showed that some extractors can also extract from Yao entropy.

4.2   Extracting from Conditional Yao Entropy

Barak et al. [BSW03] observed that extractors with the so-called reconstruction procedure can be used to extract from Yao Entropy. Thus, Theorem 1 ($\mathbf{H}^{\mathsf{Yao}}(X|Z) \gg \mathbf{H}^{\mathsf{HILL}}(X|Z)$) suggests that such a *reconstructive* extractor with a Yao-entropy-based analysis may yield more pseudorandom bits than a generic extractor with a traditional HILL-entropy-based analysis. We begin with a definition from [BSW03].

Definition 6 (Reconstruction procedure). *An (ℓ, ϵ)-reconstruction for a function $E : \{0,1\}^n \times \{0,1\}^d \to \{0,1\}^m \times \{0,1\}^d$ (where the last d outputs are equal to the last d inputs bits) is a pair of machines C and D, where $C : \{0,1\}^n \to \{0,1\}^\ell$ is a randomized Turing machine, and $D^{(\cdot)} : \{0,1\}^\ell \to \{0,1\}^n$ is a randomized oracle Turing machine which runs in time polynomial in n. Furthermore, for every x and T, if $|\mathbf{Pr}[T(E(x, U_d)) = 1] - \mathbf{Pr}[T(U_m \times U_d) = 1]| > \epsilon$, then $\mathbf{Pr}[D^T(C^T(x)) = x] > 1/2$ (the probability is over the random choices of C and D).*

Trevisan [Tre99] showed, implicitly, that any E with an (ℓ, ϵ)-reconstruction is an $(\ell + \log \frac{1}{\epsilon}, 3\epsilon)$-extractor, and Barak et al. [BSW03] showed that such extractors can be used to extract pseudorandom bits from distributions with Yao entropy. We extend the proof of Barak et al. so that their result holds for the conditional version of Yao entropy.

Lemma 5. *Let X be a distribution with $\mathbf{H}^{\mathsf{Yao}}_{\epsilon,s}(X|Z) \geq k$, and let E be an extractor with a $(k - \log \frac{1}{\epsilon}, \epsilon)$-reconstruction (C, D). Then $\mathrm{cdist}_{s'}((E(X, U_d), Z), U_m \times U_d \times Z) \leq 5\epsilon$, where $s' = s/(size(C) + size(D))$.*

Proof. Assume, for the purpose of contradiction, that there is a distinguisher T of size s' such that $\mathbf{Pr}[T(E(X, U_d), Z) = 1] - \mathbf{Pr}[T(U_m \times U_d \times Z) = 1] > 5\epsilon$. By the Markov inequality, there is a subset S in the support of (X, Z) such that $\mathbf{Pr}[(X, Z) \in S] \geq 4\epsilon$, and $\forall (x, z) \in S$, $\mathbf{Pr}[T(E(x, U_d), z) = 1] - \mathbf{Pr}[T(U_m \times U_d, z) = 1] > \epsilon$. For every pair $(x, z) \in S$, $\mathbf{Pr}[D^{T(\cdot, z)}(C(x)) = x] > 1/2$, where the probability is over the random choices of C and D. Thus, there is a fixing of the random choices of C and D, denoted by circuits $\bar{C}, \bar{D}$, such that $\mathbf{Pr}_{(x,z) \leftarrow (X,Z)}[\bar{D}^{T(\cdot, z)}(\bar{C}(x)) = x] > 2\epsilon$. Let $c(x, z) = \bar{C}(x)$ and $d(y, z) = \bar{D}^{T(\cdot, z)}(y)$ be the compression and decompression circuits, respectively. Then $\mathbf{Pr}_{(x,z) \leftarrow (X,Z)}[d(c(x, z), z) = x] > 2\epsilon = 2^{\ell - k} + \epsilon$, a contradiction. $\square$

The above lemma does not yield more pseudorandom bits when given a distribution that has high Yao but low HILL entropy, unless we have a reconstructive extractor with long output length (compared to generic extractors, which work for HILL entropy). Fortunately, there is a simple way to increase the output length of a reconstructive extractor, at the expense of increasing the seed length; namely, by applying the extractor multiple times on the same input distribution but each time with an independent fresh seed. Furthermore, there do exist reconstructive extractors; e.g., the Goldreich-Levin extractor: $GL(x, y) \stackrel{\text{def}}{=} (x \cdot y) \circ y$, where $\circ$ denotes concatenation and $\cdot$ denotes inner product. Below, we describe more precisely how to increase the output length. For a proof, we refer the readers to Section 3.5 in the survey by Shaltiel [Sha02].

Proposition 1. *Let $GL : \{0,1\}^n \times \{0,1\}^n \to \{0,1\} \times \{0,1\}^n$ be an extractor with (ℓ, ϵ)-reconstruction. Then $E : \{0,1\}^n \times \{0,1\}^{mn} \to \{0,1\}^m \times \{0,1\}^{mn}$ defined below is an extractor with $(m + \ell, m\epsilon)$-reconstruction. Let $\circ$ denote componentwise concatenation (i.e., to agree syntactically with the definition of extractor, we concatenate the 1-bit outputs and the n-bit seeds separately)*

$$E(x, y_1, \ldots, y_m) \stackrel{\text{def}}{=} GL(x, y_1) \circ \cdots \circ GL(x, y_m).$$

For the Goldreich-Levin extractor, $\ell = O(\log \frac{1}{\epsilon})$. Then Lemma 5 implies that E extracts m pseudorandom bits out of any distribution that has Yao entropy $m + \ell + \log \frac{1}{\epsilon} = m + O(\log \frac{1}{\epsilon})$. This shows that it is possible to extract almost all of Yao entropy (e.g., if the negligible $\epsilon = 2^{-\text{polylog}(n)}$ suffices, then all but a polylogarithmic amount of entropy can be extracted).

Using the distribution of Theorem 1, we can set $\epsilon = 2^{-n}$ to extract $\lambda(n) - O(n)$ bits from X that are pseudorandom even given Z. This is more than the linear number of bits extractable from X using the analysis based on conditional HILL entropy.

4.3   Unconditional Extraction

In this subsection, let $(X, Z) = ((G(U_n), \Pi), R) = \{\alpha \leftarrow U_n \; ; \; r \leftarrow U_{q(n)} \; ; \; \pi \leftarrow \mathsf{P}(G(\alpha), \alpha, r) : ((G(\alpha), \pi), r)\}$ as defined in Section 3.2. The question is: how many pseudorandom bits can we extract from the unconditional distribution (X, Z)? Surprisingly, analysis based on conditional entropy yields more bits than unconditional analysis, demonstrating that the notion of conditional entropy may be a useful tool even in the analysis of pseudorandomness of unconditional distributions.

Analysis based on unconditional entropy. The straightforward way is to apply an extractor on (X, Z). This gives us almost k pseudorandom bits provided that $\mathbf{H}^{\mathsf{HILL}}(X, Z) \geq k$, or $\mathbf{H}^{\mathsf{Yao}}(X, Z) \geq k$ for reconstructive extractors (see previous subsections). However, the best we can show is that $\mathbf{H}^{\mathsf{HILL}}(X, Z) = \lambda(n) + q(n) + O(n)$ (the analysis appears in Appendix B), and hence we cannot prove, using HILL entropy, that more than $\lambda(n) + q(n) + O(n)$ bits can be extracted. On the other hand, we do not know if $\mathbf{H}^{\mathsf{Yao}}(X, Z)$ is higher; this is closely related to the open problem of whether HILL entropy is equivalent to Yao entropy, and appears to be difficult.[6] Thus, analysis based on unconditional entropy does not seem to yield more than $\lambda(n) + q(n) + O(n)$ bits.

More bits from conditional Yao entropy. Analysis based on conditional HILL entropy seems to yield even fewer bits (see Lemma 2). But using conditional Yao entropy, we get the following result.

Lemma 6. *It is possible to extract $4\lambda(n) + q(n) - O(n)$ pseudorandom bits out of (X, Z).*

Proof (Sketch). According to Remark 1 following Lemma 3, we can show that the conditional Yao entropy $\mathbf{H}^{\mathsf{Yao}}(X|Z) \geq \lambda(n) + 3\gamma(n)$, where $\gamma(n)$ is the number

---

[6] To show that $\mathbf{H}^{\mathsf{Yao}}(X, Z)$ is high, one would have to show that the pair (X, Z) cannot be compressed; the same indistinguishability argument as in Lemma 3 does not work for the pair (X, Z), because in the simulated distribution, Z is simulated and thus has less entropy. It is thus possible that both the real distribution (where Z is random and ϕ in X is pseudorandom) and the simulated distribution (where ϕ is random and Z is pseudorandom), although indistinguishable, can be compressed with the help of the proof π.

of clauses in the 3-CNF formula produced from ϕ in the reduction from L to 3-SAT. Since $\gamma(n) \geq \lambda(n)$, we can extract $4\lambda(n) - O(n)$ bits from X that are pseudorandom even given Z, by the last paragraph of Section 4.2. Noting that Z is simply a uniform string[7], we can append it to the pseudorandom bits extracted from X and obtain an even longer pseudorandom string. Thus, we get $4\lambda(n) + q(n) - O(n)$ pseudorandom bits using the analysis based on conditional Yao entropy.   □

5   Unpredictability Entropy

In this section, we introduce a new computational entropy, which we call unpredictability entropy. Analogous to min-entropy, which is the logarithm of the maximum predicting probability, unpredictability entropy is the logarithm of the maximum predicting probability where the predictor is restricted to be a circuit of polynomial size. Note that in the unconditional setting, unpredictability entropy is just min-entropy; a small circuit can have the most likely value hardwired. In the conditional setting, however, this new definition can be very different from min-entropy, and in particular, allows us to talk about the entropy of a value that is unique, such as g^{xy} where g^x and g^y are known to the observer, and possibly even verifiable, such as the preimage x of a one-way permutation f, where $y = f(x)$ is known to the observer.

Definition 7 (Unpredictability entropy). *For a distribution (X, Z), we say that X has* **unpredictability entropy** *at least k conditioned on Z, denoted by $\mathbf{H}^{unp}_{\epsilon,s}(X|Z) \geq k$, if there exists a collection of distributions Y_z (giving rise to a joint distribution (Y, Z)) such that $\mathrm{cdist}_s((X, Z), (Y, Z)) \leq \epsilon$, and for all circuits C of size s,*

$$\mathbf{Pr}[C(Z) = Y] \leq 2^{-k}.$$

Remark 3. The parameter ϵ and the variable Y do not seem to be necessary in the definition; we can simply require $\mathbf{Pr}[C(Z) = X] \leq 2^{-k}$. However, they make this definition *smooth* [RW04] and easier to compare with existing definitions of HILL and Yao entropy.

Remark 4. Note that our entropy depends primarily on the predicting probability, as opposed to on the size of the predicting circuit or the combination of both (see e.g., [TSZ01, HILL99]). We choose to have s fixed, in order to accommodate distributions with nonzero information-theoretic entropy; otherwise the computational entropy of such distribution would be infinite because the predicting probability doesn't increase no matter how big the predicting circuit grows. For the case of one-way function, unpredictability entropy is what is often called "hardness." This notion is more general, and provides a simple language for pseudorandomness extraction: namely, a distribution with computational entropy k contains k pseudorandom bits that can be extracted (see below).

---

[7] In case Z is not uniform but contains some amount of entropy, we can apply another extractor on it.

5.1   Relation to Other Notions and Bit Extraction

In this subsection we show that high conditional HILL entropy implies high unpredictability entropy, which in turn implies high conditional Yao entropy. Note that, assuming exponentially strong one-way permutations f exist, unpredictability entropy does not imply conditional HILL entropy: simply let $(X, Z) = (x, f(x))$.

Lemma 7. $\mathbf{H}_{\epsilon,s}^{\mathsf{HILL}}(X|Z) \geq k \Rightarrow \mathbf{H}_{\epsilon,s}^{\mathsf{unp}}(X|Z) \geq k$.

Proof. $\mathbf{H}_{\epsilon,s}^{\mathsf{HILL}}(X|Z) \geq k$ means that there is a Y such that $\tilde{\mathbf{H}}_{\infty}(Y|Z) \geq k$ and $\mathtt{cdist}_s((X, Z), (Y, Z)) \leq \epsilon$. And $\tilde{\mathbf{H}}_{\infty}(Y|Z) \geq k$ means that $\mathbb{E}_{z \leftarrow Z}[\max_y \Pr[Y = y|Z = z]] \leq 2^{-k}$, which implies that for all circuits C of size s, $\Pr[C(Z) = Y] \leq 2^{-k}$. □

Lemma 8. $\mathbf{H}_{\epsilon,s}^{\mathsf{unp}}(X|Z) \geq k \Rightarrow \mathbf{H}_{\epsilon,s}^{\mathsf{Yao}}(X|Z) \geq k$.

Proof. $\mathbf{H}_{\epsilon,s}^{\mathsf{unp}}(X|Z) \geq k$ means that there is a collection of $\{Y_z\}_{z \in Z}$ such that $\mathtt{cdist}_s((X, Z), (Y, Z)) \leq \epsilon$, and for all circuits C of size s, $\Pr[C(Z) = Y] \leq 2^{-k}$. We will show that $\mathbf{H}_{0,s}^{\mathsf{Yao}}(Y|Z) \geq k$, which in turn implies $\mathbf{H}_{\epsilon,s}^{\mathsf{Yao}}(X|Z) \geq k$.

Suppose for contradiction that $\mathbf{H}_{0,s}^{\mathsf{Yao}}(Y|Z) < k$. Then there exists a pair of circuits c, d of total size s with the outputs of c having length ℓ, such that $\Pr_{(y,z) \leftarrow (Y,Z)}[d(c(y, z), z) = y] > 2^{\ell-k}$. Because $|c(y, z)| = \ell$, guessing the correct value is at least $2^{-\ell}$, so $\Pr_{(a,y,z) \leftarrow (U_\ell, Y, Z)}[d(a, z) = y] > 2^{\ell-k} \cdot 2^{-\ell} = 2^{-k}$, a contradiction since $d(a, \cdot)$ (with some fixing of a) is a circuit of size at most s. So $\mathbf{H}_{0,s}^{\mathsf{Yao}}(Y|Z) \geq k$.

Next, suppose for contradiction that $\mathbf{H}_{\epsilon,s}^{\mathsf{Yao}}(X|Z) < k$. Then there exists a pair of circuits c, d of total size s with the outputs of c having length ℓ, such that $\Pr_{(x,z) \leftarrow (X,Z)}[d(c(x, z), z) = x] > 2^{\ell-k} + \epsilon$. But $\Pr_{(y,z) \leftarrow (Y,Z)}[d(c(y, z), z) = y] \leq 2^{\ell-k}$, which means that $d(c(\cdot, \cdot), \cdot)$ can be used to distinguish (X, Z) from (Y, Z) with advantage more than ϵ, a contradiction to $\mathtt{cdist}_s((X, Z), (Y, Z)) \leq \epsilon$. Hence $\mathbf{H}_{\epsilon,s}^{\mathsf{Yao}}(X|Z) \geq k$. □

From Section 4, we know how to extract almost k bits from distributions with Yao entropy k, by using reconstructive extractors. Lemma 8 implies that the same method works for unpredictability entropy. Thus, the notion of unpredictability entropy allows for more general statements of results on hardcore bits (such as, for example, [GL89, TSZ01]), which are usually formulated in terms of one-way functions. Most often these results generalize easily to other conditionally unpredictable distributions, for instance, the Diffie-Hellman distribution $(g^{xy} \mid g, g^x, g^y)$. However, such generalization is not automatic, because a prediction of a one-way function inverse is verifiable (namely, knowing y, one can check if the guess for $f^{-1}(y)$ is correct), while a guess of a value of a conditionally unpredictable distribution in general is not (indeed, the Diffie-Hellman distribution does not have it unless the decisional Diffie-Hellman problem is easy). Thus, it would be beneficial if results were stated for the more general case of unpredictable distributions whenever such verifiability is not crucial. Unpredictability entropy provides a simple language for doing so.

Acknowledgments

We thank anonymous referees for their helpful comments, and Moni Naor for pointing out related work. This work supported was in part by the US National Science Foundation grants CCR-0311485, CCF-0515100 and CNS-0546614, the Taiwan National Science Council grants NSC95-2218-E-001-001, NSC95-2218-E-011-015 and NSC95-3114-P-001-002-Y02, and the Institute for Pure and Applied Mathematics at UCLA.

References

[BDMP91] Manuel Blum, Alfredo De Santis, Silvio Micali, and Giuseppe Persiano. Noninteractive zero-knowledge. *SIAM Journal on Computing*, 20(6):1084–1118, December 1991.

[BFM88]   Manuel Blum, Paul Feldman, and Silvio Micali. Non-interactive zero-knowledge and its applications (extended abstract). In *Proceedings of the Twentieth Annual ACM Symposium on Theory of Computing*, pages 103–112, Chicago, Illinois, 2–4 May 1988.

[BSW03]   Boaz Barak, Ronen Shaltiel, and Avi Wigderson. Computational analogues of entropy. In Sanjeev Arora, Klaus Jansen, José D. P. Rolim, and Amit Sahai, editors, *RANDOM-APPROX 2003*, volume 2764 of *LNCS*, pages 200–215. Springer, 2003.

[CW79]   J.L. Carter and M.N. Wegman. Universal classes of hash functions. *Journal of Computer and System Sciences*, 18:143–154, 1979.

[DH76]   Whitfield Diffie and Martin E. Hellman. New directions in cryptography. *IEEE Transactions on Information Theory*, IT-22(6):644–654, 1976.

[DORS06] Yevgeniy Dodis, Rafail Ostrovsky, Leonid Reyzin, and Adam Smith. Fuzzy extractors: How to generate strong keys from biometrics and other noisy data. Technical Report 2003/235, Cryptology ePrint archive, http://eprint.iacr.org, 2006. Previous version appeared at *EURO-CRYPT 2004*.

[GKR04]   Rosario Gennaro, Hugo Krawczyk, and Tal Rabin. Secure hashed Diffie-Hellman over non-DDH groups. In Christian Cachin and Jan Camenisch, editors, *Advances in Cryptology—EUROCRYPT 2004*, volume 3027 of *LNCS*, pages 361–381. Springer-Verlag, 2004.

[GL89]   O. Goldreich and L. Levin. A hard-core predicate for all one-way functions. In *Proceedings of the Twenty First Annual ACM Symposium on Theory of Computing*, pages 25–32, Seattle, Washington, 15–17 May 1989.

[GM84]   S. Goldwasser and S. Micali. Probabilistic encryption. *Journal of Computer and System Sciences*, 28(2):270–299, April 1984.

[HILL99] J. Håstad, R. Impagliazzo, L.A. Levin, and M. Luby. Construction of pseudorandom generator from any one-way function. *SIAM Journal on Computing*, 28(4):1364–1396, 1999.

[Imp99]   Russell Impagliazzo. Remarks in open problem session at the dimacs workshop on pseudorandomness and explicit combinatorial constructions, 1999.

[LMS05]   Matt Lepinski, Silvio Micali, and Abhi Shelat. Fair-zero knowledge. In Joe Kilian, editor, *TCC*, volume 3378 of *LNCS*, pages 245–263. Springer-Verlag, 2005.

[Nao96]   Moni Naor. Evaluation may be easier than generation. In *Proceedings of the Twenty-Eighth Annual ACM Symposium on the Theory of Computing*, pages 74–83, Philadelphia, Pennsylvania, 22–24 May 1996.

[NZ96]   Noam Nisan and David Zuckerman. Randomness is linear in space. *Journal of Computer and System Sciences*, 52(1):43–53, 1996.

[RT00]   Jaikumar Radhakrishnan and Amnon Ta-Shma. Bounds for dispersers, extractors, and depth-two superconcentrators. *SIAM Journal on Computing*, 13(1):2–24, 2000.

[RW04]   Renato Renner and Stefan Wolf. Smooth rényi entropy and applications. In *Proceedings of IEEE International Symposium on Information Theory*, page 233, June 2004.

[RW05]   Renato Renner and Stefan Wolf. Simple and tight bounds for information reconciliation and privacy amplification. In Bimal Roy, editor, *Advances in Cryptology—ASIACRYPT 2005*, LNCS, Chennai, India, 4–8 December 2005. Springer-Verlag.

[Sha48]   Claude E. Shannon. A mathematical theory of communication. *Bell System Technical Journal*, 27:379–423 and 623–656, July and October 1948. Reprinted in D. Slepian, editor, *Key Papers in the Development of Information Theory*, IEEE Press, NY, 1974.

[Sha02]   Ronen Shaltiel. Recent developments in explicit constructions of extractors. *Bulletin of the EATCS*, 77:67–95, 2002.

[Tre99]   Luca Trevisan. Construction of extractors using pseudo-random generators (extended abstract). In *STOC*, pages 141–148, 1999.

[TSZ01]   Amnon Ta-Shma and David Zuckerman. Extractor codes. In *STOC*, pages 193–199, 2001.

[TVZ05]   Luca Trevisan, Salil P. Vadhan, and David Zuckerman. Compression of samplable sources. Technical Report TR05-012, Electronic Colloquium on Computational Complexity (ECCC), 2005.

[Wee04]   Hoeteck Wee. On pseudoentropy versus compressibility. In *IEEE Conference on Computational Complexity*, pages 29–41. IEEE Computer Society, 2004.

[Yao82]   A. C. Yao. Theory and applications of trapdoor functions. In *23rd Annual Symposium on Foundations of Computer Science*, pages 80–91, Chicago, Illinois, 3–5 November 1982. IEEE.

A   Modifications to the Proof of [LMS05]

The proof of Theorem 1 in [LMS05] requires the n-bit modulus x chosen by the prover (and, in our case, included as part of the proof) to be a Blum integer, i.e., a product of two primes that are each congruent to 3 modulo 4. However, the proof π (using the techniques from [BDMP91]) guarantees only that x is "Regular(2)," i.e., is square-free and has exactly two distinct odd prime divisors. In other words, we are assured only that x is of the form $p^i q^j$ for some odd primes p, q and some i, j not simultaneously even. Soundness does not suffer if a prover maliciously chooses such an x that is not a Blum integer, but the uniqueness property does: there may be more than one valid proof π, because π consists of square roots s of values in $\mathbb{Z}_x^*$ such that the Jacobi symbol $\left(\frac{s}{x}\right) = 1$ and $s < x/2$, and there may be more than one such square root if x is not a Blum integer.

One approach to remedy this problem is to use the technique proposed in countable zero-knowledge of Naor [Nao96, Theorem 4.1]: to include into π the proof that x is a Blum integer. Another, simpler, approach (which does not seem to work for the problem in [Nao96], because the length of the primes is important there) is to require the verifier to check that $x \equiv 1 \pmod 4$. This guarantees that either $p \equiv q \equiv 3 \bmod 4$ and i, j are odd, in which case uniqueness of a square root $r < x/2$ with $\left(\frac{r}{x}\right) = 1$ is guaranteed, or $p^i \equiv q^j \equiv 1 \bmod 4$, in which case simple number theory (case analysis by the parity of i, j) shows that half the quadratic residues in $\mathbb{Z}_x^*$ have *no* square root r with $\left(\frac{r}{x}\right) = 1$. Thus, such an x that allows for non-unique proofs is very unlikely to work for a shared random string r, and we can simply add strings r for which such an x exists to the set of bad strings (which will remain of negligible size).

B   Unconditional HILL Entropy of (X, Z)

Recall that $(X, Z) = ((G(U_n), \Pi), R) = \{\alpha \leftarrow U_n \; ; \; r \leftarrow U_{q(n)} \; ; \; \pi \leftarrow \mathsf{P}(G(\alpha), \alpha, r) : ((G(\alpha), \pi), r)\}$. Below, we show that $\mathbf{H}^{\mathsf{HILL}}(X, Z) \geq \lambda(n) + q(n) + O(n)$; it is unclear if higher HILL entropy can be shown. The discussion assumes some familiarity with the NIZK system for 3-SAT, by Lepinski, Micali, and shelat [LMS05].

By the zero-knowledgeness, the output distribution $(X_{\mathsf{SIM}}, Z_{\mathsf{SIM}})$ of the simulator is indistinguishable from (X, Z). So $\mathbf{H}^{\mathsf{HILL}}(X, Z)$ is no less than the min-entropy of $(X_{\mathsf{SIM}}, Z_{\mathsf{SIM}})$. We count how many choices the simulator SIM has: there are,

- $2^{\lambda(n)}$ theorems to prove,
- fewer than 2^{2n} proving pairs to choose from (a proving pair is an n-bit Blum integer x and an n-bit quadratic residue $y \in \mathbb{Z}_x^*$),
- $2^{q(n)-\kappa(n)}$ choices for shared "random" string r, where $\kappa(n)$ is the number of Jacobi symbol 1 elements of $\mathbb{Z}_x^*$ included in r (because in the simulated r, these elements must be quadratic residues in $\mathbb{Z}_x^*$),
- $2^{\kappa(n)}$ choices for claiming, in the simulated proof, whether each of the Jacobi symbol 1 elements in r is a quadratic residue or a quadratic nonresidue (the simulator gets to make false claims about that, because in the simulated r, they are all residues).

Taking the logarithm of the number of choices, we have $\mathbf{H}^{\mathsf{HILL}}(X, Z) > \lambda(n) + q(n) + O(n)$. This seems to be the best we can do, as we do not know whether there are other distribution that is indistinguishable from (X, Z).

Zero Knowledge and Soundness Are Symmetric[*]

Shien Jin Ong and Salil Vadhan

School of Engineering and Applied Sciences
Harvard University
Cambridge, Massachusetts, USA
{shienjin,salil}@eecs.harvard.edu

Abstract. We give a complexity-theoretic characterization of the class of problems in **NP** having zero-knowledge argument systems. This characterization is symmetric in its treatment of the zero knowledge and the soundness conditions, and thus we deduce that the class of problems in **NP** ∩ **coNP** having zero-knowledge arguments is closed under complement. Furthermore, we show that a problem in **NP** has a *statistical* zero-knowledge argument system if and only if its complement has a computational zero-knowledge *proof* system. What is novel about these results is that they are *unconditional*, i.e., do not rely on unproven complexity assumptions such as the existence of one-way functions.

Our characterization of zero-knowledge arguments also enables us to prove a variety of other unconditional results about the class of problems in **NP** having zero-knowledge arguments, such as equivalences between honest-verifier and malicious-verifier zero knowledge, private coins and public coins, inefficient provers and efficient provers, and non-black-box simulation and black-box simulation. Previously, such results were only known unconditionally for zero-knowledge *proof systems*, or under the assumption that one-way functions exist for zero-knowledge argument systems.

1 Introduction

Zero-knowledge protocols are interactive protocols whereby one party, the *prover*, convinces another party, the *verifier*, that some assertion is true with the remarkable property that the verifier "learns nothing" other than the fact that the assertion being proven is true. Since their introduction by Goldwasser, Micali, and Rackoff [GMR], zero-knowledge protocols have played a central role in the design and study of cryptographic protocols.

Zero-knowledge protocols come in several flavors, depending on how one formulates the two security conditions: (1) the zero-knowledge condition, which says that the verifier "learns nothing" other than the fact the assertion being proven is true, and (2) the soundness conditions, which says that the prover

---

[*] A preliminary version of this paper appeared in the *Electronic Colloquium on Computational Complexity* [OV]. Both the authors were supported by NSF grant CNS-0430336 and ONR grant N00014-04-1-0478.

cannot convince the verifier of a false assertion. In *statistical zero knowledge*, the zero-knowledge condition holds regardless of the computational resources the verifier invests into trying to learn something from the interaction. In *computational zero knowledge*, we only require that a probabilistic polynomial-time verifier learn nothing from the interaction.[1] Similarly, for soundness, we have *statistical soundness*, giving rise to *proof systems*, where even a computationally unbounded prover cannot convince the verifier of a false statement (except with negligible probability), and *computational soundness*, giving rise to *argument systems* [BCC], where we only require that a polynomial-time prover cannot convince the verifier of a false statement. Using a prefix of **S** or **C** to indicate whether the zero knowledge is statistical or computational and a suffix of **P** or **A** to indicate whether we have a proof system or argument system, we obtain four complexity classes corresponding to the different types of zero-knowledge protocols: **SZKP**, **CZKP**, **SZKA**, **CZKA**. More precisely, these are the classes of *decision problems* Π having the correponding type of zero-knowledge protocol. In such a protocol, the prover and verifier are given as common input an instance x of Π, and the prover is trying convince the verifier that x is a YES instance of Π.

These two security conditions seem to be of very different flavors; zero knowledge is a 'secrecy' condition, whereas soundness is more like an 'unforgeability' condition. However, in a remarkable paper, Okamoto [Oka] showed that they are actually symmetric in the case of statistical security.

Theorem 1 ([Oka, GSV][2]**).** *The class* **SZKP** *of problems having statistical zero-knowledge proofs is closed under complement. That is,* $\Pi \in$ **SZKP** *if and only if* $\overline{\Pi} \in$ **SZKP**.

In a zero-knowledge protocol for proving that a string x is a YES instance of a problem Π, zero knowledge is required only when x is a YES instance (that is, when the statement being proven is true) and soundness is required only when x is a NO instance (that is, when the statement is false). Thus, by showing that **SZKP** is closed under complement, Okamoto established a symmetry between zero knowledge and soundness, in the case when both security conditions are statistical.

We ask whether an analogous theorem holds when the security conditions are *computational*, namely when considering computational zero-knowledge arguments. If we make complexity assumptions, then the answer is yes. Indeed, the classical results of Goldreich, Micali, and Wigderson [GMW], and Brassard, Chaum, and Crépeau [BCC] show that every problem in **NP** has computational

---

[1] More precisely, in statistical zero knowledge, we require that the verifier's view of the interaction can be efficiently simulated up to negligible statistical distance, whereas in computational zero knowledge, we only require that the simulation be computationally indistinguishable from the verifier's view.

[2] Okamoto's result was actually for the class of languages having *honest-verifier* statistical zero-knowledge proofs, but in [GSV] it was shown this is the same as the class of languages having general statistical zero-knowledge proofs.

zero-knowledge argument systems under widely believed complexity assumptions, and in fact either one of the security conditions can be made statistical. Moreover, it is known that the existence of one-way functions (OWF) suffices for the construction of computational zero-knowledge proof systems and statistical zero-knowledge argument systems for every problem in **NP** [Nao, HILL, NOV]. Thus, the existence of one-way functions implies that computational zero knowledge and computational soundness are symmetric for problems in $\mathbf{NP} \cap \mathbf{coNP}$, by implying that all problems in $\mathbf{NP} \cap \mathbf{coNP}$ and their complements have computational zero-knowledge arguments. We note that here, and throughout the paper, we usually restrict attention to problems in **NP**, because argument systems are mainly of interest when the prover can be implemented in polynomial time given a witness of membership, which only makes sense for problems in $\mathbf{NP}$.[3]

In this paper, we establish an *unconditional* symmetry between computational zero knowledge and computational soundness.

Theorem 2 (Symmetry Theorem)

1. *(**CZKA** versus **co-CZKA**) A problem $\Pi \in \mathbf{NP} \cap \mathbf{coNP}$ has a computational zero-knowledge argument system if and only if $\overline{\Pi}$ has a computational zero-knowledge argument system.*
2. *(**SZKA** versus **CZKP**) A problem $\Pi \in \mathbf{NP}$ has a statistical zero-knowledge argument system if and only if $\overline{\Pi}$ has a computational zero-knowledge proof system.*

Observe how the quality of the zero-knowledge condition for Π translates to the quality of the soundness condition for $\overline{\Pi}$ and vice-versa.

1.1 The SZKP–OWF Characterization

The Symmetry Theorem is obtained by new characterizations of the classes of problems having zero-knowledge protocols, and moreover these characterizations treat zero knowledge and soundness symmetrically. These characterizations are a generalization of the "SZK/OWF Characterization Theorem" of [Vad], which says that any problem having a computational zero-knowledge *proof* system can be described as a problem having a statistical zero-knowledge proof plus a set of YES instances from which we can construct a one-way function. To characterize zero-knowledge *argument* systems, we will also allow some additional NO instances from which we can construct a one-way function.

To formalize this, we will need the notion of a *promise problem*, which is simply a decision problem with some inputs excluded. More precisely, a promise problem Π consists of two disjoint sets of strings (Π_Y, Π_N), corresponding to YES and NO instances respectively. All of the complexity classes that we consider—for

---

[3] Actually polynomial-time provers also make sense for problems in **MA**, which is a variant of **NP** where the verification of witnesses is probabilistic. All of our results easily extend to **MA**, but we state them for **NP** for simplicity.

instance, **SZKP**, **CZKP**, **SZKA**, and **CZKA**—generalize to promise problems in a natural way; completeness and zero knowledge are required for YES instances, and soundness is required for NO instances.

Definition 1 (SZKP–OWF CONDITION). *We say that promise problem* $\Pi = (\Pi_Y, \Pi_N)$ *satisfies the* SZKP–OWF CONDITION *if there exists a set of instances* $I \subseteq \Pi_Y \cup \Pi_N$ *such that the following two conditions hold:*

- *The promise problem* $(\Pi_Y \setminus I, \Pi_N \setminus I)$ *is in* **SZKP**.
- *There exists a polynomial-time computable function* $f_x \colon \{0,1\}^{n(|x|)} \to \{0,1\}^{m(|x|)}$, *with* $n(\cdot)$ *and* $m(\cdot)$ *being polynomials and instance* x *given as an auxiliary input, such that for every nonuniform probabilistic polynomial-time adversary* A, *and for every constant* $c > 0$, *we have*

$$\Pr_{y \leftarrow \{0,1\}^{n(|x|)}} \left[A(f_x(y)) \in f_x^{-1}(f_x(y)) \right] \leq |x|^{-c} ,$$

for every sufficiently long $x \in I$.

We call I *the set of* OWF *instances,* $I \cap \Pi_Y$ *the set of* OWF YES *instances, and* $I \cap \Pi_N$ *the set of* OWF NO *instances.*

We use the SZKP–OWF CONDITION to characterize the classes of problems having zero-knowledge protocols.

Theorem 3 (SZKP–OWF Characterization of Zero Knowledge)

1. (**SZKP** [trivial]) *A problem* $\Pi \in$ **IP** *has a statistical zero-knowledge proof system if and only if* Π *satisfies the* SZKP–OWF CONDITION *without* OWF *instances, namely* $I = \emptyset$.
2. (**CZKP** [Vad]) *A problem* $\Pi \in$ **IP** *has a computational zero-knowledge proof system if and only if* Π *satisfies the* SZKP–OWF CONDITION *without* OWF NO *instances, namely* $I \cap \Pi_N = \emptyset$.
3. (**SZKA** [new]) *A problem* $\Pi \in$ **NP** *has a statistical zero-knowledge argument system if and only if* Π *satisfies the* SZKP–OWF CONDITION *without* OWF YES *instances, namely* $I \cap \Pi_Y = \emptyset$.
4. (**CZKA** [new]) *A problem* $\Pi \in$ **NP** *has a computational zero-knowledge argument system if and only if* Π *satisfies the* SZKP–OWF CONDITION.

Theorem 2, our Symmetry Theorem between computational zero knowledge and computational soundness, follows directly from: (i) Theorem 3 above, (ii) Okamoto's Theorem that **SZKP** is closed under complement (Theorem 1), and (iii) the symmetric role played by the set of OWF instances I in the SZKP–OWF CONDITION.

The advantage of the SZKP–OWF Characterization Theorem is that it reduces the study of the various forms of zero-knowledge protocols to the study of **SZKP** together with the study of the consequences of one-way functions, both of which are by now quite well-developed. Indeed, we also use these characterizations to prove many other unconditional theorems about the classes of problems in **NP** possessing zero-knowledge arguments, such as equivalences between honest-verifier and malicious-verifier zero knowledge, private coins and public coins, inefficient provers and efficient provers, and non-black-box simulation and black-box simulation. Previously, such results were only known unconditionally for the case of zero-knowledge *proof* systems [Oka, GSV, Vad, NV], or were known under the complexity assumptions like the existence of one-way functions for the case of zero-knowledge argument systems [GMW, Nao, HILL, NOV].

While our characterizations of **SZKA** and **CZKA** (Items 3 and 4) are similar in spirit to the **CZKP** characterization of [Vad] (Item 2), both directions of the implications require new ingredients that were not present in [Vad].

In the forward direction, going from **CZKA** or **SZKA** to an SZKP–OWF CONDITION, we combine the work of [Vad] with an idea of Ostrovsky [Ost] to construct a one-way function on NO instances in $I \cap \Pi_N$. Ostrovsky showed that if a *hard-on-average* problem has a statistical zero-knowledge argument system, then (standard) one-way functions exist.[4] (This was later generalized to computational zero knowledge in [OW].) We use the same construction, but with a slightly different analysis. In Ostrovsky's work, the hardness of inverting the one-way function is derived from the assumed (average-case) hardness of the problem having the zero-knowledge protocol, and it is shown to be hard to invert on YES instances. In our proof, the hardness of inverting the one-way function is instead derived from a gap between between statistical soundness and computational soundness, and it is analyzed on NO instances.

In the reverse direction, going from an SZKP–OWF CONDITION to **CZKA** or **SZKA**, there were more fundamental obstacles in extending the work of [Vad]. First, the construction of [Vad] made use of a computationally unbounded prover in an essential way (as did the previous work on **SZKP**, such as [Oka]), whereas argument systems are rather unnatural with unbounded provers and hence are typically defined with respect to efficient provers. Second, at the time we did not know of a construction of statistical zero-knowledge arguments for **NP** from any one-way function, which is necessary to make use of the one-way functions constructed from instances in $I \cap \Pi_N$—this is clear when trying to characterize **SZKA**, but it also turns out to be important for characterizing **CZKA**. Fortunately, both of these obstacles have been recently overcome in [NV] and [NOV], respectively.

In more detail, the way the reverse direction is proved is to show that for any problem Π satisfying the SZKP–OWF CONDITION, we can construct an *instance-dependent* commitment scheme,[5] and then we use the instance-dependent commitment scheme to construct a zero-knowledge protocol for Π. In the original version of this paper [OV], our instance-dependent commitment scheme inherited a certain "1-out-of-2" binding property from [NV] and [NOV]. This property is weaker and more complicated than the standard binding

---

[4] Ostrovky's theorem is only stated in terms of statistical zero-knowledge proofs, but it immediately extends to arguments.

[5] Informally, instance-dependent commitment schemes for a problem Π are commitment schemes where the hiding and binding properties are required to hold only on the YES and NO instances of Π, respectively. A formal definition is given in Sect. 2.1.

property of commitments, but sufficed for establishing our main theorems (Theorems 2 and 3). Subsequently, the results of [NV] and [NOV] have been improved to yield standard-binding commitments, the latter by Haitner and Reingold [HR] and the former by [HORV]. Thus in this version, we use standard-binding instance-dependent commitments, as it simplifies our presentation.

2   Preliminaries

If X is a random variable taking values in a finite set $\mathcal{U}$, then we write $x \leftarrow X$ to indicate that x is selected according to X. If S is a subset of $\mathcal{U}$, then $x \leftarrow S$ means that x is selected according to the uniform distribution on S. We adopt the convention that when the same random variable occurs several times in an expression, they refer to a single sample. For example, $\Pr[f(X) = X]$ is defined to be the probability that when $x \leftarrow X$, we have $f(x) = x$. We write U_n to denote the random variable distributed uniformly over $\{0,1\}^n$.

A function $\varepsilon : \mathbb{N} \to [0,1]$ is called *negligible* if $\varepsilon(n) = n^{-\omega(1)}$. We let $\mathrm{neg}(n)$ denote an arbitrary negligible function (i.e., when we say that $f(n) < \mathrm{neg}(n)$ we mean that *there exists* a negligible function $\varepsilon(n)$ such that for every n, $f(n) < \varepsilon(n)$). Likewise, $\mathrm{poly}(n)$ denotes an arbitrary function $f(n) = n^{O(1)}$.

PPT refers to probabilistic algorithms (i.e., Turing machines) that run in *strict* polynomial time. A *nonuniform* PPT algorithm is a pair $(A, \bar{z})$, where $\bar{z} = z_1, z_2, \ldots$ is an infinite sequence of strings where $|z_n| = \mathrm{poly}(n)$, and A is a PPT algorithm that receives pairs of inputs of the form $(x, z_{|x|})$. (The string z_n is the called the *advice string* for A for inputs of length n.) Nonuniform PPT algorithms are equivalent to (nonuniform) families of polynomial-sized Boolean circuits.

Statistical Difference. The *statistical difference* (a.k.a. *variation distance*) between random variables X and Y taking values in $\mathcal{U}$ is defined to be $\Delta(X,Y) = \max_{S \subseteq \mathcal{U}} |\Pr[X \in S] - \Pr[Y \in S]|$. We say that X and Y are ε-*close* if $\Delta(X,Y) \leq \varepsilon$. Conversely, we say that X and Y are ε-*far* if $\Delta(X,Y) > \varepsilon$. For basic facts about this metric, see [SV, Sec 2.3].

Promise problems. Roughly speaking, a *promise problem* [ESY] is a decision problem where some inputs are excluded. Formally, a promise problem is specified by two disjoint sets of strings $\Pi = (\Pi_Y, \Pi_N)$, where we call Π_Y the set of YES *instances* and Π_N the set of NO *instances*. Such a promise problem is associated with the following computational problem: given an input that is "promised" to lie in $\Pi_Y \cup \Pi_N$, decide whether it is in Π_Y or in Π_N. Note that languages are a special case of promise problems (namely, a language L over alphabet Σ corresponds to the promise problem $(L, \Sigma^* \setminus L)$). Thus working with promise problems makes our results more general. Moreover, even to prove our results just for languages, it turns out to be extremely useful to work with promise problems along the way. We refer the reader to the recent survey of Goldreich [Gol2] for more on the utility and subtleties of promise problems.

2.1 Instance-Dependent Cryptographic Primitives

It will be very useful for us to work with cryptographic primitives that may depend on an instance x of a problem $\Pi = (\Pi_Y, \Pi_N)$, and where the security condition will hold only if x is in some particular set $I \subseteq \{0,1\}^*$. Indeed, recall that the SZKP–OWF CONDITION (Definition 1) refers to such a variant of of one-way functions, as captured by Definition 3 below.

Instance-Dependent One-Way Functions. To define instance-dependent one-way functions, we will need to define what it means for a function to be *instance dependent*.

Definition 2. *An* instance-dependent function *is a family* $\mathcal{F} = \{f_x \colon \{0,1\}^{n(|x|)} \to \{0,1\}^{m(|x|)}\}_{x \in \{0,1\}^*}$, *where* $n(\cdot)$ *and* $m(\cdot)$ *are polynomials. We call* $\mathcal{F}$ polynomial-time computable *if there is a deterministic polynomial-time algorithm* F *such that for every* $x \in \{0,1\}^*$ *and* $y \in \{0,1\}^{n(|x|)}$, *we have* $F(x,y) = f_x(y)$.

To simplify notation, we often write $f_x \colon \{0,1\}^{n(|x|)} \to \{0,1\}^{m(|x|)}$ to mean the family $\{f_x \colon \{0,1\}^{n(|x|)} \to \{0,1\}^{m(|x|)}\}_{x \in \{0,1\}^*}$.

Definition 3 (Instance-Dependent One-Way Function). *For any set* $I \subseteq \{0,1\}^*$, *a polynomial-time computable instance-dependent function* $f_x \colon \{0,1\}^{n(|x|)} \to \{0,1\}^{m(|x|)}$ *is an* instance-dependent one-way function on I *if for every nonuniform PPT adversary* A, *there exists a negligible function* ε *such that for every* $x \in I$,

$$\Pr_{y \leftarrow \{0,1\}^{n(|x|)}} \left[A(x, f_x(y)) \in f_x^{-1}(f_x(y)) \right] \leq \varepsilon(|x|) \ .$$

Next we consider an instance-dependent variant of *distributionally one-way functions*, which are functions that are hard for PPT adversaries to invert in a distributional manner—that is, given y it is hard for PPT adversaries to output a random preimage $f^{-1}(y)$. The standard definition of distributionally one-way function is given by Impagliazzo and Luby [IL]; here we give the instance-dependent analogue.

Definition 4 (Instance-Dependent Distributionally One-Way Function). *For any set* $I \subseteq \{0,1\}^*$, *a polynomial-time computable instance-dependent function* $f_x \colon \{0,1\}^{n(|x|)} \to \{0,1\}^{m(|x|)}$ *is an* instance-dependent distributionally one-way function on I *if there exists a polynomial* $p(\cdot)$ *such that for every nonuniform PPT adversary* A, *the random variables* $(U_{n(|x|)}, f_x(U_{n(|x|)}))$ *and* $(A(f_x(U_{n(|x|)})), f_x(U_{n(|x|)}))$ *are* $1/p(|x|)$-*far for all sufficiently long* $x \in I$.

Asking to invert in a distributional manner is a stronger requirement that just finding a preimage, therefore distributionally one-way functions might seem weaker than one-way functions. However, Impagliazzo and Luby [IL] proved that they are in fact equivalent. Like almost all reductions between cryptographic primitives, this result immediately extends to the instance-dependent analogue (using the same proof).

Proposition 1 (based on [IL, Lemma 1]). *For every set $I \subseteq \{0,1\}^*$, there exists an instance-dependent one-way function on I if and only if there exists an instance-dependent distributionally one-way function on I.*

Indistinguishability of Instance-Dependent Ensembles. The notions of statistical and computational indistinguishability have instance-dependent analogues. But first, we define an instance-dependent analogue of probability ensembles.

Definition 5. *An* instance-dependent probability ensemble *is a collection of random variables $\{A_x\}_{x \in \{0,1\}^*}$, where A_x takes values in $\{0,1\}^{p(|x|)}$ for some polynomial p. We call such an ensemble* samplable *if there is a probabilistic polynomial-time algorithm M such that for every x, the output $M(x)$ is distributed according to A_x.*

Definition 6. *Two instance-dependent probability ensembles $\{A_x\}_{x \in \{0,1\}^*}$ and $\{B_x\}_{x \in \{0,1\}^*}$ are* computationally indistinguishable on $I \subseteq \{0,1\}^*$ *if for every nonuniform PPT D, there exists a negligible function ε such that for all $x \in I$,*

$$|\Pr[D(x, A_x) = 1] - \Pr[D(x, B_x) = 1]| \leq \varepsilon(|x|) \ .$$

Similarly, we say that $\{A_x\}_{x \in \{0,1\}^*}$ and $\{B_x\}_{x \in \{0,1\}^*}$ are statistically indistinguishable on $I \subseteq \{0,1\}^*$ *if the above is required for all functions D, instead of only nonuniform PPT ones. Equivalently, $\{A_x\}_{x \in \{0,1\}^*}$ and $\{B_x\}_{x \in \{0,1\}^*}$ are statistically indistinguishable on I iff A_x and B_x are $\varepsilon(|x|)$-close for some negligible function ε and all $x \in I$. We write $\approx_c$ and $\approx_s$ to denote computational and statistical indistinguishability, respectively.*

Often, we will informally say "A_x and B_x are computationally indistinguishable when $x \in I$" to mean the ensembles $\{A_x\}_{x \in \{0,1\}^*}$ and $\{B_x\}_{x \in \{0,1\}^*}$ are computationally indistinguishable on I.

Instance-Dependent Commitment Schemes. Recall that a (standard) commitment scheme is a two-stage protocol between a sender and a receiver. In the first stage, called the *commit stage*, the sender "commits" to a private message m. In the second stage, called the *reveal stage*, the sender reveals m and "proves" that it was the message to which she committed in the first stage. We require two properties of commitment schemes. The *hiding* property says that the receiver learns nothing about m in the commit stage. The *binding* property says that after the commit stage, the sender is bound to a particular value of m; that is, she cannot successfully open the commitment to two different bits in the reveal stage. A commitment scheme is said to be *public coin* if the all messages from the receiver to the sender are random coin tosses.

Instance dependent analogues of commitments schemes are commitments schemes that are tailored specifically to a specific problem Π. More precisely, *instance-dependent commitment schemes* receive an instance x of the problem Π as auxiliary input, and are required to be hiding when $x \in \Pi_Y$ and be binding when $x \in \Pi_N$. Thus, they are a relaxation of standard commitment schemes, since we do not require that the hiding and binding properties hold at the same time. Nevertheless, this relaxation is still useful in constructing zero-knowledge protocols. The reason is that zero-knowledge protocols based on commitments (for example, the protocol of [GMW]) typically use only the hiding property in proving zero knowledge (which is required only when x is a YES instance) and use only the binding property in proving soundness (which is required only when x is a NO instance).

2.2   Interactive Protocols and Zero Knowledge

In general, we follow the standard definitions of *interactive protocols*, *interactive proofs* and *arguments*, and *zero-knowledge proofs* and *arguments*, as in [Gol1]. We provide informal definitions of completeness, soundness, and public coin properties of an interactive protocol (P, V) for a promise problem $\Pi = (\Pi_Y, \Pi_N)$; the reader is referred to [Gol1] for the formal definitions.

- The *completeness error* of (P, V) is the maximum probability of V rejecting when interacting with an honest prover P on an input $x \in \Pi_Y$; we usually insist that the completeness error of an interactive protocol be bounded by $1/3$. We say that (P, V) has *perfect completeness* if it has zero completeness error; in other word, V always accepts with probability 1 when interacting with the honest prover P on every input $x \in \Pi_Y$.
- The *statistical [resp., computational] soundness error* of (P, V) is the probability of V accepting when interacting with any [resp., nonuniform PPT] adversarial prover P^* on input $x \in \Pi_N$. Protocol (P, V) is said to be a *proof [resp., argument] system* if it has statistical [resp., computational] soundness error bounded by $1/3$.
- We say (P, V) is *public coin* if all the messages sent by verifier V to prover P are random coin tosses.

Informally, an interactive protocol is *zero knowledge* if the verifier "learns nothing" from interacting with the prover other than the fact that the assertion being proven is true. This guarantee of "learning nothing" is formalized by exhibiting a PPT algorithm, called a *simulator*, whose output is indistinguishable from the verifier's view of the interaction with the prover. (Unlike the verifier, the simulator does not have access to the prover.) Intuitively, the verifier learns nothing because it could run the simulator instead of interacting with the prover. There are various notions of zero knowledge, referring to how rich a class of verifier strategies are considered. We informally describe them as follows:

- *Honest-verifier zero knowledge* refers to interactive protocols where there exists a PPT simulator for the verifier that follows the prescribed (honest) strategy.[6] This is the weakest formulation of zero knowledge, but it is already a nontrivial and interesting notion.

---

[6] This is an instantiation of what is called an "honest-but-curious adversary" or "passive adversary" in the literature on cryptographic protocols.

- *Auxiliary-input zero knowledge* or just *zero knowledge* refers to interactive protocols where for every (nonuniform PPT) verifier V^*, even one that deviates from the prescribed strategy, there exists a PPT simulator that simulates the view of V^* in the interaction with the prover.
- *Black-box zero knowledge* refers to zero knowledge protocols where the zero knowledge property is established by exhibiting a single, universal simulator that simulates an arbitrary verifier strategy V^* by using V^* as a subroutine. In other words, the simulator does not depend on or use the code of V^* (or its auxiliary input), and instead only requires black-box access to V^*.

The complexity classes that we use are defined as follows:

- **IP** denotes the class of promise problems possessing interactive proof systems.
- **HV-SZKP** and **HV-CZKP** denote the classes of promise problems having honest-verifier statistical and computational zero-knowledge proofs, respectively. Analogously, **HV-SZKA** and **HV-CZKA** denote the classes of promise problems having honest-verifier statistical and computational zero-knowledge *arguments*, respectively.
- **SZKP** and **CZKP** are the classes of promise problems possessing statistical and computational (auxiliary-input) zero-knowledge proofs, respectively. Analogously, **SZKA** and **CZKA** are the classes of promise problems possessing statistical and computational (auxiliary-input) zero-knowledge *arguments*, respectively.

We highlight the following points:

1. (Proof vs. argument systems) Interactive argument systems refer to protocols whose *soundness* condition is *computational*. That is, only nonuniform PPT cheating provers are guaranteed not to be able to convince the verifier of false statements except with probability $1/3$; this is a weaker condition than proof systems, where the soundness condition is required of all cheating provers instead of just nonuniform PPT ones. Hence, we say that proof systems have *statistical soundness*.

2. (Prover complexity) In interactive proofs and interactive arguments, and in their zero-knowledge analogues, we allow the honest prover to be computationally unbounded, unless we specify *efficient prover*, which means a polynomial-time honest prover strategy given a witness for membership. It was shown in [NV] that for problems in **NP**, any zero-knowledge *proof* system with an unbounded prover can be transformed into one with an efficient prover; we will show the same for *argument* systems.

3   Unconditional Characterizations of Zero Knowledge

In this section, we provide *unconditional* characterizations of zero knowledge that would among other things allow us to establish our Symmetry Theorem between computational zero knowledge and computational soundness (Theorem 2). We first present our main characterization theorems in Sect. 3.1, which expands upon Theorem 3. The steps involved in proving these characterization theorems are outlined in Sect. 3.2, and lemmas needed to establish these theorems are given in Sects. 3.3, 3.4, and 3.5.

3.1 Our Main Characterization Theorems

In this subsection, we elaborate upon the SZKP–OWF Characterization of Zero Knowledge Theorem (Theorem 3). Specifically, we state four theorems giving a variety of equivalent characterizations of the classes **SZKP**, **CZKP**, **CZKA**, and **SZKA**. The ones for zero-knowledge arguments, namely **CZKA** and **SZKA**, are new; the other for zero-knowledge proofs, namely **CZKP** and **SZKP**, contain results from previous work, but are given for comparison. In addition to establishing Theorem 3 (and hence Theorem 2), these theorems show an equivalence between problems having only honest-verifier zero-knowledge protocols, problems satisfying the SZKP–OWF CONDITION, and problems with (malicious-verifier) zero-knowledge protocols having desirable properties like an efficient prover, perfect completeness, public coins, and black-box simulation. We note that these characterizations refer only to the classes of problems, and do not necessarily preserve other efficiency measures like round complexity, unless explicitly mentioned.

The following two previously known theorems give unconditional characterizations of zero-knowledge *proofs*.

Theorem 4 (SZKP Characterization Theorem [Oka, GSV, NV, HORV]).
*For every problem $\Pi \in$ **IP**, the following conditions are equivalent.*

1. $\Pi \in$ **HV-SZKP**.
2. Π *satisfies the* SZKP–OWF CONDITION *without* OWF *instances.*
3. Π *has an instance-dependent commitment scheme that is statistically hiding on the* YES *instances and statistically binding on the* NO *instances. Moreover, the scheme is public coin.*
4. $\Pi \in$ **SZKP**, *and the statistical zero-knowledge proof system for Π has a black-box simulator, is public coin, and has perfect completeness. Furthermore, if $\Pi \in$ **NP**, the proof system has an efficient prover.*

Theorem 5 (CZKP Characterization Theorem [Vad, NV, HORV]).
*For every problem $\Pi \in$ **IP**, the following conditions are equivalent.*

1. $\Pi \in$ **HV-CZKP**.
2. Π *satisfies the* SZKP–OWF CONDITION *without* OWF NO *instances.*
3. Π *has an instance-dependent commitment scheme that is computationally hiding on the* YES *instances and statistically binding on the* NO *instances. Moreover, the scheme is public coin.*
4. $\Pi \in$ **CZKP**, *and the computational zero-knowledge proof system for Π has a black-box simulator, is public coin, and has perfect completeness. Furthermore, if $\Pi \in$ **NP**, the proof system has an efficient prover.*

We give analogous characterizations for zero-knowledge *arguments*.

Theorem 6 (SZKA Characterization Theorem). *For every problem* $\Pi \in$ **NP**, *the following conditions are equivalent.*

1. $\Pi \in$ **HV-SZKA**.
2. Π *satisfies the* SZKP–OWF CONDITION *without* OWF YES *instances.*
3. Π *has an instance-dependent commitment scheme that is statistically hiding on the* YES *instances and computationally binding on the* NO *instances. Moreover, the scheme is public coin.*
4. $\Pi \in$ **SZKA**, *and the statistical zero-knowledge argument system for* Π *has a black-box simulator, is public coin, has perfect completeness, and an efficient prover.*

Theorem 7 (CZKA Characterization Theorem). *For every problem* $\Pi \in$ **NP**, *the following conditions are equivalent.*

1. $\Pi \in$ **HV-CZKA**.
2. Π *satisfies the* SZKP–OWF CONDITION.
3. Π *has an instance-dependent commitment scheme that is computationally hiding on the* YES *instances and computationally binding on the* NO *instances. Moreover, the scheme is public coin.*
4. $\Pi \in$ **CZKA**, *and the computational zero-knowledge proof system for* Π *has a black-box simulator, is public coin, has perfect completeness, and an efficient prover.*

We prove Theorems 6 and 7 using lemmas established in Sections 3.3, 3.4, and 3.5. Notice that in these theorems involving zero knowledge arguments, we have restricted Π to be in **NP** in contrast to the theorems involving zero-knowledge proofs (Theorems 4 and 5), which are naturally restricted to **IP**. The reason for this is that argument systems are mainly interesting when the honest prover runs in polynomial time given a witness for membership (otherwise the protocol would not even be sound against prover strategies with the same resources as the honest prover), and such efficient provers only make sense for problems in **NP** (or actually, **MA**, to which our results generalize easily). In fact our theorems above show that for problems in **NP**, a zero-knowledge protocol without an efficient prover can be converted into one with an efficient prover (by the equivalence of Items 1 and 4 in Theorems 4 to 6 above).

3.2 Steps of Our Proof

Having stated our main characterization theorems in the previous subsection, we now provide an outline of the steps involved in establishing these characterization theorems:

1. We show that every problem Π possessing a (honest-verifier) zero-knowledge protocol satisfies the SZKP–OWF CONDITION. Depending on the zero knowledge and soundness guarantee, the types of SZKP–OWF CONDITION that Π satisfies will differ (in whether the sets of OWF YES instances and OWF NO instances are empty or nonempty). This extends the unconditional characterization work of [Vad] for zero-knowledge proof systems to the more general zero-knowledge argument systems, and is in Section 3.3.

2. Next, we show that every problem Π satisfying the SZKP–OWF CONDITION yields an *instance-dependent commitment scheme* for Π. This is based on the techniques of [NOV, NV, HR, HORV], and is in Section 3.4.

3. Finally, we show that every problem $\Pi \in \mathbf{NP}$ having instance-dependent commitments allow us to construct zero-knowledge argument systems for Π with desirable properties like perfect completeness, black-box zero knowledge, public coins, and an efficient prover. This is done by substituting instance-dependent commitments for standard (non-instance-dependent) commitments used in existing zero-knowledge protocols like the Goldreich–Micali–Wigderson [GMW] zero-knowledge protocol for $\mathbf{NP}$, and is in Section 3.5.

3.3 From Zero-Knowledge Protocols to SZKP–OWF Characterizations

In this subsection, we show that problems possessing (honest verifier) zero-knowledge arguments satisfy the SZKP–OWF CONDITION. Specifically, we prove that for every problem Π having a zero-knowledge argument also satisfies the SZKP–OWF CONDITION. This involving establishing a set of instances $I \subseteq \Pi_Y \cup \Pi_N$ such that $(\Pi_Y \setminus I, \Pi_N \setminus I) \in \mathbf{SZKP}$, and from which instance-dependent one-way functions can be constructed. The main difference from [Vad] is that [Vad] characterizes only zero-knowledge proofs and has no OWF NO instances, namely $I \cap \Pi_N = \emptyset$. In other words, the characterizations of [Vad] satisfy the SZKP–OWF CONDITION without OWF NO instances.

Lemma 1. *If problem $\Pi \in \mathbf{HV\text{-}CZKA}$, then Π satisfies the* SZKP–OWF CONDITION. *In addition, if $\Pi \in \mathbf{HV\text{-}SZKA}$, then Π satisfies the* SZKP–OWF CONDITION *without* OWF YES *instances, namely $I \cap \Pi_Y = \emptyset$.*

Proof Idea. To show that $\Pi \in \mathbf{HV\text{-}CZKA}$ satisfies the SZKP–OWF CONDITION, we will need to establish a set I with $(\Pi_Y \setminus I, \Pi_N \setminus I) \in \mathbf{SZKP}$, and construct an instance-dependent one-way on I. We will do a separate analysis for the YES and NO instances, and therefore we first show how to define sets $I_Y \subseteq \Pi_Y$ and $I_N \subseteq \Pi_N$ such that instance-dependent one-way functions can be constructed on these sets, and that $(\Pi_Y \setminus I_Y, \Pi_N \setminus I_N) \in \mathbf{SZKP}$. Having two (different) instance-dependent one-way functions f_x and g_x on I_Y and I_N, respectively, we construct a single instance-dependent one-way function on $I \overset{\text{def}}{=} I_Y \cup I_N$ by concatenating the functions f_x and g_x.

Next, we describe, on an intuitive level, how to define the sets $I_Y \subseteq \Pi_Y$ and $I_N \subseteq \Pi_N$. Fix an instance x of the problem $\Pi \in \mathbf{HV\text{-}CZKA}$. From the simulator S on input x, we consider a simulation-based prover P_S and a simulation-based verifier V_S. On a high level, P_S replies with the same conditional probability as the prover in the output of S, and V_S sends its messages with the same conditional probability as the verifier in the output of S. We make the following observations:

1. The interaction between P_S and V_S is identical to the output of the simulator S, on every x.
2. By the zero-knowledge condition, we have that $\langle P_S, V_S \rangle$ is computationally indistinguishable from $\langle P, V \rangle$, when $x \in \Pi_Y$.
3. By assuming, without loss of generality, that the simulator always outputs accepting transcripts, it holds that P_S makes V_S accepts with probability 1, on every x.

We consider a statistical measure of how "similar" V_S is to V (on instance x, when interacting with simulation-based prover P_S). Using this statistical measure (given in the full proof below), we define sets I_Y and I_N as follows:

- I_Y contains instances $x \in \Pi_Y$ for which V_S is *statistically different* from V.
- I_N contains instances $x \in \Pi_N$ for which V_S is *statistically similar* to V.

Now the proof that this gives a SZKP–OWF CONDITION proceed as follows:

1. On I_Y, we have that V_S is statistically different from V. Nevertheless, by the zero-knowledge condition (as noted above), V_S is computationally similar to V. This enables us to construct one-way functions for instances in I_Y, as shown in [Vad].
2. On I_N, we have that V_S is statistically similar to V. Combining this with the fact that P_S will always convince V_S to accept (as noted above), we conclude that P_S convinces V to accept with high probability. By computational soundness of (P, V), it must be the case that P_S is not PPT. Using techniques from Ostrovsky [Ost], this allows us to convert the simulator S into an instance-dependent distributional one-way function g_x.[7] Then by Proposition 1, due to Impagliazzo and Luby [IL], we can obtain an instance-dependent one-way function from g_x.
3. To see that $(\Pi_Y \setminus I_Y, \Pi_N \setminus I_N) \in$ **SZKP**, we observe the following: For those YES instances not in I_Y—that is, instances in $\Pi_Y \setminus I_Y$—the simulated verifier V_S is statistically similar to V. And for those NO instances not in I_N—that is, instances in $\Pi_N \setminus I_N$—the simulated verifier V_S is statistically different from V. This gap in the statistical properties allows us to reduce promise problem $(\Pi_Y \setminus I_Y, \Pi_N \setminus I_N)$ to one of the complete problems for **SZKP** [SV, GV, Vad].

Proof of Lemma 1. Let (P, V) be a zero-knowledge argument system for Π, with simulator S. We now proceed as in the proof of [Vad] and modify our interactive protocol (P, V) to satisfy the following (standard) additional properties.

- The completeness error $c(|x|)$ and soundness error $s(|x|)$ are both negligible. This can be achieved by standard error reduction via (sequential) repetition.

---

[7] If g_x is not distributionally one-way, then P_S can be made to be efficient, hence contradicting the computational soundness of (P, V). Interestingly, Ostrovsky [Ost] uses the assumption that g_x is not distributionally one-way to invert the simulator S on the YES instances, and conclude that Π is not "hard-on-average". Although we use similar techniques as [Ost], we instead invert S on the NO instances to contradict the computational soundness of (P, V).

- On every input x, the two parties exchange $2\ell(|x|)$ messages for some polynomial ℓ, with the verifier sending even-numbered messages and sending all of its $r(|x|)$ random coin tosses in the last message. (Without loss of generality, we may assume that $r(|x|) \geq |x|$.) Having the verifier send its coin tosses at the end does not affect soundness because it is after the prover's last message, and does not affect honest-verifier zero knowledge because the simulator is anyhow required to simulate the verifier's coin tosses.
- On every input x, the simulator always outputs *accepting transcripts*, where we call a sequence τ of 2ℓ messages an accepting transcript on x if all of the verifier's messages are consistent with its coin tosses (as specified in the last message), and the verifier would accept in such an interaction.

For a transcript τ, we denote by τ_i the *prefix* of τ consisting of the first i messages. For readability, we often drop the input x from the notation, for instance using $\ell = \ell(|x|)$, $\langle P, V \rangle = \langle P, V \rangle(x)$, $r = r(|x|)$, and so forth. Thus, in what follows, $\langle P, V \rangle_i$ and S_i are random variables representing prefixes of transcripts generated by the real interaction and simulator, respectively, on a specified input x.

We define the *simulation-based prover*, denoted as $P_S(x)$, as follows: Given an execution prefix τ_{2i}, for $i = 1, 2, \ldots, \ell - 1$, prover P_S responses as follows.

1. If simulator $S(x)$ outputs a transcript that begins with τ_{2i} with probability 0, then P_S replies with a dummy message.
2. Otherwise, P_S replies according with the same conditional probability as the prover in the output of the simulator. That is, it replies with a string β with probability $p_\beta = \Pr\left[S(x)_{2i+1} = \tau_{2i} \circ \beta | S(x)_{2i} = \tau_{2i}\right]$.

Following [AH, PT, GV, Vad], we consider the following quantity:

$$h(x) = \sum_{i=1}^{\ell} [\mathrm{H}(S(x)_{2i}) - \mathrm{H}(S(x)_{2i-1})] \quad , \tag{1}$$

where $\mathrm{H}(\cdot)$ denotes the *(Shannon) entropy* measure, which is given by $\mathrm{H}(X) = \mathrm{E}_{x \leftarrow X}[\log(1/\Pr[X = x])]$.

Now, we define the sets I_Y and I_N as follows:

$$I_Y = \{x \in \Pi_Y : h(x) < r(|x|) - 1/q(|x|)\} \quad ;$$
$$I_N = \{x \in \Pi_N : h(x) > r(|x|) - 2/q(|x|)\} \quad ,$$

where the polynomial $q(|x|) = 256 \cdot \ell(|x|)$.

Having defined sets I_Y and I_N, Lemma 1 is established by the following claims, where the first three are established using techniques in [Vad].

Claim 1. *Problem* $(\Pi_Y \setminus I_Y, \Pi_N \setminus I_N) \in \mathbf{SZKP}$.

Claim 2. *There exists an instance-dependent one-way function on I_Y.*

Claim 3. *For $\Pi \in \mathbf{HV\text{-}SZKA}$, we can take $I_Y = \emptyset$.*

The main novelty in our analysis is the following claim.

Claim 4. *There exists an instance-dependent one-way function on I_N.*

Proof of Claim. To get an instance-dependent one-way function on I_N, we use the following idea of Ostrovsky [Ost]: If we can invert the simulator, then P_S's replies can be approximated efficiently. By the computational soundness of (P, V), this is impossible, so the simulator must be a one-way function. More precisely, we define the function g_x, whose purpose is to output messages of the simulator, as follows:

$$g_x(i, \omega) = (x, i, S(x; \omega)_{2i}) \ . \tag{2}$$

Note that g_x is polynomial-time computable because the simulator S runs in polynomial time. If g_x is *not* distributionally one-way (in the sense of Definition 4), then we can devise an efficient cheating prover strategy, call it $\widetilde{P}$, that *efficiently* "simulates" our simulation-based prover P_S upto negligible statistical error. The way to do this is to feed a given transcript prefix τ_{2i} after the verifier has responded in round $2i$, into the inversion algorithm of g_x to obtain the simulation-based prover response for round $2i + 1$. In doing so, we contradict the computational soundness property of (P, V). This argument is captured by following proposition.

Proposition 2 (based on [Ost, Lemma 1]). [8] *Let g_x be as in (2). For every set $K \subseteq \{0, 1\}^*$, if g_x is not an instance-dependent distributionally one-way function on K, then for every polynomial p, there exists a nonuniform PPT prover $\widetilde{P}$ such that*

$$\Delta(\langle \widetilde{P}, V \rangle(x), S(x)) \leq \ell(|x|) \cdot \left(\frac{1}{p(|x|)} + 2 \cdot \Delta(\langle P_S, V \rangle(x), S(x)) \right) \ ,$$

for infinitely many $x \in K$.

This leaves us to upper bound $\Delta(\langle P_S, V \rangle, S)$ in order to obtain an upper bound on $\Delta(\langle \widetilde{P}, V \rangle, S)$, and hence contradict the computational soundness of V (because S always outputs accepting transcripts). Recall that for every $x \in I_N$, we have $h > r - 2/q$. From [AH, PT, GV], we know that $h = r - \mathrm{KL}(\langle P_S, V \rangle, S)$, where KL is the *Kullback-Leibler* distance defined as $\mathrm{KL}(X, Y) = \mathrm{E}_{\alpha \leftarrow X} \left[\log(\Pr[X = \alpha]) - \log(\Pr[Y = \alpha]) \right]$. (See [GV, Lemma 2.2].) Hence, we get $\mathrm{KL}(\langle P_S, V \rangle, S) < 2/q$. Using the fact that for any random variables X and Y, $\mathrm{KL}(X, Y) \geq (1/2) \cdot (\Delta(X, Y))^2$ [CT, Lemma 12.6.1], we get that for all $x \in I_N$,

$$\Delta(\langle P_S, V \rangle, S) < 2/\sqrt{q} = 1/(8 \cdot \ell) \ , \tag{3}$$

since $q = 256 \cdot \ell$.

---

[8] As pointed out to us by Lilach Bien, the statement and application of this proposition in the original version of our paper [OV] erroneously neglected the dependence on $\Delta(\langle P_S, V \rangle(x), S(x))$.

Now by Proposition 2, if g_x is not distributionally one-way on I_N, we can take $I_N = K$ and choose $p(|x|) = 4 \cdot \ell(|x|)$, to get a nonuniform PPT $\widetilde{P}$ such that

$$\Delta(\langle \widetilde{P}, V \rangle, S) \leq \ell \cdot (1/p + 2 \cdot \Delta(\langle P_S, V \rangle, S))$$
$$= 1/4 + 2 \cdot \ell \cdot \Delta(\langle P_S, V \rangle, S)$$
$$< 1/2 . \qquad \qquad \text{(by (3))}$$

And since the simulator S always produce accepting transcripts, we have

$$\Pr[(\widetilde{P}, V)(x) = \texttt{accept}] \geq 1/2 ,$$

for infinitely many $x \in I_N$. This contradicts the computational soundness of (P, V). Therefore, g_x must be a distributionally one-way function on I_N. By Proposition 1 (due to Impagliazzo and Luby [IL]), g_x can be converted into an instance-dependent (standard) one-way function on I_N, as desired. $\qquad \square$

Let us see how the above five claims establish Lemma 1. Define set $I = I_Y \cup I_N$. This means that the promise problem $(\Pi_Y \setminus I, \Pi_N \setminus I) = (\Pi_Y \setminus I_Y, \Pi_N \setminus I_N)$, and Claim 1 places this problem in **SZKP**. Claims 2 and 4 give us instance-dependent one-way functions on I_Y and I_N, respectively; to obtain a single instance-dependent one-way function on $I = I_Y \cup I_N$, we use the following claim.

Claim 5. *For any sets $J, K \subseteq \{0, 1\}^*$, if there exist instance-dependent one-way functions on J and there exist instance-dependent one-way functions on K, then there exist instance-dependent one-way functions on $J \cup K$.*

Therefore, by Claim 5 above, we know that $\Pi \in$ **HV-CZKA** satisfies the SZKP–OWF CONDITION. Furthermore, if $\Pi \in$ **HV-SZKA**, Claim 3 tells us that $I_Y = \emptyset$, and hence $I \cap \Pi_Y = I_Y = \emptyset$, giving us that Π satisfies the SZKP–OWF CONDITION without OWF YES instances. $\qquad \square$

3.4 From SZKP–OWF Characterization to Instance-Dependent Commitment Schemes

In this subsection, we show that every problem Π satisfying the SZKP–OWF CONDITION yields an instance-dependent commitment scheme for Π. This is obtained by combining statistically-binding commitments from one-way functions [Nao, HILL], statistically-hiding commitments from one-way functions [NOV, HR], and instance-dependent commitments for **SZKP** [NV, HORV]. In the original version of this paper [OV], our instance-dependent commitment scheme inherited a certain "1-out-of-2" binding property from [NV, NOV]. This property is weaker and more complicated than the standard binding property of commitments, but sufficed for establishing our main theorems (Theorems 2 and 3). Due to improvements by [HR, HORV], it is now possible to construct instance-dependent commitments with the standard binding property, and hence we use standard-binding commitments to simplify our presentation.

Lemma 2. *The following conditions hold for problems* Π *satisfying the* SZKP– OWF CONDITION.

- *If* Π *satisfies the* SZKP–OWF CONDITION *without* OWF NO *instances [resp., without* OWF *instances], then it has an instance-dependent commitment scheme that is computationally [resp., statistically] hiding on the* YES *instances and statistically binding on the* NO *instances.*
- *If* Π *satisfies the* SZKP–OWF CONDITION *[resp., without* OWF YES *instances], then it has an instance-dependent commitment scheme that is computationally [resp., statistically] hiding on the* YES *instances and computationally binding on the* NO *instances.*

Furthermore, all the above instance-dependent commitment schemes are public coin.

The proof of Lemma 2, tying together all the following propositions and claims, is given at the end of this subsection. Before stating our propositions and claims, we provide an outline of what we intend to construct in the next paragraph.

Given that problem Π satisfies the SZKP–OWF CONDITION, we let the set of OWF YES instances be denoted as $I_Y = I \cap \Pi_Y$, and the set of OWF NO instances be denoted as $I_N = I \cap \Pi_N$. Our task of constructing an instance-dependent commitment scheme for Π is broken into following four steps: (1) construct an instance-dependent commitment scheme for the problem $(\Pi_Y \setminus I, \Pi_N \setminus I) \in \mathbf{SZKP}$, (2) construct an instance-dependent commitment scheme for the problem $(I_Y, \overline{I_Y})$, (3) construct an instance-dependent commitment scheme for the problem $(\overline{I_N}, I_N)$, and (4) combine all these three instance-dependent commitment schemes into a single instance-dependent commitment scheme for Π. We will explain why these four steps yield an instance-dependent commitment scheme for Π in the proof of Lemma 2, given at the end of this subsection.

Step 1: The instance-dependent commitment for the problem $(\Pi_Y \setminus I_Y, \Pi_N \setminus I_N) \in$ **SZKP** follows from [HORV] (which builds on [NV]).

Proposition 3 ([HORV]). *For any problem* Γ ∈ **SZKP**, *problem* Γ *has an instance-dependent commitment scheme that is statistically hiding on the* YES *instances and statistically binding on the* NO *instances. Moreover, the instance-dependent commitment scheme obtained is public coin.*

Step 2: Notice that the instance-dependent commitments given by the above proposition do not guarantee hiding or binding properties on the OWF instances sets I_Y and I_N. Nevertheless, we noted in [Vad], we can use the instance-dependent one-way functions on I_Y to construct instance-dependent commitment schemes that are computationally hiding on I_Y and statistically binding elsewhere, based on Naor's [Nao] commitment scheme. This is because Naor's scheme can be based on any one-way function [HILL], and the statistical binding property of the scheme does not depend on the one-way security of the function.

Proposition 4 (based on [Nao, HILL]). *For every set $K \subseteq \{0,1\}^*$, if there is an instance-dependent one-way function on K, then problem $(K, \overline{K})$ has an instance-dependent commitment scheme that is computationally hiding on the YES instances (namely, instances in K), and statistically binding on the NO instances (namely, instances in $\overline{K}$). Moreover, the instance-dependent commitment scheme obtained is public coin.*

Step 3: We construct instance-dependent commitment schemes that are computationally binding on I_N and statistically hiding elsewhere, based on the fact that statistically hiding and computationally binding commitments can be constructed from any one-way function [NOV, HR].

Proposition 5 (based on [NOV, HR]). *For every set $K \subseteq \{0,1\}^*$, if there is an instance-dependent one-way function on K, then problem $(\overline{K}, K)$ has an instance-dependent commitment that is statistically hiding on the YES instances (namely, instances in $\overline{K}$), and computationally binding on the NO instances (namely, instances in K). Moreover, the instance-dependent commitment scheme obtained is public coin.*

Step 4: Finally, we use standard methods to combine the three instance-dependent commitment schemes that we have constructed into a single instance-dependent commitment scheme for Π. The first method gives a combined scheme for the intersection of two problems.

Claim 6. *Suppose problems $\Gamma = (\Gamma_Y, \Gamma_N)$ and $\Gamma' = (\Gamma'_Y, \Gamma'_N)$ have instance-dependent commitment schemes Com_x and Com'_x, respectively. Then problem $\Gamma \cap \Gamma' = (\Gamma_Y \cap \Gamma'_Y, \Gamma_N \cup \Gamma'_N)$ has an instance-dependent commitment scheme Com''_x with the following properties:*

- Com''_x *is statistically [resp., computationally] hiding if both Com_x and Com'_x are statistically [resp., computationally] hiding.*
- Com''_x *is statistically [resp., computationally] binding if either of Com_x or Com'_x is statistically [resp., computationally] binding.*
- Com''_x *is public coin if both Com_x and Com'_x are public coin.*

Proof. In commitment scheme Com''_x, the sender commits to b by committing to b in both schemes Com_x and Com'_x, with the execution of both schemes done in parallel. The claimed properties of Com''_x follow by inspection. $\qquad\square$

The second method provides a combined scheme for the union of two problems.

Claim 7. *Suppose problems $\Gamma = (\Gamma_Y, \Gamma_N)$ and $\Gamma' = (\Gamma'_Y, \Gamma'_N)$ have instance-dependent commitment schemes Com_x and Com'_x, respectively. Then problem $\Gamma \cup \Gamma' = (\Gamma_Y \cup \Gamma'_Y, \Gamma_N \cap \Gamma'_N)$ has an instance-dependent commitment scheme Com''_x with the following properties:*

- Com_x'' is statistically [resp., computationally] hiding if either of Com_x or Com_x' is statistically [resp., computationally] hiding.
- Com_x'' is statistically [resp., computationally] binding if both Com_x and Com_x' are statistically [resp., computationally] binding.
- Com_x'' is public coin if both Com_x and Com_x' are public coin.

Proof. In commitment scheme Com_x'', the sender on input bit b, first secret shares b into two shares, b_1 and b_2, with the property that $b_1 \oplus b_2 = b$ and each b_i is uniform in $\{0, 1\}$. (This can be done by choosing a random $b_1 \leftarrow \{0, 1\}$, and setting $b_2 = b_1 \oplus b$.) The sender then commits to b by committing to bits b_1 and b_2 in schemes Com_x and Com_x', respectively. The execution of schemes Com_x and Com_x' is done in parallel.

The hiding property follows from the fact that bit b remains hidden as long as one of the bits b_1 or b_2 remains hidden. Then binding property follows from the fact that $b = b_1 \oplus b_2$, and hence b is bounded to a fixed value if both b_1 and b_2 are bounded to fixed values. The public coin property and round complexity of Com_x'' follow by inspection. □

Having established the propositions and claims that we need, we now prove Lemma 2.

Proof of Lemma 2. Given that problem Π satisfies the SZKP–OWF CONDITION, let I be the set of OWF instances, and let the OWF YES instances be $I_Y = I \cap \Pi_Y$ and the OWF NO instances be $I_N = I \cap \Pi_N$. By Propositions 3, 4, and 5, we have three instance-dependent commitment schemes, call them $\mathsf{Com}_x^{(1)}$, $\mathsf{Com}_x^{(2)}$, and $\mathsf{Com}_x^{(3)}$, for the problems $(\Pi_Y \setminus I, \Pi_N \setminus I) \in \textbf{SZKP}$, $(I_Y, \overline{I_Y})$, and $(\overline{I_N}, I_N)$, respectively. Moreover, all three schemes are public coin.

If Π satisfies the SZKP–OWF CONDITION without OWF instances, then set $I = \emptyset$, and hence $\mathsf{Com}_x^{(1)}$ suffices to be our instance-dependent commitment scheme for Π. If Π satisfies the SZKP–OWF CONDITION without OWF NO instances, then $I_N = I \cap \Pi_N = \emptyset$. Consequently, we do not need scheme $\mathsf{Com}_x^{(3)}$, and can just combine schemes $\mathsf{Com}_x^{(1)}$ and $\mathsf{Com}_x^{(2)}$ in a manner prescribed by Claim 7 to get an instance-dependent commitment scheme for Π.

Analogously, if Π satisfies the SZKP–OWF CONDITION without OWF YES instances, then $I_Y = I \cap \Pi_Y = \emptyset$. Consequently, we do not need scheme $\mathsf{Com}_x^{(2)}$, and can just combine schemes $\mathsf{Com}_x^{(1)}$ and $\mathsf{Com}_x^{(3)}$ in a manner prescribed by Claim 6 to get an instance-dependent commitment scheme for Π. Finally, if Π satisfies the SZKP–OWF CONDITION, we first combine schemes $\mathsf{Com}_x^{(1)}$ and $\mathsf{Com}_x^{(2)}$ in a manner prescribed by Claim 7 to get an instance-dependent commitment scheme for $(\Pi_Y, \Pi_N \setminus I_N)$, and then combine this scheme with $\mathsf{Com}_x^{(3)}$ in a manner prescribed by Claim 6 to get an instance-dependent commitment scheme for Π.

The hiding, binding, and public coin properties of the instance-dependent commitment scheme for Π follow by inspection. □

3.5 From Instance-Dependent Commitment Schemes to Zero-Knowledge Protocols

Having obtained instance-dependent commitments in the previous subsection, we now use these commitments to construct unconditional zero-knowledge protocols for problems $\Pi \in \mathbf{NP}$ having these instance-dependent commitments. We observe that the existing zero-knowledge protocols for $\mathbf{NP}$ require complexity assumptions because they use standard (non-instance dependent) commitments, and standard commitments are not known to exist unconditionally. Therefore, we can remove the complexity assumptions needed by substituting standard commitments for instance-dependent commitments in these existing protocols. Specifically, we do this substitution in the Goldreich–Micali–Wigderson [GMW] zero-knowledge protocol for $\mathbf{NP}$.

Lemma 3 (based on [GMW]). *If problem $\Pi \in \mathbf{NP}$ has an instance-dependent commitment scheme Com_x, then it has a zero-knowledge protocol (P, V) with the following properties:*

- *(P, V) is statistical [resp., computational] zero knowledge if Com_x is statistically [resp., computationally] hiding on the* YES *instances. Moreover, (P, V) has a black-box simulator.*
- *(P, V) is a proof [resp., argument] system if Com_x is statistically [resp., computationally] binding on the* NO *instances.*
- *(P, V) has perfect completeness and has an efficient prover.*
- *(P, V) is public coin if Com_x is public coin.*

3.6 Putting It All Together

We now show how our lemmas in Sects. 3.3, 3.4, and 3.5 imply our main characterization theorems in Sect. 3.1.

Proof of Theorems 6 and 7. The implications for both theorems are captured by the same lemmas, so we can conveniently state them together.

- **(1)** $\Rightarrow$ **(2)** is established by Lemma 1.
- **(2)** $\Rightarrow$ **(3)** is established by Lemma 2.
- **(3)** $\Rightarrow$ **(4)** is established by Lemma 3. This is the only step that requires the problem $\Pi \in \mathbf{NP}$.
- **(4)** $\Rightarrow$ **(1)** follows directly from definition. $\square$

Acknowledgements

We are grateful to Lilach Bien for pointing out an error in the statement and use of Proposition 2 in the original version of our paper [OV]. We thank Oded Goldreich and the anonymous *EUROCRYPT 2007* reviewers for their helpful comments. We also thank Iftach Haitner and Omer Reingold for allowing us to simplify the presentation of our results with their wonderful result [HR] and the follow-up [HORV].

References

[AH]   AIELLO, W., AND HÅSTAD, J. Statistical zero-knowledge languages can be recognized in two rounds. *J. Comput. Syst. Sci.*, 42(3):327–345, 1991.

[BCC]   BRASSARD, G., CHAUM, D., AND CRÉPEAU, C. Minimum disclosure proofs of knowledge. *J. Comput. Syst. Sci.*, 37(2):156–189, 1988.

[CT]   COVER, T. M., AND THOMAS, J. A. *Elements of information theory*. Wiley-Interscience, New York, NY, USA, 2 edition, 2006.

[ESY]   EVEN, S., SELMAN, A., AND YACOBI, Y. The complexity of promise problems with applications to public-key cryptography. *Inform. Control*, 61(2):159–173, 1984.

[GMR]   GOLDWASSER, S., MICALI, S., AND RACKOFF, C. The knowledge complexity of interactive proof systems. *SIAM J. Comput.*, 18(1):186–208, 1989.

[GMW]   GOLDREICH, O., MICALI, S., AND WIGDERSON, A. Proofs that yield nothing but their validity or all languages in NP have zero-knowledge proof systems. *J. ACM*, 38(1):691–729, 1991.

[Gol1]   GOLDREICH, O. *Foundations of Cryptography: Basic Tools*. Cambridge University Press, 2001.

[Gol2]   GOLDREICH, O. On promise problems (a survey in memory of Shimon Even [1935-2004]). Technical Report TR05–018, ECCC, 2005.

[GSV]   GOLDREICH, O., SAHAI, A., AND VADHAN, S. Honest verifier statistical zero-knowledge equals general statistical zero-knowledge. In *Proc. 30th STOC*, pages 399–408, 1998.

[GV]   GOLDREICH, O., AND VADHAN, S. Comparing entropies in statistical zero-knowledge with applications to the structure of SZK. In *Proc. 14th Comput. Complex.*, pages 54–73, 1999.

[HILL]   HÅSTAD, J., IMPAGLIAZZO, R., LEVIN, L., AND LUBY, M. A pseudorandom generator from any one-way function. *SIAM J. Comput.*, 28(4):1364–1396, 1999.

[HORV]   HAITNER, I., ONG, S., REINGOLD, O., AND VADHAN, S. Instance-dependent commitments for statistical zero-knowledge proofs. In preparation, March 2007.

[HR]   HAITNER, I., AND REINGOLD, O. Statistically-hiding commitment from any one-way function. Technical Report 2006/436, Cryptol. ePrint Arch., 2006.

[IL]   IMPAGLIAZZO, R., AND LUBY, M. One-way functions are essential for complexity based cryptography. In *Proc. 30th FOCS*, pages 230–235, 1989.

[Nao]   NAOR, M. Bit commitment using pseudorandomness. *J. Cryptol.*, 4(2):151–158, 1991.

[NOV]   NGUYEN, M., ONG, S., AND VADHAN, S. Statistical zero-knowledge arguments for NP from any one-way function. In *Proc. 47th FOCS*, pages 3–14, 2006.

[NV]   NGUYEN, M., AND VADHAN, S. Zero knowledge with efficient provers. In *Proc. 38th STOC*, pages 287–295, 2006.

[Oka]   OKAMOTO, T. On relationships between statistical zero-knowledge proofs. *J. Comput. Syst. Sci.*, 60(1):47–108, 2000.

[Ost]   OSTROVSKY, R. One-way functions, hard on average problems, and statistical zero-knowledge proofs. In *Proc. 6th Annual Structure in Complexity Theory Conference*, pages 133–138, 1991.

[OV]   ONG, S., AND VADHAN, S. Zero knowledge and soundness are symmetric. Technical Report TR06-139, ECCC, 2006.

[OW]   OSTROVSKY, R., AND WIGDERSON, A. One-way functions are essential for non-trivial zero-knowledge. In *Proc. 2nd Israel Symposium on Theory of Computing Systems*, pages 3–17, 1993.

[PT]   PETRANK, E., AND TARDOS, G. On the knowledge complexity of NP. *Combinatorica*, 22(1):83–121, 2002.

[SV]   SAHAI, A., AND VADHAN, S. A complete problem for statistical zero knowledge. *J. ACM*, 50(2):196–249, 2003.

[Vad]   VADHAN, S. An unconditional study of computational zero knowledge. *SIAM J. Comput.*, 36(4):1160–1214, 2006.

Mesh Signatures
How to Leak a Secret with Unwitting and Unwilling Participants

Xavier Boyen

Voltage Inc., Palo Alto
xb@boyen.org

Abstract. We define the mesh signature primitive as an anonymous signature similar in spirit to ring signatures, but with a much richer language for expressing signer ambiguity. The language can represent complex access structures, and in particular allows individual signature components to be replaced with complete certificate chains. Because withholding one's public key from view is no longer a shield against being named as a possible cosignatory, mesh signatures may be used as a ring signature with compulsory enrollment.

We give an efficient construction based on bilinear maps in the common random string model. Our signatures have linear size, achieve everlasting perfect anonymity, and reduce to very efficient ring signatures without random oracles as a special case. We prove non-repudiation from a mild extension of the SDH assumption, which we introduce and justify meticulously.

1 Introduction

We introduce mesh signatures, which are similar in spirit and purpose to the ring signatures of Rivest, Shamir, and Tauman [27], but overcome some of their crucial limitations.

Ring signatures are pseudonymous signatures that are issued in the name of a "ring" of users, and created by one of them without the participation of the others, in a way that preserves the instigator's anonymity. The canonical application is for an individual "to leak a secret" non-repudiably on behalf of a crowd. Technically, ring signatures can thus be viewed as a witness-indistinguishable disjunction of regular signatures, but because of this, only people who have previously published a verification key are eligible to be enrolled in such a crowd, ring signatures can only ever implicate individuals who, by the very act of publishing their key, are proclaiming their consent.

Mesh signatures generalize this notion to mononote access structures representable as a tree, whose interior nodea are And, Or, and Threshold gates, and whose leaves are regular signatures. The access structure can be satisfied using different subsets of the regular signatures; once created, the mesh signature will not reveal what particular subset was used. The regular signatures at the leaves can be "static", and thus PKI certificates are eligible if the mesh signer does

M. Naor (Ed.): EUROCRYPT 2007, LNCS 4515, pp. 210–227, 2007.

not have the CA's signing key. Since furthermore the monotone tree structure is powerful enough to express disjunctions of certificate chains, we are no longer dependent on individual ring members publishing their keys. As a toy example, suppose that *Alice* wants to implicate *Bob*, who may or may not have a verification key on record. *Alice* can still produce the following mesh signature,

$$\sigma = [VK_{Alice}: Msg_1] \text{ or } ([VK_{CertAuth}: VK_{Bob}] \text{ and } [VK_{Bob}: Msg_2]) ,$$

All that *Alice* needs to create σ is her own private key and *CertAuth*'s certificate verification key. Even if *Bob* has published no verification key, the mesh signature σ implicates him via the certificate $[VK_{CertAuth}: VK_{Bob}]$ that binds his name to the string VK_{Bob}; the certificate can be real or a fake. Conversely, *Bob* could have created σ himself, using the real certificate and his own private key, to implicate *Alice*; although in this case her public key would have to be available to the verifier since her certificate is not part of σ. Another feature of mesh signatures is that they provide threshold gates, which makes it easy to scale constructs like,

$$\sigma = \text{ 2-out-of-3 in } \{[\text{CEO}: skrt\text{-}memo], [\text{CFO}: skrt\text{-}memo], [\text{COO}: skrt\text{-}memo]\} .$$

Threshold gates like this can feed or be fed from other gates as in the earlier example. The unconditional anonymity of mesh signatures guarantees that, as long as the signature σ is valid, there is no way to tell the true and false clauses apart in the formula expressed by σ.

We can immediately see how much more practical mesh signatures are than ring signatures: instead of requiring that each and everyone generate and publish their public key in a ring scheme, here we just need one trustworthy certificate authority (or preferably a few) to publish their keys in the mesh scheme—a natural demand to place on certificate authorities, though not on individuals.

To make the use of certificate chains truly believable, it is important that mesh signatures be constructible non-interactively from reusable constituent atomic signatures (in our case, these are Boneh-Boyen short signatures [4]).

1.1   Related Work

The original ring signature primitive was defined in [27], to enable secret leaking that is at once authenticated (by a crowd) and anonymous (within the crowd). Whereas that construction [27] was based on trapdoor permutations, a number of alternatives have subsequently been proposed, based on bilinear pairings [6], discrete logarithms [21], factoring (Strong-RSA specifically) [16], or hybrids [1]; all these constructions are set in the random oracle model. Most have linear size in the ring membership count, except [16] which squeezes it all in constant size using accumulators in the random oracle model.

A number of general protocols bear similarities with our new primitive. Perhaps the first such scheme is an anonymous authentication protocol of [15] that supports access structures and can be turned into a signature using the Fiat-Shamir heuristic. Another is an interactive anonymous authentication protocol, called deniable ring authentication [26], that combines the anonymity of ring signatures with the non-transferability of deniable authentication [18], and supports threshold and access structures. Among specific constructions in the random oracle model, we note the distributed ring signatures of [22] which lets coalitions of users cooperate in an interactive signing protocol, and the hierarchical identity-based ring signatures of [32], which adds signer ambiguity to the notion of hierarchical identity-based signature. Additionally, we mention that mesh signatures could in principle be realized using signatures of knowledge [11], which allow the knowledge of a witness to an NP statement to serve as a signing key, in the common random string model.

Another related notion that has received much attention is that of group signatures, originally introduced in [12], and which also provides for the anonymous creation of signatures on behalf of a crowd. The main difference is that group signatures require the anonymity to be revocable by a group manager, who also controls enrollment into the group. Group membership is often immutable although this restriction has been relaxed in [9]. There exists efficient constant-size group signature schemes, with random oracles [5], from interactive assumptions [2], and in the standard model [8].

Efficient ring signatures constructions without random oracles have also been proposed recently, such as [14], [3], and [28]. The construction of [14] uses bilinear groups and is efficient but relies on a cumbersome assumption stated without justification. The results of [3] include an impractical scheme from non-interactive Zaps [17], but also two efficient constructions (based on [10] or [31] signatures) for rings of size two, and a discussion of security models for ring signatures.

Probably the most closely related to our work is the very recent ring scheme of [28] which can efficiently creates linear-size ring signatures in the "trusted parameters" model; unforgeability is based on computational Diffie-Hellman, and anonymity on the decisional Subgroup [7] assumption. Because of the latter, the scheme requires a bilinear map in a group of composite order with a hidden factorization; such a group is set up explicitly by a central authority, which afterwards must erase the factorization to ensure anonymity. It may be possible to use ideas from [20] and base anonymity on the decisional Linear [5] assumption, which would no longer require secret-coin *trusted parameters* (TP) but only a public-coin *common random string* (CRS), as in our scheme; however anonymity would still remain computational. The main advantage of [28] over our ring scheme is that unforgeability rests on a weaker assumption.

2   Definitions and Security Models

Intuitively, a mesh signature is a non-interactive witness-indistinguishable proof that some monotone boolean expression Υ is true, where each input of Υ is notionally labeled with a key & message pair and is true only if the mesh signer is in possession of a valid atomic signature for the stated message and key.

A mesh signature scheme should satisfy two security properties. First, it should be anonymous (ideally, unconditionally so), *i.e.*, it should not reveal what assignment to the inputs of Υ caused it to be satisfied. Second, it should be unforgeable, *i.e.*, the creation of a valid mesh signature must be predicated on the possession of a set of valid atomic signatures sufficient to satisfy Υ.

2.1   Recursive Mesh Signature Specification

We use ℓ to denote the number of atomic clauses in a mesh structure (in a ring signature, this would be equal to the number of users in the ring). Let Υ be the expression generated by the following grammar, with propositional-logic semantics, under the restriction that, for each $i = 1, ..., \ell$, the production EXPR $::= L_i$ corresponding to the symbol L_i be used at most once (in other words, no L_i may appear more than once in the written expression of Υ):

$$
\begin{array}{lll}
\text{EXPR} ::= & L_1 \mid ... \mid L_\ell & \text{single-use input symbols} \\
\mid & \geq_t\{\text{EXPR}_1, ..., \text{EXPR}_m\} & t\text{-out-of-}m \text{ threshold, with } 1 < t < m \\
\mid & \wedge\{\text{EXPR}_1, ..., \text{EXPR}_m\} & m\text{-wise conjunction, with } 1 < m \\
\mid & \vee\{\text{EXPR}_1, ..., \text{EXPR}_m\} & m\text{-wise disjunction, with } 1 < m
\end{array}
$$

Equivalently, we call Υ an "arborescent monotone threshold circuit" with ℓ Boolean inputs $L_1, ..., L_\ell$ and one Boolean output denoted $\Upsilon(L_1, ..., L_\ell)$. It is apparent by induction that Υ is always a non-trivial monotone function of its inputs, and in particular $\Upsilon(\bot, ..., \bot) = \bot$ and $\Upsilon(\top, ..., \top) = \top$.

We use expressions of this form to state the meaning of mesh signatures. The signer specifies the circuit Υ, and assigns to each symbol L_j an atomic proposition $[VK: Msg]$ to convey the meaning: "This is Msg signed under VK." The mesh signature then simply expresses that $\Upsilon(L_1, ..., L_\ell) = \top$ holds for the stated interpretation of the L_i (without revealing their individual truth values). For the example in the introduction, $\Upsilon = L_1 \vee (L_2 \wedge L_3)$ where L_1 denotes $[VK_{Alice}: Msg_1]$, etc.

We emphasize that two distinct symbols L_i and L_j can express the same sentence and yet have opposite truth values, since the signer is free to use a valid atomic signature for one and not for the other. The current construction does not support cloning truth values without losing the original, just as it cannot express the negation of a truth value.

2.2   Anonymity Model

The strongest notion of anonymity defined in [3], "anonymity against full key exposure", in the context of ring signatures, requires that the signer remain anonymous following full exposure of all the private keys, after their use. It is however a constrained notion of anonymity because the keys are not chosen by the adversary, and are only revealed *a posteriori*. We contend that, since the motivating application of ring and mesh schemes is to leak secrets, it is crucial that anonymity be *unconditional* and *everlasting*, subsequently to the exposure of all secrets, for the long-term peace of mind of the signer. We thus insist on perfect (*i.e.*, information theoretic) anonymity, even upon prior disclosure of the signer's and every user's secret keys.

Precisely, we require that the identity of the signer be statistically independent, conditionally on all public keys and the mesh formula, of any long-term secret held by any party in the system. We exclude ephemeral randomness from the above requirement, for the reason that there is no way to prevent the signer to prove willingly that she herself created a particular signature: revealing the ephemerals used to create a signature is but one way to do this. By contrast, the signer must be protected againt coerced disclosure, which is why independence from her long-term keys is crucial.

2.3   Unforgeability Model

The strongest notion of unforgeability defined in [3], "unforgeability with respect to insider corruption", for ring signatures, gives the adversary the ability to corrupt users dynamically, and include its own public keys when making ring signature queries. Since the point of mesh signatures is to implicate uncooperative users, it is judicious to allow them to choose their keys maliciously.

However, as a compromise for unconditional anonymity, we relax the fully dynamic corruption model into an enhanced static one, in which the honest users are static and created ahead of time by a challenger, and the corrupted users are under the full control of an adversary who can bring them to life dynamically. We also need to specify what constitutes a valid forgery. For ring signatures, a forgery is any signature by a ring without adversarially controlled users. For mesh signatures, this is overly restrictive, since it excludes forgeries such as,

$$\Upsilon = ([U_1 : m_1] \wedge [U_3 : m_3]) \vee ([U_2 : m_2] \wedge [U_4 : m_4]) ,$$

where U_1 and U_2 are honest users, and U_3 and U_4 are corrupted. Since Υ nominally entails $\Upsilon' = [U_1 : m_1] \vee [U_2 : m_2]$, a forger who signs Υ lacking the imprimatur of both U_1 and U_2 should be deemed successful. We capture these circumstances by deeming admissible any forgery on a statement Υ if there exists a well-formed Υ' that involves only honest users and such that $\Upsilon \Rightarrow \Upsilon'$.

To see where this comes from, for all corrupted users let us set the corresponding literal $L_i \leftarrow \top$, which is the most that they can supposedly do. If Υ evaluates to $\top$, the forgery is inadmissible; otherwise, Υ reduces to some well-formed formula Υ' which involves honest users, exclusively. Hence, the condition demands that Υ be unsatisfiable by the volition of the adversarial users alone. We distill all of this into the following existential unforgeability game, and define the adversary's advantage as the probability of outputting an admissible valid forgery.

Challenger setup: the challenger designates a number ℓ of public keys, corresponding to the honest target users under the challenger's control.

Interaction: the following occurs interactively, in any order, driven by the adversary.

Adversary setup: the adversary reveals polynomially many public keys, one at a time, corresponding to the users under the adversary's control.

Signature queries: the adversary makes up to q mesh signature queries, one at a time, on specifications Υ_j whose satisfiability involves the challenger's users.

The adversary may also query q atomic signatures to each of
the users controlled by the challenger (since atomic signatures
should be usable instead of signing keys for mesh signing.)

The challenger processes each request before accepting the next one.

Signature forgery: the adversary produces a forged signature whose
specification Υ contains no clause $[VK_i : Msg_i]$ from an atomic query,
and is such that $\forall j, \Upsilon \neq \Upsilon_j$ and $\exists \Upsilon', \Upsilon(L_1, ..., L_\ell, ...) \Rightarrow \Upsilon'(L_1, ..., L_\ell)$
where Υ' is a well-formed formula with honest user clauses only.

3   Framework and Computational Assumption

We write $\mathbb{F}_p$ for the finite field of prime order p, and $\mathbb{F}_p^\times = \mathbb{F}_p \setminus \{0\}$ for its multiplicative group of order $p - 1$. Let a bilinear context $\mathbf{G} = (p, \mathbb{G}, \hat{\mathbb{G}}, \mathbb{G}_t, g, \hat{g}, \mathbf{e})$,
where $\mathbf{e} : \mathbb{G} \times \hat{\mathbb{G}} \to \mathbb{G}_t$ is a pairing [25]. We use the "hat-notation" (as in $\hat{g}$) to
indicate that an element belongs to $\hat{\mathbb{G}}$ rather than $\mathbb{G}$.

3.1   Review of the SDH Assumption

The complexity assumption we shall need is inspired from the Strong Diffie-
Hellman assumption proposed in [4], which we now review. The q-SDH problem
in a (bilinear) group $\mathbb{G}$ is stated:

(Original SDH) Given elements $g, g^\alpha, g^{\alpha^2}, ..., g^{\alpha^q} \in \mathbb{G}$, choose $w \in \mathbb{F}_p$ and
output $(w, g^{1/(\alpha+w)})$.

The SDH assumption then posits that the q-SDH problem above is intractable
for $q = \mathrm{O}(\mathrm{poly}(\kappa))$. What makes this assumption special is that the problem
admits not one but exponentially many "independent" solutions, which are all
equally hard to find. Hence the modified q-SDH problem:

(Modified SDH) Given $g, g^\alpha \in \mathbb{G}$ and $q - 1$ pairs $(w_j, g^{1/(\alpha+w_j)})$, output
another $(w, g^{1/(\alpha+w)})$.

It is known from [4] that if the original q-SDH problem is hard, then so is the
modified problem.

Although the SDH problem statement does not require a bilinear group, it is
because the bilinear map provides an efficient Decision Diffie-Hellman procedure
[23] that the correctness of an SDH solution can be decided openly. Specifically,
given g and g^α, deciding whether $(w, u) = (c, g^{1/(\alpha+w)})$ amounts to checking the
equality $\mathbf{e}(u, \hat{g}^\alpha \hat{g}^w) = \mathbf{e}(g, \hat{g})$, basically a DDH a test that anyone can perform
from public information. The short signature scheme of [4] relies on this.

3.2   Poly-SDH: For Better Use of the Pairing

The verifiability of SDH solutions with a simple DDH test suggests that more
general assumptions could be made, based on the observation that the pairing is

a powerful tool that can be used to decide more complex relations that are not efficiently reducible to DDH. For example, a natural generalization of the SDH problem is that of finding ℓ pairs $(w_i, u_i = g^{r_i/(\alpha+w_i)})$ for $i = 1, ..., \ell$, such that $\sum_{i=1}^{\ell} r_i = 1 \pmod{p}$. Purported solutions can then be verified by checking,

$$\prod_{i=1}^{\ell} \mathbf{e}(u_i, \hat{g}^{\alpha}\,\hat{g}^{w_i}) = \mathbf{e}(g, \hat{g}) \ . \tag{1}$$

Clearly, when $\ell = 1$, this is identical to the SDH problem. For larger values of ℓ, the adversary is given to spread the exponent inversion task across multiple pairs, by means of linear combination.

Unfortunately, for $\ell > 1$, the problem is in fact trivial, because Equation (1) admits spurious solutions that do not require the solver to know the secret α and invert the exponent: for example, for $\ell = 2$ the solution $w_1 = 1$, $u_1 = g$, $w_2 = 0$, $u_2 = g^{-1}$ satisfies the equality regardless of α.

To remedy the preceding problem, we change the solver's task slightly, and ask that the ℓ pairs to be output involve ℓ independent secrets $\alpha_1, ..., \alpha_\ell$ that appear once each, $i.e.$, find,

$$\left(w_i, u_i = g^{\frac{r_i}{\alpha_i+w_i}} \right) : i = 1, ..., \ell \ , \qquad \text{s.t.} \qquad \sum_{i=1}^{\ell} r_i = 1 \pmod{p} \ .$$

To decide whether a solution $((w_1, u_1), ..., (w_\ell, u_\ell))$ to the new problem is correct, one also needs, besides the generators g and $\hat{g}$, the ℓ group elements $(\hat{g}_1, ..., \hat{g}_\ell) = (\hat{g}^{\alpha_1}, ..., \hat{g}^{\alpha_\ell})$. The verification equation is then,

$$\prod_{i=1}^{\ell} \mathbf{e}(u_i, \hat{g}_i\,\hat{g}^{w_i}) = \mathbf{e}(g, \hat{g}) \ . \tag{2}$$

Notice that (1) is a special case of (2) where $\alpha_1 = ... = \alpha_\ell = \alpha$; however, for the security of the assumption it is important that the α_i be independently and uniformly distributed. Despite the added variables, we stress that Equation (2) is no more expensive to verify.

Based on the previous observations, the (q, ℓ)-Poly-SDH problem can be informally stated as:

(Poly-SDH) Given g, g^{α_1}, ..., $g^{\alpha_\ell} \in \mathbb{G}$ and $q\ell$ pairs $(w_{i,j}, g^{1/(\alpha_i+w_{i,j})})$ for $1 \le i \le \ell$ and $1 \le j \le q$, choose fresh w_1, ..., $w_\ell \in \mathbb{F}_p$ and output ℓ pairs $(w_i, g^{r_i/(\alpha_i+w_i)})$ such that $\sum_{i=1}^{\ell} r_i = 1$.

The α_i and $w_{i,j}$ in the instance are drawn from a uniform distribution. The w_i and r_i are chosen by the respondent. We require that $\forall i, \forall j, w_i \ne w_{i,j}$, lest the task be easy. The exponents r_i need not be revealed, since Equation (2) can establish that a solution is correct, and thus $\sum_i r_i = 1$, without seeing the r_i.

We have chosen to state the (q, ℓ)-Poly-SDH problem in a form analogue to Modified SDH, rather than Original SDH. There are a few justifications for this:

- the modified form results in a weaker assumption (by analogy to the implication from Original SDH to Modified SDH);
- the input/output symmetry simplifies the security reductions;
- its instances are more concisely stated when more than one iterator is needed;
- the modified problem form is impervious to a (benign) generic analysis described in [13], which relies on the availability of g, g^α, and g^{α^d} for certain d, as in Original SDH instances.

The reason why there are no undesirably easy solutions to the (q, ℓ)-Poly-SDH problem will become apparent as we prove generic hardness in Section 3.3.

3.3   Generic Hardness of Poly-SDH

We now take some time to explain why the Poly-SDH assumption based on Equation (2) is plausible, unlike our first attempt from Equation (1) that was so easily broken. We give a heuristic argument based on the impossibility of efficient generic attacks. Specifically, we show that finding a solution to the (q, ℓ)-Poly-SDH problem will require, on expectation, $\Omega(\sqrt{p/q\,\ell})$ generic group operations.

The generic group model [29] assumes the lack of any structure beyond that of an (Abelian) cyclic group, restricting all manipulations on group elements to the group operation and its inverse (i.e., multiplication and division if the group is written multiplicatively). In the bilinear version of the model [4], one can also compute a pairing $\mathbf{e} : \mathbb{G} \times \hat{\mathbb{G}} \to \mathbb{G}_t$, as well as an isomorphism $\psi : \hat{\mathbb{G}} \to \mathbb{G}$ (for "type-1" and "type-2" contexts) and its inverse $\psi^{-1} : \mathbb{G} \to \hat{\mathbb{G}}$ (for "type-1" only).

Let us assume that $\mathbb{G} = \hat{\mathbb{G}}$, which only makes the attack easier. Recall that the Poly-SDH instance furnishes g, g^{α_1}, ..., g^{α_ℓ}, and a large number of pairs $(w_{i,j}, u_{i,j} = g^{1/(\alpha_i + w_{i,j})})$. Based on this information, the attacker must output ℓ pairs $(w_i, u_i = g^{r_i/(\alpha_i + w_i)})$ such that $\sum_i r_i = 1$, and where w_i is distinct from all $w_{i,j}$ with the same index i.

First, notice that the pairing $\mathbf{e}$ is useful to verify a solution, but not really to find one. This is because $\mathbf{e}$ ranges into $\mathbb{G}_t$, and once we have landed in $\mathbb{G}_t$ we can never leave it. Also, ψ and ψ^{-1} just model the identity function since we have already assumed that $\mathbb{G} = \hat{\mathbb{G}}$. We can thus focus on multiplication and division in the multiplicative group $\mathbb{G}$ of prime order p.

Next, observe that all the group elements that can be created from g, $\{g^{\alpha_i}\}$, and $\{g^{1/(\alpha_i + w_{i,j})}\}$ are of the form $g^{\frac{\pi(\alpha_1,...,\alpha_\ell)}{\Delta}}$, where $\pi \in \mathbb{F}_p[\alpha_1, ..., \alpha_\ell]_{q\ell+1}$ is any multivariate polynomial in $\alpha_1, ..., \alpha_\ell$ of total degree at most $q\,\ell + 1$, and where Δ is the common denominator $\Delta = \prod_{i=1}^{\ell} \prod_{j=1}^{q} (\alpha_i + w_{i,j})$. We need to produce ℓ elements $u_i = g^{r_i/(\alpha_i + w_i)}$ and the corresponding w_i. Our task is thus to find ℓ polynomials π_1, ..., $\pi_\ell \in \mathbb{F}_p[\alpha_1, ..., \alpha_\ell]_{q\ell+1}$ such that $\pi_i/\Delta = r_i/(\alpha_i + w_i)$ for some $\sum_i r_i = 1$, i.e., such that,

$$\sum_{i=1}^{\ell} (\alpha_i + w_i)\, \pi_i = \Delta = \prod_{i=1}^{\ell} \prod_{j=1}^{q} (\alpha_i + w_{i,j}) \,.$$

We show that there can be no such polynomials π_i using a linear change of variable. For all $i = 1, ..., \ell$ and $j = 1, ..., q$, we define $\alpha'_i = \alpha_i + w_i$ and $w'_{i,j} = w_{i,j} - w_i$. Notice that all $w'_{i,j} \neq 0$. Our new task becomes to find ℓ polynomials $\pi'_1, ..., \pi'_\ell$ of degree $\leq q\,\ell + 1$ in the variables $\alpha'_1, ..., \alpha'_\ell$, such that,

$$\sum_{i=1}^{\ell} \alpha'_i \pi'_i = \Delta = \prod_{i=1}^{\ell} \prod_{j=1}^{q} (\alpha'_i + w'_{i,j}) \ .$$

Clearly, all the monomials in the left-hand side have degree in $\alpha'_1, ..., \alpha'_\ell$ at least 1. On the other hand, all $w'_{i,j}$ are non-zero, so the right-hand side yields a non-vanishing independent degree-0 term equal to $\prod_i \prod_j w'_{i,j} = \prod_i \prod_j (w_{i,j} - w_i) \neq 0$, which is a contradiction.

The contradiction shows that the equations above cannot be satisfied identically in $\mathbb{F}_p[\alpha'_1, ..., \alpha'_\ell]$ or $\mathbb{F}_p[\alpha_1, ..., \alpha_\ell]$, which proves that the polynomials π'_i and thus π_i cannot exist. A standard argument then shows that the equations can only be satisfied in $\mathbb{F}_p$ for certain assignments of $\alpha_1, ..., \alpha_\ell \in \mathbb{F}_p$: the polynomial roots. Since the α_i are chosen at random, we can bound the probability of hitting those roots. We find that, if $q\,\ell < \mathrm{O}(\sqrt[3]{p})$, it takes $q_G = \Omega(\sqrt{\epsilon\,p/q\,\ell})$ operations to solve (q, ℓ)-Poly-SDH with probability ϵ in generic groups of order p.

4   Special Case: Ring Signatures

We first describe a ring signature based on the Poly-SDH assumption as a special case of our technique. It is more efficient than other provably secure ring signature schemes without random oracles, and is set in the common random string model without trusted parameters.

Initialization: Given a security parameter κ and a public random string $K \in \{0,1\}^{\mathrm{poly}(\kappa)}$, the parties generate from K a common bilinear instance $\mathbf{G} = (p, \mathbb{G}, \hat{\mathbb{G}}, \mathbb{G}_t, g, \hat{g}, \mathbf{e}) \leftarrow \mathcal{G}(1^\kappa; K)$ and a collision-resistant hash function $H : \{0,1\}^* \rightarrow \mathbb{F}_p$ shared by all. Since $\mathbf{G}$ has prime order and no hidden structure, it can safely be generated from public coins.

　　The string K is also used to generate three random elements $\hat{A}_0$, $\hat{B}_0$, and $\hat{C}_0$ in $\hat{\mathbb{G}}$. These elements define a public verification key "in the sky" whose matching signing key is undefined.

　　For notational convenience, we suppose for now that the isomorphism $\psi : \hat{\mathbb{G}} \rightarrow \mathbb{G}$ is efficiently computable in the instance $\mathbf{G}$, and we let $A_0 = \psi(\hat{A}_0)$, $B_0 = \psi(\hat{B}_0)$, and $C_0 = \psi(\hat{C}_0)$ in $\mathbb{G}$. This temporary restriction will be lifted later in this section.

Key generation: User #i draws a signing key $(a_i, b_i, c_i) \in (\mathbb{F}_p^\times)^3$, and publishes $(A_i, B_i, C_i, \hat{A}_i, \hat{B}_i, \hat{C}_i) = (g^{a_i}, g^{b_i}, g^{c_i}, \hat{g}^{a_i}, \hat{g}^{b_i}, \hat{g}^{c_i}) \in \mathbb{G}^3 \times \hat{\mathbb{G}}^3$. In case $\psi : \hat{\mathbb{G}} \rightarrow \mathbb{G}$ is easy to compute, users publish only $(\hat{A}_i, \hat{B}_i, \hat{C}_i)$.

Ring signature: To create a ring signature on a message $m \in \mathbb{F}_p$ attributed to a ring of ℓ users, any member of the ring would proceed as follows. W.l.o.g., suppose that the signer is User #ℓ in the ring $R = (1, ..., \ell)$. The signer selects $2\ell + 1$ random integers $s_0, s_1, ..., s_{\ell-1}, t_0, t_1, ..., t_\ell \in \mathbb{F}_p$, and outputs the signature $\sigma = (S_0, ..., S_\ell, t_0, ..., t_\ell) \in \mathbb{G}^{\ell+1} \times \mathbb{F}_p^{\ell+1}$, given by,

$$\sigma = \left(g^{s_0}, ..., g^{s_{\ell-1}}, \left(g \cdot \prod_{i=0}^{\ell-1} (A_i\, B_i^{m_i}\, C_i^{t_i})^{-s_i} \right)^{\overline{a_\ell + b_\ell\, m_\ell + c_\ell\, t_\ell}}, t_0, ..., t_\ell \right) ,$$

with $m_1, ..., m_\ell$ the messages to be signed, and $m_0 = H((1, m_1), ..., (\ell, m_\ell))$, a collision-resistant hash of the statement expressed by the signature.

Ring verification: To verify a signature $\sigma = (S_1, ..., S_\ell, t_1, ..., t_\ell)$ with respect to a message m and a ring $R = (1, ..., \ell)$, it suffices to set $m_1 = ... = m_\ell = m$ and $m_0 = H((1, m_1), ..., (\ell, m_\ell))$, and test the equality,

$$\prod_{i=0}^{\ell} \mathbf{e}(S_i,\ \hat{A}_i\, \hat{B}_i^{m_i}\, \hat{C}_i^{t_i}) = \mathbf{e}(g, \hat{g}) .$$

Consistency of the algorithms is readily verified. Note that the scheme is trivially modified to force all messages $m_1, ..., m_\ell$ to be the same, as in traditional ring signatures.

The purpose of signing a hash of the message and ring composition under the public key "in the sky" is to prevent outsiders from appending new components to an existing signature, which would otherwise give an easy forgery. It also helps in the security proof.

We emphasize that the public string K has no hidden structure, and can be drawn publicly at random as long as it is not chosen to grant anyone undue advantage to compute discrete logarithms in $\mathbf{G}$ or sign under the key "in the sky". The absence of secret coins is the main difference between a common random string (CRS) and the much more demanding trusted parameters (TP) model: in the former the parameters can be drawn in the open; in the latter they must be crafted in a special way from secret coins by some trusted setup authority, who must then voluntarily give up the secret knowledge it has (and convince everyone that it did not cheat). No trusted setup agent, either centralized or distributed, is needed in our system.

Furthermore, we have irrevocable, or everlasting, unconditional anonymity of the signatures (*i.e.*, with forward security against coerced disclosure of the long-term signing keys), as stated by the following theorem.

Theorem 1. *The ring signature has everlasting perfect anonymity.*

The second security theorem states that the scheme is existentially unforgeable in the model of Section 2.3. The proofs will appear in the full version.

Theorem 2. *The ring signature is existentially unforgeable under an adaptive attack, against a static adversary that makes no more than q ring signature queries, and q atomic signature queries to each one of the ℓ honest users, adaptively, provided that the $(q, \ell + 1)$-Poly-SDH assumption holds in $\mathbf{G}$, in the common random string model.*

Withholding the Isomorphism. Since the most general types of bilinear instance $\mathbf{G}$ may fail to provide both an efficient isomorphism $\psi : \hat{\mathbb{G}} \to \mathbb{G}$ and an efficient sampling procedure in $\hat{\mathbb{G}}$, it is useful to modify the ring scheme in order to relax both requirements [19]. This is done as follows.

- First, we redefine the random key "in the sky" to consist just of A_0, B_0, and C_0, to be sampled directly in $\mathbb{G}$ from the common random seed K (skipping $\hat{\mathbb{G}}$ altogether).
- Next, we modify the group element of index 0 in the signature, $\hat{g}^{s_0} \in \hat{\mathbb{G}}$ replacing $g^{s_0} \in \mathbb{G}$. The signature becomes, *e.g.*, with User $\#\ell$ as the signer: $\sigma = (\hat{S}_0, ..., S_\ell, t_0, ..., t_\ell) \in \hat{\mathbb{G}} \times \mathbb{G}^\ell \times \mathbb{F}_p^{\ell+1}$, given by,

$$\left(\hat{g}^{s_0}, g^{s_1}, ..., g^{s_{\ell-1}}, \left(g \cdot \prod_{i=0}^{\ell-1} (A_i B_i^{m_i} C_i^{t_i})^{-s_i} \right)^{\frac{1}{a_\ell + b_\ell m_\ell + c_\ell t_\ell}}, t_0, ..., t_\ell \right) ,$$

- Last, we exchange the arguments under the pairing of index 0 and amend the verification equation into,

$$\mathbf{e}(A_0 B_0^{m_0} C_0^{t_0}, \hat{S}_0) \cdot \prod_{i=1}^{\ell} \mathbf{e}(S_i, \hat{A}_i \hat{B}_i^{m_i} \hat{C}_i^{t_i}) = \mathbf{e}(g, \hat{g}) .$$

It is easy to see that the security theorems continue to hold in the modified ring signature scheme. On the one hand, anonymity is unconditional and thus insensitive to the existence of some efficient algorithm for ψ or for sampling in $\hat{\mathbb{G}}$. On the other hand, unforgeability relies no more on the presence of such algorithms than on their absence, as an inspection of the proof would show.

5   General Case: Mesh Signatures

We now describe our mesh signature scheme, based on the Poly-SDH assumption. We proceed in stages: we first define a few useful notions, which we then use to describe the actual system.

5.1   Flattened Mesh Representation

Recall that a mesh signature is characterized by an expression Υ generated by the grammar: $\Upsilon ::= N$ and,

$$N ::= L_1 \mid ... \mid L_\ell \mid \geq_t \{N_1, ..., N_m\} \mid \wedge \{N_1, ..., N_m\} \mid \vee \{N_1, ..., N_m\} .$$

To harmonize the notation with the scheme description, we need to consider an extra literal L_0 whose meaning is unimportant for now, and let $\tilde{\Upsilon}$ be as above with $\ell + 1$ input literals $L_0, ..., L_\ell$.

We show how to convert the recursive expression of $\tilde{\Upsilon}$ into a representation as a list of $\ell+1$ polynomials in $\ell+1$ variables (or fewer, depending on the structure of $\tilde{\Upsilon}$), akin to Linear Secret Sharing Structures [24,30].

The principle is as follows. To each input symbol L_i we associate a degree-1 homogeneous polynomial $\pi_i = \sum_{j=0}^{\ell} y_{i,j} Z_j$, where the variables $Z_0, ..., Z_\ell$ are common to all polynomials and the integer coefficients $y_{i,j}$ are constant. The polynomials are such that, if the formula $\tilde{\Upsilon}$ is satisfied by setting some subset of symbols to $\top$, then the span of the corresponding polynomials will contain the pure monomial Z_0; conversely, any set of polynomials whose span contains the monomial Z_0 indicates a satisfying assignment.

The following algorithm computes such a representation from $\tilde{\Upsilon}$. Proceeding recursively, it assigns temporary polynomials to the interior nodes as it walks down the tree from the root to the leaves (*i.e.*, from the output gate to the input symbols):

1. Initialize a counter $k_c \leftarrow 0$.
   The counter k_c is used for allocating new variables, so that each Z_{k+k_c} is always a "fresh" variable that is never used before or after in the algorithm.
2. Label the root node N_0 with the polynomial $\pi_{N_0} \leftarrow Z_0$.
3. Select a non-leaf node N with non-empty label $\pi_N \neq \emptyset$.
   (a) Denote by $N_1, ..., N_m$ the $m \geq 2$ children of N.
   (b) If N is $\vee\{N_1, ..., N_m\}$, then $\forall i = 1, ..., m$ let $\pi_{N_i} = \pi_N$.
   (c) If N is $\wedge\{N_1, ..., N_m\}$, then $\forall i = 1, ..., m$ let $\pi_{N_i} = \pi_N + \sum_{k=1}^{m-1} l_{i,k} Z_{k+k_c}$
       where $l_{i,k} \in \mathbb{Z}$. The selection of $l_{i,k}$ is explained below.
   (d) If N is $\geq_t \{N_1, ..., N_m\}$, then $\forall i = 1, ..., m$ let $\pi_{N_i} = \pi_N + \sum_{k=1}^{t-1} l_{i,k} Z_{k+k_c}$
       where $l_{i,k} \in \mathbb{Z}$.
   (e) Label each child N_i with the polynomial π_{N_i}.
   (f) Unlabel node N, *i.e.*, set $\pi_N \leftarrow \emptyset$.
   (g) Increment $k_c \leftarrow k_c + t - 1$ (using $t = 1$ and $t = m$ for $\vee$- and $\wedge$-gates).
   (h) Continue at Step 3 if an eligible node remains, otherwise skip to Step 4.
4. Let $\vartheta \leftarrow k_c$ and output the polynomials $(\pi_0, ..., \pi_\ell)$ associated with the leaf nodes $L_0, ..., L_\ell$. Each polynomial π_i is represented as a vector of coefficients $(y_{i,0}, ..., y_{i,\vartheta}) \in \mathbb{F}_p^{\vartheta+1}$ such that $\pi_i = \sum_{k=0}^{\vartheta} y_{i,k} Z_k$ is the result of the sequence of operations in Steps 3b, 3c and 3d.

We note that the only variables with non-zero coefficients in the output polynomials are $Z_0, ..., Z_\vartheta$, where $\vartheta = k_c$ is the final counter value and may be equal to or lesser than ℓ.

In Steps 3c and 3d, the coefficients $l_{i,k}$ must ensure that no linear relation exists in any set of π_i of size $< m$ or $< t$. (By construction, m or t of them will always be linearly dependent.) To ensure this property, we let $(l_{i,k})$ form a Vandermonde matrix in $\mathbb{F}_p^{m \times (m-1)}$ or $\mathbb{F}_p^{m \times (t-1)}$, *i.e.*, set $l_{i,k} = a_i^k$ for distinct $a_i \in \mathbb{F}_p$; independence follows from the existence of polynomial interpolation. We also require that $(l_{i,k})$ be constructed deterministically, so that anyone can verify that the π_i faithfully encode $\tilde{\Upsilon}$ simply by reproducing the process.

The following lemma shows the equivalence between the recursive specification of $\tilde{\Upsilon}$ and its flattened representation. It is adapted from a classic result [24] for Linear Secret Sharing Structures, and proven by induction on the structure of $\tilde{\Upsilon}$. We refer to the literature [30] for further details.

Lemma 1. [24] *Let $\tilde{\Upsilon}$ be an arborescent monotone threshold circuit as defined, and $(\pi_0, ..., \pi_\ell)$ a flattened representation of it per the above algorithm. A minimal truth assignment $\chi : \{L_0, ..., L_\ell\} \to \{\bot, \top\}$ satisfies $\tilde{\Upsilon}(\chi(L_0), ..., \chi(L_\ell)) = \top$ if and only if there exist integer coefficients $(\nu_0, ..., \nu_\ell)$ such that,*

$$\sum_{i=0}^{\ell} \nu_i \pi_i = Z_0 , \qquad \text{and} \quad \forall i : \nu_i = 0 \iff \chi(L_i) = \bot .$$

5.2   Information-Theoretic Blinding

In the signature scheme (yet to be described), we use both the polynomials $(\pi_0, ..., \pi_\ell)$ and the linear combination $(\nu_0, ..., \nu_\ell)$ from Lemma 1: the latter to create a signature, and the former to indicate how to verify it. However, since the linear coefficients ν_i reveal which of the L_i are true, they must be kept secret. In the actual signature, these coefficients appear not as integers but as exponents of elements of $\mathbb{G}$, and are thus already computationally hidden; however, this is not enough and we need to take an extra step to ensure perfect hiding.

By Lemma 1 we know that $\sum_{i=0}^{\ell} \nu_i \pi_i = Z_0$, where each $\nu_i \in \mathbb{F}_p$ and each $\pi_i \in \mathbb{F}_p[Z_0, ..., Z_\vartheta]_1$. We hide the linear coefficients ν_i using random blinding terms $(h_0, ..., h_\ell)$ such that $\sum_{i=0}^{\ell} h_i \pi_i = 0$. Since $\sum_{i=0}^{\ell} (\nu_i + h_i) \pi_i = Z_0$, the blinded coefficients $\nu_i + h_i$ still bear witness that $\tilde{\Upsilon}(L_0, ..., L_\ell) = \top$. However, these witnesses have been rendered information-theoretically indistinguishable, because the distribution of $(\nu_0 + h_0, ..., \nu_\ell + h_\ell)$ is conditionally independent of the truth values of the L_i given that $\tilde{\Upsilon}(L_0, ..., L_\ell) = \top$.

The difficulty is that no scalar h_i will satisfy $\sum_{i=0}^{\ell} h_i \pi_i = 0$ when the π_i contain uninstantiated variables. However, given a specific set of π_i, it is easy to build h_i that have polynomial values.

1. Draw a random vector $s = (s_1, ..., s_\ell) \in \mathbb{F}_p^\ell$ of scalar coefficients.
2. For $i = 1, ..., \ell$, define $h_i = -s_i \pi_0$, and set the remaining term $h_0 = \sum_{j=1}^{\ell} s_j \pi_j$.

In the actual scheme, these polynomials are evaluated "in the exponent" for unknown assignments to the Z_k, but regardless of their values we have $\sum_{i=0}^{\ell} h_i \pi_i = (\sum_{j=1}^{\ell} s_j \pi_j) \pi_0 + \sum_{i=1}^{\ell} (-s_i \pi_0) \pi_i = 0$, and so the blinding terms $(h_0, ..., h_\ell)$ meet our requirements.

Remark that the random vector s can be chosen independently of the π_i. This is important for the actual signature scheme, where the relevant polynomials will have coefficients that involve discrete logarithms not known explicitly (in addition to the Z_k being instantiated as discrete logarithms of random group elements). In spite of this, we will be able to select a suitable vector s and compute the blinding terms h_i "in the exponent".

5.3   Construction

The full mesh signature scheme can now be described as follows.

Initialization: Given a security parameter κ and a public random string $K \in \{0,1\}^{\text{poly}(\kappa)}$, all participants generate the common bilinear instance $\mathbf{G} = (p, \mathbb{G}, \hat{\mathbb{G}}, \mathbb{G}_t, g, \hat{g}, \mathbf{e}) \leftarrow \mathcal{G}(1^\kappa; K)$. Here, we require that the accompanying isomorphism $\psi : \hat{\mathbb{G}} \to \mathbb{G}$ be efficiently computable.

The string K also indicates a hash function $H : \{0,1\}^* \to \mathbb{F}_p$ from a collision-resistant family.

Given a mesh size parameter λ, the string K then specifies $\lambda + 1$ elements $\hat{g}_0, \hat{g}_1, ..., \hat{g}_\lambda$ in $\hat{\mathbb{G}}$, on which the efficient algorithm for ψ can be applied to obtain the images $g_0, g_1, ..., g_\lambda$ in $\mathbb{G}$.

Additionally, K defines $\lambda + 1$ random triples $(\hat{A}_{0,k}, \hat{B}_{0,k}, \hat{C}_{0,k}) \in \hat{\mathbb{G}}^3$ for $k \in \{0, ..., \lambda\}$; these elements together constitute a public verification key "in the sky" with no known signing key. We define $A_{0,k} = \psi(\hat{A}_{0,k})$, $B_{0,k} = \psi(\hat{B}_{0,k})$, $C_{0,k} = \psi(\hat{C}_{0,k})$, in $\mathbb{G}$, again easy to compute.

Key generation: To create a key pair, User #i draws a triple $(a_i, b_i, c_i) \in (\mathbb{F}_p^\times)^3$ as signing key. User #i computes for each $k \in \{0, ..., \lambda\}$ the triple $(\hat{A}_{i,k}, \hat{B}_{i,k}, \hat{C}_{i,k}) = (\hat{g}_k^{a_i}, \hat{g}_k^{b_i}, \hat{g}_k^{c_i}) \in \hat{\mathbb{G}}^3$, and lets these $3(\lambda+1)$ group elements constitute his or her verification key.

For simplicity, we write $(A_{i,k}, B_{i,k}, C_{i,k}) = (\psi(\hat{A}_{i,k}), \psi(\hat{B}_{i,k}), \psi(\hat{C}_{i,k})) = (g_k^{a_i}, g_k^{b_i}, g_k^{c_i}) \in \mathbb{G}^3$, which anyone can compute from the verification key of User #i thanks to ψ.

Mesh signature: On input the following mesh signature specification:

- ℓ atomic signature specifications $[VK_i : Msg_i]$, not necessarily all distinct, and ℓ boolean flags L_i, for $i = 1, ..., \ell$;
- a well-formed formula Υ with ℓ boolean inputs; and an assignment $\chi : \{L_1, ..., L_\ell\} \to \{\bot, \top\}$ that satifies $\Upsilon(L_1, ..., L_\ell) = \top$;
- $\forall i = 1, ..., \ell$ such that $\chi(L_i) = \top$, a valid Boneh-Boyen signature in $\mathbf{G}$ on the statement $[VK_i : Msg_i]$, given as a pair,

$$(u_i = g^{\frac{1}{a+b\,w+c\,t_i}}, \; t_i), \qquad \text{for random} \quad t_i \in \mathbb{F}_p ,$$

where $w = Msg_i$ and (a, b, c) is the signing key for the clause $[VK_i : Msg_i]$. The signer firsts extends Υ into Υ' that involves the public key "in the sky":

1. Compute $Msg_0 = H([VK_1 : Msg_1], ..., [VK_\ell : Msg_\ell], \Upsilon)$ by hashing the mesh specification, and associate the literal L_0 to the clause $[VK_0 : Msg_0]$.
2. Construct $\tilde{\Upsilon} = L_0 \vee \Upsilon$, which is well-formed per the definition.
3. Extend χ so that $\chi(L_0) = \bot$, as we lack an atomic signature for L_0.

The signer then builds the mesh signature from the circuit $\tilde{\Upsilon}$, the assignment χ, and the atomic signatures (u_i, t_i) known for such i that $\chi(L_i) = \top$, as:

4. Create a flattened representation of $\tilde{\Upsilon}$ and χ as discussed in Section 5.1. Accordingly, let $\pi_0, ..., \pi_\ell \in \mathbb{F}_p[Z_0, ..., Z_\vartheta]$ be public degree-1 multivariate polynomials that encode $\tilde{\Upsilon}$, and $\nu_0, ..., \nu_\ell \in \mathbb{F}_p$ the secret scalar coefficients of a linear combination that expresses χ. Explicitly determine all the coefficients $y_{j,k} \in \mathbb{F}_p$ in all polynomials $\pi_j = \sum_{k=0}^\vartheta y_{j,k} Z_k$.
5. Create a random blinding vector $s = (s_1, ..., s_\ell) \in \mathbb{F}_p^\ell$ as in Section 5.2.
6. $\forall i \in \{0, ..., \ell\} : \chi(L_i) = \bot$, pick $t_i \in \mathbb{F}_p$ and fix $u_i = g^0 = 1 \in \mathbb{G}$.

7. For all $j = 0, ..., \ell$ and $k = 0, ..., \vartheta$, let $m_j = Msg_j$ and calculate,

$$v_{j,k} = \left(A_{j,k} \, B_{j,k}^{m_j} \, C_{j,k}^{t_j} \right)^{y_{j,k}} \, , \qquad v_j = \prod_{k=0}^{\vartheta} v_{j,k} \, .$$

8. Compute, for $i = 1, ..., \ell$, and $k = 0, ..., \vartheta$, respectively,

$$S_i = u_i^{\nu_i} \, v_0^{-s_i} \, , \qquad P_k = \prod_{j=1}^{\ell} v_{j,k}^{s_j} \, .$$

(The value of any intervening u_i such that $\chi(L_i) = \bot$ is unimportant since then $\nu_i = 0$; this is true in particular for the 0-th user "in the sky".)

9. Output the mesh signature, consisting of the statement Υ and the tuple,

$$\sigma = (\, t_0, \ ..., \ t_\ell, \ S_1, \ ..., \ S_\ell, \ P_0, \ ..., \ P_\vartheta \,) \ \in \ \mathbb{F}_p^{\ell+1} \times \mathbb{G}^{\ell+\vartheta+1} \, .$$

Mesh verification: A fully qualified mesh signature package consists of:
- $\ell + 1$ propositions $[VK_0 : Msg_0], ..., [VK_\ell : Msg_\ell]$ viewed as inputs to,
- an arborescent monotone threshold circuit $\tilde{\Upsilon} : \{\bot, \top\}^{\ell+1} \to \{\bot, \top\}$,
- a mesh signature $\sigma = (t_0, ..., t_\ell, S_1, ..., S_\ell, P_0, ..., P_\vartheta) \in \mathbb{F}_p^{\ell+1} \times \mathbb{G}^{\ell+\vartheta+1}$.

To verify such a signature, the verifier proceeds as follows:

1. Ascertain that $\tilde{\Upsilon}(\top, \star, ..., \star) = \top$, extract from $\tilde{\Upsilon}(L_0, ..., L_\ell)$ the subcircuit $\Upsilon(L_1, ..., L_\ell)$ such that $\tilde{\Upsilon} = \Upsilon \vee L_0$, and verify that $Msg_0 = H([VK_1 : Msg_1], ..., [VK_\ell : Msg_\ell], \Upsilon)$.

2. Compute the representation $(\pi_0, ..., \pi_\ell)$ of the formula $\tilde{\Upsilon}$ by reproducing the deterministic conversion of Section 5.1.

3. For $i = 0, ..., \ell$, determine the coefficients $y_{i,k} \in \mathbb{F}_p$ of the polynomials $\pi_i = \sum_{k=0}^{\vartheta} y_{i,k} Z_k$.

4. For $i = 0, ..., \ell$ and $k = 0, ..., \vartheta$, retrieve $(\hat{A}_{i,k}, \hat{B}_{i,k}, \hat{C}_{i,k})$ from the key VK_i, let $m_i = Msg_i$, and calculate,

$$\hat{v}_{i,k} = \left(\hat{A}_{i,k} \, \hat{B}_{i,k}^{m_i} \, \hat{C}_{i,k}^{t_i} \right)^{y_{i,k}} \, , \qquad \hat{v}_i = \prod_{k=0}^{\vartheta} \hat{v}_{i,k} \, .$$

5. Using the pairing, verify the equalities, for all $k = 0, ..., \vartheta$,

$$\mathbf{e}\,(P_k, \, \hat{v}_0) \cdot \prod_{i=1}^{\ell} \mathbf{e}\,(S_i, \, \hat{v}_{i,k}) = \begin{cases} \mathbf{e}(g, \, \hat{g}_0) & \text{for } k = 0 \\ 1 & \text{otherwise} \end{cases} \, .$$

6. Accept the signature if and only if all $\vartheta + 1$ equalities hold in $\mathbb{G}_t$.

(Optional) **Probabilistic check:** Mesh signatures can be verified using fewer total pairings, at the cost of some additional random bits and exponentiations. In the same setting as above, replace Step 5 onward by the following:

5'. Set $d_0 = 1$, pick random $d_1, ..., d_\vartheta \in \mathbb{F}_p$, and verify the single equality,

$$\mathbf{e}(\prod_{k=0}^{\vartheta} P_k^{d_k}, \, \hat{v}_0) \cdot \prod_{i=1}^{\ell} \mathbf{e}(S_i, \, \prod_{k=0}^{\vartheta} \hat{v}_{i,k}^{d_k}) = \mathbf{e}(g, \, \hat{g}_0) \, .$$

6'. Accept the signature as valid if and only if the equality holds in $\mathbb{G}_t$.

Theorem 3. *The mesh signature is consistent.*

Proof. For any list of public polynomials $\pi_0, ..., \pi_\ell$ and secret coefficients $\nu_0, ..., \nu_\ell$ that respectively encode per Lemma 1 a well-formed mesh specification $\tilde{\Upsilon}$ and an assignment χ that satisfies it, we need to show that a signature created by the above algorithm will be accepted by the same. A straightforward sequence of substitutions in the scheme description shows this to be the case.

Theorem 4. *The mesh signature has everlasting perfect anonymity.*

Theorem 5. *The mesh signature is existentially unforgeable under an adaptive attack, against a static adversary that makes no more than q mesh signature queries, and no more than q atomic signature queries to each of the ℓ honest users, adaptively, provided that the $(q, \ell + 1)$-Poly-SDH assumption holds in $\mathbf{G}$, in the common random string model.*

Optimization. We note that the user keys and the key "in the sky" can be shortened significantly. It turns out that in the proofs the b_i are always known to the simulators and are thus superfluous: we can set $b_i = 1$ and omit the $\hat{B}_{i,k} = \hat{g}_k^{b_i} = \hat{g}_k$ from the keys. This holds in the ring scheme, too.

We can independently compress the key "in the sky" to just two elements of $\hat{\mathbb{G}}$, if we observe that for $\tilde{\Upsilon} = \Upsilon \vee L_0$ the encoding algorithm of Section 5.1 always gives $\pi_0 = Z_0$, *i.e.*, $y_{0,0} = 1$ and $y_{0,k} = 0$ for $k \neq 0$, meaning that the tuples $(\hat{A}_{0,k}, ..., \hat{C}_{0,k})$ for $k \neq 0$ are in fact never used. Furthermore, it is safe to set $\hat{B}_{0,0} = \hat{g}$, which leaves just the pair $(\hat{A}_{0,0}, \hat{C}_{0,0})$.

6   Conclusion

We have introduced mesh signatures as a generalization of ring signatures with a richer language for expressing signer ambiguity. Mesh signatures scale to large crowds with many co-signers and independent certificate authorities; they can even implicate unwilling individuals who, by withholding their ring public key, would have otherwise remained out of reach. Because in principle mesh signatures require neither trusted setup nor centralized authorities, they provide a credible answer to the question of how to leak a secret authoritatively.

We have constructed a simple and practical mesh signature scheme in prime order bilinear groups, that achieves everlasting unconditional anonymity, and existential unforgeability in the common random string model, without trusted setup authority. To obtain this result, we introduced a new complexity assumption, which we prove sound in the generic model; it is in the spirit of the SDH assumption, but better exploits the group structure of the values computed by pairing. Incidentally, we obtain an efficient ring signature without random oracles as a special case of our construction.

Acknowledgements

The author wishes to express his gratitude to Anna Lysyanskaya and the anonymous referees of Eurocrypt 2007 for many valuable comments.

References

1. Masayuki Abe, Miyako Ohkubo, and Koutarou Suzuki. 1-out-of-n signatures from a variety of keys. In *Proceedings of AsiaCrypt 2002*, volume 2501 of *LNCS*, pages 415–32. Springer, 2002.
2. Giuseppe Ateniese, Jan Camenisch, Susan Hohenberger, and Breno de Medeiros. Practical group signatures without random oracles. Cryptology ePrint Archive, Report 2005/385, 2005. http://eprint.iacr.org/.
3. Adam Bender, Jonathan Katz, and Ruggero Morselli. Ring signatures: Stronger definitions, and constructions without random oracles. In *Proceedings of TCC 2006*, LNCS. Springer, 2006.
4. Dan Boneh and Xavier Boyen. Short signatures without random oracles. In *Advances in Cryptology—EUROCRYPT 2004*, volume 3027 of *LNCS*, pages 56–73. Springer, 2004.
5. Dan Boneh, Xavier Boyen, and Hovav Shacham. Short group signatures. In *Advances in Cryptology—CRYPTO 2004*, volume 3152 of *LNCS*, pages 41–55. Springer, 2004.
6. Dan Boneh, Craig Gentry, Ben Lynn, and Hovav Shacham. Aggregate and verifiably encrypted signatures from bilinear maps. In *Advances in Cryptology—EUROCRYPT 2003*, volume 2656 of *LNCS*, pages 416–32. Springer, 2003.
7. Dan Boneh, Eu-Jin Goh, and Kobbi Nissim. Evaluating 2-DNF formulas on ciphertexts. In *Proceedings of TCC 2005*, Lecture Notes in Computer Science. Springer-Verlag, 2005.
8. Xavier Boyen and Brent Waters. Full-domain subgroup hiding and constant-size group signatures. In *Public Key Cryptography—PKC 2007*, volume 4450 of *LNCS*, pages 1–15. Springer, 2007.
9. Jan Camenisch and Anna Lysyanskaya. Signature schemes with efficient protocols. In *Proceedings of SCN 2002*, LNCS. Springer, 2002.
10. Jan Camenisch and Anna Lysyanskaya. Signature schemes and anonymous credentials from bilinear maps. In *Advances in Cryptology—CRYPTO 2004*, volume 3152 of *LNCS*. Springer, 2004.
11. Melissa Chase and Anna Lysyanskaya. Signature of knowledge. In *Advances in Cryptology—CRYPTO 2006*, volume 4117 of *LNCS*. Springer, 2006.
12. David Chaum and Eugène van Heyst. Group signatures. In *Advances in Cryptology—EUROCRYPT 1991*, volume 547 of *LNCS*, pages 257–65. Springer, 1991.
13. Jung Hee Cheon. Security analysis of the strong Diffie-Hellman problem. In *Advances in Cryptology—EUROCRYPT 2006*, volume 4004 of *LNCS*, pages 1–13. Springer, 2006.
14. Sherman S. M. Chow, Victor K.-W. Wei, Joseph K. Liu, and Tsz Hon Yuen. Ring signatures without random oracles. In *Proceedings of AsiaCCS 2006*, pages 297–302. ACM Press, 2006.
15. Ronald Cramer, Ivan Damgård, and Berry Schoenmakers. Proofs of partial knowledge and simplified design of witness hiding protocols. In *Advances in Cryptology—CRYPTO 1994*, volume 839 of *LNCS*, pages 174–87. Springer, 1994.
16. Yevgeniy Dodis, A. Kiayias, Antonio Nicolosi, and Victor Shoup. Anonymous identification in ad hoc groups. In *Advances in Cryptology—EUROCRYPT 2004*, volume 3027 of *LNCS*, pages 609–26. Springer, 2004.
17. Cynthia Dwork and Moni Naor. Zaps and their applications. In *Proceedings of FOCS 2000*, pages 542–552. IEEE Press, 2000.

18. Cynthia Dwork, Moni Naor, and Amit Sahai. Concurrent zero-knowledge or the timing model for designing concurrent protocols. *Journal of the ACM*, 51(6):851–98, 2004.
19. Steven D. Galbraith, Kenneth G. Paterson, and Nigel P. Smart. Pairings for cryptographers. Cryptology ePrint Archive, Report 2006/165, 2006. http://eprint.iacr.org/2006/165/.
20. Jens Groth, Rafail Ostrovsky, and Amit Sahai. Non-interactive Zaps and new techniques for NIZK. In *Advances in Cryptology—CRYPTO 2006*, LNCS. Springer, 2006.
21. Javier Herranz and Germán Sáez. Forking lemmas for ring signature schemes. In *Proceedings of IndoCrypt 2003*, volume 2904 of *LNCS*, pages 266–79. Springer, 2003.
22. Javier Herranz and Germán Sáez. New distributed ring signatures for general families of signing subsets. Cryptology ePrint Archive, Report 2004/377, 2004. http://eprint.iacr.org/.
23. Antoine Joux and Kim Nguyen. Separating decision Diffie-Hellman from computational Diffie-Hellman in cryptographic groups. *Journal of Cryptology*, 16(4), 2003.
24. Mauricio Karchmer and Avi Wigderson. On span programs. In *Annual Conference on Structure in Complexity Theory*, 1993.
25. Victor Miller. The Weil pairing, and its efficient calculation. *Journal of Cryptology*, 17(4), 2004.
26. Moni Naor. Deniable ring authentication. In *Advances in Cryptology—CRYPTO 2002*, volume 2442 of *LNCS*, pages 481–98. Springer, 2002.
27. Ron Rivest, Adi Shamir, and Yael Tauman. How to leak a secret. In *Proceedings of AsiaCrypt 2001*, volume 2248 of *LNCS*, pages 552–65. Springer, 2001.
28. Hovav Shacham and Brent Waters. Efficient ring signatures without random oracles. In *Public Key Cryptography—PKC 2007*, volume 4450 of *LNCS*. Springer, 2007.
29. Victor Shoup. Lower bounds for discrete logarithms and related problems. In *Advances in Cryptology—EUROCRYPT 1997*, volume 1233 of *LNCS*. Springer, 1997.
30. Marten van Dijk. A linear construction of secret sharing schemes. *Designs, Codes and Cryptography*, 12(2):161–201, 1997.
31. Brent Waters. Efficient identity-based encryption without random oracles. In *Advances in Cryptology—EUROCRYPT 2005*, volume 3494 of *LNCS*. Springer, 2005.
32. Victor K. Wei and Tsz Hon Yuen. (Hierarchical identity-based) threshold ring signatures. Cryptology ePrint Archive, Report 2006/193, 2006. http://eprint.iacr.org/.

The Power of Proofs-of-Possession: Securing Multiparty Signatures against Rogue-Key Attacks

Thomas Ristenpart and Scott Yilek

Dept. of Computer Science & Engineering 0404, University of California San Diego
9500 Gilman Drive, La Jolla, CA 92093-0404, USA
{tristenp,syilek}@cs.ucsd.edu
http://www-cse.ucsd.edu/users/{tristenp,syilek}

Abstract. Multiparty signature protocols need protection against *rogue-key attacks*, made possible whenever an adversary can choose its public key(s) arbitrarily. For many schemes, provable security has only been established under the *knowledge of secret key* (KOSK) assumption where the adversary is required to reveal the secret keys it utilizes. In practice, certifying authorities rarely require the strong proofs of knowledge of secret keys required to substantiate the KOSK assumption. Instead, *proofs of possession* (POPs) are required and can be as simple as just a signature over the certificate request message. We propose a general *registered key* model, within which we can model both the KOSK assumption and in-use POP protocols. We show that simple POP protocols yield provable security of Boldyreva's multisignature scheme [11], the LOSSW multisignature scheme [28], and a 2-user ring signature scheme due to Bender, Katz, and Morselli [10]. Our results are the *first* to provide formal evidence that POPs can stop rogue-key attacks.

Keywords: Proofs of possession, PKI, multisignatures, ring signatures, bilinear maps.

1 Introduction

We refer to any scheme that generates signatures bound to a group of parties as a *multiparty signature scheme*. We focus on schemes that are both adaptive and decentralized: the set of potential signers is dynamic and no group manager is directly involved in establishing eligibility of participants. Examples include multisignatures, ring signatures, designated-verifier signatures, and aggregate signatures. These schemes require special care against *rogue-key attacks*, which can be mounted whenever adversaries are allowed to choose their public keys arbitrarily. Typical attacks have the adversary use a public key that is a function of an honest user's key, allowing him to produce forgeries easily. Rogue-key attacks have plagued the development of multiparty signature schemes [26,20,22,30,32,33,25,11,28,38,31].

M. Naor (Ed.): EUROCRYPT 2007, LNCS 4515, pp. 228–245, 2007.

One method for preventing rogue-key attacks is to require, during public key registration with a certificate authority (CA), that a party proves knowledge of its secret key. This setting has typically been formalized as the *knowledge of secret key* (KOSK) assumption [11]: schemes are analyzed in a setting where adversaries must reveal their secret keys directly. This abstraction has lead to simple schemes and straightforward proofs of security. To name a few: Boldyreva's multisignature scheme [11] (we call it BMS), the LOSSW multisignature scheme [28] (we call it WMS for brevity and its basis on Waters signatures [40]), the LOSSW sequential aggregate signature scheme [28], and many designated-verifier signature schemes [23,39,27,24]. Since simple rogue-key attacks against these schemes are known, it might appear that the security of these schemes actually depends on parties performing proofs of knowledge during registration.

DRAWBACKS OF THE KOSK ASSUMPTION. Unfortunately, there are substantial drawbacks to using the KOSK assumption. Bellare and Neven discuss this in detail [7]; we briefly recall some of their discussion. First and foremost, the KOSK assumption is not realized by existing public key infrastructures (PKI). Registration protocols specified by the most widely used standards (RSA PKCS#10 [36], RFC 4210 [1], RFC 4211 [37]) do not specify that CA's should require proofs of knowledge. Thus, to use schemes proven secure under the KOSK assumption, one would be faced with the daunting task of upgrading existing (and already complex) PKI. This would likely require implementing clients and CA's that support zero-knowledge (ZK) proofs of knowledge that have extraction guarantees in fully concurrent settings [4]. Non-interactive ZK proofs of knowledge [17,16,19] could also be utilized, but these are more computationally expensive.

THE PLAIN SETTING. In the context of multisignatures, Bellare and Neven [7] show that it is possible to dispense with the KOSK assumption. They provide a multisignature scheme which is secure, even against rogue-key attacks, in the *plain public-key setting*, where registration with a CA ensures nothing about a party's possession or knowledge of a secret key. Here we are interested in something different, namely investigating the security of schemes (that are *not* secure in the plain setting) under more realistic key registration protocols, discussed next.

PROOFS OF POSSESSION. Although existing PKIs do not require proofs of knowledge, standards mandate the inclusion of a *proof of possession* (POP) during registration. A POP attests that a party has access to the secret key associated with his/her public key, which is typically accomplished using the functionality of the key pair's intended scheme. For signature schemes, the simplest POP has a party sign its certificate request message and send both the message and signature to the CA. The CA checks that the signature verifies under the public key being registered. In general, such proofs of possession (POPs) are clearly not sufficient for substantiating the KOSK assumption. In fact, POPs have not (previously) lead to any formal guarantees of security against rogue key attacks, even though intuitively they might appear to stop adversaries from picking arbitrary public keys. This logical gap has contributed to contention regarding the need for POPs in PKI standards [2].

OUR CONTRIBUTIONS. We suggest analyzing the security of multiparty signature schemes in a registered key model, which allows modeling a variety of key registration assumptions including those based on POPs. Using the new model, we analyze the security of the BMS and WMS multisignature schemes under POP protocols. We show that, interestingly, requiring the in-use and standardized POP protocol described above *still* admits rogue-key attacks. This implies the intuition mentioned above is flawed. On the positive side, we show how a slight change to the standardized POP protocol admits proofs of security for these schemes. We also investigate the setting of ring signatures. We describe how the key registration model can be utilized to result in improved unforgeability guarantees. In particular we show that the Bender, Katz, and Morselli 2-user ring signature scheme based on Waters signatures [10] is secure against rogue-key attacks under a simple POP protocol. We now look at each of these contributions in more detail.

THE REGISTERED KEY MODEL. A key registration protocol is a pair of interactive algorithms (RegP, RegV), the former executed by a registrant and the latter executed by a certifying authority (CA). We lift security definitions to the registered key model by giving adversaries an additional key registration oracle, which, when invoked, executes a new instance of RegV. The security game can then restrict adversarial behavior based on whether successful registration has occurred. Security definitions in the registered key model are thus parameterized by a registration protocol. This approach allows us to straightforwardly model a variety of registration assumptions, including the KOSK assumption, the plain setting and POP-based protocols.

MULTISIGNATURES UNDER POP. A multisignature scheme allows a set of parties to jointly generate a compact signature for some message. These schemes have numerous applications, e.g. contract signing, distribution of a certificate authority, or co-signing. The BMS and WMS schemes are simple multisignature schemes that are based directly on the short signature schemes of Boneh, Lynn, and Shacham (BLS) [14] and Waters [40]. (That is, a multisignature with group of size one is simply a BLS or Waters signature.) These schemes give short multisignatures (just 160 bits for BMS). Moreover, multisignature generation is straightforward: each party produces its BLS or Waters signature on the message, and the multisignature is just the (component-wise) product of these signatures. Both schemes fall prey to straightforward rogue-key attacks, but have proofs of security under the KOSK assumption [11,28].

We analyze these schemes when key registration requires POPs. We show that the standardized POP mechanism described above, when applied to these schemes, *does not* lead to secure multisignatures. Both schemes fall to rogue-key attacks despite the use of the standardized POPs. We present a straightforward and natural fix for this problem: simply use separate hash functions for POPs and multisignatures. We prove the security of BMS and WMS multisignatures under such POP mechanisms, giving the first formal justification that these desirable schemes can be used in practice. Both proofs reduce to the same computational assumptions used in previous KOSK proofs and the reductions are just as tight.

RING SIGNATURES UNDER POP. Ring signatures allow a signer to choose a group of public keys and sign a message so that it is verifiable that some party in the group signed it, but no adversary can determine which party it was. The canonical application of ring signatures is leaking secrets [35]. Bender, Katz, and Morselli (BKM) have given a hierarchy of anonymity and unforgeability definitions for ring signature schemes [10]. For κ-user schemes, where only rings of size κ are allowed, we point out that the ability to mount rogue-key attacks (as opposed to the ability to corrupt honest parties) is a crucial distinguisher of the strength of unforgeability definitions. We introduce new security definitions that facilitate a formal analysis of this fact. BKM also propose two 2-user ring signature schemes that do not rely on random oracles, and prove them to meet the weaker unforgeability guarantee. As pointed out by Shacham and Waters, these schemes do not meet the stronger definition due to rogue-key attacks [38].

We show that the KOSK assumption provably protects against rogue-key attacks for a natural class of ring signature schemes (both the BKM 2-user schemes fall into this class). We go on to prove the security of the BKM 2-user scheme based on Waters signatures under a simple POP-based registration protocol.

SCHEMES IN THE PLAIN SETTING. We briefly overview some schemes built for the plain setting. The Micali, Ohta, and Reyzin multisignature scheme [31] was the first to be proven secure in the plain setting, but it requires a dedicated key setup phase after which the set of potential signers is necessarily static. The multisignature scheme of Bellare and Neven [7] does not require a key setup phase, and is proven secure in the plain setting. While computationally efficient, it requires several rounds of communication between all co-signers, which is more than the "non-interactive" BMS and WMS schemes.

Bender, Katz, and Morselli introduced the first ad-hoc ring signature scheme that provably resists rogue-key attacks [10]. Their scheme is not efficient, requiring semantically-secure encryption for each bit of a message. The ring signature scheme of Shacham and Waters [38] is more efficient but still not as efficient as the BKM schemes for rings of size two. Particularly, their ring signatures are at least three times as long as those given by the BKM scheme based on Waters signatures and they require more computational overhead. Of course, their solution works on rings with size greater than two.

Finally, aggregate signature schemes due to Boneh et al. [13] and Lysyanskaya et al. [29] are secure in the plain setting.

RELATED WORK AND OPEN PROBLEMS. Boldyreva et al. [12] investigate certified encryption and signature schemes. They utilize a POP-based protocol to show the security of traditional certified signatures. They do not consider multiparty signatures. Many schemes beyond those treated here rely on the KOSK assumption and finding POP-based protocols for such schemes, if possible, constitutes an important set of open problems. A few examples are the LOSSW sequential aggregate signature scheme [28], the StKD encryption scheme due to Bellare, Kohno, and Shoup [5], and various designated-verifier signature schemes [23,39,27,24].

2 Preliminaries

BASIC NOTATION. We denote string concatenation by $||$. Let S be any set. Then we define $S \overset{\cup}{\leftarrow} s$ for any appropriate value s as $S \leftarrow S \cup \{s\}$. For a multiset $\mathcal{S}$, let $\mathcal{S} - \{s\}$ denote the multiset $\mathcal{S}$ with one instance of element s removed. For multisets $\mathcal{S}$ and R, let $\mathcal{S} \backslash \mathcal{R}$ be the multiset formed by repeatedly executing $\mathcal{S} \leftarrow \mathcal{S} - \{r\}$ for each $r \in \mathcal{R}$ (including duplicates). We define $s \overset{\$}{\leftarrow} S$ as sampling uniformly from S and $s \overset{\$}{\leftarrow} A(x_1, x_2, \ldots)$ assigns to s the result of running A on fresh random coins and the inputs $x_1, x_2, \ldots$. For any string M, let $M[i]$ denote the ith bit of M. For a table H, let $\mathsf{H}[s]$ denote the element associated with s. We write $\mathsf{Time}(A) = \max\{t_1, t_2, \ldots\}$ where $A = (A_1, A_2, \ldots)$ is a tuple of algorithms and $t_1, t_2, \ldots$ are their worst case running times.

BILINEAR MAPS AND CO-CDH. The schemes we consider use bilinear maps. Let $\mathbb{G}_1$, $\mathbb{G}_2$, and $\mathbb{G}_T$ be groups, each of prime order p. Then $\mathbb{G}_1^*$, $\mathbb{G}_2^*$, and $\mathbb{G}_T^*$ represent the set of all generators of the groups (respectively). Let $\mathbf{e} \colon \mathbb{G}_1 \times \mathbb{G}_2 \to \mathbb{G}_T$ be an efficiently computable bilinear map (also called a pairing). For the multisignature schemes we consider, we use the asymmetric setting [14,13] where $\mathbb{G}_1 \neq \mathbb{G}_2$ and there exists an efficiently computable isomorphism $\psi \colon \mathbb{G}_2 \to \mathbb{G}_1$. The asymmetry allows for short signatures, while ψ is needed in the proofs. For the ring signature schemes we consider, we instead use the symmetric setting [10,38] where $\mathbb{G}_1 = \mathbb{G}_2$. Let n represent the number of bits needed to encode an element of $\mathbb{G}_1$; for the asymmetric setting n is typically 160. Finally let g be a generator in $\mathbb{G}_2$. For the rest of the paper we treat $\mathbb{G}_1, \mathbb{G}_2, \mathbb{G}_T, p, g, \mathbf{e}$ as fixed, globally known parameters. Then we define the advantage of an algorithm A in solving the Computational co-Diffie-Hellman (co-CDH) problem in the groups $(\mathbb{G}_1, \mathbb{G}_2)$ as

$$\mathbf{Adv}_{(\mathbb{G}_1, \mathbb{G}_2)}^{\text{co-cdh}}(A) = \Pr\left[A(g, g^x, h) = h^x : x \overset{\$}{\leftarrow} \mathbb{Z}_p; h \overset{\$}{\leftarrow} \mathbb{G}_1\right]$$

where the probability is over the random choices of x and h and the coins used by A. Here $\mathbb{Z}_p$ is the set of integers modulo p. Note that in the symmetric setting this is just the CDH problem.

For a group element g, we write $\langle g \rangle$ to mean some canonical encoding of g as a bit string of the appropriate length. We write $\langle g \rangle_n$ to mean the first n bits of $\langle g \rangle$. We use the shorthand $\vec{u}$ (resp. $\vec{w}$) to mean a list of group elements $u_1, \ldots, u_n$ (resp. $w_1, \ldots, w_n$). Let t_E, t_ψ, and $t_{\mathbf{e}}$ be the maximum times to compute an exponentiation in $\mathbb{G}_1$, compute ψ on an element in $\mathbb{G}_2$, and compute the pairing.

SIGNATURE SCHEMES. A signature scheme $\mathsf{S} = (\mathsf{Kg}, \mathsf{Sign}, \mathsf{Ver})$ consists of a key generation algorithm, a signing algorithm that outputs a signature given a secret key and a message, and a verification algorithm that outputs a bit given a public key, message, and signature. We require that $\mathsf{Ver}(pk, M, \mathsf{Sign}(sk, M)) = 1$ for all allowed M and valid pk, sk. Following [18], we define the advantage of an adversary A in forging against S in a chosen message attack as

$$\mathbf{Adv}_{\mathsf{S}}^{\text{uf}}(A) = \Pr\left[\mathsf{Ver}(pk, M, \sigma) = 1 : (pk, sk) \overset{\$}{\leftarrow} \mathsf{Kg}; (M, \sigma) \overset{\$}{\leftarrow} A^{\mathsf{Sign}(sk, \cdot)}(pk)\right]$$

where the probability is over the coins used by Kg, Sign, and A.

3   The Registered Key Model

KEY REGISTRATION PROTOCOLS. Let $\mathcal{P}$ and $\mathcal{S}$ be sets and $\mathcal{K} \subseteq \mathcal{P} \times \mathcal{S}$ be a relation on the sets (representing public keys, secret keys, and valid key pairs, respectively). A *key registration protocol* is a pair of interactive algorithms (RegP, RegV). A party registering a key runs RegP with inputs $pk \in \mathcal{P}$ and $sk \in \mathcal{S}$. A certifying authority (CA) runs RegV. We restrict our attention (without loss of generality) to protocols in which the last message is from RegV to RegP and contains either a $pk \in \mathcal{P}$ or a distinguished symbol $\bot$. We require that running RegP(pk, sk) with RegV results in RegV's final message being pk whenever $(pk, sk) \in \mathcal{K}$.

We give several examples of key registration protocols. The plain registration protocol Plain = (PlainP, PlainV) has the registrant running PlainP(pk, sk) send pk to the CA. The CA running PlainV, upon receiving a public key pk, simply replies with pk. This protocol will be used to capture the plain model, where no checks on public keys are performed by a CA. To model the KOSK assumption, we specify the registration protocol Kosk = (KoskP, KoskV). Here KoskP(pk, sk) sends (pk, sk) to the CA. Upon receiving (pk, sk), the KoskV algorithm checks that $(pk, sk) \in \mathcal{K}$. (We assume that such a check is efficiently computable; this is the case for key pairs we consider.) If so, it replies with pk and otherwise with $\bot$.

We refer to registration protocols that utilize the key's intended functionality as proof-of-possession based. For example, let S = (Kg, Sign, Ver) be a signature scheme. Define the registration protocol S-Pop = (PopP, PopV) as follows. Running PopP on inputs pk, sk results in sending the message $pk \parallel \mathsf{Sign}(sk, \langle pk \rangle)$ to the CA. Upon receiving message $pk \parallel \sigma$, a CA running PopV replies with pk if $\mathsf{Ver}(pk, \langle pk \rangle, \sigma) = 1$ and otherwise replies with $\bot$. This corresponds to the simplest POPs for signature schemes specified in PKCS#10 and RFCs 4210/4211.

THE REGISTERED KEY MODEL. We consider security definitions that are captured by a game between an adversary and an environment. To lift such security definitions to the *registered key model*, we use the following general approach. Adversaries are given an additional key registration oracle OKReg that, once invoked, runs a new instance of RegV for some key registration protocol (RegP, RegV). If the last message from RegV is a public key pk, then pk is added to a table $\mathcal{R}$. This table can now be used to modify winning conditions or restrict which public keys are utilized by the adversary in interactions with the environment. Security of schemes under the new definition is therefore always with respect to some registration protocol.

The key registration protocols mentioned so far are *two round protocols*: the registrant sends a first message to the CA, which replies with a second message being either pk or $\bot$. For any two round protocol Reg = (RegP, RegV), the OKReg oracle can be simplified as follows. An adversary queries with a first message, at which point RegP is immediately run and supplied with the message. The oracle halts RegP before it sends its reply message. The message is added to $\mathcal{R}$ if it is not $\bot$. The oracle finally returns pk or $\bot$ appropriately.

Experiment $\mathbf{Exp}_{MS,Reg}^{msuf\text{-}kr}(A)$

  $par \xleftarrow{\$} \mathsf{MPg}; \; (pk^*, sk^*) \xleftarrow{\$} \mathsf{MKg}(par); \; \mathcal{Q} \leftarrow \emptyset; \; \mathcal{R} \leftarrow \emptyset$

  Run $A(par, pk^*)$ handling oracle queries as follows

  $\mathsf{OMSign}(\mathcal{V}, M)$, where $pk^* \in \mathcal{V}$: $\mathcal{Q} \xleftarrow{\cup} M$; Simulate a new instance of $\mathsf{MSign}(sk^*, \mathcal{V}, M)$, forwarding messages to and from A appropriately.

  OKReg: Simulate a new instance of algorithm RegV, forwarding messages to and from A. If the instance's final message is $pk \neq \perp$, then $\mathcal{R} \xleftarrow{\cup} pk$.

  A halts with output $(\mathcal{V}, M, \sigma)$
  If $(pk^* \in \mathcal{V}) \wedge (M \notin \mathcal{Q}) \wedge (\mathsf{MVf}(\mathcal{V}, M, \sigma) = 1) \wedge ((\mathcal{V} - \{pk^*\}) \setminus \mathcal{R} = \emptyset)$ then
      Return 1
  Return 0

Fig. 1. Multisignature security experiment in the registered key model

4 Multisignatures Using POPs

The goal of a multisignature scheme is for a group of parties, each with its own public and secret keys, to jointly create a compact signature on some message. Following the formulation in [7], a *multisignature scheme* is a tuple of algorithms $\mathsf{MS} = (\mathsf{MPg}, \mathsf{MKg}, \mathsf{MSign}, \mathsf{MVf})$. A central authority runs the (randomized) parameter generation algorithm MPg to create a parameter string par that is given to all parties and is an (usually implicit) input to the other three algorithms. The (randomized) key generation algorithm MKg, independently run by each party, outputs a key pair (pk, sk). The MSign interactive protocol is run by some group of players. Each party locally runs MSign on input being a secret key sk, a multiset of public keys $\mathcal{V}$, and a message M. It may consist of multiple rounds, though the protocols we consider here only require two rounds: a request broadcast to all parties and the response(s). Finally, the verification algorithm MVf takes as input a tuple $(\mathcal{V}, M, \sigma)$, where $\mathcal{V}$ is a multiset of public keys, M is a message, and σ is a signature, and returns a bit. We require that $\mathsf{MVf}(\mathcal{V}, M, \mathsf{MSign}(sk, \mathcal{V}, M)) = 1$ for any M and where every participant correctly follows the algorithms.

MULTISIGNATURE SECURITY. Let $\mathsf{MS} = (\mathsf{MPg}, \mathsf{MKg}, \mathsf{MSign}, \mathsf{MVf})$ be a multisignature scheme, $\mathsf{Reg} = (\mathsf{RegP}, \mathsf{RegV})$ be a key registration protocol, and A be an adversary. Figure 1 displays the security game $\mathbf{Exp}_{MS,Reg}^{msuf\text{-}kr}(A)$. The experiment simulates one honest player with public key pk^*. The goal of the adversary is to produce a *multisignature forgery*: a tuple $(\mathcal{V}, M, \sigma)$ that satisfies the following four conditions. First, the honest public key pk^* is in the multiset $\mathcal{V}$ at least once. Second, the message M was not queried to the multisignature oracle. Third, the signature verifies. Fourth, each public key in $\mathcal{V} - \{pk^*\}$ must be in $\mathcal{R}$, where $\mathcal{V} - \{pk^*\}$ means the multiset $\mathcal{V}$ with *one* occurrence of the honest key removed. We define the msuf-kr-advantage of an adversary A against a multisignature scheme MS with respect to registration protocol Reg as $\mathbf{Adv}_{MS,Reg}^{msuf\text{-}kr}(A) = \Pr\left[\mathbf{Exp}_{MS,Reg}^{msuf\text{-}kr}(A) \Rightarrow 1\right]$. The probability is taken over the random coins used in the course of running the experiment, including those used by A. The definitions can be lifted to the random oracle model [8] in the natural way. It is easy to show that our definition is equivalent to the definition in [7] when $\mathsf{Reg} = \mathsf{Plain}$ and equivalent to the definition in [11] when $\mathsf{Reg} = \mathsf{Kosk}$.

In the case of two-round multisignature schemes, the multisignature oracle can be simplified: it just computes the honest parties' share of the multisignature and outputs it. Furthermore, we assume without loss of generality that adversaries never output a forgery on a message previously queried to their signing oracle and that they always output a forgery with $\mathcal{V}$ including the trusted party's public key.

We now prove the security of the BMS and WMS multisignature schemes relative to POP-based protocols that differ from current standards only by use of a distinct hash function. In Section 4.3 we discuss attacks against the schemes when standardized registration protocols are utilized.

4.1 Multisignatures Based on BLS Signatures

BLS SIGNATURES AND MULTISIGNATURES. Let $H\colon \{0,1\}^* \to \mathbb{G}_1$ be a random oracle. Boneh, Lynn, and Shacham [14] specify a signature scheme $\mathsf{BLS} = (\mathsf{B\text{-}Kg}, \mathsf{B\text{-}Sign}, \mathsf{B\text{-}Vf})$. The algorithms work as follows:

B-Kg:	B-Sign$^H(sk, M)$:	B-Vf$^H(pk, M, \sigma)$:
$sk \xleftarrow{\$} \mathbb{Z}_p$; $pk \leftarrow g^{sk}$	Return $H(M)^{sk}$	If $\mathbf{e}(H(M), pk) = \mathbf{e}(\sigma, g)$ then
Return (pk, sk)		Return 1
		Return 0

The $\mathsf{BMS} = (\mathsf{B\text{-}MPg}, \mathsf{B\text{-}MKg}, \mathsf{B\text{-}MSign}, \mathsf{B\text{-}MVf})$ multisignature scheme [11] is a simple extension of BLS signatures. Parameter generation just selects the groups, generators, and pairings as described in Section 2. Key generation, using the global parameters, creates a key pair as in B-Kg. Multisignature generation for participants labeled $1, \ldots, v$, public keys $\mathcal{V} = \{pk_1, \ldots, pk_v\}$, and a message M proceeds as follows. Each participant i computes $\sigma_i \xleftarrow{\$} \mathsf{B\text{-}Sign}(sk_i, M)$ and broadcasts σ_i to all other participants. The multisignature is $\sigma \leftarrow \prod_{i=1}^v \sigma_i$. On input $\mathcal{V}, M, \sigma$ the verification algorithm B-MVf computes $PK = \prod_{i=1}^v pk_i$ and then runs B-Vf$^H(PK, M, \sigma)$, returning its output. Boldyreva proved the scheme secure under the KOSK assumption [11].

THE B-Pop PROTOCOL. We now specify a POP-based key registration protocol under which we can prove BMS secure. Let $H_{\mathrm{pop}}\colon \{0,1\}^* \to \mathbb{G}_1$ be a random oracle. Then we define the $\mathsf{B\text{-}Pop} = (\mathsf{B\text{-}PopP}, \mathsf{B\text{-}PopV})$ protocol as follows. Algorithm B-PopP(pk, sk) sends $pk \parallel \mathsf{B\text{-}Sign}^{H_{\mathrm{pop}}}(sk, \langle pk \rangle)$ and algorithm B-PopV, upon receiving (pk, π) computes B-Vf$^{H_{\mathrm{pop}}}(pk, \langle pk \rangle, \pi)$ and if the result is 1 replies with pk and otherwise with $\bot$. We point out that one can use the same random oracle (and underlying instantiating hash function) for both H and H_{pop} as long as domain separation is enforced. The following theorem captures the security of BMS with respect to this key registration protocol.

Theorem 1. *Let* $H, H_{\text{pop}}: \{0,1\}^* \rightarrow \mathbb{G}_1$ *be random oracles. Let* A *be an* msuf-kr-*adversary, with respect to the* B-Pop *propocol, that runs in time* t, *makes* q_h, q_{pop}, q_s, *and* q_k *queries to* H, H_{pop}, *the signing oracle, and the key registration oracle, and outputs a multisignature forgery on a group of size at most* v. *Then there exists an adversary* B *such that*

$$\mathbf{Adv}_{\text{BMS,B-Pop}}^{\text{msuf-kr}}(A) \leq e(q_s + 1) \cdot \mathbf{Adv}_{(\mathbb{G}_1,\mathbb{G}_2)}^{\text{co-cdh}}(B)$$

where B *runs in time* $t' \in \mathcal{O}(t \log t + (q_h + q_{\text{pop}} + v)t_E + (q_k + 1)t_{\mathbf{e}})$.

Proof. We wish to construct a co-CDH adversary B, which on input g, X, h utilizes an msuf-kr adversary A to help it compute h^x where $x = \log_g X$. We adapt a game-playing [9] approach due to Bellare for proving the security of BLS signatures [3]. Without loss of generality, we assume that A always queries $H(M)$ before querying B-Sign(M). Likewise we assume that A always queries $H_{\text{pop}}(\langle pk \rangle)$ before querying OKReg(pk, π) for any π. Figure 2 details a sequence of four games. The game G0, which does not include the boxed statements, represents the core of our adversary B.

The execution G0(A) proceeds as follows. First Initialize is executed, which initializes several variables, including a co-CDH problem instance (g, X, h). Then A is run with input X. Oracle queries by A are handled as shown. Game G0 programs H (lazily built using an initially empty table $\mathsf{H}[\cdot]$) to sometimes return values that include h and sometimes not, depending on a δ-biased coin (we notate flipping such a coin by $\delta_c \overset{\delta}{\leftarrow} \{0,1\}$). Intuitively, the δ_c values correspond to guessing which H query will correspond to the forgery message. G0 programs H_{pop} (lazily built using an initially empty table $\mathsf{H}_{\mathsf{p}}[\cdot]$) to always include h. This is so that the adversarially-supplied POPs can be used to help extract the co-CDH solution from a forgery. Queries to OKReg invoke an execution of B-PopV, utilizing the H_{p} table for the algorithm's random oracle. Successful registrations have the POP signature stored in the table P (which is initially set everywhere to $\perp$). Once A halts with output a potential forgery the Finalize procedure is executed. We define the subroutine CheckForgery (not explicitly shown in the games for brevity) as follows. It checks that all keys in the multiset have an entry defined in P except the honest user's key (though if there are multiple copies of the honest user's key, then $\mathsf{P}[pk^*]$ must not be $\perp$). Then it checks if the multisignature verifies under the multiset of public keys given. If either check fails it returns zero, otherwise it returns one. Note that in the game x is not used beyond defining the co-CDH problem instance.

The adversary B, when run on input (g, X, h), follows exactly the steps of G0(A), except that it uses its co-CDH problem instance to supply the appropriate values. We now must justify that $\mathbf{Adv}_{(\mathbb{G}_1,\mathbb{G}_2)}^{\text{co-cdh}}(B) = \Pr[\text{G0}(A) \not\Rightarrow \perp]$ where G0$(A) \not\Rightarrow \perp$ means that the output of G0's Finalize procedure is not $\perp$. By construction the behavior of G0(A) and B are equivalent, and thus all that remains to be shown is that if the variable G0(A) does not output $\perp$, then it outputs the co-CDH solution h^x. Let us fix some more notation related to the variables in the Finalize procedure of G0. Define $sk_i = \log_g pk_i$ and $\pi_i = \mathsf{P}[pk_i]$ for each $i \in [1..d]$. Then $\psi(pk_i) = g_1^{sk_i}$ holds for each i. Define $\beta_i = \alpha - \gamma_i = \mathsf{B}[pk_i]$. Now,

procedure Initialize
$x \xleftarrow{\$} \mathbb{Z}_p; X \leftarrow g^x; h \xleftarrow{\$} \mathbb{G}_1; c \leftarrow 0$
$g_1 \leftarrow \psi(g); X_1 \leftarrow \psi(X)$
Return X

On query $H(M)$:
$c \leftarrow c + 1; M_c \leftarrow M$
$\alpha_c \xleftarrow{\$} \mathbb{Z}_p; \delta_c \xleftarrow{\delta} \{0,1\}$
If $\delta_c = 1$ then $H[M] \leftarrow g_1^{\alpha_c}$
Else $H[M] \leftarrow hg_1^{\alpha_c}$
Return $H[M]$

On query B-Sign(M):
Let k be such that $M = M_k$
$S_k \leftarrow 1$
If $\delta_k = 1$ then $S_k \leftarrow X_1^{\alpha_k}$
Else $bad \leftarrow$ true $\boxed{; S_k \leftarrow H[M]^x}$
Return S_k

On query $H_{\mathrm{pop}}(N)$:   G0 $\boxed{\text{G1}}$
$B[N] \xleftarrow{\$} \mathbb{Z}_p$; Return $H_{\mathrm{p}}[N] \leftarrow hg_1^{B[N]}$

On query OKReg(pk, π):
If B-Vf$^{H_{\mathrm{p}}}(pk, \langle pk \rangle, \pi) = 1$ then
$\quad$P$[pk] \leftarrow \pi$; Return pk
Return $\perp$

procedure Finalize$(\{X, pk_1, \ldots, pk_d\}, M, \sigma)$
$f \leftarrow$ CheckForgery$(\{X, pk_1, \ldots, pk_d\}, M, \sigma)$
If $f = 0$ then Return $\perp$
Let k be such that $M = M_k; \alpha \leftarrow \alpha_k$
For each $i \in [1..d]$ do $\gamma_i \leftarrow \alpha - B[\langle pk_i \rangle]$
$w \leftarrow \perp$
If $\delta_k = 0$ then
$\quad w \leftarrow \sigma X_1^{-\alpha} \prod_{i=1}^{d} \left(P[pk_i]^{-1} \psi(pk_i)^{-\gamma_i}\right)$
Else $bad \leftarrow$ true $\boxed{; w \leftarrow h^x}$
Return w

---

procedure Initialize
$x \xleftarrow{\$} \mathbb{Z}_p; X \leftarrow g^x; h \xleftarrow{\$} \mathbb{G}_1^*; c \leftarrow 0$
Return X

On query $H(M)$:
$c \leftarrow c + 1; M_c \leftarrow M$
$\alpha_c \xleftarrow{\$} \mathbb{Z}_p; \delta_c \xleftarrow{\delta} \{0,1\}$
Return $H[M] \xleftarrow{\$} \mathbb{G}_1$

On query B-Sign(M):
Let k be such that $M = M_k$
if $\delta_k = 1$ then $S_k \leftarrow H[M]^x$
Else $bad \leftarrow$ true; $S_k \leftarrow H[M]^x$
Return S_k

On query $H_{\mathrm{pop}}(N)$:   G2
Return $H_{\mathrm{p}}[N] \xleftarrow{\$} \mathbb{G}_1$

On query OKReg(pk, π):
If B-Vf$^{H_{\mathrm{p}}}(pk, \langle pk \rangle, \pi) = 1$ then
$\quad$P$[pk] \leftarrow \pi$; Return pk
Return $\perp$

procedure Finalize$(\{X, pk_1, \ldots, pk_d\}, M, \sigma)$
$f \leftarrow$ CheckForgery$(\{X, pk_1, \ldots, pk_d\}, M, \sigma)$
If $f = 0$ then Return $\perp$
Let k be such that $M = M_k; \alpha \leftarrow \alpha_k$
If $\delta_k = 0$ then $w \leftarrow h^x$
Else $bad \leftarrow$ true; $w \leftarrow h^x$
Return w

---

procedure Initialize
$x \xleftarrow{\$} \mathbb{Z}_p; X \leftarrow g^x; h \xleftarrow{\$} \mathbb{G}_1^*; c \leftarrow 0$
Return X

On query $H(M)$:
Return $H[M] \xleftarrow{\$} \mathbb{G}_1$

On query B-Sign(M):
$c \leftarrow c + 1$; Return $H[M]^x$

On query $H_{\mathrm{pop}}(N)$:
Return $H_{\mathrm{p}}[N] \xleftarrow{\$} \mathbb{G}_1$

On query OKReg(pk, π):   G3
If B-Vf$^{H_{\mathrm{p}}}(pk, \langle pk \rangle, \pi) = 1$ then
$\quad$P$[pk] \leftarrow \pi$; Return pk
Return $\perp$

procedure Finalize$(\{X, pk_1, \ldots, pk_d\}, M, \sigma)$
$f \leftarrow$ CheckForgery$(\{X, pk_1, \ldots, pk_d\}, M, \sigma)$
If $f = 0$ then Return $\perp$
For each $j \in [1..c]$ do
$\quad \delta_j \xleftarrow{\delta} \{0,1\}$; If $\delta_j = 0$ then $bad \leftarrow$ true
$\quad \delta_{j+1} \xleftarrow{\delta} \{0,1\}$; If $\delta_{j+1} = 1$ then $bad \leftarrow$ true
Return h^x

Fig. 2. Games used in proof that BMS is secure using POPs

because CheckForgery returns one if $G0(A)$ does not output $\perp$, we necessarily have that $\mathbf{e}(\mathtt{H}[M], PK) = \mathbf{e}(\sigma, g)$ and that $\mathbf{e}(\mathtt{H}_{\mathtt{p}}[\langle pk_i\rangle], pk_i) = \mathbf{e}(\pi_i, g)$ for each $i \in [1\mathinner{.\,.} d]$. In turn this means that $\sigma = (hg_1^\alpha)^{x+sk_1+\ldots+sk_d}$ and $\pi_i = (hg_1^{\beta_i})^{sk_i}$ for each $i \in [1\mathinner{.\,.} d]$. Thus, we can see that $w = h^x$:

$$w = \sigma X_1^{-\alpha} \prod_{i=1}^{d} \pi_i^{-1}\, \psi(pk_i)^{-\gamma_i} = \frac{(hg_1^\alpha)^{x+sk_1+\ldots+sk_d}}{X_1^\alpha \cdot \prod (hg_1^{\beta_i})^{sk_i}(g_1^{sk_i})^{\alpha-\beta_i}} = h^x\,.$$

Now we move through a sequence of games to lower bound the probability that $G0(A)$ actually succeeds in terms of A's advantage. Let Good be the event that bad is never set to true. What we show is that

$$\begin{aligned}
\Pr\left[G0(A) \not\Rightarrow \perp\right] \geq \Pr\left[G0(A) \not\Rightarrow \perp \wedge \mathsf{Good}\right] &= \Pr\left[G1(A) \not\Rightarrow \perp \wedge \mathsf{Good}\right] &(1)\\
&= \Pr\left[G2(A) \not\Rightarrow \perp \wedge \mathsf{Good}\right] &(2)\\
&= \Pr\left[G3(A) \not\Rightarrow \perp \wedge \mathsf{Good}\right] &(3)\\
&= \Pr\left[G3(A) \not\Rightarrow \perp\right] \cdot \Pr\left[\mathsf{Good}\right] &(4)\\
&\geq \mathbf{Adv}_{\mathsf{BMS}}^{\mathrm{msuf}}(A) \cdot \frac{1}{e} \cdot \frac{1}{q_s+1} &(5)
\end{aligned}$$

which implies the theorem statement. Now to justify this sequence of equations. ▶ Game G0 and G1 are identical-until-bad. A variant [6] of the fundamental lemma of game-playing [9] justifies Equation 1. ▶ Game G2 simplifies game G1 by taking advantage of knowing x. Queries to H are always answered with values uniformly chosen from $\mathbb{G}_1$. Signature queries are always answered with $\mathtt{H}[M]^x$. The value h^x is always returned by Finalize. These changes mean we never need g_1 and X_1, so they are omitted. The only distinction between G2 and G1 then is how these values are computed; their distributions remain the same and we therefore have justified Equation 2. ▶ We now note that in game G2 the values chosen for the δ_c variables have no impact on any of the values returned by procedures in the game, and only affect the setting of bad. Furthermore, not all of the δ values can actually set bad: only those that end up being referenced during signing queries and the one extra for the forgery. With these facts in mind, we modify G2 to get game G3, in which we defer all possible settings of bad until the Finalize procedure. We only perform δ-biased coin tosses $c + 1$ times: one for each signature query and one for the forgery. Equation 3 is justified by the fact that none of these changes affect the other variables in the game. (We also make some other cosmetic changes to simplify the games, but these do not modify distributions involved.) ▶ It is clear in game G3 that the event Good and "$G3(A) \not\Rightarrow \perp$" are independent, justifying Equation 4. ▶ Lastly, we note that G3 now exactly represents the environment of $\mathbf{Exp}_{\mathsf{BMS}}^{\mathrm{msuf\text{-}pop}}(A)$ because if $G3(A)$ does not output $\perp$ then A's output is a valid forgery. The lower bound $\Pr\left[\mathsf{Good}\right] \geq (e(q_s + 1))^{-1}$ is standard (see, e.g. [6,15]).

The adversary B runs A. Additionally B must perform an exponentiation for each H and H_{pop} query and one for each key in the forgery set $\mathcal{V}$. Finally B must perform a pairing for each OKReg query and to verify the forgery. Thus B runs in time $t' \in \mathcal{O}(t \log t + (q_h + q_{\mathrm{pop}} + v)t_E + (q_k + 1)t_\mathbf{e})$ where $|\mathcal{V}| = v$.   □

4.2   Multisignatures Based on Waters Signatures

WATERS SIGNATURES AND MULTISIGNATURES. Let $H\colon \{0,1\}^n \to \mathbb{G}_1$ be a hash function and define the signature scheme $\mathsf{W} = (\mathsf{W\text{-}Kg}, \mathsf{W\text{-}Sign}, \mathsf{W\text{-}Vf})$ as shown below.

W-Kg:	W-Sign$^H(sk, M)$:	W-Vf$^H(pk, M, (\sigma, \rho))$:
$\alpha \xleftarrow{\$} \mathbb{Z}_p;\ sk \leftarrow h^\alpha$	$r \xleftarrow{\$} \mathbb{Z}_p;\ \rho \leftarrow g^r$	If $\mathbf{e}(\sigma, g) \cdot \mathbf{e}(H(M), \rho)^{-1} = pk$ then
$pk \leftarrow \mathbf{e}(h, g)^\alpha$	$\sigma \leftarrow sk \cdot H(M)^r$	Return 1
Return (pk, sk)	Return (σ, ρ)	Return 0

Although one could use a random oracle for H, we can avoid the random oracle model by using the following hash function, as done in [28]. A trusted party, in addition to picking h, chooses $u, u_1, \ldots, u_n \xleftarrow{\$} \mathbb{G}_1$ and publishes them globally. Define $H_{u, \vec{u}}\colon \{0,1\}^n \to \mathbb{G}_1$ by $H_{u, \vec{u}}(M) = u \cdot \prod_{i=1}^{n} u_i^{M[i]}$. For simplicity we restrict ourselves to the message space $\{0,1\}^n$, but in practice we can use a collision-resistant hash function to expand the domain.

The WMS $= (\mathsf{W\text{-}Pg}, \mathsf{W\text{-}MKg}, \mathsf{W\text{-}MSign}, \mathsf{W\ MVf})$ multisignature scheme [28] is a straightforward extension of the Waters' signature scheme. Parameter generation chooses $h, u, \vec{u}$ as specified above in addition to fixing all the groups, generators, and pairings as per Section 2. Key generation, using the generated parameters, computes keys as in W-Kg. To generate a multisignature for multiset $\mathcal{V} = \{pk_1, \ldots, pk_v\}$, each participant i computes $(\sigma_i, \rho_i) \xleftarrow{\$} \mathsf{W\text{-}Sign}(sk_i, M)$ and broadcasts (σ_i, ρ_i). The multisignature is $(\prod_{i=1}^{v} \sigma_i, \prod_{i=1}^{v} \rho_i)$. To verify a signature (σ, ρ) for a message M and public keys $\mathcal{V} = \{pk_1, \ldots, pk_v\}$, simply let $PK \leftarrow \prod_{i=1}^{v} pk_i$ and then return $\mathsf{W\text{-}Vf}(PK, M, (\sigma, \rho))$. This scheme was proven secure using the KOSK assumption in [28].

THE WM-Pop PROTOCOL. Let $w, w_1, \ldots, w_n \xleftarrow{\$} \mathbb{G}_1$ be global parameters with associated hash function $H_{w, \vec{w}}$. These parameters require trusted setup, particularly because the CA should *not* know their discrete logs. (One might therefore have the trusted party that runs W-Pg also generate $w, \vec{w}$.) We define the following key registration protocol WM-Pop $= (\mathsf{WM\text{-}PopP}, \mathsf{WM\text{-}PopV})$: Algorithm WM-PopP takes as input (pk, sk) and sends $pk \mathbin{\|} (\pi, \varpi)$ where $(\pi, \varpi) = \mathsf{W\text{-}Sign}^{H_{w, \vec{w}}}(sk, \langle pk \rangle_n)$. Algorithm WM-PopV receives $pk \mathbin{\|} (\pi, \varpi)$ and then runs $\mathsf{W\text{-}Vf}^{H_{w, \vec{w}}}(pk, \langle pk \rangle_n, (\pi, \varpi))$ and if the result is 1, replies with pk and else replies with $\bot$. The following theorem, proof of which is given in the full version of the paper [34], and Theorem 2 in [28] (security of WMS under the KOSK assumption) establish the security of WMS under WM-Pop.

Theorem 2. *Let A be an* msuf-kr-*adversary, with respect to the* WM-Pop *protocol, that runs in time t, makes q_s signing queries, q_k registration queries, and outputs a forgery for a group of size at most v. Then there exists an adversary B such that*

$$\mathbf{Adv}_{\mathsf{WMS},\mathsf{WM\text{-}Pop}}^{\mathrm{msuf\text{-}kr}}(A) \leq \mathbf{Adv}_{\mathsf{WMS},\mathsf{Kosk}}^{\mathrm{msuf\text{-}kr}}(B)$$

and where B runs in time $t' \in \mathcal{O}(t \log t + nt_E + (t_E + t_\psi + t_\mathbf{e})q_k)$ and makes q_s signature queries.

4.3   Attacks Against Standardized Key Registration Protocols

We show how the standardized proof-of-possession based key registration protocols (as per PKCS#10 [36] and RFCs 4210/4211 [1,37]) fail to prevent rogue key attacks. Let $\mathsf{BadPop} = (\mathsf{BadP}, \mathsf{BadV})$ be the standardized key registration protocol for BMS and let the algorithms be as follows: Algorithm BadP, on input (pk, sk) sends $pk \parallel \mathsf{B\text{-}Sign}^H(sk, \langle pk \rangle)$ and algorithm BadV, upon receiving (pk, π), runs $\mathsf{B\text{-}Vf}^H(pk, \langle pk \rangle, \pi)$ and replies with pk if the result is 1 and $\perp$ otherwise. Here H is the same hash function as used in B-MSign and B-MVf.

We define a simple msuf-kr adversary A that successfully mounts a rogue-key attack against BMS with respect to the BadPop registration protocol. Adversary A gets the honest party's public key pk^* which is equal to g^{sk^*}. It then chooses $s \overset{\$}{\leftarrow} \mathbb{Z}_p$. Its public key is set to $pk = g^s/pk^* = g^{s-sk^*}$. The forgery on any message M and multiset $\{pk^*, pk\}$ is simply $H(M)^s$, which clearly verifies under the two public keys given. Now to register its key, the adversary makes the query $\mathsf{OMSign}(\{pk^*\}, \langle pk \rangle)$, receiving $\sigma = H(\langle pk \rangle)^{sk^*}$. Then A sets $\pi \leftarrow H(\langle pk \rangle)^s/\sigma$ and registers with $pk \parallel \pi$. It is easy to see that this verifies, and thus A can always output a multisignature forgery: its msuf-kr advantage is one.

An analogous key registration protocol could be defined for WMS, and again a simple attack shows its insecurity. Both approaches fall to attacks because the signatures used for key registration and normal multisignatures are calculated in the same manner. This motivated our simple deviations from standardized registration protocols for the B-Pop and WM-Pop protocols.

4.4   Other POP Variants

Another class of POP-based registration protocols for signature schemes has the CA send a random challenge to the registrant. The registrant must then supply a signature over the challenge message. Our results apply to such protocols, also, see the full version for details.

5   Ring Signatures in the Registered Key Model

A *ring signature scheme* $\mathsf{RS} = (\mathsf{RPg}, \mathsf{RKg}, \mathsf{RSign}, \mathsf{RVf})$ consists of four algorithms. The parameter generation algorithm generates a string par given to all parties and (often implicitly) input to the other three algorithms. The key generation algorithm RKg outputs a key pair (pk, sk). The algorithm $\mathsf{RSign}_{sk}(\mathcal{V}, M) \equiv \mathsf{RSign}(sk, \mathcal{V}, M)$ generates a ring signature on input a secret key sk, a message M, and a set of public keys $\mathcal{V}$ such that there exists $pk \in \mathcal{V}$ for which (pk, sk) is a valid key pair. We further assume that $|\mathcal{V}| \geq 2$ and all keys in $\mathcal{V}$ are distinct. It outputs a ring signature. Lastly the verification algorithm $\mathsf{RVf}(\mathcal{V}, M, \sigma)$ outputs a bit. We require that $\mathsf{RVf}(\mathcal{V}, M, \mathsf{RSign}_{sk}(\mathcal{V}, M)) = 1$ for any message M, any valid set of public keys, and for any valid sk with a $pk \in \mathcal{V}$. Ring signatures that only allow rings of size κ are called κ-*user ring signatures*.

NEW ANONYMITY DEFINITION. We propose a stronger definition of anonymity than those given by Bender et al. [10]. Intuitively, our definition requires that no adversary should be able to tell what secret key was used to generate a ring signature, even if the adversary itself chooses the secret keys involved. Formally, let A be an adversary and $\mathsf{RS} = (\mathsf{RPg}, \mathsf{RKg}, \mathsf{RSign}, \mathsf{RVf})$ be a ring signature scheme. Then the experiment $\mathbf{Exp}_{\mathsf{RS}}^{\text{r-anon-ind-}b}(A)$ works as follows: it runs $par \xleftarrow{\$} \mathsf{RPg}$ and then runs $A(par)$, giving it a left-or-right oracle $\mathsf{ORSignLR}(\cdot, \cdot, \cdot)$. The oracle takes queries of the form $\mathsf{ORSignLR}(\mathcal{S}, \mathcal{V}, M)$ where $\mathcal{S} = (sk_0, sk_1)$ and $\mathcal{V} = pk_0, \ldots, pk_{v-1}$ is a set of public keys such that (pk_0, sk_0) and (pk_1, sk_1) are valid key pairs. The oracle returns $\mathsf{RSign}_{sk_b}(\mathcal{V}, M)$. Finally A outputs a bit b', and wins if $b = b'$. The r-anon-ind advantage of A is

$$\mathbf{Adv}_{\mathsf{RS}}^{\text{r-anon-ind}}(A) = \Pr\left[\mathbf{Exp}_{\mathsf{RS}}^{\text{r-anon-ind-}0}(A) \Rightarrow 1\right] - \Pr\left[\mathbf{Exp}_{\mathsf{RS}}^{\text{r-anon-ind-}1}(A) \Rightarrow 1\right].$$

We say a scheme is *perfectly r-anon-ind anonymous* if the advantage of any adversary is zero.

The r-anon-ind definition is stronger than the strongest definition given in [10] (see the full version for details). Even so, both of the BKM 2-user ring signature schemes meet it, and are, in fact, perfectly r-anon-ind anonymous.

UNFORGEABILITY DEFINITIONS. We expand the unforgeability definitions given in [10], drawing a distinction between attacks where honest parties can be corrupted and rogue-key attacks (where the adversary can choose public keys). We also lift the strongest unforgeability definition to the registered key model. Fix some number η, representing the number of trusted potential honest signers. Figure 3 gives the security experiment for the strongest definition of security lifted to the registered key model, r-uf3-kr, which represents resistance to rogue-key attacks. A weaker definition, r-uf2, is obtained by defining an experiment $\mathbf{Exp}_{\mathsf{RS}}^{\text{r-uf2}}(A)$ that is the same as $\mathbf{Exp}_{\mathsf{RS},\mathsf{Reg}}^{\text{r-uf3-kr}}(A)$ except we do not allow the adversary to choose its own public keys. We also omit the key registration oracle and remove the requirement in ORSign that all adversarily chosen keys must be in $\mathcal{R}$. Lastly, we weaken this definition one step further by defining $\mathbf{Exp}_{\mathsf{RS}}^{\text{r-uf1}}(A)$, which disallows corruption queries. We thus define the following advantages:

- $\mathbf{Adv}_{\mathsf{RS},\mathsf{Reg}}^{\text{r-uf3-kr}}(A) = \Pr[\mathbf{Exp}_{\mathsf{RS},\mathsf{Reg}}^{\text{r-uf3-kr}}(A) \Rightarrow 1]$ (rogue-key attacks, equivalent to Definition 7 in [10] when $\mathsf{Reg} = \mathsf{Plain}$)
- $\mathbf{Adv}_{\mathsf{RS}}^{\text{r-uf2}}(A) = \Pr[\mathbf{Exp}_{\mathsf{RS}}^{\text{r-uf2}}(A) \Rightarrow 1]$ (corruption attacks, similar to a definition in [21])
- $\mathbf{Adv}_{\mathsf{RS}}^{\text{r-uf1}}(A) = \Pr[\mathbf{Exp}_{\mathsf{RS}}^{\text{r-uf1}}(A) \Rightarrow 1]$ (chosen subring attacks, Definition 6 in [10])

For κ-user ring signatures that meet the strongest anonymity definition, we have that security against corruption attacks is actually implied by security against chosen subring attacks. The reduction is tighter for small κ. This stems from having to guess a particular ring out of the η potential participants in the proof. The proof is given in the full version.

Theorem 3. *Let* RS *be a* κ-*user ring signature scheme. Let* $\eta \geq \kappa$ *be some number and let* A *be an* r-uf2 *adversary that makes at most* q_s *signature queries,*

Experiment $\mathbf{Exp}_{\mathsf{RS},\mathsf{Reg}}^{\text{r-uf3-kr}}(A)$

   $par \xleftarrow{\$} \mathsf{RPg}; (pk_i, sk_i) \xleftarrow{\$} \mathsf{RKg}(par)$ for $i \in [1 .. \eta]; \mathcal{S} \leftarrow \{pk_1, \ldots, pk_\eta\}$

   $\mathcal{Q} \leftarrow \mathcal{C} \leftarrow \mathcal{R} \leftarrow \emptyset$

   Run $A(par, \mathcal{S})$ handling oracle queries as follows

   $\mathsf{ORSign}(s, \mathcal{V}, M)$, where $s \in [1 .. \eta]$ and $pk_s \in \mathcal{V}$:

      $\mathcal{Q} \xleftarrow{\cup} (\mathcal{V}, M)$; If $(\mathcal{V} \setminus \mathcal{S}) \setminus \mathcal{R} \neq \emptyset$ then Return $\perp$

      Return $\mathsf{RSign}_{sk_s}(\mathcal{V}, M)$

   $\mathsf{OCorrupt}(i)$, where $i \in [1 .. \eta]$: $\mathcal{C} \xleftarrow{\cup} pk_i$; Return sk_i

   OKReg: Simulate a new instance of algorithm RegV, forwarding messages to

      and from A. If the instance's last message is $pk \neq \perp$, then $\mathcal{R} \xleftarrow{\cup} pk$.

   A outputs $(\mathcal{V}, M, \sigma)$

   If $\mathsf{RVf}(\mathcal{V}, M, \sigma) = 1 \wedge ((\mathcal{V}, M) \notin \mathcal{Q}) \wedge (\mathcal{V} \subseteq \mathcal{S} \setminus \mathcal{C})$ then Return 1

   Return 0

Fig. 3. Ring signature unforgeability experiment in the registered key model

q_c *corruption queries, and runs in time at most* t. *Then there exists adversaries* B_a *and* B_u *such that*

$$\mathbf{Adv}_{\mathsf{RS}}^{\text{r-uf2}}(A) \leq \binom{\eta}{\kappa} \mathbf{Adv}_{\mathsf{RS}}^{\text{r-anon-ind}}(B_a) + \binom{\eta}{\kappa} \mathbf{Adv}_{\mathsf{RS}}^{\text{r-uf1}}(B_u)$$

where B_a *uses* q_s *queries and runs in time* $t_a \in \mathcal{O}(t \log t + (\eta + 1)\mathsf{Time}(\mathsf{RS}))$ *and* B_u *uses* q_s *queries and runs in time* $t_u \in \mathcal{O}(t \log t + (\eta + 1 + q_s)\mathsf{Time}(\mathsf{RS}))$.

USING KOSK. Using the key registration protocol Kosk, any scheme that is unforgeable with respect to corruption attacks (r-uf2) and meets our strong definition of anonymity is also secure against rogue-key attacks (r-uf3-kr). Note that this result (unlike the last) applies to any ring signature scheme, not just κ-user ring signature schemes.

Theorem 4. *Fix* η *and let* RS *be a ring signature scheme for which* t_κ *is the maximal time needed to validate a key pair. Let* A *be an* r-uf3-kosk *adversary that makes at most* (q_s, q_c, q_k) *signature queries, corruption queries, and registration queries, and runs in time at most* t. *Then there exists adversaries* B_a *and* B_u *such that*

$$\mathbf{Adv}_{\mathsf{RS},\mathsf{Kosk}}^{\text{r-uf3-kr}}(A) \leq \mathbf{Adv}_{\mathsf{RS}}^{\text{r-anon-ind}}(B_a) + \mathbf{Adv}_{\mathsf{RS}}^{\text{r-uf2}}(B_u)$$

where B_a *runs in time* $t_a \in \mathcal{O}(t \log t + (\eta + 1)\mathsf{Time}(\mathsf{RS}))$, *using at most* q_s *sign queries, and* B_u *runs in time* $t_u \in \mathcal{O}(t \log t + q_k t_\kappa + (\eta + 1 + q_s)\mathsf{Time}(\mathsf{RS}))$, *using at most* q_s *sign queries and* q_c *corrupt queries.*

The proof is given in the full version. We can apply Theorem 3 and then Theorem 4 to the two 2-user ring signature schemes from Bender et al., rendering them secure against rogue-key attacks when Kosk is used for key registration.

USING POPs. For all the reasons already described, we'd like to avoid the KOSK assumption wherever possible. Thus, we give a proof-of-possession based registration protocol for the 2-user scheme based on Waters signatures from Bender et al. [10]. Let WRS = (W-RPg, W-RKg, W-RSign, W-RVf). The parameter generation selects groups, generators, and a pairing in the symmetric setting as per Section 2. The key generation algorithm W-RKg chooses $\alpha \stackrel{\$}{\leftarrow} \mathbb{Z}_q$, sets $g_1 \leftarrow g^\alpha$, and chooses random elements $u, u_1, \ldots, u_n \stackrel{\$}{\leftarrow} \mathbb{G}_1^*$. Finally it outputs $pk \leftarrow g_1, u, u_1, \ldots, u_n$ and $sk \leftarrow \alpha$. Define W-RSign$_{sk}(\{pk, pk'\}, M)$ as follows. (Without loss we assume sk corresponds to pk.) Parse pk as $g_1, u, \vec{u}$ and pk' as $g_1', u', \vec{u}'$ and let $H'(M) = H_{u,\vec{u}}(M) \cdot H_{u',\vec{u}'}(M)$. Finally, return W-Sign$^{H'}(g_1'^{sk}, M)$. The verification algorithm W-RVf$(\{pk, pk'\}, M, (\sigma, \rho))$ first parses pk as $g_1, u, u_1, \ldots, u_n$ and pk' as $g_1', u', u_1', \ldots, u_m'$ and defines H' as in signature generation. Then it outputs one if $\mathbf{e}(g_1, g_1') \cdot \mathbf{e}(H'(M), \rho) = \mathbf{e}(\sigma, g)$.

In [10] the scheme is proven secure against r-uf1 adversaries, but as shown in [38] the scheme is *not* secure against rogue-key attacks without key registration. We now show a simple proof-of-possession based registration protocol to render the scheme secure. Choose global parameters $h_0, h_1, w, w_1, \ldots, w_n \stackrel{\$}{\leftarrow} \mathbb{G}_1$ (this will require trusted setup, and could be accomplished with W-RPg). Then we specify the registration protocol WR-Pop = (WR-PopP, WR-PopV). Algorithm WR-PopP takes as input (sk, pk) and sends $pk \parallel (\pi_0, \varpi_0, \pi_1, \varpi_1)$ which is simply computed by generating the two signatures W-Sign$^{H_{w,\vec{w}}}(h_0^{sk}, \langle pk \rangle_n)$ and W-Sign$^{H_{w,\vec{w}}}(h_1^{sk}, \langle pk \rangle_n)$. Algorithm WR-PopV, upon receiving the message, verifies the signatures in the natural way: get g_1 from pk and check that both $\mathbf{e}(g_1, h_0) \cdot \mathbf{e}(H_{w,\vec{w}}(\langle pk \rangle_n), \varpi_0) = \mathbf{e}(\pi_0, g)$ and $\mathbf{e}(g_1, h_1) \cdot \mathbf{e}(H_{w,\vec{w}}(\langle pk \rangle_n), \varpi_1) = \mathbf{e}(\pi_1, g)$. If both signatures verify, the algorithm replies with pk and otherwise $\perp$. The following theorem captures security of WRS with respect to the WR-Pop registration protocol. The proof is given in the full version.

Theorem 5. *Fix η. Let A be an r-uf3-kr adversary with respect to the* WR-Pop *protocol that makes at most q_s signature queries, q_c corruption queries, q_k key registration queries, and runs in time at most t. Then there exists an adversary B such that*

$$\mathbf{Adv}_{\mathsf{WRS},\mathsf{WR\text{-}Pop}}^{\mathrm{r\text{-}uf3\text{-}kr}}(A) \le \eta^2 \mathbf{Adv}_{\mathsf{W}}^{\mathrm{uf}}(B)$$

where B makes at most q_s signing queries and runs in time $t_B \in \mathcal{O}(t \log t + \eta t_E + q_s t_E + (q_k + 1) t_\mathbf{e})$.

Acknowledgements

The authors thank Mihir Bellare for suggesting that they investigate the security of multisignature schemes when the proof of knowledge is replaced by a proof of possession akin to ones currently used in PKIs, and for many useful discussions. The authors thank the anonymous reviewers for their helpful comments. The first author is supported by NSF grant CNS–0524765. The second author is supported by NSF grant CNS–0430595 and a Jacobs School Fellowship.

References

1. C. Adams, S. Farrell, T. Kause, T. Mononen. Internet X.509 public key infrastructure certificate management protocols (CMP). Request for Comments (RFC) 4210, Internet Engineering Task Force (September 2005)
2. N. Asokan, V. Niemi, P. Laitinen. On the usefulness of proof-of-possession. In *Proceedings of the 2nd Annual PKI Research Workshop.* (2003) 122–127
3. M. Bellare. CSE 208: Advanced Cryptography. UCSD course (Spring 2006).
4. M. Bellare, O. Goldreich. On Defining Proofs of Knowledge. In *CRYPTO '92.* Volume 740 of *LNCS*, Springer (1993) 390–420
5. M. Bellare, T. Kohno, V. Shoup. Stateful public-key cryptosystems: how to encrypt with one 160-bit exponentiation. ACM Conference on Computer and Communications Security. (2006) 380–389
6. M. Bellare, C. Namprempre, G. Neven. Unrestricted aggregate signatures. Cryptology ePrint Archive, Report 2006/285 (2006) http://eprint.iacr.org/.
7. M. Bellare, G. Neven. Multi-signatures in the plain public-key model and a generalized forking lemma. In *ACM Conference on Computer and Communications Security.* (2006) 390–399
8. M. Bellare, P. Rogaway. Random oracles are practical: a paradigm for designing efficient protocols. In *ACM Conference on Computer and Communications Security.* (1993) 62–73
9. M. Bellare, P. Rogaway. The security of triple encryption and a framework for code-based game-playing proofs. In *EUROCRYPT '06.* Volume 4004 of *LNCS*, Springer (2006) 409–426
10. A. Bender, J. Katz, R. Morselli. Ring signatures: Stronger definitions, and constructions without random oracles. In *TCC '06.* Volumne 3876 of *LNCS*, Springer (2006) 60–79
11. A. Boldyreva. Threshold signatures, multisignatures and blind signatures based on the gap-diffie-hellman-group signature scheme. In *PKC '03.* Volume 2567 of *LNCS*, Springer (2002) 31–46
12. A. Boldyreva, M. Fischlin, A. Palacio, B. Warinschi. A closer look at PKI: security and efficiency. Public Key Cryptography (2007), to appear.
13. D. Boneh, C. Gentry, B. Lynn, H. Shacham. Aggregate and verifiably encrypted signatures from bilinear maps. In *EUROCRYPT '03.* Volume 2656 of *LNCS*, Springer (2003) 416–432
14. D. Boneh, B. Lynn, H. Shacham. Short signatures from the weil pairing. In *ASIACRYPT '01.* Volume 2248 *LNCS*, Springer (2001) 514–532
15. J.S. Coron. On the exact security of full domain hash. In *CRYPTO '00.* Volume 1880 of *LNCS*, Springer (2000) 229–235
16. A. De Santis, G. Persiano. Zero-knowledge proofs of knowledge without interaction (extended abstract). In *FOCS '92.* IEEE (1992) 427–436
17. M. Fischlin. Communication-efficient non-interactive proofs of knowledge with online extractors. In *CRYPTO '05.* Volume 3621 of *LNCS*, Springer (2005) 152–168
18. S. Goldwasser, S. Micali, R. Rivest. A Paradoxical Solution to the Signature Problem. In *FOCS '84.* IEEE (1984) 441–449.
19. J. Groth, R. Ostrovsky, A. Sahai. Perfect non-interactive zero knowledge for NP. In *EUROCRYPT '06.* Volume 4004 of *LNCS*, Springer (2006) 339–358
20. L. Harn. Group-oriented (t,n) threshold digital signature scheme and digital multisignature. Computers and Digital Techniques, IEEE Proceedings **141**(5) (1994) 307–313

21. J. Herranz. *Some digital signature schemes with collective signers.* Ph.D. Thesis, Universitat Politècnica De Catalunya, Barcelona. April, 2005.
22. P. Horster, M. Michels, H. Petersen. Meta signature schemes based on the discrete logarithm problem. In *IFIP TC11 Eleventh International Conference on Information Security (IFIP/SEC 1995)* (1995) 128–141
23. M. Jakobsson, K. Sako, R. Impagliazzo. Designated verifier proofs and their applications. In *EUROCRYPT '96.* Volume 1070 of *LNCS*, Springer (1996) 143–154
24. F. Laguillaumie, D. Vergnaud. Designated verifier signatures: anonymity and efficient construction from any bilinear map. In *Security in Communication Networks, 4th International Conference, SCN 2004.* Volume 3352 of *LNCS*, Springer (2005) 105–119
25. S.K. Langford. Weakness in some threshold cryptosystems. In *CRYPTO '96.* Volume 1109 of *LNCS*, Springer (1996) 74–82
26. C.M. Li, T. Hwang, N.Y. Lee. Threshold-multisignature schemes where suspected forgery implies traceability of adversarial shareholders. In *EUROCRYPT '94.* Volume 950 of *LNCS*, Springer (1995) 194–204
27. H. Lipmaa, G. Wang, F. Bao. Designated verifier signature schemes: attacks, new security notions and a new construction. In *ICALP 2005.* Volume 3580 *LNCS*, Springer (2005) 459–471
28. S. Lu, R. Ostrovsky, A. Sahai, H. Shacham, B. Waters. Sequential aggregate signatures and multisignatures without random oracles. In *EUROCRYPT '06.* Volume 4004 of *LNCS*, Springer (2006) 465–485
29. A. Lysyanskaya, S. Micali, L. Reyzin, H. Shacham. Sequential Aggregate Signatures from Trapdoor Permutations. In *EUROCRPYT '04.* Volume 3027 of *LNCS*, Springer (2004) 74–90
30. M. Michels, P. Horster. On the risk of disruption in several multiparty signature schemes. In *ASIACRYPT '96.* Volume 1163 of *LNCS*, Springer (1996) 334–345
31. S. Micali, K. Ohta, L. Reyzin. Accountable-subgroup multisignatures. In *ACM Conference on Computer and Communications Security.* (2001) 245–254
32. K. Ohta, T. Okamoto. A digital multisignature scheme based on the Fiat-Shamir scheme. In *ASIACRYPT '91.* Volume 739 of *LNCS*, Springer (1993) 139–148
33. K. Ohta, T. Okamoto. Multi-signature schemes secure against active insider attacks. IEICE Transactions on Fundamentals of Electronics Communications and Computer Sciences E82-A(1) (1999) 21–31
34. T. Ristenpart, S. Yilek. The power of proofs-of-possession: securing multiparty signatures against rogue-key attacks. Full version of current paper. http://www.cse.ucsd.edu/users/tristenp/
35. R.L. Rivest, A. Shamir, Y. Tauman. How to leak a secret. In *ASIACRYPT '01.* Volume 2248 of *LNCS*, Springer (2001) 552–565
36. RSA Laboratories: RSA PKCS #10 v1.7: Certification Request Syntax Standard ftp://ftp.rsasecurity.com/pub/pkcs/pkcs-10/pkcs-10v1_7.pdf.
37. J. Schaad. Internet X.509 public key infrastructure certificate request message format (CRMF). Request for Comments (RFC) 4211, Internet Engineering Task Force (September 2005)
38. H. Shacham, B. Waters. Efficient ring signatures without random oracles. Public Key Cryptography (2007), to appear.
39. R. Steinfeld, L. Bull, H. Wang, J. Pieprzyk. Universal designated-verifier signatures. In *ASIACRYPT '03.* Volume 2894 of *LNCS*, Springer (2003) 523–542
40. B. Waters. Efficient identity-based encryption without random oracles. In *EUROCRYPT '05.* Volume 3494 of *LNCS*, Springer (2005) 114–127

Batch Verification of Short Signatures

Jan Camenisch[1], Susan Hohenberger[2,*], and Michael Østergaard Pedersen[3,*]

[1] IBM Research, Zürich Research Laboratory
jca@zurich.ibm.com
[2] The Johns Hopkins University
susan@cs.jhu.edu
[3] University of Aarhus
michael@daimi.au.dk

Abstract. With computer networks spreading into a variety of new environments, the need to authenticate and secure communication grows. Many of these new environments have particular requirements on the applicable cryptographic primitives. For instance, several applications require that communication overhead be small and that many messages be processed at the same time. In this paper we consider the suitability of public key signatures in the latter scenario. That is, we consider signatures that are 1) short and 2) where many signatures from (possibly) different signers on (possibly) different messages can be verified quickly.

We propose the first batch verifier for messages from many (certified) signers without random oracles and with a verification time where the dominant operation is independent of the number of signatures to verify. We further propose a new signature scheme with very short signatures, for which batch verification for *many* signers is also highly efficient. Prior work focused almost exclusively on batching signatures from the same signer. Combining our new signatures with the best known techniques for batching certificates from the *same* authority, we get a fast batch verifier for certificates and messages combined. Although our new signature scheme has some restrictions, it is the only solution, to our knowledge, that is a candidate for some pervasive communication applications.

1 Introduction

As the world moves towards pervasive computing and communication, devices from vehicles to dog collars will soon be expected to communicate with their environments. For example, many governments and industry consortia are currently planning for the future of *intelligent cars* that constantly communicate with each other and the transportation infrastructure to prevent accidents and to help alleviate traffic congestion [11,37]. Raya and Hubaux suggest that vehicles will transmit safety messages every 300ms to all other vehicles within a minimum range of 110 meters [36], which in turn may retransmit these messages.

---

* Research performed while at IBM Research, Zürich Research Laboratory.

M. Naor (Ed.): EUROCRYPT 2007, LNCS 4515, pp. 246–263, 2007.

For such pervasive systems to work properly, there are many competing constraints [11,37,27,36]. First, there are physical limitations, such as a limited spectrum allocation for specific types of communications and the potential roaming nature of devices, that require that messages be kept very short and (security) overhead be minimal [27]. Yet for messages to be trusted by their recipients, they need to be authenticated in some fashion, so that entities spreading false information can be held accountable. Thus, some short form of authentication must be added. Third, different messages from many different signers may need to be verified and processed quickly (e.g., every 300ms [36]). A possible fourth constraint that these authentications remain anonymous or pseudonymous, we leave as an exciting open problem.

In this work, we consider the suitability of public key signatures to the needs of pervasive communication applications. Generating one signature every 300ms is not a problem for current systems, but transmitting and/or verifying 100+ messages per second might pose a problem. Using RSA signatures for example seems attractive as they are verified quickly, however, one would need approximately 3000 bits to represent a signature on a message plus the certificate (i.e., the public key and signature on that public key) which might be too much (see Section 8.2 of [36]). While many new schemes based on bilinear maps can provide the same security with significantly smaller signatures, they take significantly more time to verify.

1.1   Our Contributions

Now, if one wants both, short signatures and short verification times, it seems that one needs to improve on the verification of the bilinear-map based schemes. In this paper we take this route and investigate the known batch-verification techniques and to what extent they are applicable to such schemes. More precisely, the main contributions of this paper are:

1. We instantiate the general batch verification definitions of Bellare, Garay, and Rabin [2] to the case of signatures from many signers. We also do this for a weaker notion of batch verification called *screening* and show the relation of these notions to the one of aggregate signatures. Surprisingly, for most known aggregate signature schemes a batching algorithm is provably *not* obtained by aggregating many signatures and then verifying the aggregate.
2. We present a batch verifier for the Waters IBS scheme [39,7]. To our knowledge, this is the *first* batch verifier for a signature scheme without random oracles. When identities are k_1 bits and messages are k_2 bits, our algorithm verifies n Waters IBS signatures using only (k_1+k_2+3) pairings. Individually verifying n signatures would cost $3n$ pairings.
3. We present a new signature scheme, CL*, derived from the Camenisch and Lysyanskaya signature scheme [8]. We show that CL* can be realized without random oracles when the message space is polynomial. CL* signatures require only one-third the space of the original CL signatures– on par with the shortest signatures known [5] –, but users may only issue one signature

per period (e.g., users might only be allowed to sign one message per 300ms). We present a batch verifier for these signatures from many different signers that verifies n signatures using only three total pairings, instead of the $5n$ pairings required by n original CL signatures. Yet, our batch verifier has the restriction that it can only batch verify signatures made during the same period. CL* signatures form the core of the only public key authentication, known to us, that is extremely short and highly efficient to verify in bulk.

4. Often signatures and certificates need to be verified together. This happens implicitly in IBS schemes, such as Waters. To achieve this functionality with CL* signatures, we use a known batch verifier for the Boneh, Lynn, and Shacham signatures in the random oracle model [5,4] that can batch verify n signatures from the *same* signer using only two pairings.

1.2   Batch Verification Overview

Batch cryptography was introduced in 1989 by Fiat [17] for a variant of RSA. Later, in 1994, Naccache, M'Raïhi, Vaudenay and Raphaeli [35] gave the first efficient batch verifier for DSA signatures, however an interactive batch verifier presented in an early version of their paper was broken by Lim and Lee [31]. In 1995 Laih and Yen proposed a new method for batch verification of DSA and RSA signatures [29], but the RSA batch verifier was broken five years later by Boyd and Pavlovski [6]. In 1998 Harn presented two batch verification techniques for DSA and RSA [22,23] but both were later broken [6,25,26]. The same year, Bellare, Garay and Rabin took the first systematic look at batch verification [2] and presented three generic methods for batching modular exponentiations, called the *random subset test*, the *small exponents test* and the *bucket test* which are similar to the ideas from [35,29]. They showed how to apply these methods to batch verification of DSA signatures and also introduced a weaker form of batch verification called *screening*. In 2000 some attacks against different batch verification schemes, mostly ones based on the small exponents test and related tests, were published [6]. These attacks do not invalidate the proof of security for the small exponents test, but rather show how the small exponents test is often used in a wrong way. However, they also describe methods to repair some broken schemes based on this test. In 2001 Hoshino, Masayuki and Kobayashi [24] pointed out that the problem discovered in [6] might not be critical for batch verification of signatures, but when using batch verification to verify for example zero-knowledge proofs, it would be. In 2004 Yoon, Cheon and Kim proposed a new ID-based signature scheme with batch verification [15], but their security proof is for aggregate signatures and does not meet the definition of batch verification from [2]; hence their title is somewhat misleading. Of course not all aggregate signature schemes claim to do batch verification. For example Gentry and Ramzan present a nice aggregate signature scheme in [19] that does not claim to be, nor is, a batch verification scheme. Other schemes for batch verification based on bilinear maps were proposed [12,40,41,42] but all were later broken by Cao, Lin and Xue [10]. In 2006, a method was proposed for identifying invalid signatures in RSA-type batch signatures [30], but Stanek [38] showed that this method is flawed.

1.3   Efficiency of Prior Work and Our Contributions

Efficiency will be given as an abstract cost for computing different functions. We begin by discussing prior work on RSA, DSA, and BLS signatures mostly for single signers, and then discuss our new work on Waters, BLS, and CL signatures for many signers. Note that Lim [32] provides a number of efficient methods for doing m-term exponentiations and Granger and Smart [21] give improvements over the naive method for computing a product of pairings, which is why we state them explicitly.

$m\text{-MultPairCost}_{\mathbb{G},\mathbb{H}}^s$ s m-term pairings $\prod_{i=1}^m \mathbf{e}(g_i, h_i)$ where $g_i \in \mathbb{G}$, $h_i \in \mathbb{H}$.
$m\text{-MultExpCost}_{\mathbb{G}}^s(k)$ s m-term exponentiations $\prod_{i=1}^m g^{a_i}$ where $g \in \mathbb{G}$, $|a_i| = k$.
$\text{PairCost}_{\mathbb{G},\mathbb{H}}^s$ s pairings $\mathbf{e}(g_i, h_i)$ for $i = 1 \ldots s$, where $g_i \in \mathbb{G}$, $h_i \in \mathbb{H}$.
$\text{ExpCost}_{\mathbb{G}}^s(k)$ s exponentiations g^{a_i} for $i = 1 \ldots s$ where $g \in \mathbb{G}$, $|a_i| = k$.
$\text{GroupTestCost}_{\mathbb{G}}^s$ Testing whether or not s elements are in the group $\mathbb{G}$.
$\text{HashCost}_{\mathbb{G}}^s$ Hashing s values into the group $\mathbb{G}$.
MultCost^s s multiplications in one or more groups.

If $s = 1$ we will omit it. Throughout this paper we assume that n is the number of message/signature pairs and ℓ_b is a security parameter such that the probability of accepting a batch that contains an invalid signature is at most $2^{-\ell_b}$.

RSA* is a modified version of RSA by Boyd and Pavlovski [6]. The difference to normal RSA is that the verification equation accepts a signature σ as valid if $\alpha\sigma^e = m$ for some element $\alpha \in \mathbb{Z}_m^*$ of order no more than 2, where m is the product of two primes. The signatures are usually between $1024 - 2048$ bits and the same for the public key. A single signer batch verifier for this signature scheme with cost $n\text{-MultExpCost}_{\mathbb{Z}_m}^2(\ell_b) + \text{ExpCost}_{\mathbb{Z}_m}(k)$, where k is the number of bits in the public exponent e, can be found in [6]. Note that verifying n signatures by verifying each signature individually only costs $\text{ExpCost}_{\mathbb{Z}_m}^n(k)$, so for small values of e ($|e| < 2\ell_b/3$) the naive method is a faster way to verify RSA signatures and it can also handle signatures from multiple signers. Bellare et al. [2] presents a screening algorithm for RSA that assumes distinct messages from the same signer and costs $2n + \text{ExpCost}_{\mathbb{Z}_m}(k)$.

DSA** is a modified version of DSA from [35] compatible with the *small exponents test* from [6]. There are two differences to normal DSA. First there is no reduction modulo q, so the signatures are 672 bits instead of 320 bits and second, individual verification should check both a signature σ and $-\sigma$ and accept if one of them holds. Messages and public keys are both 160 bits long. Using the small exponents test the cost is $n\text{-MultExpCost}_{\mathbb{G}}(\ell_b) + \text{ExpCost}_{\mathbb{G}}^2(160) + \text{HashCost}_{\mathbb{G}}^n + \text{MultCost}^{2n+1}$ multiplications. This method works for a single signer only.

Waters IBS is the Waters 2-level hierarchical signature scheme from [7] for which we provide a batch verifier without random oracles in Section 4. An interesting property of this scheme is that the identity does not need to be verified separately. Identities are k_1 bits, messages are k_2 bits and a signature is three group elements in a bilinear group. The computational effort required depends on the number of messages and the security parameters. Let $\mathsf{M} = n\text{-MultExpCost}_{\mathbb{G}_T}(\ell_b) + n\text{-MultExpCost}_{\mathbb{G}}^3(\ell_b) + \mathsf{PairCost}_{\mathbb{G},\mathbb{G}}^3 + \mathsf{GroupTestCost}_{\mathbb{G}}^{3n} + \mathsf{MultCost}^4$ and refer to the table below for efficiency of the scheme. We assume that $k_1 < k_2$.

$$n < k_1 : \mathsf{M} + n\text{-MultPairCost}_{\mathbb{G},\mathbb{G}}^2 + \mathsf{ExpCost}_{\mathbb{G}}^{2n}(\ell_b) + \mathsf{MultCost}^{k_1+k_2}$$
$$k_1 \leq n \leq k_2 : \mathsf{M} + k_1\text{-MultPairCost}_{\mathbb{G},\mathbb{G}} + n\text{-MultPairCost}_{\mathbb{G},\mathbb{G}}$$
$$+ n\text{-MultExpCost}_{\mathbb{G}}^{k_1}(\ell_b) + \mathsf{ExpCost}_{\mathbb{G}}^{n}(\ell_b) + \mathsf{MultCost}^{k_2}$$
$$n > k_2 : \mathsf{M} + k_1\text{-MultPairCost}_{\mathbb{G},\mathbb{G}} + k_2\text{-MultPairCost}_{\mathbb{G},\mathbb{G}}$$
$$+ n\text{-MultExpCost}_{\mathbb{G}}^2(\ell_b)$$

The naive application of Waters IBS to verify n signatures costs $\mathsf{PairCost}_{\mathbb{G},\mathbb{G}}^{3n} + \mathsf{MultCost}^{n(k_1+k_2+3)}$. Also note that in many applications we do not need to transmit the identity as a separate parameter, as it is given to us for free. For example as the hardware address of the network interface card.

BLS is the signature scheme by Boneh, Lynn and Shacham [5,4]. We discuss batch verifiers for BLS signatures based on the small exponents test. For a screening algorithm, aggregate signatures by Boneh, Gentry, Lynn and Shacham [3] can be used. The signature is only one group element in a bilinear group and the same for the public key. For different signers the cost of batch verification is $n\text{-MultPairCost}_{\mathbb{G},\mathbb{G}} + n\text{-MultExpCost}_{\mathbb{G}}(\ell_b) + \mathsf{PairCost}_{\mathbb{G},\mathbb{G}} + \mathsf{ExpCost}_{\mathbb{G}_T}^n(\ell_b) + \mathsf{GroupTestCost}_{\mathbb{G}}^n + \mathsf{HashCost}_{\mathbb{G}}^n$, but for single signer it is only $n\text{-MultExpCost}_{\mathbb{G}}^2(\ell_b) + \mathsf{PairCost}_{\mathbb{G},\mathbb{G}}^2 + \mathsf{GroupTestCost}_{\mathbb{G}}^n + \mathsf{HashCost}_{\mathbb{G}}^n$.

CL* is a new variant of Camenisch and Lysyanskaya signatures [8] presented in Section 5. The signature is only one bilinear group element and the same for the public key. Batch verification costs $n\text{-MultExpCost}_{\mathbb{G}}^2(\ell_b) + n\text{-MultExpCost}_{\mathbb{G}}(|w| + \ell_b) + \mathsf{PairCost}_{\mathbb{G},\mathbb{G}}^3 + \mathsf{GroupTestCost}_{\mathbb{G}}^n + \mathsf{HashCost}_{\mathbb{G}}^n$, where w is the output of a hash function. However, the scheme has some additional restrictions.

Bucket Test. Bellare, Garay and Rabin [2] provide a method called the *bucket test* which is even more efficient than the small exponents test for large values of n. We note that one can use the tests we outline in this paper as subroutines to the *bucket test* to further speed up verification.

2   Definitions

Recall that a *digital signature scheme* is a tuple of algorithms (Gen, Sign, Verify) that also is *correct* and *secure*. The correctness property states that for all Gen$(1^\ell) \rightarrow (pk, sk)$, the algorithm Verify$(pk, m, \mathsf{Sign}(sk, m)) = 1$.

There are two common notions of security. Goldwasser, Micali, and Rivest [20] defined a scheme to be *unforgeable* as follows: Let $\mathsf{Gen}(1^\ell) \rightarrow (pk, sk)$. Suppose (m, σ) is output by a p.p.t. adversary $\mathcal{A}$ with access to a signing oracle $\mathcal{O}_{sk}(\cdot)$ and input pk. Then the probability that m was *not* queried to $\mathcal{O}_{sk}(\cdot)$ and yet $\mathsf{Verify}(pk, m, \sigma) = 1$ is negligible in ℓ. An, Dodis, and Rabin [1] proposed the notion of *strong unforgeability*, where if $\mathcal{A}$ outputs a pair (m, σ) such that $\mathsf{Verify}(pk, m, \sigma) = 1$, then except with negligible probability at some point oracle $\mathcal{O}_{sk}(\cdot)$ was queried on m and outputted signature σ exactly. In other words, an adversary cannot create a new signature even for a previously signed message. Our batch verification definitions work with either notion. The signatures used in Section 4 meet the GMR [20] definition, while those in Section 5 meet the ADR [1] definition.

Now, we consider the case where we want to quickly verify a set of signatures on (possibly) different messages by (possibly) different signers. The input is $\{(t_1, m_1, \sigma_1), \ldots, (t_n, m_n, \sigma_n)\}$, where t_i specifies the verification key against which σ_i is purported to be a signature on message m_i. We extend the definitions of Bellare, Garay and Rabin [2] to deal with multiple signers. And this is an important point that wasn't a concern with only a single signer: *one or more of the signers may be maliciously colluding.*

Definition 1 (Batch Verification of Signatures). *Let ℓ be the security parameter. Suppose $(\mathsf{Gen}, \mathsf{Sign}, \mathsf{Verify})$ is a signature scheme, $n \in \mathrm{poly}(\ell)$, and $(pk_1, sk_1), \ldots, (pk_n, sk_n)$ are generated independently according to $\mathsf{Gen}(1^\ell)$. Then we call probabilistic Batch a batch verification algorithm when the following conditions hold:*

- *If $\mathsf{Verify}(pk_{t_i}, m_i, \sigma_i) = 1$ for all $i \in [1, n]$, then $\mathsf{Batch}((pk_{t_1}, m_1, \sigma_1), \ldots, (pk_{t_n}, m_n, \sigma_n)) = 1$.*
- *If $\mathsf{Verify}(pk_{t_i}, m_i, \sigma_i) = 0$ for any $i \in [1, n]$, then $\mathsf{Batch}((pk_{t_1}, m_1, \sigma_1), \ldots, (pk_{t_n}, m_n, \sigma_n)) = 0$ except with probability negligible in k, taken over the randomness of Batch.*

Note that Definition 1 does not require verification keys *to belong* to honest users, only to keys that were *generated* honestly (and are perhaps now held by an adversary). In practice, users could register their keys and prove some necessary properties of the keys at registration time.

Confusion between Batch Verification, Aggregate Signatures, and Screening. As we discussed in the introduction, several works (e.g., [15,16]) claim to do batch verification when, in fact, they often meet a weaker guarantee called *screening* [2]. However, in most cases the confusion is about words, e.g. when the words *batch verification* are used to describe an aggregate signature scheme.

Definition 2 (Screening of Signatures). *Let ℓ be the security parameter. Suppose $(\mathsf{Gen}, \mathsf{Sign}, \mathsf{Verify})$ is a signature scheme, $n \in \mathrm{poly}(\ell)$ and $(pk^*, sk^*) \leftarrow \mathsf{Gen}(1^\ell)$. Let $\mathcal{O}_{sk^*}(\cdot)$ be an oracle that on input m outputs $\sigma = \mathsf{Sign}(sk^*, m)$. Then* *for all p.p.t. adversaries $\mathcal{A}$, we call probabilistic Screen a screening algorithm when $\mu(\ell)$ defined as follows is a negligible function:*

$$\Pr[(pk^*, sk^*) \leftarrow \mathsf{Gen}(1^\ell), (pk_1, sk_1) \leftarrow \mathsf{Gen}(1^\ell), \dots, (pk_n, sk_n) \leftarrow \mathsf{Gen}(1^\ell),$$
$$D \leftarrow \mathcal{A}^{\mathcal{O}_{sk^*}(\cdot)}(pk^*, (pk_1, sk_1), \dots, (pk_n, sk_n)) :$$
$$\mathsf{Screen}(D) = 1 \ \wedge \ (pk^*, m_i, \sigma_i) \in D \ \wedge \ m_i \notin Q] = \mu(\ell),$$

where Q is the set of queries that $\mathcal{A}$ made to $\mathcal{O}$.

The above definition is generalized to the multiple-signer case from the single-signer screening definition of Bellare et al. [2].

Interestingly, screening is the (maximum) guarantee that most aggregate signatures offer if one were to attempt to batch verify a group of signatures by first aggregating them together and then executing the aggregate-verification algorithm. Consider the aggregate signature scheme of Boneh, Gentry, Lynn and Shacham [3] based on the BLS signatures [5,4]. First, we describe the BLS signatures. Let $\mathbf{e} : \mathbb{G} \times \mathbb{G} \to \mathbb{G}_T$, where g generates the group $\mathbb{G}$ of prime order q. Gen chooses a random $sk \in \mathbb{Z}_q$ and outputs $pk = g^{sk}$. A signature on message m is $\sigma = H(m)^{sk}$, where H is a hash function. To verify signature σ on message m, one checks that $\mathbf{e}(\sigma, g) = \mathbf{e}(H(m), pk)$. Given a group of message-signature pairs $(m_1, \sigma_1), \dots, (m_n, \sigma_n)$ (all purportedly from the same signer), BGLS aggregates them as $A = \prod_{i=1}^n \sigma_i$. Then all signatures can be verified in aggregate (i.e., screened) by testing that $\mathbf{e}(A, g) = \mathbf{e}(\prod_{i=1}^n H(m_i), pk)$. This scheme is *not*, however, a batch verification scheme since, for any $a \neq 1 \in \mathbb{G}$, the two *invalid* message-signature pairs $P_1 = (m_1, a \cdot H(m_1)^{sk})$ and $P_2 = (m_2, a^{-1} \cdot H(m_2)^{sk})$ will verify under Definition 2 (as BGLS prove [3]), but will not verify under Definition 1. Indeed, for some pervasive computing applications only guaranteeing screening would be disastrous, because only P_1 may be relevant information to forward to the next entity – and it won't verify once it arrives! To be fair, batch verification is not what aggregate schemes were designed to do, but it is a common misuse of them.

Let $D = \{(t_1, m_1, \sigma_1), \dots, (t_n, m_n, \sigma_n)\}$. We note that while $\mathsf{Screen}(D) = 1$ does not guarantee that $\mathsf{Verify}(pk_{t_i}, m_i, \sigma_i)$ for all i; it does guarantee that the holder of sk_{t_i} authenticated m_i. Thus, we can always prove this by first creating a *new* signature scheme $(\mathsf{Gen}, \mathsf{Sign}, \mathsf{Verify}')$ where the verification algorithm Verify' is modified w.r.t. the original scheme as follows. Apart from the original signatures, it also accepts signatures σ_i' derived from D such that if and only if for all $(t_i, m_i, \sigma_i) \in D$, $\mathsf{Verify}'(pk_{t_i}, m_i, \sigma_i') = 1$ we have $\mathsf{Screen}(D) = 1$. One method to construct σ_i' would be to give a zero-knowledge proof of knowledge of D such that $\mathsf{Screen}(D) = 1$, although (using the naive solution) these new signatures σ_i' will require $O(n)$ space and Verify' will run in $O(n)$ time.

3   Algebraic Setting and Group Membership

Bilinear Groups. Let BSetup be an algorithm that, on input the security parameter 1^ℓ, outputs the parameters for a bilinear map as $(q, g, \mathbb{G}, \mathbb{G}_T, \mathbf{e})$, where

$\mathbb{G}, \mathbb{G}_T$ are of prime order $q \in \Theta(2^\ell)$. The efficient mapping $\mathbf{e} : \mathbb{G} \times \mathbb{G} \to \mathbb{G}_T$ is both: (*Bilinear*) for all $g \in \mathbb{G}$ and $a, b \leftarrow \mathbb{Z}_q$, $\mathbf{e}(g^a, g^b) = \mathbf{e}(g, g)^{ab}$; and (*Non-degenerate*) if g generates $\mathbb{G}$, then $\mathbf{e}(g, g) \neq 1$. Following prior work, we write $\mathbb{G}$ and $\mathbb{G}_T$ in multiplicative notation, although $\mathbb{G}$ is actually an additive group. Our constructions from Section 5 also work in the setting $\mathbf{e} : \mathbb{G}_1 \times \mathbb{G}_2 \to \mathbb{G}_T$, where $\mathbb{G}_1$ and $\mathbb{G}_2$ are distinct groups, possibly without efficient isomorphisms between them, but it is more tedious to write. However, this later implementation allows for the shortest group elements. We note that if the Waters IBS scheme also works in this setting, so will our proposed batch verifier in Section 4.

Testing Membership in $\mathbb{G}$. In a *non-bilinear* setting, Boyd and Pavlovski [6] observed that the proofs of security for many previous batch verification or screening schemes *assumed* that the signatures (potentially submitted by a malicious adversary) were elements of an appropriate subgroup. For example, it was common place to assume that signatures submitted for batch DSA verification contained an element in a subgroup $\mathbb{G}$ of $\mathbb{Z}_p^*$ of prime order q. Boyd and Pavlovski [6] pointed out efficient attacks on many batching algorithms via exploiting this issue. Of course, group membership cannot be *assumed*, it must be *tested* and the work required by this test might well obliterate all batching efficiency gains. E.g., verifying that an element y is in $\mathbb{G}$ by testing if $y^q \mod q = 1$; easily obliterates the gain of batching DSA signatures. Boyd and Pavlovski [6] suggest methods for overcoming this problem through careful choice of q.

In this paper, we will work in a bilinear setting, and we must be careful to avoid this common mistake in batch verification. To do so, we must say more about the groups in which we are working. Let E be an elliptic curve over a finite field $\mathbb{F}_p$ and let $\mathcal{O}$ denote the point at infinity. We denote the group of points on E defined over $\mathbb{F}_p$ as $E(\mathbb{F}_p)$. Then, a prime subgroup $\mathbb{G} \subseteq E(\mathbb{F}_p)$ of order q is chosen appropriately for our mapping. Our proofs will require that elements of purported signatures are members of $\mathbb{G}$ and *not* $E(\mathbb{F}_p) \setminus \mathbb{G}$. The question is: how efficiently can this fact be verified? Determining whether some data represents a point on a curve is easy. The question is whether it is in the correct subgroup. Assuming we have a bilinear map $\mathbf{e} : \mathbb{G}_1 \times \mathbb{G}_2 \to \mathbb{G}_T$. In all the schemes we use, signatures are in $\mathbb{G}_1$, so this is the group we are interested in testing membership of. Elements in $\mathbb{G}_1$ will always be in $\mathbb{F}_p$ and have order q, so we can use cofactor multiplication: The curve has hq points over $\mathbb{F}_p$, so if an element y satisfies the curve equation and $hy \neq \mathcal{O}$ (here $\mathbb{G}_1$ is expressed in additive notation), then that element is in $\mathbb{G}_1$. If one chooses a curve with $h = 1$ then this test is trivial, but even if $h > 1$, but still much smaller than q, this test is efficient. Chen, Cheng and Smart discuss this and ways to test membership in $\mathbb{G}_2$ in [14].

4   Batch Verification Without Random Oracles

In this section, we present a method for batch verifying signatures together with their accompanying certificates. We propose using Waters Two-Level Hierarchical Signatures [39,7] with the first level corresponding to the certificate and the second level used for signing messages. We assume all certificates originate from the same authority. The scheme is secure under the Computational Diffie-Hellman assumption in the plain model. This batch verification method can execute in different modes, optimizing for the lowest runtime. Let n be the number of certificate/signature pairs, let 2^{k_1} be the number of users, and let k_2 be the bits per message. Then our batch verifier will verify n certificate/signature pairs with asymptotic complexity of the dominant operations roughly $\text{MIN}\{(2n+3)\,,\ (k_1+n+3)\,,\ (n+k_2+3)\,,\ (k_1+k_2+3)\}$. Suppose there are one billion users ($k_1 = 30$) and RIPEMD-160 is used to hash all the messages ($k_2 = 160$), then when $n \geq 64$ batching becomes faster than individual verification. However, one can imagine many usage scenarios where devices send predefined messages requiring less than 160 bits, e.g., ISO defined error messages. For example, if $k_1 = 30$ and $k_2 = 32$, then when $n \geq 22$ batching should be used.

4.1   Batch Verification for Waters IBS

We describe a batch verification algorithm for the Waters IBS scheme from [7], where the number of pairings depends on the security parameter and not on the number of signatures. We assume that the identities are bit strings of length k_1 and the messages are bit strings of length k_2. First we describe the signature scheme. Let $\mathsf{BSetup}(1^\ell) \to (q, g, \mathbb{G}, \mathbb{G}_T, \mathbf{e})$.

Setup: First choose a secret $\alpha \in \mathbb{Z}_q$ and calculate $A = \mathbf{e}(g, g)^\alpha$. Then pick two random integers $y', z' \in \mathbb{Z}_q$ and two random vectors $y = (y_1, \ldots, y_{k_1}) \in \mathbb{Z}_q^{k_1}$ and $z = (z_1, \ldots, z_{k_2}) \in \mathbb{Z}_q^{k_2}$. The master secret key is $MK = g^\alpha$ and the public parameters are given as: $PP = g, A, u' = g^{y'}, u_1 = g^{y_1}, \ldots, u_{k_1} = g^{y_{k_1}}, v' = g^{z'}, v_1 = g^{z_1}, \ldots, v_{k_2} = g^{z_{k_2}}$.

Extract: To create a private key for a user with identity $ID = (\kappa_1, \ldots, \kappa_{k_1}) \in \{0,1\}^{k_1}$, select $r \in \mathbb{Z}_q$ and return $K_{ID} = \left(g^\alpha(u' \prod_{i=1}^{k_1} u_i^{\kappa_i})^r, g^{-r}\right)$.

Sign: To sign a message $m = (m_1, \ldots, m_{k_2}) \in \{0,1\}^{k_2}$ using private key $K = (K_1, K_2)$, select $s \in \mathbb{Z}_q$ and return $S = \left(K_1(v' \prod_{j=1}^{k_2} v_j^{m_j})^s, K_2, g^{-s}\right)$.

Verify: To verify a signature $S = (S_1, S_2, S_3)$ from identity $ID = (\kappa_1, \ldots, \kappa_{k_1})$ on message $M = (m_1, \ldots, m_{k_2})$, check that $\mathbf{e}(S_1, g) \cdot \mathbf{e}(S_2, u' \prod_{j=1}^{k_1} u_j^{\kappa_j}) \cdot \mathbf{e}(S_3, v' \prod_{j=1}^{k_2} v_j^{m_j}) = A$. If this equation holds, output *accept*; otherwise output *reject*.

We now introduce a batch verifier for this signature scheme. The basic idea is to adopt the small exponents test from [2] and to take advantage of the peculiarities of bilinear maps. Let *KeyGen*, *Sign* and *Verify* be as before.

Batch Verify: Let κ_j^i and m_k^i denote the j'th bit of the identity of the i'th signer respectively the k'th bit in the message signed by the i'th signer, and let $S^i = (S_1^i, S_2^i, S_3^i)$ denote the signature from the i'th signer. First check if $S_1^i, S_2^i, S_3^i \in \mathbb{G}$ for all i. If not; output *reject*. Otherwise generate a vector

$\Delta = (\delta_1, \ldots, \delta_n)$ where each δ_i is a random element of ℓ_b bits from $\mathbb{Z}_q$ and set $P = \mathbf{e}(\prod_{i=1}^{n} S_1^{i \, \delta_i}, g) \cdot \mathbf{e}(\prod_{i=1}^{n} S_2^{i \, \delta_i}, u') \cdot \mathbf{e}(\prod_{i=1}^{n} S_3^{i \, \delta_i}, v')$. Depending on the values of k_1, k_2 and n (c.f. below), pick and check one of the following equations:

$$\prod_{i=1}^{n} A^{\delta_i} = P \cdot \prod_{i=1}^{n} \mathbf{e}(S_2^{i \, \delta_i}, \prod_{j=1}^{k_1} u_j^{\kappa_j^i}) \cdot \prod_{i=1}^{n} \mathbf{e}(S_3^{i \, \delta_i}, \prod_{j=1}^{k_2} v_j^{m_j^i}) \tag{1}$$

$$\prod_{i=1}^{n} A^{\delta_i} = P \cdot \prod_{j=1}^{k_1} \mathbf{e}(\prod_{i=1}^{n} S_2^{i \, \delta_i \kappa_j^i}, u_j) \cdot \prod_{i=1}^{n} \mathbf{e}(S_3^{i \, \delta_i}, \prod_{j=1}^{k_2} v_j^{m_j^i}) \tag{2}$$

$$\prod_{i=1}^{n} A^{\delta_i} = P \cdot \prod_{j=1}^{k_1} \mathbf{e}(\prod_{i=1}^{n} S_2^{i \, \delta_i \kappa_j^i}, u_j) \cdot \prod_{j=1}^{k_2} \mathbf{e}(\prod_{i=1}^{n} S_3^{i \, \delta_i m_j^i}, v_j) \tag{3}$$

Output *accept* if the chosen equation holds; otherwise output *reject*.

Let us discuss which equation should be picked. Assume that $k_1 < k_2$ (i.e., that fewer bits are used for the identities of the users than for the messages). If $n < k_1$ use equation 1, if $k_1 \leq n \leq k_2$ use equation 2; otherwise use equation 3.

Theorem 1. *The above algorithm is a batch verifier for the Waters IBS.*

Proof. First we prove that $\mathsf{Verify}(ID_{t_1}, M_1, S_1) = \cdots = \mathsf{Verify}(ID_{t_n}, M_n, S_n) = 1 \Rightarrow \mathsf{Batch}((ID_{t_1}, M_1, S_1), \ldots, (ID_{t_n}, M_n, S_n)) = 1$. This follows from the verification equation for the Waters IBS scheme:

$$\prod_{i=1}^{n} A^{\delta_i} = \prod_{i=1}^{n} \left(\mathbf{e}(S_1^i, g) \cdot \mathbf{e}(S_2^i, u' \prod_{j=1}^{k_1} u_j^{\kappa_j^i}) \cdot \mathbf{e}(S_3^i, v' \prod_{j=1}^{k_2} v_j^{m_j^i}) \right)^{\delta_i}$$

$$= \mathbf{e}(\prod_{i=1}^{n} S_1^{i \, \delta_i}, g) \cdot \prod_{i=1}^{n} \mathbf{e}(S_2^{i \, \delta_i}, u' \prod_{j=1}^{k_1} u_j^{\kappa_j^i}) \cdot \prod_{i=1}^{n} \mathbf{e}(S_3^{i \, \delta_i}, v' \prod_{j=1}^{k_2} v_j^{m_j^i})$$

$$= P \cdot \prod_{i=1}^{n} \mathbf{e}(S_2^{i \, \delta_i}, \prod_{j=1}^{k_1} u_j^{\kappa_j^i}) \cdot \prod_{i=1}^{n} \mathbf{e}(S_3^{i \, \delta_i}, \prod_{j=1}^{k_2} v_j^{m_j^i}) \tag{4}$$

Since for all i, $\mathsf{Verify}(ID_{t_i}, M_i, S_i) = 1$, S_2^i is a valid part of a signature, and hence $S_2^i \in \mathbb{G}$. This means that we can write $S_2^i = g^{b_i}$ for some $b_i \in \mathbb{Z}_q$ and get $\prod_{i=1}^{n} \mathbf{e}(S_2^{i \, \delta_i}, \prod_{j=1}^{k_1} u_j^{\kappa_j^i}) = \prod_{i=1}^{n} \mathbf{e}(g^{\delta_i b_i}, g^{\sum_{j=1}^{k_1} y_j \kappa_j^i}) = \mathbf{e}(g, g)^{\sum_{i=1}^{n} \sum_{j=1}^{k_1} \delta_i b_i y_j \kappa_j^i} = \prod_{j=1}^{k_1} \mathbf{e}(g^{\sum_{i=1}^{n} \delta_i b_i \kappa_j^i}, g^{y_j}) = \prod_{j=1}^{k_1} \mathbf{e}(\prod_{i=1}^{n} S_2^{i \, \delta_i \kappa_j^i}, u_j)$. We can do the same with $\prod_{i=1}^{n} \mathbf{e}(S_3^{i \, \delta_i}, \prod_{j=1}^{k_2} v_j^{m_j^i})$, so correctness of the different verification equations follows from this and equation 4.

We must now show the other direction. This proof is an application of the technique for proving the small exponents test in [2]. Since $S_1^i, S_2^i, S_3^i \in \mathbb{G}$ we can write $S_1^i = g^{a_i}, S_2^i = g^{b_i}$ and $S_3^i = g^{c_i}$ for some $a_i, b_i, c_i \in \mathbb{Z}_q$. This means that the verification equation for Waters IBS can be rewritten as:

$$\mathbf{e}(g,g)^\alpha = \mathbf{e}(g^a, g) \cdot \mathbf{e}(g^b, g^{y'} g^{\sum_{j=1}^{k_1} y_j \kappa_j}) \cdot \mathbf{e}(g^c, g^{z'} g^{\sum_{j=1}^{k_2} z_j m_j})$$

$$= \mathbf{e}(g,g)^{a + y'b + z'c + b \sum_{j=1}^{k_1} y_j \kappa_j + c \sum_{j=1}^{k_2} z_j m_j} \tag{5}$$

Using 5 we can rewrite equation 1 as:

$$\mathbf{e}(g,g)^{\sum_{i=1}^n \delta_i \alpha} = \mathbf{e}(g,g)^{\sum_{i=1}^n \delta_i \left(a_i + y'b_i + z'c_i + b_i \sum_{j=1}^{k_1} y_j \kappa_j + c_i \sum_{j=1}^{k_2} z_j m_j\right)} \tag{6}$$

Setting $\beta_i = \alpha - \left(a_i + y'b_i + z'c_i + b_i \sum_{j=1}^{k_1} y_j \kappa_j^i + c_i \sum_{j=1}^{k_2} z_j m_j^i\right)$ and rewriting equation 6 we get:

$$\mathbf{e}(g,g)^{\sum_{i=1}^n \delta_i \alpha - \sum_{i=1}^n \delta_i \left(a_i + y'b_i + z'c_i + b_i \sum_{j=1}^{k_1} y_j \kappa_j + c_i \sum_{j=1}^{k_2} z_j m_j\right)} = 1$$

$$\Rightarrow \sum_{i=1}^n \delta_i \alpha - \sum_{i=1}^n \delta_i \left(a_i + y'b_i + z'c_i + b_i \sum_{j=1}^{k_1} y_j \kappa_j + c_i \sum_{j=1}^{k_2} z_j m_j\right) \equiv 0 \pmod{q}$$

$$\Rightarrow \sum_{i=1}^n \delta_i \beta_i \equiv 0 \pmod{q} \tag{7}$$

Assume that $\mathsf{Batch}((ID_{t_1}, M_1, S_1), \ldots, (ID_{t_n}, M_n, S_n)) = 1$, but for at least one i it is the case that $\mathsf{Verify}(ID_{t_i}, M_i, S_i) = 0$. Assume wlog that this is true for $i = 1$, which means that $\beta_1 \neq 0$. Since q is a prime then β_1 has an inverse γ_1 such that $\beta_1 \gamma_1 \equiv 1 \pmod{q}$. This and equation 7 gives us:

$$\delta_1 \equiv -\gamma_1 \sum_{i=2}^n \delta_i \beta_i \pmod{q} \tag{8}$$

Given (ID_{t_i}, M_i, S_i) where $i = 1 \ldots n$, let E be an event that occurs if $\mathsf{Verify}(ID_{t_1}, M_1, S_1) = 0$ but $\mathsf{Batch}((ID_{t_1}, M_1, S_1), \ldots, (ID_{t_n}, M_n, S_n)) = 1$, or in other words that we break batch verification. Note that we do not make any assumptions about the remaining values. Let $\Delta' = \delta_2, \ldots, \delta_n$ denote the last $n - 1$ values of Δ and let $|\Delta'|$ be the number of possible values for this vector. Equation 8 says that given a fixed vector Δ' there is exactly one value of δ_1 that will make event E happen, or in other words that the probability of E given a randomly chosen δ_1 is $\Pr[E|\Delta'] = 2^{-\ell_b}$. So if we pick δ_1 at random and sum over all possible choices of Δ' we get $\Pr[E] \leq \sum_{i=1}^{|\Delta'|} (\Pr[E|\Delta'] \cdot \Pr[\Delta'])$. Plugging in the values, we get: $\Pr[E] \leq \sum_{i=1}^{2^{\ell_b(n-1)}} \left(2^{-\ell_b} \cdot 2^{-\ell_b(n-1)}\right) = 2^{-\ell_b}$. $\qquad \square$

5   Faster Batch Verification with Restrictions

In this section, we present a second method for batch verifying signatures together with their accompanying certificates. We propose using the BLS signature scheme [5] for the certificates and a modified version of the CL signature scheme [8] for signing messages. This method requires only two pairings to verify n certificates (from the same authority) and three pairings to verify n signatures (from possibly different signers). The cost for this significant efficiency gain is some usage restrictions, although as we will discuss, these restrictions may not be a problem for some of the applications we have in mind.

Certificates: We use a batch verifier for BLS signatures from the same authority as described in Section 5.1. The scheme is secure under CDH in the random oracle model. To verify n BLS certificates costs $n\text{-MultExpCost}_{\mathbb{G}}^2(\ell_b) + \text{PairCost}_{\mathbb{G},\mathbb{G}}^2 + \text{GroupTestCost}_{\mathbb{G}}^n + \text{HashCost}_{\mathbb{G}}^n$, using the Section 1.2 notation.

Signatures: We describe a new signature scheme CL* with a batch verifier in Section 5.2. The scheme is secure under the LRSW assumption in the plain model when the message space is a polynomial and in the random oracle model when the message space is super-polynomial. We assume that there are discrete time or location identifiers $\phi \in \Phi$. A user can issue at most one signature per ϕ (e.g., this might correspond to a device being allowed to broadcast at most one message every 300ms) and only signatures from the same ϕ can be batch verified together. To verify n CL* signatures, costs $n\text{-MultExpCost}_{\mathbb{G}}^2(\ell_b) + n\text{-MultExpCost}_{\mathbb{G}}(|w| + \ell_b) + \text{PairCost}_{\mathbb{G},\mathbb{G}}^3 + \text{GroupTestCost}_{\mathbb{G}}^n + \text{HashCost}_{\mathbb{G}}^n$, where w is the output of a hash function.

5.1 Batch Verification of BLS Signatures

We describe a batch verifier for *many signers* for the Boneh, Lynn, and Shacham signatures [5,4] described in Section 2, using the small exponents test [2].

Batch Verify: Given purported signatures σ_i from n users on messages M_i for $i = 1 \ldots n$, first check that $\sigma_i \in \mathbb{G}$ for all i and if not; output *reject*. Otherwise compute $h_i = H(M_i)$ and generate a vector $\delta = (\delta_1, \ldots, \delta_n)$ where each δ_i is a random element of ℓ_b bits from $\mathbb{Z}_q$. Check that $\mathbf{e}(\prod_{i=1}^n \sigma_i^{\delta_i}, g) = \prod_{i=1}^n \mathbf{e}(h_i, pk_i)^{\delta_i}$. If this equation holds, output *accept*; otherwise output *reject*.

Theorem 2. *The algorithm above is a batch verifier for BLS signatures.*

Proof. The proof is similar to proof 4.1 and omitted for space reasons.

Single Singer for BLS. However, BLS [5,4] previously observed that if we have a single signer with public key v, the verification equation can be written as $\mathbf{e}(\prod_{i=1}^n \sigma_i^{\delta_i}, g) = \mathbf{e}(\prod_{i=1}^n h_i^{\delta_i}, v)$ which reduces the load to only two pairings.

Theorem 3 ([5,4]). *The algorithm above is a single-signer, batch verifier for BLS signatures.*

5.2 A New Signature Scheme CL*

In this section we introduce a new signature scheme secure under the LRSW assumption [33], which is based on the Camenisch and Lysyanskaya signatures [8].

Assumption 4 (LRSW Assumption). *Let* $\mathsf{BSetup}(1^\ell) \to (q, g, \mathbb{G}, \mathbb{G}_T, \mathbf{e})$. *Let* $X, Y \in \mathbb{G}$, $X = g^x$, *and* $Y = g^y$. *Let* $\mathcal{O}_{X,Y}(\cdot)$ *be an oracle that, on input a value* $m \in \mathbb{Z}_q^*$, *outputs a triple* $A = (a, a^y, a^{x+mxy})$ *for a randomly chosen* $a \in \mathbb{G}$. *Then for all probabilistic polynomial time adversaries* $\mathcal{A}^{(\cdot)}$, $\nu(\ell)$ *defined as follows is a negligible function:*

$$\Pr[(q, g, \mathbb{G}, \mathbb{G}_T, \mathbf{e}) \leftarrow \mathsf{BSetup}(1^\ell); x \leftarrow \mathbb{Z}_q; y \leftarrow \mathbb{Z}_q; X = g^x; Y = g^y;$$
$$(m, a, b, c) \leftarrow \mathcal{A}^{\mathcal{O}_{X,Y}}(q, g, \mathbb{G}, \mathbb{G}_T, \mathbf{e}, X, Y) \ : \ m \notin Q \ \wedge \ m \in \mathbb{Z}_q^* \ \wedge$$
$$a \in \mathbb{G} \wedge \ b = a^y \ \wedge \ c = a^{x+mxy}] = \nu(\ell) \ ,$$

where Q *is the set of queries that* $\mathcal{A}$ *made to* $\mathcal{O}_{X,Y}(\cdot)$.

The Original CL Scheme. Recall the Camenisch and Lysyanskaya signatures [8]. Let $\mathsf{BSetup}(1^\ell) \to (q, g, \mathbb{G}, \mathbb{G}_T, \mathbf{e})$. Choose the secret key $(x, y) \in \mathbb{Z}_q^2$ at random and set $X = g^x$ and $Y = g^y$. The public key is $pk = (X, Y)$. To sign a message $m \in \mathbb{Z}_q^*$, choose a random $a \in \mathbb{G}$ and compute $b = a^y$, $c = a^x b^{xm}$. Output the signature (a, b, c). To verify, check whether $\mathbf{e}(X, a) \cdot \mathbf{e}(X, b)^m = \mathbf{e}(g, c)$ and $\mathbf{e}(a, Y) = \mathbf{e}(g, b)$ holds.

CL*: A version of the CL Scheme Allowing Batch Verification. Our goal is to batch-verify CL signatures made by different signers. That is we need to consider how to verify equations of the form $\mathbf{e}(X, a) \cdot \mathbf{e}(X, b)^m = \mathbf{e}(g, c)$ and $\mathbf{e}(a, Y) = \mathbf{e}(g, b)$. The fact that the values X, a, b, and c are different for each signature seems to prevent efficient batch verification. Thus, we need to find a way such that many different signers share some of these values. Obviously, X and c need to be different. Now, depending on the application, all the signers can use the same value a by choosing a as the output of some hash function applied to, e.g., the current time period or location. We then note that all signers can use the same b in principle, i.e., have all of them share the same Y as it is sufficient for each signer to hold only one secret value (i.e., $sk = x$). Indeed, the only reason that the signer needs to know Y is to compute b. However, it turns out that if we define b such that $\log_a b$ is not known, the signature scheme is still secure. So, for instance, we can derive b in a similar way to a using a second hash function. Thus, all signers will virtually sign using the same Y per time period (but a different one for each period).

Let us now describe the resulting scheme. Let $\mathsf{BSetup}(1^\ell) \to (q, g, \mathbb{G}, \mathbb{G}_T, \mathbf{e})$. Let $\phi \in \Phi$ denote the current time period or location, where $|\Phi|$ is polynomial. Let $\mathcal{M}$ be the message space, for now let $\mathcal{M} = \{0, 1\}^*$. Let $H_1 : \Phi \to \mathbb{G}$, $H_2 : \Phi \to \mathbb{G}$, and $H_3 : \mathcal{M} \times \Phi \to \mathbb{Z}_q$ be different hash functions.

KeyGen: Choose a random $x \in \mathbb{Z}_q$ and set $X = g^x$. Set $sk = x$ and $pk = X$.

Sign: If this is the first call to Sign during period $\phi \in \Phi$, then on input message $m \in \mathcal{M}$, set $w = H_3(m\|\phi)$, $a = H_1(\phi)$, $b = H_2(\phi)$ and output the signature $\sigma = a^x b^{xw}$. Otherwise, abort.

Verify: On input message-period pair (m, ϕ) and purported signature σ, compute $w = H_3(m\|\phi)$, $a = H_1(\phi)$ and $b = H_2(\phi)$, and check that $\mathbf{e}(\sigma, g) = \mathbf{e}(a, X) \cdot \mathbf{e}(b, X)^w$. If true, output *accept*; otherwise output *reject*.

Theorem 5. *Under the LRSW assumption in* $\mathbb{G}$, *the CL* signature scheme is existentially unforgeable in the random oracle model for message space* $\mathcal{M} = \{0,1\}^*$.

Proof. We show that if there exists a p.p.t. adversary $\mathcal{A}$ that succeeds with probability ε in forging CL* signatures, then we can construct a p.p.t. adversary $\mathcal{B}$ that solves the LRSW problem with probability $\varepsilon \cdot |\Phi|^{-1} \cdot q_H^{-1}$ in the random oracle model, where q_H is the maximum number of oracle queries $\mathcal{A}$ makes to H_3 during any period $\phi \in \Phi$. Recall that $|\Phi|$ is a polynomial. Adversary $\mathcal{B}^{\mathcal{O}_{X,Y}(\cdot)}$ against LRSW operates as follows on input $(q, g, \mathbb{G}, \mathbb{G}_T, \mathbf{e}, X, Y)$. Let ℓ be the security parameter. We assume that Φ is pre-defined. Let q_H be the maximum number of queries $\mathcal{A}$ makes to H_3 during any period $\phi \in \Phi$.

1. *Setup:* Send the bilinear parameters $(q, g, \mathbb{G}, \mathbb{G}_T, \mathbf{e})$ to $\mathcal{A}$. Choose a random $w' \in \mathcal{M}$ and query $\mathcal{O}_{X,Y}(w')$ to obtain an LRSW instance (w', a', b', c'). Choose a random $\phi' \in \Phi$. Treat H_1, H_2, H_3 as random oracles. Allow $\mathcal{A}$ access to the hash functions H_1, H_2, H_3.
2. *Key Generation:* Set $pk^* = X$. For $i = 1$ to n, choose a random $sk_i \in \mathbb{Z}_q$ and set $pk_i = g^{sk_i}$. Output to $\mathcal{A}$ the keys pk^* and all (pk_i, sk_i) pairs.
3. *Oracle queries:* $\mathcal{B}$ responds to $\mathcal{A}$'s hash and signing queries as follows. Choose random r_i and s_i in $\mathbb{Z}_q$ for each time period (except ϕ'). Set up H_1 and H_2 such that:

$$H_1(\phi_i) = \begin{cases} g^{r_i} & \text{if } \phi_i \neq \phi' \\ a' & \text{otherwise} \end{cases} \qquad (9)$$

and

$$H_2(\phi_i) = \begin{cases} g^{s_i} & \text{if } \phi_i \neq \phi' \\ b' & \text{otherwise} \end{cases} \qquad (10)$$

Pick a random j in the range $[1, q_H]$. Choose random $t_{l,i} \in \mathbb{Z}_q$, such that $t_{l,i} \neq w'$, for $l \in [1, q_H]$ and $i \in [1, |\Phi|]$. Set up H_3 such that:

$$H_3(m_l \| \phi_i) = \begin{cases} t_{l,i} & \text{if } \phi_i \neq \phi' \text{ or } l \neq j \\ w' & \text{otherwise} \end{cases} \qquad (11)$$

$\mathcal{B}$ records $m^* := m_j$. Finally, set the signing query oracle such that on the lth query involving period ϕ_i:

$$\mathcal{O}_{sk^*}(m_l \| \phi_i) = \begin{cases} \text{abort} & \text{if } \phi_i = \phi' \text{ and } l \neq j \\ c' & \text{else if } \phi_i = \phi' \text{ and } l = j \\ X^{r_i} X^{(s_i) t_{l,i}} & \text{otherwise} \end{cases} \qquad (12)$$

4. *Output:* At some point $\mathcal{A}$ stops and outputs a purported forgery $\sigma \in \mathbb{G}$ for some (m_l, ϕ_i). If $\phi_i \neq \phi'$, $\mathcal{B}$ did not guess the correct period and thus $\mathcal{B}$ outputs a random guess for the LRSW game. If $m_l = m^*$ or the CL* signature does not verify, $\mathcal{A}$'s output is not a valid forgery and thus $\mathcal{B}$ outputs a random guess for the LRSW game. Otherwise, $\mathcal{B}$ outputs $(t_{l,i}, a', b', \sigma)$ as the solution to the LRSW game.

We now analyze $\mathcal{B}$'s success. If $\mathcal{B}$ is not forced to abort or issue a random guess, then we note that $\sigma = H_1(\phi_i)^x H_2(\phi_i)^{xH_3(m_l||\phi_i)}$. In this scenario $\phi_i = \phi'$ and $t_{l,i} \neq w'$. We can substitute as $\sigma = (a')^x (b')^{x(t_{l,i})}$. Thus, we see that $(t_{l,i}, a', b', \sigma)$ is indeed a valid LRSW instance. Thus, $\mathcal{B}$ succeeds at LRSW whenever $\mathcal{A}$ succeeds in forging CL* signatures, except when $\mathcal{B}$ is forced to abort or issue a random guess. First, when simulating the signing oracle, $\mathcal{B}$ is forced to abort whenever it incorrectly guesses which query to H_3, during period ϕ', $\mathcal{A}$ will eventually query to $\mathcal{O}_{sk^*}(\cdot, \cdot)$. Since all outputs of H_3 are independently random, $\mathcal{B}$ will be forced to abort at most q_H^{-1} probability. Next, provided that $\mathcal{A}$ issued a valid forgery, then $\mathcal{B}$ is only forced to issue a random guess when it incorrectly guesses which period $\phi \in \Phi$ that $\mathcal{A}$ will choose to issue its forgery. Since, from the view of $\mathcal{A}$ conditioned on the event that $\mathcal{B}$ has not yet aborted, all outputs of the oracles are perfectly distributed as either random oracles (H_1, H_2, H_3) or as a valid CL* signer ($\mathcal{O}_{sk^*}$). Thus, this random guess is forced with probability at most $|\Phi|^{-1}$. Thus, if $\mathcal{A}$ succeeds with ε probability, then $\mathcal{B}$ succeeds with probability $\varepsilon \cdot |\Phi|^{-1} \cdot q_H^{-1}$. $\square$

Theorem 6. *Under the LRSW assumption in $\mathbb{G}$, the CL* signature scheme is existentially unforgeable in the plain model when $|\mathcal{M}|$ is polynomial.*

Proof sketch. If there exists a p.p.t. adversary $\mathcal{A}$ that succeeds with probability ε in forging CL* signatures when $|\mathcal{M}| = \text{poly}(\ell)$, then we can construct a p.p.t. adversary $\mathcal{B}$ that solves the LRSW problem with probability $\varepsilon \cdot |\Phi|^{-1} \cdot |\mathcal{M}|^{-1}$. Canetti, Halevi, and Katz [9] described one method of constructing a universal one-way hash function that satisfies a polynomial number of input/output constraints, i.e., pairs (x_i, y_i) such that $H(x_i) = y_i$. Furthermore, we note that H_1, H_2 and H_3 have $|\Phi|$, $|\Phi|$, and $|\Phi| \cdot |\mathcal{M}|$ constraints, respectively. Since these are all polynomials, $\mathcal{B}$ can efficiently construct the appropriate hash functions. The analysis follows the proof with random oracles.

Efficiency Note. First, we observe that the CL* signatures are *very* short, requiring only one element in $\mathbb{G}$. Since the BLS signatures also require only one element in $\mathbb{G}$, and since a public key for the CL* scheme is also only one group element, the entire signature plus certificate could be transmitted in three $\mathbb{G}$ elements. In order to get the shortest representation for these elements, we need to use asymmetric bilinear maps $\mathbf{e} : \mathbb{G}_1 \times \mathbb{G}_2 \to \mathbb{G}_T$, where $\mathbb{G}_1 \neq \mathbb{G}_2$, which will allow elements in $\mathbb{G}_1$ to be 160 bits and elements of $\mathbb{G}_2$ to be 1024 bits for a security level comparable to RSA-1024 [28,18]. For BLS this means that the public key will be around 1024 bits, but since we use it for single signer, the public key of the certifying authority is probably embedded in the systems at production time. For CL* signatures we need to hash into $\mathbb{G}_1$ which according to Galbraith, Paterson and Smart [18] can be done efficiently. To summarize; using BLS and CL* we can represent the signature plus certificate using approximately 1344 bits with security comparable to RSA-1024, compared to around 3072 bits for actually using RSA-1024. We note that this is based on current state of the art for pairings, and might improve in the future.

Second, suppose one uses the universal one-way hash functions described by Canetti, Halevi, and Katz [9] to remove the random oracles from CL*. These hash functions require one exponentiation per constraint. In our case, we may require as many as $|\Phi| \cdot |\mathcal{M}|$ constraints. Thus, the cost to compute the hashes may dampen the efficiency gains of batch verification. However, our scheme will benefit from improvements in the construction of universal one-way hash functions with constraints. To keep $|\Phi|$ small in practice, users might need to periodically change their keys.

Batch Verification of CL* Signatures. Batch verification of n signatures $\sigma_1, \ldots, \sigma_n$ on messages $m_1, \ldots, m_n$ for the same period ϕ can be done as follows. Assume that user i with public key X_i signed message m_i. Set $w_i = H(m_i||\phi)$. First check if $\sigma_i \in \mathbb{G}$ for all i. If not; output *reject*. Otherwise pick a vector $\Delta = (\delta_i, \ldots, \delta_n)$ with each element being a random ℓ_b-bit number and check that $\mathbf{e}(\prod_{i=1}^{n} \sigma_i^{\delta_i}, g) = \mathbf{e}(a, \prod_{i=1}^{n} X_i^{\delta_i}) \cdot \mathbf{e}(b, \prod_{i=1}^{n} X_i^{w_i \delta_i})$. If this equation holds, output *accept*; otherwise output *reject*.

Theorem 7. *The algorithm above is a batch verifier for CL* signatures.*

Proof. The proof is similar to proof 4.1 and omitted for space reasons.

6   Conclusions and Open Problems

In this paper we focused on batch verification of signatures. We overviewed the large body of existing work, almost exclusively dealing with single signers. We extended the general batch verification definition of Bellare, Garay and Rabin [2] to the case of multiple signers. We then presented, to our knowledge, the first efficient and practical batch verification scheme for signatures without random oracles. We focused on solutions that comprehended the time to verify the signature *and* the corresponding certificate for the verification key. First, we presented a batch verifier for the Waters IBS that can verify n signatures using only $(k_1 + k_2 + 3)$ pairings (the dominant operation), where identities are k_1 bits and messages are k_2 bits. This is a significant improvement over the $3n$ pairings required by individual verification. Second, we presented a solution in the random oracle model that batch verifies n certificates and n CL* signatures using only 5 pairings. Here, CL* is a variant of the Camenisch-Lysyanskaya signatures that is much shorter, allows for efficient batch verification from many signers, but where only one signature can be safely issued per period.

It is an open problem to find a fast batch verification scheme for short signatures without the period restrictions from Section 5. Perhaps this can be achieved by improving the efficiency of our scheme in Section 4, using some of the techniques applied to the Waters IBE by Naccache [34] and Chatterjee and Sarkar [13]. Another exciting open problem is to develop fast batching schemes for various forms of anonymous authentication such as group signatures, e-cash, and anonymous credentials.

Acknowledgments

We thank Jean-Pierre Hubaux and Panos Papadimitratos for helpful discussions about the practical challenges of vehicular networks. We are also grateful to Paulo Barreto, Steven Galbraith, Hovav Shacham, and Nigel P. Smart for their comments on testing membership in bilinear groups. Finally we thank Ivan Damgård and the anonymous reviewers for their valuable input. Jan Camenisch was supported by the EU projects ECRYPT and PRIME, contracts IST-2002-507932 and IST-2002-507591. Michael Østergaard Pedersen was supported by the eu-DOMAIN IST EU project, contract no. IST-004420.

References

1. J. H. An, Y. Dodis, and T. Rabin. On the security of joint signature and encryption. In *EUROCRYPT '02*, volume 2332, p. 83–107, 2002.
2. M. Bellare, J. A. Garay, and T. Rabin. Fast batch verification for modular exponentiation and digital signatures. In *EUROCRYPT '98*, LNCS vol. 1403.
3. D. Boneh, C. Gentry, B. Lynn, and H. Shacham. Aggregate and verifiably encrypted signatures from bilinear maps. In *EUROCRYPT '03*, LNCS vol. 2656.
4. D. Boneh, B. Lynn, and H. Shacham. Short signatures from the Weil pairing. *J. Cryptology*, 17(4):297–319, 2004.
5. D. Boneh, H. Shacham, and B. Lynn. Short signatures from the Weil pairing. In *ASIACRYPT '01*, volume 2248 of LNCS, p. 514–532, 2001.
6. C. Boyd and C. Pavlovski. Attacking and repairing batch verification schemes. In *ASIACRYPT '00*, volume 1976 of LNCS, p. 58–71, 2000.
7. X. Boyen and B. Waters. Compact group signatures without random oracles. In *EUROCRYPT '06*, volume 4004 of LNCS, p. 427–444, 2006.
8. J. Camenisch and A. Lysyanskaya. Signature schemes and anonymous credentials from bilinear maps. In *CRYPTO '04*, volume 3152 of LNCS, p. 56–72, 2004.
9. R. Canetti, S. Halevi, and J. Katz. A forward-secure public-key encryption scheme. In *EUROCRYPT '03*, volume 2656 of LNCS, p. 255–271, 2003.
10. T. Cao, D. Lin, and R. Xue. Security analysis of some batch verifying signatures from pairings. *International Journal of Network Security*, 3(2):138–143, 2006.
11. Car 2 Car. Communication consortium. `http://car-to-car.org`.
12. J. C. Cha and J. H. Cheon. An identity-based signature from gap Diffie-Hellman groups. In *PKC '03*, p. 18–30, 2003.
13. S. Chatterjee and P. Sarkar. Trading time for space: Towards an efficient IBE scheme with short(er) public parameters in the standard model. In *ICISC '05*, volume 3935 of LNCS, p. 257–273, 2005.
14. L. Chen, Z. Cheng, and N. Smart. Identity-based key agreement protocols from pairings, 2006. Cryptology ePrint Archive: Report 2006/199.
15. J. H. Cheon, Y. Kim, and H. J. Yoon. A new ID-based signature with batch verification, 2004. Cryptology ePrint Archive: Report 2004/131.
16. S. Cui, P. Duan, and C. W. Chan. An efficient identity-based signature scheme with batch verifications. In *InfoScale '06*, p. 22, 2006.
17. A. Fiat. Batch RSA. In *CRYPTO '89*, volume 435, p. 175–185, 1989.
18. S. D. Galbraith, K. G. Paterson, and N. P. Smart. Pairings for cryptographers, 2006. Cryptology ePrint Archive: Report 2006/165.

19. C. Gentry and Z. Ramzan. Identity-based aggregate signatures. In *Public Key Cryptography 2005*, volume 3958 of LNCS, p. 257–273, 2006.
20. S. Goldwasser, S. Micali, and R. L. Rivest. A digital signature scheme secure against adaptive chosen-message attacks. *SIAM J. Computing*, 17(2), 1988.
21. R. Granger and N. Smart. On computing products of pairings, 2006. Cryptology ePrint Archive: Report 2006/172.
22. L. Harn. Batch verifying multiple DSA digital signatures. In *Electronics Letters*, volume 34(9), p. 870–871, 1998.
23. L. Harn. Batch verifying multiple RSA digital signatures. In *Electronics Letters*, volume 34(12), p. 1219–1220, 1998.
24. F. Hoshino, M. Abe, and T. Kobayashi. Lenient/strict batch verification in several groups. In *ISC '01*, p. 81–94, 2001.
25. M.-S. Hwang, C.-C. Lee, and Y.-L. Tang. Two simple batch verifying multiple digital signatures. In *ICICS '01*, p. 233–237, 2001.
26. M.-S. Hwang, I.-C. Lin, and K.-F. Hwang. Cryptanalysis of the batch verifying multiple RSA digital signatures. *Informatica, Lith. Acad. Sci.*, 11(1):15–19, 2000.
27. IEEE. 5.9 GHz Dedicated Short Range Communications. http://grouper.ieee.org/groups/scc32/dsrc.
28. N. Koblitz and A. Menezes. Pairing-based cryptography at high security levels, 2005. Cryptology ePrint Archive: Report 2005/076.
29. C.-S. Laih and S.-M. Yen. Improved digital signature suitable for batch verification. *IEEE Trans. Comput.*, 44(7):957–959, 1995.
30. S. Lee, S. Cho, J. Choi, and Y. Cho. Efficient identification of bad signatures in RSA-type batch signature. *IEICE Trans. on Fundamentals of Electronics, Communications and Computer Sciences*, E89-A(1):74–80, 2006.
31. C. Lim and P. Lee. Security of interactive DSA batch verification. In *Electronics Letters*, volume 30(19), p. 1592–1593, 1994.
32. C. H. Lim. Efficient multi-exponentation and application to batch verification of digital signatures, 2000. http://dasan.sejong.ac.kr/~chlim/english_pub.html.
33. A. Lysyanskaya, R. L. Rivest, A. Sahai, and S. Wolf. Pseudonym systems. In *SAC*, volume 1758 of *LNCS*, p. 184–199, 1999.
34. D. Naccache. Secure and practical identity-based encryption, 2005. Cryptology ePrint Archive: Report 2005/369.
35. D. Naccache, D. M'Raïhi, S. Vaudenay, and D. Raphaeli. Can D.S.A. be improved? complexity trade-offs with the digital signature standard. In *EUROCRYPT '94*, volume 0950 of LNCS, p. 77–85, 1994.
36. M. Raya and J.-P. Hubaux. Securing vehicular ad hoc networks. *Journal of Computer Security*, 15:39–68, 2007.
37. SeVeCom. Security on the road. http://www.sevecom.org.
38. M. Stanek. Attacking LCCC batch verification of RSA signatures, 2006. Cryptology ePrint Archive: Report 2006/111.
39. B. Waters. Efficient identity-based encryption without random oracles. In *EUROCRYPT '05*, volume 3494 of LNCS, p. 320–329, 2005.
40. H. Yoon, J. H. Cheon, and Y. Kim. Batch verifications with ID-based signatures. In *ICISC*, p. 233–248, 2004.
41. F. Zhang and K. Kim. Efficient ID-based blind signature and proxy signature from bilinear pairings. In *ACISP '03*, volume 2727 of LNCS, p. 312–323, 2003.
42. F. Zhang, R. Safavi-Naini, and W. Susilo. Efficient verifiably encrypted signature and partially blind signature from bilinear pairings. In *Indocrypt 2003*, volume 2904 of LNCS, p. 191–204, 2003.

Cryptanalysis of SFLASH with Slightly Modified Parameters

Vivien Dubois, Pierre-Alain Fouque, and Jacques Stern

École normale supérieure
DI, 45 rue d'Ulm, 75230 Paris cedex 05, France
{vivien.dubois,pierre-alain.fouque,jacques.stern}@ens.fr

Abstract. SFLASH is a signature scheme which belongs to a family of multivariate schemes proposed by Patarin *et al.* in 1998 [9]. The SFLASH scheme itself has been designed in 2001 [8] and has been selected in 2003 by the NESSIE European Consortium [6] as the best known solution for implementation on low cost smart cards. In this paper, we show that slight modifications of the parameters of SFLASH within the general family initially proposed renders the scheme insecure. The attack uses simple linear algebra, and allows to forge a signature for an arbitrary message in a question of minutes for practical parameters, using only the public key. Although SFLASH itself is not amenable to our attack, it is worrying to observe that no rationale was ever offered for this "lucky" choice of parameters.

1   Introduction

Multivariate Cryptography is an area of research which attempts to build asymmetric primitives, based on hard computational problems related to multivariate quadratic polynomials over a finite field. Multivariate schemes have recently received much attention, for several reasons. First, the hard problems of reference are not known to be polynomial in the quantum model, unlike integer factorization and the discrete logarithm problems. More importantly, Multivariate Cryptography offers a large collection of primitives and problems of a new flavor. In general, multivariate schemes require modest computational resources and can be implemented on low cost smart cards. Moreover, these schemes benefit from several nice properties such as providing very short or very fast signatures. Also, they are quite versatile: a number of generic non-exclusive variations can be derived from a few basic schemes. Even when the original schemes are weak, variations are often considered to avoid structural attacks.

One of the more elaborate outcomes of Multivariate Cryptography is probably the SFLASH signature scheme. Designed by Patarin *et al.* [8], it is among the fastest signatures schemes known, with NTRUSign and TTS [4,11]. Although initial tweaks in the first version of SFLASH were shown inappropriate [3], the second version of SFLASH is currently considered secure, as testified from the recent acceptance of this primitive by the NESSIE European Consortium [6].

M. Naor (Ed.): EUROCRYPT 2007, LNCS 4515, pp. 264–275, 2007.

The structure of SFLASH is among the simplest in Multivariate Cryptography. Roughly speaking, SFLASH is a truncated C^* scheme. The C^* scheme was invented by Matsumoto and Imai in 1988 [5], and was shown to be insecure by Patarin in 1995 [7]. Later, Patarin *et al.* considered the simple variation of C^* consisting in deleting from the public key a large number of coordinates [9]. Schemes derived from C^* by this principle are called C^{*-} schemes; they are well suited for signature. As soon as the number of deleted coordinates is large enough, C^{*-} schemes are considered secure. SFLASH belongs to the C^{*-} family and has been chosen as a candidate for the NESSIE selection, and finally accepted.

Our Results. We argue that the security of the C^{*-} schemes remains insufficiently understood. In particular, one may rightfully question the reasons for the particular choice of parameters opted for in SFLASH. Might other parameters yield the same security ?

In this paper, we show that many choices of parameters for C^{*-} schemes are insecure. Our approach uses basic properties of the differential as introduced in [2]. Since the differential is bilinear and symmetric, it seems natural to consider skew-symmetric linear maps with respect to this function. This property is so specific and overdefined that the space of skew-symmetric maps is left unchanged when we replace the full public key of C^* by its truncated version C^{*-}, even when the number of deleted coordinates is very large. Skew-symmetric maps can be recovered from their defining equation in terms of the differential of a C^{*-} public key, using only linear algebra. Once this has been achieved, compositions of these maps with the public key can be used to recover a full C^* public key, which can then be inverted using the original attack by Patarin [7].

The schemes under attack are those for which the internal C^* parameter and the number of variables are not coprime. Such parameters are perfectly acceptable for practical realizations of C^{*-} schemes in the current state of cryptanalysis. SFLASH with the recommended set of parameters escapes this attack. However, this shows that the elements underlying the security of C^{*-} schemes and their relations with parameters are not well identified. To illustrate this point, we show that changing the parameters of SFLASH by one renders the scheme breakable in a few minutes.

Organization of the Paper. In Section 2, we recall the definition of C^* and C^{*-} schemes. Then, in Section 3, we characterize skew-symmetric maps with respect to the differential of C^*. In Section 4, we show that the same maps can be recovered from a truncated version C^{*-} of the original C^* public key. Finally, in section 5, we show how their use allows us to restore a full C^* public key.

2 C^* and C^{*-}

Before we describe the C^* and C^{*-} schemes, we recall the generic construction of multivariate schemes.

2.1   The Generic Construction of Multivariate Schemes

We denote by $\mathbb{F}_q^n$ the n-dimensional vector space over the finite field $\mathbb{F}_q$ with q elements. A function from $\mathbb{F}_q^n$ to $\mathbb{F}_q^m$ is defined by m coordinate-polynomials in n variables. When these polynomials have multivariate degree 2, the function is termed quadratic. Finding a preimage of a quadratic function involves solving a multivariate quadratic system of equations, an NP-hard problem in general. Nevertheless, some classes of easily invertible quadratic functions are known and can form the basis of a multivariate asymmetric scheme. More precisely, the generic construction of multivariate schemes is the following. The key generation algorithm hides an easily invertible quadratic function F by two linear (or affine) changes of coordinates U and T into a function P defined by

$$P = T \circ F \circ U$$

P is the public key and U, T are the secret key. The proponents of multivariate cryptography argue that the function P is a random-looking quadratic function, which is expected to be hard to invert by general purpose techniques. An encrypted message can be decrypted by using the secret key (T, U) to undo the hiding process and by solving the easy internal quadratic system.

2.2   The C^* Scheme

The C^* scheme was proposed by Matsumoto and Imai in 1988 [5]. In the C^* scheme, the internal easy-to-invert function is defined from a monomial over the degree n extension field of $\mathbb{F}_q$, denoted $\mathbb{F}_{q^n}$, of the form

$$F(x) = x^{1+q^\theta}$$

where θ is a positive integer. The function F is isomorphic to a quadratic function from $\mathbb{F}_q^n$ into itself and provided q is even, the integer θ can be chosen so that F is a permutation. This happens if and only if $\gcd(q^\theta + 1, q^n - 1) = 1$. In Appendix A, we show that, denoting by d the gcd of θ and n, this is equivalent to the condition that $\frac{n}{d}$ is odd.

The C^* scheme, as previously described, was shown to be insecure by Patarin [7]. It was observed that, for any x, y such that $F(x) = y$, we have

$$y^{q^\theta}.x - y.x^{q^{2\theta}} = 0$$

It follows that there exist n bilinear relations between a ciphertext and the corresponding plaintext. These bilinear relations can be found from the public key with a few plaintext-ciphertext pairs. Using these bilinear relations allows us to recover the plaintext from any ciphertext, by linear algebra.

Several ways to withstand the attack by Patarin were later considered. Among the most promising, are the C^{*-} schemes. In the next section, we recall these schemes in detail.

2.3 C^{*-} Schemes

A C^{*-} scheme is derived from a C^* scheme by simply deleting from the C^* public key some of the quadratic polynomials. More precisely, for some additional parameter r, the key generation builds a C^* scheme and then deletes from the public key the last r coordinates. In the sequel, Π will denote the projection on the first $(n - r)$ coordinates, P the C^* public key, and P_Π the resulting C^{*-} public key.

To find a preimage under P_Π of a string y of $(\mathbb{F}_q)^{n-r}$, the user first has to pad y with some string k of $(\mathbb{F}_q)^r$, and then has to find the preimage of (y, k) by P using its secret key U, T. Using a C^{*-} scheme for encryption is therefore quite awkward: to recover the plaintext, the user has to review all possible paddings k, compute for each k the corresponding preimage, and identify the plaintext among these preimages by using some message redundancy. However, C^{*-} schemes are well-suited for signature, even for large q and r, since in this setting any of the q^r preimages of y by P_Π is a valid signature of y. To sign the message y, the user chooses an arbitrary k and the signature consists in the preimage of (y, k) by P. In the sequel, we only consider the C^{*-} scheme in the signature setting.

C^{*-} schemes were first introduced by Patarin $et\ al.$ [9], but the idea of enhancing the security of multivariate schemes by deleting a few coordinates from the public key first appeared in Shamir [10]. In [9], Patarin $et\ al.$ describe a technique for reconstructing a C^* public key from a C^{*-} public key with complexity of the order of q^r. Accordingly, the parameters q and r must be chosen such that $q^r \simeq 2^{80}$ for practical instantiations of C^{*-} schemes. The illustrative notation C^{*--} is sometimes used in this case. No condition is specified in the literature for choosing the parameter θ besides the obvious requirement that the corresponding monomial should be invertible and, as seen before, all values of θ whose gcd d with n is such that $\frac{n}{d}$ is odd can be chosen. In fact, choosing a large d allows a faster inversion of the C^* monomial, as observed by Ding [1], and can be an attractive choice.

SFLASH. Practical instantiations of C^{*-} schemes were proposed by Patarin $et\ al.$ as candidates to the European call for primitives NESSIE in 2001. These instantiations were called FLASH and SFLASH. Initially, some tweak was added to SFLASH to decrease the size of the public key, however this tweak rendered the scheme insecure, as shown by Gilbert and Minier in 2002 [3], and discarded. Without this tweak, FLASH and SFLASH are very similar and therefore, only SFLASH was later considered by the NESSIE evaluation process, and finally accepted in 2003. The recommended parameters of SFLASH are $q = 2^7$, $n = 37$, $\theta = 11$ and $r = 11$; signatures are 239 bits long. Until now, no weakness was reported in either SFLASH or the general design principle of C^{*-} schemes.

In the sequel, we will show that many C^{*-} schemes are insecure. The C^{*-} schemes under attack are those for which the gcd d of θ and n is not 1. Note that this is different than the condition that $\frac{n}{d}$ is odd, which is needed to make the mapping invertible. The attack makes it possible to forge a signature in a matter of minutes for practical parameters. The attack does not apply to SFLASH for which the recommended parameters θ and n are coprime. However this "lucky" choice appears to be accidental since no rationale was offered for it.

3   Skew-Symmetric Maps w.r.t the Differential of C^*

In this section, we consider some properties of the differential of the internal C^* monomial. Implications of these properties to C^{*-} schemes will be addressed in the next section.

The differential, defined as follows, can be considered for any quadratic function F. For any element a, the difference function $x \mapsto F(x + a) - F(x)$ is affine and its constant term is $F(a) - F(0)$. Its linear part is called the differential of F at a and is denoted $DF(a, x)$:

$$DF(a, x) = F(x + a) - F(x) - F(a) + F(0)$$

$DF(a, x)$ is actually bilinear and symmetric when considered as a function of a and x. Our attack is based on considering skew-symmetric maps with respect to this bilinear function *i.e.* linear maps M such that for all choices of x and a

$$DF(a, M(x)) + DF(M(a), x) = 0$$

This is a very strong condition, and when F is defined by a random collection of quadratic polynomials, only trivial solutions M are expected to satisfy this condition. However, when $F(x) = x^{1+q^\theta}$, its differential is

$$DF(a, x) = a^{q^\theta} x + ax^{q^\theta} \tag{1}$$

The skew-symmetric maps with respect to the differential of such a C^* monomial are given by the following theorem.

Theorem 1. *Let M be a linear map; M is skew-symmetric with respect to $DF(a, x)$ if and only if M is the multiplication by some element ξ satisfying $\xi^{q^\theta} + \xi = 0$.*

Proof. A linear map M over $\mathbb{F}_{q^n}$ is a sum of q-powerings : $M(x) = \sum_{i=0}^{n-1} \lambda_i x^{q^i}$. When DF is the differential of the C^* monomial given by (1), we get for any elements a, x in $\mathbb{F}_{q^n}$

$$\sum_{i=0}^{n-1} \lambda_i a^{q^\theta} x^{q^i} + \sum_{i=0}^{n-1} \lambda_i^{q^\theta} a \, x^{q^{i+\theta}} + \sum_{i=0}^{n-1} \lambda_i a^{q^i} x^{q^\theta} + \sum_{i=0}^{n-1} \lambda_i^{q^\theta} a^{q^{i+\theta}} x = 0$$

Since the monomials $a^{q^u} x^{q^v}$ are a basis of the space of bilinear maps over $\mathbb{F}_{q^n}$, we obtain the following equations corresponding to the various elements of the basis

$$\lambda_0 + \lambda_0^{q^\theta} = 0 \qquad \text{(coefficient of } ax^{q^\theta})$$

$$\lambda_i = 0, \, i \neq 0, \theta \qquad \text{(coefficient of } a^{q^i} x^{q^\theta}, \, i \neq 0, \theta)$$

$$(\lambda_\theta)^{q^\theta} = 0 \qquad \text{(coefficient of } ax^{q^{2\theta}})$$

Conversely, it is straightforward to see that multiplications by an element ξ satisfying $\xi^{q^\theta} + \xi = 0$ are skew-symmetric with respect to DF :

$$DF(a, \xi.x) + DF(\xi.a, x) = \xi^{q^\theta} a^{q^\theta} x + \xi a x^{q^\theta} + a^{q^\theta} \xi x + a \xi^{q^\theta} x^{q^\theta} = 0$$

which concludes the proof. $\qquad\qquad\qquad\qquad\qquad\qquad\qquad\qquad\qquad\qquad\qquad\quad$ $\square$

We denote by $\mathcal{K}_\theta$ the set of the elements ξ such that $\xi^{q^\theta} + \xi = 0$. By the linearity of q-powerings, this is a linear space. The non-zero elements of $\mathcal{K}_\theta$ are the $(q^\theta - 1)$-th roots of the unity and the number of these elements is $\gcd(q^\theta - 1, q^n - 1) = q^d - 1$ where d is the gcd of θ and n. Consequently, $\mathcal{K}_\theta$ is a linear space of dimension d.

For any element ξ in $\mathcal{K}_\theta$, we denote by M_ξ multiplication by ξ. As stated by the theorem, the maps M_ξ are the skew-symmetric applications with respect to the differential of the C^* monomial. They form a linear space isomorphic to $\mathcal{K}_\theta$. When $d = 1$, $\mathcal{K}_\theta$ is generated by 1, and all the maps M_ξ are colinear to the identity. This case is trivial since scalar multiples of the identity are skew-symmetric with respect to any bilinear product. Accordingly, non-trivial maps M_ξ only exist when $d > 1$.

4   Recovering the Skew-Symmetric Maps from a C^{*-} Public Key

Let P be a C^* public key and let P_Π be the C^{*-} public key obtained from P by deleting the last r coordinates. Since P is a composition $T \circ F \circ U$ where F is the internal C^* monomial and U, T are secret changes of coordinates, P_Π is the composition $T_\Pi \circ F \circ U$ where T_Π is obtained from T by removing the last r rows. The differential of P_Π is

$$DP_\Pi(a, x) = T_\Pi \left(DF(U(a), U(x)) \right)$$

Since $DF(U(a), U(x))$ is isomorphic by U to $DF(a, x)$, the skew-symmetric maps with respect to $DF(U(a), U(x))$ are the maps denoted N_ξ defined by

$$N_\xi = U^{-1} \circ M_\xi \circ U$$

By the linearity of T_Π, all the maps N_ξ are also skew-symmetric with respect to the truncated DP_Π:

$$DP_\Pi(a, N_\xi(x)) + DP_\Pi(N_\xi(a), x) = 0$$

We argue that they are likely to be the only ones, even when the number r of deleted coordinates is very close to n.

For any pair (a, x), the equation

$$DP_\Pi(a, L(x)) + DP_\Pi(L(a), x) = 0 \tag{2}$$

gives us $n - r$ linear equations in the n^2 coefficients of the unknown L. Since Equation (2) is bilinear and symmetric in (a, x) and trivial when $a = x$, taking

n^2 linearly independent choices for a and x, we construct a system of $(n-r)n(n-1)/2$ linear equations in the n^2 coefficients of L. The kernel of these equations must contain the d-dimensional space formed by the maps N_ξ. Assuming that all the generated linear equations are otherwise independent, the kernel does not contain other solutions up to r satisfying

$$(n-r)\frac{n(n-1)}{2} \geq n^2 - d$$

According to this heuristic, the maps N_ξ are likely to be the only solutions of our greatly overdefined system of linear equations provided that $r \leq r_{max}$ where

$$r_{max} = n - \left\lceil 2\frac{n^2 - d}{n(n-1)} \right\rceil = n - 3$$

which is very close to n. Consequently, we expect to find the same linear subspace of solutions even if we delete from the C^* public key almost all the quadratic polynomials, in order to generate the C^{*-} public key.

Though this analysis is rather naive, it provides a good estimate of the actual value of r_{max} as observed from some computer experiments. In the table below, we report on the actual value of r_{max} found for several parameters, to be compared with the heuristic value $n - 3$.

n	36	36	38	39	39	40	42	42	44
θ	8	12	10	13	9	8	12	14	12
d	4	12	2	13	3	8	6	14	4
r_{max}	33	32	35	35	36	37	39	38	41

The parameters chosen for these experiments are very close to the recommended parameters $n = 37$ and $\theta = 11$ for SFLASH, with the same value of $q = 2^7$. Note that in practice r would be chosen to be much smaller than n – about $\frac{n}{3}$ in SFLASH – and thus our approach could be easily applied even if not all the equations happen to be sufficiently independent.

Using Equation (2) with n^2 independent choices for a and x, we find all maps N_ξ by linear algebra. This takes a few seconds for practical parameters.

5   Recomposing a C^* Public Key Using Skew-Symmetric Maps

At this point, we assume that the linear space of skew-symmetric maps N_ξ has been computed. Non-trivial N_ξ are those which are not colinear to the identity. For any non-trivial N_ξ, we can now generate two C^{*-} public keys P_{Π} and $P_{\Pi} \circ N_\xi$. We next show that, provided r is at most $\frac{n}{2}$, completing P_{Π} with r arbitrary polynomials from $P_{\Pi} \circ N_\xi$ creates a valid C^* public key with high probability. Though higher values of r are not of practical interest, the technique can be generalized to $r \leq n(1 - \frac{1}{d})$ using $d - 1$ linearly independent non-trivial maps N_ξ, as shown in Appendix B.

Let us recall that the function P_Π is a composition $T_\Pi \circ F \circ U$, where U is a secret isomorphism, F is the C^* monomial and T_Π consists of $n - r$ linearly independent rows. Besides, N_ξ equals $U^{-1} \circ M_\xi \circ U$, where M_ξ denotes multiplication by ξ. The composition of P_Π and N_ξ is

$$P_\Pi \circ N_\xi = T_\Pi \circ F \circ M_\xi \circ U$$

Since F is multiplicative, multiplying the input by ξ results in multiplying the output by $F(\xi)$. Therefore

$$P_\Pi \circ N_\xi = T_\Pi \circ M_{F(\xi)} \circ F \circ U$$

where $M_{F(\xi)}$ denotes the multiplication by $F(\xi)$. Since N_ξ is non-trivial, ξ is not colinear to 1, and since the inverse of F is a power function, $F(\xi)$ is not colinear to 1 either. Hence, $M_{F(\xi)}$ is non-trivial and the matrices T_Π and $T_\Pi \circ M_{F(\xi)}$ are distinct.

The $n - r$ quadratic polynomials defining P_Π are linear combinations encoded by the rows of T_Π of the n quadratic polynomials defining $F \circ U$, whereas the $n - r$ quadratic polynomials defining $P_\Pi \circ N_\xi$ are different linear combinations encoded by the rows of $T_\Pi \circ M_{F(\xi)}$ of the same n quadratic polynomials defining $F \circ U$. Adding r polynomials of $P_\Pi \circ N_\xi$ to P_Π recomposes a valid C^* public key if and only if the corresponding rows of $T_\Pi \circ M_{F(\xi)}$ added to the rows of T_Π form a full rank system. Let us select, for instance, the r first rows of $T_\Pi \circ M_{F(\xi)}$. The rows of T_Π generate a subspace of dimension $n - r$ of $(\mathbb{F}_q)^n$. A random vector lies in a subspace of dimension $n - k$ of $(\mathbb{F}_q)^n$ with probability q^{-k}. Therefore, if we assume that the selected rows of $T_\Pi \circ M_{F(\xi)}$ are random vectors, the probability that they form with the rows of T_Π a full rank system is

$$\left(1 - \frac{1}{q^r}\right)\left(1 - \frac{1}{q^{r-1}}\right) \cdots \left(1 - \frac{1}{q}\right) \simeq 1 - \frac{1}{q}$$

With this probability, adding the r first polynomials of $P_\Pi \circ N_\xi$ to P_Π will recover a valid C^* public key (which is not necessarily identical to the C^* key we started with). This public key corresponds to a secret key T obtained by adding to T_Π the first r rows of $T_\Pi \circ M_{F(\xi)}$. We then apply Patarin's attack and recover n message-signature bilinear relations. If adding the r first polynomials fails to recover a C^* public key (which can be detected by the failure of Patarin's attack), we can retry with a different set of r polynomials of $P_\Pi \circ N_\xi$, or try a different value of ξ. The probability of success in at most t independent trials is expected to be $1 - q^{-t}$.

The table below provides some timings (in seconds) for an actual implementation of our attack on a single PC. We successfully recovered a C^* public key from a C^{*-} public key for all the listed values of the parameters n, θ which are close to those of SFLASH and with the same value of $q = 2^7$.

n	36	36	38	39	39	40	42	42	44
θ	8	12	10	13	9	8	12	14	12
d	4	12	2	13	3	8	6	14	4
r	11	11	11	12	12	12	13	13	13
$C^{*-} \mapsto C^*$	$57s$	$57s$	$94s$	$105s$	$90s$	$105s$	$141s$	$155s$	$155s$

6   Forging Signatures Using Patarin's Attacks

Our attack makes it possible to recover a C^* public key from a C^{*-} public key in a few seconds for practical parameters. Then, it remains to apply Patarin's attack to this public key and this is the "expensive" step of the attack.

As shown in [7], once Patarin's bilinear relations have been computed, we get for any message a subspace of dimension d containing at least one valid signature. Finding this signature requires trying all the q^d elements of this subspace. When d is large, additional linear equations can be generated to avoid exhaustive search using another attack also described in [7] which takes advantage of a small value of $\frac{n}{d}$.

The first attack, involving a precomputation in time $(\log_2 q)^2 n^6$ and then $q^d (\log_2 q)^2 n^3$ for each signature, is efficient when d is small. The second attack, involving a precomputation in time $(\log_2 q)^2 n^{3\frac{k+1}{2}}$ where $k = \frac{n}{d}$ and then $(\log_2 q)^2 n^3$ for each signature, is efficient when $\frac{n}{d}$ is small.

We summarize in the table below the complexities of Patarin's attacks for several choices of parameters which are close to those of SFLASH (and with the same value of $q = 2^7$). The star symbol at parameter d or $\frac{n}{d}$ specifies which of the two attacks devised by Patarin is considered.

n	36	36	38	39	39	40	42	42	44
θ	8	12	10	13	9	8	12	14	12
d	4*	12	2*	13	3*	8	6*	14	4*
n/d	9	3*	19	3*	13	5*	7	3*	11
r	11	11	11	12	12	12	13	13	13
Precomputation	2^{36}	2^{36}	2^{36}	2^{36}	2^{36}	2^{51}	2^{36}	2^{36}	2^{36}
Signature forgery	2^{49}	2^{21}	2^{35}	2^{21}	2^{36}	2^{21}	2^{57}	2^{21}	2^{49}

7   Conclusion

We have demonstrated a very simple but very powerful attack against a large class of C^{*-} schemes, namely those for which the number of variables n and the C^* parameter θ are not coprime. This attack transforms any such C^{*-} scheme into a full C^* scheme in a few seconds, even when the number of deleted coordinates is much larger than encountered for practical purposes. This is a major discovery since it was currently believed that even a weak scheme such as C^* can be made secure by simply deleting a sufficiently large number of coordinates from the public key. We have shown that this design fails for some choices of parameters. This shows that the security of C^{*-} schemes relies on mechanisms which are more subtle than anticipated, and does not necessarily improve when we increase the parameters. In particular, it is quite worrying to observe that no rationale was ever offered for the parameters recommended for SFLASH. It should be added that further unpublished work performed by the authors together with Adi Shamir has shown that the weakness of C^{*-} schemes was not only a matter of parameter choice, since they were able to mount a practical attack against the actual SFLASH schemes.

Acknowledgements. We are very grateful to Adi Shamir for interesting discussions and helpful remarks. Part of this work is supported by the Commission of the European Communities through the IST program under contract IST-2002-507932 ECRYPT.

References

1. J. Ding. A New Variant of the Matsumoto-Imai Cryptosystem through Perturbation. In *PKC '04*, LNCS 2947, pages 305–318. Springer-Verlag, 2004.
2. P. A. Fouque, L. Granboulan, and J. Stern. Differential Cryptanalysis for Multivariate Schemes. In *Eurocrypt '05*, LNCS 3494, pages 341–353. Springer-Verlag, 2005.
3. H. Gilbert and M. Minier. Cryptanalysis of SFLASH. In *Eurocrypt '02*, LNCS 2332, pages 288–298. Springer-Verlag, 2002.
4. J. Hoffstein, N. Howgrave-Graham, J. Pipher, J. H. Silverman, and W. Whyte. NTRUSIGN : Digital Signatures Using the NTRU Lattice. In *CT-RSA '03*, LNCS 2612, pages 122–140. Springer-Verlag, 2003.
5. T. Matsumoto and H. Imai. Public Quadratic Polynomial-tuples for Efficient Signature-Verification and Message-Encryption. In *Eurocrypt '88*, LNCS 330, pages 419–453. Springer-Verlag, 1988.
6. NESSIE. New European Schemes for Signatures Integrity and Encryption. Portfolio of recommended cryptographic primitives. http://www.nessie.eu.org/index.html
7. J. Patarin. Cryptanalysis of the Matsumoto and Imai Public Key Scheme of Eurocrypt'88. In *Crypto '95*, LNCS 963, pages 248–261. Springer-Verlag, 1995.
8. J. Patarin, N. Courtois, and L. Goubin. FLASH, a Fast Multivariate Signature Algorithm. In *CT-RSA '01*, LNCS 2020, pages 297–307. Springer-Verlag, 2001.
9. J. Patarin, L. Goubin, and N. Courtois. C^*_{-+} and HM : Variations Around Two Sechemes of T. Matsumoto and H. Imai. In *Asiacrypt '98*, LNCS 1514, pages 35–49. Springer-Verlag, 1998.
10. A. Shamir. Efficient Signature Scheme Based on Birational Permutations. In *Crypto '93*, LNCS 773, pages 1–12. Springer-Verlag, 1993.
11. B. Y. Yang and J. M. Chen. Building Secure Tame-like Multivariate Public-Key Cryptosystems: The New TTS. In *ACISP '05*, LNCS 3574, pages 518–531. Springer-Verlag, 2005.

A   Constructing a Bijective C^* Monomial

The internal C^* monomial x^{1+q^θ} is bijective in the field $\mathbb{F}_{q^n}$ if and only if $q^\theta + 1$ and $q^n - 1$ are coprime.

When q is odd, both $q^\theta + 1$ and $q^n - 1$ are even, and their gcd is a multiple of 2. Therefore, q odd never yields a bijective C^* monomial.

When q is even, then $q^\theta - 1$ and $q^\theta + 1$ are coprime and therefore

$$\gcd(q^{2\theta} - 1, q^n - 1) = \gcd(q^\theta - 1, q^n - 1). \gcd(q^\theta + 1, q^n - 1)$$

We denote by A, B and C the above gcds. We determine A and B and then deduce C. Denoting by d the gcd of θ and n, B equals $q^d - 1$. On the other hand, A equals $q^{\gcd(2\theta,n)} - 1$. We have

$$\gcd(2\theta, n) = d. \gcd(2\frac{\theta}{d}, \frac{n}{d})$$

and since $\frac{\theta}{d}$ and $\frac{n}{d}$ are coprime, the right-hand gcd is 2 when $\frac{n}{d}$ is even and 1 otherwise. Hence, A equals $q^{2d} - 1$ when $\frac{n}{d}$ is even, and $q^d - 1$ when $\frac{n}{d}$ is odd. Finally, C equals $q^d + 1$ when $\frac{n}{d}$ is even, and 1 when $\frac{n}{d}$ is odd.

The choices of θ and n yielding a bijective C^* monomial are therefore those for which $\frac{n}{d}$ is odd.

B   Recovering a Full C^* When r Is over $\frac{n}{2}$

In Section 5, we have shown how to recover a C^{*-} public key into a full C^* public key, using one single non-trivial skew-symmetric map N_ξ, when $r \leq \frac{n}{2}$. In this appendix, we show that this technique can be generalized up to

$$r = \min \left\{ r_{max} ; n \left(1 - \frac{1}{d}\right) \right\}$$

Let us recall that r_{max} is the maximal value of r allowing to recover the d-dimensional space of skew-symmetric maps. This value can be found experimentally and is given in Section 4 for some parameters. For r smaller than r_{max}, let $N_\xi^1, \ldots, N_\xi^{d-1}$ form with the identity a basis of the space of skew-symmetric maps. Aside from P_{II}, we get $d - 1$ independent C^{*-} public keys $P_{II} \circ N_\xi^1, \ldots, P_{II} \circ N_\xi^{d-1}$. We use coordinates of these additional C^{*-} public key to complete P_{II} into a full C^* public key. The overall number of coordinates available is $d(n - r)$, so that there is no hope to recover a full C^* if $r > n(1 - \frac{1}{d})$. When all coordinates are linearly independent, we can recover a full C^* up to $r = n(1 - \frac{1}{d})$. This has never failed to work in practice. The table below provides timings for some parameters and the largest value of r allowing the attack. The star symbol at parameter r indicates that the value considered corresponds to r_{max}.

n	36	36	38	39	39	40	42	42	44
θ	8	12	10	13	9	8	12	14	12
d	4	12	2	13	3	8	6	14	4
$r = \min\{r_{max}, n(1 - \frac{1}{d})\}$	27	32*	19	35*	26	35	35	38*	33
$C^{*-} \mapsto C^*$	65s	51s	112s	79s	107s	95s	134s	117s	202s

Differential Cryptanalysis of the Stream Ciphers Py, Py6 and Pypy[*]

Hongjun Wu and Bart Preneel

Katholieke Universiteit Leuven, ESAT/SCD-COSIC
Kasteelpark Arenberg 10, B-3001 Leuven-Heverlee, Belgium
{wu.hongjun,bart.preneel}@esat.kuleuven.be

Abstract. Py and Pypy are efficient array-based stream ciphers designed by Biham and Seberry. Both were submitted to the eSTREAM competition. This paper shows that Py and Pypy are practically insecure. If one key is used with about 2^{16} IVs with special differences, with high probability two identical keystreams will appear. This can be exploited in a key recovery attack. For example, for a 16-byte key and a 16-byte IV, 2^{23} chosen IVs can reduce the effective key size to 3 bytes. For a 32-byte key and a 32-byte IV, the effective key size is reduced to 3 bytes with 2^{24} chosen IVs. Py6, a variant of Py, is more vulnerable to these attacks.

Keywords: Differential Cryptanalysis, Stream Cipher, Py, Py6, Pypy.

1 Introduction

RC4 has inspired the design of a number of fast stream ciphers, such as ISAAC [8], Py [2], Pypy [3] and MV3 [10]. RC4 was designed by Rivest in 1987. Being the most widely used software stream cipher, RC4 is extremely simple and efficient. At the time of the invention of RC4, its array based design was completely different from the previous stream ciphers mainly based on linear feedback shift registers.

There are two main motives to improve RC4. One motive is that RC4 is byte oriented, so we need to design stream ciphers that can run more efficiently on today's 32-bit microprocessors. Another motive is to strengthen RC4 against various attacks [7,11,16,5,6,12,15,17,13,14]. Two of these attacks affect the security of RC4 in practice: the broadcast attack which exploits the weakness that the first few keystream bytes are heavily biased [12], and the key recovery attack using related IVs [6] which results in the practical attack on RC4 in WEP [13]. These two serious weaknesses are caused by the imperfection in the initialization of RC4.

Recently Biham and Seberry proposed the stream cipher Py [2] which is related to the design of RC4. Py is one of the fastest stream ciphers on 32-bit

---

[*] This work was supported in part by the Concerted Research Action (GOA) Ambiorics 2005/11 of the Flemish Government and in part by the European Commission through the IST Programme under Contract IST-2002-507932 ECRYPT.

M. Naor (Ed.): EUROCRYPT 2007, LNCS 4515, pp. 276–290, 2007.

processors (about 2.5 times faster than RC4). A distinguishing attack against Py was found by Paul, Preneel and Sekar [18]. In that attack, the keystream can be distinguished from random with about 2^{88} bytes. Later, the attack was improved by Crowley [4], and the data required in the attack is reduced to 2^{72}. In order to resist the distinguishing attack on Py, the designers of Py decided to discard half of the outputs, i.e., the first output of the two outputs at each step is discarded. The new version is called Pypy [3]. Py and Pypy are selected as focus ciphers in the Phase 2 of the ECRYPT eSTREAM project.

The initializations of Py and Pypy are identical. In this paper, we show that there are serious flaws in the initialization of Py and Pypy, thus these two ciphers are vulnerable to differential cryptanalysis [1]. Two keystreams can be identical if a key is used with about 2^{16} IVs with special differences. It is a practical threat since the set of IVs required in the attack may appear with high probability in applications. Then we show that part of the key of Py and Pypy can be recovered with chosen IVs. For a 16-byte key and a 16-byte IV, 2^{23} chosen IVs can reduce the effective key size to 3 bytes.

Py6 [2] is a variant of Py with reduced internal state size. We show that Py6 is more vulnerable to the attacks against Py and Pypy.

This paper is organized as follows. In Sect. 2, we illustrate the Key and IV setups of Py and Pypy. Section 3 describes the attack of generating identical keystreams. The key recovery attack is given in Sect. 4. In Sect. 5, we outline the attacks against Py6. Section 6 concludes this paper.

2   The Specifications of Py and Pypy

Py and Pypy are two synchronous stream ciphers supporting key and IV sizes up to 256 bytes and 64 bytes, respectively. The initializations of Py and Pypy are identical. The initialization consists of two stages: key setup and IV setup.

In the following descriptions, P is an array with 256 8-bit elements. Y is an array with 260 32-bit elements, s is a 32-bit integer. $YMININD = -3$, $YMAXIND = 256$. The table 'internal_permutation' is a constant permutation table with 256 elements. '$\wedge$' and '$\&$' in the pseudo codes denote binary XOR and AND operations, respectively. 'u8' and 'u32' mean 'unsigned 8-bit integer' and 'unsigned 32-bit integer', respectively. 'ROTL32(a,n)' means that the 32-bit a is left rotated over n bits.

2.1   The Key Setup

The key setups of Py and Pypy are identical. In the key setup, the key is used to initialize the array Y. The description is given below.

```
keysizeb=size of key in bytes;
ivsizeb=size of IV in bytes;
YMININD = -3; YMAXIND = 256;
s = internal_permutation[keysizeb-1];
s = (s<<8) | internal_permutation[(s ^ (ivsizeb-1))&0xFF];
```

```
s = (s<<8) | internal_permutation[(s ^ key[0])&0xFF];
s = (s<<8) | internal_permutation[(s ^ key[keysizeb-1])&0xFF];
for(j=0; j<keysizeb; j++)
{
    s = s + key[j];
    s0 = internal_permutation[s&0xFF];
    s = ROTL32(s, 8) ^ (u32)s0;
}
/* Again */
for(j=0; j<keysizeb; j++)
{
    s = s + key[j];
    s0 = internal_permutation[s&0xFF];
    s ^= ROTL32(s, 8) + (u32)s0;
}
/* Algorithm C is the following 'for' loop */
for(i=YMININD, j=0; i<=YMAXIND; i++)
{
    s = s + key[j];
    s0 = internal_permutation[s&0xFF];
    Y(i) = s = ROTL32(s, 8) ^ (u32)s0;
    j = (j+1) mod keysizeb;
}
```

2.2   The IV Setup

The IV setups of Py and Pypy are identical. In the IV setup, the IV is used to affect every bit of the internal state. EIV is a temporary byte array with the same size as the IV. The IV setup is given below.

```
/* Create an initial permutation */
u8 v= iv[0] ^ ((Y(0)>>16)&0xFF);
u8 d=(iv[1 mod ivsizeb] ^ ((Y(1)>>16)&0xFF))|1;
for(i=0; i<256; i++)
{
    P(i)=internal_permutation[v];
    v+=d;
}
/* Now P is a permutation */
/* Initial s */
s = ((u32)v<<24)^((u32)d<<16)^((u32)P(254)<<8)^((u32)P(255));
s ^= Y(YMININD)+Y(YMAXIND);

/* Algorithm A is the following 'for' loop */
for(i=0; i<ivsizeb; i++)
```

```
{
    s = s + iv[i] + Y(YMININD+i);
    u8 s0 = P(s&0xFF);
    EIV(i) = s0;
    s = ROTL32(s, 8) ^ (u32)s0;
}
/* Again, but with the last words of Y, and update EIV */
/* Algorithm B is the following 'for' loop */
for(i=0; i<ivsizeb; i++)
{
    s = s + iv[i] + Y(YMAXIND-i);
    u8 s0 = P(s&0xFF);
    EIV(i) += s0;
    s = ROTL32(s, 8) ^ (u32)s0;
}

/*updating the rolling array and s*/
for(i=0; i<260; i++)
{
    u32 x0 = EIV(0) = EIV(0)^(s&0xFF);
    rotate(EIV);
    swap(P(0),P(x0));
    rotate(P);
    Y(YMININD)=s=(s^Y(YMININD))+Y(x0);
    rotate(Y);
}
s=s+Y(26)+Y(153)+Y(208);
if(s==0)
    s=(keysizeb*8)+((ivsizeb*8)<<16)+0x87654321;
```

2.3   The Keystream Generation

After the key and IV setup, the keystream is generated. One step of the keystream generation of Py is given below. Note that the first output at each step is discarded in Pypy.

```
/* swap and rotate P */
swap(P(0), P(Y(185)&0xFF));
rotate(P);

/* Update s */
s+=Y(P(72)) - Y(P(239));
s=ROTL32(s, ((P(116) + 18)&31));

/* Output 8 bytes (least significant byte first) */
output ((ROTL32(s, 25) ^ Y(256)) + Y(P(26)));
```

```
output (( s ^ Y(-1)) + Y(P(208)));
/* Update and rotate Y */
Y(-3)=(ROTL32(s, 14) ^ Y(-3)) + Y(P(153));
rotate(Y);
```

3   Identical Keystreams

We notice that the IV appears only in the IV setup algorithm described in Sect. 2.2. At the beginning of the IV setup, only 15 bits of the IV ($iv[0]$ and $iv[1]$) are applied to initialize the array P and s (the least significant bit of $iv[1]$ is not used). For an IV pair, if those 15 bits are identical, then the resulting P are the same. Then we notice that the IV is applied to update s and EIV as follows.

```
for(i=0; i<ivsizeb; i++)
{
    s = s + iv[i] + Y(YMININD+i);
    u8 s0 = P(s&0xFF);
    EIV(i) = s0;
    s = ROTL32(s, 8) ^ (u32)s0;
}
for(i=0; i<ivsizeb; i++)
{
    s = s + iv[i] + Y(YMAXIND-i);
    u8 s0 = P(s&0xFF);
    EIV(i) += s0;
    s = ROTL32(s, 8) ^ (u32)s0;
}
```

We call the first 'for' loop Algorithm A, and the second 'for' loop Algorithm B. In the following, we give two types of IV pairs that result in identical keystreams.

3.1   IVs Differing in Two Bytes

We illustrate the attack with an example. Suppose that two IVs, iv_1 and iv_2, differing in only two consecutive bytes with $iv_1[i] \oplus iv_2[i] = 1$, the least significant bit of $iv_1[i]$ is 1, $iv_1[i+1] \neq iv_2[i+1]$ ($1 \leq i \leq ivsizeb - 1$), and $iv_1[j] = iv_2[j]$ for $0 \leq j < i$ and $i + 1 < j \leq ivsizeb - 1$. We trace how the difference in IV affects s and EIV in Algorithm A. At the ith step in Algorithm A,

```
s = s + iv[i] + Y(YMININD+i);
u8 s0 = P(s&0xFF);
EIV(i) = s0;
s = ROTL32(s, 8) ^ (u32)s0;
```

At the end of the ith step, $EIV_1[i] \neq EIV_2[i]$. Let $\beta_1 = EIV_1[i]$, and $\beta_2 = EIV_2[i]$. We obtain that $s_1 - s_2 = 256 + \delta_1$, where $\delta_1 = (\beta_1 \oplus x) - (\beta_2 \oplus x)$, and $x = ROTL32(s, 8)$. Then we look at the next step.

```
s = s + iv[i+1] + Y(YMININD+i+1);
u8 s0 = P(s&0xFF);
EIV(i+1) = s0;
s = ROTL32(s, 8) ^ (u32)s0;
```

Because $iv_1[i+1] \neq iv_2[i+1]$, if $iv_2[i+1] - iv_1[i+1] = \delta_1$, then s_1 and s_2 become identical with high probability. Let $s_1 = s_2$ with probability p_1. Based on the simulation, we obtain that $p_1 = 2^{-10.6}$. If $s_1 = s_2$, then $EIV_1[i+1] = EIV_2[i+1]$, and in the following steps $i+2, i+3, \cdots, i + ivsizeb - 1$ in Algorithm A, s_1 and s_2 remain the same, and $EIV_1[j] = EIV_2[j]$ for $j \neq i$.

After Algorithm A, the $iv[i]$ and $iv[i+1]$ are used again to update s and EIV in Algorithm B. At the ith step in Algorithm B,

```
s = s + iv[i] + Y(YMAXIND-i);
u8 s0 = P(s&0xFF);
EIV(i) += s0;
s = ROTL32(s, 8) ^ (u32)s0;
```

At the end of this step, $EIV_1[i] = EIV_2[i]$ with probability $\frac{1}{255}$. Let $\gamma_1 = s0_1$, and $\gamma_2 = s0_2$. If $EIV_1[i] = EIV_2[i]$, we know that $\gamma_2 - \gamma_1 = \beta_1 - \beta_2$. At the end of this step, $s_1 - s_2 = 256 + \delta_2$, where $\delta_2 = (\gamma_1 \oplus y) - (\gamma_2 \oplus y)$, and y is ROTL32(s,8). Note that δ_1 and δ_2 are correlated since $\gamma_2 - \gamma_1 = \beta_1 - \beta_2$. Then we look at the next step.

```
s = s + iv[i+1] + Y(YMAXIND-i-1);
u8 s0 = P(s&0xFF);
EIV(i+1) += s0;
s = ROTL32(s, 8) ^ (u32)s0;
```

At the end of this step, if $iv_2[i + 1] - iv_1[i + 1] = \delta_2$, then s_1 and s_2 become identical with high probability. Note that $iv_2[i + 1] - iv_1[i + 1] = \delta_1$, and δ_1 and δ_2 are correlated, so $iv_2[i + 1] - iv_1[i + 1] = \delta_2$ with probability larger than 2^{-8}. Let $s_1 = s_2$ with probability p_1'. Based on a simulation, we obtain that $p_1' = 2^{-5.6}$. Once the two s values are identical, $EIV_1[i + 1] = EIV_2[i + 1]$, and in the following steps $i + 2, i + 3, \cdots, i + ivsize - 1$ in Algorithm B, s_1 and s_2 remain the same, and $EIV_1[i + 2] = EIV_2[i + 2]$, $EIV_1[i + 3] = EIV_2[i + 3]$, $\cdots$, $EIV_1[i + ivsize - 1] = EIV_2[i + ivsize - 1]$.

Thus after introducing the IV to update s and EIV, $s_1 = s_2$ and $EIV_1 = EIV_2$ with probability $p_1 \times \frac{1}{255} \times p_1' \approx 2^{-24.2}$.

Note that once an IV has been introduced in Algorithm A and B, the IV is not used in the rest of the IV setup. Thus once $s_1 = s_2$ and $EIV_1 = EIV_2$ at the end of Algorithm B, we know that those two keystreams will be the same.

Experiment 1. We use 2^{14} random 128-bit keys in the attack. For each key, we randomly generate 2^{16} pairs of 128-bit IV that differ in only two bytes: $iv_1[6] \oplus iv_2[6] = 1$, $iv_1[7] \neq iv_2[7]$. We found that 111 pairs of those 2^{30} keystream pairs are identical. For example, for the key (08 da f2 35 a3 d5 94 e2 85 cc 68 ba 7e 10 8a b4), and the IV pair (6e e7 09 b1 35 85 2f 07 1a fe 3f 50 a8 84 30 11) and (6e e7 09 b1 35 85 2e 80 1a fe 3f 50 a8 84 30 11), the two keystreams are identical, and the first 16 keystream bytes of Pypy are (6f eb ca 18 54 3f 59 96 b6 17 8a 54 6e bd 45 1f).

From the experiment, we deduce that for an IV pair with the required difference, the two keystreams are identical with probability about $\frac{111}{2^{30}} = 2^{-23.2}$, about twice the theoretical value.

The IV difference at two bytes. In the above analysis, the difference is chosen as $iv_1[i] \oplus iv_2[i] = 1$, $iv_1[i+1] \neq iv_2[i+1]$ $(i \geq 1)$. We can generalize this type of IV difference so that $iv_1[i]$ and $iv_2[i]$ can take other differences. **As long as $(iv_1[i] - iv_2[i]) \bmod 256 = 1$ or 255, $iv_1[i+1] \neq iv_2[i+1]$ $(i \geq 2)$, there is a non-zero probability that the two keystreams can be identical.**

For example, if $iv_1[i] \oplus iv_2[i] = 3$, the two least significant bits of $iv_1[i]$ are 01 or 10, and $iv_1[i+1] \neq iv_2[i+1]$ $(i \geq 2)$, then two identical keystreams appear with probability $2^{-23.2}$. On average, if $iv_1[i] - iv_2[i] = 1$, and $iv_1[i+1] \neq iv_2[i+1]$ $(i \geq 2)$, then two identical keystreams appear with probability $2^{-26.4}$.

3.2   IVs Differing in Three Bytes

In the above attack, we deal with the ith and $(i+1)$th bytes of the IV, and use the difference at $iv[i+1]$ to eliminate the difference introduced by $iv[i]$ in s. In the following, we introduce another type of difference to deal with the situation when the difference at $iv[i+1]$ cannot eliminate the difference introduced by $iv[i]$ in s. The solution is to introduce a difference in $iv[i+4]$.

We illustrate the attack with an example. Suppose that two IVs, iv_1 and iv_2, differ in only three bytes $iv_1[i] \oplus iv_2[i] = 0x80$, the most significant bit of $iv_1[i]$ is 1, $iv_1[i+1] \neq iv_2[i+1]$, $iv_1[i+4] \oplus iv_2[i+4] = 0x80$, and the most significant bit of $iv_1[i+4]$ is 0, where $i \geq 2$. We trace how the difference affects s and EIV. At the ith step in Algorithm A,

```
s = s + iv[i] + Y(YMININD+i);
u8 s0 = P(s&0xFF);
EIV(i) = s0;
s = ROTL32(s, 8) ^ (u32)s0;
```

At the end of this step, $EIV_1[i] \neq EIV_2[i]$, and $s_1 - s_2 = 0x8000 + \delta_1$, where δ_1 is the difference of two different 8-bit numbers. Then we look at the next step.

```
s = s + iv[i+1] + Y(YMININD+i+1);
u8 s0 = P(s&0xFF);
EIV(i+1) = s0;
s = ROTL32(s, 8) ^ (u32)s0;
```

Because $iv_1[i+1] \neq iv_2[i+1]$, $s_1 - s_2 = 0x8000$ with probability $p_2 = 2^{-8}$. If $s_1 - s_2 = 0x8000$, then $EIV_1[i+1] \oplus EIV_2[i+1] = 0$.

Since $v_1[i+2] = v_2[i+2]$, at the end of the $(i+2)$th step of Algorithm A, $EIV_1[i+2] = EIV_2[i+2]$, and $s_1 - s_2 = 0x800000$ with probability close to 1.

Since $v_1[i+3] = v_2[i+3]$, at the end of the $(i+3)$th step of Algorithm A, $EIV_1[i+3] = EIV_2[i+3]$, and $s_1 - s_2 = 0x80000000$ with probability close to 1. Now consider the $(i+4)$th step.

```
s = s + iv[i+4] + Y(YMININD+i+4);
u8 s0 = P(s&0xFF);
EIV(i+4) = s0;
s = ROTL32(s, 8) ^ (u32)s0;
```

At the end of this step, the probability that $EIV_1[i+4] = EIV_2[i+4]$, and $s_1 = s_2$ is 1. So for the above 5 steps, $s_1 = s_2$ with probability p_2. Once $s_1 = s_2$, in the following steps $i+5, i+6, \cdots, i+ivsize-1$ in Algorithm A, the s_1 and s_2 remain the same, and $EIV_1[i+5] = EIV_2[i+5]$, $EIV_1[i+6] = EIV_2[i+6]$, $\cdots$, $EIV_1[i+ivsize-1] = EIV_2[i+ivsize-1]$.

Then $iv[i]$ and $iv[i+1]$ are used again to update s and EIV. With a similar analysis, we can show that at the end of the updating, $EIV_1 = EIV_2$, $s_1 = s_2$ with probability about $(p_2)^2 \times \frac{1}{255} \approx 2^{-24}$. (As shown in the experiment in the next subsection, this probability is about $2^{-22.9}$.)

The IV difference at three bytes. In the above analysis, the difference is chosen at only three bytes, $iv_1[i] \oplus iv_2[i] = 0x80$, the most significant bit of $iv_1[i]$ is 1, $iv_1[i+1] \neq iv_2[i+1]$, $iv_1[i+4] \oplus iv_2[i+4] = 0x80$, and the most significant bit of $iv_1[i+4]$ is 0 ($i \geq 2$). For this type of IV difference, we can generalize it so that $iv_1[i]$ and $iv_2[i]$ can choose other differences instead of 0x80. In fact, **once we set the difference as** $iv_1[i] - iv_2[i] = iv_2[i+4] - iv_1[i+4]$, $iv_1[i+1] \neq iv_2[i+1]$ ($i \geq 2$), **then the two keystreams are identical with probability close to** 2^{-23}. For two IVs different only at three bytes, if $iv_1[1] \oplus iv_2[1] = 1$, $iv_1[2] \neq iv_2[2]$, and $iv_1[1] - iv_2[1] = iv_2[5] - iv_1[5]$, then this IV pair is also weak.

3.3   Improving the Attack

The number of IVs required to generate identical keystreams can be reduced in practice. The idea is to generate more IV pairs from a group of IVs. For the IV pair with a two-byte difference $iv_1[i] \oplus iv_2[i] = 1$, $iv_1[i+1] \neq iv_2[i+1]$, if $iv[2]$ takes all the 256 values, then we can obtain $255 \times 255 = 2^{15.99}$ IV pairs with the required differences from 512 IVs. Thus with 512 chosen IVs, the probability that there is one pair of identical keystreams becomes $2^{15.99} \times 2^{-23.2} \approx 2^{-7.2}$. With about $2^{7.2} \times 512 = 2^{16.2}$ IVs, identical keystreams can be obtained.

Experiment 2. We use 2^{16} random 128-bit keys in the improved attack. For each key, we generate 512 128-bit IVs with the values of the least significant bit of $iv[4]$ and the eight bits of $iv[5]$ choosing all the 512 possible values, while all the other 119 IV bits remain unchanged for each key (but those 119 IV bits are random from key to key). Then we obtain $255 \times 255 = 2^{15.99}$ IV pairs with the required difference. Among these $2^{16} \times 2^{15.99} \approx 2^{32}$ IV pairs, 447 IV pairs result in identical keystreams.

The above experiment shows that with $2^{16} \times 512 = 2^{25}$ selected IVs, 447 IVs result in identical keystreams. It shows that two identical keystreams appear for every $\frac{2^{25}}{447} = 2^{16.2}$ IVs.

For the IV pair with three-byte difference, a similar improvement can also be applied.

Experiment 3. We use 2^{16} random 128-bit keys in the improved attack. For each key, we generate 512 128-bit IVs with the values of the most significant bit of $iv[4]$ and the eight bits of $iv[5]$ choosing all the 512 possible values, and the most significant bit of $iv[8]$ is different from the most significant bit of $iv[4]$, while all the other 118 IV bits remain unchanged for each key (but those 118 IV bits are randomly generated for each key). Then we obtain $255 \times 255 = 2^{15.99}$ IV pairs with the required difference. Among these $2^{16} \times 2^{15.99} \approx 2^{32}$ IV pairs, 570 IV pairs result in identical keystreams.

The above experiment shows that with $2^{16} \times 512 = 2^{25}$ selected IVs, 570 IVs result in identical keystreams. It means that two identical keystreams appear for every $\frac{2^{25}}{570} = 2^{15.9}$ IVs.

Remarks. The attacks show that the Py and Pypy are practically insecure. In the application, if the IVs are generated from a counter, or if the IV is short (such as 3 or 4 bytes), then the special IVs (with the differences as illustrated above) appear with high probability, and identical keystreams can be obtained with high probability.

4   Key Recovery Attack on Py and Pypy

In this section, we develop a key recovery attack against Py and Pypy by exploiting the collision in the internal state. The key recovery attack consists of two stages: recovering part of the array Y in the IV setup and recovering the key information from Y in the key setup.

4.1   Recovering Part of the Array Y

We use the following IV differences to illustrate the attack (the other IV differences can also be used). Let two IVs iv_1 and iv_2 differ only in two bytes, $iv_1[i] \oplus iv_2[i] = 1$, $iv_1[i+1] \neq iv_2[i+1]$ ($i \geq 1$), and the least significant bit of $iv_1[i]$ be 1. This type of IV pair results in identical keystreams with probability $2^{-23.2}$.

We first recover part of Y from Algorithm A in the IV setup (more information of Y will be recovered from Algorithm B).

Note that the permutation P in Algorithm A is unknown. According to the IV setup algorithm, there is 15 bits of secret information in P, i.e., there are at most 2^{15} possible permutations. During the recovery of Y, we assume that P is known (the effect of the 15-bit secret information in P will be analyzed in Sect. 4.2). For iv_m, denote the s at the end of the jth step of Algorithm A as s_j^m, and denote the least and most significant bytes of s_j^m as $s_{j,0}^m$ and $s_{j,3}^m$, respectively. Denote the least and most significant bytes of $Y(j)$ with $Y_{j,0}$ and $Y_{j,3}$, respectively. Note that in Algorithm A, Y remains the same for all the IVs. Denote ξ as a binary random variable with value 0 with probability 0.5. Denote with $B(x)$ a function that gives the least significant byte of x. If the keystreams for iv_1 and iv_2 identical, then from the analysis given in Sect. 3.1, we know that $s_{i+1}^1 = s_{i+1}^2$, i.e.,

$$s_i^1 + iv_1[i+1] = s_i^2 + iv_2[i+1]. \tag{1}$$

From Algorithm A, we know

$$s_i = \text{ROTL32}(s_{i-1} + iv[i] + Y(-3+i), 8)$$
$$\oplus P(B(s_{i-1} + iv[i] + Y(-3+i))) \tag{2}$$

Thus we obtain

$$s_{i,0} = P(B(s_{i-1,0} + iv[i] + Y(-3+i))) \oplus B(s_{i-1,3} + Y(-3+i) + \xi_i), \tag{3}$$
$$(s_i^1 - s_{i,0}^1) - (s_i^2 - s_{i,0}^2) = (iv_1[i] - iv_2[i]) \ll 8 = 256, \tag{4}$$

where ξ_i is caused by the carry bits at the 24th least significant bit position when $iv[i]$ and $Y(-3+i)$ are introduced, and (4) holds with probability $1 - 2^{-15}$. From (1), (3) and (4), we obtain

$$(P(B(s_{i-1,0}^1 + iv_1[i] + Y_{-3+i,0})) \oplus B(s_{i-1,3}^1 + Y_{-3+i,3} + \xi_{i,1})) + 256 + iv_1[i+1]$$
$$= (P(B(s_{i-1,0}^2 + iv_2[i] + Y_{-3+i,0})) \oplus B(s_{i-1,3}^2 + Y_{-3+i,3} + \xi_{i,2}) + iv_2[i+1], \tag{5}$$

where $\xi_{i,1} = \xi_{i,2}$ with probability $1 - 2^{-15}$ since the $iv[i]$ has a negligible effect on the value of ξ_1 and ξ_2. In the following, we use ξ_i to represent $\xi_{i,1}$ and $\xi_{i,2}$.

Denote iv_θ as a fixed IV with the first i bytes being identical to all the IVs with differences only at $iv[i]$ and $iv[i+1]$. Thus $s_{i-1,0}^\theta = s_{i-1,0}^1 = s_{i-1,0}^2$, and $s_{i-1,3}^\theta = s_{i-1,3}^1 = s_{i-1,3}^2$. (5) becomes

$$(P(B(s_{i-1,0}^\theta + iv_1[i] + Y_{-3+i,0})) \oplus B(s_{i-1,3}^\theta + Y_{-3+i,3} + \xi_i)) + 256 + iv_1[i+1]$$
$$= (P(B(s_{i-1,0}^\theta + iv_2[i] + Y_{-3+i,0})) \oplus B(s_{i-1,3}^\theta + Y_{-3+i,3} + \xi_i)) + iv_2[i+1]. \tag{6}$$

Using another IV pair different at $iv[i]$ and $iv[i+1]$, and the first i bytes being the same as iv_θ, another equation (6) can be obtained if there is collision in their internal states. Suppose that several equations (6) are available. We consider that the value of ξ_i is independent of $iv[i]$ in the following attack since ξ_i is affected by $iv[i]$ with small probability 2^{-15}. We can recover the values of $B(s_{i-1,0}^\theta + Y_{-3+i,0})$

and $B(s_{i-1,3}^{\theta} + Y_{-3+i,3} + \xi_i)$. From the experiment, we find that if there are two equations (6), on average the correct values can be recovered together with 5.22 wrong values. If there are three, four, five, six, seven equations (6), in average the correct values can be recovered together with 1.29, 0.54, 0.25, 0.12, 0.06 wrong values, respectively. It shows that the values of $B(s_{i-1,0}^{\theta} + Y_{-3+i,0})$ and $B(s_{i-1,3}^{\theta} + Y_{-3+i,3} + \xi_i)$ can be determined with only a few equations (6).

After recovering several consecutive $B(s_{i-1,0}^{\theta}+Y_{-3+i,0})$ and $B(s_{i-1,3}^{\theta}+Y_{-3+i,3} +\xi_i)$ ($i \geq 1$), we proceed to recover part of the information of the array Y. From the values of $B(s_{i-1,0}^{\theta} + Y_{-3+i,0})$, $B(s_{i-1,3}^{\theta} + Y_{-3+i,3} + \xi_i)$ and (3), we determine the value of $s_{i,0}^{\theta}$. From the values of $B(s_{i,0}^{\theta} + Y_{-3+i+1,0})$ and $s_{i,0}^{\theta}$, we know the value of $Y_{-3+i+1,0}$.

Generating the equations (6). The above attack can only be successful if we can find several equations (6) with the same $s_{i-1,0}^{\theta}$ and $s_{i-1,3}^{\theta}$. In the following, we illustrate how to obtain these equations for $2 \leq i \leq ivsizeb - 3$. At the beginning of the attack, we set a fixed iv_θ. For all the IVs different at only $iv[i]$ and $iv[i+1]$, we require that their first i bytes are identical to that of iv_θ. Let the least significant bit of $iv[i]$ and the 8 bits of $iv[i+1]$ choose all the 512 values, and the other 119 bits remain unchanged, then we obtain a $255 \times 255 \approx 2^{16}$ desired IV pairs. We call these 512 IVs a desired IV group. According to Experiment 2, this type of IV pair results in identical keystreams with probability $2^{-23.2}$, we thus obtain $\frac{2^{-23.2}}{2^{16}} = 2^{-7.2}$ identical keystream pairs from one desired IV group. It means that we can obtain $2^{-7.2}$ equations (1) from one desired IV group. We modify the values of the 7 most significant bits of $iv_1[i]$ and $iv_2[i]$, and 3 bits of $iv_1[i+2]$ and $iv_2[i+2]$, then we obtain $2^7 \times 2^3 = 2^{10}$ desired IV groups. From these desired IV groups, we obtain $2^{10} \times 2^{-7.2} = 7$ equations (1). There are $2^7 \times 2^3 \times 2^9 = 2^{19}$ IVs being used in the attack. To find all the $s_{i,0}$ for $2 \leq i \leq ivsizeb - 3$, we need $(ivsizeb - 4) \times 2^{19}$ IVs in the attack.

We are able to recover $s_{i,0}^{\theta}$ for $2 \leq i \leq ivsizeb - 3$, which implies that we can recover the values of $Y_{-3+i,0}$ for $3 \leq i \leq ivsizeb - 3$. Then we proceed to recover more information of Y by considering Algorithm B. Applying an attack similar to the above attack and reusing the IVs, we can recover the values of $Y_{256-i,0}$ for $3 \leq i \leq ivsizeb - 3$.

Thus with $(ivsizeb - 4) \times 2^{19}$ IVs, we are able to recover $2 \times (ivsizeb - 6)$ bytes of Y: $Y_{-3+i,0}$ and $Y_{256-i,0}$ for $3 \leq i \leq ivsizeb - 3$.

4.2   Recovering the Key

In the above analysis, we recovered the values of $Y_{-3+i,0}$ and $Y_{256-i,0}$ for $3 \leq i \leq ivsizeb - 3$ by exploiting the difference elimination in s. Next, we will recover the 15-bit secret information in P by exploiting the difference elimination in EIV. Denote s_i^{θ} in Algorithm A and B as $s_i^{A,\theta}$ and $s_i^{B,\theta}$, respectively. Denote $EIV_1[i]$ at the end of Algorithm A and B as $EIV_1^A[i]$ and $EIV_1^B[i]$, respectively. For two IVs differing in only $iv[i]$ and $iv[i+1]$ and generating identical keystreams, $EIV_1^A[i]$, $EIV_2^A[i]$, $EIV_1^B[i]$ and $EIV_2^B[i]$ are computed as:

$$EIV_1^A[i] = P(B(s_{i-1,0}^{A,\theta} + iv_1[i] + Y_{-3+i,0})) \tag{7}$$

$$EIV_2^A[i] = P(B(s_{i-1,0}^{A,\theta} + iv_2[i] + Y_{-3+i,0})) \tag{8}$$

$$EIV_1^B[i] = EIV_1^A[i] + P(B(s_{i-1,0}^{B,\theta} + iv_1[i] + Y_{256-i,0})) \tag{9}$$

$$EIV_2^B[i] = EIV_2^A[i] + P(B(s_{i-1,0}^{B,\theta} + iv_2[i] + Y_{256-i,0})) \tag{10}$$

Since the two keystreams are identical, it is required that

$$EIV_1^B[i] = EIV_2^B[i] . \tag{11}$$

Note that the values of $B(s_{i-1,0}^{A,\theta} + Y_{-3+i,0})$ and $B(s_{i-1,0}^{B,\theta} + Y_{256-i,0})$ are determined when we recover part of Y from Algorithm A and Algorithm B, respectively. Eight bits of information on P is revealed from (7),(8),(9),(10) and (11). In Sect. 4.1, there are about 7 pairs of IVs resulting in identical keystreams for each value of i. Thus P can be recovered completely.

We proceed to recover the key information. We consider the last part of the key schedule:

```
for(i=YMININD, j=0; i<=YMAXIND; i++)
{
    s = s + key[j];
    s0 = internal_permutation[s&0xFF];
    Y(i) = s = ROTL32(s, 8) ^ (u32)s0;
    j = (j+1) mod keysizeb;
}
```

We call the above algorithm Algorithm C. From Algorithm C, we obtain the following relation:

$$B(Y_{-3+i,0} + key[i+1 \bmod keysizeb] + \xi_i')$$
$$\oplus P'(B(Y_{-3+i+3,0} + key[i+4 \bmod keysizeb])) = Y_{-3+i+4,0} , \tag{12}$$

where P' indicates the 'internal_permutation', ξ_i' indicates the carry bit noise introduced by $key[i+2]$ and $key[i+3]$; it is computed as $\xi_i' \approx (key[i+2] + Y_{-3+i+1,0}) \gg 8$. The value of the binary ξ_i' is 0 with probability about 0.5.

Once the values of $Y_{-3+i,0}$ ($3 \leq i \leq ivsizeb - 3$) are known, we find a relation (12) linking $key[i+1 \bmod keysizeb]$ and $key[i+4 \bmod keysizeb]$ for $3 \leq i \leq ivsizeb - 7$. Each relation leaks at least 7 bits of $key[i+1 \bmod keysizeb]$ and $key[i+4 \bmod keysizeb]$. The values of $Y_{256-i,0}$ ($3 \leq i \leq ivsizeb - 3$) are also known, thus we can find a relation (12) linking $key[i+1 \bmod keysizeb]$ and $key[i+4 \bmod keysizeb]$ for $262 - ivsizeb \leq i \leq 252$. Thus there are $2 \times (ivsizeb - 9)$ relations (12) linking the key bytes.

For the 16-byte key and 16-byte IV, 14 relations (12) can be obtained: 7 relations linking $key[i]$ and $key[i+3]$ for $4 \leq i \leq 10$, and another 7 relations (12) linking $key[i]$ and $key[i+3 \bmod 16]$ for $7 \leq i \leq 13$. There are 13 key bytes in these 14 relations (12). Note that the randomness of ξ_i' does not affect the overall attack (once we guess the values of $key[4]$, $key[5]$ and $key[6]$, then we obtain the other key bytes $key[j]$ ($7 \leq j \leq 15$), $key[0]$, and all the ξ'_j ($3 \leq j \leq 9$ and $247 \leq j \leq 249$). Thus these 14 relations are sufficient to recover the 13 key bytes. The effective key size is reduced to 3 bytes and these three bytes can be found easily with brute force search.

For the 32-byte key and 32-byte IV, 46 relations (12) can be obtained: 23 relations linking $key[i]$ and $key[i + 3]$ for $4 \leq i \leq 26$, and another 23 relations (12) linking $key[i]$ and $key[i + 3 \bmod 32]$ for $7 \leq i \leq 29$. There are 29 key bytes in these 46 relations. The effective key size is again reduced to 3 bytes.

5   The Security of Py6

Py6 is a variant of Py with reduced internal state size. The array P is a permutation with only 64 elements, and the array Y has 68 entries. Py6 was proposed to achieve fast initialization, but it is weaker than Py. Paul and Preneel has developed distinguishing attack against Py6 with data complexity $2^{68.6}$ [19]. In the following, we show that identical keystreams are genereated from Py6 with high probability. There is no detailed description of the key and IV setups of Py6. Thus we use the source code of Py6 submitted to eSTREAM as reference. In our experiment, the following IV differences are used: $iv_1[i] - iv_2[i] = 32$, $iv_1[i+1] \neq iv_2[i+1]$, $iv_1[i+1] \gg 6 = iv_2[i+1] \gg 6$, and $iv_2[i+5] - iv_1[i+5] = 8$ ($i \geq 2$). After testing 2^{30} pairs with the original Py6 source code, we found that identical keystreams appear with probability $2^{-11.45}$. This probability is much larger than the probability 2^{-23} for Py and Pypy. It shows that Py6 is much weaker than Py and Pypy.

6   Conclusion

In this paper, we developed practical differential attacks against Py, Py6 and Pypy: the identical keystreams appear with high probability, and the key information can be recovered when the IV size is more than 9 bytes. To resist the attacks given in this paper, we suggest that the IV setup be performed in an invertible way.

Several ciphers in the eSTREAM competition have been broken due to the flaws in their IV setups: DECIM [20], WG [21], LEX [21], Py, Pypy and VEST [9]. We should pay great attention to the design of the stream cipher IV setup.

Acknowledgements

The authors would like to thank the anonymous reviewers of Eurocrypt 2007 for their helpful comments.

References

1. E. Biham, A. Shamir, "Differential Cryptanalysis of DES-like Cryptosystems." *Advances in Cryptology – Crypto'90*, LNCS 537, A. J. Menezes and S. A. Vanstone (Eds.), pp. 2–21, Springer-Verlag, 1991.

2. E. Biham, J. Seberry, "Py (Roo): A Fast and Secure Stream Cipher Using Rolling Arrays." The ECRYPT eSTREAM project Phase 2 focus ciphers. Available at http://www.ecrypt.eu.org/stream/ciphers/py/py.ps .

3. E. Biham, J. Seberry, "Pypy (Roopy): Another Version of Py." The ECRYPT eSTREAM project Phase 2 focus ciphers. Available at http://www.ecrypt.eu.org/stream/p2ciphers/py/pypy_p2.ps

4. P. Crowley, "Improved Cryptanalysis of Py." Available at http://www.ecrypt.eu.org/stream/papersdir/2006/010.pdf .

5. S. R. Fluhrer, D. A. McGrew, "Statistical Analysis of the Alleged RC4 Keystream Generator," *Fast Software Encryption – FSE 2000*, LNCS 1978, B. Schneier (Ed.), pp. 19–30, Springer-Verlag, 2000.

6. S. R. Fluhrer, I. Mantin, A. Shamir, "Weaknesses in the Key Scheduling Algorithm of RC4," *Selected Areas in Cryptography – SAC 2001*, LNCS 2259, S. Vaudenay and A.M. Youssef (Eds.), pp. 1–24, Springer-Verlag, 2001.

7. J. Golić, "Linear statistical weakness of alleged RC4 keystream generator," *Advances in Cryptology – Eurocrypt'97*, LNCS 1233, W. Fumy (Ed.), pp. 226–238, Springer-Verlag, 1997.

8. R. J. Jenkins Jr., "ISAAC," *Fast Software Encryption – FSE 1996*, LNCS 1039, D. Gollmann (Ed.), pp. 41–49, Springer-Verlag, 1996.

9. A. Joux, J. Reinhard, "Overtaking VEST." *Fast Software Encryption – FSE 2007*, LNCS, A. Biryukov (Ed.), Springer-Verlag, to appear.

10. N. Keller, S. D. Miller, I. Mironov, and R. Venkatesan, "MV3: A new word based stream cipher using rapid mixing and revolving buffers," *Topics in Cryptology – CT-RSA 2007, The Cryptographers' Track at the RSA Conference 2007*, LNCS 4377, M. Abe (Ed.), pp. 1–19, Springer-Verlag, 2006.

11. L. R. Knudsen, W. Meier, B. Preneel, V. Rijmen and S. Verdoolaege, "Analysis Methods for (Alleged) RC4," *Advances in Cryptology – ASIACRYPT'98*, LNCS 1514, K. Ohta and D. Pei (Eds.), pp. 327–341, Springer-Verlag, 1998.

12. I. Mantin, A. Shamir, "A Practical Attack on Broadcast RC4," *Fast Software Encryption – FSE 2001*, LNCS 2355, M. Matsui (Ed.), pp. 152–164, Springer-Verlag, 2001.

13. I. Mantin, "A Practical Attack on the Fixed RC4 in the WEP Mode." *Advances in Cryptology – ASIACRYPT 2005*, LNCS 3788, B. Roy (Ed.), pp. 395–411, Springer-Verlag, 2005.

14. I. Mantin, "Predicting and Distinguishing Attacks on RC4 Keystream Generator." *Advances in Cryptography – EUROCRYPT 2005*, LNCS 3494, R. Cramer (Ed.), pp. 491–506, Springer-Verlag, 2005.

15. I. Mironov, "(Not so) random shuffles of RC4," *Advances in Cryptology – CRYPTO'02*, LNCS 2442, M. Yung (Ed.), pp. 304–319, Springer-Verlag, 2002.

16. S. Mister and S. E. Tavares, "Cryptanalysis of RC4-like Ciphers," *Selected Areas in Cryptography – SAC'98*, LNCS 1556, S. Tavares, H. Meijer (Eds.), pp. 131–143, Springer-Verlag, 1998.

17. S. Paul, B. Preneel, "A NewWeakness in the RC4 Keystream Generator and an Approach to Improve the Security of the Cipher," *Fast Software Encryption – FSE 2004*, LNCS 3017, B. Roy (Ed.), pp. 245–259, Springer-Verlag, 2004.

18. S. Paul, B. Preneel, S. Sekar, "Distinguishing Attack on the Stream Cipher Py." *Fast Software Encryption – FSE 2006*, LNCS 4047, M. J. Robshaw (Ed.), pp. 405–421, Spring-Verlag, 2006.

19. S. Paul, B. Preneel, "On the (In)security of Stream Ciphers Based on Arrays and Modular Addition." *Advances in Cryptology – ASIACRYPT 2006*, LNCS 4284, K. Chen, and X. Lai (Eds.), pp. 69–83, Spring-Verlag, 2006.

20. H. Wu, B. Preneel, "Cryptanalysis of the Stream Cipher DECIM." *Fast Software Encryption – FSE 2006*, LNCS 4047, M. J. Robshaw (ed.), pp. 30–40, Springer-Verlag, 2006.
21. H. Wu, B. Preneel, "Resynchronization Attacks on WG and LEX." *Fast Software Encryption – FSE 2006*, LNCS 4047, M. J. Robshaw (ed.), pp. 422–432, Springer-Verlag, 2006.

Secure Computation from Random Error Correcting Codes

Hao Chen[1,*], Ronald Cramer[2,**], Shafi Goldwasser[3], Robbert de Haan[4,***],
and Vinod Vaikuntanathan[5]

[1] Department of Computing and Information Technology, School of Information
Science and Engineering, Fudan University, Shanghai, China
chenhao@fudan.edu.cn
[2] CWI, Amsterdam & Mathematical Institute, Leiden University, The Netherlands
http://www.cwi.nl/~cramer
[3] MIT, Cambridge, Massachusetts, USA & Weizmann Institute of Science, Rehovot,
Israel
http://theory.lcs.mit.edu/~shafi
[4] CWI, Amsterdam, The Netherlands
http://www.cwi.nl/~haan
[5] MIT, Cambridge, Massachusetts, USA
http://www.mit.edu/~vinodv

Abstract. Secure computation consists of protocols for secure arithmetic: secret values are added and multiplied securely by networked processors. The striking feature of secure computation is that security is maintained even in the presence of an adversary who corrupts a quorum of the processors and who exercises full, malicious control over them. One of the fundamental primitives at the heart of secure computation is secret-sharing. Typically, the required secret-sharing techniques build on Shamir's scheme, which can be viewed as a cryptographic twist on the Reed-Solomon error correcting code. In this work we further the connections between secure computation and error correcting codes. We demonstrate that threshold secure computation in the secure channels model can be based on arbitrary codes. For a network of size n, we then show a reduction in communication for secure computation amounting to a multiplicative logarithmic factor (in n) compared to classical methods for small, e.g., constant size fields, while tolerating $t < (\frac{1}{2} - \epsilon)n$ players to be corrupted, where $\epsilon > 0$ can be arbitrarily small. For large networks this implies considerable savings in communication. Our results hold in the broadcast/negligible error model of Rabin and Ben-Or, and complement results from CRYPTO 2006 for the zero-error model of Ben-Or, Goldwasser and Wigderson (BGW). Our general theory can be extended so as to encompass those results from CRYPTO 2006 as well. We also present a new method for constructing high information rate ramp schemes based on arbitrary codes, and in particular we give a new construction based on algebraic geometry codes.

* Hao Chen's research has been supported by NSFC grants 10225106 and 90607005.
** Ronald Cramer's research has been partially supported by NWO VICI.
*** Robbert de Haan's research has been partially funded by the Dutch BSIK/BRICKS project PDC1.

M. Naor (Ed.): EUROCRYPT 2007, LNCS 4515, pp. 291–310, 2007.

1   Introduction

Secure computation consists of protocols for secure arithmetic: secret values are added and multiplied securely by networked processors. The striking feature of secure computation is that security is maintained even in the presence of an adversary who corrupts a quorum of the processors and who exercises full, malicious control over them. A crowning achievement of cryptography in the late '80s was the following result (stated informally):

> *Any function that can be computed, can be computed securely.*

This statement (appropriately formalized) was shown in the computational setting by Goldreich, Micali and Wigderson [16] and in the information-theoretic setting by Ben-Or, Goldwasser and Wigderson [2] and Chaum, Crépeau and Damgaard [5]. Our focus in this paper will be on the information-theoretic setting.

One of the fundamental primitives at the heart of information-theoretic secure computation is secret-sharing. Typically, the required secret-sharing techniques build on Shamir's scheme, which can be viewed as a cryptographic twist on the Reed-Solomon error correcting code. In this work we further the study on the connections between secure computation and error correcting codes. We demonstrate that threshold secure computation in the secure channels model can be based on arbitrary codes, in two steps.

First we identify sufficient, specialized conditions on a secret sharing scheme in order that it can serve as an essentially seamless replacement of Shamir's scheme in the context of secure computation. Second, we show how arbitrary error correcting codes give rise to such dedicated secret sharing schemes, and we prove various bounds on the relevant achievable parameters. We also analyze high information rate ramp schemes based on arbitrary codes, and in particular we give a new construction based on algebraic geometry codes.

A t-threshold secret-sharing scheme among n players typically has the following complementary pair of guarantees: (1) *Privacy:* The shares of any set of at most t players reveal no information about the secret, and (2) *Reconstruction:* The shares of $t+1$ players, together, reveal the entire secret. Linear threshold secret sharing schemes are known to be equivalent to maximum-distance-separable (MDS) codes. By known lower bounds on MDS codes (or equivalently, on matroids), the smallest possible field K on which the shares can lie is of size at least $\max\{n-t, t+2\} \geq \frac{n+2}{2}$ [1]. We show that this obstacle can be circumvented by bounding corruption tolerance an arbitrary constant fraction of n away from its maximal value $\lfloor \frac{n-1}{2} \rfloor$.

In turn, we use this result to improve the existing results on information-theoretic secure computation. The existing approaches, which use variants of Shamir's threshold secret-sharing scheme, incur a communication overhead as the size of the working field is larger than n due to Shamir's scheme. This can

---

[1] In fact, the so-called Main Conjecture on MDS codes implies that $|K|$ is at least n minus a constant.

amount to a multiplicative factor of a (large) power of $\log n$ bits. Our results alleviate this and allow, for instance, constant size fields K as opposed to linear size, while corruption tolerance t is at most an (arbitrary) constant fraction of n away from optimal. Such a (small) loss is unavoidable over sub-linear size fields due to (the above-mentioned) impossibility results from combinatorics.

Concretely, by using Gilbert-Varshamov type of arguments, we show that for each ϵ there is a constant size field K and an infinite family of quasi-threshold (i.e., ramp) parameters (t_i, n_i) such that for each of them there is an ideal (or information rate $1/2$) linear secret sharing scheme over K that has multiplication, t_i-privacy and $(n_i - t_i)$-reconstruction and $(\frac{1}{2} - \epsilon)n_i \leq t_i < \frac{1}{2}n_i$. Other interesting examples include schemes over $\mathbb{F}_2$ where corruption tolerance t is about $\frac{n}{10}$, or in fact, $t \approx \frac{n}{5}$ for $n \leq 100$.

Trading corruption tolerance for small fields was first used in [6] where a class of algebraic geometric secret sharing schemes was introduced that are ideal, linear, offer t-privacy and $(n - 2t)$-reconstruction and satisfy the strong multiplication property rather than only the multiplication property. It was shown there how this enables low-communication threshold multi-party computation over small (e.g. constant) size fields in the zero-error/perfect security/active adversary model of Ben-Or, Goldwasser and Wigderson (BGW) [2]. This result owes to the special multi-linear algebraic structure induced by rational function evaluation (for the strong multiplication property, which also implies efficient error correction algorithms), the existence of families of algebraic curves with many rational points (to enable a small field), and reductions from secure computation to these dedicated secret sharing schemes. Of course, the techniques from [6] can be adapted to obtain the quasi-threshold schemes of the type we consider in this work (at least when $|K|$ is a square); their properties are different but similar enough to facilitate easy adaptation.

However, our first point is that quasi-threshold schemes of the type we consider here are much easier to design. In fact, they can be constructed from arbitrary (or even randomly chosen) error correcting codes. Our second point is that, although these quasi-threshold schemes cannot be used as the basis for BGW type of secure computation (as opposed to the schemes from [6]), they can serve as an essentially seamless replacement of Shamir's scheme in known secure computation protocols in the broadcast model of Rabin and Ben-Or (such as [26,9,12]) supplemented with preprocessing. In this model a broadcast primitive is given and small, non-zero errors are tolerated, but corruption tolerance is greater, i.e., up to $\frac{1}{2}n$ instead of $\frac{1}{3}n$ as in the BGW model. An important advantage of the use of our quasi-threshold schemes here is that they can lead to much more communication-efficient protocols. More concretely, when operating in Beaver's preprocessing model [1], we can obtain a reduction in communication amounting to a multiplicative logarithmic factor (in n), while tolerating a number of corrupted players that is arbitrarily close to the optimal value of $n/2$. Note that this may offer a considerable gain in case of very large networks. For similar results in the zero-error BGW model, see [6].

We also consider high information rate ramp schemes based on arbitrary codes. These are schemes where the secret is a vector of field elements, but shares consist of a single field element (or at least a shorter vector than the secret). This of course is impossible in perfect secret sharing schemes, which necessarily have shares of size at least the size of the secret. In ramp schemes one has t-privacy and r-reconstruction, and one does care if there are sets of size in between these bounds whose joint shares reveal partial information about the secret. The earliest example of such a scheme we are aware of is the one by Blakley and Meadows [4] (see also [19,24] in the references therein), which is a variation on Shamir's scheme. We give a full treatment of linear ramp schemes from arbitrary error correcting codes, and show various bounds. As an application we give a new scheme based on algebraic geometry that improves the high information rate scheme given at the end of [6].

1.1   Organization of the Paper

This paper is organized as follows. In Section 3 we study linear quasi-threshold secret sharing schemes with multiplication and show how these can be constructed from codes. Additionally, we prove several bounds on the achievable parameters. We also argue there how these schemes can essentially seamlessly replace Shamir's scheme in secure computation in the Rabin/Ben-Or model with preprocessing and indicate what savings can be achieved due to our results.

In Section 4.1 and Section 4.2, we describe a general approach for constructing high information rate ramp schemes from linear codes. Finally, in Section 4.3, we present a new high information rate ramp scheme based on algebraic geometry that improves the one presented in [6] and demonstrate that we can obtain high information rate ramp schemes from randomly generated codes and can predict bounds on their parameters with high probability.

2   Preliminaries and Definitions

2.1   Basic Definitions from Coding Theory

We establish notational conventions that we will use throughout this paper. Let K be a finite field.

DEFINITION 1. The Hamming weight $w_H(c)$ of a vector $c \in K^n$ is the number of non-zero positions in c. For a subspace $C \subset K^n$, the minimum distance $d_{min}(C)$ is defined as $\min\{w_H(c) \mid c \in C \backslash \{0\}\}$.

An $[n, k, d]$-code C over K is defined to be a k-dimensional subspace of K^n with $d_{\min}(C) = d$.

DEFINITION 2. The dual code $C^\perp$ for a code C consists of all vectors $c^* \in K^n$ such that $\langle c^*, c \rangle = 0$ for all $c \in C$, where $\langle \cdot, \cdot \rangle$ denotes the standard inner product. Whenever d is used to denote the minimum distance of C, $d^\perp$ is used to denote the minimum distance of $C^\perp$.

2.2   Threshold and Ramp Secret Sharing Schemes

In what follows, the reader is assumed to be familiar with linear secret sharing schemes (For details, see [10,11,6]). However, we give a brief survey of the most relevant properties below.

A secret-sharing scheme with t-privacy and r-reconstruction over a field K is an algorithm that, on input a secret $s_0 \in K^{d_0}$, outputs a vector $(s_1, \ldots, s_n)$ of shares, where $s_i \in K^{d_i}$ for certain $d_i > 0$, such that for any $A \subset \{1, 2, \ldots, n\}$ the following properties hold:

1. If $|A| \geq r$, then the shares $(s_i)_{i \in A}$ jointly determine the value s_0.
2. If $|A| \leq t$, then the shares $(s_i)_{i \in A}$ jointly give no information about s_0.

Such a scheme is called a *t-threshold secret-sharing scheme* when $r = t + 1$. In general (that is, when this is not the case), the scheme is called a *ramp (quasi-threshold) scheme with t-privacy and r-reconstruction*.

The sets A for which the shares allow for reconstruction are referred to as the *accepted* sets, whereas the sets for which the shares give no information are called the *rejected* sets. The *information rate* of a secret sharing scheme is $d_0/\max\{d_1, \ldots, d_n\}$. A secret sharing scheme with information rate 1, which is maximal for threshold secret sharing schemes, is said to be *ideal*.

A secret sharing scheme is said to be *linear* if for any two secrets s and s' and respective share vectors $(s_1, s_2, \ldots, s_n)$ and $(s'_1, s'_2, \ldots, s'_n)$, the vectors $(s_1 + s'_1, s_2 + s'_2, \ldots, s_n + s'_n)$ and $(\lambda s_1, \lambda s_2, \ldots, \lambda s_n)$ are valid share vectors for the secrets $s + s'$ and λs respectively. It is said to have the *multiplication property* if given any two full share vectors $(s_1, s_2, \ldots, s_n)$ and $(s'_1, s'_2, \ldots, s'_n)$ for secrets s and s', there is a vector r such that $\langle r, (s_1 s'_1, s_2 s'_2, \ldots, s_n s'_n) \rangle = ss'$, where $\langle \cdot, \cdot \rangle$ denotes the standard inner product. It has *strong multiplication* with respect to a t-adversary structure if the multiplication property holds with respect to any combination of $n - t$ shares. The latter property allows for reconstruction of the secret after a pooling of all shares, even when the shares for up to t indices are replaced by random values.

3   Linear Ramp Schemes with Multiplication from Codes

3.1   Massey's Secret Sharing from Codes

Massey [22,23] gave the following construction of a secret sharing scheme from an error correcting code. Let C be an $[n + 1, k, d]$-code over a finite field K. We use coordinates $(c_0, c_1, \ldots, c_n)$ for codewords. The dual code $C^\perp$ is then an $[n + 1, n + 1 - k, d^\perp]$-code. We tacitly assume in this section that C is non-degenerate, i.e., that the minimum distances of both C and $C^\perp$ are greater than 1.

Let $s \in K$ be a secret value. Select a codeword $c = (c_0, c_1, \ldots, c_n) \in C$ uniformly at random such that $c_0 = s$, and define the share-vector as $(c_1, \ldots, c_n)$. Let LSSS(C) denote this linear secret sharing scheme. The access structure $\Gamma(C)$, i.e., the collection of accepted sets, is as follows. For a vector x, define $\sup(x) =$

$\{i : x_i \neq 0\}$. Consider the set V_0 of all $c^* \in C^\perp$ such that $c_0^* = 1$. Then $\Gamma(C) = \{\sup(c^*) \setminus \{0\} : c^* \in V_0\}$.

We now extend this idea in several ways in order to obtain the claimed quasi-threshold schemes, and we prove bounds on their existence.

3.2 Extensions of Massey's Idea

We first report the following consequence (which appears to be part of folklore) about the ramp parameters of this scheme and include a proof.

THEOREM 1. *Let C be an $[n+1, k, d]$-code over a finite field K. Then $LSSS(C)$ offers linearity, $(d^\perp - 2)$-privacy and $(n - d + 2)$-reconstruction.*

PROOF. Linearity is clear; the sum of two code-words is a share-vector for the sum of the secrets, and likewise for scalar multiplication. First, we argue that $\Gamma(C) = (\Gamma(C^\perp))^*$, i.e., the access structure of $LSSS(C)$ is the dual of the access structure of $LSSS(C^\perp)$, and vice versa.[2] Indeed, $A \in \Gamma(C)$ if and only if there is $c^* \in C^\perp$ with $c_0^* = 1$ and $c_i = 0$ for all $i \in \{1, \ldots, n\} \setminus A$ $(:= \overline{A})$. The latter is a share vector with secret equal to 1 in $LSSS(C^\perp)$, with shares equal to 0 for $\overline{A}$. The existence of such a share vector is equivalent to $\overline{A} \notin \Gamma(C^\perp)$. Now, from the characterization of $\Gamma(C)$ it is immediate that $LSSS(C)$ rejects all sets of size $d^\perp - 2$. Since $LSSS(C^\perp)$ rejects all sets of size $d - 2$ and since $\Gamma(C) = (\Gamma(C^\perp))^*$, it must be that $LSSS(C)$ accepts all sets of size $n - d + 2$. △

The exact privacy threshold $t_{\max}$ is equal to $-2 + \min\{w_H(c^*) : c^* \in C^\perp : c_0^* = 1\}$, i.e., this is the largest cardinality such that the joint shares of any set of this cardinality give no information on the secret. The exact reconstruction threshold $r_{\min}$ is equal to $n + 2 - \min\{w_H(c) : c \in C : c_0 = 1\}$.

For $A \subset \{1, \ldots, n\}$, let $\phi_A(C)$ denote the code restricted to the coordinates from the set $i \in A \cup \{0\}$, i.e., consisting of all codewords of C stripped of the coordinates not in $A \cup \{0\}$.

DEFINITION 3. *A self-dual code C is one for which $C = C^\perp$. A code is* weakly self-dual *if it there is a diagonal matrix $W \in K^{n+1,n+1}$ such that $w_{00} = 1$ and $Wc \in C^\perp$ for all $c \in C$. A code C is t-locally weakly self-dual if for all sets $B \subset \{1, \ldots, n\}$ with $|B| = n - t$ the code $\phi_B(C)$ is weakly self-dual.*

The definition of self-dual is standard in the coding literature, while our definition for weakly self-dual codes is a slight relaxation of the notion of quasi self-orthogonal[3] codes. The t-local variation appears to be novel. Simple examples are the following: the $[n + 1, t + 1, n - t + 1]$-Reed Solomon code is weakly self-dual if $t < \frac{n}{2}$ and t-locally weakly self-dual if $t < \frac{n}{3}$. The following theorem demonstrates the relevance of these notions in secure computation.

---

[2] The dual Γ^* is defined as $A \in \Gamma^*$ if and only if $\{1, \ldots, n\} \setminus A \notin \Gamma$. It holds that $(\Gamma^*)^* = \Gamma$.

[3] For quasi self-orthogonal codes, the matrix W is required to be regular.

THEOREM 2. *If C is a self-dual code of length $n+1$ with minimum distance d, then LSSS(C) offers linearity, t-privacy and $(n-t)$-reconstruction with $t = d-2$, and it has the multiplication property. If C is weakly self-dual, then C has the multiplication property and $t = d^{\perp} - 2$ if the matrix W is regular and otherwise $t = min\{d - 2, d^{\perp} - 2\}$. If C is t-locally weakly self-dual then LSSS(C) has the strong multiplication property with respect to the t-adversary structure.*

PROOF. Since $d = d^{\perp}$ for self-dual codes, the privacy and reconstruction claims follow from Theorem 1. From $\langle c, c' \rangle = 0$ for all $c, c' \in C$ we get $c_0 c'_0 = -c_1 c'_1 - \cdots - c_n c'_n$. This implies the multiplication property (see [10,11,6] for the definition). For weakly self-dual codes, if W is regular then the minimum distance of WC is the same as that of C. Since $WC \subset C^{\perp}$, we must have $d^{\perp} \leq d$, and we apply Theorem 1. As to multiplication, we now have $\langle Wc, c' \rangle = 0$, so $c_0 c'_0 = -w_1 c_1 c'_1 - \cdots - w_n c_n c'_n$. The claim about the strong multiplication property is now obvious from the definition. $\triangle$

We can generalize this as follows, using a twist on an idea from [10]. Let C be a code of length $n + 1$ and minimum distance d. Consider the linear secret sharing scheme LSSS$^{\dagger}(C)$ defined as follows. Take the secret s, and generate random shares $(c_1, \ldots, c_n)$ according to LSSS(C), and generate independently random shares $(c_1^*, \ldots, c_n^*)$ according to LSSS($C^{\perp}$). The share vector is then defined as $((c_1, c_1^*), \ldots, (c_n, c_n^*))$.

THEOREM 3. *Let C be a code of length $n + 1$ and minimum distance d. Define $t(C) = min\{d - 2, d^{\perp} - 2\}$. Then: LSSS$^{\dagger}(C)$ offers $t(C)$-privacy and $(n - t(C))$-reconstruction and it has the multiplication property. In particular, $t(C) < n/2$.*

The claim that $t(C) < n/2$ can for instance be verified by applying the Singleton-bound to C as well as to $C^{\perp}$. Note however that this scheme has information rate $1/2$.

Strong multiplication is much more elusive and is not achieved by the construction above. In fact, the only way known to ensure strong multiplication (with respect to the t-adversary structure) for LSSS(C) is when C is an algebraic geometry code defined by the Riemann-Roch space of a divisor of degree $2g + t$ on a genus g algebraic curve over a finite field, where $3t < n - 4g$ [6]. If $2t < n - 4g$ it is weakly self-dual. For the special case where $g = 0$, these correspond to the well-known Reed-Solomon codes with the appropriate parameters.

3.3 Existence and Bounds

Our main objective in this section is to prove several lower bounds on the maximal value T taken over all values $t = min\{d - 2, d^{\perp} - 2\}$ as C ranges over all K-linear codes of length $n + 1$. In the following, an $[n + 1, k]$-*code* C is simply a k-dimensional subspace of $\mathbb{F}_q^{n+1}$ and q is some fixed prime power. Where the parameters n and k are clear, $[n + 1, k]$-code is simply abbreviated to code.

General lower bounds on T. In Theorem 5 we give a general lower bound on the maximal t. In Corollary 2 we treat the general case when $K = \mathbb{F}_2$. In Theorem 6 we show that one can asymptotically get arbitrarily close to $\frac{1}{2}n$, over some constant size field. We also treat in that same corollary the parameterized case where C is randomly selected and a security parameter regulates the error probability that t is below a certain bound.

DEFINITION 4. *Let $n \in \mathbb{Z}_{>0}$ be fixed. Then $T(n+1, q) := \max_C t(C)$, where C ranges over all subcodes of $\mathbb{F}_q^{n+1}$. Similarly, $T'(n+1, q) := \max_C t(C)$, where C ranges over all weakly self-dual subcodes of $\mathbb{F}_q^{n+1}$.*

DEFINITION 5. *Let $\mathcal{C}_k$ have the uniform distribution over the set of $[n+1, k]$-subcodes of $\mathbb{F}_q^{n+1}$. Then we define*

$$T(n+1, q, m, k) := \max\{d - 2 : P(\min\{d_{min}(\mathcal{C}_k), d_{min}(\mathcal{C}_k^{\perp})\} < d) < 2^{-m}\}$$

and $T(n+1, q, m) := \max_k T(n+1, q, m, k)$.

It is easy to see that $T(n+1, q) \geq T(n+1, q, 0)$. The following lemma is trivial.

LEMMA 1. *Suppose $k \leq n$. For each pair (x, y) with $x \in \mathbb{F}_q^k \setminus \{0\}$ and $y \in \mathbb{F}_q^n \setminus \{0\}$ there exists an $n \times k$ matrix M of rank k such that $Mx = y$.*

The following theorem bounds the probability that a randomly chosen code has a minimum distance less than some fixed value d. It is used for most of the bounds that follow later.

THEOREM 4. *Let $\mathcal{C}$ have the uniform distribution over the set of $[n, k]$-subcodes of $\mathbb{F}_q^n$. Furthermore assume that $d = \alpha n \in \mathbb{Z}$, where $0 < \alpha < \frac{1}{2}$. Then*

$$P(\exists y \in \mathcal{C} : w_H(y) < d) < q^{k+n(H_q(\alpha)-1)},$$

where $H_q(\lambda) = \lambda \log_q(q-1) - \lambda \log_q \lambda - (1-\lambda)\log_q(1-\lambda)$.

PROOF. Let $\mathcal{H}$ have the uniform distribution over the set of $n \times k$ matrices of rank k over $\mathbb{F}_q$. Every such matrix corresponds to an ordered basis for a subcode V of $\mathbb{F}_q^n$. Since there is a one-to-one correspondence between the ordered bases for V and the linear isomorphisms between V and $\mathbb{F}_q^k$, each such subcode has the same number of ordered bases. Therefore, the variable $\mathcal{H}$ induces a uniformly random selection of an $[n, k]$-subcode of $\mathbb{F}_q^n$.

Fix some non-zero $x \in \mathbb{F}_q^k$. The variable $\mathcal{H}x$ then corresponds to a uniformly random selection from $\mathbb{F}_q^n$, which can be seen as follows: First, by Lemma 1 for any non-zero $y \in \mathbb{F}_q^n$ there exists an $n \times k$ matrix M of rank k such that $Mx = y$. Now fix some $y \in \mathbb{F}_q^n$ and assume that $Mx = y$ for some $n \times k$-matrix M of rank k. Then $\#\{M' : M'x = y\} = \#\{M' : (M - M')x = 0\} = \#\{M' : M'x = 0\}$, so for every $y \in \mathbb{F}_q^n$ there are the same number of matrices of rank k such that $Mx = y$.

Now let x range over the elements of $\mathbb{F}_q^k$. It follows that

$$P(\exists y \in \mathcal{C} : w_H(y) < d) = P(\exists x \in \mathbb{F}_q^k : w_H(\mathcal{H}x) < d) \leq \sum_{x \in (\mathbb{F}_q^k)^*} P(w_H(\mathcal{H}x) < d)$$

$$= \frac{q^k - 1}{q^n - 1} \cdot \sum_{i=1}^{d-1} \binom{n}{i}(q-1)^i < \frac{q^k}{q^n} \cdot (q-1)^d \sum_{i=1}^{d-1} \binom{n}{i}$$

$$< \frac{q^k}{q^n} \cdot q^{\alpha n \log_q(q-1)} \cdot 2^{n H_2(\alpha)} = q^{k+n(H_q(\alpha)-1)}. \qquad \triangle$$

Since there is a one-to-one correspondence between subcodes C of $\mathbb{F}_q^n$ and their dual codes $C^\perp$, the random variable $\mathcal{C}^\perp$ corresponds to a uniformly random selection from the set of $[n, n-k]$-subcodes of $\mathbb{F}_q^n$. Therefore, we immediately obtain the following corollary.

COROLLARY 1. *Let $\mathcal{C}$ have the uniform distribution on the set of $[n,k]$-subcodes of $\mathbb{F}_q^n$. Furthermore assume that $d^* = \alpha n \in \mathbb{Z}$, where $0 < \alpha < \frac{1}{2}$. Then*

$$P(\exists y \in \mathcal{C}^\perp : w_H(y) < d^*) < q^{n H_q(\alpha)-k}.$$

Using the fact that $-\lambda \ln \lambda - (1-\lambda)\ln(1-\lambda) < 3.3\lambda$ for $1/10 \leq \lambda \leq 1/2$, we obtain that

$$H_q(\lambda) < \lambda \log_q(q-1) - \frac{3.3}{\ln q}\lambda \qquad (1)$$

for $1/10 \leq \lambda \leq 1/2$. This gives rise to the following theorem.

THEOREM 5 $T(n+1, q, m) \geq \lfloor \beta(n+1, q, m) \rfloor - 2$ *with*

$$\beta(n+1, q, m) = \frac{(n+1)\ln q - 2(m+1)\ln 2}{2\ln(q-1) + 6.6},$$

provided that $\lfloor \beta(n+1, q, m) \rfloor \geq n/10$.

PROOF. Set $k = (n+1)/2$ and let $\mathcal{C}$ be as in Theorem 4. By Theorem 4 and Corollary 1,

$$P(\min\{d_{\min}(\mathcal{C}), d_{\min}(\mathcal{C}^\perp)\} < d) \leq P(d_{\min}(\mathcal{C}) < d) + P(d_{\min}(\mathcal{C}^\perp) < d)$$

$$< 2 \cdot q^{(n+1)H_q(\alpha)-(n+1)/2}.$$

We want $P(\min\{d_{\min}(\mathcal{C}), d_{\min}(\mathcal{C}^\perp)\} < d) < 2^{-m}$. Filling in (1) and rewriting, we see that this is the case if

$$d \leq \frac{(n+1)\ln q - 2(m+1)\ln 2}{2\ln(q-1) + 6.6}. \qquad \triangle$$

COROLLARY 2. *If $n \geq 21$, then $T(n+1, 2) \geq \lfloor 0.1n \rfloor - 2$.*

THEOREM 6. *Fix any arbitrarily small $\epsilon > 0$ and any $m \in \mathbb{Z}_{>0}$. Then there exists a fixed finite field $\mathbb{F}_q$ over which for infinitely many n there exist $[n+1, k]$-codes $C \subset \mathbb{F}_q^{n+1}$ with $(1/2 - \epsilon)n \le t(C) \le n/2$ where such a code can be selected with probability at least $1 - 2^{-m}$ using a random selection among the $[n, k]$-subcodes of $\mathbb{F}_q^{n+1}$.*

PROOF. Let d be the minimum distance of C and $d^\perp$ the minimum distance of $C^\perp$. By Theorem 3, $t(C) < n/2$. Therefore, it suffices to show that $(d - 2)$ and $(d^\perp - 2)$ can simultaneously get arbitrarily close to $n/2$ (relative to n) with probability at least $1 - 2^{-m}$.

By Theorem 5,

$$T(n + 1, q, m) \ge \beta(n + 1, q, m) - 2 = \frac{(n + 1)\ln q - 2(m + 1)\ln 2}{2\ln(q - 1) + 6.6} - 2$$

and we have that

$$\lim_{q \to \infty} \frac{(n + 1)\ln q - 2(m + 1)\ln 2}{2\ln(q - 1) + 6.6} - 2 = \lim_{q \to \infty} \frac{(n + 1)\ln q}{2\ln(q - 1) + 6.6} - 2$$

$$\ge \lim_{q \to \infty} \frac{(n + 1)\ln q}{2\ln q + 6.6} - 2.$$

Since $\lim_{x \to \infty} \frac{x}{x + 3.3} = \lim_{y \to \infty} \frac{y - 3.3}{y} = \lim_{y \to \infty} (1 - \frac{3.3}{y}) = 1$, the final term converges to $(n + 1)/2 - 2$ as $q \to \infty$. We can therefore for any $\delta > 0$ select a q large enough such that $T(n, q, m) \ge n/2 - 3/2 - \delta$. For large enough n, $(3/2 + \delta)/n < \epsilon$ and the claim follows. $\triangle$

So far we have assumed a random selection from the set of $[n, k]$-subcodes of $\mathbb{F}_q^n$. The lemma below demonstrates, together with the proof of Theorem 4, that we can in fact perform this random selection by selecting $n \times k$ matrices at random, where we obtain a matrix of rank k with probability at least $1/4$.

LEMMA 2 *The probability that a randomly selected $n \times k$-matrix over $\mathbb{F}_q$ has full rank is larger than $1 - 1/q - 1/q^2$.*

Bounds from (Weakly) Self-Dual Codes. In Corollary 3 we prove a general lower bound on T for binary self-dual codes, and Theorem 8 shows that for $n < 100$ the situation is much better than the bound indicates. We are especially interested in self-dual codes, because secret sharing schemes based on self-dual codes do not suffer from the $1/2$ information rate loss that occurs in the general case. Finally, in Theorem 9 we prove a much better lower bound for weakly self-dual codes based on algebraic geometry, and not random codes. Note that the results based on algebraic geometry are only known to hold if the size of the field is a square.

THEOREM 7. *Let n be any positive integer and let d_{GV} be the largest integer such that*

$$\sum_{\substack{0 < i < d \\ 2|i}} \binom{n}{i} < 2^{n/2-1} + 1.$$

Then there exists a self-dual binary code of length n and minimum distance at least d_{GV}.

PROOF. See [21,29,27]. △

COROLLARY 3. *Fix $\epsilon > 0$. For large enough n, $T'(n,2) \geq \lfloor(\delta - \epsilon)n\rfloor - 2$, where $\delta \approx 0.11002786$ is any truncated approximation of the unique solution less than $1/2$ of $H_2(\delta) = 1/2$.*

PROOF. ([21,29,27]) Let $d = \alpha(n + 1)$. Since for $\alpha < 1/2$, $\sum_{0<i<d} \binom{n+1}{i} \leq 2^{(n+1)H(\alpha)}$, the conditions of Theorem 7 are met if

$$(n + 1)H(\alpha) \leq \frac{n+1}{2} - 1 \Leftrightarrow H(\alpha) \leq \frac{1}{2} - \frac{1}{n+1}.$$

The solution for α then comes arbitrarily close to δ as n increases. △

THEOREM 8. *There exist self-dual binary codes C of length $n + 1 < 100$ for which $d_{min}(C) > n/5$. In particular, there exist self-dual binary codes C with the following parameters:*

$n+1$	$d_{min}(C)$
12	4
22	6
24	8
46	10
48	12

PROOF. See [14]. △

THEOREM 9. *When we take the maximum over algebraic geometry codes, then*

$$T(n + 1, q^2) > \left(\frac{1}{2} - \frac{1}{q-1}\right) n.$$

PROOF. This follows from a suitable choice of parameters for algebraic geometry codes and their duals and the existence of Garcia-Stichtenoth curves, using techniques similar to those in [6]. △

For a corresponding result that ranges over t-locally weakly self-dual codes, see [6].

3.4   Application to VSS and Secure Computation

Using the results from Sections 3.2 and 3.3, we are now ready to discuss the fact that our specialized secret sharing schemes can essentially seamlessly replace Shamir's scheme in the broadcast model of Rabin/Ben-Or, yielding significant reductions in communication when working over a small field. More concretely, when operating in Beaver's preprocessing model [1] with a network of size n, this results in a reduction in communication amounting to a multiplicative logarithmic factor (in n) in the on-line phase, while tolerating $(\frac{1}{2} - \epsilon)n$ corrupted players, where $\epsilon > 0$ is arbitrarily small. Note that this may offer a considerable gain in case of very large networks. For similar results in the zero-error BGW model, see [6].

As an illustration, Theorem 6 together with Theorem 3 implies that for any $\epsilon > 0$, there exists a (fixed) finite field K and an infinite family of specialized secret sharing schemes tolerating a $(\frac{1}{2} - \epsilon)n$-fraction of corrupted players. We now focus on the communication-efficient protocol of Cramer, Damgaard and Fehr [12] and outline the main changes necessary to enable the use of these specialized secret-sharing schemes. The CDF protocol is stated in the broadcast model of Rabin and Ben-Or [26] supplemented with a preprocessing phase as introduced by Beaver [1]. The claimed reduction in communication will be achieved in the on-line phase of the adapted CDF protocol.

The model of Rabin and Ben-Or assumes the presence of a broadcast channel and induces a non-zero (negligible) error probability. In Beaver's model, an independent preprocessing phase is implemented, which can take place even before the selection of the type of computation, that is used to compute VSSes of random values and secret-shared "multiplication tables" of random values. The attractive feature of this model is that, during the subsequent *on-line phase* when the actual computation is performed, players only need to open a constant number of VSSes for every secure multiplication (which saves a lot of communication). Moreover, no secure channels are required at all during this on-line phase, as all communication is by broadcast. [4]

Briefly, the main changes are as follows. First, in VSS we modify the usual bivariate Shamir-sharing by using a technique from [10] for extending a linear secret sharing scheme so as to enable the pair-wise checking protocols for VSS. This is by having the fixed secret sharing matrix operating on random symmetric matrices, rather than on random vectors. This can trivially be adapted to our scenario here. Exactly as in the CDF protocol, the resulting two-level secret-sharings are then augmented with unconditionally secure Information Checking (IC) signatures. This completes the basis for VSS with a two-level sharing, where all shares and sub-shares are signed. Multiplication of VSS'ed values can be performed based on the linearity of the scheme and the multiplication property, while addition essentially comes for free due to linearity of the VSS itself.

The preprocessing in the CDF protocol is a secure multi-party computation that prepares VSSes of random multiplication tables, as well as VSSes of random

---

[4] In some implementations broadcast isn't even necessary in the on-line phase, but in our case it is.

inputs of players. The point however, is that by a specialized secure multi-party computation the CDF preprocessing strips off one layer of shares, resulting in VSSes with just a single layer of signed shares. This makes an on-line phase possible that is much more communication-efficient. We assume now that the security parameters are set so that these signatures in these one-level sharings are correct except with negligible probability. This can be done by repeating the information checking step sufficiently many times; the total amount of communication in this preprocessing phase would be the same as in CDF though, since our field is small.

In the on-line phase each player first VSSes his real inputs, by broadcasting the difference of this input with the random VSSed input that he has been given in the pre-processing. The corresponding VSS is accordingly updated (noninteractively). Secure computation in the on-line phase can subsequently take place. Note that, as opposed to CDF, we are working here over a constant size field. This means that, though the signatures themselves are correct with high probability as a result of the CDF preprocessing as instructed above, they are "so small" (as a matter of fact, equal to field elements) that successful forgeries can be constructed with high probability. Thus, when opening such a (stripped) VSS, a corrupted player could in principle make an individual honest player accept a false share with high probability, by guessing the "small signature value" for this individual player. An additional concern would be the following. For their use in secure addition and secure multiplication, these signatures enjoy a certain linearity property [9]. This requires, for each ordered pair of players, a secret key part held by one of those players. This part remains fixed throughout the protocol. Now, this fixed key part can be extracted from an honest player in a single successful forgery, which, as we have seen above, has a high probability of success. So, at first sight, there seems to be a risk that security might degrade fatally over time, if there was any in the first place.

What saves the day completely is the ϵ-gap with $n/2$ in the number of corrupted players, in combination with a simple elimination strategy regarding corrupted players. Consider a corrupted player, and focus on his very first attempt at cheating in the on-line phase. It is easy to see that if he doesn't modify his correct share, he can predict the behavior all of all honest players; rejection if the corresponding correct signature was modified and acceptance otherwise. This is due to the fact that the signature is deterministic given all secret information held by the receiver and the purported share. So, he cannot gain advantage unless he modifies the correct share. In our adaptation of the CFD protocol, we instruct that he broadcasts that purported share. Thus, if the correct share is modified, he must also modify the corresponding correct signatures for many honest players individually. More precisely, we instruct that a purported share is accepted only if a majority of the players individually accept it. This is done by local verification of individual signatures followed by majority voting using broadcast. [5] This means that he must guess the signatures for roughly ϵn

---

[5] There is a slightly more sophisticated strategy involving error correction that gives still better error probabilities.

honest players, so as to get a majority (assuming that the adversary appropriately coordinates this with the actions of the other $t-1$ corrupted players). Now, if the field size $|K|$ is, say, about $2/\epsilon$, then this probability is exponentially small in n. Note that we can always replace our original fixed finite field K with a large enough fixed extension field so that this condition holds, without changing the other parameters and properties of the underlying specialized secret sharing scheme. Thus, if a corrupted player makes his first attempt, he will be caught in the voting phase with very high probability, and he is subsequently eliminated from the network. This also means that the entropy of the fixed secret keys of all honest players remains essentially intact, so the error probability analysis is essentially the same throughout the on-line phase if n is indeed very large. The network then moves to the next computation with the remaining players, applying the same strategy as above. All in all, this reduces the communication by a multiplicative factor $\log n$, due to the fact that in the stripped VSS each of the n shares now carries a signature for each individual receiving player that is a $\log n$ factor smaller.

A Concrete Example. The case $K = \mathbb{F}_2$ is especially interesting, since the algebraic geometry results have no known strong bearing on this case. Our results show that in the secure channels model (passive case), secure multiplication over $\mathbb{F}_2$ can be done with just n^2 *bits* communication, with corruption tolerance of a constant fraction of n. This saves a multiplicative factor of $O(\log n)$ bits compared to the standard approach based on Shamir's scheme. For n below 100, about 20 percent of the network may be corrupted, while the underlying scheme is ideal due to the use of a self-dual code. For instance, with $n = 48 - 1 = 47$, an adversary corrupting $t = 12 - 2 = 10$ players can be tolerated. In the active adversary case (with preprocessing, as in [12]), the savings also amount to a multiplicative factor of $O(\log n)$ bits. For large networks these savings in communication can be rather substantial.

4   Ramp Schemes with High Information Rate

In a secret sharing scheme each subset of the player set is either rejected, which means that the shares held by the players in the given set jointly do not give any information about the underlying secret-shared value, or it is accepted, which means that those shares jointly determine that secret uniquely. In other words, there is no way in between. As a consequence (by an argument very similar to the one used to show that the key is at least the size of the plain-text in the perfectly secure one-time pad encryption scheme), the size of a share is at least the size of the secret.

In what is sometimes called a non-perfect secret sharing scheme, there is a third category of subsets, consisting of subsets whose joint shares gives some partial (but not full) information about the secret. In such schemes it is possible to have high information rate, i.e., the size of a share may be much smaller than the size of the secret.

Ramp schemes are a special case, and a variation on Shamir's threshold secret sharing scheme constitutes a well-known example [13]. This goes as follows. Let K be a finite field with $|K| > n + \ell$, let $x_1, \ldots, x_\ell, y_1, \ldots, y_n \in K$ be distinct and let the y_i's be non-zero. Let τ, ℓ be positive integers with $1 \le \ell \le \tau$. Consider a secret vector $\alpha \in K^\ell$ of length ℓ. Sample a polynomial $f(X) \in K[X]$ uniformly at random such that its degree is at most τ and such that $f(x_1) = \alpha_1, \ldots, f(x_\ell) = \alpha_\ell$, and define the shares as $s_1 = f(y_1), \ldots, s_n = f(y_n)$. This is a scheme on n players, and using Lagrange interpolation one proves that all player sets of size at least $\tau + 1$ are accepted, while all player sets of size at most $\tau + 1 - \ell$ are rejected. Note that the scheme has information rate ℓ, i.e., each player gets one element of K as a share while in fact the secret is a K-vector of length ℓ. In other words, this is an $(n, \tau + 1, \tau + 1 - \ell, \ell)$-ramp scheme over K. It is also linear in that each share is a K-linear combination of the coordinates of the secret vector and (random) field elements.

An alternative [7] is to encode the secret vector in the first ℓ lower order coefficients of the polynomial f instead. This yields a ramp scheme with the same parameters, except that the requirement on the size of the field K can be relaxed, namely, $|K| > n$ suffices here. Later we analyze this scheme in terms of our general results from Section 4.1 and in Section 4.3 we generalize this result in terms of algebraic geometry codes.

Interestingly, these two schemes give rise to complementary applications in secure computation. The first one to parallel secure multi-party computation with good amortized communication complexity [13], and the second to secure atomic multiplication with low communication [7].

We generalize Massey's scheme from Section 3.1 to high information rate ramp schemes in Section 4.1. In Section 4.2, we give a completely general construction that does not consume codelength (which corresponds to the number of players in the scheme) for an increased information rate. As an application we use this theory to analyse the alternative high information rate ramp scheme based on Shamir presented above. Also, our general method gives rise to a new high information rate ramp scheme based on algebraic geometry code which we introduce in Section 4.3.

4.1 A High Information Rate Ramp Scheme

Let C be an $[n + \ell, k, d]$-code over a finite field K. We now extend Massey's scheme from Section 3.1 in the direction of high information rate as follows. Let ℓ be a non-negative integer such that $\ell < d^\perp$.

Let $s \in K^\ell$. Select a codeword $c = (c'_0, \ldots, c'_{\ell-1}, c_1, \ldots, c_n) \in C$ at random such that $s = (c'_0, \ldots, c'_{\ell-1})$. Such c always exists. Define the coefficients of $(c_1, \ldots, c_n)$ to be the shares. We claim that this is a linear ramp scheme with information rate ℓ that has $(d^\perp - \ell - 1)$-privacy and $(n + l - d + 1)$-reconstruction. This can be verified from the following facts.

Reconstruction follows from the fact that if there would exist two codewords in C that agreed on $n + l - d + 1$ share locations, their difference would give a codeword in C with Hamming weight less than d. As for privacy, note that in a generator matrix for C, any collection of $m < d^\perp$ rows (the code is generated by the columns) are linearly independent. So the corresponding columns span K^m. Therefore, for each $j \in \{0, \ldots, \ell - 1\}$ and for each $A \subset \{1, \ldots, n\}$ with $|A| \leq d^\perp - \ell - 1$ there exists a codeword c such that $c'_j = 1$ and $c'_i = 0$ for all $i \in \{0, \ldots, \ell - 1\} \setminus \{j\}$ and $c_u = 0$ for all $u \in A$. This implies privacy as claimed.

4.2   A More Fruitful Approach

A disadvantage of the scheme above is that it consumes code-length in exchange for secret-length. Below we describe an entirely general approach that doesn't have this disadvantage, and by means of which one can prove the existence of improved ramp schemes (see Section 4.3).

Let $\hat{C}$ and C be linear codes of length n over K, i.e., they are subspaces of the vector space K^n. Assume that C has dimension greater than 0 and that it is a proper subspace of $\hat{C}$. Choose an arbitrary linear code S such that

$$\hat{C} = S + C \text{ and } S \cap C = \{0\},$$

i.e., a direct sum. This is always possible of course, for instance by completing a basis of C to one of $\hat{C}$. Write

$$\ell = \dim_K \hat{C} - \dim_K C \ (= \dim_K S)$$

and fix an arbitrary isomorphism $\psi : K^\ell \longrightarrow S$.

We now define the following linear ramp scheme. Let $s \in K^\ell$ be the secret vector. Sample uniformly at random $c \in C$ and define the share vector $\hat{c}$ as $\hat{c} = \psi(s) + c$. [6]

Note that this is a generalization of a scheme used by Ozarow and Wyner [25], who considered the case $\hat{C} = K^n$. In fact, all possible linear ramp schemes are captured by this general scheme we consider here.

For $A \subset \{1, \ldots, n\}$, let ϕ_A denote the function $\phi_A : K^n \longrightarrow K^{|A|}$ where $(x_1, \ldots, x_n) \mapsto (x_i)_{i \in A}$, i.e., restriction to the coordinates labeled with A. Given A, consider the restriction of ϕ_A to $\hat{C}$. The set A is said to offer privacy if the collection of shares $\{\hat{c}_i\}_{i \in A}$ give no information on the secret vector, and reconstruction if those shares always determine the secret vector uniquely.

THEOREM 10 *Let $\ell = \dim \hat{C} - \dim C$. The set A offers privacy if and only if $\dim \phi_A(\hat{C}) - \dim \phi_A(C) = 0$. The set A offers reconstruction if and only if $\dim \phi_A(\hat{C}) - \dim \phi_A(C) = \ell$. More generally, the uncertainty about the secret vector s, given the shares of A, is equal to r elements of K, where r is such that $\ell - r = \dim \phi_A(\hat{C}) - \dim \phi_A(C)$.*

PROOF. Privacy (for the set A) is equivalent to saying that for each possible secret vector $s \in K^\ell$, there is a share vector $\hat{c}$ that "encodes" s and that satisfies $\phi_A(\hat{c}) = 0$. This is the same as saying that for each $z \in S$, there exists

---

[6] Equivalently, one can say that we fixed an arbitrary isomorphism from K^ℓ to $\hat{C}/C$, and that the share vector is selected by mapping s to the residue-class of $\psi(s)$ modulo C, and that $\hat{c}$ is chosen uniformly at random from that residue-class.

$c \in C$ such that $0 = \phi_A(z + c) = \phi_A(z) + \phi_A(c)$. Thus, $\phi_A(\hat{C}) \subset \phi_A(C)$. Since the other inclusion holds regardless of A, the privacy claim follows. As for unique reconstruction (for the set A), this is equivalent to saying that there are no two distinct $z, z' \in S$ so that $\phi_A(z + c) = \phi_A(z' + c')$ for some $c, c' \in C$. This is equivalent to saying that $\dim \phi_A(S) = \ell$ and $\phi_A(S) \cap \phi_A(C) = \{0\}$. Since $\dim \phi_A(\hat{C}) - \dim \phi_A(C) = \dim \phi_A(S) - \dim \phi_A(S) \cap \phi_A(C)$, the reconstruction claim follows. The cases in between these two extremes should now be obvious. $\triangle$

We give the following estimate with respect to privacy and reconstruction (which, as one can prove by giving counter-examples, is not always sharp).

COROLLARY 4. *The set A offers privacy if $|A| < d_{min}(C^{\perp})$. The set A offers reconstruction if $|A| > n - d_{min}(\hat{C})$.*

PROOF. As for privacy, if $|A| < d_{\min}(C^{\perp})$, then $\phi_A(C)$ clearly has rank $|A|$, since otherwise we could construct a codeword in $C^{\perp}$ whose weight is smaller than $d_{\min}(C^{\perp})$. Since $\phi_A(C) \subset \phi_A(\hat{C}) \subset K^{|A|}$, we must have $\phi_A(C) = \phi_A(\hat{C})$, and privacy follows from the theorem. As for reconstruction, if $|A| > n - d_{\min}(C)$, then $\phi_A(\hat{c}) = 0$ if and only if $\hat{c} = 0$, since otherwise C would contain a codeword whose weight is smaller than $d_{\min}(C)$. Thus, ϕ_A is injective when restricted to $\hat{C}$, and $\hat{c}$ follows uniquely from $\phi_A(\hat{c})$. Since $S \cap C = \{0\}$, $\psi(s)$ and c follow uniquely from $\hat{c}$. The secret vector s now follows uniquely from $\psi(s)$ since ψ is bijective. $\triangle$

Note that from the Singleton-bound, we have $\dim_K \hat{C} \leq n - d_{\min}(\hat{C}) + 1$ and $d_{\min}(C^{\perp}) - 1 \leq n - \dim_K C^{\perp} = \dim_K C$. Thus, $r - t \geq \dim_K \hat{C} - \dim_K C$ in any linear ramp scheme.

Before presenting constructive results, we argue as an example that the Shamir ramp scheme discussed earlier can be easily analyzed with this theory. Suppose $n > |K|$, and let $x_1, \ldots, x_n$ be distinct non-zero elements of K. Consider the Vandermonde matrix M with n rows and t columns whose i-th row is $(1, x_i, \ldots, x^t)$. Let $\hat{C}$ be the code generated by all the columns. This is an $(n, t + 1, n - t)$-MDS code. So its dual is an $(n, n - t - 1, t + 2)$-code. Let C be the code generated by the last $t + 1 - \ell$ columns. Clearly $C \subset \hat{C}$. By appropriately scaling the rows of C it is immediate that C is equivalent to an $(n, t + 1 - \ell, n - t + \ell)$-code. This is an MDS code, so its dual is an $(n, n - t - 1 + \ell, t + 2 - \ell)$-code. So by our theorem the resulting ramp scheme rejects all sets of size $t + 1 - \ell$, and accepts all sets of size $t + 1$. Note that the gap between the two bounds here is ℓ, so that is optimal.

4.3 High Information Rate Ramp Schemes: Existence and Bounds

In this section we demonstrate two methods for constructing high information rate ramp schemes. First, we present a new high information rate ramp scheme that improves the one presented in [6], where $\hat{C}$ will be an algebraic geometry code and C will be a carefully selected algebraic geometry subcode of $\hat{C}$. Then, we demonstrate that high information rate ramp schemes can be obtained from random codes and bound the error probabilities on their predicted parameters.

Algebraic Geometry Codes. Select an absolutely irreducible smooth projective curve over a finite field K, write g for its genus and let $\{Q, P_1, P_2, \ldots, P_n\}$ denote distinct points on the curve. Consider the rational divisor $\hat{D} = (2g+t) \cdot Q$, and let $\mathcal{L}(\hat{D})$ denote the corresponding Rieman-Roch space of rational functions. Write $\hat{C}$ for the Goppa-code consisting of the codewords $(f(P_1), \ldots, f(P_n))$, where f ranges over $\mathcal{L}(\hat{D})$. Also define the rational divisor $D = (2g + t - \ell) \cdot Q$, and let $\mathcal{L}(D)$ denote the corresponding Rieman-Roch space of rational functions. Write C for the Goppa-code consisting of the codewords $(f(P_1), \ldots, f(P_n))$, where f ranges over $\mathcal{L}(D)$.

By the Riemann-Roch Theorem the dimension of $\hat{C}$ is $g + t + 1$, whereas the dimension of C is $g + t + 1 - \ell$. Since $\hat{D} \geq D$, we have $\mathcal{L}(D) \subset \mathcal{L}(\hat{D})$, and hence $C \subset \hat{C}$. It is fact that the minimum distance of $C^\perp$ is at least $\deg D - 2g + 2 = t - \ell + 2$. Furthermore, it has been proven in [6] that we have reconstruction for $\deg \hat{D} + 1 = 2g + t + 1$ shares. Thus, by our theorem, we have a linear ramp scheme over K with $t - \ell + 1$ privacy, $2g + t + 1$ reconstruction and information rate ℓ. Note that the improvement consists in the fact that the scheme above does not use up any points on the curve in order to encode the secret vector. Also note that by taking the projective line (i.e., $g = 0$) we recover the earlier Shamir ramp scheme example. Using Garcia-Stichtenoth towers [15] our ramp scheme can be defined over constant size fields. See [6] for more details.

Random Codes. Finally, the results in Section 3.3 demonstrate that we can also obtain high information rate ramp schemes from randomly selected codes $\hat{C}$ and C, provided that $C \subset \hat{C}$. Theorem 10 demonstrates that for such codes C and $\hat{C}$, the corresponding ramp scheme provides privacy for any subset consisting of at most $d_{\min}(C^\perp) - 1$ players and reconstruction for any subset consisting of at least $n - d_{\min}(\hat{C}) + 1$ players.

One method of obtaining the appropriate distribution for C and $\hat{C}$, as demonstrated in the proof of Theorem 4, is to randomly select a matrix M from the set of $n \times \hat{k}$-matrices of rank $\hat{k}$ and let $\hat{C}$ be the code spanned by the columns. It is easy to see that if we now look at the last k columns of M, these columns in turn span a random $[n, k]$-subcode C of K^n that is furthermore contained in $\hat{C}$. Clearly, the corresponding scheme allows for a secret vector of length $\ell = \hat{k} - k$.

Suppose that we want the scheme to provide privacy for up to t players and reconstruction for at least $n - \hat{t}$ players. Using a similar argument as in Theorem 4 and using the fact that $-\lambda \ln \lambda - (1 - \lambda) \ln(1 - \lambda) < 1.2\sqrt{\lambda}$ for $0 \leq \lambda \leq 1/2$, the following theorem is now straightforward to obtain. It provides, for many different parameters and with arbitrarily high probability, a lower bound on t and $\hat{t}$ when we select the codes C and $\hat{C}$ at random.

THEOREM 11. *Select an $[n, k]$-code C and an $[n, \hat{k}]$-code $\hat{C}$ over $\mathbb{F}_q$ at random under the restriction that $C \subset \hat{C}$. Then*

$$P(d_{min}(\mathcal{C}^{\perp}) < t) < q^{-(k - t \log_q(q-1) - \frac{1.2\sqrt{tn}}{\ln q})}$$

and

$$P(d_{min}(\hat{C}) < \hat{t}) < q^{-(n - \hat{k} - \hat{t} \log_q(q-1) - \frac{1.2\sqrt{tn}}{\ln q})}.$$

References

1. D. Beaver. Efficient multiparty protocols using circuit randomization. In *Proceedings of CRYPTO '91*, volume 576, pages 420–432. Springer Verlag LNCS, 1992.

2. M. Ben-Or, S. Goldwasser, and A. Wigderson. Completeness theorems for non-cryptographic fault-tolerant distributed computation. In *Proceedings of STOC 1988*, pages 1–10. ACM Press, 1988.

3. G. R. Blakley. Safeguarding cryptographic keys. In *Proceedings of National Computer Conference '79*, volume 48 of *AFIPS Proceedings*, pages 313–317, 1979.

4. G. R. Blakley and C. Meadows. Security of ramp schemes. In *Proceedings CRYPTO '85*, volume 196, pages 242–269. Springer Verlag LNCS, 1985.

5. D. Chaum, C. Crépeau, and I. Damgaard. Multi-party unconditionally secure protocols. In *Proceedings of STOC 1988*, pages 11–19. ACM Press, 1988.

6. H. Chen and R. Cramer. Algebraic Geometric Secret Sharing Schemes and Secure Multi-Party Computation over Small Fields. In *Proceedings of 26th Annual IACR CRYPTO*, volume 4117, pages 516–531, Santa Barbara, Ca., USA, August 2006. Springer Verlag LNCS.

7. R. Cramer, I. Damgaard, and R. de Haan. Atomic Secure Multi-Party Multiplication with Low Communication. In *Proceedings of EUROCRYPT 2007*, May 2007.

8. R. Cramer, I. Damgaard, and S. Dziembowski. On the complexity of verifiable secret sharing and multi-party computation. In *Proceedings of STOC 2000*, pages 325–334. ACM Press, 2000.

9. R. Cramer, I. Damgaard, S. Dziembowski, M. Hirt, and T. Rabin. Efficient Multi-Party Computations with Dishonest Minority. In *Proceedings of 18th Annual IACR EUROCRYPT*, volume 1592, pages 311–326, Prague, Czech Republic, May 1999. Springer Verlag LNCS.

10. R. Cramer, I. Damgaard, and U. Maurer. General secure multi-party computation from any linear secret sharing scheme. In *Proceedings of EUROCRYPT 2000*, volume 1807 of *LNCS*, pages 316–334. Springer Verlag, 2000.

11. R. Cramer, V. Daza, I. Gracia, J. Jimenez Urroz, G. Leander, J. Martí-Farré, and C. Padró. On codes, matroids and secure multi-party computation from linear secret sharing schemes. In *Proceedings of CRYPTO 2005*, volume 3621 of *LNCS*, pages 327–343. Springer-Verlag, 2005.

12. R. Cramer, I. Damgård, and S. Fehr. On the Cost of Reconstructing a Secret– Or: VSS with Optimal Reconstruction. In *Proceedings of 21th Annual IACR CRYPTO*, volume 2139, pages 503–523, Santa Barbara, Ca., USA, August 2001. Springer Verlag LNCS.

13. M. Franklin and M. Yung. Communication complexity of secure computation. In *Proceedings of STOC 1992*, pages 699–710. ACM Press, 1992.

14. P. Gaborit and A. Otmani. Experimental constructions of self-dual codes. Manuscript. Available from http://www.unilim.fr/pages_perso/philippe.gaborit/SD/, 2002.

15. A. García and H. Stichtenoth. On the asymptotic behavior of some towers of function fields over finite fields. *J. Number Theory*, 61:248–273, 1996.

16. O. Goldreich, S. Micali, and A. Wigderson. How to Play Any Mental Game. In *Proceedings of STOC 1987*, pages 218–229. ACM Press, 1987.

17. V. D. Goppa. Codes on algebraic curves. *Soviet Math. Dokl*, 24:170–172, 1981.

18. M. Karchmer and A. Wigderson. On span programs. In *Proceedings of the Eight Annual Structure in Complexity Theory Conference*, pages 102–111. IEEE, 1993.

19. K. Kurosawa, K. Okada, K. Sakano, W. Ogata, and S. Tsujii. Nonperfect Secret Sharing Schemes and Matroids. In *Proceedings EUROCRYPT 1993*, pages 126–141. Springer Verlag, 1993.

20. S. Lang. *Algebra*. Addison-Wesley Publishing Company, 1997.

21. F. J. MacWilliams, N. J. A. Sloane, and J. G. Thompson. Good self-dual codes exist. *Discrete Math.*, 3:153–162, 1972.

22. J. L. Massey. Minimal codewords and secret sharing. In *Proceedings of the 6-th Joint Swedish-Russian Workshop on Information Theory*, pages 269–279, Molle, Sweden, August 1993.

23. J. L. Massey. Some applications of coding theory in cryptography. *Codes and Ciphers: Cryptography and Coding IV*, pages 33–47, 1995.

24. W. Ogata and K. Kurosawa. Some Basic Properties of General Nonperfect Secret Sharing Schemes. *J. UCS*, 4(8):690–704, 1998.

25. L. H. Ozarow and A. D. Wyner. "Wire-tap-channel II". *AT&T Bell Labs Tech. J.*, 63:2135–2157, 1984.

26. T. Rabin and M. Ben-Or. Verifiable secret sharing and multiparty protocols with honest majority. In *Proceedings of ACM STOC 1989*, pages 73–85, 1989.

27. E. M. Rains and N. J. A. Sloane. Self-Dual Codes. A long survey article written for the Handbook of Coding Theory. Available from http://www.research.att.com/~njas/, 1998.

28. A. Shamir. How to share a secret. *Communications of the ACM*, 22(11):612–613, 1979.

29. J. G. Thompson. Weighted averages associated to some codes. *Scripta Math.*, 29:449–452, 1973.

30. J. H. van Lint. *Introduction to Coding Theory*. Graduate Texts in Mathematics. Springer Verlag, 1999.

31. V. K. Wei. Generalized Hamming Weights for Linear Codes. *IEEE Transactions on Information Theory*, 37(5):1412–1418, 1991.

Round-Efficient Secure Computation in Point-to-Point Networks*

Jonathan Katz** and Chiu-Yuen Koo

Dept. of Computer Science, University of Maryland, College Park, USA
{jkatz,cykoo}@cs.umd.edu

Abstract. Essentially all work studying the round complexity of secure computation assume broadcast as an atomic primitive. Protocols constructed under this assumption tend to have very poor round complexity when compiled for a point-to-point network due to the high overhead of emulating each invocation of broadcast. This problem is compounded when broadcast is used in more than one round of the original protocol due to the complexity of handling sequential composition (when using round-efficient emulation of broadcast).

We argue that if the goal is to optimize round complexity in point-to-point networks, then it is preferable to design protocols — assuming a broadcast channel — minimizing the *number of rounds in which broadcast is used* rather than minimizing the *total number of rounds*. With this in mind, we present protocols for secure computation in a number of settings that use only a *single* round of broadcast. In all cases, we achieve optimal security threshold for adaptive adversaries, and obtain protocols whose round complexity (in a point-to-point network) improves on prior work.

1   Introduction

The round complexity of cryptographic protocols — and, in particular, protocols for secure multi-party computation of general functionalities — has been the subject of intense study. Establishing bounds on the round complexity of various tasks is, of course, of fundamental theoretical importance. Moreover, reducing the round complexity of existing protocols is crucial if we ever hope to use these protocols in the real world. If the best known protocol for a given task requires hundreds of rounds, it will never be used; on the other hand, if we know (in principle) that round-efficient solutions are possible, we can then turn our attention to improving other aspects (such as computation) in an effort to obtain a protocol that can be used in practice.

Previous research investigating the round complexity of protocols for secure multi-party computation (MPC) has almost exclusively focused on optimizing the round complexity *under the assumption that a broadcast channel is available*. (We survey some of this work in Section 1.2.) In most settings where MPC

---

* Work done in part while the authors were visiting IPAM.
** This research was supported by NSF CAREER award #0447075 and US-Israel Binational Science Foundation grant #2004240.

might potentially be used, however, only point-to-point channels are likely to be available and a broadcast channel is not expected to exist. Nevertheless, the use of a broadcast channel is justified in previous work by the fact that the broadcast channel can always be *emulated* by having the parties run a broadcast protocol over the point-to-point network.

We argue that if the ultimate goal is to optimize round complexity for point-to-point networks (i.e., where the protocol will actually be run), then the above may be a poor approach due to the high overhead introduced by the final step of emulating the broadcast channel. Specifically:

- If the broadcast channel is emulated using a deterministic protocol [16,11], then a lower bound due to Fischer and Lynch [13] shows that $\Omega(t+R)$ rounds are needed to emulate R rounds of broadcast in the original protocol (this is true regardless of how many parties broadcast during the same round). Here and in the rest of the paper, t denotes the number of malicious parties and may be linear in the total number of parties n. In particular, this will not lead to sub-linear-round protocols with optimal security threshold $t = \Theta(n)$.

- Using randomized protocols, each round of broadcast in the original protocol can be emulated in an expected constant number of rounds [12,14,22]. Nevertheless, the exact constant is rather high. More problematic is that if broadcast is used in *more than one* round of the original protocol, then it is necessary to explicitly handle sequential composition of protocols without simultaneous termination [6,24,22]. (This is not an issue if broadcast is used in only a *single* round.) This leads to a substantial increase in round complexity; we refer the reader to Appendix A for details.

To illustrate the point, consider the protocols of Micali and Rabin [25] and Fitzi, et al. [15] (building on [17]) for verifiable secret sharing (VSS) with $t < n/3$. The Micali-Rabin protocol uses 16 rounds but only a single round of broadcast; the protocol of Fitzi et al. uses three rounds, two of which involve broadcast. Compiling these protocols for a point-to-point network using the most round-efficient randomized broadcast protocol known, the Micali-Rabin protocol runs in an expected 31 rounds while the protocol by Fitzi et al. requires an expected 55 rounds! The conclusion is that optimizing round complexity using broadcast does not, in general, lead to round-optimal protocols in the point-to-point model.

This suggests that if the ultimate goal is a protocol for a point-to-point network, then it is preferable to focus on minimizing the number of rounds *in which broadcast is used* rather than on minimizing the total number of rounds. This raises in particular the following question:

*Is it possible to construct **constant-round** (or even sub-linear-round) protocols for secure computation that use only a **single** round of broadcast?*

Note that for $t = \Theta(n)$ at least one round of broadcast is necessary if the protocol uses a strict constant number of rounds, since broadcast itself cannot be achieved over point-to-point channels in a strict constant number of rounds.

We resolve the above question in the affirmative in a number of settings. Specifically, we show:

1. A constant-round protocol using a single round of broadcast that is secure for $t < n/3$ and assumes only the existence of one-way functions.
2. A constant-round protocol using a single round of broadcast that is secure for $t < n/2$ and assumes only a public-key infrastructure (PKI) along with secure signatures.
3. A protocol using a single round of broadcast and achieving information-theoretic security for $t < n/3$. Here, the round complexity is linear in the depth of the circuit being computed.

All protocols are secure even for adaptive adversaries.

Of course, the fact that a protocol uses broadcast in only a single round does not necessarily imply that it yields the most round-efficient protocol in a point-to-point setting. For the protocols we construct, however, this is indeed the case (at least given the most round-efficient known techniques for emulating broadcast over point-to-point channels). For example, the first protocol mentioned above requires 41 rounds (in expectation) when compiled for a point-to-point network. In contrast, *any* protocol for $t < n/3$ that uses broadcast in *two* rounds (even if that is all it does!) will require at least 55 rounds (in expectation) when run in a point-to-point network (see Appendix A). Similarly, any protocol for $t < n/2$ that uses broadcast in two rounds will require at least 96 rounds (in expectation) in a point-to-point network. We stress again that the main issue in moving from one broadcast to two (or more) broadcasts is the significant overhead in the latter case needed to deal with sequential composition of protocols that do not terminate in the same round.

1.1   Overview of Our Techniques

We give a high-level overview of the main techniques we use in constructing the protocols outlined above. Call (a, b, c), where a, b, and c are elements of some field, a *random multiplication triple* if a and b are uniformly distributed, each of a, b, c is shared among the players,[1] and $c = ab$. Beaver [3] shows that if, in a "setup phase," the parties share their inputs along with sufficiently-many multiplication triples — in particular, one multiplication triple for each multiplication gate of the circuit being evaluated — then the parties can evaluate the circuit in a round-efficient manner *without using any further invocations of broadcast*. Our task is thus reduced to showing how to perform the necessary setup using only a single round of broadcast.

To achieve this, we use the concept of *moderated protocols* as introduced in [22]. In such protocols, there is a distinguished party P_m known as the *moderator*. Given a protocol Π, designed under the assumption of a broadcast channel, the moderated version of Π is a protocol Π' that runs in a point-to-point network and has the following properties (roughly speaking):

- At the end of Π', each party P_i outputs a binary value $\mathsf{trust}_i(m)$.
- If the moderator P_m is honest, then each honest P_i outputs $\mathsf{trust}_i(m) = 1$. This represents the fact that an honest party P_i "trusts" the moderator P_m.

---

[1] For now, we do not specify the exact manner in which sharing is done.

- If any honest party P_i outputs $\text{trust}_i(m) = 1$, then Π' achieves the functionality of Π.

In our prior work [22], we have shown[2] how to compile any protocol Π into its moderated version Π', while increasing the round complexity of Π by at most a constant multiplicative factor (the exact effect on the round complexity depends on the number of invocations of broadcast in Π). For $t < n/3$, the compilation does not require any assumptions; for $n/3 \leq t < n/2$, the compilation assumes a PKI and digital signatures.

Let Π_i denote some protocol, designed assuming a broadcast channel, that shares the input value of party P_i as well as sufficiently-many multiplication triples. Such protocols are constructed in, e.g., [7,29,1,19,9,10]. We compile Π_i into a moderated protocol Π'_i where P_i itself acts as the moderator. Now consider the following protocol that uses broadcast in only a single round:

1. Run protocols $\{\Pi'_i\}_{i=1}^n$ in parallel.[3] Recall that P_i is the moderator in Π'_i.
2. Each party P_i broadcasts $\{\text{trust}_i(1), \ldots, \text{trust}_i(n)\}$.
3. A party P_i is disqualified if $|\{j : \text{trust}_j(i) = 1\}| \leq t$; i.e., if t or fewer players broadcast $\text{trust}_j(i) = 1$. If P_i is disqualified, then a default value is used as the input for P_i.
4. Let i^* be the minimum value such that P_{i^*} is not disqualified. The set of random multiplication triples that the parties will use is taken to be the set that was generated in Π'_{i^*}.

Analyzing the above, note that if P_i is honest and there exists an honest majority, then at least $t + 1$ parties broadcast $\text{trust}_j(i) = 1$. Hence an honest P_i is never disqualified. On the other hand, at least one of the parties that broadcast $\text{trust}_j(i^*) = 1$ must be honest. The properties of moderated protocols discussed earlier thus imply that Π'_{i^*} achieves the functionality of Π_{i^*}. Since Π_{i^*} is assumed to securely share sufficiently-many multiplication triples, it follows that the above protocol securely shares sufficiently-many multiplication triples. A similar argument shows that the inputs of all non-disqualified parties are shared appropriately. We conclude that the above protocol implements the necessary setup phase using only one round of broadcast.

In a naive compilation of Π_i to Π'_i (following [22]), each round of broadcast in Π_i is replaced by six rounds in Π'_i (for the case $t < n/3$). Proceeding directly thus yields secure MPC protocols with relatively high round complexity: after all, existing constructions of protocols Π_i achieving the needed functionality do not attempt to minimize the number of rounds of broadcast. We present instead a new set of protocols that minimize their use of broadcast. Furthermore, our implementation of the setup phase deviates from the above simplified approach in order to further optimize the round complexity of the final protocol. Along the way, we construct round-efficient protocols for VSS that use broadcast only

---

[2] Although our prior work only claims the result when Π is a VSS protocol, it is not hard to verify that the proof extends for more general classes of functionalities.

[3] In fact, only protocols $\Pi'_1, \ldots, \Pi'_{t+1}$ need to share multiplication triples; the remaining protocols only need to share the input of the appropriate player.

once; these in turn yield the most round-efficient VSS and broadcast protocols for point-to-point networks. For $t < n/3$ we show a 7-round VSS protocol using broadcast once (the best previous VSS protocol using broadcast once, obtained by combining [15,22], requires 14 rounds), and for $t < n/2$ we obtain a 5-round VSS protocol using broadcast once (the best previous protocol required 34 rounds [22]). The latter implies an expected 36-round broadcast protocol for the same threshold (improved from 58 rounds in [22]).

1.2   Prior Work

There is a vast amount of work in the cryptographic and distributed computing literature studying the round complexity of various tasks; here, we summarize the work most relevant to our own.

Broadcast/Byzantine agreement. For $t < n/2$, broadcast and Byzantine agreement (BA) have essentially the same round complexity (to within one round); therefore, we freely interchange between the two. In a synchronous network with pairwise authenticated channels and no additional setup, BA is achievable iff $t < n/3$ [26,23]. In this setting, a lower bound of $t + 1$ rounds for any deterministic protocol is known [13]. A protocol with this round complexity (but exponential message complexity) was shown by Pease, et al. [26,23]. Following a long sequence of works, Garay and Moses [16] show a fully-polynomial BA protocol with optimal resilience and round complexity.

To obtain protocols with sub-linear round complexity, researchers explored the idea of using randomization [28,5]. This culminated in the work of Feldman and Micali [12], who show a randomized BA protocol with optimal resilience running in an expected *constant* number of rounds.

To achieve resilience $t \geq n/3$, additional assumptions are needed; the most common assumptions are digital signatures and a PKI. Under these assumptions, linear-round deterministic broadcast protocols are known for $t < n$ [26,23,11]. For $t < n/2$, randomized protocols with expected constant-round complexity exist [14,22], the latter without any additional computational assumptions.

VSS. Gennaro, et al. [17] show a 2-round VSS protocol for $t < n/4$ and a 4-round protocol for $t < n/3$. They also give a 3-round protocol for $t < n/3$ with exponential complexity. Fitzi, et al. [15] determine the exact round complexity of VSS by showing a fully-polynomial 3-round VSS protocol for $t < n/3$. Their work also shows how to run many sequential VSS protocols at an amortized cost of only $(1 + \epsilon)$ rounds.

We stress that the above consider the round complexity of VSS *under the assumption that a broadcast channel is available*. (In particular, the VSS protocol from [15] is only optimal in this setting.) While of theoretical interest, this appears to be a poor approach (as explained in the Introduction) if one is ultimately interested in round-efficient protocols for point-to-point networks.

General secure MPC. Unconditionally-secure MPC protocols in point-to-point networks exist for $t < n/3$ (combining [7,8] with [26]), or for $t < n/2$ assuming a broadcast channel is available [2,30]. The broadcast channel can be removed for $t < n/2$ by relying on a PKI and digital signatures [11] or information-theoretic pseudo-signatures [27].

Beaver, Micali, and Rogaway [4] gave a constant-round (computationally-secure) protocol for secure MPC with $t < n/2$, assuming a broadcast channel and one-way functions. Damgård and Ishai [10] showed a constant-round protocol under the same assumptions that is secure even for adaptive adversaries. These can both be converted to *expected* constant-round protocols in point-to-point networks by using the broadcast protocols mentioned above [12,22]. We stress that the constant is rather high, on the order of hundreds of rounds.

The work of Gennaro et al. [17] mentioned earlier implies a 3-round MPC protocol with resilience $t < n/4$, assuming the existence of one-way functions. The resulting protocol uses broadcast in only a single round, and so yields a very round-efficient protocol in point-to-point networks; the drawback is that the resilience is not optimal. In subsequent work [18], the same authors show that 2-round MPC is not possible (in general) for $t \geq 2$. However, they show that certain functionalities can be securely computed in 2 rounds for $t < n/6$.

Hirt, Nielsen, and Przydatek [21] show a protocol for asynchronous secure MPC that uses only one round of broadcast. Their result is not directly comparable to ours due to differences in the way rounds are counted in the synchronous and asynchronous settings. (In particular, their protocol requires a linear number of rounds when directly adapted to the synchronous setting.) They assume $t < n/3$ and a global setup assumption (stronger than a PKI).

Goldwasser and Lindell [20] show various round-efficient MPC protocols for point-to-point networks; however, the point of their work is to consider weaker security definitions in which fairness and output delivery are not guaranteed (even when an honest majority exists).

1.3   Outline of the Paper

We review and formalize some standard notions in Section 2. In Section 3 we focus on the case $t < n/3$ in both the computational and information-theoretic settings. Section 4 discusses the case of $t < n/2$ (with computational security). Due to lack of space, we defer some of the details to the full version.

2   Model and Preliminaries

We use the standard synchronous communication model where parties communicate using pairwise private/authenticated channels. In addition, we assume a broadcast channel with the understanding that it will be emulated using a round-efficient broadcast sub-routine. As a convenient shorthand, we say that a protocol has round complexity (r, r') if it uses r rounds in total and $r' \leq r$ of these rounds invoke broadcast (possibly by all parties). We emphasize that since our aim is to minimize the eventual round complexity in point-to-point networks, we will construct protocols that access the broadcast channel in only a single round (i.e., $(\star, 1)$-round protocols).

When we say a protocol tolerates t malicious parties, we always mean that it is secure against a *rushing* adversary who may *adaptively* corrupt up to t parties during execution of the protocol and coordinate the actions of these parties as they deviate from the protocol in an arbitrary manner. Parties not corrupted by the adversary are called *honest*. In our protocol descriptions, we implicitly assume that parties send a properly-formatted message at all times; this is without loss of generality, as an improper or missing message can always be interpreted as some default message.

For $t < n/3$ we do not assume any setup, but for $t < n/2$ we assume a PKI. Note that, since we are assuming a broadcast channel, the additional assumption of a PKI may not be necessary; nevertheless, we see no harm in assuming it since a PKI will be needed anyway once we compile our protocols to run in a point-to-point network. We leave open the question of constructing an $(O(1), 1)$-round secure MPC protocol for $t < n/2$ that uses a broadcast channel but no PKI.

2.1 Gradecast

Gradecast was introduced by Feldman and Micali [12].

Definition 1. *(Gradecast): A protocol for parties* $\mathcal{P} = \{P_1, \ldots, P_n\}$, *where a distinguished* dealer $P^* \in \mathcal{P}$ *holds initial input* M, *is a* gradecast *protocol tolerating* t malicious parties *if the following conditions hold for any adversary controlling at most t parties:*

- *Each honest party* P_i *outputs a* message m_i *and a* grade $g_i \in \{0, 1, 2\}$.
- *If the dealer is honest, then the output of every honest party* P_i *satisfies* $m_i = M$ *and* $g_i = 2$.
- *If there exists an honest party* P_i *who outputs message* m_i *and grade* $g_i = 2$, *then the output of every honest party* P_j *satisfies* $m_j = m_i$ *and* $g_j \geq 1$.

Lemma 1 ([12,22]). *There exists a* $(3, 0)$-*round gradecast protocol tolerating* $t < n/3$ *malicious parties and, assuming a PKI, a* $(4, 0)$-*round gradecast protocol tolerating* $t < n/2$ *malicious parties.*

2.2 Generalized Secret Sharing and VSS

Throughout, we assume a finite field $\mathbb{F}$ whose order is a power of 2 and which contains $[n]$ as a subset.

Definition 2. *(1-level sharing): We say a value* $s \in \mathbb{F}$ *has been* 1-level shared *if there exists a degree-t polynomial* $F_s(x)$ *such that (1)* $F_s(0) = s$ *and (2) player* P_i *holds the share* $s_i \overset{\text{def}}{=} F_s(i)$. *In this case, we say that* $F_s(x)$ shares s.

When $t < n/3$ the parties can reconstruct s by having all parties send their shares to all other parties, and then having each party use Reed-Solomon decoding to recover s. If s, s' are 1-level shared then for any publicly-known $\alpha, \beta \in \mathbb{F}$ the value $\alpha s + \beta s'$ has been 1-level shared as well. We recall the following technical lemma concerning multiplication of shares [19]:

Lemma 2. *Let A, B be degree-t polynomials over $\mathbb{F}$, and $\alpha_1, \ldots, \alpha_{2t+1} \in \mathbb{F}$ distinct elements. Then $A(0) \cdot B(0) = \sum_{i=1}^{2t+1} \beta_i \cdot A(\alpha_i) \cdot B(\alpha_i)$ for some constants $\beta_1, \ldots, \beta_{2t+1} \in \mathbb{F}$.*

Definition 3. *(2-level sharing):* We say a value $s \in \mathbb{F}$ has been 2-level shared if (1) there exists a degree-t polynomial $F_s(x)$ that shares s and (2) for $i \in [n]$, there exists a degree-t polynomial $F_{s_i}(x)$, known to P_i, that shares $s_i \overset{\text{def}}{=} F_s(i)$ (i.e., each party P_j holds a share $s_{j,i}$ of s_i).

Definition 4. *(3-level sharing):* We say a value $s \in \mathbb{F}$ has been 3-level shared if (1) there exists a degree-t polynomial $F_s(x)$ that shares s; (2) for $i \in [n]$, the value $s_i \overset{\text{def}}{=} F_s(i)$ has been 2-level shared; and (3) each party P_i knows the polynomial $F_{s_i}(x)$ that shares s_i.

Note that if s is 3-level (resp., 2-level) shared, then it is 2-level (resp., 1-level) shared as well.

Definition 5. *(VSS with 2-level (resp., 3-level) sharing):* A protocol for parties $\mathcal{P} = \{P_1, \ldots, P_n\}$, where a distinguished dealer $P^* \in \mathcal{P}$ holds an initial input s, is a VSS protocol with 2-level (resp., 3-level) sharing tolerating t malicious parties if the following conditions hold for any adversary controlling at most t parties by the end of the protocol:

Secrecy: If the dealer is honest, then the joint view of the malicious parties is independent of the dealer's input s.

Commitment: At the end of the protocol, some value s' is 2-level (resp., 3-level) shared. Moreover, if the dealer is honest then $s' = s$. On the other hand, if the dealer is dishonest, then s' can be efficiently computed from the messages sent from the malicious parties to the honest parties during the protocol execution. We refer to this latter property as **extraction.**

3   Secure Multiparty Computation for $t < n/3$

3.1   Outline of the Construction

We first construct a $(7, 1)$-round VSS protocol with 2-level sharing; this protocol will be used by parties to share their inputs. This VSS protocol is based on the $(4, 3)$-round VSS protocol due to Gennaro et al. [17].

Based on the above VSS protocol with 2-level sharing, we can construct an $(8, 1)$-round VSS protocol with *3-level* sharing. We sketch the protocol below:

1. The dealer shares the secret s using the VSS protocol with 2-level sharing. In parallel, the dealer shares $g_1 \overset{\text{def}}{=} F_s(1), \ldots, g_n \overset{\text{def}}{=} F_s(n)$ using n invocations of the 2-level VSS protocol.

2. The parties reconstruct $g_1 - F_s(1), \ldots, g_n - F_s(n)$ and check if all the values are equal to 0. If this condition does not hold, the dealer is disqualified.

Using the above as a building block, we construct a $(17,3)$-round protocol for sharing a random multiple triple $(a,b,c=ab)$. On a high level, our protocol consists of the following steps:

1. Each party P_i shares two random values $a^{(i)}$ and $b^{(i)}$ using VSS with 3-level sharing.
2. Set $a = \sum a^{(i)}$ and $b = \sum b^{(i)}$. Note that a and b have been 3-level shared. Let $F_a(x)$ and $F_b(x)$ be the polynomials sharing a and b respectively. Using the sharing of product of shares protocol from [7], each party P_i shares $F_a(i) \cdot F_b(i)$ (using VSS with 2-level sharing) and proves that the right value is being shared; all parties can identify the set of parties that are not sharing the correct value.
3. Since $t < n/3$, there exist $2t+1$ parties P_i that correctly share $F_a(i) \cdot F_b(i)$ in step 2. Following Lemma 2, each party can compute its share of c non-interactively.

The above protocol runs VSS twice sequentially. Using the amortization technique from [15], the round complexity can be reduced to $(11,3)$. The idea is as follows: Suppose a party P_i needs to share two values a and b using VSS in two consecutive steps. P_i can do the following instead:

1. P_i picks a random value r and shares a and r using VSS.
2. P_i broadcasts the value $b-r$. Since the value r has been shared and $b-r$ has been made public, each party can compute its share of b non-interactively.

By running the above protocols in parallel, we obtain a $(11,3)$-round protocol Π_i that allows party P_i to both share its input and generate sufficiently-many random multiplication triples (cf. the overview in Section 1.1). In Section 3.3, we show how to use the ideas described in Section 1.1 (in particular, the idea of using moderated protocols) to implement the needed setup for all parties via a $(21,1)$-round protocol. (Our implementation of this protocol does not exactly follow the description in Section 1.1 for the reason described there). Based on this setup, we then show MPC protocols using only one round of broadcast in both the information-theoretic and computational settings (based on [7] and [10], respectively).

3.2   A $(7,1)$-Round VSS Protocol with 2-Level Sharing

Our protocol is based on the $(4,3)$-round VSS protocol of Gennaro et al. [17]. For readers who are already familiar with their protocol, we describe the two main modifications we make:

- Instead of using a "random pad" technique to detect inconsistent shares and resolve the inconsistencies in the next round — which requires two rounds of broadcast — we use a different method that requires only one round of broadcast. This gives us a $(6,2)$-round protocol.
- After the above, two rounds of broadcast still remain in the protocol of [17]. We devise a way for parties to postpone the first broadcast (and then combine it with the second) without affecting the progress of the protocol. This gives the $(7,1)$-round protocol as claimed.

We start by describing a $(6,2)$-round protocol. When we say the dealer P^* is *disqualified* we mean that execution of the protocol halts, and a default value s' is 2-level shared (using some default polynomials).

Round 1. The dealer chooses a random bivariate polynomial $F \in \mathbb{F}[x,y]$ of degree t in each variable with $F(0,0) = s$. The dealer sends to P_i the polynomials $g_i(x) \overset{\text{def}}{=} F(x,i)$ and $h_i(y) \overset{\text{def}}{=} F(i,y)$.

Round 2. P_i sends $h_i(j)$ to P_j.

Round 3. Let $h'_{j,i}$ be the value P_i received from P_j. If $h'_{j,i} \neq g_i(j)$, then P_i sends "complain$_i(j)$" to the dealer.

Round 4. If the dealer receives "complain$_i(j)$" from P_i in the last round, then the dealer sends "complain$_i(j)$" to P_j.

Round 5. For every ordered pair (i,j), parties P_i, P_j, and the dealer do the following:

- If P_i sent "complain$_i(j)$" to the dealer in round 3, then P_i broadcasts "$(P_i, i, j) : g_i(j)$" else P_i broadcasts "(P_i, i, j): no complaint".
- If P_j received "complain$_i(j)$" from the dealer in round 4, then P_j broadcasts "$(P_j, i, j) : h_j(i)$" else P_j broadcasts "(P_j, i, j): no complaint".
- If the dealer received "complain$_i(j)$" from P_i in round 3, then the dealer broadcasts "$(P^*, i, j) : F(j,i)$" else the dealer broadcasts "(P^*, i, j): no complaint".

We say party P_i is *unhappy* if P_i broadcasted a message of the form "$(P_i, i, j) : Y$," the dealer broadcasted a message of the form "$(P^*, i, j) : X$," and[4] $X \neq Y$. Similarly, P_i is unhappy if P_i broadcasted a message of the form "$(P_i, j, i) : Y$," the dealer broadcasted a message of the form "$(P^*, j, i) : X$," and $X \neq Y$.

Round 6. For each unhappy party P_j, the dealer broadcasts the polynomials $g_j(x)$ and $h_j(y)$, and each party P_i who is not unhappy broadcasts $b'_{i,j} \overset{\text{def}}{=} h_i(j)$ and $c'_{i,j} \overset{\text{def}}{=} g_i(j)$.

A party P_i that is not unhappy becomes *accusatory* if, in round 6, for some unhappy party P_j, the dealer broadcasts polynomial $g_j(x)$ and $h_j(y)$ but $b'_{i,j} \neq g_j(i)$ and $c'_{i,j} \neq h_j(i)$.

A party that is neither unhappy nor accusatory is said to be *happy*. The dealer is disqualified if the number of happy parties is less than $n - t$.

Output determination. If the dealer has not been disqualified, then a happy party P_i keeps the polynomials $g_i(x)$ and $h_i(y)$ it received from the dealer in the first round. An unhappy party P_i takes the polynomials broadcasted by the dealer in the final round as $g_i(x)$ and $h_i(y)$. (We do not define what accusatory players do, since if the dealer is not disqualified then all such parties are malicious.) The share P_i holds with respect to s is $s_i \overset{\text{def}}{=} g_i(0)$, and the share P_i holds with respect to s_i is $s_{i,j} \overset{\text{def}}{=} h_i(j)$.

We briefly argue that the requirements of Definition 5 hold. Consider an honest dealer P^*. For any pair of honest parties (P_i, P_j), the parties P^*, P_i, and P_j

---

[4] Note that X or Y can be field elements or the string "no complaint."

will always broadcast "no complaint" with respect to the ordered pair (i, j) in round 5. Hence secrecy will not be violated. It is easy to see that an honest party P_i will never become unhappy or accusatory. Therefore P^* will not be disqualified as dealer.

Next consider a malicious dealer P^*. Suppose two honest parties P_i and P_j are holding inconsistent shares (i.e., $h'_{j,i} \neq g_i(j)$ or $h'_{i,j} \neq g_j(i)$). In round 5, they will broadcast different messages with respect to the ordered pair (i, j) (or the ordered pair (j, i)). Hence the inconsistency is made known to all parties. The commitment property then follows the argument in [17].

Call the above protocol $\Pi_{(6,2)}$. We now show how to transform $\Pi_{(6,2)}$ into a $(7, 1)$-round protocol $\Pi_{(7,1)}$. The first four rounds of $\Pi_{(7,1)}$ are the same as $\Pi_{(6,2)}$. Then the parties carry out the following instructions:

Round 5. If P_i is instructed to broadcast a message m in round 5 of $\Pi_{(6,2)}$, then P_i sends the message m to all parties (via point-to-point links).

Round 6. Parties forward all the messages received in last round to all parties.

Round 7. The dealer does the following:
- For every ordered pair (i, j), if in round 6 the dealer received messages of the form "$(P_i, i, j) : X$" and "$(P^*, i, j) : Y$," with $X \neq Y$, each from $t + 1$ different parties, then the dealer broadcasts the polynomials $g_i(x)$ and $h_i(y)$.
- Similarly, for every ordered pair (i, j), if in round 6 the dealer received messages of the form "$(P_j, i, j) : X$" and "$(P^*, i, j) : Y$," with $X \neq Y$, each from $t + 1$ different parties, then the dealer broadcasts the polynomials $g_j(x)$ and $h_j(y)$.

In parallel with the above, all parties P_k do the following:
- A party broadcasts all the messages it received in round 5 (note that the round-5 messages include the identity of the sender).
- For every ordered pair (i, j), if in round 6 party P_k received messages of the form "$(P_i, i, j) : X$" and "$(P^*, i, j) : Y$," with $X \neq Y$, each from $t + 1$ different parties, then P_k broadcasts $b'_{k,i} \overset{\text{def}}{=} h_k(i)$ and $c'_{k,i} \overset{\text{def}}{=} g_k(i)$.
- For every ordered pair (i, j), if in round 6 party P_k received messages of the form "$(P_j, i, j) : X$" and "$(P^*, i, j) : Y$," with $X \neq Y$, each from $t + 1$ different parties, then P_k broadcasts $b'_{k,j} \overset{\text{def}}{=} h_k(j)$ and $c'_{k,j} \overset{\text{def}}{=} g_k(j)$.

Output determination. Parties decide on their output as follows:
1. A party P_i is said to *announce* a message m if, in round 7, at least $n - t$ parties broadcast that they received m from P_i in round 5.
2. A party P_i is *unhappy* if P_i announced a message of the form "$(P_i, i, j) : Y$," the dealer announced a message of the form "$(P^*, i, j) : X$," and $X \neq Y$. Similarly, P_i is unhappy if P_i announced a message of the form "$(P_i, j, i) : Y$," the dealer announced a message of the form "$(P^*, j, i) : X$," and $X \neq Y$.
3. A party P_i that is not unhappy becomes *accusatory* if, in round 7, for some unhappy party P_j, the dealer broadcasts polynomials $g_j(x)$ and $h_j(y)$, and P_i broadcasts $b'_{i,j}$ and $c'_{i,j}$ with $g_j(i) \neq b'_{i,j}$ or $h_j(i) \neq c'_{i,j}$.

We remark that because broadcast is used in round 7, all parties agree on which parties are unhappy or accusatory.

4. The dealer is disqualified if any of the following conditions hold:

   (DQ.1) There exists an ordered pair (i, j) such that the dealer does not announce a message of the form "$(P^*, i, j) : X$."

   (DQ.2) There exists an unhappy party P_i such that the dealer does not broadcast $g_i(x)$ or $h_i(y)$ in round 7.

   (DQ.3) The number of unhappy and accusatory parties exceeds t.

5. If the dealer has not been disqualified, then the parties determine their output the same way as in $\Pi_{(6,2)}$.

We now make two observations regarding the protocol $\Pi_{(7,1)}$:

- If an honest party P_i sends a message m to all parties in round 5, then P_i will be considered as announcing m by the end of round 7.
- If a (possibly malicious) party P_i announces a message m, then every honest party received m from at least $t + 1$ different parties in round 6.

If an honest party P_i sends a message m to all parties in round 5, then all honest parties receive it. Since all honest parties broadcast this information in round 7 and there are at least $n - t$ of them, the first condition above holds. If a party P_i announces a message m in round 5, then, by definition, in round 7 at least $n - t$ parties broadcast that they received m from P_i in round 5. At least $n - t - t \geq t + 1$ of them are honest. These honest parties will forward m to all parties in round 6. Hence the second condition above holds.

With the above observations, it is not hard to see that $\Pi_{(7,1)}$ will preserve the commitment property of $\Pi_{(6,2)}$. Now we argue that secrecy is preserved as well. The only issue is that, if P_i is malicious, then in round 7 an honest party P_k may broadcast $h_k(i)$ and $g_k(i)$ (or an honest dealer may broadcast $g_i(x)$ and $h_i(y)$) even if P_i is not considered unhappy by the end of the protocol. However, this does not affect secrecy since the malicious P_i already knows these values.

We defer the full description of the protocol $\Pi_{(7,1)}$ and the proof of correctness to the full version.

3.3   Secure Multiparty Computation Using One Round of Broadcast

In this section, we describe how parties can share their inputs and generate random multiplication triples using only one round of broadcast. As discussed in Section 1.1, following such a "setup" phase the parties can then compute their respective outputs *without using any additional invocations of broadcast* using the techniques of Beaver [3]. For completeness, this too is discussed below.

As stated in Section 3.1, we can construct an $(11, 3)$-round protocol Π_i that simultaneously allows a party P_i to share its input and generate sufficiently-many random multiplication triples. In the resulting protocol, broadcast is invoked in the $7^{\text{th}}, 9^{\text{th}}$, and 10th rounds. We now show how to transform Π_i into a $(21, 1)$-round protocol Π_i' with the following properties: (1) By the end of the protocol, all honest parties output a common bit $\mathsf{trust}(i)$; (2) if P_i is honest, then $\mathsf{trust}(i) = 1$. Moreover, the view of the adversary remains independent of P_i's input; (3) if $\mathsf{trust}(i) = 1$, then P_i's input as well as all the random multiplication triples have been 2-level shared. Furthermore, given the view of the adversary, the first two components of each multiplication triple (a, b, c) are uniformly distributed in the field $\mathbb{F}$.

Π_i' proceeds as follows: Each party P_j initializes a binary flag f_j to 1. Roughly speaking, the flag f_j indicates whether P_j "trusts" P_i or not. The parties then run an execution of Π_i. When a party P is directed by Π_i to send message m to another party over a point-to-point channel, it simply sends this message. When a party P is directed to broadcast a message m in the 7th or 9th round of Π_i, all parties run the following "simulated broadcast" sub-routine:

- P gradecasts the message m.
- Each party P_i gradecasts the message it output in the previous step.
- Let (m_j, g_j) and (m_j', g_j') be the output of party P_j in steps 1 and 2, respectively. Within the underlying execution of Π_i, party P_j will use m_j' as the message "broadcast" by P. Furthermore, P_j sets $f_j := 0$ if either (or both) of the following conditions hold: (1) $g_j' \neq 2$, or (2) $m_j' \neq m_j$ and $g_j = 2$.

In the 10th round of Π_i, when a party P_j is directed to broadcast a message m, it simply broadcasts this message along with the flag f_j. If fewer than $2t + 1$ parties broadcast $f_j = 1$, then all parties set $\mathsf{trust}(i) = 0$; otherwise, all parties set $\mathsf{trust}(i) = 1$.

The transformation from Π_i to Π_i' is similar to the compilation of VSS to *moderated* VSS in [22], except that we retain the last invocation of broadcast in Π_i. The proof of correctness is similar and is omitted due to space limitations. We now describe how to use the above to obtain a secure MPC protocol.

The information-theoretic setting. Suppose the given circuit has K multiplication gates. Following the approach in [7], we can construct the following error-free multiparty computation protocol:

1. For $1 \leq i \leq n$, protocol Π_i' is executed in parallel (i.e., P_i shares its input values and generates[5] K random multiplication triples). At the end of this step, all parties agree on values $\mathsf{trust}(i)$ for $i \in [n]$.
2. For each i, if $\mathsf{trust}(i) = 0$, then P_i is disqualified and default values are used as the input values of P_i. Let i^* be the minimum value such that $\mathsf{trust}(i^*) = 1$. In the next step, parties use the multiplication triples generated in Π_{i^*}'.
3. The parties evaluate the circuit gate by gate. Suppose that values x and y, representing the values on the two input-wires of some gate in the circuit, have been 1-level shared. The value of the output wire can be 1-level shared as follows:

Addition gate: This is easy to do non-interactively.

---

[5] An optimization is to have each party generate $K/(n - t)$ random multiplication triples, and then use (in step 3, below) the multiplication triples generated by the first $n - t$ non-disqualified parties.

Multiplication gate: Using the method suggested in [3], this can be reduced to one round of value reconstruction while consuming one random multiplication triple (a, b, c). Specifically: the parties publicly reconstruct $d_x = x - a$ and $d_y = y - b$. Then, parties non-interactively compute shares of $d_x d_y + d_x b + d_y a + c$ (using their shares of a, b, c). Note that if $c = ab$, then $d_x d_y + d_x b + d_y a + c = xy$ (recall that calculations are performed in a field of characteristic 2).

3. Output values are reconstructed by having all parties send appropriate shares to the appropriate parties and using error correction.

The above protocol invokes one round of broadcast. The total number of rounds required is equal to the depth of the circuit being computed plus 22.

The computational setting. Assuming the existence of one-way functions, Damgård and Ishai [10] give a multiparty computation protocol with round complexity $(O(1), O(1))$. Roughly speaking, they transform evaluations of a given circuit into evaluations of degree-3 polynomials. Using the approach described in the last section, we can obtain an $(O(1), 1)$-round MPC protocol. (Details are omitted due to space constraints.) The end result is that we obtain a $(26, 1)$-round MPC protocol. Using the expected constant-round broadcast protocol from [22], the (expected) round complexity of the MPC protocol becomes 41.[6]

4 Secure Multiparty Computation for $t < n/2$

For $t < n/2$, we assume a PKI and a secure digital signature scheme. Our protocol is based on the protocols in [9,10]. On a high level, the construction is similar to the case of $t < n/3$ with the following differences:

1. Since t may be greater than $n/3$, we can no longer apply Reed-Solomon decoding to reconstruct shared values. Instead, we use the linear information checking tool from [9]. Unfortunately, their protocol as described requires additional invocations of broadcast. We show how to eliminate this usage of broadcast by utilizing the PKI.

2. The presence of a PKI enables us to "catch" a malicious party who cheats more easily. For instance, if a malicious party P_i sends two contradicting messages (with valid signatures) to P_j and P_k, then the latter two parties can conclude that P_i is cheating upon exchange of messages. This allows us to construct more round-efficient protocols.

3. As in the case of $t < n/3$ (see Section 3.1), in the protocol for sharing a random multiplication triple (a, b, c), the parties first share two random field elements a and b and then each party shares $a_i b_i$ (where a_i and b_i are the shares held by P_i with respect to a and b) and proves that the correct value has been shared. In order for the parties to compute their shares of c (by

---

[6] Even though broadcast is not used in the final round of the resulting protocol, we do not need to introduce special techniques to deal with issues of non-simultaneous termination since only secrets are reconstructed after broadcast is invoked.

applying Lemma 2), there should exist a set of $2t + 1$ parties P_i that have correctly shared $a_i b_i$. For $t < n/3$, this condition is always satisfied since there are at least $2t + 1$ honest parties. However, for $t < n/2$, we can only guarantee that $t + 1$ (honest) parties will correctly share the product of their shares. Hence the protocol needs to be designed in such a way that if a party does not share the product of its shares correctly, then its shares will be made public.

Due to lack of space, we defer the actual details of the construction to the full version. The round complexity of the final MPC protocol we construct is $(34, 1)$. When the protocol is compiled for a point-to-point network, the round complexity is 64 (in expectation).

5   Conclusion

Previous work on round complexity has (for the most part) aimed to minimize *the total number of rounds* for a given task, but under the assumption of a broadcast channel "for free". In fact, broadcast is not for free since emulating broadcast over point-to-point channels is rather expensive. We argue here that if one is ultimately interested in round-efficient protocols for point-to-point networks (which is where most protocols would eventually be run), then it is more productive to focus on minimizing *the number of rounds in which broadcast is used*. With this motivation, we have shown here protocols for secure multi-party computation in a number of settings that use broadcast in a *single* round.

A number of interesting open questions are suggested by our work:

1. The work of [17,15] characterizes the round complexity of VSS (for $t < n/3$) when a broadcast channel is available. Can we obtain a similar characterization of the round complexity of VSS in a point-to-point network? As a step toward this goal, one might start by establishing the optimal round complexity of VSS when the broadcast channel is used only *once*.

2. Our $(O(1), 1)$-round MPC protocol for $t < n/2$ assumes the existence of a PKI. Does there exist a constant-round MPC protocol using a single round of broadcast that does *not* rely on a PKI? Although a PKI will anyway be needed to implement broadcast, the question is of theoretical interest. Furthermore, such a protocol may be useful in settings (such as a small-scale wireless network) when a broadcast channel *is* available, or when it is desirable to minimize the usage of digital signatures for reasons of efficiency.

References

1. D. Beaver. Multiparty protocols tolerating half faulty processors. In *Advances in Cryptology — Crypto '89*, pages 560–572. Springer-Verlag, 1989.

2. D. Beaver. Secure multi-party protocols and zero-knowledge proof systems tolerating a faulty minority. *Journal of Cryptology*, 4(2):75–122, 1991.

3. D. Beaver. Efficient multiparty protocols using circuit randomization. In *Advances in Cryptology — Crypto '91*, pages 420–432. Springer-Verlag, 1992.
4. D. Beaver, S. Micali, and P. Rogaway. The round complexity of secure protocols. In *22nd Annual ACM Symposium on Theory of Computing*, pages 503–513, 1990.
5. M. Ben-Or. Another advantage of free choice: Completely asynchronous agreement protocols. In *2nd Annual ACM Symposium on Principles of Distributed Computing (PODC)*, 1983.
6. M. Ben-Or and R. El-Yaniv. Resilient-optimal interactive consistency in constant time. *Distributed Computing*, 16(4):249–262, 2003.
7. M. Ben-Or, S. Goldwasser, and A. Wigderson. Completeness theorems for non-cryptographic fault-tolerant distributed computation. In *Proc. 20th Annual ACM Symposium on Theory of Computing*, pages 1–10. ACM Press, 1988.
8. D. Chaum, C. Crepeau, and I. Damgård. Multiparty unconditionally secure protocols. In *Proc. 20th Annual ACM Symposium on Theory of Computing*, pages 11–19. ACM Press, 1988.
9. R. Cramer, I. Damgård, S. Dziembowski, M. Hirt, and T. Rabin. Efficient multiparty computations secure against an adaptive adversary. In *Advances in Cryptology — Eurocrypt '99*, volume 1592 of *LNCS*, pages 311–326. Springer-Verlag, 1999.
10. I. Damgård and Y. Ishai. Constant-round multiparty computation using a black-box pseudorandom generator. In *Adv. in Cryptology — Crypto 2005*, pages 378–394. Springer-Verlag, 2005.
11. D. Dolev and H. Strong. Authenticated algorithms for Byzantine agreement. *SIAM J. Computing*, 12(4):656–666, 1983.
12. P. Feldman and S. Micali. An optimal probabilistic protocol for synchronous Byzantine agreement. *SIAM J. Comput.*, 26(4):873–933, 1997.
13. M. J. Fischer and N. A. Lynch. A lower bound for the time to assure interactive consistency. *Info. Proc. Lett.*, 14(4):183–186, 1982.
14. M. Fitzi and J. A. Garay. Efficient player-optimal protocols for strong and differential consensus. In *22nd Annual ACM Symposium on Principles of Distributed Computing*, pages 211–220, 2003.
15. M. Fitzi, J. A. Garay, S. Gollakota, C. P. Rangan, and K. Srinathan. Round-optimal and efficient verifiable secret sharing. In *3rd Theory of Cryptography Conference*, pages 329–342, 2006.
16. J. A. Garay and Y. Moses. Fully polynomial Byzantine agreement for $n > 3t$ processors in $t + 1$ rounds. *SIAM J. Comput.*, 27(1):247–290, 1998.
17. R. Gennaro, Y. Ishai, E. Kushilevitz, and T. Rabin. The round complexity of verifiable secret sharing and secure multicast. In *33rd Annual ACM Symposium on Theory of Computing*, pages 580–589, 2001.
18. R. Gennaro, Y. Ishai, E. Kushilevitz, and T. Rabin. On 2-round secure multiparty computation. In *Advances in Cryptology — Crypto 2002*, pages 178–193. Springer-Verlag, 2002.
19. R. Gennaro, M. O. Rabin, and T. Rabin. Simplified VSS and fast-track multiparty computation with applications to threshold cryptography. In *Proc. 17th Annual ACM Symposium on Principles of Distributed Computing*, pages 101–111. ACM Press, 1998.
20. S. Goldwasser and Y. Lindell. Secure computation without agreement. In *16th Intl. Conf. on Distributed Computing (DISC)*, pages 17–32. Springer-Verlag, 2002.

21. M. Hirt, J. B. Nielsen, and B. Przydatek. Cryptographic asynchronous multi-party computation with optimal resilience. In *Adv. in Cryptology — EUROCRYPT 2005*, volume 3494 of *Lecture Notes in Computer Science*, pages 322–340. Springer-Verlag, 2005.

22. J. Katz and C.-Y. Koo. On expected constant-round protocols for Byzantine agreement. In *Adv. in Cryptology — Crypto 2006*. Full version available at http://eccc.hpi-web.de/eccc-reports/2006/TR06-028/index.html.

23. L. Lamport, R. Shostak, and M. Pease. The Byzantine generals problem. *ACM Trans. Program. Lang. Syst.*, 4(3):382–401, 1982.

24. Y. Lindell, A. Lysyanskaya, and T. Rabin. Sequential composition of protocols without simultaneous termination. In *Proc. 21st Annual ACM Symposium on Principles of Distributed Computing*, pages 203–212, 2002.

25. S. Micali and T. Rabin. Collective coin tossing without assumptions nor broadcasting. In *Adv. in Cryptology — Crypto '90*, pages 253–266. Springer-Verlag, 1991.

26. M. Pease, R. Shostak, and L. Lamport. Reaching agreement in the presence of faults. *J. ACM*, 27(2):228–234, 1980.

27. B. Pfitzmann and M. Waidner. Information-theoretic pseudosignatures and Byzantine agreement for $t \geq n/3$. Technical Report RZ 2882 (#90830), IBM Research, 1996.

28. M. Rabin. Randomized Byzantine generals. In *Proc. 24th IEEE Symposium on Foundations of Computer Science*, pages 403–409, 1983.

29. T. Rabin. Robust sharing of secrets when the dealer is honest or cheating. *J. ACM*, 41(6):1089–1109, 1994.

30. T. Rabin and M. Ben-Or. Verifiable secret sharing and multiparty protocols with honest majority. In *Proc. 21st Annual ACM Symposium on Theory of Computing*, pages 73–85. ACM Press, 1989.

A   Round Complexity of Emulating Broadcast

In this section, we discuss the round complexity of emulating broadcast by the most round-efficient randomized protocol known [22], The randomized broadcast protocols of [22] allow all parties to broadcast a message simultaneously.[7] Roughly speaking, the protocols consist of two phases. The first phase is a "setup" phase that is independent of the messages being broadcast, and (only) consists of parallel executions of moderated VSS. Using the VSS protocol developed in this work, for $t < n/3$, this initial phase can be implemented in (strict) 12 rounds.

The execution of the second phase depends on the messages being broadcast, and terminates in 16 rounds (in expectation) assuming $t < n/3$. For a single invocation of broadcast, the first 5 rounds of the second phase can be executed in parallel with the last 5 rounds of the first phase, and hence the round complexity of the entire broadcast protocol (in expectation) is 23. (See [22, Appendix C] for

---

[7] Note, however, that the protocols do not achieve the "simultaneous broadcast" functionality (i.e., with all broadcast messages being independent of each other). Instead, they simply emulate a round of broadcast with rushing.

further details. Note that the numbers computed there do not take into account the more efficient VSS protocol constructed here.)

However, if broadcast is invoked multiple times sequentially the round complexity does not simply scale linearly. The reason is that the second phase does not guarantee simultaneous termination and so sequential executions do not compose directly; instead, additional steps (which increase the round complexity) are needed. Without going into the details (see [22, Appendix C]), it is possible to show using the techniques of [6,24,22] that emulation of multiple rounds of broadcast requires 32 rounds (in expectation) per additional broadcast, in addition to an initial 23 rounds.

As stated in the Introduction, the Micali-Rabin VSS protocol uses 16 rounds but only a single round of broadcast. In fact, the broadcast is used in the final round. If the first 15 rounds of the Micali-Rabin protocol are executed in parallel to the first phase of the broadcast protocol, then the Micali-Rabin protocol takes $31 = 15 + 16$ rounds (in expectation) when compiled for a point-to-point network. On the other hand, the protocol by Fitzi et al. requires $55 = 32 + 23$ rounds since they use two rounds of broadcast. (Messages sent during the one round of their protocol that does not use broadcast can be "piggy-backed" on messages sent as part of the broadcast protocols.)

For the case of $t < n/2$, the numbers are even worse: a single invocation of broadcast takes 36 rounds (in expectation), while 58 rounds are needed per additional broadcast when broadcast is used in two or more rounds.

Atomic Secure Multi-party Multiplication with Low Communication

Ronald Cramer[1,*], Ivan Damgård[2], and Robbert de Haan[3,**]

[1] CWI, Amsterdam & Mathematical Institute, Leiden University, The Netherlands
cramer@cwi.nl
http://www.cwi.nl/~cramer, http://www.math.leidenuniv.nl/~cramer
[2] Comp. Sc. Dept., Aarhus University & BRICS, Denmark
ivan@daimi.au.dk
[3] CWI, Amsterdam, The Netherlands
R.de.Haan@cwi.nl
http://www.cwi.nl/~haan

Abstract. We consider the standard secure multi-party multiplication protocol due to M. Rabin. This protocol is based on Shamir's secret sharing scheme and it can be viewed as a practical variation on one of the central techniques in the foundational results of Ben-Or, Goldwasser, and Wigderson and Chaum, Crépeau, and Damgaard on secure multi-party computation. Rabin's idea is a key ingredient to virtually all practical protocols in threshold cryptography.

Given a passive t-adversary in the secure channels model with synchronous communication, for example, secure multiplication of two secret-shared elements from a finite field K based on this idea uses one communication round and has the network exchange $O(n^2)$ field elements, if $t = \Theta(n)$ and $t < n/2$ and if n is the number of players. This is because each of $O(n)$ players must perform Shamir secret sharing as part of the protocol. This paper demonstrates that under a few restrictions much more efficient protocols are possible; even at the level of a single multiplication.

We demonstrate a twist on Rabin's idea that enables one-round secure multiplication *with just $O(n)$ bandwidth* in certain settings, thus reducing it from quadratic to linear. The ideas involved can additionally be employed in the evaluation of arithmetic circuits, where under appropriate circumstances similar efficiency gains can be obtained.

1 Introduction

Given a passive t-adversary in the secure channels model with synchronous communication, secure multiplication of two secret-shared elements from a finite field K based on Rabin's idea uses one communication round and has the network exchange $O(n^2)$ field elements, if $t = \Theta(n)$ and $t < n/2$ and if n is the number of

---

* Ronald Cramer's research has been partially supported by NWO VICI.
** Robbert de Haan's research has been partially funded by the Dutch BSIK/BRICKS project PDC1.

M. Naor (Ed.): EUROCRYPT 2007, LNCS 4515, pp. 329–346, 2007.

players. This is because each of $O(n)$ players must perform Shamir secret sharing as part of the protocol.

We demonstrate a twist on Rabin's idea that enables one-round secure multiplication *with just $O(n)$ bandwidth*, thus reducing it from quadratic to linear. However, to obtain this efficiency we need to decrease the maximal corruption tolerance, but still $t = \Theta(n)$, i.e., a number of corruptions is tolerated that is still a constant fraction of n. Furthermore, we require the finite field L to have a certain property; it should contain a subfield K over which it has an extension degree linear in n.

For this result we emphasize that, unlike in previous approaches (such as [10]), the mentioned costs analysis is not amortized, as we consider "single-shot" (or "atomic") secure multiplication only. The techniques involved can provide considerable efficiency gain in certain secure linear algebra computations, such as securely computing the determinant of a matrix and securely solving a linear system of equations, where the chosen field is typically large in order to ensure a small error probability [5], [6].

A main handle that enables the result mentioned above is a theorem that demonstrates that when certain values can be extracted from the shares in a ramp scheme by means of a linear function, several linear functions on these values can be securely computed at the cost of only a single multiplication and using only a single round of communication. We demonstrate how this theorem, together with a technique due to Franklin and Yung [7], can be used to speed up computation over arithmetic circuits.

After discussing the theorem and the main idea behind our variation, we detail some further handles for trade-offs between communication efficiency and corruption tolerance. We also demonstrate similar reductions in communication complexity for secure computation in the presence of an active adversary.

2   Rabin's Secure Multiplication Protocol

We consider Rabin 's idea (as explained in [9]) for secure multiplication. This protocol is a key ingredient to virtually all practical protocols in threshold cryptography.

For the moment we focus on the secure channels model with synchronous communication, in the presence of a passive t-adversary where t is maximal such that $t < n/2$ and where n is the number of players in the network. Assuming that the network has (t, n)-Shamir-sharings of two secret values a and b, the protocol allows the network to securely generate a (t, n)-Shamir-sharing of the product $a \cdot b$. The technical idea behind this protocol is a simple and elegant reduction from secure multiplication to secure linear computation.

Concretely, let K be a finite field with $|K| > n$. Let $x_1, \ldots, x_n$ be distinct non-zero elements from K. Each player P_i has a share a_i in the secret a and a share b_i in the secret b. Let f denote the polynomial of degree at most t such that $f(0) = a$ and such that $f(x_i) = a_i$ for all i. Similarly, g is the polynomial defining the secret sharing of b.

Now note that the values $(a_1 \cdot b_1, \ldots, a_n \cdot b_n)$ are consistent with the polynomial $f \cdot g$, i.e., $(fg)(x_i) = a_i b_i$ for all i. Since fg has degree at most $2t$ and since $2t < n$, these values uniquely determine $f \cdot g$, by Lagrange interpolation. Concretely, there exists a (public) linear map $\phi : K^n \longrightarrow K$ such that $\phi(a_1 b_1, \ldots, a_n b_n) = ab$ always.

This reduces secure multiplication to secure linear computation: it is sufficient to compute ϕ securely on the secret inputs $a_i b_i$, where $a_i b_i$ is the input of player P_i. These inputs can of course be computed locally. So, first the players perform input sharing, i.e., each player P_i (t, n)-Shamir-shares $a_i b_i$ among the network, using a polynomial h_i. Then each player P_j simply computes locally $\phi(h_1(x_j), \ldots, h_n(x_j))$ as his share in ab. The overall result is clearly a (t, n)-Shamir-sharing of ab, defined by the polynomial $h = \phi(h_1, \ldots, h_n)$. This protocol takes a singe round of communication, and it involves the exchange of $O(n^2)$ elements from K.

3   Prior Work: Parallel Secure Computation

Franklin and Yung [7] have shown that interesting advantages can be offered in secure computation by relaxing the corruption tolerance level by just a constant fraction of the number of players. They showed an *amortized* cost reduction in communication complexity. More precisely, they assume that the number of corrupted parties t satisfies $t < cn$ where c is a constant less than the standard maximum that can be tolerated in the given scenario (typically $1/2$ or $1/3$). The same secure evaluation can now be performed on several different inputs in parallel, while the total communication amounts to that of a single secure evaluation.

Although our goals and techniques substantially differ from [7], we do use some of the ideas. We recall their techniques below. Consider for simplicity the secure channels model with a passive adversary, just as in the description of Rabin's idea, though with the following differences.

Let $\hat{t}$ be a positive integer with $\hat{t} < n/2$, and let k be an integer with $1 \leq k \leq \hat{t}$. Define $t = \hat{t} - k$. The finite field K is chosen such that $|K| > n + k$.

First consider the following variation on Shamir's secret sharing scheme. Let the sets $\{x_1, \ldots, x_n\}$ and $\{e_0, \ldots, e_k\}$ be two disjoint sets of distinct elements from K.

- Let $a = (u_0, \ldots, u_k)$ be a vector of secret elements from K.
- Choose a random polynomial $f(X) \in K[X]$ of degree at most $\hat{t}$ such that

$$f(e_0) = u_0, \ldots, f(e_k) = u_k.$$

- Define the shares as

$$a_1 = f(x_1), \ldots, a_n = f(x_n).$$

Clearly, $\hat{t} + 1$ shares or more jointly determine f and hence the secret vector a. As to privacy, it is a straightforward consequence of Lagrange-interpolation that t or fewer shares jointly give no information on the secret vector. So it is a $(t, \hat{t} + 1)$-ramp scheme, with secrets of length $\hat{t} - t + 1$.

Now, $k + 1$ secure multiplications of $(u_0 v_0, \ldots, u_k v_k)$ can be performed in a very compact manner. Suppose that vectors $a = (u_0, \ldots, u_k)$ and $b = (v_0, \ldots v_k)$ have been secret-shared. Say that the shares in a are $(a_1, \ldots, a_n)$ (with defining polynomial f) and the shares in b are $(b_1, \ldots, b_n)$ (with defining polynomial g). The network may now obtain a secret-sharing according to the scheme above (and with the same parameters) of the vector $a * b := (u_0 v_0, \ldots, u_k v_k)$ as follows.

First we note that for $j = 0, 1, \ldots, k$, it holds that $(fg)(e_j) = u_j v_j$. For a reason similar to the one used in the description of Rabin's idea, there exists linear maps $\phi_j : K^n \longrightarrow K$ such that $u_j v_j = \phi_j(a_1 b_1, \ldots, a_n b_n)$ $(j = 0 \ldots k)$.

Each player P_i now simply secret-shares (according to the scheme above, with the same parameters) the vector $(\phi_0(\epsilon_i) a_i b_i, \ldots, \phi_k(\epsilon_i) a_i b_i)$, where $\epsilon_i \in K^n$ is the i-th unit vector. Define the polynomial $h(X) = \sum_{i=1}^n h_i(X)$, where h_i is the polynomial used by P_i in the sharing step above $(i = 1 \ldots n)$. This polynomial is consistent with the parameters of the scheme, the secret encoded by it is the vector $a * b$ and each player P_i can locally compute his share as $\sum_{j=1}^n h_j(x_i)$. We will demonstrate later that there is a more general way to look at this last resharing step (see Theorem 1).

4   Ramp Schemes and Share Conversion

We now present a formal definition of (linear) ramp schemes, which can be seen as a generalization of threshold secret sharing schemes.

Definition 1. *Let M_i be a $d_i \times e$ matrix for $i = 1, 2, \ldots, n$. For every set $A \subset \{1, 2, \ldots, n\}$, let M_A be the matrix defined by stacking the matrices $(M_i)_{i \in A}$ on top of each other. The scheme defined as such is called a (linear) $(t, \hat{t} + 1)$-ramp scheme of embedding degree $k + 1$ if the following two properties hold:*

- *For any $A \subset \{1, 2, \ldots, n\}$ with $|A| \geq \hat{t} + 1$, there are vectors $r_0, r_1, \ldots, r_k$ such that $r_i M_A = u_i$, where u_i is the ith unit vector.*
- *For any $A \subset \{1, 2, \ldots, n\}$ with $|A| \leq t$ and any vector $v = (v_0, v_1, \ldots, v_k)$ there is a vector $\kappa \in Ker M_A$ where the first $k+1$ coordinates of κ correspond with the coordinates of v.*

Ramp schemes are used for secret sharing as follows. Let $s = (s_0, s_1, \ldots, s_k) \in K^{k+1}$ be a secret vector and choose $b = (b_0, b_1, \ldots, b_{e-1}) \in K^e$ at random under the restriction that $b_i = s_i$ for $i = 0, 1, \ldots, k$. Now define $s_i := M_i b \in K^{d_i}$ as the share for the ith player. Note that the embedding degree of the ramp scheme defines the dimension of the secret space over K.

The first condition for ramp schemes is now equivalent to the statement for the corresponding secret sharing scheme that $\hat{t} + 1$ or more players can compute every coordinate of the secret vector via a linear combination of their shares. Furthermore, the second condition is equivalent to the statement that for any subset consisting of at most t players every possible secret vector is equally consistent with their shares. Another key point to note is that ramp schemes allow for a "gray zone" between the unqualified and the qualified number of players, which allows the size of the shares to be smaller than the size of the secret.

There is a way to rewrite the scheme due to Franklin and Yung to the notation used above by applying appropriate operations on the columns of a Vandermonde matrix. However, since the representation using polynomials is rather convenient for both their scheme and our scheme from Section 5, we will stick to a polynomial notation for these schemes in the sequel. Naturally, there is also a straightforward way of rewriting our scheme to the formal notation above, which boils down to the elimination of a number of columns from a Vandermonde matrix.

One of the key ingredients of our results is the following theorem, which allows us to convert shares between different types of linear ramp schemes, while at the same time computing a number of linear functions on secret values in the ramp scheme in parallel.

Theorem 1. *Assume that the players hold shares $c_1, .., c_n$ in a linear ramp scheme of the secret vector $(s_1, .., s_m)$ – which means there exist linear maps $\phi_j : K^n \longrightarrow K$ such that $s_j = \phi_j(c_1, \ldots, c_n)$ $(j = 0 \ldots m)$ and that the set of all players is qualified in this scheme. Furthermore, let arbitrary linear functions $F_1, .., F_\ell$, $F_i : K^m \to K$, be given. Then in a single round of communication, the shares in this scheme can be transformed into shares in any other linear ramp scheme with secret space of dimension at least ℓ with secret vector*

$$(F_0(s_0, \ldots, s_m), \ldots, F_\ell(s_0, \ldots, s_m)),$$

Furthermore, privacy is maintained for any subset of players for which privacy holds in both of the ramp schemes involved.

Proof. Assume that the functions F_j are $F_j(x_0, \ldots, x_m) := \sum_{w=0}^m \mu_w^{(j)} x_w$ for some $\mu_w^{(j)} \in K$ and define $\beta_i^{(j)} := \sum_{w=0}^m \mu_w^{(j)} \phi_w(\epsilon_i) c_i$. Note that

$$s_j = \phi_j(c_1, \ldots, c_n) = \sum_{i=1}^n \phi_j(\epsilon_i) c_i,$$

so that

$$F_j(s_0, \ldots, s_m) = \sum_{w=0}^m \mu_w^{(j)} s_w = \sum_{w=0}^m \mu_w^{(j)} \left(\sum_{i=1}^n \phi_w(\epsilon_i) c_i \right)$$
$$= \sum_{i=1}^n \left(\sum_{w=0}^m \mu_w^{(j)} \phi_w(\epsilon_i) c_i \right) = \sum_{i=1}^n \beta_i^{(j)},$$

and that player i can ramp share the vector $(\beta_i^{(0)}, \ldots, \beta_i^{(\ell)})$ in the target scheme, as the coefficients $\beta_i^{(j)}$ only depend on its share c_i and public information. After all players have reshared their shares in this way and the players locally sum up their new shares, they obtain shares in the target scheme with secret vector $(F_0(s_0, \ldots, s_m), \ldots, F_\ell(s_0, \ldots, s_m))$. The privacy claim is straightforward to verify and the result follows.   □

In particular, Theorem 1 demonstrates that we can in a single round of communication securely compute any list of linear functions (up to a certain size) on the ramp shared secret values. Combined with the techniques of Franklin and Yung, this is used in Section 8 to enable more efficient evaluation of certain arithmetic circuits. Theorem 1 is later also used in combination with the ramp scheme from Section 5, where the resulting scheme allows to compute products of values in an extension field of K using only shares and communication consisting of elements in K.

5   Atomic Secure Multiplication: The Main Idea

In [7], the amortized communication complexity of a secure computation is reduced by performing a linear number of multiplications in parallel. The more general techniques described in this section alternatively allow to reduce the *atomic* communication complexity, i.e., the minimum communication complexity required to perform a *single* secure multiplication. In particular, we demonstrate how a decreased maximum corruption tolerance, while still a constant fraction of n, allows one to gain a linear factor in communication complexity for a single multiplication. However, for this we require that the finite field that is used in the computation has some additional structure. These techniques can provide considerable efficiency gain, for instance when used as a building block in secure linear algebra computations over large finite extension fields [5,6].

The technical idea behind our result can be summarized as follows. We use a dedicated ramp scheme, different from the one in [7]. It is defined using an extension field L over K, but each share is just a single element from K. The secret is an element in L, which is represented as a vector of elements from K by fixing a basis of L over K and interpreting L as a vector space over K. This way, the information rate of the scheme improves as the degree of L over K increases, but we pay for this by having to decrease the corruption tolerance appropriately.

This approach is additive in the sense that sums of sharings of two elements from L give a sharing of their sum. The relative difficulty lies in the product. We show a variation on Rabin's idea that allows the network to securely compute, in a single round, the vector-representation over K of the product of two elements from L, using just $O(n)$ bandwidth. [1] Our idea depends crucially on the properties of our dedicated ramp scheme.

---

[1] Our results here should be contrasted with those of [4], which deals with low communication secure computation over very small fields, and uses an entirely different technique.

Definition 2. *For each integer d with $0 \leq d \leq 2k$ the polynomial H_d is defined as*

$$H_d(X_0, \ldots, X_k, Y_0, \ldots, Y_k) = \sum_{0 \leq q, r \leq k \ : \ q+r=d} X_q \cdot Y_r.$$

Definition 3. *Let k be a non-negative integer and let $\hat{t}$ be an integer with $2k < \hat{t}$. The linear subspace $V_{k,\hat{t}}(K)$ of the vector space of polynomials of degree at most $\hat{t}$ consists of all polynomials $f(X) \in K[X]$ of the form*

$$f(X) = a(X) + R(X) \cdot X^{2k+1},$$

where $a(X) \in K[X]$ is a polynomial of (formal) degree k and where $R(X) \in K[X]$ is a polynomial of (formal) degree $\hat{t} - 2k - 1$.

Note the presence of a "gap" in the polynomials. It ensures that after local multiplication of shares none of the higher-term random coefficients in the corresponding product polynomial interferes with the coefficients that results from the lower-term coefficients (which contain the secret vectors). Furthermore, the degree of the polynomials is chosen large enough to ensure that the higher-term coefficients provide sufficient privacy.

Now assume that $2k < \hat{t}$. Thus, $a(X)$ has degree at most k as a polynomial, but its coefficient vector will be taken of length $k + 1$ in all the cases. We will sometimes "identify $a(X)$ with its coefficient vector a." Similar for $R(X)$. We have the following trivially verified property.

Lemma 1. *If $f(X) = a(X) + R(X) \cdot X^{2k+1}$ and $g(X) = b(X) + R'(X) \cdot X^{2k+1} \in V_{k,\hat{t}}(K)$, then*

$$f(X) \cdot g(X) = H_0(a, b) + H_1(a, b) \cdot X + \ldots + H_{2k}(a, b) \cdot X^{2k} + S(X) \cdot X^{2k+1},$$

where a, b are taken as the coefficient vectors (of length $k+1$) of the corresponding polynomials and where $S(X)$ is a polynomial of degree at most $2\hat{t} - 2k - 1$.

Now let L be an extension field of K of degree $k + 1$, and let θ be such that

$$L = K(\theta).$$

The fact that $1, \theta, \ldots, \theta^k$ is a basis for the field L as a $k + 1$-dimensional K-vector space implies the following lemma. Let $a = u_0 + u_1 \cdot \theta + \ldots + u_k \cdot \theta^k \in L$ and $b = v_0 + v_1 \cdot \theta + \ldots + v_k \cdot \theta^k \in L$, with the u_i and v_j elements from K.

Lemma 2. *With K, θ and L fixed as above, the following holds. There exist linear maps $\chi_j : K^{2k+1} \longrightarrow K$ $(j = 0 \ldots k)$ such that for all $a, b \in L$*

$$ab = \sum_{j=0}^{k} \chi_j(H_0(a, b), \ldots, H_{2k}(a, b)) \cdot \theta^j,$$

where a and b are given by their respective coordinate vectors $(u_0, \ldots, u_k)$ and $(v_0, \ldots, v_k)$.

This lemma is easily verified by multiplying everything out, rewriting the powers θ^j with $j > k$ with respect to the basis chosen and making the substitutions.

Now consider the following secret sharing scheme. It is assumed that θ is fixed (and public), as well as the other parameters introduced above. A secret can be any element $a \in L$, *represented as a $k + 1$-vector of elements from K*: $a = u_0 + u_1\theta + \ldots + u_k\theta^k$, with the u_j in K. Each share will be an element of K however. Define

$$t = \hat{t} - 2k.$$

1. Let

$$a = u_0 + u_1 \cdot \theta + \ldots + u_k \cdot \theta^k \in L$$

   be the secret value.
2. Choose $f(X) \in V_{k,\hat{t}}(K)$ at random such that

$$f(X) = a(X) + R(X) \cdot X^{2k+1},$$

   where $a(X) \in K[X]$ is the polynomial of degree at most k whose coefficient vector is $(u_0, \ldots, u_k)$ and where $R(X) \in K[X]$ is a polynomial of degree at most $\hat{t} - 2k - 1$.
3. Set

$$a_1 = f(x_1) \in K, \ldots, a_n = f(x_n) \in K$$

   as the shares.
4. For any set $A \subset \{1, \ldots, n\}$ with $|A| \geq \hat{t} + 1$, the reconstruction of $a \in L$ from the shares $\{a_i\}_{i \in A}$ is by standard Lagrange Interpolation.

As for privacy, we note the following. If $|A| \leq t \ (= \hat{t} - 2k)$, then the collection of shares $\{a_i\}_{i \in A}$ gives no information on the secret a. Indeed, for each such set A and for each $z \in L$ there exists a $\kappa(X) \in V_{k,\hat{t}}(K)$ such that

$$\kappa(X) = z(X) + T(X) \cdot X^{2k+1},$$

where $T(X)$ is a polynomial of degree at most $\hat{t} - 2k - 1$, and such that

$$\kappa(x_i) = 0 \text{ for all } i \in A,$$

and this implies the privacy claim, for instance by a simple argument similar to the one used in the analysis of general linear secret sharing. The existence of $\kappa(X)$ per se follows from the lemma below, an immediate consequence of Lagrange's Interpolation Theorem.

Lemma 3. *Let $x_1, x_2, \ldots, x_e$ be distinct non-zero elements of K. Let d be an integer with $d \geq e$. For any $z_0, \ldots, z_{d-e} \in K$ and for any $y_1, \ldots, y_e$ there exists a polynomial $\kappa(X) \in K[X]$ of degree at most d such that*

$$\kappa(X) = z_0 + z_1 \cdot X + \ldots + z_{d-e} \cdot X^{d-e} + \text{ higher order terms },$$

and

$$\kappa(x_1) = y_1, \ldots, \kappa(x_e) = y_e.$$

Proof. Define $f_1(X) = \sum_{j=0}^{d-e} z_j X^j$ and let $f_2(X)$ be the polynomial of degree at most $e - 1$ through the e points $(y_i - f_1(x_i))/x_i^{d-e+1}$. Then the polynomial $\kappa(X) = f_1(X) + f_2(X) \cdot X^{d-e+1}$ is the unique polynomial that has the required properties. □

Thus, the dedicated scheme above is a $(t, \hat{t} + 1)$ ramp scheme with shares in K and the secret in L (as a vector of length $\frac{\hat{t}-t}{2} + 1$ over K).

In order to state the claimed secure multiplication protocol we need the following lemma, which can easily be verified using arguments similar to the ones used in standard proofs of Langrange's Interpolation Theorem, or by using the properties of Vandermonde determinants.

Lemma 4. *Let* $x_1, \ldots, x_{\ell+1}$ *be fixed distinct elements of* K. *Then there exist linear maps* $\phi_j : K^{\ell+1} \longrightarrow K$ $(j = 0 \ldots \ell)$ *such that the following holds. Let* $y_1, \ldots, y_{\ell+1}$ *be any elements of* K. *Let* $f \in K[X]$ *be the unique polynomial of degree at most* ℓ *such that* $f(x_1) = y_1, \ldots, f(x_{\ell+1}) = y_{\ell+1}$. *Then*

$$f(X) = \phi_0(y_1, \ldots, y_{\ell+1}) + \phi_1(y_1, \ldots, y_{\ell+1}) \cdot X + \ldots + \phi_\ell(y_1, \ldots, y_{\ell+1}) \cdot X^\ell.$$

Still in the secure channels model as before, assume that $\hat{t} < n/2$. Suppose that values $a = u_0 + u_1 \cdot \theta + \ldots + u_k \cdot \theta^k \in L$ and $b = v_0 + v_1 \cdot \theta + \ldots + v_k \cdot \theta^k \in L$, with coefficients in K, have been secret-shared according to the dedicated scheme explained above. Write $f \in K[X]$ for the polynomial defining the sharing of $a \in L$, with respective shares $a_1, \ldots, a_n \in K$, and write g defining that of $b \in L$, with respective shares $b_1, \ldots, b_n \in K$.

It now follows immediately from the fact that $\hat{t} < n/2$ and from Lemmas 1, 2, and 4 that there exist linear maps $\psi_j : K^n \longrightarrow K$ such that

$$ab = \sum_{j=0}^{k} \psi_j(a_1 b_1, \ldots, a_n b_n) \cdot \theta^j \in L.$$

The coefficients defining these linear maps can be computed efficiently. We can now use Theorem 1 to convert the local products of the shares of the players into a sharing of ab.

If the degree $[L : K] = k + 1$ of the extension field L satisfies the conditions detailed below, we can now achieve $O(n)$ communication.

We have

$$t + 2k = \hat{t} \text{ and } \hat{t} < n/2.$$

So if we set, say,

$$2\hat{t} + 1 = n,$$

and

$$k = cn,$$

for some real constant c, then we can achieve t maximal such that

$$t < \frac{(1-\delta)n}{2}, \text{ where } \delta = 4c.$$

If the parameters are such, secure multiplication of two elements from the field L is done with communication $O(n^2)$ elements from K, which is equivalent to $O(n)$ elements from L. This is as claimed.

5.1 A More General View

It is possible to look at the secure multiplication protocols in a more general way, that contain both our results and those of Franklin and Yung [7] as special cases.

Both in the protocol of Franklin and Yung and our protocol from Section 5, the protocols start out with two sets of shares, defining secret vectors $(s_0, ..., s_m)$, $(s'_0, ..., s'_m)$ respectively. We then compute locally the pairwise products of shares in the two vectors and these pairwise products can be seen as shares in a new ramp scheme, different from the original one.

For instance, in the scheme by Franklin and Yung the secret vector defined by the "local products" of the shares is $(s_0 s'_0, s_1 s'_1, \ldots, s_m s'_m)$ according to the new scheme defined On the other hand, in the protocol from Section 5, assuming the same initial secret vectors, we obtain secret vector $(\sum_{i+j=0} s_i \cdot s'_j, \sum_{i+j=1} s_i \cdot s'_j, \ldots, \sum_{i+j=2m} s_i \cdot s'_j)$ consisting of all homogeneous sums of the secret coefficients. This is why we can obtain different results after the application of Theorem 1.

In general, we can start from any ramp scheme $\mathcal{R}$, do the local multiplications and obtain a sharing of some quadratic function of the two original secret vectors in a new ramp scheme $\mathcal{R}'$ that depends on $\mathcal{R}$. This is not always useful – for instance, it is not always the case that the set of all players is qualified in $\mathcal{R}'$. Franklin/Yung and our scheme are two nicely structured examples, where useful results are indeed obtained.

We note that one can also obtain the homogeneous sums we use by multiple applications of Franklin and Yung's scheme, but since this would require $O(n)$ applications of their scheme (in order to obtain the required cross-products) this would be much less efficient.

6   Further Trade-Offs

In Section 5, we presented a scheme which is secure against a t-adversary. We now show a variation that is secure against a (stronger) t'-adversary with $t' > t$, where $t' - t$ is a constant fraction of n. Given again a finite field L with extension degree $k + 1$ over a subfield K, the bandwidth requirement remains $O(n)$, but there is a larger hidden constant.

The idea is to introduce a slightly modified version of the dedicated ramp scheme from Section 5. Basically, the coefficients of a secret element $a \in L$ are distributed over two polynomials f_1 and f_2 with smaller gaps than the polynomial that was used before, and the secure multiplication is then performed with these two polynomials by exploiting cross-products. This doubles the size of the shares and the required bandwidth. There is also a natural generalization of this idea involving more then two polynomials and cross-products of shares.

In Section 5, a t-adversary was defined where $t = \hat{t} - 2k$ for some integer $\hat{t} < n/2$ and where $k + 1$ is the degree of L over K. In this section, we fix the value $\hat{k} = \lceil (k-1)/2 \rceil$, and define the t'-adversary by $t' = \hat{t} - 2\hat{k}$. We now explain the details of the variation.

For an arbitrary value $a = u_0 + u_1 \cdot \theta + \ldots + u_k \cdot \theta^k \in L$, with coefficients in K, we denote $a^{(1)} = u_0 + u_1 \cdot \theta + \ldots + u_{\hat{k}} \cdot \theta^{\hat{k}}$ and $a^{(2)} = a - a^{(1)}$. Furthermore, we define $a^{(1)}(X) = u_0 + u_1 \cdot X + \ldots + u_{\hat{k}} \cdot X^{\hat{k}}$ and $a^{(2)}(X) = u_{\hat{k}+1} + u_{\hat{k}+2} \cdot X + \ldots + u_k \cdot X^{k-\hat{k}-1}$.

For $i \in \{1, 2\}$, choose $f_i(X) \in V_{\hat{k}, \hat{t}}(K)$ at random such that

$$f_i(X) = a^{(i)}(X) + R_i(X) \cdot X^{2\hat{k}+1},$$

where $a^{(1)}(X)$ is the polynomial of formal degree $\hat{k}$ with the initial coefficients $(u_0, u_1, \ldots, u_{\hat{k}})$, $a^{(2)}(X)$ is the polynomial of formal degree $k - \hat{k} - 1$ with the remaining coefficients $(u_{\hat{k}+1}, u_{\hat{k}+2}, \ldots, u_k)$ and where $R_1(X), R_2(X) \in K[X]$ are polynomials of formal degree $\hat{t} - 2\hat{k} - 1$. Then f_1 and f_2 both encode exactly half of the coefficients of a (if k is odd) or f_1 encodes one more coefficient of a than f_2 (if k is even). These polynomials are used in this section to perform the secure multiplication.

Assume that a value $b = v_0 + v_1 \cdot \theta + \ldots + v_k \cdot \theta^k \in L$ has likewise been encoded, resulting in polynomials $b^{(1)}(X)$, $b^{(2)}(X)$, $g_1(X)$ and $g_2(X)$, and that every player P_i received the values $a_i^{(1)} = f_1(x_i)$, $a_i^{(2)} = f_2(x_i)$, $b_i^{(1)} = g_1(x_i)$ and $b_i^{(2)} = g_2(x_i)$. By Lemma 3, no subset of $t - 2\hat{k}$ players can obtain any information about $a^{(1)}$, $a^{(2)}$, $b^{(1)}$ or $b^{(2)}$, and therefore the players in such a subset also cannot obtain any information about a or b.

We now make use of the observation that

$$(ab)(X) = (a^{(1)}b^{(1)})(X) + (a^{(1)}b^{(2)} + a^{(2)}b^{(1)})(X) \cdot X^{\hat{k}+1} + (a^{(2)}b^{(2)})(X) \cdot X^{2\hat{k}+2},$$

with as coefficients the values $H_0(a, b), H_1(a, b), \ldots, H_{2k}(a, b)$. This is straightforward to verify using the discussion from the last section. Since by Lemma 4 there exists a linear map ϕ_ℓ such that for $i, j \in \{1, 2\}$ the ℓth coefficient of $(f_i g_j)(X)$ can be computed as $\phi_\ell(a_1^{(i)} b_1^{(j)}, a_2^{(i)} b_2^{(j)}, \ldots, a_n^{(i)} b_n^{(j)})$, the same holds for $a^{(i)} b^{(j)}(X)$. In particular there exist linear maps $\psi_\ell : K^n \longrightarrow K$ such that

$$ab = \sum_{\ell=0}^{k} \psi_\ell(C_{11}, C_{12}, C_{21}, C_{22}) \cdot \theta^\ell \in L,$$

where $C_{ij} = (a_1^{(i)} b_1^{(j)}, \ldots, a_n^{(i)} b_n^{(j)})$ for $i, j \in \{1, 2\}$. Therefore, the techniques from the previous section can be used to construct a multiplication protocol that leads to two polynomials $h_1(X)$ and $h_2(X)$ of the proper form that encode the coefficients of $ab \in L$.

7   Secure MPC Against an Active Adversary (Overview)

Using the new techniques, we construct a protocol for secure multiplication in the presence of an active t-adversary that requires only $O(n^2)$ bandwidth when the multiplication is performed in a field L with extension degree $k + 1$ over a subfield K. Again, the corruption tolerance is not maximal, as we require $t = \hat{t} - 3k$ with $t < \hat{t} < n/4$, but it is still a constant fraction of n. Below we sketch the underlying ideas of the protocol. A more detailed description can be found in the appendix.

The obvious weakness of the protocol described in Section 5 is that the outcome completely depends on the polynomials h_i that the players select. Even if only one of these polynomials is not selected according to the protocol specification, the final outcome can encode any arbitrary element of L or not even be of the correct form. However, a closer inspection of the protocol reveals that the values of the leading coefficients of every polynomial h_i mainly depend on the corresponding value $a_i b_i$. Therefore, we can use VSS to let the players secret share their value $a_i b_i$, and then let the players locally compute their shares in polynomials h_i that are guaranteed to be of the proper form. We now sketch the key ingredients of the protocol.

Dedicated VSS. We use an adaptation of the four-round VSS protocol by Gennaro et al. [8] that allows the players to verify the presence of a gap in a secret sharing polynomial. In particular, we show that it is sufficient if the polynomials that the honest players receive as their shares using this scheme contain the desired gap.

Resharing step. Every player P_i reshares the value $a_i b_i$ using an instance of the dedicated VSS scheme by embedding it in a secret sharing polynomial v_i of formal degree $\hat{t} - k$ that has a gap of size $2k$ following the constant coefficient. Furthermore, player P_i uses VSS to distribute evaluations on a random polynomial of formal degree $2\hat{t}$ that has a zero constant coefficient. The value $a_i b_i$ is the constant coefficient of a polynomial of formal degree $2\hat{t}$ in which all the players have a share due to the VSS scheme. Therefore, the players can jointly subtract the polynomial v_i from this polynomial and mask the result by adding the random polynomial. These operations can all be performed locally on the shares and lead to shares in the resulting polynomial. The players then publicly reconstruct this polynomial by pooling their shares and verify whether it has a zero constant coefficient. This ensures that player P_i indeed reshared the value $a_i b_i$.

Local computation. Since the polynomial $h_i(X)$ should contain the element

$$\sum_{j=0}^{k} \psi_j(\epsilon_i) a_i b_i \theta^j \in L,$$

the polynomial

$$h_i(X) = \sum_{j=0}^{k} \psi_j(\epsilon_i) X^j v_i(X)$$

is of the correct form and every player P_m can locally compute a share $h_i(x_m)$ in this polynomial using the share $v_i(x_m)$. The sum of these shares then gives a share in a polynomial of the proper form that encodes the product ab.

8 Efficient Circuit Evaluation

This section shows another application of Theorem 1. Consider any arithmetic circuit C and a set of inputs to C and suppose that we evaluate the circuit by repeating the following two steps until all the gates have been evaluated:

1. Evaluate all linear gates for which we have both inputs, i.e., the addition gates and gates that perform multiplication by a constant.
2. Evaluate all multiplication gates for which we have both inputs.

Now let $S(C)$ be the minimum number of multiplication gates that are handled in one instance of step 2. We will refer to this value $S(C)$ as the *multiplicative speedup* of C.[2] Arithmetic circuits with large multiplicative speedup occur frequently in settings related to secure linear algebra [5]. For instance, constant-round protocols for secure unbounded fan-in multiplication and secure matrix multiplication require many parallel secure multiplications in a single step.

It is a natural idea to apply the scheme of Franklin and Yung here to perform these multiplications in parallel, but in order to do this it is required that the values that are to be multiplied are "aligned" in the corresponding instances of the ramp scheme.

Theorem 1 enables us to perform this aligning and more. If the inputs to the multiplication are available as secrets of some ramp scheme, or even merely available via a linear function on the shares that the players hold in a number of (potentially different) ramp schemes, a single resharing round can be used in order to correctly align the inputs to the parallel multiplications. This also implies that the same resharing step can simultaneously perform the operations required in step 1 before the multiplications are performed, and after local multiplication of the new shares we can continue with the preparations for the next multiplication round. We formulate this consequence of Theorem 1 a bit more precisely below.

Theorem 2. *Consider an arithmetic circuit C over the field $\mathbb{F}_q$ with multiplicative speedup m. Then there exists a passively secure protocol for n players that securely evaluates C having communication complexity $O(|C|n^2k/m+C')$, where C' is the complexity of sharing the inputs and $k = \log(q)$. The protocol is secure against at most $n/2 - m$ passive corruptions.*

Proof. (Sketch) Assume for simplicity that each multiplication layer in C consists of exactly m gates. Then to perform one set of multiplications, the protocol of Franklin and Yung requires ramp sharings, say in ramp scheme $\mathcal{R}$ of two blocks A

---

[2] This term is inspired by [11], where the speedup is defined to be the factor you save in runtime due to parallelism.

and B of m values each, where A contains all the left inputs to the multiplication gates and B contains all the right inputs in matching order. Local multiplication of the shares of A and B then produces a linear secret sharing (in a new scheme $\mathcal{R}'$) of all the outputs from the multiplication gates.

Now note that we can assume that as input to an instance of Step 1 above, we have a linear sharing of all values going into Step 1. This is either obtained because the inputs are shared initially, or we have a sharing in $\mathcal{R}'$ which was output from a previous instance of Step 2. All we need is that the set of all players is qualified in the scheme that occurs here. We now need to subject these values to a linear function and place the results in the blocks A and B. Using Theorem 1, we can do exactly this in one round and communication complexity $n^2 k$. Clearly, there can be no more than $|C|/m$ multiplication layers, and the scheme of Franklin and Yung that we start from is private as long as there are at most $n/2 - m$ corruptions. □

References

1. M. Ben-Or, S. Goldwasser, and A. Wigderson. Completeness theorems for non-cryptographic fault-tolerant distributed computation. In *Proceedings of STOC 1988*, pages 1–10. ACM Press, 1988.
2. G. R. Blakley. Safeguarding cryptographic keys. In *Proceedings Proceedings of National Computer Conference '79*, volume 48 of *AFIPS Proceedings*, pages 313–317, 1979.
3. D. Chaum, C. Crépeau, and I. Damgaard. Multi-party unconditionally secure protocols. In *Proceedings of STOC 1988*, pages 11–19. ACM Press, 1988.
4. H. Chen and R. Cramer. Algebraic Geometric Secret Sharing Schemes and Secure Multi-Party Computation over Small Fields. In *Proceedings of 26th Annual IACR CRYPTO*, volume 4117, pages 516–531, Santa Barbara, Ca., USA, August 2006. Springer Verlag LNCS.
5. R. Cramer and I. Damgaard. Secure Distributed Linear Algebra in Constant Number of Rounds. In *Proceedings of CRYPTO 2001*, volume 2139, pages 119–136. Springer LNCS, 2001.
6. R. Cramer, E. Kiltz, and C. Padró. A Note on Secure Computation of the Moore-Penrose and Its Application to Secure Linear Algebra. Manuscript, 2006.
7. M. Franklin and M. Yung. Communication complexity of secure computation. In *Proceedings of STOC 1992*, pages 699–710. ACM Press, 1992.
8. R. Gennaro, Y. Ishai, E. Kushilevitz, and T. Rabin. The Round Complexity of Verifiable Secret Sharing and Secure Multicast. In *Proceedings of STOC 2001*, pages 580–589. ACM Press, 2001.
9. R. Gennaro, M. O. Rabin, and T. Rabin. Simplified VSS and fasttrack multi-party computations with applications to threshold cryptography. In *Proceedings of PODC 1997*, pages 101–111, 1998.
10. M. Hirt and U. Maurer. Robustness for Free in Unconditional Multi-Party Computation. In *Proceedings of CRYPTO 2001*, volume 2139, pages 101–118. Springer LNCS, 2001.
11. C. P. Kruskal, L. Rudolph, and M. Snir. A complexity theory of efficient parallel algorithms. *Theoretical Computer Science*, 71(1):95–132, 1990.
12. A. Shamir. How to share a secret. *Communications of the ACM*, 22(11):612–613, 1979.

A   Secure MPC Against an Active Adversary

In this section we describe in detail the key ingredients of the protocol secure against an active adversary, as described in Section 7. Throughout this section, we assume that $t = \hat{t} - 3k$ with $\hat{t} < n/4$.

A.1   VSS

We start by describing the adaptation of the four-round VSS protocol by Gennaro at al. [8] that allows the players to verify the presence of a gap in the secret sharing polynomial.

Let f be the polynomial defining the sharing of $a \in L$, as described in Section 5. The dealer D randomly selects a symmetric bivariate polynomial $F(X,Y) = \sum_{i,j=0}^{\hat{t}} e_{ij} X^i Y^j \in K[X,Y]$ under the restriction that $F(X,0) = f(X)$ and that $e_{ij} = e_{ji} = 0$ for $j = k+1, k+2, \ldots, 2k$ and $i = 0, 1, \ldots, n$. The dealer D and the players now execute the following steps:

1. D privately sends to every player P_i the polynomial $f_i(X) := F(X, x_i)$ by transmitting the $\hat{t} - k$ coefficients that are not equal to zero by default. In every subset of two players $\{i, j\}$ one of the players (which one can be fixed before execution of the protocol) selects a random pad $r_{ij} = r_{ji}$ and transmits this value privately to the other player in this set.
2. Player P_i broadcasts for every player P_j the value $a_{ij} = f_i(x_j) + r_{ij}$.
3. For every pair $a_{ij} \neq a_{ji}$, the dealer, P_i and P_j each broadcast the value $f_i(x_j) = f_j(x_i) = F(x_i, x_j)$.
   A player is called *unhappy* if his value does not match the dealer's value. If there are more than t unhappy players, the dealer is disqualified and the protocol stops.
4. For every unhappy player P_i the dealer broadcasts $f_i(X)$ and every player P_j that is not unhappy broadcasts the value $f_j(x_i)$.
5. Every player checks for every broadcast polynomial $f_i(X)$ whether at least $3\hat{t} + 1$ happy players P_j broadcast a value $f_j(x_i)$ such that $f_i(x_j) = f_j(x_i)$. If this is not the case, the dealer is disqualified. The broadcast polynomials are from here on (publicly) used as the shares of the corresponding players.

As in [8], this protocol has the properties that when the dealer is honest, no new information is disclosed to the adversary after the first round and that when the protocol completes all honest players have obtained consistent polynomials $f_i(X)$. Therefore, the main properties to be verified here are that in the case of an honest dealer no information is disclosed about a in the first round and that in the case of a dishonest dealer the polynomial $f(X)$ that is fixed by the resulting polynomials $f_i(X)$ is of the proper form. Note that, since $\hat{t} < n/4$, this protocol can easily be adjusted so that polynomials of formal degree $2\hat{t}$ are distributed.

Below we present a security proof for a setting in which none of the initial $2k + 1$ coefficients has a fixed value. The security for the case where some of the coefficients are fixed to zero, but only $k + 1$ of the coefficients need to remain secret, then follows as a straightforward application of this result.

Lemma 5. *Let $F(X,Y)$ be a random symmetric bivariate polynomial of formal degree $\hat{t}$ in each variable and define $f_i(X) := F(X, x_i)$ for $i = 1, 2, \ldots, n$. If $0 \le d \le \hat{t}$, then any subset of $\hat{t} - d$ polynomials $f_i(X)$ gives no information about the first $d + 1$ coefficients of $f(X) := F(X, 0)$.*

Proof. Assume wlog that the given polynomials are $\{f_i(X)\}_{i=1}^{\hat{t}-d}$. We need to show that for any selection for the first $d + 1$ coefficients of $f(X)$, there is a symmetric bivariate polynomial $F(X, Y)$ that is consistent with the given polynomials and the selected coefficients. We show the equivalent statement that there exist symmetric bivariate polynomials $F_j(X, Y)$ for $j = 0, 1, \ldots, d$ such that $F_j(X, x_i) = 0$ for $i = 1, 2, \ldots, \hat{t} - d$ and all the $d + 1$ lower coefficients of $F_j(X, 0)$ are zero except for the jth one, which is equal to one.

By Lemma 3, any selection $c_0, c_1, \ldots, c_d$ for the first $d + 1$ coefficients of f leads to a polynomial $f'(X)$ that is consistent with the selection and for which $f'(x_i) = 0$ for $i = 1, 2, \ldots, \hat{t} - d$. Let C_j be the selection where all selected coefficients are zero, except for the first and the jth one which are equal to one, and let f_{C_j} be the corresponding polynomial with those first coefficients for which $f_{C_j}(x_i) = 0$ for $i = 1, 2, \ldots, \hat{t} - d$.

Define a number of symmetric bivariate polynomials $F_{C_j}(X, Y)$ by setting $F_{C_j}(X, Y) := f_{C_j}(X) f_{C_j}(Y)$ for $j = 0, 1, \ldots, d$. Then we have that $F_{C_j}(X, x_i) = f_{C_j}(X) f_{C_j}(x_i) = 0$ for $i = 1, 2, \ldots, \hat{t} - d$ and $F_{C_j}(X, 0) = f_{C_j}(X) f_{C_j}(0) = f_{C_j}(X)$. The polynomials $F_0(X, Y) := F_{C_0}(X, Y)$ and $F_j(X, Y) := F_{C_j}(X, Y) - F_{C_0}(X, Y)$ for $j = 1, 2, \ldots, d$ are now of the desired form.   □

We now show that the default zeros in the polynomials $f_i(X)$ that the players receive as their share ensure that the required gap is present in the polynomial $f(X)$.

Lemma 6. *Take $x_0 = 0$. For $i = 0, 1 \ldots, n$, let $f_i(X) := F(X, x_i) = c_{i0} + c_{i1}X + \cdots + c_{i\hat{t}}X^{\hat{t}}$ for certain $c_{ij} \in K$. If $c_{ik} = 0$ for at least $\hat{t} + 1$ values of i, then the coefficient c_{0k} of the polynomial $f_0(X)$ is zero.*

Proof. Since $F(X,Y) = \sum_{i,j=0}^{\hat{t}} e_{ij} X^i Y^j$, $f_v(X) = \sum_{i=0}^{\hat{t}} (\sum_{j=0}^{\hat{t}} e_{ij} v^j) X^i$ and in particular $f_0(X) = \sum_{i=0}^{\hat{t}} e_{i0} X^i$. Now assume that $c_{i_l k} = 0$ for distinct $i_1, i_2, \ldots, i_{\hat{t}+1}$. This amounts to saying that $\sum_{j=0}^{\hat{t}} e_{kj} i_l^{\,j} = 0$ for $l = 1, \ldots, \hat{t} + 1$ and therefore the polynomial $\sum_{j=0}^{\hat{t}} e_{kj} Y^j$ has to be the zero polynomial. We conclude that $e_{kj} = 0$ for $j = 0, 1 \ldots, \hat{t}$ so that in particular $e_{k0} = c_{0k} = 0$.   □

A.2   Multiplication/Resharing Step

Suppose that both $a \in L$ and $b \in L$ have been secret-shared according to the dedicated VSS scheme described above, resulting in distributed polynomials $f_i(X)$ and $g_i(X)$. The aim is to let the players execute a secure resharing protocol that results in a secret-sharing of ab according to the dedicated VSS scheme. The resharing protocol proceeds as follows for every player P_i:

1. Player P_i selects a polynomial of the form $v_i(X) = a_i b_i + \sum_{l=2k+1}^{\hat{t}-k} r_l X^l$, where r_l is chosen at random from K for $l = 2k+1, 2k+2, \ldots, \hat{t} - k$ and embeds it in a random symmetric bivariate polynomial $V_i(X,Y) = \sum_{i,j=0}^{\hat{t}} e_{ij} X^i Y^j \in K[X,Y]$ under the restriction that $V_i(X,0) = v_i(X)$ and that $e_{ij} = e_{ji} = 0$ for $j = 1, 2, \ldots, 2k$ and $i = 0, 1, \ldots, n$. This bivariate polynomial is then used for VSS, leading to shared polynomials $v_{ij}(X) := V_i(X, x_j)$.

2. Player P_i selects at random a symmetric bivariate polynomial $R_i(X,Y)$ of formal degree $2\hat{t} - 1$ in each variable and distributes using VSS polynomials $r_{ij}(X) := R_i(X, x_j)$, where the evaluations $r_{ij}(0)$ determine the polynomial $r_i(X) := R_i(X,0)$ of formal degree $2\hat{t} - 1$.

3. All players P_j broadcast the value $f_j(x_i) g_j(x_i) - v_{ij}(0) + x_j r_{ij}(0)$ and use error correction to reconstruct a polynomial of degree $2\hat{t}$. If the first coefficient of the reconstructed polynomial is not zero, player P_i is disqualified.

First note that $f_j(x_i) g_j(x_i) - v_{ij}(0) + x_j r_{ij}(0) = (f_i g_i - v_i)(x_j) + x_j r_i(x_j)$, so that the players reconstruct the sum of two polynomials where one of the polynomials is random under the restriction that the first coefficient is equal to zero. Since the VSS-schemes have the property that all honest players have consistent shares at the end of the procedure, the polynomials $r_i(X)$, $v_i(X)$, $f_i(X)$ and $g_i(X)$ are uniquely determined when all players pool their shares in these polynomials. Since $\hat{t} < n/4 < n/3$, the same holds for the polynomials $X r_i(X)$ and $(f_i g_i)(X)$ and therefore also for the polynomial $(f_i g_i - v_i)(X) + X r_i(X)$. Furthermore, this polynomial has an initial coefficient equal to zero if and only if the first coefficient of $v_i(X)$ is equal to $a_i b_i$. Note also that the additional zero's in the bivariate polynomial $V_i(X,Y)$ ensure to the players that the polynomial $v_i(X)$ is of the proper form.

We need to show that the n polynomials $(f_i g_i - v_i)(X) + X r_i(X)$ together with t evaluations on the points $x_1, x_2, \ldots, x_t$ for every polynomial r, v_i, f_i and g_i do not give any information about a, b or ab. First, we can conclude by the following lemma that the sum of two arbitrary polynomials of degree $2\hat{t}$ together with t evaluations for these polynomials give no information about the first $\hat{t} - t + 1$ first coefficients of one of these two polynomials.

Lemma 7. *Let f and g be polynomials of formal degree $\hat{t}$ and let the polynomial $f + g$ and evaluations $f(x_i)$ and $g(x_i)$ be given for $i = 1, 2, \ldots, d$. Then $f + g$ together with the given evaluations $f(x_i)$, $g(x_i)$ give no information about the first $\hat{t} - d + 1$ coefficients of f.*

Proof. By Lemma 3, for any selection $C = (c_0, c_1, \ldots, c_{\hat{t}-d+1})$ there exists a polynomial with these values as the first $\hat{t}-d+1$ coefficients that evaluates to zero in the points $x_1, x_2, \ldots, x_d$. Then adding this polynomial to f and subtracting it from g leads to consistent polynomials f' and g' with different initial coefficients, while the sum $f' + g'$ remains the same. This works for every arbitrary selection C, and therefore the given information is consistent with any selection for the first coefficients of f. $\qquad\square$

As a consequence of the lemma, given the evaluations of t players we can choose polynomials of formal degree $\hat{t}$ with arbitrary first $k+1$ coefficients that evaluate to zero in the given points and add them to the polynomials f_i and g_i to give polynomials f_i' and g_i'. Then, the polynomial $f_i'g_i' - f_ig_i$ can be subtracted from r_i, which gives a polynomial r_i' that is consistent with the given points on r_i, but for which the sum $f_i'g_i' + r_i'$ is equal to $f_ig_i + r_i$. Therefore, no information about a, b or ab is leaked during the protocol.

A.3   Local Computation

In order to obtain the desired polynomials $h_i(X)$, every player P_i now locally computes the polynomial

$$h_i(X) = \sum_{j=1}^{n} \left(\sum_{l=0}^{k} \psi_k(\epsilon_i)(X^l + x_i{}^l) \right) v_{ji}(X),$$

where $\psi_l : K^n \to K$ for $l = 1, 2, \ldots, n$ have been defined in Section 5.

Define $h(X) := \sum_{i=1}^{n} (\sum_{l=0}^{k} \psi_l(\epsilon_i)X^l)v_i(X)$. Then it is easy to verify that $h(X)$ has degree $\hat{t}$ and we can write it in the form

$$\left(\sum_{l=0}^{k} \left(\sum_{i=1}^{n} \psi_l(\epsilon_i)a_ib_i \right) X^l \right) + \sum_{l=2k+1}^{\hat{t}} r_l''X^l$$

for certain $r_{2k+1}'', r_{2k+2}'', \ldots, r_{\hat{t}}'' \in K$. In particular, the first $k + 1$ coefficients are the coefficients of ab. Below, we show that the evaluations $h_i(0)$ all give evaluations on $h(X)$ and that for all $i, j \in \{1, 2, \ldots, n\}$ we have that $h_i(x_j) = h_j(x_i)$, so that there exists a symmetric bivariate polynomial $H(X, Y)$ such that $H(X, 0) = h(X)$ and $H(X, x_i) = h_i(X)$. Therefore, the resulting sharing is of the desired form.

The following two, easy to verify lemmas show that the polynomials $h_i(X)$ that the players obtain are part of a proper sharing of the polynomial $h(X)$. Therefore, the protocol described above gives us a proper sharing of the new (product) secret.

Lemma 8. $\forall 1 \leq i \leq n : h_i(0) = h(i)$.

Lemma 9. $\forall 1 \leq i, j \leq n : h_i(j) = h_j(i)$.

Cryptanalysis of the Sidelnikov Cryptosystem

Lorenz Minder[*] and Amin Shokrollahi

Laboratoire de mathématiques algorithmiques (LMA), EPFL

Abstract. We present a structural attack against the Sidelnikov cryptosystem [8]. The attack creates a private key from a given public key. Its running time is subexponential and is effective if the parameters of the Reed-Muller code allow for efficient sampling of minimum weight codewords. For example, the length 2048, 3rd-order Reed-Muller code as proposed in [8] takes roughly an hour to break on a stock PC using the presented method.

Keywords: Sidelnikov cryptosystem, McEliece cryptosystem, error-correcting codes, structural attack.

1 Introduction

The McEliece cryptosystem [6] is one of the oldest known public-key cryptosystems. The fact that it has not been broken in more than a quarter of a century and that the best known attacks today are still exponential speaks for itself.

Despite its impressive security record, it plays a rather marginal role in practice, being far less popular than other systems, such as RSA. The principal reason for this is that the McEliece cryptosystem is not as efficient as the alternatives. The main problems are its large public keys, the fact that the message is subject to expansion (the cryptogram is longer than the plaintext message), and its potentially high decryption complexity.

To understand the tradeoffs, we recall how the McEliece cryptosystem works. Let $\mathcal{C}$ be a linear binary Goppa code of block length n, dimension k, and having a decoding algorithm correcting up to t errors. Let G be a $k \times n$ generator matrix for the code. Let P be a random $n \times n$ permutation matrix. Then there is an efficient decoder for the code generated by $G \cdot P$. Let A be a $k \times k$ invertible matrix. The code generated by

$$G_{\mathrm{pub}} := AGP$$

is the same as the code generated by GP. The public key is the pair (G_{pub}, t). To encrypt a message vector $x := (x_1, \ldots, x_k) \in \mathbb{F}_2^k$, we first compute xG_{pub} and then add t errors at random positions. The resulting vector y is the cryptogram.

The decrypting problem is to recover the value of x given y. The receiver can decrypt this message, since he knows a decoding algorithm for $\mathcal{C}$. An attacker has to either resort to general decoding techniques, and attempt to solve a problem

---

[*] Supported by the Swiss National Fund, grant 200021-103683.

M. Naor (Ed.): EUROCRYPT 2007, LNCS 4515, pp. 347–360, 2007.

which appears to be intractable, or recover the structure of the code given by G_{pub}. For further details, see [6].

The McEliece cryptosystem can be generalized to codes other than Goppa codes. A priori, any family of linear codes having decoders which allow for efficient correction of a large number of errors with high probability could be used.

The efficiency and security of such a cryptosystem depend on several factors. First, the number t of correctable errors has to be very large to render general linear decoding algorithms inefficient, which is a security requirement. In addition, the McEliece cryptosystem can be modified in a manner so that the expansion factor to which a message is subjected depends on t, see the paper by Niederreiter [7] for such modifications. At the limit, if capacity-achieving codes could be used, t could be made so large that the expansion factor would converge to 1.

Secondly, the difficulty of recovering the structure of a code given by an arbitrary, permuted generator matrix is highly dependent on the code in question. An interesting dichotomy can be observed here: While modern, graph-based codes (like LDPC-, expander-, LT- or turbo-codes) are all unsafe because of the sparse parity checks revealing their structure, classical algebraic codes have proved widely resistant to structural attacks. The most notable exception is given by Sidelnikov and Shestakov [9], showing that generalized Reed-Solomon codes are unsafe.

In 1994, the first author of [9] proposed a variant of the McEliece system, basically replacing the Goppa codes with Reed-Muller codes [8]. The advantage of using Reed-Muller codes is that very efficient decoding algorithms are known for these codes. Thus, using these codes allows simultaneously for faster decryption, smaller key sizes and expansion factors close to 1, if the Niederreiter variant is used.

While all those properties sound very promising, we will show in this paper that Reed-Muller codes are a bad choice, too. More specifically, we present a method to find a private key for a given public key. The most costly step of this procedure is that of finding minimum weight codewords in the code. In the low-rate setting of Reed-Muller codes, this is feasible even for fairly long block lengths. This attack is, to our knowledge, the first known effective attack against this cryptosystem, and it breaks in particular Sidelnikov's original proposed parameters $(m = 11, r = 3)$ in less than an hour on a stock PC.

The key observation that makes the attack work, is the fact that minimum weight words in the r-th Reed-Muller code of length 2^m (this code is denoted $\mathcal{R}(r, m)$) are products of r minimum weight words in $\mathcal{R}(1, m)$. The attack uses this fact to reduce the order: First, minimum weight codewords in the given, permuted $\mathcal{R}(r, m)$ are found, and then a statistical test is applied to find factors of those words which lie in the accordingly permuted $\mathcal{R}(r - 1, m)$. By iterating this procedure, ultimately the permuted version of $\mathcal{R}(1, m)$ is found, which allows easy identification of a suitable permutation.

The fact that there is only a single Reed-Muller code for a given block length and dimension has been noticed as a cryptographic weakness before. The best known previous attack is using the support splitting algorithm[11], an algorithm to find a permutation between two equivalent codes. While this algorithm is generally very fast, it is ineffective against Reed-Muller codes: Its running time is exponential in the dimension of the hull (i.e., the intersection of the code with its dual), and Reed-Muller codes have a hull as big as the code itself.

This paper is organized as follows. First, we give a short summary of Reed-Muller codes. Second, we present our results on the structure of these codes which form the basis for the attack. Third, we present the attack with a running time analysis.

2   Reed-Muller Codes

To recall the notation, we start by presenting the construction of these codes. For further details the reader is referred to [5] or [3].

Reed-Muller codes can be constructed by using Boolean functions. A Boolean function of m variables can be evaluated on 2^m different positions. So to each Boolean function we can associate a binary word of length 2^m. The code $\mathcal{R}(r, m)$ is the set of words obtained by evaluating all the Boolean functions of degree $\leq r$ in this way. We will subsequently call the variables evaluated $v_1, \ldots, v_m$.

We denote by $\mathcal{B}(r, \{v_1, \ldots, v_m\})$ the set of Boolean functions in the variables $v_1, \ldots, v_m$ of degree at most r.

Note that since the base field is $\mathbb{F}_2$, the term v_i^2 can be simplified to v_i, which implies that the degree of any variable in any term of these Boolean functions is at most 1.

The fact that all functions generating words in $\mathcal{B}(r-1, \{v_1, \ldots, v_m\})$ are also in $\mathcal{B}(r, \{v_1, \ldots, v_m\})$ implies the following observation.

Proposition 2.1. *For any m, we have $\mathcal{R}(0, m) \subset \mathcal{R}(1, m) \subset \cdots \subset \mathcal{R}(m, m)$.*

In what follows, we frequently switch back and forth between $\mathcal{B}(r, \{v_1, \ldots, v_m\})$ and $\mathcal{R}(r, m)$. Doing so in the most explicit manner would make the reasonings a lot harder to read, and for this reason we decided to treat codewords and Boolean functions as interchangeable. Note, however, that a codeword has a fixed length, while a function does not. If $x \in \mathcal{R}(r, m)$ is a codeword, its *extension* to $\mathcal{R}(r, m+1)$ is the codeword (x, x), i.e., the codeword obtained by evaluating the function $f \in \mathcal{B}(r, \{v_1, \ldots, v_m\})$ at all the possible values of $(v_1, \ldots, v_m, v_{m+1})$. Similarly, if $x \in \mathcal{R}(r, m)$ is a codeword whose corresponding function does not depend on v_m, then we can *reduce* x to $\mathcal{R}(r, m-1)$, by evaluating the function f corresponding to x on all the possible values of $(v_1, \ldots, v_{m-1})$. Note that in a Reed-Muller code, a position (coordinate) within a codeword can be specified by the value of $(v_1, \ldots, v_m)$.

The block length n, dimension k and minimum distance d of $\mathcal{R}(r, m)$ are

$$n = 2^m, \qquad k = \sum_{i=0}^{r} \binom{m}{i}, \qquad d = 2^{m-r}.$$

The *support* of a codeword $x \in \mathcal{R}(r, m)$, noted by $\mathrm{supp}(x)$, is the set of positions i, for which $x_i \neq 0$.

3   Minimum-Weight Codewords

We will now present the structural property of Reed-Muller codes which constitutes the theoretical foundation of our cryptanalysis of the Sidelnikov cryptosystem.

The fact that products of r linearly independent first-order codewords are minimum weight in $\mathcal{R}(r, m)$ is well-known. The following proposition states the converse, namely, that minimum weight codewords in Reed-Muller codes can always be written as a (pointwise) product of suitable words in the corresponding first order code. In other words, the only functions giving rise to minimum weight codewords are products of functions in $\mathcal{B}(1, \{v_1, \ldots, v_m\})$.

Proposition 3.1. *Let* $f \in \mathcal{R}(r, m)$ *be a word of minimum weight. Then there exist* $f_1, f_2, \ldots, f_r \in \mathcal{R}(1, m)$, *such that*

$$f = f_1 \cdot f_2 \cdots f_r,$$

as functions. The f_i *are of minimum weight in* $\mathcal{R}(1, m)$.

Proposition 3.1 is proved in [4]. The same paper also gives more precise formulas for the weight distribution, which can be used, in particular, to estimate the number of minimum-weight words:

Proposition 3.2. *There are at least*

$$2^{mr - r(r-1)}.$$

minimum weight codewords in $\mathcal{R}(r, m)$.

We will make use of this fact in the analysis of the running time of our algorithm.

4   Cryptanalysis of the Sidelnikov Cryptosystem

The Sidelnikov variant of the McEliece cryptosystem [8] uses Reed-Muller codes in combination with powerful decoding algorithms.

Reed-Muller codes are low-rate if any interesting error-correction capability is to be obtained, which makes it easy to apply algorithms such as the Canteaut-Chabaud-algorithm [1] to find low weight words, and also to decode if the number of errors is less than $d/2$ (half the minimum distance). However, there are decoding-algorithms for Reed-Muller codes which decode many more errors (with high probability) than $d/2$, and thus the low weight word finding algorithms cannot be directly used for decoding.

Such algorithms can still be used to find minimum weight words in codes with suitable parameters, though. In this section, we show how to exploit this fact to invert trapdoors from Reed-Muller codes.

4.1   Outline of the Attack

We now present an algorithm which, given a permuted, scrambled Reed-Muller code $\mathcal{C}$, constructs a permutation σ such that if the positions of $\mathcal{C}$ are permuted accordingly, the resulting code is a Reed-Muller code.

Let σ be any permutation on $\{1, \ldots, n\}$. For any code $\mathcal{C}$ of length n, we denote by $\mathcal{C}^\sigma$ the code obtained from $\mathcal{C}$ with the positions permuted according to σ, i.e., a word $(x_0, x_1, \ldots, x_n)$ will be a codeword in $\mathcal{C}^\sigma$ if and only if $(x_{\sigma^{-1}(1)}, \ldots, x_{\sigma^{-1}(n)}) \in \mathcal{C}$.

The sketch of the attack is as follows. Let $\mathcal{C} = \mathcal{R}(r, m)^\sigma$ for some unknown σ, given by an arbitrary generator matrix.

1. Find codewords in $\mathcal{C}$ which with very high probability also belong to $\mathcal{R}(r - 1, m)^\sigma$. Find enough such vectors to build a basis of $\mathcal{R}(r - 1, m)^\sigma$.
2. Iterate the previous step (with decreasing r) until obtaining $\mathcal{R}(1, m)^\sigma$.
3. Determine a permutation τ such that $\mathcal{R}(1, m)^{\tau \circ \sigma} = \mathcal{R}(1, m)$. Then $\mathcal{R}(r, m)^{\tau \circ \sigma} = \mathcal{R}(r, m)$, and this fact can then be used to decode.

The meat of the attack lies in the first step, which is based on the properties of Reed-Muller codes stated in the previous section.

4.2   Finding the Subcode $\mathcal{R}(r - 1, m)^\sigma \subseteq \mathcal{R}(r, m)^\sigma$

The basic idea of this step is to find a codeword for which we know that it is a product of other codewords, and then to split off a factor lying in the $\mathcal{R}(r-1, m)^\sigma$ subcode.

By proposition 3.1, a minimum weight codeword is actually a product of several codewords of $\mathcal{R}(1, m)^\sigma$. Hence, we do the following: We find a minimum weight codeword x and split off a factor of this word.

To this end, we shorten the code on $\mathrm{supp}(x)$, and use the structure of the shortened code to find a factor of x which lies in $\mathcal{R}(r - 1, m)^\sigma$.

Finding enough words in $\mathcal{R}(r - 1, m)^\sigma$ will result in a basis of $\mathcal{R}(r - 1, m)^\sigma$.

Finding factors of minimum weight words. We drop the permutation σ in this section, since our ideas do not depend on σ.

Let $x \in \mathcal{R}(r, m)$ be a minimum weight codeword. Using proposition 3.1, and changing the basis, we can assume that $x = v_1 v_2 \cdots v_r$. Let $\mathcal{C}_{\mathrm{supp}(x)}$ be the code $\mathcal{R}(r, m)$ shortened on the support of x. (In other words, $\mathcal{C}_{\mathrm{supp}(x)}$ is the subcode of $\mathcal{R}(r, m)$ containing only the words which are zero on $\mathrm{supp}(x)$, and with these positions punctured afterwards.)

Write $\bar{v} = (v_{r+1}, \ldots, v_m)$, and let f be a codeword in $\mathcal{C}_{\mathrm{supp}(x)}$. Then we can write f as

$$f(v_1, \ldots, v_r, \bar{v}) = \sum_{I \in 2^{\{1, \ldots, r\}}} f_I(\bar{v}) \cdot \prod_{i \in I} v_i,$$

where for each $I \subseteq \{1, \ldots, r\}$, we have $f_I \in \mathcal{B}(r - |I|, \{v_{r+1}, \ldots, v_m\})$. The condition that f be 0 on $\{v_1 = v_2 = \cdots = v_r = 1\}$ implies

$$0 = \sum_{I \in 2^{\{1, \ldots, r\}}} f_I(\bar{v}), \tag{1}$$

and shows in particular that $f_\emptyset(\bar{v}) \in \mathcal{B}(r-1, \{v_{r+1}, \ldots, v_m\})$. Therefore, if we take any codeword in the shortened code, fix a value for $(v_1, \ldots, v_r)$, and look at the positions determined by this value, we get a codeword in $\mathcal{R}(r-1, m-r)$. In other words, the shortened code is a concatenated code[1] with the inner codewords being on the disjoint sets of positions determined by the value of $(v_1, \ldots, v_r)$.

We shorten on $\mathrm{supp}(x)$, i.e., the set $\{v_1 = \cdots = v_r = 1\}$, so there are actually $2^r - 1$ such sets, and each is of length 2^{m-r}. We apply the algorithm of the next section (Algorithm 1.) to find the sets and then construct a word y of length 2^m that has ones exactly on the points $\{v_1 = \cdots = v_r = 1\} \cup S$, where S is one of the determined sets, say $\{v_1 = v_2 = \cdots = v_\ell = 0, v_{\ell+1} = v_{\ell+2} = \cdots = v_r = 1\}$. The set $\mathrm{supp}(x) \cup S$ can also be written as $\{v_1 = v_2 = \cdots = v_\ell, v_{\ell+1} = \cdots = v_r = 1\}$, and hence we can write

$$y = (1 + v_1 + v_2)(1 + v_2 + v_3) \cdots (1 + v_{\ell-1} + v_\ell)v_{\ell+1} \cdots v_r,$$

which shows that $y \in \mathcal{B}(r-1, \{v_1, \ldots, v_r\})$. Note that one can write $x = v_i y$ for any $1 \le i \le \ell$, showing that y is indeed a factor of x.

Finding inner words in the shortened code. To solve the problem of distinguishing the sets with different values of $(v_1, \ldots, v_r)$, we use the fact that the code is a concatenated code, with an inner codeword on each of these sets.

The problem of recovering concatenated codes has previously been studied by Sendrier; the algorithm presented in [10] could possibly be applied in our case, if one showed that the code in question verifies the assumptions of this algorithm, namely that the most lightweight parity checks all have their support within one inner word.

Another possibility is to use a similar method which acts on the code itself, rather than on its dual, and works well in our setting. The method is based on a statistical analysis, and we start by describing the relevant random experiment. Let $\mathcal{C}$ be a concatenated code, i.e.,

$$\mathcal{C} \subseteq \underbrace{\mathcal{C}_i \times \cdots \times \mathcal{C}_i}_{n \text{ times } \mathcal{C}_i},$$

where $\mathcal{C}_i$ is a non-trivial code of length n_i and relative minimum distance δ, called the *inner code*.

For our analysis, we need the following assumption: If $Y \in \mathcal{C}$ is sampled randomly in the low weight words of $\mathcal{C}$, we assume that the events $\{Y_i = 1\}$ and $\{Y_j = 1\}$ are independent if the positions i and j do not belong to the same inner block. (Note that this is almost universally true for linear codes with Y sampled from all the words, and not just the low weight ones.)

Now we randomly sample words of relative weight $< \delta$ from $\mathcal{C}$. Call these samples $X_0, X_1, \ldots$, and denote by $(X_\ell)_k$ the k-th position of X_ℓ. For two indexes $1 \le i < j \le n_i \cdot n$, we define the random variable

---

[1] A *concatenated code* is a subspace of the Cartesian product of several nontrivial codes.

$$I_{ij,k} := \begin{cases} 1 & \text{if } (X_k)_i = 1 \text{ and } (X_k)_j = 1, \\ 0 & \text{otherwise.} \end{cases}$$

The punch line will be that the behaviour of $I_{ij,k}$ depends on whether i and j lie within the same inner code or not.

We first assume that i and j are not in the same inner block; then $(X_k)_i$ and $(X_k)_j$ will be independent random variables, and we get

$$E[I_{ij,k}] = \text{Prob}((X_k)_i = 1 \wedge (X_k)_j = 1)$$
$$= \text{Prob}((X_k)_i = 1)\text{Prob}((X_k)_j = 1)$$
$$\approx \delta^2,$$

assuming the relative weight of X_k is very likely close to δ.

The situation is different if i and j are in the same inner block. Let ϵ_k denote the fraction of zero inner codewords of X_k, and let $T_{k,i}$ be the indicator variable being one whenever the inner block of X_k containing the position i (and also position j) is nonzero. Then we get the following estimate for the case where i and j are in the same inner block:

$$E[I_{ij,k} \mid \epsilon_k] = \text{Prob}((X_k)_i = 1 \wedge (X_k)_j = 1 \mid T_{k,i} = 1, \epsilon_k) \cdot \text{Prob}(T_{k,i} = 1 \mid \epsilon_k)$$
$$\approx \left(\frac{\delta}{1 - \epsilon_k}\right)^2 \epsilon_k$$
$$= \delta^2 \cdot \frac{\epsilon_k}{(1 - \epsilon_k)^2}.$$

Since the relative weight of X_k is less than δ, this means that the average relative weight of the inner blocks is less than δ. Knowing that the relative distance of the inner code is δ, we get the combinatorial guarantee that at least one of the inner code blocks contains the zero codeword. We therefore know that $\epsilon_k \geq n^{-1}$. (In reality, we expect a constant fraction of them to be zero.)

Now, if for each pair of indices (i, j), we compute

$$S_{ij} := \sum_{k=1}^{N} I_{ij,k},$$

then, if N is large enough, those random variables can be used to determine the inner codewords: Just declare (i, j) as belonging to the same set whenever S_{ij} is large enough.

After the sampling, the values S_{ij} can then be used to recover the sets, using a greedy algorithm, for example. Algorithm 1. illustrates this approach.

Note that the behaviour of ϵ_k has an impact on the complexity of the algorithm. The bound $\epsilon_k \geq n^{-1}$ guarantees that only a polynomial number of low weight codewords has to be sampled, but larger values cause much faster convergence. (In practice, choosing the number of observations linear in the number of sets works well over a wide parameter range, although this is significantly less than what we can prove to be sufficient.)

Algorithm 1. Decompose inner sets of $\mathcal{C}$

$\mathcal{C}$ is a concatenated code of block length $N = n \cdot n_i$. The inner code $\mathcal{C}_i$ has distance d_i and length n_i. M is the number of samples deemed sufficient.

Let $S_{ij} \leftarrow 0$, $1 \leq i, j \leq N$.
for $i = 1, \ldots, M$ **do**
    Sample a word $(x_1, \ldots, x_N) \in \mathcal{C}$ of weight $< N(d_i/n_i)$.
    for each (i, j) with $x_i = x_j = 1$ and $i \neq j$ **do**
        Increment S_{ij}.
    end for
end for
for $e = 1, \ldots, n$ **do**
    Let i be such that S_{ij} is maximal for some j, i.e., $i \leftarrow \arg \max_{1 \leq i \leq N} \max_{1 \leq j \leq N} S_{ij}$

    $T_e \leftarrow \{i\}$
    while $|T_e| < n_i$ **do**
        Let $1 \leq i \leq N$ be a vertex such that $\sum_{j \in T_e} S_{ij}$ is maximal.
        $T_e \leftarrow T_e \cup \{i\}$
        Let $S_{ji} \leftarrow -\infty$ and $S_{ij} \leftarrow -\infty$ for all $1 \leq j \leq N$.
    end while
end for

We close this section by noting that according to our definition, Reed-Muller codes themselves are concatenated codes, so one could think of applying this method directly, rather than first finding minimum weight words and shortening. This does not work, since the minimum distance is in this case just large enough to prevent any codewords from lying in the space we want to sample from.

4.3   The Case $r = 1$

Consider the matrix A formed by the rows corresponding to the codewords v_m, $v_{m-1}, \ldots, v_1$ of the (unpermuted) $\mathcal{R}(1, m)$. By construction, the i-th column of this matrix is just the number $i - 1$, if we read the vector as a binary number. Any possible binary vector of length m appears exactly once among the columns of this matrix, and if we add the all-one row, we get a generator matrix for a first-order Reed-Muller code.

Now, let $f_1, f_2, \ldots, f_m, f_{m+1}$ be a random basis of $\mathcal{R}(1, m)^\sigma$. If the all-one codeword is not linearly dependent on $f_1, \ldots, f_m$, then in the matrix A^σ formed by the rows $f_1, \ldots, f_m$, each column-vector is distinct. Thus, we can just reorder the columns by moving the zero-vector to the first position, etc., and thus obtain the matrix A. The same permutation applied to the positions of $\mathcal{R}(1, m)^\sigma$ will then yield $\mathcal{R}(1, m)$.

This suggests a simple method to find a suitable permutation: Pick any basis $f_1, \ldots, f_{m+1}$ of $\mathcal{R}(1, m)^\sigma$, check if the columns of the corresponding matrix A^σ are distinct, repeat if not, identify the corresponding permutation otherwise.

What is the success-probability of such an iteration? Since the f_i are linearly independent, the following estimate of this probability holds:

$$\frac{(2^{m+1} - 2)(2^{m+1} - 2^2) \cdots (2^{m+1} - 2^m)}{(2^{m+1} - 1)(2^{m+1} - 2) \cdots (2^{m+1} - 2^{m-1})} = \frac{2^m}{2^{m+1} - 1} > \frac{1}{2}.$$

In other words, we need merely two trials on the average.

4.4   Running Time Analysis

In the analysis, we will take the quantity $n = 2^m$ (the block length) as the input length, and we will assume r to be small with respect to m which leads to a low-rate setting. This assumption is based on the fact that Reed-Muller codes behave very poorly when r is large, and are therefore practically useless in these instances. For this reason, we will assume $r/m \to 0$ and $r < m/2$. In practice, r is usually a small constant. See [2] for tradeoffs between r, m and decoding thresholds.

The only computationally hard operation of the attack is the one of finding low weight words in a code, everything else is polynomial time. Thus, in order to determine the running time up to a polynomial factor, it is sufficient to verify that only a polynomial number of low weight words is needed, and then to restrict attention to the low weight word finding algorithm.

Checking that only a polynomial number of low weight words has to be found is straightforward: In order to find a single vector in $\mathcal{R}(r - 1, m)^\sigma$, a minimum weight word in the original code has to be found, and then the statistical test has to be performed to recover the concatenated structure of the shortened code. Since the bias in the statistical test is at least $(2^r - 1)^{-2}$ per observation, we have to collect $O(2^{2r}) = O(2^m) = O(n)$ vectors to get good estimates.

Thus, finding a single vector of $\mathcal{R}(r - 1, m)^\sigma$ needs the sampling of a polynomial number of low weight words. But then, since $O(k) = O(n)$, so does clearly the sampling for a complete basis of $\mathcal{R}(r - 1, m)^\sigma$. And given that $r \ll n$, the reduction to $\mathcal{R}(1, m)^\sigma$ requires still only a polynomial number of samples.

Because of this, we conclude that, in the exponent, only the complexity of the low weight word finding algorithm matters asymptotically, and we restrict our attention to this algorithm. In practice, those polynomial factors do of course matter to some extent, but notice that the degree of the polynomial is not very large.

Finding very low weight codewords. The problem of finding very low weight words is generally intractable for linear codes. For example, if the rate is kept fixed and the relative weight of the sought word is fixed and small enough, then even the best known algorithms are exponential in the block length.

However, finding low weight words is much easier if the rate is low, and if it is not actually fixed but converges to 0 with the block length.

Good methods for finding low weight words are based on the following (information set decoding) algorithm: Take a random $k \times n$ generator matrix G of the code, pick a random set I of k columns of G, and diagonalize the matrix G on the set I. Check the rows of G to see if any of them is low weight. If not, try again with another random set I.

The condition for a specific word in G of weight w to pop up as a row in such a diagonalized matrix is that exactly one of its bits is inside the information set and the other ones are outside. To simplify, we instead compute the probability that none of its bits are in the information set, a probability which is a bit smaller. We can approximate this by noting that if k is small compared to n, the probability that none of the positions of I match with the support of the word of weight w is roughly

$$\left(1 - \frac{w}{n}\right)^k. \tag{2}$$

This probability becomes large if k is very small with respect to n (i.e., the rate is very low), or if w is very small.

Note that (2) estimates the probability of finding a *single* word of the given weight given a random set I. If many words of the desired weight exist, the probability has to be multiplied with the number of such words. The above estimate decreases with w, but if such an algorithm is applied to find any word of weight $\leq w_0$, then the larger w_0 is, the easier the task becomes. The reason for this apparent contradiction is simply that the number of acceptable words increases dramatically with w_0.

Finite-length analysis. The goal of this section is to specialize (2) to the case of Reed-Muller codes, and to devise a crude bound which allows to estimate the feasibility of the low weight word finding problems (and thus the attack) for different values of r and m.

We first study the hardness of the minimum weight word finding procedure for Reed-Muller codes. In this case, we have $w = 2^{m-r}$ and $k = \sum_{i=0}^{r} \binom{m}{i} \leq \frac{m-r+1}{m-2r+1} \cdot \frac{m^r}{r!}$. If we plug this into (2), we get the hit probability of at least

$$\exp\left\{ \frac{m-r+1}{m-2r+1} \cdot \frac{m^r}{r!} \cdot \ln\left(1 - 2^{-r}\right) \right\} \tag{3}$$

for a single codeword per information set. By Proposition 3.2, there are at least $2^{mr-r(r-1)}$ such words, and so the cost for finding any one of them can be estimated to be at most

$$2^{-\frac{m-r+1}{m-2r+1} \cdot \frac{m^r}{r!} \cdot \log_2\left(1 - 2^{-r}\right) - mr + r(r-1)} \tag{4}$$

diagonalizations of the generator matrix. This rough estimate predicts, for example, that finding a minimum weight word in $\mathcal{R}(3, 11)$ would cost roughly 2^{37} diagonalizations, and thus finding such words is feasible in that case.

As expected and easily seen by comparing to real running times, the bound (4) is somewhat pessimistic, i.e., it overestimates the running time. For example, finding a minimum weight word in $\mathcal{R}(3, 11)$ needs only about 2^{17} diagonalizations in practice. More precise estimates are of course possible, but result in uglier formulas.

The other low weight finding instance operates on the shortened code. In practice the sampling turns out to be much easier, because of the lower rate and the weakened condition on the weight. A conservative estimate is easy to find. For example, one can show that there are at least $2^{mr-r(r-1)-(m+r^2-4r+2)}$ minimum weight words in the shortened code, and then apply (3) to get a bound similar to (4). The obtained bound is even weaker than (4), though: It does not take into account the lower rate of the code, nor the fact that words do not have to be strictly minimum weight in this case.

Asymptotic analysis. Asymptotically, the running time for the algorithm is

$$O(\mathrm{poly}(n)) \cdot e^{O(\mathrm{poly}(\log(n)))} \tag{5}$$

for any fixed value of r.

To see this, we start again with (3). Using the assumption that $r/m \to 0$, and writing the expression in terms of the block length $n = 2^m$ instead of m, we get that this probability behaves like

$$\exp\left\{-\log_2(n)^r C_r(1 + o(1))\right\},$$

where C_r is a constant depending only on r. This time, we assume there is just a single minimum weight codeword, and thus a conservative estimate for the number of trials to find a minimum weight word is

$$C_{\mathrm{lw}} := \exp\left\{\log_2(n)^r C_r(1 + o(1))\right\}. \tag{6}$$

We take C_{lw} as the cost for both of the low weight sampling instances, deferring justification of this to A.1. Using the fact that only a polynomial (in n) number of samplings is needed, we conclude that (5) is is indeed a bound for the running time of the algorithm.

For large r, the numbers get very large. That is not an artifact: If the code is not sufficiently low-rate, then finding minimum weight becomes a very hard problem, rendering the attack infeasible.

4.5   Experimental Running Time

To check the real-life behaviour, we ran our algorithm for different parameters on a 2.4GHz PC. Our implementation uses rather simple low weight word finding algorithms, and not elaborate ones like, e.g., the ones described in [1]. The average running-times for ten runs were:[2]

	$r = 2$	$r = 3$	$r = 4$
$m = 5$ $(n = 32)$	< 0.01s		
$m = 6$ $(n = 64)$	< 0.01s		
$m = 7$ $(n = 128)$	0.02s	5.261s	
$m = 8$ $(n = 256)$	0.081s	2.059s	
$m = 9$ $(n = 512)$	0.448s	3.462s	176.914s
$m = 10$ $(n = 1024)$	2.46s	26.6s	82197.4s
$m = 11$ $(n = 2048)$	18.34s	1192.71s	no try

---

[2] We only look at $m > 2r$; since in the other case, the attack can be carried out more efficiently on the dual code.

As predicted by the analysis, the performance degrades quickly with larger r. This does indeed exhibit a limit of our attack, but note that since the performance of Reed-Muller codes degrades with large r, choosing such values would very likely open the doors to other attacks.

The $(r = 3, m = 7)$-case is an anomaly of our implementation; we have decided to leave the high numbers for consistency reasons.

Acknowledgement

We would like to thank Gérard Maze, Arjen Lenstra and Martijn Stam for helpful discussions and reviewing early draft versions of this paper; their help and support has been invaluable in the process of writing up this paper.

References

1. A. Canteaut, F. Chabaut, *A new algorithm for finding minimum-weight words in a linear code: application to primitive narrow-sense BCH-codes of length 511*, 1998, IEEE Transactions on Information Theory, 44(1):367-378
2. I. Dumer, K. Shabunov, *Soft-decision decoding of Reed-Muller codes: a simplified algorithm*, 2006, IEEE Transactions on Information Theory 52(3): 954-963
3. W. Cary Huffman, V. Pless, *Fundamentals of Error-Correcting Codes*, 2003, Cambridge University Press
4. T. Kasami, N. Tokura, *On the Weight Structure of Reed-Muller Codes*, 1970, IEEE Transactions on Information Theory, 16(6): 752-759
5. F. J. MacWilliams, N. J. A. Sloane, *The Theory of Error-Correcting Codes*, 1978, North-Holland
6. R. J. McEliece, *A public key cryptosystem based on algebraic coding theory*, DSN progress report, 42-44:114-116, 1978
7. H. Niederreiter, *Knapsack-Type Cryptosystems and Algebraic Coding Theory*, Problems of Control and Information Theory, 15(2):159–166, 1986.
8. V. M. Sidelnikov, *A public-key cryptosystem based on binary Reed-Muller codes*, Discrete Mathematics and Applications, 4 No. 3, 1994
9. V. M. Sidelnikov, S. O. Shestakov, *On insecurity of cryptosystems based on generalized Reed-Solomon codes*, Discrete Mathematics and Applications, 2, No. 4:439–444, 1992
10. N. Sendrier, *On the Structure of a randomly permuted concatenated code*, EUROCODE 94, October 1994.
11. N. Sendrier, *Finding the permutation between equivalent codes: the support splitting algorithm*, IEEE Transactions on Information Theory, 46(4):1193-1203, 2000

A   Appendix

This appendix contains detailed proofs that have been omitted in the paper, as well as some other comments we do not consider vital for the understanding of the paper.

A.1 The Low Weight Word Problem in the Shortened Code

In the running-time analysis, we based our running-time estimates on estimates on the difficulty of the low weight word finding problem.

It should be noted that two different low weight word finding problems have to be solved; the minimum weight word finding problem in the Reed-Muller code, and the low weight word finding problem in the shortened Reed-Muller code.

We assumed that the low weight word finding problem in the shortened code is easier than the minimum weight word finding algorithm in the original code. The reason for this is that first the weight restriction is relieved, and second, the shortened code has lower rate, as we will now show.

The shortened code has lower rate. The correctness of our running time analysis depends on the fact that the shortened code has lower rate. This is not an obvious fact, since, even though the dimension clearly has to decrease, the length does so too. We prove the assertion in this section.

We write $P_{r,m}$ the number of linearly independent parity checks that $\mathcal{R}(r,m)$ has, i.e., $P_{r,m} = \dim(\mathcal{R}(r,m)^{\perp})$.

We can use (1) to deduce the number its number of linearly independent parity checks in the shortened code, and get that there are

$$\sum_{i=1}^{r} \binom{r}{i} P_{r-i,m-r}$$

of them. We can deduce a similar formula for $P_{r,m}$ itself using an induction on the equality $P_{r,m} = P_{r,m-1} + P_{r-1,m-1}$. We then get that for any $\ell \leq r$, we have

$$P_{r,m} = \sum_{i=0}^{\ell} \binom{\ell}{i} P_{r-i,m-\ell}.$$

Proposition A.1. *The shortened code (constructed in the section on the attack) has lower rate than the original code.*

Proof. We have to show that

$$\frac{\sum_{i=1}^{r} \binom{r}{i} P_{r-i,m-r}}{2^m - 2^{m-r}} > \frac{\sum_{i=0}^{r} \binom{r}{i} P_{r-i,m-r}}{2^m}.$$

Rearranging the terms, we get that this is equivalent to showing that

$$\frac{1}{1 - 2^{-r}} > \frac{\sum_{i=0}^{r} \binom{r}{i} P_{r-i,m-r}}{\sum_{i=1}^{r} \binom{r}{i} P_{r-i,m-r}},$$

or yet,

$$2^{-r} > \frac{P_{r,m-r}}{\sum_{i=0}^{r} \binom{r}{i} P_{r-i,m-r}}.$$

Let μ be the weighted average of the $P_{r-i,m-r}$, i.e.,

$$\mu = 2^{-r} \sum_{i=0}^{r} \binom{r}{i} P_{r-i,m-r}.$$

Then we see that we have to show

$$\mu > P_{r,m-r}.$$

Now this last equation is true because

$$P_{r,m-r} < P_{r-1,m-r} < \cdots < P_{0,m-r},$$

as implied by proposition 2.1.   □

A.2 A Brief Note on the Generalized Sidelnikov System

The paper [8] also proposes to use more than a single generator. The proposition is to juxtapose several differently scrambled generators, and then intermingle the separate blocks with a right-hand permutation matrix. If u is some small integer, R is a generator matrix of some $\mathcal{R}(r,m)$-matrix of dimension $k \times n$, $E_1, \ldots, E_u$ are random invertible matrices, and Γ is a $un \times un$ random permutation matrix, then the public key is of the form

$$|E_1 R, E_2 R, \ldots, E_u R|\Gamma,$$

which is the generator matrix of some $[un, k]$-code.

In fact, there is no added security using this when compared to the case $u = 1$. To see this, note that on the positions corresponding to $E_i R$, all the parity checks for $\mathcal{R}(r,m)$ are valid parity checks. So to recover the independent code blocks, it is enough to sample low weight parity checks and to mark the bits in their support as belonging to the same inner block. Doing this for not to many codewords should be enough to recover the block decomposition.

In general, it is hard to find low weight words, but not if the sought words are very low weight. Since $\mathcal{R}(r,m)^{\perp} = \mathcal{R}(m-r-1,m)$, the lowest-weight in the dual code is 2^{r+1}, which is indeed very low weight for the values of r of interest in practice.

So, in summary, breaking the general Sidelnikov system is roughly equivalent to recovering a single Reed-Muller code.

Toward a Rigorous Variation of Coppersmith's Algorithm on Three Variables

Aurélie Bauer[2] and Antoine Joux[1,2]

[1] DGA
[2] Université de Versailles Saint-Quentin-en-Yvelines
Laboratoire PRISM, 45, Avenue des Etats-Unis
78035 Versailles cedex, France
aurelie.bauer@prism.uvsq.fr, antoine.joux@m4x.org

Abstract. In 1996, Coppersmith introduced two lattice reduction based techniques to find small roots in polynomial equations. One technique works for modular univariate polynomials, the other for bivariate polynomials over the integers. Since then, these methods have been used in a huge variety of cryptanalytic applications. Some applications also use extensions of Coppersmith's techniques on more variables. However, these extensions are heuristic methods. In the present paper, we present and analyze a new variation of Coppersmith's algorithm on three variables over the integers. We also study the applicability of our method to short RSA exponents attacks. In addition to lattice reduction techniques, our method also uses Gröbner bases computations. Moreover, at least in principle, it can be generalized to four or more variables.

Keywords: Lattice reduction, Coppersmith's algorithms, Gröbner basis.

1   Introduction

In 1996, Coppersmith introduced two methods for finding small roots of polynomial equations using lattice reduction, one for the univariate modular case and another one for the bivariate case over the integers [6,5,7]. These algorithms are based on the same idea: using lattice reduction (e.g. LLL) in order to create a second polynomial that has the same root as the first one. In both cases, this construction leads to a rigorous method to recover the root. In particular, in the bivariate case, the use of orthogonal lattice guarantees the independence of the two polynomials and ensures that the root can be recovered. In order to simplify and help understand Coppersmith's methods, Howgrave-Graham [13] and Coron [8] revisited his ideas and proposed alternative constructions.

Since 1996, many cryptanalytic applications have been based on these methods, for example the factorization of $N = pq$ knowing a fraction of the most significant bits on each factor. Another well-known example is the cryptanalysis of RSA with small private key [4,2].

The applications of these algorithms for finding small roots of polynomial equations can roughly be divided into two parts. On the one hand some researchers try to generalize the original Coppersmith's methods. For example, in

M. Naor (Ed.): EUROCRYPT 2007, LNCS 4515, pp. 361–378, 2007.

[3], Blömer and May present new results using Coppersmith's method for polynomials whose shapes are more complicated than those originally considered in Coppersmith's articles. Another example is the paper of Howgrave-Graham [14] in which he explains how to cast the problem of finding roots for particular polynomials in the more general context of approximate GCD computations. On the other hand, there are researchers trying to adapt all these methods for more than two variables. As an example, several new attacks on RSA are proposed in [9], using variants of the original method on three variables.

However, with more than two variables, one encounters a major obstruction. Indeed, one can not guarantee any more that the polynomials outputted by LLL reduction are algebraically independent. Still in some practical applications, the approach continues to work. Despite this, more and more articles mention problematic cases. For example, in [2] the authors analyze in details one of the heuristic multivariate attacks proposed by Boneh and Durfee in [4]. In [11,12], Hinek analyzes the problem of algebraic independence of the polynomials. He focuses on the fact that in experiments algebraic dependency often leads to difficulties and he says "in light of the observations in this work, it might be the case that this lack of reported instances is simply due to a lack of experimental observations". In this paper, in order to avoid these difficulties we propose a new generalization of Coppersmith's method in three variables, using a new lattice construction to find a third independent polynomial. Our construction uses Gröbner basis in addition to lattice reduction.

This paper is organized as follows. In section 2, we recall a few facts about lattice reduction and known heuristic variations of Coppersmith's method on three variables over the integers. We discuss the issue of polynomials independence. In section 3, we present an overview of our main idea which generalizes the method by using LLL reduction on a different lattice. To construct it, we show that the use of Gröbner bases and their properties are essential. In section 4 we describe a criterion on the input polynomials that when satisfied allows to develop a rigorous version. In section 5, we focus on one of the RSA attacks proposed in [9] and we show some results of experiments made with our method. The two approaches can then be compared. Finally, in section 6 we discuss the possibility of generalizing to four or more variables.

2   Preliminaries

2.1   Lattices

Since lattices are an essential tool for Coppersmith's attack, let us recall a few facts about lattices and reduced basis. A lattice L is a discrete subgroup of $\mathbb{R}^n$. If L is a non-empty subset of $\mathbb{R}^n$, L is a lattice if and only if there exists r linearly independent vectors over $\mathbb{R}$ (with $r \leq n$) such that

$$L = \mathbb{Z}b_1 \oplus \cdots \oplus \mathbb{Z}b_r$$

The set $\mathcal{B} = (b_1, \ldots, b_r)$ is called a basis of L. In this paper, as in many cryptographic applications, we focus on integer lattices $L \subset \mathbb{Z}^n$.

Let L be a lattice generated by the vectors $\mathcal{B}=(b_1,\ldots,b_r)$ and $(b_1^\star,\ldots,b_r^\star)$, the vectors from Gram-Schmidt's orthogonalization of $\mathcal{B}$. Let B be the $r \times n$-matrix whose rows are the b_i's. The determinant of L is defined as

$$\det L = \sqrt{\det(B^t B)} = \prod_{i=1}^{r} \|b_i^\star\|^2$$

where $\|\|$ denotes the Euclidean norm. When L is a full-rank lattice (i.e. when $n = r$) the formula can be simplified to $\det L = |\det B|$.

In 1982, Lenstra, Lenstra and Lovasz [15] introduced the LLL reduction algorithm. Using this algorithm, one can obtain a reduced basis of a lattice L. To analyze Coppersmith's algorithm, we need to know that

$$\|b_r^\star\| \geq (\det L)^{1/r} 2^{-(r-1)/4} \tag{1}$$

for any LLL reduced basis $(b_1,\ldots,b_r)$.

2.2   Gröbner Basis on Three Variables

Let $\mathbb{Z}[x,y,z]$ be the polynomial ring in three variables x,y,z over $\mathbb{Z}$. A monomial is an elementary polynomial $x^{\alpha_1}y^{\alpha_2}z^{\alpha_3}$ with $(\alpha_1,\alpha_2,\alpha_3) \in \mathbb{N}^3$ and a term is $\lambda x^{\alpha_1}y^{\alpha_2}z^{\alpha_3}$ with $\lambda \in \mathbb{Z}$. In the following, when we refer to a monomial in a set, we use both the notations $(\alpha_1,\alpha_2,\alpha_3)$ and $x^{\alpha_1}y^{\alpha_2}z^{\alpha_3}$. If p is a polynomial defined over $\mathbb{Z}$, the Newton polygon of p refers to the convex hull of all monomials (viewed as points in $\mathbb{N}^3$) that appear with a non zero coefficient in p.

A monomial ordering $<$ on $\mathbb{Z}[x,y,z]$ is a total ordering on the set of monomials which is compatible with multiplication. Among all existing orderings, a frequently encountered one is called *deglex* and is defined as:

$$x^{\alpha_1}y^{\alpha_2}z^{\alpha_3} < x^{\beta_1}y^{\beta_2}z^{\beta_3} \Leftrightarrow \begin{cases} \alpha < \beta \\ \text{or} \\ \alpha = \beta \text{ and } \exists i \in \{1,2,3\}, \alpha_i < \beta_i \\ \qquad \forall j < i, \alpha_j = \beta_j \end{cases}$$

where $\alpha = (\alpha_1 + \alpha_2 + \alpha_3)$ and $\beta = (\beta_1 + \beta_2 + \beta_3)$. If a monomial ordering is chosen, the initial term of a polynomial p, denoted by $in(p)$, refers to its greatest term. Let I be an ideal of $\mathbb{Z}[x,y,z]$, $in(I)$ is the set of all initial terms of the polynomials which belong to I. If the set $\{q_1,\ldots,q_l\}$ is composed by polynomials of I such that $(in(q_1),\ldots,in(q_l)) = in(I)$, we call it a Gröbner basis of I. In practice, a Gröbner basis can be computed using F4 algorithm [10] implemented in Magma. For a system of generators having d as its maximal degree, the theoretical complexity is polynomial in d when the number of variables is fixed.[1]

---

[1] According to M. Bardet [1], the complexity in this case is upper bounded by d^{72}. In practice, the computation is very fast under Magma.

2.3 Primary Decomposition

Let I be an ideal of $\mathbb{Z}[x, y, z]$. It is said to be prime if the condition $fg \in I$ implies that either f or g belongs to I. The radical of I, denoted by $\sqrt{I}$, refers to the set $\{f \in I, \exists n \in \mathbb{N}, f^n \in I\}$. A primary ideal J satisfies the following condition: if fg belongs to J with $f \notin J$, then g belongs to $\sqrt{J}$. If I is a primary ideal, then $\sqrt{I}$ is a prime one. In a noetherian ring, each ideal can be written as an intersection of primary ideals. In practice, with the help of Magma, a few seconds are needed to compute the primary decomposition of an ideal I or to obtain its radical.

The set defined as $\{(x_1, y_1, z_1) \in \mathbb{Z}^3, \forall p \in I, p(x_1, y_1, z_1) = 0\}$ is denoted as $V(I)$. In the following, we say that (x_1, y_1, z_1) is a root of I if it belongs to $V(I)$. If I has $I_1 \cap \cdots \cap I_r$ as a primary decomposition, then $V(I) = V(I_1) \cup \cdots \cup V(I_r)$. The following property holds: $V(I) = V(\sqrt{I})$.

2.4 Coppersmith's Method, a Basic Variation on 3 Variables

Let $p_1(x, y, z)$ be an irreducible polynomial of $\mathbb{Z}[x, y, z]$ having (x_0, y_0, z_0) as root over the integers satisfying $|x_0| < X, |y_0| < Y$ and $|z_0| < Z$. As usual when working with Coppersmith's method, we denote by W_1 the quantity $\|\tilde{p}_1\|_\infty$ where $\|p(x, y, z)\|_\infty$ is the maximum of the absolute values of the coefficients of p and $\tilde{p}_1(x, y, z)$ represents $p_1(xX, yY, zZ)$. Our goal is to recover the root (x_0, y_0, z_0) in polynomial time.

As in [3], we use the notion of admissible sets. Let M be a non-empty set of three variables monomials. A polynomial $p(x, y, z)$ is said to be defined over M if p can be written as linear combination of monomials in M. Let S be another non-empty set and f, g be two polynomials such that $g = fp_1$. The ordered pair (S, M) is said to be admissible for p_1 if the property "g defined over M" is equivalent with "f defined over S". The cardinality of M and S are denoted by m and s.

Coppersmith's algorithm works by finding a second polynomial p_2 algebraically independent from p_1, which has the same root over the integers. When working with two variables, the resultant of p_1 and p_2 is non zero and the root can easily be recovered. However, in our case, since we work with three variables, two polynomials are not enough to recover the root. Still, it is a first important step. Thus, we now describe how Coppersmith's algorithm can be adapted in three variables to find p_2. We start by introducing the notation $(x_0{}^f y_0{}^g z_0{}^h)_M$ that refers to the vector $(1, x_0, y_0, z_0, \ldots, x_0{}^f y_0{}^g z_0{}^h, \ldots)$ with $(f, g, h) \in M$, where the order of the coordinates depends on the monomial ordering. Then, let us take the vector $r_0 = (x_0^f y_0^g z_0^h)_M$ and the lattice L_1 generated by the rows of the matrix M_1 (see figure 1).

The right hand part of M_1 is denoted by P_1 and the left hand one by D_M. As $s < m$, there exists a sublattice $L_1' \subset L_1$ of dimension $(m - s)$ such that its vectors have their s last coordinates equal to zero. As (x_0, y_0, z_0) is a root of p_1, the product $s_0 = r_0 M_1$ gives a short vector of L_1' defined by $s_0 = ((\frac{x_0}{X})^f (\frac{y_0}{Y})^g (\frac{z_0}{Z})^h)_M | (0, \ldots, 0)$ where the symbol $|$ refers to the concatenation of the two vectors. Assume that $(b_1, \ldots, b_r)$ is an LLL reduced basis of

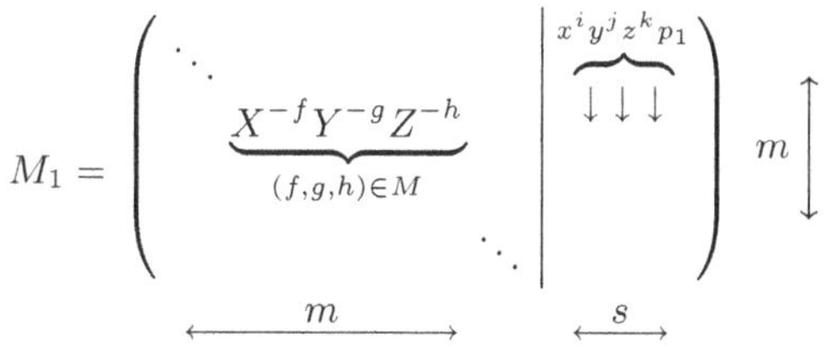

Fig. 1.

L'_1 (with $r = m - s$), then when $\|s_0\| < \|b^\star_r\|$ we know that the inner product $< s_0 | b^\star_r >$ is equal to zero. That leads to a new polynomial p_2 that has the same root as p_1.

As in [5], we can show that p_2, as all polynomials obtained from L'_1, is by construction independent from p_1. The crucial point of this proof is based on the fact that, in this case, algebraic independence relies on linear independence.

In order to prove in advance that the inequality $\|s_0\| < \|b^\star_r\|$ holds, we need to compute $|\det L'_1|$. This can be done by adapting the method of [5], see appendix A. From the determinant computation, we derive the conditions on the bounds X, Y, Z:

$$X^{s_x} Y^{s_y} Z^{s_z} < W_1^s 2^{-(6+c)s(d_x^2 + d_y^2 + d_z^2)} \qquad (2)$$

with c a well-chosen constant. In this formula d_x, d_y and d_z denote the maximum degree of p_1 in x, y, z and s_x refers to $\sum_{(f,g,h)\in M\setminus S} f$. The corresponding sums on y and z are denoted by s_y and s_z.

2.5   Recovering the Root

With this method, we have two polynomials p_1 and p_2 that have (x_0, y_0, z_0) as common root over the integers, and are algebraically independent. Two approaches have been proposed to recover the root. The first idea is to compute the (provably non-zero) resultant of p_1 and p_2 in one of the variables. This leads to a polynomial in two variables, on which Coppersmith's algorithm can be reused. However, this polynomial usually has a very high degree. As a consequence, the conditions on the bounds are too restrictive to make the method useful. Another idea is to reuse Coppersmith's method in three variables trying to find another polynomial p_3. The difficulty here is to ensure that p_3 is algebraically independent from p_1 and p_2. Many authors use this approach together with the heuristic hypothesis that p_3 happens to be independent from $\{p_1, p_2\}$.

2.6   The Notion of Independence

As this works focuses on the problem of algebraic independence, this notion has to be rigorously defined. Three polynomials p_1, p_2, p_3 are algebraically independent if and only if $P(p_1, p_2, p_3) = 0$ implies $P = 0$ for a polynomial P defined over $\mathbb{Q}[x, y, z]$. In general, showing this property is rather difficult. In our case, knowing that p_1 is irreducible and that p_2 does not belong to (p_1), it can be reduced to a simpler problem. When the ideal $I = (p_1, p_2)$ is prime, whenever p_3 does not belong to I, then p_1, p_2 and p_3 are algebraically independent. The proof of this result can be found in appendix B. It uses the fact that (x_0, y_0, z_0) is a common root of these three polynomials.

In the sequel, when we refer to I, it implicitly means a prime ideal. As a consequence, showing that p_3 does not belong to I is a sufficient condition to obtain the independence. Let us now discuss on what happens if I is not a prime ideal. In this case, the analysis is more complicated. Two behaviors are possible depending on the fact that I is a primary ideal or not. If I is primary, it is sufficient to replace it by its radical, which is prime. In the other case, I can be written as an intersection of primary ideals $I_1 \cap \cdots \cap I_r$, such that at least one of the I_j has (x_0, y_0, z_0) as root. One has just to replace I by the well-chosen ideal and to take its radical if it is primary.

3   A New Lattice to Find a Third Independent Polynomial

Having recovered p_1 and p_2, we now want a method to create a third polynomial p_3 that has again the same root as p_1 and p_2 and moreover does not belong to the ideal $I = (p_1, p_2)$. The main idea is to construct a new lattice very similar to Coppersmith's one that can produce this third independent polynomial.

3.1   Overview of the Main Idea

Let start with analyzing the first step of Coppersmith's algorithm. The proof concerning the independence of p_2 from p_1 uses the fact that, in this case, algebraic independence relies on linear independence. In three variables, a third polynomial has to be found. As explained before, the main difficulty is less its construction than the proof of its independence from I. Our goal is to adapt the previous construction and to keep information both from p_1 and p_2. If I_M is the set of all polynomials belonging to I that are defined over M, one possible idea would be to create a new lattice by using generators of I_M as a $\mathbb{Z}$-module. Thus, any polynomial belonging to I_M is generated by the columns of this new matrix.

Finding these generators is quite complicated as it is strongly linked to the shape of the set M. For this reason, in the rest of this section, we only focus on a pair $(M, <)$ such that there exists an equivalence between belonging to M and being smaller than a given monomial ($<$ is compatible with the shape of M). As an example, one can consider the set M defined as all $(f, g, h) \in \mathbb{N}^3$ satisfying $(f + g + h) \leq n$ (with n an integer) and the *deglex* ordering. In order to construct these generators, we need an additional tool. In the sequel, we show that the use of truncated Gröbner basis gives us the right tool.

3.2   Truncated Gröbner Basis

As explained before, our goal here is to find linear generators of I (up to some degree). Moreover, we want these polynomials to be defined over M in order to preserve the dimension of the lattice. Let us denote by $\mathcal{F}$, the set we are looking for. To sum up the property we require for the construction, one can say that if p is a polynomial defined over M and belonging to I, we want it to be written as linear combination of the polynomials r_i where the set $\{r_1, \ldots, r_t\}$ refers to $\mathcal{F}$. To construct such a set, we need the use of "truncated Gröbner basis" whose definition is given as follows:

Definition 1. *Let $G = \{q_1, \ldots, q_r\}$ be a minimal Gröbner basis of I. A truncated Gröbner basis of I related to M is the set of polynomials $\{q_{i_1}, \ldots, q_{i_l}\}$ of G such that for each $j \in \{i_1, \ldots, i_l\}$, q_{i_j} is defined over M. The corresponding set is denoted by G_M.*

The idea is just to keep among all polynomials that generate the ideal I, those which are defined over M. Then, to obtain the set $\mathcal{F}$, it is sufficient to multiply the q_{i_j} by monomials and keep products which remain defined over M. Thus, we have a system of generators of the vector space I_M. The creation of the set G_M from G has a complexity equal to $O(rm)$ whereas those of $\mathcal{F}$ has one of $O(rm^2)$. With this construction, the set $\mathcal{F}$ is not necessarily minimal. As a consequence, one could improve this construction by deleting the extra polynomials. However, keeping them does not increase the dimension of the lattice, and does not change the proofs.

Note: When the set M cannot be described directly by a monomial ordering, finding a set of generators $\mathcal{F}$ is more difficult. However, the examples of section 5 show that it can still be done in practice. The difficulty here is to give a theoretic construction that works for all cases.

3.3   A Second Coppersmith's Iteration

Knowing how to find the set $\mathcal{F}$, we are able to construct a new lattice to recover a polynomial p_3 having (x_0, y_0, z_0) as root over the integers. The positive point of this construction is that we can now prove that p_3 does not belong to the ideal I. Let us explain in more details, how the lattice is constructed.

Let start by considering the $m \times t$ matrix P_2 whose columns represent the polynomials $\{r_1, \ldots, r_t\}$. From P_2, one can construct the lattice L_I generated by the rows of the following matrix:

$$M_I = \begin{pmatrix} \ddots & & \overbrace{\begin{matrix} & r_1, \ldots, r_t \\ \downarrow \downarrow \downarrow \end{matrix}} & \\ & \underbrace{X^{-f} Y^{-g} Z^{-h}}_{(f,g,h) \in M} & & m \\ & & \ddots & \end{pmatrix} \Big\updownarrow m$$
$$\underset{\xleftarrow{\hspace{2cm}} m \xrightarrow{\hspace{2cm}}}{} \quad \underset{\xleftarrow{} t \xrightarrow{}}{}$$

If we assume that $t < m$, there exists a sublattice $L_I' \subset L_I$ whose dimension is $(m - t)$ such that its vectors have their t last coordinates equal to zero. Let see again the vector $r_0 = (x_0^f y_0^g z_0^h)_M$. As (x_0, y_0, z_0) is a root of all polynomials in I, the product $t_0 = r_0 M_I$ satisfies $t_0 = ((\frac{x_0}{X})^f (\frac{y_0}{Y})^g (\frac{z_0}{Z})^h)_M | (0, \ldots, 0)$. This is a short vector of L'_I. Assume that $(c_1, \ldots, c_r)$ is an LLL reduced basis of L'_I (with $r = m - t$). When $\|t_0\| < \|c_r^\star\|$, the inner product $< t_0 | c_r^\star >$ is equal to zero, that leads to a new polynomial p_3 that has the same common root as p_1 and p_2.

Let focus on the most important point which is the independence of p_3 from the ideal I. By construction, the vector which refers to p_3 is orthogonal to all polynomials of the set $\mathcal{F}$. Knowing that in each vector space E, if there exists a vector x such that for all $y \in E$, $< x | y > = 0$, then $x = 0$, we necessarily have $p_3 \notin I$. Indeed, if p_3 is assumed to belong to I, it would be equal to zero, which is not the case.

Then, from p_1 and p_2, we construct a polynomial p_3 that again has (x_0, y_0, z_0) as a root over the integers and that does not belong to I. The resultant computation of the three polynomials leads to a non-zero result and the root can be recovered easily. When trying to check if $\|t_0\| < \|c_r^\star\|$ is verified, some technical difficulties related to the evaluation of the determinant of M_I, are encountered. As the considered lattice L'_I is much more complicated than the initial one used in the first iteration of Coppersmith's algorithm on three variables, it makes the analysis more difficult. In the general case, as we are not able to evaluate the determinant of M_I precisely, this can not give explicit bounds.

4   A Criterion That Guarantees Rigorous Success

Starting with the ideal $I = (p_1, p_2)$, we give in this section a criterion on the input polynomials that guarantees that p_3 can be found with no further restrictions on X, Y, Z. Let us consider the set $\mathcal{F}$ related to the ideal I. In the sequel, we use the following criterion: $\mathcal{F}$ should be equal to $\{\{x^i y^j z^k p_1\}_{(i,j,k) \in S}, p_2\}$. The monomial $x^a y^b z^c$ refers to those which verifies $|\tilde{p}_{2,(a,b,c)}| = \|\tilde{p}_2(xX, yY, zZ)\|_\infty = W_2$. The gcd of the coefficients of p_2 is denoted by d. In the sequel, we show that p_3 can be found with no further restrictions than what was required to obtain p_2. The proof relies on a variation of the method explained in appendix A and is written using the same notations.

4.1   Some Preliminary Results

Consider $\bar{P}_2$ the $(m \times (s+1))$ matrix whose s first columns are multiples of $\tilde{p}_1$ and the last one represent the polynomial $\tilde{p}_2$. Thus, $\bar{P}_2$ is just composed by the matrix $\bar{P}_1$ and one additional column that is $\tilde{p}_2$. Using all results of the appendix A, there exists a subset $\hat{M} \subset M$ of size s such that if $\hat{P}_1$ is the matrix composed by the rows of P_1 related to $\hat{M}$, we are able to evaluate $|\det \hat{P}_1|$. As p_2 is defined over $M \setminus \hat{M}$, we know that (a, b, c) can not belong to $\hat{M}$. Then, let take the set $\dot{M} = \hat{M} \cup \{(a, b, c)\}$. We have $|\dot{M}| = (s+1)$.

If we select from $\bar{P}_2$, the rows related to monomials in $\dot{M}$, we obtain the following matrix $\dot{P}_2$:

$$\hat{P}_2 = \left(\begin{array}{c|c} & 0 \\ \hat{P}_1 & \vdots \\ & 0 \\ \hline \times \ \dots \ \times & \pm W_2 \end{array} \right)$$

That leads to

$$|\det \hat{P}_2| \geq W_2 W_1^s 2^{-6s(d_x^2 + d_y^2 + d_z^2)}$$

4.2   Construction of the Lattice $\mathcal{L}_I$

Let P_2 be the $(m \times (s+1))$ matrix constructed as follows: the s first columns represent $x^i y^j z^k p_1$ for all $(i, j, k) \in S$ and the last one is p_2. $\mathcal{L}_I$ is the lattice generated by the rows of the following matrix:

$$N_I = \left(D_{M \setminus \dot{M}} \mid P_2 \right)$$

where $D_{M \setminus \dot{M}}$ is the resulting matrix coming from deletion in D_M of the columns related to monomials in $\dot{M}$. This definition of $\mathcal{L}_I$ is different from those of L_I, which has been announced in the previous section. However using this construction does not change the explanation, moreover it gives an easier analysis. Multiplying the rows of N_I related to $(f, g, h) \in M$ by $X^f Y^g Z^h$ and the s first columns of P_2 related to $(i, j, k) \in S$ by $X^{-i} Y^{-j} Z^{-k}$ leads to the matrix $\bar{N}_I$ satisfying:

$$|\det \bar{N}_I| = |\det N_I| X^{s_x} Y^{s_y} Z^{s_z}$$

Making some elementary row operations on $\bar{N}_I$ leads to:

$$\begin{pmatrix} Id & A' \\ 0 & \hat{P}_2 \end{pmatrix}$$

whose determinant is equal to $(\det \hat{P}_2)$. Thus, we obtain

$$|\det N_I| \geq X^{-s_x} Y^{-s_y} Z^{-s_z} W_2 W_1^s 2^{-6s(d_x^2 + d_y^2 + d_z^2)}$$

4.3   Using LLL-Reduction to Construct p_3

The demonstration follows the same idea as in the previous case. r_0 is the vector defined by $r_0 = (x_0^f y_0^g z_0^h)_M$ and $t_0 = r_0 N_I$. We have

$$t_0 = ((\frac{x_0}{X})^f (\frac{y_0}{Y})^g (\frac{z_0}{Z})^h \underset{M \setminus \dot{M}}{}) | \underbrace{(0, \dots, 0)}_{s+1}$$

The vector t_0 has its $(s+1)$ last coordinates equal to zero. Moreover, its norm is less than $\sqrt{m - s - 1}$. As the polynomial $p_1(x, y, z)$ is irreducible, and the gcd of the coefficients of p_2 is d, some elementary row operations on N_I leads to the following matrix:

$$N_I' = \left(\begin{array}{c|c} A_1 & B \\ \hline A_2 & 0 \end{array} \right) \quad \begin{array}{l} \updownarrow (s+1) \\ \updownarrow (m - s - 1) \end{array}$$

where B is a diagonal matrix having 1 on its s first coefficients and d for the last one. If we call $\mathcal{L}'_I$ the lattice generated by the $(m-s-1)$ last rows of the previous matrix, we have $|d \cdot \det \mathcal{L}'_I| = |\det N_I|$. Moreover, the vector t_0 belongs to $\mathcal{L}'_I$. Let take $r = m - s - 1$, and assume that $(c_1, \ldots, c_r)$ is an LLL-reduced basis of $\mathcal{L}'_I$. Thus $\|c^\star_r\| \geq 2^{-(r-1)/4} |\det \mathcal{L}'_I|^{1/r}$. As t_0 belongs to $\mathcal{L}'_I$, when $\|t_0\| < \|c^\star_r\|$, the inner product $< t_0 | c^\star_r >$ is equal to zero that leads to a polynomial $p_3(x, y, z)$ having the same common root as p_1 and p_2. This condition has to be satisfied:

$$\sqrt{m-s-1} < 2^{-\frac{m-s-2}{4}} |\det \mathcal{L}'_I|^{\frac{1}{m-s-1}}$$

Then we can construct p_3 if the following one is verified:

$$\sqrt{m-s}^{(m-s)} 2^{(m-s-1)/4} < |\det \mathcal{L}'_I|$$

$$\sqrt{m-s}^{(m-s)} 2^{(m-s-1)/4} X^{s_x} Y^{s_y} Z^{s_z} < \frac{W_2}{d} \left(W_1^s 2^{-6s(d_x^2 + d_y^2 + d_z^2)} \right)$$

As X, Y, Z already verify the equation (2), we obtain that if $d < W_2$ (this is always the case), then we can construct p_3. In this case, no further restrictions on the bounds are needed to construct p_3. This polynomial is independent from the ideal $I = (p_1, p_2)$, as explained in section 3.3.

For a well-chosen pair $(M, <)$ (see section 3.2), the previous condition on $\mathcal{F}$ can be stated in terms of the truncated Gröbner basis. More precisely, we should have $G_M = \{p_1, p'_2\}$ and no multiples of p'_2 should be defined over M.[2] In practice, when $G_M = \{p_1, p'_2\}$, the other condition is often true.

5   Application to "Partial Key Exposure Attack on RSA"

In this section, in order to better understand the way the algorithm works in practice, we apply it to one of the partial key exposure attacks on RSA which have been proposed in [9]. We start by describing the basis of this attack in section 5.1 and 5.2.

5.1   The RSA Equation

Let $N = pq$ be a RSA modulus. The RSA encryption exponent e and decryption exponent d both satisfy the well-known equation $ed \equiv 1 \mod \phi(N)$ which can be rewritten into $ed = 1 + k(N - (p+q-1))$. We focus on the particular case of a small exponent d but without any restrictions on e except that $e < \phi(N)$. In addition, part of the high order bits of d ($\tilde{d}$) are known to an attacker. As a consequence, d can be rewritten as $\tilde{d} + d_0$ such that $|d| \leq N^\beta$ and $|d_0| = |d - \tilde{d}| \leq N^\delta$. The values of the two parameters β and δ will be used later. Putting these entries into the RSA equation leads to the following polynomial:

$$f_{MSB1}(x, y, z) = ex - yN + yz + R \text{ with } R = e\tilde{d} - 1$$

---

[2] The polynomial p'_2 is obtained by replacing in p_2 all multiples of the initial term of p_1 by multiples of $p_1^{(1)}$ where $p_1^{(1)} = p_1 - in(p_1)$.

The problem remains to find the root $(x_0, y_0, z_0) = (d_0, k, p + q - 1)$ of the polynomial $p_1 = f_{MSB1}(x, y, z)$ with $|x_0| < X, |y_0| < Y$ and $|z_0| < Z$ knowing that $X = N^\delta, Y = N^\beta$ and $Z = 3\sqrt{N}$.

5.2   A Heuristic Attack

We only sketch here the general idea of the attack proposed in [9], for further details, the reader can refer to it. Let m and t be two small integers which are taken in $\{0, 1, 2\}$ for the experiments. Let S and M be two sets of $\mathbb{N}^3$ defined as:

$$S = \{(i, j, k) | (i + j) \leq \mathsf{m}, k \leq j + \mathsf{t}\} \quad M = \{(f, g, h) | (f + g) \leq \mathsf{m} + 1, h \leq g + \mathsf{t}\}$$

By multiplying p_1 by monomials in S and n (a well-chosen integer) by monomials in M, a collection of polynomials is obtained whose Newton polygons are included in M and that have (x_0, y_0, z_0) as root modulo n. A lattice is then constructed with the coefficients of all these polynomials and an LLL reduction is performed. By taking the two shortest vectors of the lattice (under some conditions on the bounds, see [9]) two polynomials p_2 and p_3 can be constructed such that they have (x_0, y_0, z_0) as a root over the integers. If the three resulting polynomials p_1, p_2, p_3 are algebraically independent, it leads to the root by resultant computations. Unfortunately, one can not guarantee the independence, which makes this attack be a heuristic one.

5.3   Our Attack

Let us now explain our attack, that is the way we manage to recover the root of $p_1 = f_{MSB1}(x, y, z)$ by using the construction exposed in section 3.3. Starting with two independent polynomials, our construction allows us to construct a third one having the same common root and algebraically independent from the two others. Constructing the first two polynomials from a single one is simply done by using the construction of section 5.2. Indeed, while this construction is heuristic for the third polynomial, it rigorously yields the second one. We denote by p_2 the second polynomial thus found.

   Let us consider the ideal $I = (p_1, p_2)$ which has to be prime for our construction. If this is not the case, some preliminary computations have to be performed in order to replace I by another prime ideal which still has (x_0, y_0, z_0) as a root. If I is primary, it is sufficient to replace it by its radical. If I is not primary, we can compute its primary decomposition $I = I_1 \cap \cdots \cap I_r$ and replace it by the corresponding ideal I_j (or $\sqrt{I_j}$ if necessary). In practice, testing each I_j to find the correct one is very fast since there is a small number of such ideals in this decomposition. Finally, from I_j we construct a lattice L_{I_j} using an auxiliary set $\mathcal{F}$ as in section 3.2. After reducing L'_{I_j}, we obtain a third independent polynomial p_3.

   The polynomial p_2 is derived from p_1 using Coron's and Howgrave-Graham's variations [13,8] instead of the original Coppersmith method. As a consequence, we cannot apply the criterion of section 4 to ensure that the construction of p_3 is always easier than the construction of p_2. Nevertheless, it works extremely well in practice.

5.4   Experiments

Let us take N as a 256-bit modulus for the experiments. The following tables show the results we obtained with some fixed values of the parameters m, t and β for both the attack proposed in [9] (which we refer to as "Method 1") and ours ("Method 2"). One hundred polynomials p_1 are created for each value of δ. The first column gives the number of times the original attack only provides one polynomial, the second column refers to the number of times it provides two polynomials. In this case, the number of (p_1, p_2, p_3) really independent is given in column 3. This number is counted too with our method (column 4). The value of δ in bold corresponds to the best bound obtained in practise in [9].

Table 1. $\mathsf{m} = 1, \mathsf{t} = 1, \beta = 0.35$

	Method 1			Method 2
δ	p_2	(p_2, p_3)	Indep.	OK
0.09	0	100	98	100
0.10	0	100	92	100
0.11	0	100	95	100
0.12	0	100	92	100
0.13	0	100	80	100
0.132	0	100	86	100
0.134	0	100	77	100
0.136	0	100	71	100
0.138	1	99	75	100
0.140	0	100	71	100
0.142	1	99	72	100
0.144	0	100	55	100
0.146	4	95	57	99
0.148	7	89	50	96
0.150	6	91	43	97

Table 2. $\mathsf{m} = 2, \mathsf{t} = 0, \beta = 0.3$

	Method 1			Method 2
δ	p_2	(p_2, p_3)	Indep.	OK (Root Pb.)
0.14	0	100	100	100 (0)
0.15	0	100	97	100 (0)
0.16	0	100	97	100 (0)
0.17	0	100	82	100 (1)
0.18	0	100	60	100 (8)
0.182	0	100	47	100 (13)
0.184	0	100	47	100 (13)
0.186	0	100	33	100 (26)
0.188	0	100	18	100 (36)
0.190	0	100	16	100 (50)
0.192	0	100	6	100 (79)
0.194	7	82	0	89 (89)
0.196	14	49	0	63 (63)
0.198	4	42	0	46 (46)
0.20	4	25	0	29 (29)

The first table really show that there are no more problems due to independence. Thus, our method can be applied beyond that of [9]. In the second table, a different problem seems to appear during the computation. This behavior can be explained quite simply. Indeed, as we noticed before, in this application, we can not predict in advance the restrictions on the bounds in order to obtain p_3 such that $p_3(x_0, y_0, z_0) = 0$. Surprisingly, this is not a problem to recover the root (see the next section).

5.5   Special Cases of Interest

We saw in the previous section that even if there are sometimes root problems, it does not prevent us to recover the root. The reason why is that, in all cases, the Gröbner basis of the ideal $I = (p_1, p_2)$ is so simple that it allows to recover the root, without needing a third polynomial. Let us show some examples where the "root problem" appears for I prime, primary and non-primary. These toys examples use the tiny parameter $N \simeq 2^{50}$ with $\beta = 0.3$, $\mathsf{m} = 2$, $\mathsf{t} = 0$, and $\delta = 0.190$.

I is prime The initial parameters are:

$$p_1 = 9450886190201x + ((z - 155155341747587)y + 72582805940743679)$$
$$(x_0 = 233, y_0 = 482, z_0 = 25517171) \rightarrow (X = 496, Y = 18080, Z = 37368409)$$

After using the polynomial p_2, which is too large to print, coming from the attack of [9], we have the following Gröbner basis:

$$\begin{cases} q_1 = xz - 39521501447/12x + 46079/6z + 6785552382017/12 \\ q_2 = y - 12/197x - 92158/197 \end{cases}$$

In particular, the polynomial q_2 has (x_0, y_0, z_0) as a root. By multiplying all its coefficients by 197 and taking the equation modulo 197, we find $x_0 \equiv 36$ mod 197. By testing then $36, 233$, we find the root.

I is primary The initial parameters are:

$$p_1 = -32390526593433x + ((z - 96130883093383)y - 215591345005890049)$$
$$(x_0 = 87, y_0 = -2272, z_0 = 20056623) \rightarrow (X = 453, Y = 15661, Z = 29413906)$$

The polynomial p_2 is taken to construct $I = (p_1, p_2)$. As this ideal is not prime, it is replaced by its radical, what gives:

$$\begin{cases} r_1 = xz - 128929299037/31x + 206327/31z + 7024533450267/31 \\ r_2 = y + 31/92x + 206327/92 \end{cases}$$

The polynomial r_2 has (x_0, y_0, z_0) as a root. By multiplying it by 92 and taking the equation modulo 92, we obtain that $x_0 \equiv 87$ mod 92. We find the root $x_0 = 87$.

I is non-primary The initial parameters are:

$$p_1 = 1581190442669x + ((z - 3199926510559)y + 7690910313142015)$$
$$(x_0 = 165, y_0 = 2485, z_0 = 4282719) \rightarrow (X = 237, Y = 5642, Z = 5366501)$$

By taking p_2, we consider $I = (p_1, p_2)$ which is not a primary ideal. Its primary decomposition gives the two following prime ideals:

$$\begin{cases} q_1 = xz + 4274183387/42x + 29185/6z - 268309596605/7 \\ q_2 = y - 42/85x - 40859/17 \end{cases}$$

$$\begin{cases} q_1' = xz + 4274183387/42x + 34049/7z + 1590068930929/42 \\ q_2' = y - 42/85x - 204294/85 \end{cases}$$

By checking which of the two previous ideals has (x_0, y_0, z_0) as root, we find that it is the first one. In particular, the polynomial q_2 has the right root. By taking the equation modulo 85, we obtain that $x_0 \equiv 80$ mod 85. This gives the right root $x_0 = 165$.

Some comments. First of all, it seems for the previous examples to work very well because of the size of the parameters, but in fact we have the same behaviors with $N \simeq 2^{256}$. As the previous equations only give the modular value of x_0 and not its integer value, some tests have to be performed to recover the root.

In almost cases, it has to be tested less than five times (we even often recover the root directly). There are nevertheless some cases where the value of x_0 is more difficult to find. Another important point to notice is that in almost cases, the ideals have the shape of those studied previously. It means that we can recover the root with only two polynomials instead of three, except for very rare cases.

6   Possible Generalizations in More Variables

We expose here a method to replace the heuristic that appears in all articles concerning small roots of polynomial equations in three variables by weaker conditions. We show that the method can, in principle, be generalized to more variables. Starting with an irreducible polynomial p_1 having $(x_{0,1}, \ldots, x_{0,n})$ as root, the classical Coppersmith's method provides a second polynomial p_2 that has the same root and that is independent from p_1. For each $j \in \{3, \ldots, n\}$, considering the ideal $I_{j-1} = (p_1, \ldots, p_{j-1})$, the polynomial p_j can be constructed such that $p_j \notin I_{j-1}$. If the ideal I_{j-1} is prime, the polynomials $p_1, \ldots, p_j$ are algebraically independent. As a consequence, using a successive sequence of prime ideals, we can obtain n polynomials algebraically independent and which have the same common root, that leads to $(x_{0,1}, \ldots, x_{0,n})$.

7   Conclusion

The main result of this paper is a new variation of Coppersmith's algorithm on three variables, that uses both lattice reduction and Gröbner bases computations. In general, the success of this method is controlled by the shape of the Gröbner basis of the ideal $I = (p_1, p_2)$ produced by a straight adaption of Coppersmith's algorithm to the trivariate case. This is a first important step toward rigorous applications of Coppersmith's method with more than two variables. We also show how variations on our technique can improve applications of cryptographic interest.

Open problems are to generalize the method to more applications and to determine general criteria yielding rigorous variants of Coppersmith's algorithm with a wide range of applicability.

References

1. M. Bardet. *Etude de sytèmes algébriques surdéterminés. Applications aux codes correcteurs et à la cryptographie.* PhD thesis, University of Paris 6, 2004.
2. J. Blömer and A. May. Low Secret Exponent RSA Revisited. In *CaLC '01: Revised Papers from the International Conference on Cryptography and Lattices*, pages 4–19, London, UK, 2001. Springer-Verlag.
3. J. Blömer and A. May. A Tool Kit for Finding Small Roots of Bivariate Polynomials over the Integers. *Proceedings of Eurocrypt 2005, Lecture Notes in Computer Science*, 3494:251–257, 2005.

4. D. Boneh and G. Durfee. Cryptanalysis of RSA with Private Key Less Than $N^{0.292}$. *IEEE Transactions on Information Theory*, 46:1339–1349, July 2000.
5. D. Coppersmith. Finding a Small Root of a Bivariate Integer Equation; Factoring with high bits known. In *Advances in Cryptology-Eurocrypt '96, Lecture Notes in Computer Science*, volume 1070, pages 178–189. Springer-Verlag, 1996.
6. D. Coppersmith. Finding a Small Root of a Univariate Modular Equation. In *Advances in Cryptology-Eurocrypt '96, Lecture Notes in Computer Science*, volume 1070, pages 155–165. Springer Verlag, 1996.
7. D. Coppersmith. Finding Small Solutions to Small Degree Polynomials. In *Cryptography and Lattice Conference, Lecture Notes in Computer Science*, volume 2146. Springer-Verlag, 2001.
8. J.-S. Coron. Finding Small Roots of Bivariate Integer Polynomial Equations Revisited. In *Advances in Cryptology-Eurocrypt '04, Lecture Notes in Computer Science*, pages 492–505. Springer-Verlag, 2004.
9. M. Ernst, E. Jochemsz, A. May, and B.de Weger. Partial Key Exposure Attacks on RSA up to Full Size Exponents. *In Advances in Cryptology (Eurocrypt 2005), Lecture Notes in Computer Science Volume 3494, pages 371-386, Springer-Verlag*, 2005.
10. J.-C. Faugère. A New Efficient Algorithm for Computing Gröbner Bases (F4). *Journal of Pure and Applied Algebra*, (139):61–88, 1999.
11. M. J. Hinek. New partial key exposure attacks on RSA revisited. Technical report, CACR, Centre for Applied Cryptographic Research, University of Waterloo, 2004.
12. M. J. Hinek. Small Private Exponent Partial Key-Exposure Attacks on Multiprime RSA. Technical report, CACR, Centre for Applied Cryptographic Research, University of Waterloo, 2005.
13. N. Howgrave-Graham. Finding Small Roots of Univariate Modular Equations Revisited. In *Proceedings of the 6th IMA International Conference on Cryptography and Coding*, pages 131–142, London, UK, 1997. Springer-Verlag.
14. N. Howgrave-Graham. Approximate Integer Common Divisor. In *CaLC '01: Lecture Notes in Computer Science*, volume 2146, pages 51–66. Springer-Verlag, 2001.
15. A.K. Lenstra, Jr. H.W. Lenstra, and L. Lovasz. Factoring polynomials with rational coefficients. *Mathematische Annalen*, 261:513–534, 1982.

A   First Iteration Using a Basic Variation of Coppersmith's Method on Three Variables

Let $p_1(x, y, z)$ be an irreducible polynomial of $\mathbb{Z}[x, y, z]$ having (x_0, y_0, z_0) as root over the integers such that $|x_0| < X$, $|y_0| < Y$ and $|z_0| < Z$. Let S and M be sets of monomials over $\mathbb{N}^3$.

Theorem 1. *If S and M are admissible sets for p_1, we can find in polynomial time $p_2(x, y, z)$ which has (x_0, y_0, z_0) as a root over the integers and is algebraically independent from p_1, provided that*

$$X^{s_x} Y^{s_y} Z^{s_z} < W_1^s 2^{-(6+c)s(d_x^2 + d_y^2 + d_z^2)} \tag{3}$$

where we assume that $(m - s)^2 \le cs(d_x^2 + d_y^2 + d_z^2)$ for some constant c.

A.1   Preliminaries

We denote by $\bar{P}_1$ the $(m \times s)$ matrix defined in section 2.4 whose columns refer to the coefficients of $x^i y^j z^k \tilde{p}_1$ for all $(i, j, k) \in S$. The following result holds:

Lemma 1. *There exists a subset $\hat{M} \subset M$ of size s such that if $\hat{P}_1$ is the matrix composed by the rows of $\bar{P}_1$ corresponding to monomials in $\hat{M}$, we have*

$$|\det \hat{P}_1| \geq W_1^s 2^{-6s(d_x^2 + d_y^2 + d_z^2)}$$

We omit this proof because it follows the same idea as in [5], the point is that we work on three variables instead of two.

A.2   Construction of the Lattice $\mathcal{L}_1$

The $(m \times s)$ matrix whose columns represent the polynomials $x^i y^j z^k p_1$ for all $(i, j, k) \in S$ is denoted by P_1. Moreover, we call D_M the $(m \times m)$ diagonal matrix whose entries are $X^{-f} Y^{-g} Z^{-h}$ with $(f, g, h) \in M$. $\mathcal{L}_1$ is the lattice generated by the rows of the following matrix:

$$N_1 = \left(D_{M \setminus \hat{M}} \mid P_1 \right)$$

where $D_{M \setminus \hat{M}}$ is the resulting matrix coming from deletion in D_M of the columns related to monomials in $\hat{M}$. One can observe that the definition of $\mathcal{L}_1$ is different from those of L_1, which has been introduced in section 2.4. In fact, the same explanation holds with this definition, however, in this case, the conditions on the bounds are easier to determine. By multiplying the rows of N_1 related to $(f, g, h) \in M$ by $X^f Y^g Z^h$ and the columns of P_1 related to $(i, j, k) \in S$ by $X^{-i} Y^{-j} Z^{-k}$, a matrix $\bar{N}_1$ is constructed and satisfies:

$$|\det \bar{N}_1| = |\det N_1| X^{s_x} Y^{s_y} Z^{s_z}$$

Making some elementary row operations on $\bar{N}_1$ leads to:

$$\begin{pmatrix} Id & A \\ 0 & \hat{P}_1 \end{pmatrix}$$

whose determinant is equal to $(\det \hat{P}_1)$. Thus, we obtain that

$$|\det N_1| \geq X^{-s_x} Y^{-s_y} Z^{-s_z} W_1^s 2^{-6s(d_x^2 + d_y^2 + d_z^2)}$$

A.3   Using LLL-Reduction to Construct p_2

Let consider the vector $r_0 = (x_0^f y_0^g z_0^h)_M$ and $s_0 = r_0 N_1$. We have

$$s_0 = ((\frac{x_0}{X})^f (\frac{y_0}{Y})^g (\frac{z_0}{Z})^h)_{M \setminus \hat{M}} \mid \underbrace{(0, \ldots, 0)}_{s}$$

This vector satisfies the two following conditions :

- $\|s_0\|_2 \leq \sqrt{m-s}$
- Its s last coordinates are equal to 0.

As the polynomial $p_1(x, y, z)$ is irreducible, some elementary row operations on N_1 leads to the following matrix:

$$N_1' = \left(\begin{array}{c|c} A_1 & Id \\ \hline A_2 & 0 \end{array} \right) \begin{array}{l} \updownarrow s \\ \updownarrow m-s \end{array}$$

If we call $\mathcal{L}_1'$ the lattice generated by the $(m-s)$ last rows of the previous matrix, we obtain $|\det \mathcal{L}_1'| = |\det N_1|$. Moreover, s_0 belongs to $\mathcal{L}_1'$. Let take $r = m - s$, and assume that $(b_1, \ldots, b_r)$ is an LLL-reduced basis of $\mathcal{L}_1'$. We know that $\|b_r^\star\| \geq 2^{-(r-1)/4}|\det \mathcal{L}_1'|^{1/r}$. As s_0 is a vector belonging to the lattice $\mathcal{L}_1'$, when $\|s_0\| < \|b_r^\star\|$, the inner product $< s_0 | b_r^\star >$ is equal to zero, that leads to a polynomial $p_2(x, y, z)$ which has the same root as $p_1(x, y, z)$. The following condition has to be satisfied:

$$\sqrt{m-s} < 2^{-\frac{m-s-1}{4}}|\det \mathcal{L}_1'|^{\frac{1}{m-s}}$$

to allow the construction of p_2. Let see the more restrictive condition:

$$\sqrt{m-s} < 2^{-\frac{m-s-1}{4}}(2^{-6s(d_x^2+d_y^2+d_z^2)}W_1^s X^{-s_x}Y^{-s_y}Z^{-s_z})^{\frac{1}{m-s}}$$

Finally, if $X^{s_x}Y^{s_y}Z^{s_z} < W_1^s 2^{-(6+c)s(d_x^2+d_y^2+d_z^2)}$ is verified (for c a constant such that $(m-s)^2 \leq cs(d_x^2 + d_y^2 + d_z^2)$), the polynomial p_2 can be constructed in polynomial time. With this construction, the monomials of p_2 belong to $M \setminus \hat{M}$. It remains to prove that the polynomial p_2 is independent from p_1. In fact, if this is not the case, the vector related to p_2 is a linear combination of the columns of P_1. Knowing that p_2 is orthogonal to all multiples of p_1, this can not be possible.

B   Algebraic Independence Between p_1, p_2 and p_3

Here is the proof of the result given in section (2.6). If the ideal $I = (p_1, p_2)$ is prime and $p_3 \notin I$, we show that p_1, p_2 and p_3 are algebraically independent. Assume there exists a polynomial P defined over $\mathbb{Q}[x, y, z]$ such that $P(p_1, p_2, p_3) = 0$, our goal is to prove that $P = 0$. In the following, we denote by $\Delta(P)$ the set of all points $(a, b, c) \in \mathbb{N}^3$ such that $x^a y^b z^c$ appears in P with a non-zero coefficient.

Starting with $\sum_{(a,b,c) \in \Delta(P)} \lambda_{(a,b,c)} p_1^a p_2^b p_3^c = 0$, the polynomial $Q(x)$ can then be defined as $\sum_{c=0}^n \mu_c x^c$ with $\mu_c = \sum_{(a,b)|(a,b,c) \in \Delta(P)} \lambda_{(a,b,c)} p_1^a p_2^b$. This polynomial also has p_3 as a root and we can assume that $\mu_0 \neq 0$, otherwise the polynomial $Q(x)$ can be replaced by $Q(x)/x$. As p_1 and p_2 are already algebraically independent, proving that $Q = 0$ implies that $P = 0$.

The first step of the proof is to show that $\mu_0, \ldots, \mu_n \in I$. Indeed, let us take $Q(p_3)$ evaluated in (x_0, y_0, z_0). Knowing that p_3 has (x_0, y_0, z_0) as a root, it implies that $\mu_0(x_0, y_0, z_0) = 0$. As a consequence, its constant coefficient $\lambda_{(0,0,0)}$ is equal to zero and then $\mu_0 \in I$. Using the equation $Q(p_3) = 0$, we obtain $p_3(\mu_1 + \cdots + \mu_n p_3^{n-1}) \in I$. As the ideal I is prime and as $p_3 \notin I$, we have $\mu_1 + \cdots + \mu_n p_3^{n-1} \in I$. Evaluating again this quantity in (x_0, y_0, z_0) leads to $\mu_1(x_0, y_0, z_0) = 0$, that implies $\mu_1 \in I$. We can then go on the proof by doing the same for $\mu_2, \ldots, \mu_n$.

The previous results allow us to rewrite each μ_c as $\mu_c = p_1 F_c(p_1, p_2) + p_2 G_c(p_2)$ with $F_c \in \mathbb{Q}[x, y]$ such that $deg_x(F_c) < deg_x(\mu_c)$ and G_c a polynomial of $\mathbb{Q}[x]$ defined by $G_c(x) = \sum_{i=0}^{g_c} l_{c,i} x^i$. The equation $Q(p_3) = 0$ becomes then:

$$p_1 (F_0(p_1, p_2) + \cdots + F_n(p_1, p_2)p_3^n) = \underbrace{-p_2(G_0(p_2) + \cdots + G_n(p_2)p_3^n)}_{\in (p_1)} \quad (4)$$

As the ideal (p_1) is prime and as p_2 does not belong to (p_1), it implies that $G_0(p_2) + \cdots + G_n(p_2)p_3^n \in (p_1) \subset (p_1, p_2)$. As before, by evaluating this expression in (x_0, y_0, z_0), we can show that $l_{0,0}, \ldots, l_{n,0} = 0$. It implies that the polynomials $G_c(x)$ can be expressed as $G_c(x) = p_2 G_{c,2}(x)$ with $G_{c,2}(x)$ defined as $\sum_{i=0}^{g_c-1} l_{c,i+1} x^i$. As a consequence, the right hand part of the equation (4) which belongs to (p_1), can be rewritten into $-p_2^2(G_{0,2}(p_2) + \cdots + G_{n,2}(p_2)p_3^n)$. By the same explanation, we finally show that $G_c(x) = 0$ for all $c \in \{0, \ldots, n\}$.

Using the previous result, we can then rewrite μ_c as $p_1 F_c(p_1, p_2)$. The equation $Q(p_3) = 0$ becomes:

$$p_1(\sum_{c=0}^{n} F_c(p_1, p_2)p_3^c) = 0 \Rightarrow R(p_3) = \sum_{c=0}^{n} \nu_c p_3^c = 0$$

with $\nu_c = F_c(p_1, p_2)$. The polynomial $R(x)$ satisfies $Q(x) = p_1 R(x)$ and the coefficients ν_c are such that $deg_x(\nu_c) < deg_x(\mu_c)$. We then separate again ν_c as $p_1 H_c(p_1, p_2) + p_2 I_c(p_2)$ and we show that $I_c = 0$ for all $c \in \{0, \ldots, n\}$. By recurrence, we finally obtain that $Q(x) = p_1^k V(x)$ with $V(x)$ a polynomial defined over $\mathbb{Q}[x]$ which has p_3 as a root. It implies that $V = 0$, and thus $P = 0$. This concludes the proof.

An $L(1/3 + \varepsilon)$ Algorithm for the Discrete Logarithm Problem for Low Degree Curves

Andreas Enge[1] and Pierrick Gaudry[2]

[1] INRIA Futurs & Laboratoire d'Informatique (CNRS/UMR 7161)
École polytechnique, 91128 Palaiseau Cedex, France
[2] LORIA (CNRS/UMR 7503), Campus Scientifique, BP 239
54506 Vandœuvre-lès-Nancy Cedex, France

Abstract. The discrete logarithm problem in Jacobians of curves of high genus g over finite fields $\mathbb{F}_q$ is known to be computable with subexponential complexity $L_{q^g}(1/2, O(1))$. We present an algorithm for a family of plane curves whose degrees in X and Y are low with respect to the curve genus, and suitably unbalanced. The finite base fields are arbitrary, but their sizes should not grow too fast compared to the genus. For this family, the group structure can be computed in subexponential time of $L_{q^g}(1/3, O(1))$, and a discrete logarithm computation takes subexponential time of $L_{q^g}(1/3 + \varepsilon, o(1))$ for any positive ε. These runtime bounds rely on heuristics similar to the ones used in the number field sieve or the function field sieve algorithms.

1 Introduction

The discrete logarithm problem in algebraic curves over finite fields has been receiving particular attention since elliptic curves and subsequently Jacobian groups of further algebraic curves have been proposed for discrete logarithm based public key cryptosystems. Although it is now clear that high genus curves are unsuitable for cryptographical use, it remains crucial to study algorithms for solving the discrete logarithm problem in those curves for several reasons. The first reason is that having a better understanding of the situation for high genus curves might lead to algorithmic improvements also in the small genus case. The second reason is that the Weil descent strategy of attacking the discrete logarithm problem in elliptic curves defined over extension fields leads to a discrete logarithm problem in the Jacobian of a high genus curve. Therefore a better algorithm for high genus discrete logarithms becomes naturally a potential threat for some elliptic curves.

It turned out very early that the discrete logarithm problem in high genus hyperelliptic curves (for instance in the sense that the size q of the base field is fixed, while the genus g tends to infinity) can be solved by a subexponential algorithm of complexity $L_{q^g}(1/2, O(1))$. The first such algorithm was proposed in [1]. As other subexponential algorithms, it consists of fixing a factor base of small prime elements (here, prime divisors) and of creating relations that correspond to the zero element modulo an equivalence relation (here, equivalence

M. Naor (Ed.): EUROCRYPT 2007, LNCS 4515, pp. 379–393, 2007.

of divisors modulo principal divisors). After collecting sufficiently many relations and somehow introducing the base of the discrete logarithm and the element whose logarithm is sought, linear algebra yields the desired result. Assuming that smooth elements, that are elements decomposing over the factor base, have the same density as for instance smooth integers or polynomials, such algorithms usually end up with a complexity of $L_{q^g}(1/2, O(1))$.

The algorithm in [1] creates relations by randomly taking low degree functions (that are linear in Y for the curve $Y^2 = f(X)$), whose divisors are relations. Its analysis is only heuristic. The first proven algorithms are given in [15] for the infrastructure of real-quadratic hyperelliptic function fields and in [5] for Jacobians of hyperelliptic curves. Relations are obtained in a process similar to that of [11] by taking random linear combinations of factor base elements, reducing modulo the equivalence relation and checking for smoothness. A rigorous analysis is derived from the lower bound on the density of smooth divisors in [7]. A generic description of a similar algorithm can be found in [6]; it applies to all class groups in which a smoothness result is known. Heuristically, it obtains a running time of $L_{q^g}(1/2, O(1))$ for the discrete logarithm problem in arbitrary high genus curves, the smoothness result needed for a proof of the complexity is however only available for hyperelliptic curves.

A proven algorithm of complexity $L_{q^g}(1/2 + \varepsilon, O(1))$ for very general curves over a fixed field $\mathbb{F}_q$ and with genus g tending to infinity (with the only restriction that the curves contain a rational point and that the cardinality of the Jacobian group is bounded by $q^{g+O(\sqrt{g})}$) is given in [3]. Unlike previous algorithms, it appears to be specific to algebraic curves and relies on a double randomisation, taking random combinations of factor base elements and a random function in a Riemann–Roch space. A relation is obtained whenever the divisor of this function is smooth. A more general algorithm is proposed in [13] that yields a proven $L_{q^g}(1/2, O(1))$ complexity without any restriction on the input curve.

Another line of research on the discrete logarithm problem for algebraic curves, started in [8] and not pursued in this article, consists of fixing g and having q tend to infinity. This leads to algorithms that are exponential, but faster than generic algorithms of square root complexity as soon as $g \geq 3$, see [9, 4].

In the light of algorithms of complexity $L(1/3)$ for the discrete logarithm problem in finite fields as well as for factoring integers, it has been an open problem to determine whether this complexity can be achieved also for algebraic curves. In this article, we present the first probabilistic algorithm of heuristic complexity $L_{q^g}(1/3, O(1))$ to compute the group structure of certain curves whose total degree is relatively small compared to their genus. When introducing the two elements of the Jacobian for which the discrete logarithm problem is to be solved, some sacrifice has to be made; we obtain an algorithm of complexity bounded by $L_{q^g}(1/3 + \varepsilon, o(1))$ for any positive constant ε.

The relation collection phase is the same as in [1] and consists of looking for smooth divisors of functions linear in Y. By applying it to the curves of our special family, one readily obtains a lower degree of the affine part of the intersection divisor than in the general case, from which a complexity of $L_{q^g}(1/3, O(1))$ is derived. For smoothing the two divisors involved in the discrete logarithm problem, a process is employed that is similar to the one used in the number field sieve or in the function field sieve. This is the general *special-Q descent* strategy (also related to the so-called lattice sieving). Each divisor is partially smoothed into prime divisors of degree less than the starting divisor. Then each such prime divisor Q is smoothed again into smaller prime divisors, and we iterate until every divisor is rewritten in terms of elements of the factor base. However, in our case it is necessary to add an arbitrarily small constant ε to the $1/3$ parameter to obtain a proper descent phenomenon; otherwise, the process would get stuck after one step.

Let us mention that subsequently to our algorithm, Diem has presented at the 10th Workshop on Elliptic Curve Cryptography (ECC 2006) an algorithm based on similar ideas, but with a quite different point of view. He manages to obtain a complexity of $L(1/3, O(1))$ for the discrete logarithm phase, for which our algorithm takes $L(1/3 + \varepsilon, o(1))$. We will show how to reach a complexity of $L(1/3, O(1))$ for discrete logarithms in our setting in the long, journal version.

2   Main Idea

Before describing our algorithm with all its technical details on a general class of curves, we sketch in this section the main idea yielding a complexity of $L_{q^g}(1/3, O(1))$ for the relation collection phase for a restricted class of curves. We provide a simplified analysis by hand waving; Section 3 is devoted to a more precise description of the heuristics used and of the smoothness properties needed for the analysis.

Let $\mathbb{F}_q$ be a fixed finite field. We consider a family of C_{ab} curves over $\mathbb{F}_q$, that is, curves of the form

$$\mathcal{C} : Y^n + X^d + f(X, Y)$$

without affine singularities such that $\gcd(n, d) = 1$ and any monomial $X^i Y^j$ occurring in f satisfies $ni + dj < nd$. Such a curve has genus $g = \frac{(n-1)(d-1)}{2}$; we assume that g tends to infinity, and that $n \approx g^{1/3}$ and $d \approx g^{2/3}$ (we use the symbol $\approx$, meaning "about the same size" with no precise definition). The non-singular model of a C_{ab} curve has a unique point at infinity, and it is $\mathbb{F}_q$-rational; so there is a natural bijection between degree zero divisors and affine divisors, and in the following, we shall only be concerned with effective affine divisors. Choose as factor base $\mathcal{F}$ the $L_{q^g}(1/3, O(1))$ prime divisors of smallest degree (that is, the prime divisors up to a degree of $B \approx \log_q L_{q^g}(1/3, O(1))$). To obtain relations, consider functions linear in Y of the form

$$\varphi = a(X) + b(X)Y$$

with $a, b \in \mathbb{F}_q[X]$, $\gcd(a, b) = 1$ and $\deg a, \deg b = \delta \approx g^{1/3}$. Whenever the affine part $\operatorname{div}(\varphi)$ of the divisor of φ is smooth with respect to the factor base, it yields a relation, and we have to estimate the probability of this event.

Let N be the norm of the function field extension $\mathbb{F}_q(\mathcal{C}) = \mathbb{F}_q(X)[Y]/(Y^n + X^d + f(X,Y))$ relative to $\mathbb{F}_q(X)$. The norm of φ is computed as

$$N(\varphi) = N(b)\, N\left(Y + \frac{a}{b}\right)$$
$$= b^n\left(\left(-\frac{a}{b}\right)^n + X^d + f\left(X, -\frac{a}{b}\right)\right)$$
$$= (-a)^n + b^n X^d + f^*(X),$$

where each monomial $X^i Y^j$ occurring in f is transformed into a monomial $X^i(-a)^j b^{n-j}$ in f^*.

Since φ is linear in Y, all prime divisors it contains are totally split over $\mathbb{F}_q(X)$, and φ is B-smooth if and only if its norm is. We have

$$\deg_X N(\varphi) \le \max(n \deg a, n \deg b + d) = n\delta + d \approx g^{2/3}.$$

Heuristically, we assume that the norm behaves like a random polynomial of degree about $g^{2/3}$. Then it is B-smooth with probability $1/L_{q^g}(1/3, O(1))$ (this is the same theorem as the one stating that a random polynomial of degree g is $\log_q L_{q^g}(1/2, O(1))$-smooth with probability $1/L_{q^g}(1/2, O(1))$, cf., for instance, Theorem 2.1 of [2]). Equivalently, we may observe that $\deg(\operatorname{div}(\varphi)) = \deg_X(N(\varphi))$ and assume heuristically that $\operatorname{div}(\varphi)$ behaves like a random effective divisor of the same degree. Then the standard results on arithmetic semigroups (cf. Section 3) yield again that $\operatorname{div}(\varphi)$ is smooth with probability $1/L_{q^g}(1/3, O(1))$.

Thus, the expected time for obtaining $|\mathcal{F}| = L_{q^g}(1/3, O(1))$ relations is $L_{q^g}(1/3, O(1))$, which is also the complexity of the linear algebra step for computing the Smith normal form and thus the group structure of the Jacobian. The complexity of the discrete logarithm problem is not considered here, an analysis for the full algorithm is given in Section 5.

It remains to show that the search space is sufficiently large to yield the required $L_{q^g}(1/3, O(1))$ relations, or otherwise said, that the number of candidates for φ is at least $L_{q^g}(1/3, O(1))$. The number of φ is about

$$q^{2\delta} = q^{2g^{1/3}} = \exp(2 \log q g^{1/3})$$
$$< \exp(2(g^{1/3}(\log q)^{1/3})(\log(g \log q))^{2/3}) = L_{q^g}(1/3, O(1)).$$

The previous inequality in the place of the desired equality shows that a more rigorous analysis requires a more careful handling of the $\log q$ factors; in particular, δ has to be slightly increased. Moreover, the constant exponent in the subexponential function needs to be taken into account. This motivates the following section, in which we examine in more detail the smoothness heuristics and results that are needed for the algorithm.

3   Smoothness

The algorithm presented in this article relies on finding relations as smooth divisors of random polynomial functions of low degree. We suppose that all curves are given by an absolutely irreducible plane affine model

$$\mathcal{C} : F(X, Y)$$

with $F \in \mathbb{F}_q[X, Y]$, where $\mathbb{F}_q$ is the exact constant field of the function field of $\mathcal{C}$. The factor base $\mathcal{F}$ consists essentially of the places of degree bounded by some parameter μ, with a few technical modifications. Precisely, $\mathcal{F}$ is composed of the following places:

- the places corresponding to the resolution of singularities, regardless of their degrees, whose number is bounded by $\frac{(d-1)(d-2)}{2}$ with $d = \deg F$. By including them in $\mathcal{F}$, the algorithm can be described as if the curves were non-singular.
- the infinite places corresponding to non-singularities, regardless of their degrees, whose number is bounded by d by Bézout's theorem. By adding them, it becomes sufficient to only examine the affine part of any divisor.
- places of degree bounded by some parameter μ and of inertia degree 1 with respect to the function field extension $\mathbb{F}_q(X)[Y]/(F)$ over $\mathbb{F}_q(X)$. Otherwise said, places corresponding to prime ideals of the form $(u, Y - v)$ with $u \in \mathbb{F}_q[X]$ irreducible of degree at most μ and $v \in \mathbb{F}_q[X]$ of degree less than $\deg u$; the inertia degree is in fact the degree of the second generator in Y. Due to the way relations are obtained in the algorithm, no places of higher inertia degree may occur.

A divisor is called $\mathcal{F}$-smooth if it can be decomposed over the factor base; thus only its affine part plays a role, and for polynomial functions, this is an effective (i.e. non-negative) divisor. An effective divisor is called μ-smooth if it is composed only of places of degree up to μ. To be able to analyse the smoothness probability, we need the following reasonable assumption.

Heuristic 1. *Let D be the divisor of a uniformly randomly chosen polynomial of the form $b(X)Y - a(X)$ and ν the degree of its affine part. Then the probability of D to be $\mathcal{F}$-smooth is the same as that of a random effective divisor of degree ν to be μ-smooth.*

Heuristic 1 covers the relation collection phase. For computing discrete logarithms, arbitrary non-principal divisors need to be smoothed, and another assumption is needed.

Heuristic 2. *The probability of a uniformly randomly chosen effective divisor of degree ν to be $\mathcal{F}$-smooth is essentially the same as that of being μ-smooth.*

Heuristic 2 claims in fact that places of inertia degree larger than 1 do not play a role for smoothness considerations. In the analogous case of number fields this is justified by the observation that these places have a Dirichlet density of 0, and the situation is completely analogous for function fields: A place of degree μ and inertia degree f dividing μ corresponds to a closed point on $\mathcal{C}$ with X-coordinate in $\mathbb{F}_{q^{\mu/f}}$ and Y-coordinate in $\mathbb{F}_{q^\mu}$, of which there are on the order of $q^{\mu/f}$. Clearly, places with $f \geq 2$ are completely negligible.

The probability of μ-smoothness is ruled by the usual results on smoothness probabilities in arithmetic semigroups such as the integers or polynomials over a finite field, cf. [14].

Unfortunately, most results in the literature assume a fixed semigroup and give asymptotics for μ and ν tending to infinity, whereas we need information that is uniform over an infinite family of curves. Theorem 13 of [13] provides such a result:

Theorem 3 (Heß). *Let* $0 < \varepsilon < 1$, $\gamma = \frac{3}{1-\varepsilon}$ *and* ν, μ *and* $u = \frac{\nu}{\mu}$ *such that* $3 \log_q(14g + 4) \leq \mu \leq \nu^\varepsilon$ *and* $u \geq 2\log(g + 1)$. *Denote by* $\psi(\nu, \mu)$ *the number of* μ-*smooth effective divisors of degree* ν. *Then for* μ *and* ν *sufficiently large (with an explicit bound depending only on* ε, *but not on* q *or* g),

$$\frac{\psi(\nu, \mu)}{q^\nu} \geq e^{-u \log u \left(1 + \frac{\log \log u + \gamma}{\log u}\right)} = e^{-u \log u (1 + o(1))}.$$

Notice that the proof of Theorem 3, similar in spirit to that for hyperelliptic curves in [7], is entirely combinatorial and relies on the fact that there are essentially q^μ/μ places of degree μ. So we expect the result to hold even if one restricts to places of inertia degree 1.

Denote by

$$L(\alpha, c) = L_{q^g}(\alpha, c) = e^{c(g \log q)^\alpha (\log(g \log q))^{1-\alpha}}$$

for $0 \leq \alpha \leq 1$ and $c > 0$ the subexponential function with respect to $g \log q$, and let

$$\mathcal{M} = \mathcal{M}_{q^g} = \log_q(g \log q) = \frac{\log(g \log q)}{\log q}.$$

The parameter $g \log q$ will be the input size for the class of curves we consider; more intrinsically, this is the logarithmic size of the group in which the discrete logarithm problem is defined.

Proposition 4. *Let* $\nu = \lfloor \log_q L(\alpha, c) \rfloor = \lfloor cg^\alpha \mathcal{M}^{1-\alpha} \rfloor$ *and* $\mu = \lceil \log_q L(\beta, d) \rceil = \lceil dg^\beta \mathcal{M}^{1-\beta} \rceil$ *with* $0 < \beta < \alpha \leq 1$ *and* c, $d > 0$. *Assume that there is a constant* $\delta > \frac{1-\alpha}{\alpha - \beta}$ *such that* $g \geq (\log q)^\delta$. *Then for* g *sufficiently large,*

$$\frac{\psi(\nu, \mu)}{q^\nu} \geq L\left(\alpha - \beta, -\frac{c}{d}(\alpha - \beta) + o(1)\right),$$

where $o(1)$ *is a function that is bounded in absolute value by a constant (depending on* α, β, c, d *and* δ) *times* $\frac{\log \log(g \log q)}{\log(g \log q)}$.

Proof. One computes

$$u = \frac{\nu}{\mu} \leq \frac{c}{d}\left(\frac{g \log q}{\log(g \log q)}\right)^{\alpha - \beta}$$

(the inequality being due only to the rounding of ν and μ),

$$\log u = (\alpha - \beta) \log(g \log q)(1 + o(1))$$

and
$$\frac{\log \log u}{\log u} = o(1),$$

with both $o(1)$ terms being of the form stipulated in the proposition. Applying Theorem 3 yields the desired result. Its prerequisites are satisfied since

$$\overline{\lim}_{g \to \infty} \frac{\log \mu}{\log \nu} = \overline{\lim}_{g \to \infty} \frac{\beta \log g - (1 - \beta) \log \log q}{\alpha \log g - (1 - \alpha) \log \log q}$$

$$\leq \overline{\lim}_{g \to \infty} \frac{\beta \log g}{\alpha \log g - \frac{1-\alpha}{\delta} \log g}$$

$$= \frac{\beta}{\alpha - \frac{1-\alpha}{\delta}} =: \varepsilon < 1$$

because of the definition of δ. Notice further that $g \to \infty$ is equivalent to $g \log q \to \infty$, and that also μ and ν tend to infinity when g does. $\square$

The choice of μ shall insure that the factor base size, that is about q^μ, becomes subexponential. But the necessary rounding of μ, which may increase q^μ by a factor of almost q, may result in more than subexponentially many elements in the factor base when q grows too fast compared to g.

Proposition 5. *Let $0 < \beta < 1$ and $\delta > \frac{1-\beta}{\beta}$. If $g \geq (\log q)^\delta$, then $q = L(\beta, o(1))$ for $g \to \infty$. In particular, $\delta > \max\left(\frac{1-\alpha}{\alpha-\beta}, \frac{1-\beta}{\beta}\right)$ in Proposition 4 implies that $q^\mu = L(\beta, d + o(1))$.*

Proof. To verify the first assertion, one computes

$$q = e^{\log q} = e^{(\log q)^{1-\beta}(\log q)^\beta}$$

$$\leq e^{g^{(1-\beta)/\delta}(\log q)^\beta(\log(g \log q))^{1-\beta}}$$

$$= e^{(g \log q)^\beta(\log(g \log q)^{1-\beta})g^{\frac{1-\beta}{\delta}-\beta}},$$

and $g^{\frac{1-\beta}{\delta}-\beta} \to 0$ since $\frac{1-\beta}{\delta} - \beta < 0$. The second assertion is obvious. $\square$

4   Computing the Group Structure

This section is concerned with the relation collection phase of the discrete logarithm algorithm; an immediate application is the computation of the cardinality and the group structure of the Jacobian of the curve. Relation collection is virtually identical to the process described for hyperelliptic curves in [1]; the running time of $L(1/3, O(1))$ is obtained by applying it to a particular class of curves that are of relatively low degree with respect to their genus and for which the degrees in X and Y of a plane model are balanced in a certain way.

We consider absolutely irreducible curves over finite fields $\mathbb{F}_q$ of characteristic p of the form

$$\mathcal{C} : Y^n + F(X, Y)$$

with $F(X, Y) \in \mathbb{F}_q[X]$ of degree d in X and at most $n - 1$ in Y. The function field extension $\mathbb{F}_q(\mathcal{C}) = \mathbb{F}_q(X)[Y]/(Y^n + F(X, Y))$ over $\mathbb{F}_q(X)$ is supposed to be separable (which is for instance the case if $p \nmid n$).

Most importantly, the degrees n and d are related to the genus g by

$$n \leq n_0 g^{1/3} \mathcal{M}^{-1/3} \text{ and } d \leq d_0 g^{2/3} \mathcal{M}^{1/3}$$

where $\mathcal{M} = \frac{\log(g \log q)}{\log q}$ and n_0, d_0 are some positive constants.

For instance, $\mathcal{C}$ may be a C_{ab} curve of degree $n \sim g^{1/3} \mathcal{M}^{-1/3}$ in Y and $d \sim 2g^{2/3} \mathcal{M}^{1/3}$ in X.

For the running time analysis, we will want to apply Propositions 4 and 5 with $\alpha = 2/3$ and $\beta = 1/3$; so we have to assume that the curves belong to a family satisfying $g \geq (\log q)^\delta$ for some $\delta > 2$.

Algorithm 6 (Group structure)

Input: a curve $\mathcal{C}$ as above
Output: $h = |J_{\mathcal{C}}(\mathbb{F}_q)|$ and divisors $D_1, \ldots, D_r$ with their orders $h_1, \ldots, h_r$ s.t. $J_{\mathcal{C}}(\mathbb{F}_q) = \langle D_1 \rangle \times \cdots \times \langle D_r \rangle$

1. Compute an approximation of h within a factor of 2, that is, h_- and h_+ s.t.

$$h_- < h < h_+ \text{ and } h_+ \leq 2h_-.$$

2. Fix a smoothness bound $B = \lceil \log_q L(1/3, \rho) \rceil$ (with a parameter ρ to be determined later) and compute the factor base $\mathcal{F}$ consisting of all affine prime divisors of $\mathcal{C}$ of degree at most B as well as all infinite prime divisors and prime divisors corresponding to singularities regardless of their degrees. Let $t = |\mathcal{F}|$ and $\mathcal{F} = \{P_1, \ldots, P_t\}$.

3. Start with an empty matrix of relations R and repeat the following step until $s \geq 2t$ relations are obtained (in practice, s slightly larger than t should suffice):
   Draw uniformly at random a function

$$\varphi = b(X)Y - a(X) \in \mathbb{F}_q(\mathcal{C})$$

   with $a, b \in \mathbb{F}_q[X]$ of degree at most

$$m = \lfloor \sigma g^{1/3} \mathcal{M}^{2/3} \rfloor$$

   (with a parameter σ to be determined later). If its divisor is $\mathcal{F}$-smooth, that is,

$$\operatorname{div} \varphi = \sum_{i=1}^{t} e_i P_i,$$

   add a column $(e_1, \ldots, e_t)^T$ to the matrix R.

4. Compute the rank of R; if it is less than t, declare failure and stop.

5. *Compute the Smith normal form* $S = \mathrm{diag}(h_r, \ldots, h_1, 1, \ldots, 1)$ *of* R, *where*
   $1 \neq h_1 | h_2 | \cdots | h_r$, *and unimodular transformation matrices* $T \in \mathbb{Z}^{t \times t}$ *and*
   $U \in \mathbb{Z}^{s \times s}$ *s.t.* $TRU = (S|0)$.
   Let $h = h_1 \cdots h_r$. *If* $h \geq h_+$, *declare failure and stop.*
   Otherwise return h, $D_1, \ldots, D_r$ *s.t.*

   $$(D_1, \ldots, D_r, 0, \ldots, 0) = (P_1, \ldots, P_t) T^{-1}$$

   and $h_1, \ldots, h_r$.

That the algorithm is correct follows from standard arguments such as given
in [1, 5, 6]. It remains to prove its failure probability and running time. We also
have to show that there actually are subalgorithms to carry out the different
steps; these are given together with the following running time analysis.

1. An approximation $\tilde{h}$ of h can be obtained by appropriately truncating the
   L-series of the curve as in [13, Section 6]. The necessary counting of the
   number of points on the curve over a small number of extension fields is
   shown in [13] to be polynomial in g and $\log q$ for curves of degree in $O(g)$.
   The bounds on h are then given by $h_- = \tilde{h}/\sqrt{2}$ and $h_+ = \sqrt{2}\tilde{h}$.
2. The affine prime divisors of degree up to B are obtained by enumerating all
   irreducible monic polynomials $f \in \mathbb{F}_q[X]$ of degree up to B and factoring
   $Y^n + F(X, Y)$ over $\mathbb{F}_q[X]/(f)[Y]$. Each factor of degree w yields a prime
   divisor of degree $w \deg f$. Altogether, these factorisations can be carried out
   by $O(q^B)$ repetitions of a randomised algorithm with an expected running
   time that is polynomial in n, B and $\log q$, and thus ultimately in $g \log q$.
   Since polynomial terms are in $L(1/3, o(1))$, they can be neglected, and we
   retain only the term $O(q^B)$ for the remainder of the analysis.

   The number of singular places is bounded by $O((nd)^2) = O(g^2)$ using the
   genus formula for a plane curve. They can be fully described in polynomial
   time, by computing the desingularisation trees of the singular points (see for
   instance [10]).

   The non-singular places at infinity are included in the intersection of the
   projective curve with the line $Z = 0$, which has at most $O(nd) = O(g)$
   elements by Bézout's theorem, and these are also computable in polynomial
   time.

   So this step terminates with a factor base of size

   $$t = O\left(nq^B\right) = L(1/3, \rho + o(1))$$

   that is computed in time $L(1/3, \rho + o(1))$.
3. To estimate the smoothness probability of $\mathrm{div}\, \varphi$ under Heuristic 1, we need
   to compute the degree of its affine part. Denote the affine degree of a divisor
   by $\deg_{\mathrm{aff}}$. Let $\sigma_1, \ldots, \sigma_n$ be the different embeddings of $\mathbb{F}_q(\mathcal{C})$ into its Galois
   closure (that exists because the function field extension is assumed to be
   separable). The σ_i fixing $\mathbb{F}_q(X)$, they send affine to affine and infinite to
   infinite prime divisors. Hence, all the $\deg_{\mathrm{aff}}(\varphi^{\sigma_i})$ are the same and given by

   $$\deg_{\mathrm{aff}} \varphi = \frac{1}{n} \deg_{\mathrm{aff}} N_{\mathbb{F}_q(\mathcal{C})/\mathbb{F}_q(X)}(\varphi) = \deg_X N(\varphi).$$

The norm of φ is computed as $N(\varphi) = \mathrm{Res}_Y(\varphi, Y^n + F(X, Y))$, and its degree in X is bounded from above by

$$\deg_X \varphi \cdot \deg_Y \mathcal{C} + \deg_Y \varphi \cdot \deg_X \mathcal{C} = nm + d.$$

The divisor of φ is B-smooth if and only if its norm is; this test as well as the decomposition of a smooth div φ into prime divisors boils down to a factorisation of the norm in $\mathbb{F}_q[X]$ and takes random polynomial time.

Let $\tau = (n_0\sigma + d_0)/3$. Applying Propositions 4 and 5 under Heuristic 1 with $nm + d \leq 3\tau g^{2/3}\mathcal{M}^{1/3}$ in the place of ν and $B = \lceil \rho g^{1/3}\mathcal{M}^{2/3} \rceil$ in the place of μ shows that a relation is obtained on average in time $L\left(1/3, \frac{\tau}{\rho} + o(1)\right)$, so that this step takes overall

$$L\left(1/3, \frac{\tau}{\rho} + \rho + o(1)\right).$$

4. and 5. Since all entries of the matrix are of bit size polynomial in $g \log q$, its rank and Smith normal form can be computed in quartic time according to [16, Proposition 8.10], that is in

$$L(1/3, 4\rho + o(1)).$$

The total running time of the algorithm thus becomes

$$L\left(1/3, \max\left(\frac{\tau}{\rho} + \rho, 4\rho\right) + o(1)\right)$$

with $\tau = (n_0\sigma + d_0)/3$.

For any fixed σ (and thus τ), the value of ρ that minimises the running time is $\rho = \sqrt{\tau/3}$ and we get a complexity of $L\left(1/3, \frac{4\sqrt{\tau}}{\sqrt{3}} + o(1)\right)$.

Now τ is not a completely free parameter; it is connected to the success probability of the algorithm. It is in fact not clear whether the algorithm has a non-zero success probability at all; as in [1], it is already unknown whether the principal divisors of the special form considered in Step 3. generate the full relation lattice. The analysis of the proven subexponential algorithm in [5], for instance, exploits the fact that the created relations are essentially uniformly distributed among all possible relations in a hypercube of side length about $|J_\mathcal{C}(\mathbb{F}_q)|$. Since all our relations are sparse, this line of argumentation definitely cannot be applied; as in [1], the non-negligible success probability of the algorithm can only be conjectured (and notice also that it does not follow from a smoothness assumption such as Heuristic 1).

A necessary condition for the success of the algorithm is nonetheless that the number of potential functions φ tested for smoothness in Step 3. must be at least as large as the number of tests, since otherwise the matrix is filled with redundant multiple relations. Thus we need $q^{2m} \geq L\left(1/3, \frac{4\sqrt{\tau}}{\sqrt{3}}\right)$ or, taking logarithms,

$$2\sigma \geq \frac{4}{\sqrt{3}}\sqrt{\tau} = \frac{4}{3}\sqrt{n_0\sigma + d_0},$$

which holds asymptotically for $\sigma \to \infty$. Precisely, the optimal value of σ is the positive solution of the quadratic equation $\sigma^2 - \frac{4}{9}n_0\sigma - \frac{4}{9}d_0 = 0$.

5   Computing Discrete Logarithms

In order to smooth the basis of the discrete logarithm and the element whose logarithm is sought, we are going to perform a special-Q descent with a slightly larger subexponentiality parameter $1/3 + \varepsilon$. Let us first describe an algorithm that does one step of the special-Q descent and that will be used as a building block by the final algorithm.

Heuristic Result 7. *Let Q be an affine prime divisor of the curve $\mathcal{C}$ of the form $\mathrm{div}(u(X), Y - v(X))$, with $\deg u(X) \leq \log_q L(1/3 + t, c)$ for some constants $c > 0$ and $\varepsilon < t \leq 1/3 - \varepsilon$. There is an algorithm that finds a divisor R equivalent to Q such that all prime divisors of R are either in $\mathcal{F}$ or have a degree bounded by $\log_q L(1/3 + t - \varepsilon, c')$, and such that all these prime divisors are of the form $\mathrm{div}(u_i(X), Y - v_i(X))$. The heuristic expected running time is bounded by $L(1/3 + \varepsilon, \frac{cn_0}{c'}(1/3 + \varepsilon + o(1)))$.*

Justification. Let us consider the set $\mathcal{L}_Q$ of functions of the form $a(X) + b(X)Y$ whose divisors contain Q in their support. In other words, this is the $\mathbb{F}_q[X]$-lattice

$$\mathcal{L}_Q = \{a(X) + b(X)Y \; : \; u(X)|a(X) + v(X)b(X)\}.$$

A basis of this lattice is given by the two vectors $b_1 = u(X)$ and $b_2 = -v(X) + Y$. Hence,

$$\mathcal{L}_Q = \{\lambda(X)b_1 + \mu(X)b_2 \; : \; \lambda, \mu \in \mathbb{F}_q[X]\}.$$

When λ and μ are taken of degree at most $\delta = \log_q L(1/3 + t, c)$, the function φ corresponding to $\lambda(X)b_1 + \mu(X)b_2$ has the form $a(X) + b(X)Y$ with a and b of degree $\Delta \leq 2 \log_q L(1/3 + t, c)$. The degree of the norm of φ is then $\Delta n + d$, which is dominated by $\log_q L(2/3 + t, cn_0)$.

We rely now on Heuristic 1 that says that the zero divisor of the function has the same smoothness properties as a random effective divisor of the same degree, and apply Proposition 4. Therefore the expected number of functions one has to try before having found one whose divisor is $\log_q L(1/3 + t - \varepsilon, c')$-smooth is

$$L\left(1/3 + \varepsilon, \frac{cn_0}{c'}(1/3 + \varepsilon + o(1))\right).$$

The fact that the prime divisors that we obtain are of the same form as Q comes from the shape of the function we have chosen.

It remains to check that the number of functions we can test in the lattice is large enough compared to this expected number of tests. With our choice of δ, the size of the sieving space is $L(1/3 + t, 2c)$, which is larger than any $L(1/3 + \varepsilon)$ since t is greater than ε. □

This result suffices to carry out a full descent if one can initialise the process and finish it once smoothness is reached up to a $t < \varepsilon$. The next two heuristic results explain these steps.

Heuristic Result 8. *Assume that $\rho > (\frac{1}{3} + \varepsilon)\frac{n_0}{2}$. Let Q be an affine prime divisor of $\mathcal{C}$ of the form $\mathrm{div}(u(X), Y - v(X))$, with $\deg u(X) \leq \log_q L(1/3 + t, c)$, for some constants $c > 0$ and $0 < t \leq \varepsilon$. There is an algorithm that finds a divisor R equivalent to Q such that all prime divisors of R are in $\mathcal{F}$ (defined with this value of ρ), and such that all these prime divisors are of the form $\mathrm{div}(u_i(X), Y - v_i(X))$. The heuristic expected running time is bounded by $L\left(1/3 + t, (1/3 + t)\frac{cn_0}{\rho} + o(1)\right).$*

Justification. Let us consider the same lattice $\mathcal{L}_Q$ as in the proof of Proposition 7. Assume that λ and μ are taken of degree at most $\delta = \log_q L(1/3 + t, c)$, then, as before, the norm of the corresponding functions are of degree bounded by $\log_q L(2/3 + t, cn_0)$. Using again Heuristic 1, one gets by Proposition 4 that a $\log_q L(1/3, \rho)$-smooth divisor can be obtained in heuristic expected time

$$L\left(1/3 + t, (1/3 + t)\frac{cn_0}{\rho} + o(1)\right).$$

One has to check that we have enough possibilities for λ and μ to cover this search. The sieving space is $q^{2\delta} = L(1/3 + t, 2c)$. Therefore it is large enough if $2c > (1/3 + t)\frac{cn_0}{\rho}$, that is if $\rho > (1/3 + t)\frac{n_0}{2}$. Since $\varepsilon > t$, this is guaranteed by our hypothesis on ρ. $\square$

Heuristic Result 9. *Let D be a degree 0 divisor and $\sum_P e_P P$ its decomposition into prime divisors such that $\sum_P |m_P| \in O(g)$. Then there is an algorithm that finds a divisor R equivalent to D such that all prime divisors of R are of the form $\mathrm{div}(u_i(X), Y - v_i(X))$ with $\deg u_i(X) \leq \log_q L(2/3 - \varepsilon, c)$. The heuristic expected running time is bounded by $L(1/3 + \varepsilon, (1/3 + \varepsilon)\frac{1}{c} + o(1))$.*

Justification. In order to smooth D, we apply the classical Hafner-McCurley strategy: a random linear combination of elements of the factor base is added to D, and the obtained divisor is tested for smoothness. Each test takes polynomial time since the effective group law in the Jacobian reduces to computing Riemann-Roch spaces as in [12].

Following Heuristic 2, the additional restriction on the form of the prime divisors has no influence on the running time, and the desired result follows from Proposition 4. $\square$

Armed with these heuristic partial smoothing results, we can now derive a full special-Q descent algorithm. Let us fix a constant $\varepsilon > 0$, a parameter of the algorithm. This ε is to be thought of as small (and of course $\varepsilon < 1/6$). The algorithm assumes that Algorithm 6 has been run as a precomputation, with a value of ρ that is larger than a bound given below. Similarly, the constants c_0 and c_K are made explicit below.

Algorithm 10 (Discrete logarithm)

1. *Use Heuristic Result 9 to build a list L of prime divisors of degree at most $\log_q L(2/3-\varepsilon, c_0)$, such that if we know their discrete logarithms, the discrete logarithm of D is implied.*
2. *While there is a Q in L of degree more than $\log_q L(1/3+\varepsilon, c_K)$, use Heuristic Result 7 to replace Q in L by a list of prime divisors of degree bounded by a subexponential function with parameter reduced by ε.*
3. *For each Q in L that is not in $\mathcal{F}$, use Heuristic Result 8 to decompose Q in $\mathcal{F}$.*

In order to analyse the algorithm, let us model it by a tree: the root is the divisor D, its sons are the prime divisors coming from its decomposition using Heuristic Result 9, then each internal node corresponds to a prime divisor and its sons are the prime divisors obtained using Heuristic Result 7 or Heuristic Result 8. The depth of the tree is bounded by $1/(3\varepsilon)$ since at each intermediate step the subexponential parameter is reduced by at least ε and one has to cover a range of $1/3$. The number of sons of each node is bounded by g. Hence the total number of nodes is bounded by $g^{1/(3\varepsilon)}$. Since ε is a fixed constant, this is a polynomial in $g \log q$ and therefore contributes only for a $o(1)$ in the subexponential complexity.

Let us allow a computation time of $L(1/3 + \varepsilon, \nu + o(1))$, for fixed positive constants ε and ν. Then the first step that uses Heuristic Result 9 can decompose D in prime divisors of degree at most $\log_q L(2/3-\varepsilon, c_0)$ in time $L(1/3+\varepsilon, \nu+o(1))$ for $c_0 = (1/3 + \varepsilon)/\nu$. Going one step down the tree, one can decompose these primes using Heuristic Result 7 in primes of degrees at most $\log_q L(2/3 - 2\varepsilon, c_1)$ in the same time, for $c_1 = c_0 n_0 (1/3 + \varepsilon)/\nu$. Going from level k to level $k + 1$ in the tree will decompose in primes of degree at most $\log_q L(2/3 - (k + 2)\varepsilon, c_{k+1})$ in the same time, for $c_{k+1} = c_k n_0 (1/3 + \varepsilon)/\nu$. Finally, each last step will be feasible in the same running time if $\rho > c_K n_0 (1/3 + \varepsilon)/\nu$, where K is the depth of the tree.

This value of ρ is feasible and does not affect the overall complexity. It only changes the exponent in the $L(1/3)$ runtime of the group structure algorithm, whose complexity remains negligible compared to the $L(1/3 + \varepsilon)$ of the present algorithm. Therefore, a suitable choice of ρ, c_0 and c_K in Algorithm 10 results in a running time of $L(1/3 + \varepsilon, \nu + o(1))$ for any given ε and ν.

Choosing $\varepsilon/2$ in the place of ε (and an arbitrary ν) shows that even a complexity of $L(1/3 + \varepsilon, o(1))$ is achievable.

Remark. In the analysis, we have remained silent about the exact nature of the $o(1)$ terms. As long as a fixed number of them is involved, this does not pose any problem. But at first sight, since Heuristic Result 7 is used a non-constant number of times, one apparently needs to make the $o(1)$ terms explicit to check that they do not sum up to something that is not tending to zero. However, although the number of nodes in the tree of Algorithm 10 is in $g^{1/(3\varepsilon)}$, the $o(1)$ term is the same for any given level in the tree, so that actually only the depth of the tree is important for these $o(1)$-terms considerations. The depth of the tree is in $1/(3\varepsilon)$, which is a constant, so that we actually consider a constant number of $o(1)$ terms and need not make them explicit.

6   Extensions to Wider Families of Curves

6.1   Highly Singular Curves

Consider the case where the curve has an equation of the appropriate form, but with a genus that is much smaller than nd. Then letting $g' = nd$, one may apply the exact same algorithms yielding an $L(1/3 + \varepsilon)$ complexity. However, the subexponential function is now taken with respect to $q^{g'}$. This may still result in a subexponential complexity in q^g, depending on the relation between q, g and g'.

6.2   Different Balancing Between n and d

Here we consider the case where $n \approx g^\alpha$ and $d \approx g^{1-\alpha}$ for $\alpha \in \left[\frac{1}{3}, \frac{1}{2}\right]$. We shall just give an informal description of an algorithm that yields an $L(1/3)$ complexity for the group structure. Note that to obtain the claimed complexity without ε, the bounds on n and d should resemble the ones we have in Section 4. For instance, bounds of the form $n \leq n_0 g^\alpha \mathcal{M}^{-\alpha}$ and $d \leq d_0 g^{1-\alpha} \mathcal{M}^\alpha$ would suffice. For the sake of better readability, we content ourselves with approximate bounds.

Let us restrict to C_{ab} curves for simplicity, and let us call P_∞ the unique place at infinity. We proceed as in Algorithm 6, but the functions we consider are of the more general form:

$$\varphi = a_0(X) + a_1(X)Y + \cdots + a_k(X)Y^k,$$

where the $a_i(X)$ have a degree bounded by g^β and k is taken of the form g^γ, for some β and γ to be determined. Then the divisor of φ is of the form $E - (\deg E)P_\infty$, with E effective of degree bounded by $g^{\gamma+1-\alpha} + g^{\beta+\alpha}$.

Fix a smoothness bound of $g^{\beta+\gamma}$; with the usual heuristic, one can find E that is smooth in time about $g^{\max(\alpha-\gamma,(1-\alpha)-\beta)}$. The consistency check that the sieving space must be larger than the factor base yields the condition

$$\beta + \gamma \geq \max(\alpha - \gamma, (1-\alpha) - \beta),$$

which gives $\beta + 2\gamma \geq \alpha$ and $\gamma + 2\beta \geq 1 - \alpha$. This in turn imposes that $\beta + \gamma \geq 1/3$. Therefore, in this setting we can not hope to get something better than an $L(1/3)$ complexity. We now show that this complexity is achievable: taking $\beta = 2/3 - \alpha$ and $\gamma = \alpha - 1/3$, all the conditions are verified, and the complexity is as announced.

In the particular case of $\alpha = 1/3$, we recover $\beta = 1/3$ and $\gamma = 0$, which corresponds to Algorithm 6. In the other extremal case $\alpha = 1/2$, we get $\beta = \gamma = 1/6$.

If α gets smaller than $1/3$, then the $L(1/3)$ complexity is not achievable with this algorithm. In fact, for each value of $\alpha \in [0, 1/3]$, there is an $L(x)$ complexity with $x \in [1/3, 1/2]$, and finally, for hyperelliptic curves one essentially recovers Adleman-Demarrais-Huang's $L(1/2)$ algorithm.

All of this concerns only the group structure. For the special-Q descent however, things get more complicated and the $L(1/3 + \varepsilon)$ complexity is lost when α is bigger than $1/3$. More precisely, the same kind of computations as above yields a complexity of $L(\alpha + \varepsilon)$ for $\alpha \in [1/3, 1/2]$.

Acknowledgement. We thank Claus Diem for his careful reading of our article and many useful remarks.

References

[1] L. M. Adleman, J. DeMarrais, and M.-D. Huang. A subexponential algorithm for discrete logarithms over the rational subgroup of the jacobians of large genus hyperelliptic curves over finite fields. In L. Adleman and M.-D. Huang, editors, *ANTS-I*, volume 877 of *Lecture Notes in Comput. Sci.*, pages 28–40. Springer–Verlag, 1994.

[2] R. L. Bender and C. Pomerance. Rigorous discrete logarithm computations in finite fields via smooth polynomials. In D. A. Buell and J. T. Teitelbaum, editors, *Computational Perspectives on Number Theory: Proceedings of a Conference in Honor of A.O.L. Atkin*, volume 7 of *Studies in Advanced Mathematics*, pages 221–232. American Mathematical Society, 1998.

[3] J.-M. Couveignes. Algebraic groups and discrete logarithm. In *Public-key cryptography and computational number theory*, pages 17–27. de Gruyter, 2001.

[4] C. Diem. An index calculus algorithm for plane curves of small degree. In F. Heß, S. Pauli, and M. Pohst, editors, *ANTS-VII*, volume 4076 of *Lecture Notes in Comput. Sci.*, pages 543–557. Springer–Verlag, 2006.

[5] A. Enge. Computing discrete logarithms in high-genus hyperelliptic Jacobians in provably subexponential time. *Math. Comp.*, 71:729–742, 2002.

[6] A. Enge and P. Gaudry. A general framework for subexponential discrete logarithm algorithms. *Acta Arith.*, 102:83–103, 2002.

[7] A. Enge and A. Stein. Smooth ideals in hyperelliptic function fields. *Math. Comp.*, 71:1219–1230, 2002.

[8] P. Gaudry. An algorithm for solving the discrete log problem on hyperelliptic curves. In B. Preneel, editor, *Advances in Cryptology – EUROCRYPT 2000*, volume 1807 of *Lecture Notes in Comput. Sci.*, pages 19–34. Springer–Verlag, 2000.

[9] P. Gaudry, E. Thomé, N. Thériault, and C. Diem. A double large prime variation for small genus hyperelliptic index calculus. *Math. Comp.*, 76:475–492, 2007.

[10] G. Haché. *Construction effective de codes géométriques.* PhD thesis, Université de Paris VI, 1996.

[11] J. L. Haffner and K. S. McCurley. A rigorous subexponential algorithm for computation of class groups. *J. Amer. Math. Soc.*, 2(4):837–850, 1989.

[12] F. Heß. Computing Riemann-Roch spaces in algebraic function fields and related topics. *J. Symbolic Comput.*, 33:425–445, 2002.

[13] F. Heß. Computing relations in divisor class groups of algebraic curves over finite fields. Preprint, 2004.

[14] E. Manstavičius. Semigroup elements free of large prime factors. In F. Schweiger and E. Manstavičius, editors, *New Trends in Probability and Statistic*, pages 135–153, 1992.

[15] V. Müller, A. Stein, and C. Thiel. Computing discrete logarithms in real quadratic congruence function fields of large genus. *Math. Comp.*, 68(226):807–822, 1999.

[16] A. Storjohann. *Algorithms for Matrix Canonical Forms.* PhD thesis, Eidgenössische Technische Hochschule Zürich, 2000.

General *Ad Hoc* Encryption
from Exponent Inversion IBE

Xavier Boyen

Voltage Inc.
Palo Alto
xb@boyen.org

Abstract. Among the three broad classes of Identity-Based Encryption schemes built from pairings, the exponent inversion paradigm tends to be the most efficient, but also the least extensible: currently there are no hierarchical or other known extension of IBE based on those schemes. In this work, we show that such extensions can be realized from IBE systems that conform to a certain abstraction of the exponent inversion paradigm. Our method requires no random oracles, and is simple and efficient.

1 Introduction

Since the first practical constructions of the identity-based encryption (IBE) primitive appeared a few years ago [18,9,4], a large body of work has been devoted to creating better realizations of the basic primitive, and to extending it in many interesting ways. With the notable exception of Cocks' basic IBE scheme [9], virtually all IBE-like constructions known to date make more or less extensive use of bilinear pairings on elliptic curves.

The many extensions that have been proposed in the recent years have the common goal to extend the notion of identity from its original atomic meaning, to complex constructs of identity components on which certain operations can be performed. In particular, we mention hierarchical identities [13], fuzzy identities [16], and identities as attributes [12] among the most significant of these extensions. Fortunately, and unlike the original idea of IBE [19] which remained without construction for many years, most of the IBE extensions that have been suggested also have a known construction. However, to temper this optimism, we should note that for many of these extensions, the only realizations we know of all derive from the same basic IBE paradigm, despite the availability of alternatives. In particular, an entire family of very efficient IBE constructions does not seem to support any of the extensions afforded by other families.

Our current knowledge of pairing-based IBE schemes can be partitioned in three broad families: (1) full-domain hash, (2) exponent inversion, and (3) commutative blinding—with little doubt that others will be invented in the future. The connotations behind this taxonomy shall be explicited later on. Each of these

M. Naor (Ed.): EUROCRYPT 2007, LNCS 4515, pp. 394–411, 2007.

categories defines a general construction template, by which encryption and key derivation are matched in an identity-based manner using a bilinear pairing. The one thing that these families have in common is their use of a pairing—but not how they use it. Indeed, the shape of the template greatly affects how the schemes can be extended, and their security proved.

Among the three families, the commutative blinding method originated with BB_1-IBE [2] has distinguished itself as the most fertile ground for generalizing IBE, based on the number of extensions that it currently supports, such as forward secure hierarchies [3], partial-match or fuzzy identities [16], and complex attribute-based policies [12]. It is followed rather distantly by the full-domain hash family, defined by BF-IBE [4], which contains fewer but nevertheless interesting extensions, including hierarchies [11] also with forward security [21]. In stark contrast, based on our current state of knowledge, the exponent inversion family does not seem to have any useful extension, despite the fact that the basic IBE functionality performs more efficiently in this family, based on BB_2-IBE [2] and SK-IBE [7,17], than in the other two. This situation strikes us as odd, as there is no obvious reason why the exponent inversion family should be less accommodating than the other two.

The aim of this paper is to show that the exponent inversion paradigm is more flexible than has been previously recognized. To this end, we first give an abstraction of exponent inversion schemes such as BB_2-IBE and SK-IBE, that captures functional properties such as linearity in the exponent, and which we call Linear IBE. We also define certain security properties that such schemes should satisfy depending on the final goal of the construction; these properties have to do with simultaneous or parallel instances of the IBE running at once, which is a general technique we use in all our constructions. We then apply the method to transform any black-box Linear IBE with suitable security properties into a hierarchical, fuzzy, attribute-based, or distributed system, under generic security reductions to the underlying base IBE abstraction.

The transformations are syntactically black-box, but their security requires the parallel simulation of several base instances, hence our requirement that the underlying scheme be secure in such conditions. In general, the transformations preserve the gist of the security properties of the underlying scheme, *e.g.*, in the random oracle or standard model, and under selective or adaptive security, but keeping in mind that it requires (and consumes) the supplemental notion of parallel IBE security already mentioned. The method is quite simple and preserves the efficiency of the underlying scheme, with a multiplier that depends on the particulars of what the transformation seeks to achieve. In practice, this new approach seems appealing, as it allows the very efficient but bare-bones SK and BB_2 schemes to become more flexible and thus we hope more useful.

We call *ad hoc cryptosystem* any such public-key system that supports private sub-keys with designated restricted capabilities. This includes IBE and its extensions.

2   A Classification of IBE Schemes

The following is a rough classification of the known identity-based encryption schemes. All of them support at least a basic security reduction to a well-formulated complexity assumption, either in the standard model or in the random oracle model.

"Quadratic Residuosity" IBE (without pairings). We mention Cocks' [9] scheme as the only known example of IBE based on quadratic residuosity in RSA groups; it is inefficient in terms of bandwidth and has no known extension.

"Full Domain Hash" IBE. This is the class of the Boneh-Franklin identity-based encryption [4], and to which the earlier Sakai-Ohgishi-Kasahara identity-based key exchange [18] also belongs.

In BF encryption and the constructions that are based on it, such as [11,21], the session keys are of the form $\mathbf{e}(H(\mathsf{Id}), \hat{g}^\alpha)^s$ where Id is the recipient identity, α is the master secret, and H is a full-domain hash function into the bilinear group, viewed as a random oracle. In SOK key exchange, the session key $\mathbf{e}(H(\mathsf{Id}_A), H(\mathsf{Id}_B)^\alpha)^s$ is computed interactively from the identities of both parties, but also involves the master key α and a random oracle as in BF encryption.

"Exponent Inversion" IBE. This approach to IBE can be traced to an idea of Mitsunary, Sakai, and Kasahara in the context of traitor tracing [14]. For IBE, the principle is to obtain a session key of the form $\mathbf{e}(g, \hat{g})^s$ based on a ciphertext $(g^{f(\mathsf{Id})})^s$ and a private key $g^{1/f(\mathsf{Id})}$, where $f(\mathsf{Id})$ is a secret function of the recipient identity but $g^{f(\mathsf{Id})}$ is computable publicly. A benefit of this type of construction is that there is no need to hash directly on the curve. Notice also that the master key cancels out completely from the session key.

This category includes the Sakai-Kasahara scheme originally described in [17] and later proven secure in [7] in the random oracle model. The category also includes the second of two IBE schemes proposed by Boneh and Boyen [2], which has a selective-identity proof of security in the standard model. All these schemes rely on the fairly strong BDHI complexity assumption [2], which was first used in another context by Mitsunary, Sakai, and Kasahara [14]. This assumption, called Bilnear Diffie-Hellman Inversion (BDHI), has been further analyzed in [8].

Recently, Gentry [10] proposed another construction that has superficial similarities to the others in this category, but with a proof of security in the adaptive-identity model (based on an even stronger assumption). Gentry's IBE scheme appears to belong in the exponent inversion category, although the case is not clear-cut because the session key is not of the form $\mathbf{e}(g, \hat{g})^s$, but of the form $\mathbf{e}(g, \hat{h})^s$, where $\hat{h}$ is created by the initial setup procedure. Although $\hat{h}$ remains statistically independent of the secret key, it is not intended to be constant from one instance of the system to the next, and Gentry's security proof no longer applies if $\hat{h}$ and thus $\mathbf{e}(g, \hat{h})$ is fixed.

"Commutative Blinding" IBE. The last category of IBE systems descends from BB_1, the first scheme given in the Boneh-Boyen paper [2]. These systems are

based on the same BDH assumption as the Boneh-Franklin scheme [4], but use a mechanism that avoids random oracles. Very roughly, the general principle is to create blinding factors from two secret coefficients in a way that makes them "commute" (*i.e.*, not depend on the application order), thanks to the pairing.

The algebraic versatility exhibited by the BB$_1$ approach has given rise to a fair number of extensions to the original scheme; see for example [3,16,20,15,1]. Virtually all constructions in the commutative blinding paradigm have session keys of the form $\mathbf{e}(g, \hat{g}^{\alpha})^s$, where α is part of the master key, and s is chosen by the sender.

It is likely that the coming years will see the emergence of additional families of schemes. In this paper, we are concerned with the Exponent Inversion family, which tends to be the most computationally efficient and arguably requires the least bandwidth, but currently lacks the flexibility of the other pairing-based families (such as Commutative Blinding especially).

3   Exponent Inversion Abstractions

We now describe an abstraction of IBE that captures the properties of the exponent inversion paradigm that we need. Our abstraction is sufficiently powerful to support a wide variety of generic constructions, and sufficiently general to encompass all IBE schemes known to date that do not "obviously" fall outside of the exponent inversion paradigm.

3.1   Linear IBE Schemes

Based on the properties that our semi-generic construction will require, we define the following abstraction of IBE schemes that use the exponent inversion principle. Two basic schemes mentioned earlier (BB$_2$ and SK) fit particularly nicely within this abstraction.

Intuitively, we exploit two facets of the "linearity" exhibited by exponent inversion IBE. All such schemes construct their identity-based trapdoor from a secret polynomial $\theta(\mathsf{Id})$, and publish enough information to allow anyone to compute $g^{\theta(\mathsf{Id})}$ but not $\hat{g}^{1/\theta(\mathsf{Id})}$. The latter can serve as private key for Id, and the trapdoor arises from the cancellation of the exponents on both sides of the pairing: $\mathbf{e}(g^{\theta(\mathsf{Id})}, \hat{g}^{1/\theta(\mathsf{Id})}) = \mathbf{e}(g, \hat{g})$. To get an IBE scheme, the encryptor needs to pick a randomization exponent s; the ciphertext becomes $g^{\theta(\mathsf{Id})\,s}$ and the session key $\mathbf{e}(g, \hat{g})^s$. Because session keys constructed this way are linear in both the private key and the ciphertext, it will be easy to construct secret sharing schemes in the exponent either in the ciphertext or on the private key side. This is the first property we need (we shall precise and generalize it momentarily).

Our second property is the independence of session keys with respect to the master secret. As in any IBE scheme, the master secret is needed to construct the private keys, but here it need not affect the choice of session keys. Indeed, if the generators g and $\hat{g}$ are imposed externally, the only degree of freedom in the session key $\mathbf{e}(g, \hat{g})^s$ is the exponent s chosen by the encryptor. (This is very much unlike full-domain hash and commutative blinding IBE schemes, in which session keys are respectively of the form $\mathbf{e}(H(\mathsf{Id}), \hat{g}^\alpha)^s$ and $\mathbf{e}(g, \hat{g}^\alpha)^s$ and necessarily involve the master key α.)

As already mentioned, Gentry's IBE scheme uses session keys of the form $\mathbf{e}(g, \hat{h})^s$ rather than $\mathbf{e}(g, \hat{g})^s$, where $\hat{h}$ is created at random by the initial setup procedure. Although our template requires $\hat{h}$ to be fixed, the current proof of Gentry's IBE does not tolerate it, and so we provisionally include Gentry-IBE as a "syntactic" Linear IBE scheme until the question can be settled.

A Template for Exponent Inversion IBE. Toward formalizing the requirements above, we first define the particular template that candidate IBE schemes must obey.

Setup$(\mathbf{e}, g, \hat{g}, v, \omega)$ on input a pairing $\mathbf{e} : \mathbb{G} \times \hat{\mathbb{G}} \to \mathbb{G}_t$, generators $g \in \mathbb{G}$, $\hat{g} \in \hat{\mathbb{G}}$, $v \in \mathbb{G}_t$, and a random seed ω, outputs a master key pair (Msk, Pub) where Pub $= (\mathbf{e}, g, \hat{g}, v, ...)$.

We require key pairs generated from independent random seeds $\omega_1, \omega_2, ...$ to be mutually independent. We allow key pairs generated from the same inputs $\mathbf{e}, g, \hat{g}, v, \omega$ to be mutually independent, as the setup algorithm is permitted to use its own source of randomness.

Extract(Msk, Id) on input Msk and an identity Id, outputs a private key $\mathsf{Pvk}_{\mathsf{Id}} = (\mathsf{Id}, R, \boldsymbol{d})$, which can be deterministic or randomized.

Here, $\mathsf{Id} \in \mathcal{I}d$, the domain of identities; $R \in \mathcal{R}d$, some non-empty auxiliary domain; and $\boldsymbol{d} = (d_1, ..., d_n) \in \mathcal{D}$, a vector space of n coordinates, each a copy of one of $\mathbb{F}_p$, $\mathbb{G}$, $\hat{\mathbb{G}}$, $\mathbb{G}_t$.

Encrypt(Pub, Id, Msg, s) on input Pub, a recipient Id, a plaintext Msg, and a randomization exponent $s \in \mathbb{F}_p$, outputs a ciphertext Ctx $= (\mathsf{Id}, S, c_0, \boldsymbol{c})$.

Here we require that $\mathsf{Msg} \in \mathbb{G}_t$, that $c_0 = \mathsf{Msg} \cdot v^s$, and that $\boldsymbol{c} = (c_1, ..., c_m) \in \mathcal{C}$, where $\mathcal{C}$ is a vector space of m coordinates, each being a copy of $\mathbb{F}_p$, $\mathbb{G}$, $\hat{\mathbb{G}}$, or $\mathbb{G}_t$. Finally, we assume that $S \in \mathcal{S}d$, with $\mathcal{S}d$ some non-empty auxiliary domain.

Decrypt(Pub, $\mathsf{Pvk}_{\mathsf{Id}}$, Ctx) on input Pub, a private key $\mathsf{Pvk}_{\mathsf{Id}} = (\mathsf{Id}, R, \boldsymbol{d})$, and a ciphertext Ctx $= (\mathsf{Id}, S, \mathsf{Msg} \cdot v^s, \boldsymbol{c})$, outputs Msg provided the inputs are well-formed and the identities match.

The purpose of ω given to setup is to allow the creation of multiple instances of a single scheme with related keys; this may enable certain schemes (potentially Gentry's) to fit the template, provided that other security conditions are met. Normally, ω is ignored by the underlying scheme and all key pairs are independent.

Based on this template, we define the notion of Linear IBE to capture the intuitive linearity properties of the session keys that we discussed.

Definition 1. *A Linear IBE scheme,* (Setup, Extract, Encrypt, Decrypt), *is a quadruple of algorithms that follows the template above, and further satisfies the two properties below.*

1. *There exists a (publicly) efficiently computable function,* $f_{Pub} : \mathcal{I}d \times \mathcal{R}d \times \mathcal{S}d \times \mathcal{C} \times \mathcal{D} \to \mathbb{G}_t$, *linear in each of its last two arguments, such that, for all well-formed* $Pvk_{Id} = (Id, R, \boldsymbol{d})$ *and* $Ctx = (Id, S, c_0, \boldsymbol{c})$,

$$f_{Pub}\,(Id,\ R,\ S,\ \boldsymbol{c},\ \boldsymbol{d}) = v^{-s}\ ,$$

   where we recall that v *is the generator of* $\mathbb{G}_t$ *given as input to the* Setup *function, and thus independent of the choice of* Msk.
   Note that the decryption algorithm reduces to: $Decrypt(Pvk_{Id}, Ctx) \leftarrow c_0 \cdot f_{Pub}(Id, R, S, \boldsymbol{c}, \boldsymbol{d})$.

2. *For any two possibly identical public keys* Pub_1 *and* Pub_2 *derived from the same parameters* $(\boldsymbol{e}, g, \hat{g}, v, \omega)$, *for any auxiliary values* R_1' *and* R_2', *and for any identities* Id_1 *and* Id_2 *such that* $Pub_1 \neq Pub_2 \vee Id_1 \neq Id_2$, *one can publicly and efficiently find two "reciprocal private keys"* $\boldsymbol{d}_1' = (\hat{d}_{1,1}', ..., \hat{d}_{1,n}')$ *and* $\boldsymbol{d}_2' = (\hat{d}_{2,1}', ..., \hat{d}_{2,n}')$ *such that:*

   (a) *For* $i, j = 1, 2$, *let* $[\boldsymbol{d}_{ij} : (Id_j, R, \boldsymbol{d}_{ij}) \leftarrow Extract(Msk_j, Id_j) \mid R = R_i]$ *be the conditional distribution induced by sampling the extraction algorithm and retaining outcomes with the stated auxiliary value* R_i. *There must exist a non-trivial linear combination with coefficients* $t_{ij} \in \mathbb{F}_p$, *allowed to depend on the* R_i *and* R_j', *that renders these random variables statistically indistinguishable,*

   $$[\boldsymbol{d}_1'] \sim [(\boldsymbol{d}_{11})^{t_{11}}\,(\boldsymbol{d}_{12})^{t_{12}}]\ ,$$
   $$[\boldsymbol{d}_2'] \sim [(\boldsymbol{d}_{21})^{t_{21}}\,(\boldsymbol{d}_{22})^{t_{22}}]\ .$$

   (b) *For any two well-formed ciphertexts* $Ctx_1 = (Id_1, S_1, Msg_1 \cdot v^s, c_1)$ *and* $Ctx_2 \doteq (Id_2, S_2, Msg_2 \cdot v^s, c_2)$, *for identities* Id_1 *and* Id_2, *and built with the same randomization exponent* s, *we have,*

   $$f_{Pub}(Id_1,\ R_1',\ S_1,\ c_1,\ \boldsymbol{d}_1') \cdot f_{Pub}(Id_2,\ R_2',\ S_2,\ c_2,\ \boldsymbol{d}_2') = v^0 = 1\ .$$

Property 1 expresses our two earlier requirements: first, that the session keys be bilinear functions of both the private keys and the ciphertexts (represented by $\boldsymbol{c}$ and $\boldsymbol{d}$); and second, that session keys be of the form v^{-s} for externally fixed v, and thus independent of the master key.

Property 2 asks that anyone be able to produce $\boldsymbol{d}_1'$ and $\boldsymbol{d}_2'$ that cancel out when used as private keys. The private keys Pvk_1 and Pvk_2 and the linear coefficients $t_{11}, ..., t_{22}$ must provably exist, but they need not and should not be efficiently computable from public information (as this would be incompatible with IBE security). Requirement 2a serves to ensures that $\boldsymbol{d}_1'$ and $\boldsymbol{d}_2'$ are properly randomized and compatible with the function f_{Pub}. Requirement 2b implies a generalization to arbitrary linear combinations of keys $\boldsymbol{d}_1', ..., \boldsymbol{d}_k'$ for any number k of identities (and auxiliary values): cancellation would then occur in a k-wise product under the chosen linear combination. We shall see this in action in the HIBE scheme of Section 5.1.

3.2   Parallel IBE Security

The preceding notion of Linear IBE must be strengthened slightly in order to be useful. What we need is a weak notion of parallelism for the IBE scheme that extends to the simulation proofs, but that does not necessarily entail full concurrency.

Essentially, we want the ability to run multiple instances of the IBE at once, in a way that the session keys be all the same (though the identities might be different). For this, we need all the instances to use the same target group generator $v \in \mathbb{G}_t$ (which need not be specified externally), and allow them to use the same random exponent s to create the common session key v^s.

We define the notion of parallel semantic security under selective-identity chosen plaintext attack using the following game played against an attacker $\mathcal{A}$.

Target: $\mathcal{A}$ announces the identities $\mathsf{Id}_1^*, ..., \mathsf{Id}_\ell^*$ it intends to attack.

Setup: The challenger generates a set of public bilinear parameters $(\mathbf{e}, g, \hat{g}, v)$ and a secret random seed ω, and makes ℓ independent calls to the IBE setup algorithm $(\mathsf{Msk}_i, \mathsf{Pub}_i) \leftarrow \mathsf{Setup}(\mathbf{e}, g, \hat{g}, v, \omega)$ using these inputs, but with different internal randomness if Setup uses any. $\mathcal{A}$ is given $(\mathbf{e}, g, \hat{g}, v)$ and the ℓ public keys $\mathsf{Pub}_1, ..., \mathsf{Pub}_\ell$, which may or may not be the same.

Queries I: $\mathcal{A}$ adaptively submits private key extraction queries on each IBE scheme. For any query Id made with respect to the i-th IBE public key Pub_i, we require that $\mathsf{Id} \neq \mathsf{Id}_i^*$. The challenger answers such a query with $\mathsf{Pvk}_{\mathsf{Id},i} \leftarrow \mathsf{Extract}(\mathsf{Msk}_i, \mathsf{Id})$, recalling $\mathsf{Pvk}_{\mathsf{Id},i}$ from storage if it has been computed already.

Challenge: $\mathcal{A}$ then outputs two messages Msg_1 and Msg_2 on which it wishes to be challenged. The challenger selects $b \in \{1, 2\}$ at random, draws a random exponent $s \in \mathbb{F}_p$, and creates ℓ ciphertexts $\mathsf{Ctx}_i \leftarrow \mathsf{Encrypt}(\mathsf{Pub}_i, \mathsf{Id}_i^*, \mathsf{Msg}_b, s)$ using the same message Msg_b. The challenge given to $\mathcal{A}$ is the ℓ ciphertexts $\mathsf{Ctx}_1, ..., \mathsf{Ctx}_\ell$.

Queries II: $\mathcal{A}$ makes additional queries under the same constraints as before, to which the challenger responds as before. The total number of queries to each IBE subsystem in phases I and II may not exceed q.

Guess: $\mathcal{A}$ eventually outputs a guess $b' \in \{1, 2\}$, and wins the game if $b' = b$.

Definition 2. *We say that an IBE scheme is $(q, \ell, \tau, \epsilon)$-Par-IND-sID-CPA secure if there is no adversary $\mathcal{A}$ that and wins the preceding game in time τ with probability at least $\frac{1}{2} + \epsilon$.*

We say that an IBE scheme is $(q, \ell, \tau, \epsilon)$-Par-IND-ID-CPA secure in the same conditions, if the Target phase is moved to the beginning of the Challenge phase.

We further strengthen the security notion by offering an additional type of key extraction query, which captures the intuition that the challenger is able to create linear relations between arbitrary private keys, including the ones on the target identities (albeit without revealing what those are). We define this security property separately because it is not needed for all generic constructions. In Query phases I and II, we add a "parallel simulation" query, which goes as follows:

> **Queries I' & II':** $\mathcal{A}$ can make adaptive "parallel simulation" queries across all IBE instances. To query, $\mathcal{A}$ outputs $k+1$ pairs (i_j, Id_{i_j}) where $\{i_0, ..., i_k\} \subseteq \{1, ..., \ell\}$. We require $\mathsf{Id}_{i_j} \neq \mathsf{Id}_{i_j}^*$ for $j = 1, ..., k$ but allow $\mathsf{Id}_{i_0} = \mathsf{Id}_{i_0}^*$. To respond, $\mathcal{B}$ picks a random $\gamma \in_{\$} \mathbb{F}_p^{\times}$; for $j = 0, ..., k$, it computes $\mathsf{Pvk}_{i_j} = (\mathsf{Id}_{i_j}, R_{i,j}, \boldsymbol{d}_{i,j}) \leftarrow \mathsf{Extract}(\mathsf{Msk}_{i_j}, \mathsf{Id}_{i_j})$, or recalls it from storage if is was computed before; it then outputs $(\mathsf{Id}_{i_j}, R_{i,j}, (\boldsymbol{d}_{i,j})^{\gamma})$ for $j = 0, ..., k$.
>
> Each new needed call to $\mathsf{Extract}$ counts toward the quota of q private key queries; no $\mathsf{Pvk}_{\mathsf{Id},i}$ is ever recomputed under different randomizations.

The above game augmented with the "parallel simulation" query defines the following security notion.

Definition 3. *We say that an IBE scheme is $(q, \ell, \tau, \epsilon)$-ParSim-IND-sID-CPA secure if there is no adversary $\mathcal{A}$ that and wins the augmented game in time τ with probability at least $\frac{1}{2} + \epsilon$.*

We similarly define adaptive-identity $(q, \ell, \tau, \epsilon)$-ParSim-IND-ID-CPA security, if the Target phase is postponed to the beginning of the Challenge phase.

We short-handedly say that an IBE scheme is Exponent Inversion Compliant (or EI-compliant) if it satisfies Definitions 1 and 3, and thus 2 (with parameters that are understood from context).

4   Concrete Instantiations

In this section, we prove that the canonical examples of IBE schemes that intuitively fall under the exponent inversion umbrella are, indeed, Linear IBE schemes per our formal definition, and also fulfil the Parallel Simulation IBE security property (albeit in different ways). For completeness, we briefly review the workings of each scheme, and refer to the literature for the details.

4.1   BB$_2$-IBE

Our first example is the second of two IBE constructions given by Boneh and Boyen in [2], or BB$_2$. It was originally proven secure against selective-identity attacks from the BDHI assumption [14,2] in the standard model.

- BB$_2$.Setup outputs the master key $\mathsf{Msk} \leftarrow (a, b)$ and the public parameters $\mathsf{Pub} \leftarrow (g, g_a = g^a, g_b = g^b, v = \mathbf{e}(g, \hat{g}))$ where $a, b \in_{\$} \mathbb{F}_p$.
- BB$_2$.Extract(Msk, Id) outputs $\mathsf{Pvk}_{\mathsf{Id}} \leftarrow \left(r_{\mathsf{Id}} = r, \hat{d}_{\mathsf{Id}} = \hat{g}^{\frac{-1}{a+\mathsf{Id}+b\,r}}\right)$ for $r \in_{\$} \mathbb{F}_p$.

- BB_2.Encrypt(Pub, Id, Msg, s) outputs Ctx $\leftarrow (c_0, c_1, c_2)$ where $c_0 = $ Msg $\cdot v^s$, $c_1 = (g_a \, g^{\mathsf{Id}})^s$, $c_2 = g_b^s$ for the given s.
- BB_2.Decrypt(Pub, Pvk$_{\mathsf{Id}}$, Ctx) outputs Msg$' \leftarrow c_0 \cdot \mathbf{e}(c_1 \, c_2^{r_{\mathsf{Id}}}, \hat{d}_{\mathsf{Id}}) \in \mathbb{G}_t$.

Note that the setup seed ω is not used; the master key (a, b) is generated from internal randomness.

Lemma 1. BB_2-*IBE is a Linear IBE scheme.*[1]

Proof. For key and ciphertext with matching identities, we find that Msg$' = $ (Msg $\cdot v^s) \cdot v^{-s} = $ Msg. Towards Property 1, if follows that,

$$\mathsf{f}_{\mathsf{Pub}}\left(\mathsf{Id}, \; R = (r_{\mathsf{Id}}), \; S = \perp, \; \boldsymbol{c} = (c_1, c_2), \; \boldsymbol{d} = (\hat{d}_{\mathsf{Id}})\right) = \mathbf{e}(c_1 \, c_2^{r_{\mathsf{Id}}}, \hat{d}_{\mathsf{Id}}) = v^{-s} \; .$$

Linearity in the last two arguments is then easy to show. In particular,

$$\mathsf{f}_{\mathsf{Pub}}\left(\mathsf{Id}, \; R = (r_{\mathsf{Id}}), \; \perp, \; \boldsymbol{c}^\alpha = (c_1^\alpha, c_2^\alpha), \; \boldsymbol{d}^\beta = (\hat{d}_{\mathsf{Id}})^\beta\right) = v^{-s\,\alpha\,\beta} \; .$$

For Property 2, given Id_1, Id_2, and any $r_1', r_2' \in \mathbb{F}_p$, set $\boldsymbol{d}_1' = (\hat{g}_{a_2} \, \hat{g}_{b_2}^{r_2'} \, \hat{g}^{\mathsf{Id}_2})$ and $\boldsymbol{d}_2' = (\hat{g}_{a_1} \, \hat{g}_{b_1}^{r_1'} \, \hat{g}^{-\mathsf{Id}_1})^{-1}$, taking (g_{a_1}, g_{b_1}) from Pub$_1$ and (g_{a_2}, g_{b_2}) from Pub$_2$, which are not necessarily distinct. Then, for actual private keys Pvk$_{\mathsf{Id}_1} = (r_1, \boldsymbol{d}_1)$ and Pvk$_{\mathsf{Id}_2} = (r_2, \boldsymbol{d}_2)$, we have,

$$\boldsymbol{d}_1' = (\boldsymbol{d}_1)^{(a_1 + b_1 \, r_1 + \mathsf{Id}_1)\,(a_2 + b_2 \, r_2' + \mathsf{Id}_2)} \cdot (\boldsymbol{d}_2)^0 \; ,$$
$$\boldsymbol{d}_2' = (\boldsymbol{d}_1)^0 \cdot (\boldsymbol{d}_2)^{-(a_2 + b_2 \, r_2 + \mathsf{Id}_2)\,(a_1 + b_1 \, r_1' + \mathsf{Id}_1)} \; ,$$

and, for any $c_1 = ((g_{a_1} \, g^{\mathsf{Id}_1})^s, g_{b_1}^s)$ and $c_2 = ((g_{a_2} \, g^{\mathsf{Id}_2})^s, g_{b_2}^s)$, we have that, $\mathsf{f}_{\mathsf{Pub}}(\mathsf{Id}_1, r_1', \perp, c_1, \boldsymbol{d}_1') \cdot \mathsf{f}_{\mathsf{Pub}}(\mathsf{Id}_2, r_2', \perp, c_2, \boldsymbol{d}_2') = 1, \forall s$, as required. $\square$

The following lemma generalizes the BB_2 security theorem from [2] to the notion of parallel IBE semantic security defined in Section 3.2.

Lemma 2. BB_2-*IBE is* $(q, \ell, \tau, \epsilon)$-*ParSim-IND-sID-CPA secure in any bilinear context that satisfies the Decision* (q', τ', ϵ)-*BDHI assumption with* $q' > q\,\ell$ *and* $\tau' < \tau - \Theta(q^2\,\ell^2)$.

In other words, BB_2 is secure under a selective-identity, parallel simulation attack, in the standard model, provided that the BDHI assumption holds in the relevant bilinear context.

4.2 SK-IBE

The second scheme we describe is adapted from the identity-based key encapsulation mechanism (IBKEM) given in [7] and attributed to Sakai and Kasahara [17]. Its security proof is set in the random oracle model. For consistency with our definitions, we present an IBE version of the scheme, and call it SK.

---

[1] See Remark 1 concerning implementations in asymmetric bilinear groups.

- SK.Setup outputs the master key $\mathsf{Msk} \leftarrow a \in_{\$} \mathbb{F}_p$ and the public key $\mathsf{Pub} \leftarrow (g, g_a = g^a, v = \mathbf{e}(g, \hat{g}), H : \{0,1\}^* \to \mathbb{F}_p)$.
- SK.Extract$(\mathsf{Msk}, \mathsf{Id})$ outputs the private key $\mathsf{Pvk}_{\mathsf{Id}} \leftarrow \hat{g}^{\frac{1}{a+H(\mathsf{Id})}}$.
- SK.Encrypt$(\mathsf{Pub}, \mathsf{Id}, \mathsf{Msg}, s)$ outputs $\mathsf{Ctx} \leftarrow \left(c_0 = \mathsf{Msg} \cdot v^s, \ c_1 = (g_a\, g^{H(\mathsf{Id})})^s\right)$.
- SK.Decrypt$(\mathsf{Pub}, \mathsf{Pvk}_{\mathsf{Id}}, \mathsf{Ctx})$ outputs $\mathsf{Msg}' \leftarrow c_0 / \mathbf{e}(c_1, \mathsf{Pvk}_{\mathsf{Id}}) \in \mathbb{G}_t$.

As in BB_2, the setup seed ω is not used; the master key a is generated from internal randomness.

Lemma 3. *SK-IBE is a Linear IBE scheme.*[1]

Proof. SK-IBE clearly fits the Linear IBE template with $v = \mathbf{e}(g, \hat{g})$. Property 1 is easily verified; in particular, for $\boldsymbol{c}^\alpha = (c_1^\alpha)$ and $\boldsymbol{d}^\beta = (\mathsf{Pvk}_{\mathsf{Id}}^\beta)$,

$$\mathsf{f}_{\mathsf{Pub}} (\mathsf{Id}, \perp, \perp, \boldsymbol{c}, \boldsymbol{d}) = \mathbf{e}(c_1, \mathsf{Pvk}_{\mathsf{Id}})^{-\alpha\beta} = \mathbf{e}(g, \hat{g})^{-s\alpha\beta} = v^{-s\alpha\beta} .$$

For Property 2, given Id_1 and Id_2 anyone can pick $\boldsymbol{d}_1' = (\hat{g}_{a_2}\, \hat{g}^{H(\mathsf{Id}_2)})$ and $\boldsymbol{d}_2' = (\hat{g}_{a_1}^{-1}\, \hat{g}^{-H(\mathsf{Id}_1)})$, so,

$$\boldsymbol{d}_1' = (\mathsf{Pvk}_{\mathsf{Id}_1})^{t_{11}} \cdot (\mathsf{Pvk}_{\mathsf{Id}_2})^{t_{12}} \qquad t_{11} = (a_1 + H(\mathsf{Id}_1))\,(a_2 + H(\mathsf{Id}_2)), \quad t_{12} = 0 ,$$

$$\boldsymbol{d}_2' = (\mathsf{Pvk}_{\mathsf{Id}_1})^{t_{21}} \cdot (\mathsf{Pvk}_{\mathsf{Id}_2})^{t_{22}} \qquad t_{21} = 0, \quad t_{22} = -t_{11} ,$$

and $\forall s$, $\mathsf{f}_{\mathsf{Pub}} \left(\mathsf{Id}_1, \perp, \perp, (g_{a_1}\, g^{H(\mathsf{Id}_1)})^s, \boldsymbol{d}_1'\right) \cdot \mathsf{f}_{\mathsf{Pub}} \left(\mathsf{Id}_2, \perp, \perp, (g_{a_2}\, g^{H(\mathsf{Id}_2)})^s, \boldsymbol{d}_2'\right) = \mathbf{e}(g^{(a_1 + H(\mathsf{Id}_1))\,s}, \hat{g}^{a_2 + H(\mathsf{Id}_2)}) \cdot \mathbf{e}(g^{(a_2 + H(\mathsf{Id}_2))\,s}, \hat{g}^{-a_1 - H(\mathsf{Id}_1)}) = \mathbf{e}(g, \hat{g})^0 = 1$, as required.

Lemma 4. *SK-IBE is $(q, \ell, \tau, \epsilon)$-ParSim-IND-ID-CPA secure in any bilinear context that satisfies the Decision (q', τ', ϵ')-BDHI assumption with $q' > q\ell$ and $\tau' < \tau - \Theta(q^2\,\ell^2)$, in the random oracle model, where $\epsilon'/\epsilon \geq \prod_{i=1}^\ell Q_i$, where Q_i is the number of adversarial queries to the random oracle that hashes the identities in the i-th IBE subsystem.*

Notice that the above lemma pertains to a full adaptive-identity, parallel simulation attack. The security is not tight, however, and the security losses mount exponentially with the number of IBE subsystems in the experiment.

Proof. The security proof is similar to (and a simpler version of) the proof of Lemma 2.

4.3 The Case of the Gentry IBE

The ambiguity of Gentry's IBE as an exponent inversion candidate presents an intriguing open problem. Recall from [10] that it uses a powerful security reduction that gives it tight security under adaptive-identity attacks, albeit under a strong assumption. On the one hand, the Gentry IBE has much in common with the exponent inversion family, such as the use of session keys $\mathbf{e}(g, \hat{h})^s$ that do not involve the master secret. On the other hand, the scheme uses two generators, g

and $\hat{h}$, chosen at random by the master key generator. The security proof breaks when both g and $\hat{h}$ are fixed externally, or even when chosen randomly but reused across parallel instances in the sense of Section 3.2. Thus, Gentry-IBE currently fails the exponent inversion litmus test that session keys be of the form v^s for fixed v; it remains open whether this can be remedied using a different proof.

Since the HIBE transformation we describe next preserves adaptive-identity security, extending Gentry's proof to work in the exponent inversion setting would resolve the long-standing problem of realizing fully secure HIBE for broad and deep hierarchies. Meanwhile, the very existence of such schemes remains an open problem.

Remark 1 (Asymmetric Implementations)
Lemmas 1 and 3 tacitly assume that for each element g, $g_a = g^a$, $g_b = g^b \in \mathbb{G}$ published in Pub, the corresponding element $\hat{g}$, $\hat{g}_a = \hat{g}^a$, $\hat{g}_b = \hat{g}^b \in \hat{\mathbb{G}}$ is made available for the creation of $\boldsymbol{d}_1'$ and $\boldsymbol{d}_2'$. This is automatically true if we assume that $\mathbb{G} = \hat{\mathbb{G}}$, as was the case in the original descriptions of BB_2 [2] and SK [7,17]. Otherwise, the relevant elements will need to be published explicitly, *e.g.*, in the public key, which is harmless to the security of any scheme that was already secure under the assumption that $\mathbb{G} = \hat{\mathbb{G}}$.

5   Generic Constructions

Let an abstract scheme IBE = (IBE.Setup, IBE.Extract, IBE.Encrypt, IBE.Decrypt) with "parallel" semantic security against selective-identity chosen-plaintext attacks, that has an appropriate linear structure as above. We show how to turn it into generalizations of IBE that are semantically secure against (the appropriate notion of) selective-identity chosen-plaintext attacks.

5.1   Hierarchical Identities

In the HIBE primitive [13,11], identities are arranged in a hierarchy, and the private keys can be derived per the hierarchy without involving the global master secret. HIBE is essentially a delegation mechanism with a single root (the private key generator). We construct such a scheme generically as follows.

HIBE.Setup(L). Given a security parameter and the desired number L of levels in the hierarchy:
1. Create bilinear group parameters, $\mathbf{e}, g, \hat{g}, v$, at the desired level of security. Also pick an ephemeral shared random seed ω which is kept secret.
2. Generate L sets of IBE master key pairs with common bilinear parameters, $\mathbf{e}, g, \hat{g}, v$, by making L calls to setup (IBE.Msk$_i$, IBE.Pub$_i$) $\leftarrow$ IBE.Setup($\mathbf{e}, g, \hat{g}, v, \omega$) for $i = 1, ..., L$.
3. Select L collision-resistant hash functions (or UOWHFs) from vectors of IBE identities to single identities, $H_i : \mathcal{I}^i \rightarrow \mathcal{I}$ for $i = 1, ..., L$, where $\mathcal{I}$ is the domain of IBE identities.

4. Output the HIBE master key pair:

$$\mathsf{HIBE.Msk} = (\mathsf{IBE.Msk}_1, ..., \mathsf{IBE.Msk}_L) \ ,$$
$$\mathsf{HIBE.Pub} = (\mathsf{IBE.Pub}_1, ..., \mathsf{IBE.Pub}_L, H_1, ..., H_L) \ .$$

$\mathsf{HIBE.Extract(Msk, Id)}$. Given $\mathsf{HIBE.Msk}$ and a target identity $\mathsf{Id} = (I_1, ..., I_\ell)$ at level $\ell \le L$:

1. $\forall i = 1, ..., \ell$, let $h_i = H_i(I_1, ..., I_i)$ be the hash of the first i components.
2. $\forall i = 1, ..., \ell$, extract an IBE key $(h_i, R_i, \boldsymbol{d}_i) \leftarrow \mathsf{Extract}(\mathsf{IBE.Msk}_i, h_i)$.
3. Select $r_1, ..., r_\ell \in \mathbb{F}_p$ under the constraint that $\sum_{i=1}^{\ell} r_i = 1 \pmod{p}$.
4. Output the HIBE private key:

$$\mathsf{HIBE.Pvk}_{\mathsf{Id}} = ((I_1, R_1, \boldsymbol{d}_1^{r_1}), ..., (I_\ell, R_\ell, \boldsymbol{d}_\ell^{r_\ell})) \ .$$

Observe that all the components of the private key are bound to each other via the constraint $\sum_{i=1}^{\ell} r_i = 1 \pmod{p}$. Without it, the key would be utterly random and therefore useless. The mutual binding of the components also ensures that private keys given to different users are impervious to collusion attacks.

$\mathsf{HIBE.Derive(Pvk}_{\mathsf{Id}}, I')$. Given $\mathsf{HIBE.Pvk}_{\mathsf{Id}}$ for an ℓ-level HIBE "parent" identity Id with $\ell < L$, and an IBE identity I' to act as the $(\ell + 1)$-th component of the HIBE "child" identity:

1. Decompose $\mathsf{HIBE.Pvk}_{\mathsf{Id}}$ as a list of triples $(I_i, R_i, \boldsymbol{d}_i)$ for $i = 1, ..., \ell$. Let also $I_{\ell+1} = I'$.
2. For each $i = 1, ..., \ell + 1$, let $h_i = H_i(I_1, ..., I_i)$ be the hash of the first i components.
3. For each $i = 1, ..., \ell$:
   (a) Find two vectors $\boldsymbol{d}'_{1,i}$ and $\boldsymbol{d}'_{2,i}$ that satisfy Property 2 for $\mathsf{Id}_1 = h_i$ and $\mathsf{Id}_2 = h_{i+1}$ (and the auxiliary R_i and R_{i+1}) relative to the public keys $\mathsf{IBE.Pub}_i$ and $\mathsf{IBE.Pub}_{i+1}$.
   (b) Select $r_i \in \mathbb{F}_p^\times$ and observe that $(\boldsymbol{d}'_{1,i})^{r_i}$ and $(\boldsymbol{d}'_{2,i})^{r_i}$ also satisfy Property 2.
4. For $i = 1, ..., \ell + 1$, define $\boldsymbol{d}''_i = \begin{cases} (\boldsymbol{d}'_{1,1})^{r_1} & \text{if } i = 1 \\ (\boldsymbol{d}'_{2,i-1})^{r_{i-1}} (\boldsymbol{d}'_{1,i})^{r_i} & \text{if } 2 \le i \le \ell . \\ (\boldsymbol{d}'_{2,\ell})^{r_\ell} & \text{if } i = \ell + 1 \end{cases}$

5. Output the HIBE private key:

$$\mathsf{HIBE.Pvk}_{\mathsf{Id}'} = ((I_1, R_1, \boldsymbol{d}_1 \cdot \boldsymbol{d}''_1), ..., (I_\ell, R_\ell, \boldsymbol{d}_\ell \cdot \boldsymbol{d}''_\ell), (I_{\ell+1}, R_{\ell+1}, \boldsymbol{d}''_{\ell+1}))$$

Notice that the derived private key is fully randomized (its distribution is the same as if it had been created by $\mathsf{HIBE.Extract}$), it will decrypt correctly (because of Property 2), and its creation required only the parent private key and not the master key.

$\mathsf{HIBE.Encrypt(Pub, Id, Msg)}$. Given $\mathsf{HIBE.Pub}$, an ℓ-level identity $\mathsf{Id} = (I_1, ..., I_\ell)$ where $\ell \le L$, and a message $\mathsf{Msg} \in \mathbb{G}_t$:

1. Pick a random exponent $s \in \mathbb{F}_p$.

2. $\forall i = 1, ..., \ell$, let $h_i = H_i(I_1, ..., I_i)$ be the hash of the first i components.
3. $\forall i = 1, ..., \ell$, use s to construct an IBE ciphertext $\mathsf{Ctx}_i = (h_i, S_i, c_0, \boldsymbol{c}_i) \leftarrow$ $\mathsf{Encrypt}(\mathsf{IBE.Pub}_i, h_i, \mathsf{Msg}, s)$.
4. Output the HIBE ciphertext:

$$\mathsf{HIBE.Ctx} = ((h_1, ..., h_\ell),\ c_0,\ (S_1, ..., S_\ell),\ (\boldsymbol{c}_1, ..., \boldsymbol{c}_\ell))\ .$$

Notice that $c_0 = \mathsf{Msg} \cdot v^s$ is the same in all the IBE ciphertexts.

$\mathsf{HIBE.Decrypt}(\mathsf{Pub}, \mathsf{Pvk}_{\mathsf{Id}}, \mathsf{Ctx})$. Given the public key $\mathsf{HIBE.Pub}$, a private key $\mathsf{Pvk}_{\mathsf{Id}} = (\mathsf{Pvk}_1, ..., \mathsf{Pvk}_\ell)$ for some hierarchical identity, and a ciphertext $\mathsf{Ctx} = ((h_1, ..., h_\ell), c_0, (S_1, ..., S_\ell), (\boldsymbol{c}_1, ..., \boldsymbol{c}_\ell))$ for the same identity:

1. $\forall i = 1, ..., \ell$, assemble $\mathsf{Ctx}_i = (h_i, 1, S_i, \boldsymbol{c}_i)$, using $1 \in \mathbb{G}_t$ in lieu of c_0.
2. $\forall i = 1, ..., \ell$, IBE-decrypt $v_i \leftarrow \mathsf{IBE.Decrypt}(\mathsf{IBE.Pub}_i, \mathsf{Pvk}_i, \mathsf{Ctx}_i)$.
3. Output the decrypted plaintext:

$$\mathsf{Msg} = c_0 \cdot \prod_{i=1}^{\ell} v_i\ .$$

By Property 1, we know that $v_i = v^{-s\,r_i}$ provided that the algorithm inputs are as expected. Since $\sum_i r_i = 1$, we obtain the desired result.

The collision-resistant hash functions $H_1, ..., H_L$ serve to enforce the "inheritance" requirement that identity components of higher index be dependent on the components of lower index. The hash functions do this by creating a precedence ordering over the indices in a construction that would otherwise be indifferent to it. The schemes we build next have no such requirement.

The above construction is quite efficient. If we instantiate it using BB_2 or SK, we respectively obtain two HIBE systems that only require ℓ pairings for decryption at level ℓ, which is marginally faster than most previously known HIBE systems [11,2,3]. The specialized construction from [3] offers faster decryption for identities of depth $\ell \geq 3$.

We can prove selective-identity security of the scheme if the underlying scheme meets the weaker version of "parallel" selective-identity IBE security (from Definition 2).

Theorem 1. *The generic* HIBE *scheme is* $(q, \ell, \tau, \epsilon)$-*IND-sHID-CPA secure [5] provided that the underlying* IBE *scheme is a Linear IBE that satisfies* $(q, \ell, \tau', \epsilon)$-*Par-IND-sID-CPA security for some* $\tau' \approx \tau$.

We have essentially the same result in the adaptive-identity models.

Corollary 1. *The generic* HIBE *scheme is* $(q, \ell, \tau, \epsilon)$-*IND-HID-CPA secure [11] provided that the underlying* IBE *scheme is a Linear IBE that satisfies* $(q, \ell, \tau', \epsilon)$-*Par-IND-ID-CPA security for some* $\tau' \approx \tau$.

5.2   Fuzzy Identities

In the Fuzzy IBE primitive [16], private keys and ciphertexts pertain to multiple identities (or attributes) at once, and decryption is predicated on meeting certain threshold of matching attributes. The collusion resistance property stipulates that private keys containing different sets of attributes cannot be combined to obtain a larger set than any of them provided by itself.

Two versions of the primitive are defined in [16]: a "small universe" version which supports an enumerated set of possible attributes, and a "large universe" version, where exponentially many attributes are representable but only a constant number at a time. In both versions the attributes are boolean (either present or absent), which we call "small domain".

Here, we give a "large domain" generalization of "small universe" Fuzzy IBE, where the enumerated attributes are now key/value pairs that range in all of $\mathbb{F}_p$. This could be useful in applications of Fuzzy IBE that require non-boolean attributes, such as a biometric system with attributes such as the height of a person.

The small-universe, large-domain, generic Fuzzy IBE construction is as follows.

FuzzyIBE.Setup(n). Given a security parameter, and the number n of attribute types to support:
1. Create bilinear group parameters, $\mathbf{e}, g, \hat{g}, v$, at the desired level of security, and a secret random string ω.
2. Generate n independent IBE master key pairs with shared bilinear parameters, $\mathbf{e}, g, \hat{g}, v$, by executing setup n times, (IBE.Msk$_i$, IBE.Pub$_i$) $\leftarrow$ IBE.Setup($\mathbf{e}, g, \hat{g}, v, \omega$) for $i = 1, ..., n$.
3. Output the Fuzzy IBE master key pair:

$$\text{FuzzyIBE.Msk} = (\text{IBE.Msk}_1, ..., \text{IBE.Msk}_n) \ ,$$
$$\text{FuzzyIBE.Pub} = (\text{IBE.Pub}_1, ..., \text{IBE.Pub}_n) \ .$$

FuzzyIBE.Extract(Msk, Id, t). On input a master key FuzzyIBE.Msk, a vector Id $= (I_1, ..., I_n)$ of (positionally sensitive) attributes $I_i \in \mathbb{F}_p$, and a threshold parameter t with $1 \leq t \leq n$:
1. Pick $f_1, ..., f_{t-1} \in \mathbb{F}_p$ and let $f(x) = 1 + \sum_{i=1}^{t-1} f_i x^i$ of degree $t-1$. Note that $f(0) = 1$.
2. $\forall i = 1, ..., n$, extract an IBE key $(I_i, R_i, \boldsymbol{d}_i) \leftarrow$ Extract(IBE.Msk$_i$, I_i),
3. Output the Fuzzy IBE private key:

$$\text{FuzzyIBE.Pvk}_{\text{Id}} = \left(t, \ (I_1, \ R_1, \ \boldsymbol{d}_1^{f(1)}), \ ..., \ (I_n, \ R_n, \ \boldsymbol{d}_n^{f(n)}) \right) \ .$$

FuzzyIBE.Encrypt(Pub, Id, Msg). On input a public key FuzzyIBE.Pub, a vector Id $= (I_1, ..., I_n)$ of (positionally sensitive) attributes $I_i \in \mathbb{F}_p$, and a message Msg $\in \mathbb{G}_t$:
1. Pick a random exponent $s \in \mathbb{F}_p$.
2. For all $i = 1, ..., n$, build an IBE ciphertext Ctx$_i = (I_i, S_i, c_0, \boldsymbol{c}_i) \leftarrow$ Encrypt(IBE.Pub$_i$, I_i, Msg, s).

3. Output the Fuzzy IBE ciphertext (using $c_0 = \mathsf{Msg} \cdot v^s$ common to all IBE ciphertexts):

$$\mathsf{FuzzyIBE.Ctx} = (\mathsf{Id}, \ c_0, \ (S_1, ..., S_n), \ (c_1, ..., c_n)) \ .$$

$\mathsf{FuzzyIBE.Decrypt}(\mathsf{Pub}, \mathsf{Pvk_{Id}}, \mathsf{Ctx})$. Given $\mathsf{FuzzyIBE.Pub}$, a private key $\mathsf{Pvk_{Id}}$, and a ciphertext Ctx:

1. Determine t attributes $I_{i_1}, ..., I_{i_t}$ that appear in both $\mathsf{Pvk_{Id}}$ and Ctx in matching positions.
   (a) If there are fewer than t "key/value" matches, then output $\perp$ and halt.
   (b) Else, select any t matching attributes $I_{i_1}, ..., I_{i_t}$ and define $T = \{i_1, ..., i_t\}$.
2. For $j = 1, ..., t$:
   (a) Extract the IBE private key $(I_{i_j}, R_{i_j}, d_{i_j})$ from $\mathsf{Pvk_{Id}}$ and call it Pvk_j.
   (b) Assemble the IBE ciphertext $(I_{i_j}, 1, S_{i_j}, c_{i_j})$ from Ctx and call it Ctx_j.
   (c) Let $\Lambda_{T,i}(x) = \prod_{i' \in T \setminus \{i\}} \frac{x-i'}{i-i'}$ be the Lagrange interpolation coefficients from T to x.
   (d) Perform the IBE decryption $v_j \leftarrow \mathsf{IBE.Decrypt}(\mathsf{IBE.Pub}_j, \mathsf{Pvk}_j, \mathsf{Ctx}_j)$.
3. Output the plaintext:

$$\mathsf{Msg} = c_0 \cdot \prod_{j=1}^{t} v_j^{\Lambda_{T,i_j}(0)} \ .$$

By Property 1, we know that $v_j = v^{-s\,f(i_j)}$ if the inputs to the algorithm are as expected. The result follows by using Lagrange polynomial interpolation, $\sum_j f(i_j)\,\Lambda_{T,i_j}(0) = f(0) = 1$, "in the exponent".

The efficiency of the scheme is comparable to that of (the "small universe" version of) [16] when instantiated with $\mathsf{BB_2}$ or SK, even though this is a "large domain" construction.

Theorem 2. *The generic* $\mathsf{FuzzyIBE}$ *scheme is* (q, n, τ, ϵ)-*IND-sFuzID-CPA se-cure [16] provided that the base* IBE *scheme is a Linear IBE with* (q, n, τ', ϵ)-*ParSim-IND-sID-CPA security for* $\tau' \approx \tau$.

5.3   Attribute-Based Encryption

Attribute-based encryption (ABE) is a powerful generalization of Fuzzy IBE that was recently proposed in [12]. Instead of allowing decryption conditionally on the satisfaction of a single threshold gate (whose inputs are the matching attributes in the ciphertext and the key), ABE allows the condition to be defined by a tree of threshold gates. The construction given in [12] generalizes the Fuzzy IBE construction of [16] in the commutative blinding approach, and is based on the use of not one but multiple interpolation polynomials $f(x)$, each of which applies to a subset of the input attributes. The degrees of the random polynomials and their inputs determine the access structure in the ABE scheme; in Key-Policy (KP) ABE, they are chosen by the authority.

Our generic framework can mirror the KP-ABE construction of [12], in the same way that our Fuzzy IBE construction retains the structure of the construction in [16]. The main difference is that, since our method is to build an independent instance of the underlying IBE for each attribute, we obtain a "large domain" generalization of ABE, with attributes as key/value pairs instead of booleans.

5.4 Multiple Independent Key Generators

Our generic construction immediately generalizes to the case of multiple independent key generators, which can be useful in many applications. For example, when using Fuzzy IBE for encrypting under someone's biometric readings, one may wish to use one set of attributes constructed from fingerprints and another from iris scans, and require a combination of both to decrypt. It is quite possible in this scenario that the authority issuing fingerprint-based private keys would be different than the one issuing keys based on iris scans.

Depending on the nature of the underlying IBE system, it is possible to base our generic Fuzzy IBE construction on independent subsystems that share only the bilinear groups and generators, thereby facilitating their setup. Whether independent setup is allowed (in a commonly agreed upon bilinear group), depends on the use that the IBE.Setup function makes of the common random ω. For instance, since the BB_2 and SK schemes achieve our notion of parallel simulation security without using ω, they are suitable for building a multi-authority system without shared secret.

The only remaining difficulty lies in the final assembly of private keys given to the users, because the separate authorities will have to agree on a suitable random polynomial $f(x)$ in order to create a new key. Some amount of coordination between the servers will be required (possibly mediated by the key recipient), but since the polynomial to be agreed upon is ephemeral and decoupled from the master keys, this is an orthogonal problem that can be solved in many standard ways. In particular, Chase [6] showed how to construct a multi-authority attribute-based scheme, in the commutative blinding framework, where multiple authorities can vouch for separate attributes under the auspices of a central authority that handles the sharing of ephemerals.

6   Conclusion

We have shown that the family of identity-based encryption schemes based on the exponent inversion principle can be leveraged into building more powerful systems. We first presented an abstraction to capture a number of useful properties shared by such schemes. We then showed how to use this abstraction to construct generalizions of IBE. We described Hierarchical and Fuzzy IBE as concrete examples, as each of them illustrates a specific feature of exponent inversion schemes, but many other generalizations are possible based on the same abstraction. Our approach is fairly lightweight and is also compatible with decentralized authorities.

These results have practical implications, since the few known exponent inversion IBE schemes tend to be marginally more efficient than competing constructions, although they require stronger complexity assumptions. Our formalism has no effect on these benefits and drawbacks, but it extends the range of applicability of the relevant schemes.

Acknowledgements

The author thanks Michel Abdalla and Brent Waters for discussions on some of these ideas, and anonymous referees for valuable comments.

References

1. Michel Abdalla, Dario Catalano, Alexander W. Dent, John Malone-Lee, Gregory Neven, and Nigel P. Smart. Identity-based encryption gone wild. In *Proceedings of ICALP 2006*, volume 4051 of *Lecture Notes in Computer Science*, pages 300–11. Springer-Verlag, 2006.
2. Dan Boneh and Xavier Boyen. Efficient selective-ID secure identity based encryption without random oracles. In *Advances in Cryptology—EUROCRYPT 2004*, volume 3027 of *Lecture Notes in Computer Science*, pages 223–38. Springer-Verlag, 2004.
3. Dan Boneh, Xavier Boyen, and Eu-Jin Goh. Hierarchical identity based encryption with constant size ciphertext. In *Advances in Cryptology—EUROCRYPT 2005*, volume 3494 of *Lecture Notes in Computer Science*, pages 440–56. Springer-Verlag, 2005.
4. Dan Boneh and Matthew Franklin. Identity-based encryption from the Weil pairing. *SIAM Journal of Computing*, 32(3):586–615, 2003. Extended abstract in *Advances in Cryptology—CRYPTO 2001*.
5. Ran Canetti, Shai Halevi, and Jonathan Katz. Chosen-ciphertext security from identity-based encryption. In *Advances in Cryptology—EUROCRYPT 2004*, volume 3027 of *Lecture Notes in Computer Science*, pages 207–22. Springer-Verlag, 2004.
6. Melissa Chase. Multi-authority attribute based encryption. In *Proceedings of TCC 2007*, Lecture Notes in Computer Science. Springer-Verlag, 2007.
7. Liqun Chen, Zhaohui Cheng, John Malone-Lee, and Nigel P. Smart. An efficient ID-KEM based on the Sakai-Kasahara key construction. Cryptology ePrint Archive, Report 2005/224, 2005. http://eprint.iacr.org/2005/224/.
8. Jung Hee Cheon. Security analysis of the strong Diffie-Hellman problem. In *Advances in Cryptology—EUROCRYPT 2006*, volume 4004 of *Lecture Notes in Computer Science*, pages 1–13. Springer-Verlag, 2006.
9. Clifford Cocks. An identity based encryption scheme based on quadratic residues. In *Proceedings of the 8th IMA International Conference on Cryptography and Coding*, 2001.
10. Craig Gentry. Practical identity-based encryption without random oracles. In *Advances in Cryptology—EUROCRYPT 2006*, Lecture Notes in Computer Science. Springer-Verlag, 2006.
11. Craig Gentry and Alice Silverberg. Hierarchical ID-based cryptography. In *Proceedings of ASIACRYPT 2002*, Lecture Notes in Computer Science. Springer-Verlag, 2002.

12. Vipul Goyal, Omkant Pandey, Amit Sahai, and Brent Waters. Attribute-based encryption for fine-grained access control of encrypted data. In *ACM Conference on Computer and Communications Security—CCS 2006*, 2006.
13. Jeremy Horwitz and Ben Lynn. Towards hierarchical identity-based encryption. In *Advances in Cryptology—EUROCRYPT 2002*, Lecture Notes in Computer Science, pages 466–81. Springer-Verlag, 2002.
14. Shigeo Mitsunari, Ryuichi Sakai, and Masao Kasahara. A new traitor tracing. *IEICE Transactions on Fundamentals*, E85-A(2):481–4, 2002.
15. David Naccache. Secure and practical identity-based encryption. Cryptology ePrint Archive, Report 2005/369, 2005. http://eprint.iacr.org/2005/369/.
16. Amit Sahai and Brent Waters. Fuzzy identity-based encryption. In *Advances in Cryptology—EUROCRYPT 2005*, volume 3494 of *Lecture Notes in Computer Science*. Springer-Verlag, 2005.
17. Ryuichi Sakai and Masao Kasahara. ID based cryptosystems with pairing over elliptic curve. Cryptology ePrint Archive, Report 2003/054, 2003. http://eprint.iacr.org/2003/054/.
18. Ryuichi Sakai, Kiyoshi Ohgishi, and Masao Kasahara. Cryptosystem based on pairing. In *Symposium on Cryptography and Information Security—SCIS 2000*, Okinawa, Japan, 2000.
19. Adi Shamir. Identity-based cryptosystems and signature schemes. In *Advances in Cryptology—CRYPTO 1984*, volume 196 of *Lecture Notes in Computer Science*, pages 47–53. Springer-Verlag, 1984.
20. Brent Waters. Efficient identity-based encryption without random oracles. In *Advances in Cryptology—EUROCRYPT 2005*, volume 3494 of *Lecture Notes in Computer Science*. Springer-Verlag, 2005.
21. Danfeng Yao, Nelly Fazio, Yevgeniy Dodis, and Anna Lysyanskaya. ID-based encryption for complex hierarchies with applications to forward security and broadcast encryption. In *ACM Conference on Computer and Communications Security—CCS 2004*, pages 354–63, 2004.

Non-interactive Proofs for Integer Multiplication

Ivan Damgård and Rune Thorbek

BRICS, Dept. of Computer Science, University of Aarhus

Abstract. We present two universally composable and practical protocols by which a dealer can, verifiably and non-interactively, secret-share an integer among a set of players. Moreover, at small extra cost and using a distributed verifier proof, it can be shown in zero-knowledge that three shared integers a, b, c satisfy $ab = c$. This implies by known reductions non-interactive zero-knowledge proofs that a shared integer is in a given interval, or that one secret integer is larger than another. Such primitives are useful, e.g., for supplying inputs to a multiparty computation protocol, such as an auction or an election. The protocols use various set-up assumptions, but do not require the random oracle model.

1 Introduction

Applications such as auctions, elections or benchmarking analysis all involve computing on confidential data from several parties who do not trust each other a priori. This means that solutions involving a single trusted party are typically unsatisfactory. In principle, all such problems can be solved using general secure multiparty computation [18,2,8], where all parties take part in computing the desired results. But in practice, this is often not realistic: in auctions or elections, for instance, the number of parties holding inputs can be very large, they cannot be assumed to be expert users nor can their machines be assumed to be on-line at particular times. Hence assuming that all such parties can reliably take part in a multi-round protocol is unrealistic.

It is therefore often suggested that a smaller number of servers are assigned to do the computation, acting effectively as representatives for the clients supplying inputs. Of course, this makes sense only if the complexity of supplying inputs is much smaller than the complexity of taking part in the actual computation. In particular, we would want that supplying inputs is non-interactive. This problem can be solved using a non-interactive verifiable secret sharing (VSS) scheme. Having done the VSS's, the servers hold shares of all inputs and can do the computation using any of the (numerous) known multiparty computation techniques. Several non-interactive VSS protocols are known see, e.g., [22].

However, many applications require that the inputs supplied satisfy certain constraints. These constraints are typically phrased in a natural way as relations over the integers, because the underlying application is a computation on integers. This is the case for auctions, elections and many statistical applications such as benchmarking. For instance, an auction might specify that bids have to

M. Naor (Ed.): EUROCRYPT 2007, LNCS 4515, pp. 412–429, 2007.

be in a certain interval. In other types of auctions (so called double auctions[4]), a bid consists of a sequence of numbers that must be monotonely increasing.

Standard efficient techniques for handling this would have a client commit to his input and prove in zero-knowledge that his numbers satisfy the required relations. But this solution requires interaction in its basic form. The interaction can typically be removed following the Fiat-Shamir heuristic if we are willing to assume the random oracle model. However, it is well known that the security guarantee provided by a proof in the random oracle model leaves something to be desired: we cannot instantiate the oracle with a concrete function and be sure that this always works. Hence, our goal is to avoid random oracles and still have an efficient solution.

In [5], Boudot presents an efficient technique to prove relations, as outlined above, given a primitive to prove that a committed integer is a square. Furthermore, in [1], Abe, Cramer and Fehr propose efficient and non-interactive techniques for proving multiplicative relations on secret-shared values, using distributed-verifier proofs. Unfortunately, the protocols and definition from [1] are not directly useful in the scenarios outlined above, for several reasons: First, the relations that can be proved only hold modulo some (public) prime number, and not necessarily over the integers. Second, for the case of honest majority, the protocols in [1] are only "non-interactive with complaints", that is, if a server is unhappy with the data he received privately from the dealer, he will complain, and the dealer must intervene in a second round to resolve these conflicts. It is clear that we have to avoid this in our scenario. Third, the definition of distributed verifier proofs used in [1] works with only one prover. In our scenario, we will have many provers, some of which may be corrupted. In contrast to the single-prover case, a corrupt prover may now try to exploit the information sent by honest provers in order to cheat.

In this paper, we propose two protocols that allow a client to non-interactively VSS integers among the servers, and prove in zero-knowledge, by a distributed verifier proof, that shared integers a, b, c satisfy $ab = c$. Using known reductions [5], this implies non-interactive proofs that a shared integer is in a given interval, or that shared numbers a, b satisfy $a \geq b$. Both protocols require one broadcast from the prover and one round of messages between the verifiers (servers), which is a minimal amount of interaction for a distributed verifier proof. Details on the communication complexity of the protocols follow below. We prove our protocols secure in the Universal Composability model (with static adversary), this automatically gives us a definition handling the multiple prover case.

For the first solution, we take the protocol of [1] as the point of departure, introducing new techniques to solve the problems mentioned above. We obtain our solution by replacing in the protocol from [1] Shamir secret-sharing by Linear Integer Secret Sharing (LISS) [13] – which exists for any access structure [13]. LISS schemes are basically secret sharing schemes where the secret is reconstructed by taking a integral linear combination of the shares. Also, we replace Pedersen commitments [22] by the integer commitments from [15].

While this is quite straightforward, it is not so trivial to solve the problem of handling complaints without interaction. We first observe that the reason why the dealer must resolve conflicts in the protocol by Abe et al. is that only point-to-point channels between dealer and each server are assumed, and hence servers are not a priori committed to what they received. On the other hand, a typical implementation would realize the channels using public-key encryption, so we propose to include this encryption explicitly in the protocol. One might now hope that a server can prove it received bad data by "opening" the cipher-texts it received. However, while the sender of a ciphertext can always "open" it convincingly (simply by revealing the coins used to create it), we need that the *receiver* can do so. Since ciphertexts can be adversarially generated, and un-opened ciphertexts must remain secure, it is not immediately clear how this can be done in a non-interactive and efficient way. We propose an efficient solution to the problem based on Identity-Based Encryption (IBE). To our knowledge, this is a new application of IBE, and we believe the idea is of independent interest, as the possibility of "complaining convincingly" is often useful in protocol design.

For the case of honest majority, the VSS we obtain requires the dealer to send a total of $O(n \log n(\kappa + l + k + n))$ bits, where κ is the security parameter for the public-key and commitment schemes used, n is the number of players, l is the bit length of the numbers we share and k is an "information theoretic" security parameter, controlling the statistical leakage of information.

The protocol can handle any Q2 adversary structure (honest majority in the threshold case), which is optimal in terms of the number of corruptions that can be handled at all. However, for realistic values of the parameters, the efficiency is not what we might hope for. This is because the numbers we will be computing on will be numbers specifying bids, prices, productions costs, etc., that is, numbers that are typically much smaller than those used for public-key cryptography. Realistic parameter values might be $n = 7$, $l = 32$, $k = 60$ and $\kappa = 1024$. In such a case, each 32 bit number we share is expanded to about 25.000 bits, which hardly seems desirable.

We therefore propose another solution, where we make the stronger assumption that the adversary structure is Q3 (less than $n/3$ corruptions in the threshold case). We build a solution using a generalization of the pseudorandom secret-sharing technique from [10] to the case of linear *integer* secret sharing. In the threshold case, the protocol requires the dealer to send, once and for all, $O(T(\kappa + nk))$ bits to the servers, where T is the number of maximal unqualified sets in the adversary structure. After this, any number of VSS's can be done by sending $O(l + k)$ bits to the servers for each value to be shared. Each multiplication proof requires 3 VSS invocations and in addition $O((l + k + n)n)$ bits should be sent.

The initial step is not always efficient as a function of n because T may be exponential in n, depending on the adversary structure. In the typical threshold case, T would be about $\binom{n}{n/3}$. But for a small number of servers, T is moderate. On the other hand, for fixed n and for a large number of VSS invocations we come very close to sending only $l + k$ bits for every l-bit number we share - where of course sending l bits is necessary. It is therefore ideally suited for cases, where a large number of clients need to supply large amounts of data to a small number of servers. For the example parameter values above and assuming we share, say 200 numbers, the dealer needs to send about 230 bits per number to share.

Both our protocols use a common reference string, and assume that the verifiers have public/secret key pairs set up in advance. Note that if we do not assume random oracles, we cannot get non-interactive protocols without some sort of set-up assumption. Of course, using set-up assumptions, our problem could also be solved using standard techniques for non-interactive zero-knowledge. But with current state of the art, this approach can only prove the type of statements we are after using generic techniques. This would give non-interactive proofs of size $\Omega(l\kappa|C|)$ where $|C|$ is the size of a Boolean circuit C checking the relation in question. For realistic parameter values, this will be several orders of magnitude larger than our complexity. To our knowledge, our solutions are the first non-interactive protocols for integer relations that do not use random oracles, and have communication complexity independent of the circuit complexity of the relation.

2   Preliminaries

In a Linear Integer Secret Sharing (LISS) Scheme there are n players, which are denoted by $P_1, \ldots, P_n$. Let $\mathcal{P} = \{P_1, \ldots, P_n\}$ be the set of all the players, and let the power set of $\mathcal{P}$ be denoted by $P(\mathcal{P})$. Let $s \in [-2^l..2^l]$ be the secret which a dealer D wants to secret share between the players in $\mathcal{P}$ over a LISS. Then the sets in $P(\mathcal{P})$ which are allowed to reconstruct the secret s are called *qualified* and the sets which should not be able to obtain any information about the secret s are called *forbidden*.

Definition 1. *The collection of qualified sets, $\Gamma \subseteq P(\mathcal{P})$, is called a* monotone access structure, *if for all $A \in \Gamma$ and $A \subset B \subseteq \mathcal{P}$ it holds that $B \in \Gamma$.*

We also need the notion of an *adversary structure* [19].

Definition 2. *An* adversary structure *is a monotone collection of sets, $\Delta \subseteq P(\mathcal{P})$, for which the adversary may corrupt the players of one set in the adversary structure. It is monotone in the sense that for every $A \in \Delta$ it holds that for every $B \subset A$ that $B \in \Delta$.*

Definition 3. *An adversary structure Δ is Q2 (Q3) if no two (three) sets in the structure cover the full player set $\mathcal{P}$.*

If Γ is the collection of all qualified sets of players in $\mathcal{P}$ and Γ is a monotone access structure, then the corresponding adversary structure, Δ, is the collection of all the forbidden sets. Note that, Δ is monotone as required by an adversary structure, and that $\Gamma \cup \Delta = P(\mathcal{P})$ and $\Gamma \cap \Delta = \emptyset$. That is, an adversary structure can be seen as a complement of a monotone access structure. Since the structures, Γ and Δ, are monotone, they can be uniquely represented by their minimal and maximal sets denoted by Γ^- and Δ^+, respectively. $|\Delta^+|$ will denote the number of sets in Δ^+. In this paper we use Γ and Δ interchangeably. We proceed to define what is meant by a *correct* and *private* LISS.

Definition 4. *A LISS scheme is* correct, *if the secret can be reconstructed from shares of any qualified set in $A \in \Gamma$, by taking an integer linear combination of the shares with coefficient that depends only on the index set A.*

Definition 5. *A LISS scheme is* private, *if for any forbidden set $B \in \Delta$, any two secret $s, s' \in [-2^l..2^l]$, and independent random coins r and r', the statistical distance between the distributions of the shares $\{s_i(s, r, k) \mid i \in B\}$ and $\{s_i(s', r', k) \mid i \in B\}$ is negligible in the* security *parameter k.*

A *labeled matrix* consists of a $d \times e$ matrix M and a corresponding surjective function $\psi : \{1, \ldots, d\} \rightarrow \{1, \ldots, n\}$. We say that the i-th row is *labeled* by $\psi(i)$ or *owned* by player $P_{\psi(i)}$. For any subset $A \subset \mathcal{P}$, we let M_A denote the restriction of M to the rows labeled by some $P_{\psi(i)} \in A$. For any d-vector $\boldsymbol{x}$, we similarly denote $\boldsymbol{x}_A$ to be the restriction of entries i with $P_{\psi(i)} \in A$. For any two vectors $\boldsymbol{a}$ and $\boldsymbol{b}$, let $\langle \boldsymbol{a}, \boldsymbol{b} \rangle$ denote the inner product.

Definition 6. *An* Integer Span Program (ISP) *for a monotone access structure Γ consists of a tuple $\mathcal{M} = (M, \psi, \boldsymbol{\varepsilon})$, where $M \in Z^{d,e}$ is a labeled matrix with a surjective function $\psi : \{1, \ldots, d\} \rightarrow \{1, \ldots, n\}$, and the* target vector $\boldsymbol{\varepsilon} = (1, 0, \ldots, 0)^T \in Z^e$. *Furthermore, for every $A \subseteq \mathcal{P}$ the following holds,*

- *for every $A \in \Gamma$ there exists a* reconstruction vector $\boldsymbol{\lambda} \in Z^d$ *such that $M_A^T \boldsymbol{\lambda} = \boldsymbol{\varepsilon}$.*
- *for every $A \notin \Gamma$ there exists a* sweeping vector $\boldsymbol{\kappa} \in Z^e$ *such that $M_A \boldsymbol{\kappa} = \boldsymbol{0}$ and $\langle \boldsymbol{\kappa}, \boldsymbol{\varepsilon} \rangle = 1$.*

The size of $\mathcal{M}$ is defined to be d.

In [13] it was shown how to construct a correct and private LISS scheme from any ISP. For a given ISP we define $l_0 = l + \lceil \log_2(\kappa_{\max}(e - 1)) \rceil$, where $\kappa_{\max} = \max\{|a| \mid a$ is an entry in some sweeping vector $\}$. To share a secret $s \in [-2^l..2^l]$, we use a *distribution vector* $\boldsymbol{\rho}$ which is a uniformly random vector in $[-2^{l_0+k}..2^{l_0+k}]^e$ with the restriction that $\langle \boldsymbol{\rho}, \boldsymbol{\varepsilon} \rangle = s$. The *share vector* is computed by $M\boldsymbol{\rho} = \boldsymbol{s} = (s_1, \ldots, s_d)^T$, where the *share component* s_i is given to player $P_{\psi(i)}$ for $1 \leq i \leq n$. The *share* of player P_j is the subset of share components $\boldsymbol{s}_{\{P_j\}}$.

See [13] for a proof of correctness and privacy. There, it was also shown that LISSs exist for any adversary structure, and in particular they can be constructed for threshold structures where a player's share is $O((l+k+n^2) \log n)$ bits long. It follows from results and conjectures in [11] that this can probably be improved to $O((l + k + n) \log n)$ bits.

3 Verifiable Secret Sharing (VSS) and Distributed Verifier Proofs

3.1 Model and Definition

We have a set of dealers $\{D_1, \ldots, D_m\}$ and a set of n players or verifiers $\mathcal{P} = \{P_1, \ldots, P_n\}$. We assume an active and static adversary who may corrupt any number of dealers and a set of players in a given adversary structure. All players, dealers and the adversary are polynomially bounded. We assume (for simplicity) synchronous communication. We use the Universal Composability framework [6] and define ideal functionalities as follows:

Functionality F_{VSS}

– On input s from D_j, send ("D_j, input") to all players and the adversary. Wait one round (this models the fact that our implementation takes one round to finish, after the prover has spoken). Then, if $s = \bot$ (which may be the case if D_j is corrupt), send ("D_j, Fail") to all players, else send ("D_j, OK") to all players.

Functionality $F_{ab=c}$

– On input a, b, c from D_j, send ("D_j, input") to all players and the adversary. Wait one round. Then, if a, b, c are integers satisfying $ab = c$, send ("D_j, OK") to all players, else send ("D_j, Fail") to all players.

Both functionalities need to model that a successfully shared secret can be reconstructed. To simulate this we add a command to the functionalities, where it will send the requested shared value to everyone if asked by all honest players.

For our protocols, we will need a set-up assumption, namely $D_1, \ldots, D_m$ and $P_1, \ldots, P_n$ get common input $k, pk, pk_1, \ldots, pk_n$, where k is the security parameter, pk_i is the public key of P_i, and pk is a common reference string. As private input, P_i has a secret key sk_i corresponding to pk_i. For simplicity, we assume here that the public and secret keys are generated and given to players initially by an ideal functionality T. But we stress that T can be implemented by a once-and-for-all preprocessing among the players (it is well known that any UC functionality can be securely implemented if we have honest majority, or in general $Q2$). In Section 3.4, it is even sufficient that players generate their own key pairs and broadcast the public keys. We also assume a functionality F_{BC}, allowing any dealer to broadcast information to the verifiers.[1] Communication between verifiers uses standard authenticated but non-secret channels. Note that the UC framework incorporates, in addition to the adversary Adv attacking the protocol, an environment Z that chooses inputs for and receives outputs from honest players. We will only consider environments that give integers (and not $\bot$) as input to honest players. This models the assumption that honest players would only attempt to VSS valid integers.

3.2 An Integer Commitment Scheme

A *commitment scheme* for domain S is given by a family of functions com_{pk} : $S \times \mathcal{R}_{pk} \to \mathcal{C}_{pk}$, indexed by a *public key pk*. One commits by publishing $C =$

---

[1] Note, that even if we implement the broadcast via a subprotocol, this can be done such that we maintain the non-interactive nature of our proofs, namely the dealer sends a single (signed) message to all players, who then internally agree on what he said.

$\mathsf{com}_{pk}(s, r)$, where $s \in \mathcal{S}$ is the committed value and $r \in \mathcal{R}_{pk}$ is a random value. A *homomorphic* commitment scheme is a scheme where we assume that $\mathcal{S}$ is an additive group and that for any two commitments C and C' and any number λ, anyone can compute commitments S and P such that being able to open C and C' to s and s', respectively, allows to open S to the sum $s + s'$ and P to the product λs.

We use a modified version of the Pedersen commitment scheme [22], based on a multiplicative group G of order unknown to the players. This commitment scheme first appeared in [16] and later in [15]. We will need primes p, q where $p = 2p' + 1$ and $q = 2q' + 1$ and p', q' are also prime. The computations are done in Z_n^*, where $n = pq$, and the public key is $pk = (n, g, h)$ where g, h are chosen at random in Q_n, the set of squares modulo n. Then we use com_{pk} : $(s, r) \mapsto g^s h^r \bmod n$. The scheme is *homomorphic*, since given commitments $C = \mathsf{com}_{pk}(s, r)$ and $C' = \mathsf{com}_{pk}(s', r')$ then $CC' = \mathsf{com}_{pk}(s+s', r+r')$ and $C^\lambda = \mathsf{com}_{pk}(\lambda s, \lambda r)$. Note that if we choose r uniformly random from $[0..n2^k]$, then $r \bmod \mathrm{ord}(h)$ is statistically close to being uniformly random in $[0..\mathrm{ord}(h) - 1]$.

An important advantage of this scheme is that it allows commitment to *integers*. This follows since the commitment is done in a group G of unknown order. More specifically, the following proposition holds for the above commitment scheme.

Proposition 1 ([16]). $\mathsf{com}_{pk}(s, r)$ *is a statistically hiding and computationally binding commitment scheme, i.e.:*

- *If factoring is infeasible, then given $pk = (n, g, h)$ it is infeasible to compute $s, s', r, r' \in Z$ where $s \neq s'$ such that $\mathsf{com}_{pk}(s, r) = \mathsf{com}_{pk}(s', r')$.*
- *For any two values s, s', the distributions $(pk, \mathsf{com}_{pk}(s, r)), (pk, \mathsf{com}_{pk}(s', r'))$ are statistically indistinguishable.*

3.3   Public-Key Encryption with Verifiable Opening

We introduce here a tool that we will need later. Suppose a player P has a public/secret key pair (pk, sk), and receives ciphertext from various senders, some of whom may be corrupt. We want that the cryptosystem is chosen ciphertext (CCA) secure and has the additional property that for any received ciphertext c, P can reveal the decryption result $x = D_{sk}(c)$ and prove non-interactively and efficiently that x is correct. We want, of course, that "unopened" ciphertexts remain secure, which excludes the trivial solution of revealing the secret key.

Note that if c is a valid ciphertext, the random coins used to generate c can serve as proof of what the plaintext was. But even if the receiver could compute these coins efficiently, there is still a problem if the sender is corrupt. Then c may be invalid, and "the coins used to generate c" is not even a well-defined notion.

A formal definition of the notion we are after can be phrased as a variant of the standard chosen ciphertext security game, where the oracle answers decryption queries with the result as well as the proof of correctness. We do not give it here for lack of space. Instead, we give our solution in a form tailored for direct use in our protocol below. The proof that it works is then incorporated in the proof for the overall protocol[2].

The key pair (pk, sk) for P will be the master secret and public key for an identity-based cryptosystem (IBE)[3]. Note that, under reasonable assumptions, efficient IBE's exist that do not use random oracles[24]. For the IBE we use, we need that given identity t and pk, one can easily verify if a secret key sk_t is the secret key for identity t. This can indeed be done for all known efficient IBE's, we call this IBE with verifiable secret keys (IBE-VSK). We assume that the system is used in a protocol that assigns a unique tag to each ciphertext to be sent to P. To encrypt message m, the sender treats the tag t for this ciphertext as an identity and encrypts the message to this id, i.e., he sends $c = E_t(m)$. The receiver decrypts by computing the secret key sk_t and then $m = D_{sk_t}(c)$. To reveal the result of decrypting c, P reveals sk_t. Everyone can now compute $D_{sk_t}(c)$. One must also verify that sk_t is indeed the secret key corresponding to t. From the assumption that tags are not reused and standard properties of IBE, it follows that unopened ciphertexts remain secure. A somewhat similar idea was used for a different purpose in [7].

3.4   VSS Using Integer Commitments

In this section we construct a non-interactive *verifiable secret sharing* [9] (VSS) scheme based on LISS. We use the model described in the previous sections. Specifically, the common reference string will be a public key $pk = (g, h, n)$ for the integer commitment scheme described above. Moreover, each player P_j has a key pair (pk_j, sk_j) for an IBE-VSK as described above.

Protocol $\mathsf{VSS}_{pk}(s)$

On input $s \in [-2^l..2^l]$, the dealer D makes a commitment $C = \mathsf{com}_{pk}(s, r)$ to s, and then executes the following protocol to prove that he knows how to open C to value s, and to secret share s:

Protocol $\mathsf{Proof}_{g,h}(C)$

1. Given an ISP $\mathcal{M} = (M, \psi, \varepsilon)$, the dealer D chooses a random vector $\rho \in [-2^{l_0+k}..2^{l_0+k}]^e$ with $\langle \rho, \varepsilon \rangle = s$, and commits to this *sharing vector* $\rho = (\rho_1, \ldots, \rho_e)^T$ by commitments $R_1, \ldots, R_e$ to $\rho_1, \ldots, \rho_e$, respectively, where $R_1 = C$ and all commitments use (g, h) as public parameter. The commitments $R_2, .., R_e$ to the additional randomness are included in the proof π. D computes the shares of s: $s = (s_1, \ldots, s_d)^T = M\rho$, and computes the opening information o_i for the corresponding commitment

$$C_i = \prod_{j=1}^{e} R_j^{m_{ij}}$$

---

[2] The problem could also be solved using non-interactive zero-knowledge, but this will be much too inefficient for our purposes. Using OAEP might work as well, but only assuming random oracles which we want to avoid.

using the homomorphic property, where m_{ij} is defined by $M = [m_{ij}]$. Finally, he includes $c_i = E_{pk_{\psi(i)}}(o_i)$ in his proof π, where all these ciphertexts are assigned a tag consisting of C concatenated with the name of D (see Section 3.3). Finally, D broadcasts C, π.

2. For each i, $P_{\psi(i)}$ decrypts c_i. If he finds that the resulting opening information o_i is incorrect w.r.t. C_i, then he sends o_i to all other players, along with a proof that o_i is indeed the result of decrypting c_i, this counts as an accusation against D. Otherwise he sends "accept".

3. For any accusation from $P_{\psi(i)}$, each player verifies that any o_i received is indeed the value that c_i decrypts to. If this is not the case this o_i is discarded.

4. Each player looks at all (non-discarded) o_i-values he knows. If any such o_i is inconsistent with C_i, then he rejects. Otherwise he accepts.

A successfully shared value s can be reconstructed by simply having every player P_i open every commitment C_j where $\psi(j) = i$. For some qualified set of successfully opened shares the players can then use the corresponding reconstruction vector $\boldsymbol{\lambda}$ to reconstruct the secret. We have

Theorem 1. *Given a secure IBE-VSK, the protocol* $\mathsf{VSS}_{pk}(s)$ *securely implements* F_{VSS}, *assuming any Q2 adversary structure* Γ.

Proof. To show that $\mathsf{VSS}_{pk}(s)$ securely implements F_{VSS}, we are given an adversary Adv and an environment Z, and we need to construct a simulator S. The simulator interacts with Adv to simulate its view of attacking the protocol, and on the other hand interacts with F_{VSS} on behalf of corrupt players. This game is called the *ideal process*. This is compared to the *real process*, where Z, Adv are interacting with a real instance of the protocol. In both processes, Z and Adv may communicate at any time. The goal is now to show that Z cannot distinguish the real from the ideal process. Our simulator works as follows:

1. The simulator generates the keys $pk, \{(pk_j, sk_j)\}$ following T's algorithm, and sends all public keys to Adv, along with secret keys for corrupted players.

2. The simulator S now acts whenever required, as follows:
   - If Adv sends C and a proof π to the broadcast functionality on behalf of corrupt dealer D_j, the simulator does the following: using its secret keys, it can decrypt ciphertext in π intended for honest players and follow their algorithm to compute what they would send in the second round. This also lets it decide if the proof would be accepted. If not, the simulator sends $\perp$ to F_{VSS}. If the proof is acceptable, observe first that since Γ is Q2, the set of honest players, A, is qualified, and that every honest player can open his commitment to s_i. Let $\boldsymbol{\lambda}$ be a reconstruction vector for A, that is, $\langle \boldsymbol{s}, \boldsymbol{\lambda} \rangle = s$ and $\boldsymbol{\lambda}_{A^C} = \mathbf{0}$, i.e., if $\boldsymbol{\lambda} = (\lambda_1, \ldots, \lambda_d)^T$ then

$$\sum_{i=1}^{d} s_i \lambda_i = \sum_{i=1}^{d} \lambda_i \sum_{j=1}^{e} m_{ij} \rho_j = \rho_1 = s,$$

where $\lambda_j = 0$ for $\psi(j) \notin A$. Hence, the above equation implies that $\sum_{i=1}^{d} \lambda_i m_{ij} = \delta_{1j}$, where $\delta_{ij} = 1$ if $i = j$ and 0 otherwise. Therefore, by the homomorphic property, the simulator can open commitment $C' = \prod_{i=1}^{d} C_i^{\lambda_i}$ to $s' = \sum_{i=1}^{d} \lambda_i s_i$. Now, since

$$C' = \prod_{i=1}^{d} C_i^{\lambda_i} = \prod_{i=1}^{d} \left(\prod_{j=1}^{e} R_j^{m_{ij}} \right)^{\lambda_i} = \prod_{j=1}^{e} R_j^{\sum_i \lambda_i m_{ij}} = R_1 = C,$$

we see that the simulator can extract from the proof a way to open commitment C to a value s. The simulator sends s to F_{VSS}.

- On input ("D_j, input") from F_{VSS}, where D_j is honest, the simulator simulates what D_j would send in the protocol, as follows: First, create a commitment C to an arbitrary value. By the statistical hiding property, there exists a way to open C to the correct value s used by D_j, except with negligible probability – although s is unknown to S. We therefore proceed, assuming implicitly that C "contains" s. Now, let A be the set of corrupted players. Then there exists a sweeping vector κ such that $M_A \kappa = \mathbf{0}$ and $\langle \kappa, \varepsilon \rangle = 1$. Let $\rho_0 = (r_1, \ldots, r_e)^T$ be a random distribution vector such that $\langle \rho_0, \varepsilon \rangle = 0$, i.e., a distribution vector to a random sharing of 0. Construct $R'_1, \ldots, R'_e$ as random commitments of $r_1, \ldots, r_e$, respectively, with the exception that $R'_1 = 1$ (or the commitment of $r_1 = 0$ using randomness 0). Then, by the homomorphic property of the commitment scheme, compute commitments

$$C'_i = \prod_{j=1}^{e} R_j'^{m_{ij}},$$

to shares s_i which determines the secret 0. Now, given the commitment C for the secret s, we modify the commitments so they become consistent with s: Compute the public commitments $R_i = R'_i C^{\kappa_i}$ where $\kappa = (\kappa_1, \ldots, \kappa_e)^T$ is the sweeping vector for A. Note that $R_1 = R'_1 C^{\kappa_1} = 1C^1 = C$ as required, since $\langle \kappa, \varepsilon \rangle = 1$ (i.e., $\kappa_1 = 1$). The commitments to the shares in s will be as follows:

$$C_i = \prod_{j=1}^{e} R_j^{m_{ij}} = \prod_{j=1}^{e} (R'_j C^{\kappa_j})^{m_{ij}} = \prod_{j=1}^{e} R_j'^{m_{ij}} C^{\kappa_j m_{ij}}.$$

For the players in A we have that,

$$\prod_{j=1}^{e} C^{\kappa_j m_{ij}} = C^{\sum_j \kappa_j m_{ij}} = C^0 = 1,$$

since the inner product of κ and a row in M which is owned by a player in A is 0. So for a corrupt $P_{\psi(i)}$ we have $C'_i = C_i$, and we know how to open these commitments. The simulated proof therefore consists of the commitments $R_1, \ldots, R_e$, encryptions of correct opening information for C_i when $P_{\psi(i)}$ is corrupt, and encryptions of random values for honest players.

To see that this simulation works, note the following: First, the simulation of the initial set-up stage and of the case where a corrupt dealer gives a proof is perfect. In particular, when a corrupt dealer does a VSS that would be accepted in the real protocol, the simulator can *always* extract the correct secret, and honest players will therefore output accept also in the ideal process.

In the case where an honest dealer does a VSS, this will in the ideal process simply mean that it sends integer s to F_{VSS}. The functionality will send accept to everyone, so all honest players output accept. This is also the case in the real protocol: correct opening information for each C_i is uniquely determined from the ciphertext c_i, hence no honest player will accuse D and every other accusation will be rejected by the honest players.

Hence the only possible difference between the ideal and real process is in the simulated commitment C and proof π that is shown to Adv. By the statistical hiding property of the commitment scheme and privacy of the LISS scheme, it follows that the opening information sent to corrupt players, as well as the commitments $R_1, \ldots, R_e$ have distribution statistically close the one seen in the real protocol. So the only difference is the fact that the ciphertexts intended for honest players are random in the simulation, and contain valid openings of commitments in the real protocol.

We cannot argue that the two sets of encryptions are indistinguishable based directly on the ideal process because S knows all secret keys. Instead, we construct a machine S' that acts as an adversary breaking the underlying IBE-VSK. S' will run the algorithms of Z, Adv and S, with the following modifications to S: S' receives public keys for the honest players from an oracle. Whenever S needs to decrypt a ciphertext sent to an honest player with tag t (see Section 3.3), S' will ask the oracle for the secret key for that tag, and can then decrypt. When S wants to create ciphertext for honest players in a simulated proof, S' will ask the oracle to encrypt either 1) random data or 2) genuine opening information for the relevant commitments. The latter is possible because S' also runs Z and therefore knows each secret that is shared, this allows it to create the commitment C as a genuine commitment containing the right value, and from this it can compute how to open all the other commitments in that VSS. In the case 1), we produce exactly what we get in the ideal process, in case 2) we produce something statistically close to what we get in the real process. Hence, if Z could distinguish the two processes, S' can use the output from Z to break the underlying IBE-VSK.   □

For lack of space, we do not prove formally here that the protocol for reconstruction of the committed secret works. It is quite straightforward based on the binding property of the commitment scheme.

3.5   Verifiable Commitment Multiplication Proof

We now show a (distributed verifier) proof that VSS'ed integers s, s', s'' satisfy that $s'' = ss'$:

Protocol $\mathsf{MultProof}_{pk}(s, s', s'')$

1. The prover makes commitments C, C', C'' to s, s', s'' and then executes $\mathsf{Proof}_{g,h}(C)$, $\mathsf{Proof}_{g,h}(C')$, and $\mathsf{Proof}_{g,h}(C'')$.
2. The prover executes $\mathsf{Proof}_{C',h}(C'')$ using the *same* distribution vector ρ_s as in step 1 (but with new independent randomness for the commitments).
3. Every player verifies whether his shares obtained from $\mathsf{Proof}_{g,h}(C)$ (from step 1.) and $\mathsf{Proof}_{C',h}(C'')$ (from step 2.) coincide. If this does not hold, he accuses the dealer by opening the ciphertexts he received in $\mathsf{Proof}_{g,h}(C)$ and $\mathsf{Proof}_{C',h}(C'')$. Each player verifies any accusation made.
4. The proof is accepted if all subproofs were accepted, and no valid accusations were made.

Note that the four executions of the Proof protocol can be run in parallel. A similar protocol appeared in [1], but we have here added $\mathsf{Proof}_{g,h}(C')$.[3]

Theorem 2. *Assuming the integer commitment scheme is binding and given a secure IBE-VSK, $\mathsf{MultProof}_{pk}(s, s', s'')$ securely implements $F_{ab=c}$ assuming any Q2 adversary structure Γ.*

Proof. Note that making commitments C, C', C'' and then executing the first 3 instances of Proof is equivalent to executing 3 instances of VSS_{pk}. Therefore, to simulate this, we run the simulator from the previous theorem 3 times (in parallel). To simulate the execution of $\mathsf{Proof}_{C',h}(C'')$, we run the same simulator again, with the following changes: when simulating the actions of an honest dealer, the simulator will not create its own commitment to play the role of the commitment to the secret, instead it will use C''. Also, it will use the same distribution vector that was used in the simulation of $\mathsf{Proof}_{g,h}(C)$.

To show that this simulation works, we only need to check that when we extract opening information from an acceptable proof given by a corrupt prover, we will get values s, s', s'' such that $ss' = s''$. Note, that if the proof is accepted, it follows from the proof of Theorem 1 that we can extract from step 1. pairs $(s, r), (s', r')$ and (s'', r'') such that $C = \mathsf{com}_{g,h}(s, r)$, $C' = \mathsf{com}_{g,h}(s', r')$ and $C'' = \mathsf{com}_{g,h}(s'', r'')$. Furthermore, steps 2. and 3. ensure that we can extract (s, r^*) such that $C'' = \mathsf{com}_{C',h}(s, r^*) = C'^s h^{r^*}$ [4]. Combining this with the expression for $C' = \mathsf{com}_{g,h}(s', r') = g^{s'} h^{r'}$ we get $C'' = C'^s h^{r^*} = (g^{s'} h^{r'})^s h^{r^*} = g^{ss'} h^{r's+r^*}$ In other words, we can now open C'' to both s'' and ss', which contradicts the binding property unless $s'' = ss'$.   □

---

[3] This is necessary since the order of the group of the commitments is unknown and we can therefore not prove soundness the same way as in [1] (Lemma 1).

[4] Note that the proof in step 2. uses C', which might have been adversarially generated, in place of g which comes from the common reference string. However, this is not a problem since the extraction will work for any set of values.

4   Verifiable Multiplication Proof Based on Pseudo-Random Sharing

4.1   Replicated Integer Secret-Sharing and Share Conversion

In this section we first introduce RISS, an integer version of Replicated Secret-Sharing [20], where we share an integer over a monotone access structure. Then we define share conversion, and show that shares generated by a RISS scheme can be locally converted to shares in the same secret generated by LISS schemes.

Scheme Replicated Integer Secret-Sharing (RISS)

Let Δ be an adversary structure. For each set $B \in \Delta^+$ choose a uniformly random r_B integer from the interval $[-2^{l+k}..2^{l+k}]$ and send privately r_B to each player $P_i \notin B$. Furthermore, publish $r = s + \sum_{B \in \Delta^+} r_B$, where s is the secret from the interval $[-2^l..2^l]$.

Lemma 1. *The RISS scheme is correct and (statistically) private.*

Definition 7. *Let S and S' be two secret-sharing schemes. We say that S is locally convertible to S' if there exist local conversion functions $g_1, \ldots, g_n$ such that the following holds. If $(s_1, \ldots, s_n)$ are valid shares of a secret s in S, then $(g_1(s_1), \ldots, g_n(s_n))$ are valid shares of the same secret s in S'. We denote by g the concatenation of all g_i, namely $g(s_1, \ldots, s_n) = (g_1(s_1), \ldots, g_n(s_n))$, and refer to g as a share conversion function.*

Note by the locality feature of the conversion, that converted shares cannot reveal more information about s than the original shares.

The following theorem is proved in the full version of the paper [14], using ideas similar to what was used in [10]

Theorem 3. *The RISS scheme $\mathcal{R}_\Gamma$, realizing Γ, is locally convertible to any LISS realizing an access structure $\Gamma' \subseteq \Gamma$.*

Clearly, for any prime p, a RISS sharing of integer s can be thought of as a replicated sharing over Z_p of s mod p, by reducing all shares modulo p. Furthermore, in [10] it was shown how to locally convert a replicated sharing over Z_p to any linear secret sharing (LSS) scheme over Z_p (such as Shamir's scheme). From these two observations, we immediately get

Proposition 2. *The RISS scheme $\mathcal{R}_\Gamma$, realizing Γ, is locally convertible to any LSS over Z_p realizing an access structure $\Gamma' \subseteq \Gamma$, where the original secret s after conversion will be s mod p.*

4.2   Application to VSS

We now show how the results from the previous subsection can be used to generate a series of verifiably shared secrets by broadcasting only two values per secret, at the initial cost of distributing a set of random seeds to the players. We use the model defined earlier, where each player P_i has a public and a secret key.

In this case, we assume that there is a public key pk_B defined for each $B \in \Delta^+$, and P_i's public key consists of all pk_B for those B in which P_i is *not* a member. The secret key consists of all secret keys corresponding to relevant pk_B's. As before, we assume these are keys for an IBE-VSK.

The following protocol does the intial distribution of seeds.

Protocol Random$_{\{r_B\}}(\Delta^+)$

1. For each $B \in \Delta^+$ the dealer D choose an uniformly random r_B from $[0..2^k[$.
2. For each $B \in \Delta^+$ D broadcasts r_B encrypted under pk_B. The dealer's name is used as tag for this ciphertext. Each player decrypts all the ciphertexts for which he has the secret key.

The protocol clearly ensures that players have mutually consistent shares, i.e., all honest players not in B agree on the value of r_B, for any $B \in \Delta^+$.

Given a pseudorandom function (PRF) $\varphi.(\cdot)$ with k-bit keys and inputs, and outputs in $[-2^{l+k}..2^{l+k}]$, the following protocol is realizable.

Protocol VSS$_{\{r_B\}}(s)$

It is assumed that the dealer D has run Random$_{\{r_B\}}(\Delta^+)$ on some adversary structure, Δ.

1. D broadcasts a value a, to serve as a "label" for this instance of the protocol. The only demand is that a can be used as input to φ, and that D never reuses an a-value. D computes, with his knowledge of $\{r_B\}$, $r = s + \sum_B \varphi_{r_B}(a)$ and broadcasts r.
2. Each player P_i checks that $r \in [-(|\Delta^+| + 1)2^{l+k}..(|\Delta^+| + 1)2^{l+k}]$, and rejects if this is not the case. Otherwise, he computes $\varphi_{r_B}(a)$, for every B where $P_i \notin B$.

This lemma follows easily by inspection of the protocol:

Lemma 2. *If D is honest, no honest player will reject in* VSS$_{\{r_B\}}(s)$. *No matter what the dealer does, if honest players accept, the set of values $r, \{\varphi_{r_B}(a)| \ B \in \Delta^+\}$ form a RISS sharing of some value s'. If D is honest, $s' = s$, otherwise $s' \in [-(2|\Delta^+| + 2)2^{l+k}..(2|\Delta^+| + 2)2^{l+k}]$.*

It is also quite straightforward to see that if D is honest, and the PRF is secure, a polynomially bounded adversary does not learn anything about the secret involved. A proof of this is implicit in the proof of Theorem 4 below. We discuss in the full version of this paper [14] how a secret can be reconstructed, once it has been VSS'ed as above.

4.3 Multiplication Proof

In this section we describe a protocol which non-interactively proves that a shared value is the product of two other shared values. For simplicity, we will only consider the case of a threshold adversary who corrupts $t < n/3$ of the players, so the adversary structure Δ will in this section consist of all set of cardinality at most t. The full version of this paper [14] will describe a generalization to all Q3 adversary structures.

We will need a tool from [10], called Pseudorandom Zero Sharing (PRZS). This protocol assumes that for all $B \in \Delta^+$, players not in B have been given t random seeds $r_B^1, \ldots, r_B^t$ and a prime $p > n$ is agreed in advance. Based on this, the protocol generates (by local computation only) a pseudorandom polynomial f over Z_p of degree at most $2t$ such that $f(0) = 0$ and each player P_i knows $f(i)$. The protocol is a simple generalization of the share conversion technique.

In the following $\mathsf{Random}_{\{r_B, r_B^1, \ldots, r_B^t\}}(\Delta^+)$ will denote the protocol where the dealer distributes the seeds $r_B, r_B^1, \ldots, r_B^t$ to all players not in B using encryption under pk_B. We will choose a fixed prime p, such $p > 2(4|\Delta^+| + 2)^2 2^{2(l+k)}$.

Protocol $\mathsf{MultProof}_{\{r_B, r_B^1, \ldots, r_B^t\}}(a, b, c)$

1. The dealer D executes $\mathsf{Random}_{\{r_B, r_B^1, \ldots, r_B^t\}}(\Delta^+)$.
2. D executes $\mathsf{VSS}_{\{r_B\}}(a)$, $\mathsf{VSS}_{\{r_B\}}(b)$ and $\mathsf{VSS}_{\{r_B\}}(c)$.
3. The players use Proposition 2 to locally convert the RISS sharings we now have of a, b, c to Shamir sharings of $a \bmod p, b \bmod p$ and $c \bmod p$, consistent with polynomials f_a, f_b and f_c of degree at most t, t and $2t$ respectively. The players use PRZS to generate shares in a polynomial f of degree at most $2t$ with $f(0) = 0$.
4. D uses his knowledge of all seeds to compute the polynomial $h = f + f_a f_b - f_c$ and broadcasts h.
5. Each player P_i verifies that $h(i) = f(i) + f_a(i) f_b(i) - f_c(i)$. If the verification fails then P_i broadcast "Accusation" and opens all encrypted values $r_B, r_B^1, \ldots, r_B^t$ known by him.
6. The proof is rejected if one of the following situations happen: one of the VSS protocols in Step 2 was rejected, the broadcasted polynomial h is not of degree at most $2t$, $h(0) \neq 0$, or broadcasted values by a player are consistent with the encrypted values but inconsistent with the broadcasted values by D.

Theorem 4. *When based on a secure IBE-VSK and PRF, then the protocol* $\mathsf{MultProof}_{\{r_B, r_B^1, \ldots, r_B^t\}}(a, b, c)$ *securely implements* $F_{ab=c}$, *for any threshold-t adversary structure where* $t < n/3$.

Proof. We construct a simulator S that works as follows:

1. S generates the keys $pk, \{(pk_B, sk_B)\}$ following T's algorithm, and sends all public keys to Adv, along with secret keys for corrupted players.
2. S now acts whenever required, as follows:
   - When Adv does a proof on behalf of a corrupt dealer, S can simply decrypt everything sent by the adversary, and decide if the proof would be accepted in the real process. If so, it reconstructs values a, b and c and sends them to the ideal functionality. Otherwise, it sends $\perp$ to the ideal functionality and uses the honest players' algorithm to compute the messages (complaints) they would send to corrupt players, and sends these to Adv.

– When an honest dealer does a proof, S will generate a simulated proof by simply following the prover's algorithm, using $a = b = c = 0$.

To see that this simulation works as required, note first that the simulation of the set-up phase and proofs by corrupt dealers is perfect. This is because the simulator follows the honest players algorithm to compute their reaction to the proof, so we just need to check that when the proof is accepted, the simulator can send a correct witness to the functionality. By Lemma 2, the values a, b, c that the simulator reconstructs from the proof will be in the interval $[-(2|\Delta^+| + 2)2^{l+k} \ldots (2|\Delta^+| + 2)2^{l+k}]$, so we know that $|ab|, |c|$ are less than $p/2$. Now, from Step 5, we know that h agrees with $f + f_a f_b - f_c$ in all points owned by honest players, of which there are at least $2t + 1$. This implies that $h = f + f_a f_b - f_c$, and therefore that $ab = c \bmod p$. But if $ab \neq c$, it would have to be the case that $|ab - c| \geq p$, while on the other hand we already know that $|ab - c| \leq |ab| + |c| < p$. So indeed $ab = c$.

It remains to show that the simulation of an honest dealer's proof shown to the adversary is indistinguishable from a real proof. For this, consider the real process *Real*, and assume the worst case where the adversary has corrupted a maximal set B of players. This means that when an honest dealer does a proof, the key sk_B is the only secret key the adversary does not know. We then define a new "hybrid" process Hyb_1, where we replace the broadcasted encryptions of $r_B, r_B^1, \ldots, r_B^t$ (under pk_B) by encryptions of independent random values. By an argument similar to the proof of Theorem 1, *Real* is indistinguishable from Hyb_1 if the underlying IBE-VSK is secure. Note that in Hyb_1, we can replace evaluations of the PRF using seeds $r_B, r_B^1, \ldots, r_B^t$ by oracle access to the PRF with the same seeds, and all messages sent will remain unchanged. We define Hyb_2 by replacing the PRF oracles by oracles for truly random functions. By security of the PRF, Hyb_2 is indistinguishable from Hyb_1. Finally, we define Hyb_3 as follows: we first replace the dealer's inputs (a, b, c) to the $\mathsf{VSS}_{\{r_B\}}(\cdot)$-protocols by random values in the legal interval, and second, we choose the polynomial h to broadcast as a uniformly random polynomial, subject to $h(0) = 0$, $deg(h) \leq 2t$, and that $h(i)$ agrees with the adversary's information for all corrupt players P_i. Now, Hyb_3 is statistically indistinguishable from Hyb_2: consider, for instance, the execution of $\mathsf{VSS}_{\{r_B\}}(a)$ in Hyb_2. If we subtract the randomness that the adversary already knows, we see that he can compute $R + a$, where R is a truly random value in $I_r = [-2^{l+k}..2^{l+k}]$. This is statistically indistinguishable from $R + r$ where r is a random value in $I_s = [-2^l..2^l]$, which is what the adversary would see in Hyb_3. The polynomial h is easily seen to have exactly the same distribution in Hyb_2 and Hyb_3. It follows that *Real* is indistinguishable from Hyb_3.

To finish the proof, note that in the argument we just gave, we did not use anything special about the inputs a, b, c, other than $ab = c$. Therefore, essentially the same argument shows that the ideal process is also indistinguishable from Hyb_3 since the simulator uses $a = b = c = 0$ and otherwise follows the protocol. The theorem now follows from transitivity of indistinguishability.   □

5   Interval Proofs and Application to Secure Computing

Boudot [5] observes that to prove that a number x lies in an interval $[a, b]$ it is sufficient to prove that $x - a \geq 0$ and $b - x \geq 0$. By using a homomorphic commitments scheme and a primitive to prove that a committed integer is a square, he constructs an efficient proof that a committed number is non-negative. Only a small constant number of calls to the primitive is required.

Boudot's protocols can be run in our settings by using one of the VSS protocols we have presented to play the role of commitments in Boudot's protocols. Note that both types of VSS's we construct are linear and so we have the homomorphic properties needed. In this way, we get a non-interactive proof that a shared number is in a given interval, using a constant number of invocations of our VSS protocol.

Furthermore, each number x we prove something about is verifiably shared among the players, using a LISS scheme (a RISS scheme in case of the second protocol). If we consider the shares as numbers mod q for any prime q, we obtain a linear sharing over Z_q of $x \bmod q$. We can now, possibly after local conversion using [10], do secure computing on such numbers using, e.g., the protocols from [17,4,12]. If what we really want is secure addition and multiplication over the integers, we can use the initial interval proofs to make sure the numbers are small enough to avoid modular reductions.

Acknowledgements

We thank Matthias Fitzi, Jørgen Brandt, Mikkel Krøigård, Martin Geisler, and the anonymous referees for insightful comments.

References

1. Masayuki Abe, Ronald Cramer and Serge Fehr. *Non-interactive Distributed-Verifier Proofs and Proving Relations among Commitments.* ASIACRYPT 2002, LNCS 2501. Springer 2002.
2. M. Ben-Or, S. Goldwasser, A. Wigderson: *Completeness theorems for Non-Cryptographic Fault-Tolerant Distributed Computation*, Proc. ACM STOC '88, pp. 1–10.
3. Dan Boneh, Matthew K. Franklin: *Identity-Based Encryption from the Weil Pairing.* SIAM J. Comput. 32(3): 586-615 (2003).
4. Peter Bogetoft, Ivan Damgård, Thomas Jakobsen, Kurt Nielsen, Jakob Pagter and Tomas Toft: *A Practical Implementation of Secure Auctions based on Multiparty Integer Computation.* Proc. of Financial Cryptography 2006, Springer Verlag LNCS.
5. F. Boudot. *Efficient Proofs that a Committed Number Lies in an Interval.* EUROCRYPT'00, LNCS 1807, pp 431-444, 2000.
6. Ran Canetti: *Universally Composable Security: A New Paradigm for Cryptographic Protocols.* Proc. of FOCS 2001: pp.136-145. See also updated version on the Eprint archive, www.iacr.org.

7. Ran Canetti, Shai Halevi, and Jonathan Katz. *Chosen-Ciphertext Security from Identity-Based Encryption*. Advances in Cryptology - EUROCRYPT 2004, LNCS 3027, pp 207-222, 2004.
8. D. Chaum, C. Crépeau, I. Damgård: *Multi-Party Unconditionally Secure Protocols*, Proc. of ACM STOC '88, pp. 11–19.
9. Benny Chor, Shafi Goldwasser, Silvio Micali, and Baruch Awerbuch. *Verifiable Secret Sharing and Achieving Simultaneity in the Presence of Faults (extended abstract)*. In 26th Annual Symposium on Foundations of Computer Science. IEEE, 1985.
10. Ronald Cramer, Ivan Damgård, Yuval Ishai: *Share Conversion, Pseudorandom Secret-Sharing and Applications to Secure Computation*. Proc. of TCC 2005, pp. 342-362, Springer Verlag LNCS
11. Ronald Cramer, Serge Fehr and Martijn Stam: *Black-Box Secret Sharing from Primitve Sets in Algebraic Number Fields*, Proc. of Crypto 05, Springer Verlag LNCS.
12. Ivan Damgrd, Matthias Fitzi, Eike Kiltz, Jesper Buus Nielsen, Tomas Toft: *Unconditionally Secure Constant-Rounds Multi-party Computation for Equality, Comparison, Bits and Exponentiation*. Proc. of TCC 2006, pp. 285-304, Springer Verlag LNCS.
13. Ivan Damgård and Rune Thorbek. *Linear Integer Secret-Sharing and Distributed Exponentiation*. PKC'06, LNCS 3958, pp 75-90 (2006).
14. Ivan Damgård and Rune Thorbek. *Non-Interactive Proofs for Integer Multiplication* (full version), the Eprint archive, www.iacr.org (eprint.iacr.org/2007/086).
15. Eiichiro Fujisaki and Tatsuaki Okamoto. *A Practical and Provably Secure Scheme for Publicly Verifiable Secret Sharing and Its Applications*. EUROCRYPT'98, LNCS 1403,pp 32-46, 1998.
16. Eiichiro Fujisaki and Tatsuaki Okamoto. *Statistical Zero Knowledge Protocols to Prove Modular Polynomial Relations*. CRYPTO'97, LNCS 1294,pp 16-30, 1997.
17. R. Gennaro, M. Rabin, T. Rabin, *Simplified VSS and Fast-Track Multiparty Computations with Applications to Threshold Cryptography*, Proc of ACM PODC'98.
18. O. Goldreich, S. Micali and A. Wigderson: *How to Play Any Mental Game or a Completeness Theorem for Protocols with Honest Majority*, Proc. of ACM STOC '87, pp. 218–229.
19. Martin Hirt and Ueli Maurer *Player Simulation and General Adversary Structures in Perfect Multiparty Computation*. Journal of Cryptology: the journal of the International Association for Cryptologic Research, volume 13, pages 31-60 (2000).
20. M. Ito, A. Saito, and T. Nishizeki. *Secret sharing schemes realizing general access structures*. Proc. IEEE Global Telecommunication Conf., Globecom 87: 99-102 (1987).
21. M. Karchmer and A. Wigderson. *On Span Programs*. In Proc. of 8th IEEE Structure in Complexity Theory, pages 102-111, 1993.
22. Torben P. Pedersen. *Non-interactive and Information-theoretic Secure Verifiable Secret Sharing*. In Advances in Cryptology - CRYPTO '91, volume 576 of Lecture Notes in Computer Science. Springer, 1991.
23. Adi Shamir. *How to share a secret*. Communication of the Association for Computing Machinery, 22(11), 1979.
24. Brent Waters: *Efficient Identity-Based Encryption Without Random Oracles*. Proc. of Eurocrypt 2005: pp.114-127, Springer Verlag LNCS.

Ate Pairing on Hyperelliptic Curves

R. Granger[1,*], F. Hess[2], R. Oyono[3], N. Thériault[4,**], and F. Vercauteren[5,***]

[1] Dept. Computer Science, University of Bristol
MVB, Woodland Road, Bristol, BS8 1UB, United Kingdom
granger@cs.bris.ac.uk
[2] Technische Universität Berlin,
Fakultät II, Institut für Mathematik Sekr. MA 8-1,
Strasse des 17. Juni 136, D-10623 Berlin, Germany
hess@math.tu-berlin.de
[3] University of Waterloo,
Department of Combinatorics and Optimization,
Waterloo, Ontario, N2L 3G1, Canada
royono@uwaterloo.ca
[4] Instituto de Matemática y Física,
Universidad de Talca, Casilla 747, Talca, Chile
ntheriau@inst-mat.utalca.cl
[5] Department of Electrical Engineering, Katholieke Universiteit Leuven
Kasteelpark Arenberg 10, B-3001 Leuven-Heverlee, Belgium
frederik.vercauteren@esat.kuleuven.be

Abstract. In this paper we show that the Ate pairing, originally defined for elliptic curves, generalises to hyperelliptic curves and in fact to arbitrary algebraic curves. It has the following surprising properties: The loop length in Miller's algorithm can be up to g times shorter than for the Tate pairing, with g the genus of the curve, and the pairing is automatically reduced, i.e. no final exponentiation is needed.

Keywords: Tate pairing, Ate pairing, hyperelliptic curves, finite fields.

1 Introduction

Pairings in cryptography have received a fast growing interest in the past six years and are currently a major topic in cryptologic research. Investigations are carried out regarding the use of pairings in cryptographic protocols on one side and regarding mathematical, algorithmic foundations of pairings on the other side.

The present paper conducts investigations of the latter type. Building on and generalising ideas from [5,7,4,10,17] into a common framework, the main result

---

* Funded by the EPSRC.
** This work was done in part while the author was at the Fields Institute, Toronto, Canada.
*** Postdoctoral Fellow of the Research Foundation - Flanders (FWO).

M. Naor (Ed.): EUROCRYPT 2007, LNCS 4515, pp. 430–447, 2007.

of the paper consists in providing new classes of efficient non-degenerate pairings on higher genus algebraic curves, called Ate pairings and superspecial Ate pairings, which feature some surprising properties. These pairings are different from the well known Weil and Tate pairings in that they are defined by much simpler algebraic expressions. Of course, for prime order groups any pairing can be obtained as a suitable power of any fixed non-degenerate pairing, and we also exhibit these powers for our pairings in relation to the Tate pairing.

The surprising properties of the Ate and superspecial Ate pairings are the following: Firstly, the loop length in Miller's algorithm for evaluating the pairing function is up to g times shorter than for the corresponding Tate pairing, where g is the genus of the underlying curve C. Secondly, the pairing is automatically reduced, that is, the final exponentiation required by the Tate pairing can be omitted.

There are constructive and destructive aspects regarding the relevance of our pairings to cryptography. A discussion of constructive aspects of the Tate pairing in higher genus has been carried out in [8]. The main point here is that pairings in higher genus can make use of degenerate divisors $D_2 = (Q)$, leading to more efficient evaluation and possibly some bandwidth savings due to compression. While this gives an improvement of a factor of up to g in comparison with general D_2 of degree g, the efficiency comparison with the Ate pairing in genus one is less favourable as indicated in Appendix A.

The destructive aspects of our pairings concern pairing inversion and the difficulty of the computational Diffie-Hellman problem in finite fields. In [24] it was shown that the computational Diffie-Hellman problem in the two domains of the pairing as well as in the codomain can be efficiently reduced to the problem of computing preimages of pairing values for each argument, given a fixed opposite argument. The absence of the final powering in our pairings and the fact that the degree of the pairing function is independent of the prime group order, and can hence be very small, raises questions about the hardness of pairing inversion. What can be stated at the moment is that Ate and thus Tate pairing inversion for small q, solving for degenerate divisors $D_2 = (Q)$ in the second argument, is actually efficient and straightforward (roughly as hard as computing the roots of a polynomial of degree qg over an extension of degree about gk of $\mathbb{F}_q$ where k is the embedding degree). In protocols it is hence prudent to restrict to public degenerate divisors. As of now, the precise security implications of our pairings are unknown and much more research is needed for an assessment.

Although we state most results for hyperelliptic curves only, the theory and proofs do actually not require the hyperellipticity and readily apply to general non-singular curves with a distinguished point P_∞, once the definition of "reduced divisor" has been adopted accordingly (see for example [15]). We leave these details to the interested reader.

The remainder of this paper is organised as follows: Section 2 recalls basic properties of hyperelliptic curves and the Tate-Lichtenbaum pairing. Section 3 defines the Ate pairing on all curves and proves that it is well-defined. This is then adapted in Section 4 to superspecial curves. Finally, Section 5 concludes the paper and Appendix A provides detailed performance estimates.

2   Mathematical Background

In this section, we briefly recall arithmetic on hyperelliptic curves, the definition of the Tate-Lichtenbaum pairing and Miller's algorithm to compute it.

2.1   Hyperelliptic Curves

Let C be a nonsingular hyperelliptic curve of genus g defined over a finite field $\mathbb{F}_q$ with $q = p^n$ elements. In the remainder of the paper, we will assume that C is an imaginary hyperelliptic curve and thus has only one point P_∞ at infinity and its affine part is given by

$$y^2 + h(x)y = f(x) \ ,$$

with $h, f \in \mathbb{F}_q[x]$, $\deg h \leq g$, f monic and $\deg f = 2g + 1$.

For any algebraic extension K of $\mathbb{F}_q$ consider the set

$$C(K) := \{(x, y) \in K \times K \mid y^2 + h(x)y = f(x)\} \cup \{P_\infty\} \ ,$$

called the set of K-rational points on C. The hyperelliptic involution ι defined by $\iota(x, y) = (x, -y - h(x))$ acts on the set $C(K)$. However, unlike elliptic curves, the set $C(K)$ for $g \geq 2$ does not form a group, but we can embed C into an abelian variety of dimension g called the Jacobian of C and denoted by J_C. As usual, we will represent elements of $J_C(K)$ by elements of the divisor class group of degree 0 divisors $\mathrm{Div}_C^0(K)/\mathrm{Prin}_C(K)$, the definition of which is recalled in the following paragraphs.

A divisor D on C is a formal sum of points over the algebraic closure $\overline{\mathbb{F}}_q$

$$D = \sum_{P \in C(\overline{\mathbb{F}}_q)} c_P(P)$$

with only finitely many non-zero coefficients $c_P \in \mathbb{Z}$. The set of all divisors on C is denoted Div_C and clearly forms a group under formal addition. The degree of D is defined as $\deg(D) = \sum_{P \in C(\overline{\mathbb{F}}_q)} c_P$ and the subgroup of degree 0 divisors is denoted by Div_C^0. The support $\mathrm{supp}(D)$ of a divisor D is the set of points P with $c_P \neq 0$ and we define $\mathrm{ord}_P(D) = c_P$.

Let φ be the Frobenius morphism $\varphi : C \to C$ given by $\varphi(x, y) = (x^q, y^q)$ and define

$$\varphi(D) = \sum_{P \in C(\overline{\mathbb{F}}_q)} c_P(\varphi(P)),$$

then D is called $\mathbb{F}_{q^k}$-rational if and only if $\varphi^k(D) = D$. The set of $\mathbb{F}_{q^k}$-rational divisors is denoted by $\mathrm{Div}_C(\mathbb{F}_{q^k})$ and similarly for the degree 0 divisors. To each non-constant rational function $f \in \overline{\mathbb{F}}_q(C)^*$, we can associate the divisor $\mathrm{div}(f) = \sum_{P \in C(\overline{\mathbb{F}}_q)} \mathrm{ord}_P(f)(P)$, where $\mathrm{ord}_P(f)$ denotes the order of vanishing of f at P, i.e. $\mathrm{ord}_P(f) \neq 0$ if and only if f has either a zero or pole at P and $\mathrm{ord}_P(f)$ then equals the multiplicity of f at P. One can prove that only finitely many $\mathrm{ord}_P(f)$ are non-zero and furthermore, that $\deg(\mathrm{div}(f)) = 0$. Any divisor of the form $\mathrm{div}(f)$ with $f \in \overline{\mathbb{F}}_q(C)^*$ is called a principal divisor and the set of all these divisors is denoted Princ_C. By definition we have $J_C = \mathrm{Div}_C^0/\mathrm{Princ}_C$ and $J_C(\mathbb{F}_{q^k}) = \mathrm{Div}_C^0(\mathbb{F}_{q^k})/\mathrm{Princ}_C(\mathbb{F}_{q^k})$, where $\mathrm{Princ}_C(\mathbb{F}_{q^k}) = \mathrm{Princ}_C \cap \mathrm{Div}_C^0(\mathbb{F}_{q^k})$. Given a degree 0 divisor D, we will denote by $\overline{D}$ the corresponding divisor class in J_C.

Each divisor class $\overline{D}$ can be uniquely represented by a so called reduced divisor, i.e. a divisor of the form

$$\sum_{i=1}^m (P_i) - m(P_\infty) , \qquad m \leq g$$

with $P_i = (x_i, y_i) \in C(\overline{\mathbb{F}}_q)$, $P_i \neq P_\infty$ and $P_i \neq \iota(P_j)$ for $i \neq j$. For notational convenience, we introduce two maps on J_C: given a divisor class $\overline{D}$, we define $\rho(\overline{D})$ the unique reduced divisor in $\overline{D}$ and $\epsilon(\overline{D})$ the effective part of $\rho(\overline{D})$, i.e. $\rho(\overline{D}) = \epsilon(\overline{D}) - \deg(\epsilon(\overline{D}))(P_\infty)$. Note that the sets $\rho(J_C)$ and $\epsilon(J_C)$ can be endowed with a group law $\oplus$ by defining: $\rho(\overline{D}_1) \oplus \rho(\overline{D}_2) := \rho(\overline{D}_1 + \overline{D}_2)$ and similarly, $\epsilon(\overline{D}_1) \oplus \epsilon(\overline{D}_2) := \epsilon(\overline{D}_1 + \overline{D}_2)$. Furthermore, the notion of rationality is well defined since $P_\infty \in C(\mathbb{F}_q)$.

It is not difficult to show that any reduced $\mathbb{F}_q$-rational divisor admits a Mumford representation $[u(x), v(x)]$, i.e. a pair of polynomials $u, v \in \mathbb{F}_q[x]$, with $u = \prod_{i=1}^m (x - x_i)$, $\deg v < \deg u \leq g$ and $u|v^2 + vh - f$. Cantor's algorithm [6] can be used to compute the Mumford representation of the sum of two reduced divisors or for small genera, explicit formulae exist [3,14,18].

Given a divisor D representing a divisor class $\overline{D}$ in J_C and an integer n, we denote $[n]D := \rho(n\overline{D})$, i.e. the unique reduced divisor equivalent with nD. Finally, for D an $\mathbb{F}_{q^k}$-rational divisor, we denote by $f_{n,D} \in \mathbb{F}_{q^k}(C)$ any function (determined up to non-zero constant multiple) for which $\mathrm{div}(f_{n,D}) = nD - [n]D$.

2.2 Tate-Lichtenbaum Pairing

In this section, we briefly recall the definition of the Tate-Lichtenbaum pairing as it is usually stated in the literature and discuss the various alternatives for the domain of the pairing.

Let r be a prime with $r \mid \#J_C(\mathbb{F}_q)$ and $\gcd(r, q) = 1$ and let k be the smallest integer such that $r \mid (q^k - 1)$, then k is called the embedding degree (dependent on r). Note that this implies that the r-th roots of unity μ_r are contained in $\mathbb{F}_{q^k}$ and in no strictly smaller extension of $\mathbb{F}_q$. Note that $r > k$, since k is the order of q modulo r and hence $k \mid r - 1$ holds. Denote with $J_C(\mathbb{F}_{q^k})[r]$ the r-torsion points on J_C defined over $\mathbb{F}_{q^k}$. The Tate-Lichtenbaum pairing is a well defined, non-degenerate, bilinear pairing [9,16]

$$\langle \cdot, \cdot \rangle_r : J_C(\mathbb{F}_{q^k})[r] \times J_C(\mathbb{F}_{q^k})/rJ_C(\mathbb{F}_{q^k}) \to \mathbb{F}_{q^k}^*/(\mathbb{F}_{q^k}^*)^r ,$$

which is defined as follows: let $\overline{D}_1 \in J_C(\mathbb{F}_{q^k})[r]$ and $\overline{D}_2 \in J_C(\mathbb{F}_{q^k})$ and let $\overline{D}_1$ be represented by a divisor D_1 and $\overline{D}_2$ by a divisor D_2 with $\mathrm{supp}(D_1) \cap \mathrm{supp}(D_2) = \emptyset$. Since $\overline{D}_1$ has order r, the function $f_{r,D_1} \in \mathbb{F}_{q^k}(C)^*$ has divisor $\mathrm{div}(f_{r,D_1}) = rD_1 - [r]D_1 = rD_1$. The Tate-Lichtenbaum pairing of the divisor classes $\overline{D}_1$ and $\overline{D}_2$ is then defined by

$$\langle \overline{D}_1, \overline{D}_2 \rangle_r \equiv f_{r,D_1}(D_2) = \prod_{P \in C(\overline{\mathbb{F}}_q)} f_{r,D_1}(P)^{\mathrm{ord}_P(D_2)} ,$$

where $\equiv$ means equality up to r-th powers. Note that since D_2 has degree 0, multiplying f_{r,D_1} by a non-zero constant will give the same result.

In implementations, one works with the Mumford representation, i.e. with reduced divisors D_1 and D_2, but the Tate pairing cannot be computed as $f_{r,D_1}(D_2)$, since $P_\infty \in \mathrm{supp}(D_1) \cap \mathrm{supp}(D_2)$. The following lemma shows that if the function f_{r,D_1} is properly normalised, the Tate pairing can simply be computed as $f_{r,D_1}(\epsilon(\overline{D}_2))$. To state the lemma, we need the notion of leading coefficient: let u_∞ be a fixed $\mathbb{F}_q$-rational uniformizer at P_∞, then for any function $f \in \overline{\mathbb{F}}_q(C)^*$ we define $\mathrm{lc}_\infty(f)$ to be the leading coefficient of f as a Laurent series in u_∞. Note that when f is defined at P_∞ we simply have $f(P_\infty) = \mathrm{lc}_\infty(f)$ independent of the uniformizer chosen.

Lemma 1. *Let $\overline{D}_1 \in J_C(\mathbb{F}_{q^k})[r]$, $D_1 = \rho(\overline{D}_1)$ and $\overline{D}_2 \in J_C(\mathbb{F}_{q^k})$ and assume that $\mathrm{supp}(D_1) \cap \mathrm{supp}(\epsilon(\overline{D}_2)) = \emptyset$, then*

$$\langle \overline{D}_1, \overline{D}_2 \rangle_r \equiv f_{r,D_1}(\epsilon(\overline{D}_2))$$

if and only if $\mathrm{lc}_\infty(f_{r,D_1}) \in (\mathbb{F}_{q^k}^*)^r$. Furthermore, $\mathrm{lc}_\infty(f_{r,D_1})$ being an r-th power is independent of the uniformizer chosen.

PROOF: Let $D_2 = \rho(\overline{D}_2)$ and choose $h \in \mathbb{F}_{q^k}(C)$ such that $D_1' = D_1 + \mathrm{div}(h)$ satisfies $\mathrm{supp}(D_1') \cap \mathrm{supp}(D_2) = \emptyset$, then by definition we have

$$\langle \overline{D}_1, \overline{D}_2 \rangle_r \equiv f_{r,D_1'}(D_2) .$$

Since $D_1' = D_1 + \mathrm{div}(h)$, we can take $f_{r,D_1'} = f_{r,D_1}h^r$ (in fact we could multiply $f_{r,D_1'}$ with a constant c, but this would give the same result as remarked before) and thus

$$\langle \overline{D}_1, \overline{D}_2 \rangle_r \equiv (f_{r,D_1}h^r)(D_2) \equiv \frac{(f_{r,D_1}h^r)(\epsilon(\overline{D}_2))}{\mathrm{lc}_\infty(f_{r,D_1}h^r)^{m_2}} \equiv \frac{f_{r,D_1}(\epsilon(\overline{D}_2))}{\mathrm{lc}_\infty(f_{r,D_1})^{m_2}} ,$$

with $m_2 = \deg(\epsilon(\overline{D}_2))$. Finally, $\gcd(m_2, r) = 1$ implies that $\mathrm{lc}_\infty(f_{r,D_1})^{m_2}$ is an r-th power if and only if $\mathrm{lc}_\infty(f_{r,D_1})$ is an r-th power. Furthermore, since $\mathrm{ord}_{P_\infty}(f_{r,D_1}) = -\deg(\epsilon(\overline{D}_1))r$, i.e. a multiple of r, the property of $\mathrm{lc}_\infty(f_{r,D_1})$ being an r-th power does not depend on the uniformizer chosen. □

In practice, one often requires a unique pairing value instead of a whole coset; therefore one defines the reduced Tate-Lichtenbaum pairing as

$$e(\overline{D}_1, \overline{D}_2) = \langle \overline{D}_1, \overline{D}_2 \rangle_r^{(q^k-1)/r} \in \mu_r \subset \mathbb{F}_{q^k}^* .$$

It is easy to see that for any positive integer N with $r|N$ and $N|q^k - 1$ we have

$$e(\overline{D}_1, \overline{D}_2) = \langle \overline{D}_1, \overline{D}_2 \rangle_r^{(q^k-1)/r} = \langle \overline{D}_1, \overline{D}_2 \rangle_N^{(q^k-1)/N} \ . \tag{1}$$

For $k > 1$ and $\overline{D}_1 \in J_C(\mathbb{F}_q)$, the reduced Tate-Lichtenbaum pairing can be computed as in Lemma 1, but without the need for normalisation. Indeed, since $\rho(\overline{D}_1)$ is $\mathbb{F}_q$-rational, we conclude that $f_{r,D_1} \in \mathbb{F}_q(C)$ and thus $\mathrm{lc}_\infty(f_{r,D_1}) \in \mathbb{F}_q^* \subset (\mathbb{F}_{q^k}^*)^r$. For elliptic curves, this simplification was first noticed in [5] using a more direct proof than that of Lemma 1.

For efficiency reasons, one restricts the domain of the Tate-Lichtenbaum pairing to the groups $\mathbb{G}_1 = J_C[r] \cap \mathrm{Ker}(\varphi - [1])$ and the group $\mathbb{G}_2 = J_C[r] \cap \mathrm{Ker}(\varphi - [q])$, i.e. the eigenspaces of the Frobenius endomorphism on $J_C[r]$. Note that $\mathbb{G}_1 \subset J_C(\mathbb{F}_q)$ and $\mathbb{G}_2 \subset J_C(\mathbb{F}_{q^k})$, since for $\overline{D} \in \mathbb{G}_2$ we have $\varphi^k(\overline{D}) = [q^k]\overline{D} = \overline{D}$, because $\overline{D} \in J_C[r]$ and $q^k \equiv 1 \bmod r$. This also shows that k is the smallest integer such that the q-eigenspace of the Frobenius in $J_C[r]$ is $\mathbb{F}_{q^k}$-rational.

Remark 1. In the remainder of the paper we will assume that any representative D_1 of $\overline{D}_1 \in \mathbb{G}_1$ (resp. D_2 of $\overline{D}_2 \in \mathbb{G}_2$) is chosen to be $\mathbb{F}_q$-rational (resp. $\mathbb{F}_{q^k}$-rational).

Remark 2. In general, the smallest extension degree d such that the whole r-torsion $J_C[r]$ is $\mathbb{F}_{q^d}$-rational is larger than k [9]. This is obvious for $g \geq 2$, since $J_C[r] \simeq (\mathbb{Z}/r\mathbb{Z})^{2g}$, but even for elliptic curves, this phenomenon occurs: consider an elliptic curve $E/\mathbb{F}_q$ with $r \mid \#E(\mathbb{F}_q)$ and $r \mid q - 1$, but such that $r^2 \nmid \#E(\mathbb{F}_q)$. In this case $E(\mathbb{F}_q)[r]$ is both the 1-eigenspace and q-eigenspace and the minimal d such that $E[r] \subset E(\mathbb{F}_{q^d})$ is equal to r.

Finally, we note that the group $\mathbb{G}_2$ already occurs in the original paper [9] disguised as a Galois cohomology group $H^1(G, J_C)[r]$, with G the absolute Galois group of $\mathbb{F}_q$. In fact, in [2][Section 6.3.1] one finds that the Tate-Lichtenbaum pairing has as domain $\mathbb{G}_2 \times J_C(\mathbb{F}_q)/rJ_C(\mathbb{F}_q)$, which is yet another choice of subgroups.

2.3   Miller's Algorithm

In [20] (see also [21]), Miller described a fast algorithm to compute evaluations of the form $f_{r,D_1}(D_2)$ for divisors on elliptic curves. The algorithm easily generalises to hyperelliptic curves as follows: by definition of the group law $\oplus$ on J_C, there exists a function $G_{D_a,D_b} \in \mathbb{F}_{q^k}(C)^*$ with $\mathrm{div}(G_{D_a,D_b}) = D_a + D_b - (D_a \oplus D_b)$ where $D_a \oplus D_b$ is reduced. As such we can take the function

$$f_{i+j,D} = f_{i,D} f_{j,D} G_{[i]D,[j]D} \ .$$

This immediately leads to Algorithm 1 and the more detailed version given in Algorithm 2.

Algorithm 1. Miller's algorithm for hyperelliptic curves

Inputs: $n \in \mathbb{N}$ and $D_a, D_b \in J_C$ with disjoint support
Outputs: $f_{n,D_a}(D_b)$
    Write n as $\sum_{j=0}^{s} n_j 2^j$, with $n_j \in \{0,1\}$ and $n_s = 1$.
    $D \leftarrow D_a,\ c \leftarrow 1$.
    for $j = s - 1$ down to 0 **do**
        Compute $D \leftarrow [2]D$ and extract $G_{D,D}$.
        $c \leftarrow c^2 \cdot G_{D,D}(D_b)$.
        if $n_j = 1$ **then**
            Compute $D \leftarrow D \oplus D_a$ and extract G_{D,D_a}.
            $c \leftarrow c \cdot G_{D,D_a}(D_b)$.
        end if
    end for
    Return c.

---

3   Ate Pairing on Hyperelliptic Curves

In this section, we first recall the Ate pairing for elliptic curves and then show that with a minor, but important change, it can be extended to hyperelliptic curves.

The two main ideas of the Ate pairing are that the domain of the pairing is $\mathbb{G}_2 \times \mathbb{G}_1$ and that the loop length in Miller's algorithm is much shorter than for the Tate-Lichtenbaum pairing. The result is summarised in the following theorem from [17].

Theorem 1. *Let E be an elliptic curve over $\mathbb{F}_q$, r a large prime with $r \mid \#E(\mathbb{F}_q)$ and denote the trace of Frobenius with t, i.e. $\#E(\mathbb{F}_q) = q + 1 - t$. For $T = t - 1$, $Q \in \mathbb{G}_2 = E[r] \cap \mathrm{Ker}(\varphi - [q])$ and $P \in \mathbb{G}_1 = E[r] \cap \mathrm{Ker}(\varphi - [1])$, we have the following:*

1. *$f_{T,Q}(P)$ defines a bilinear pairing, called the Ate pairing*
2. *let $N = \gcd(T^k - 1, q^k - 1)$ and $T^k - 1 = LN$, with k the embedding degree, then*

$$e(Q,P)^L = f_{T,Q}(P)^{c(q^k-1)/N}$$

   where $c = \sum_{i=0}^{k-1} T^{k-1-i} q^i \equiv k q^{k-1} \bmod r$
3. *for $r \nmid L$, the Ate pairing is non-degenerate*

The reason why this construction works is the compatibility of the scalar $T = t - 1$ and the action of the Frobenius on $\mathbb{G}_2$. Indeed, by definition of $\mathbb{G}_2$ we have $\varphi(Q) = [q]Q$, and since $r \mid \#E(\mathbb{F}_q) = q + 1 - t$ it follows that $\varphi(Q) = [T]Q$. This last equality also determines the loop length in Miller's algorithm, i.e. $\lceil \log_2 |T| \rceil$.

For a hyperelliptic curve C with $g > 1$, the situation is somewhat different. Indeed, in this case $r \mid \#J_C(\mathbb{F}_q) = q^g + a_1(q^{g-1} + 1) + a_2(q^{g-2} + 1) + \cdots + a_g$, so in general q cannot be replaced by a smaller equivalent. However, note that for $g > 1$ and $r \approx \#J_C(\mathbb{F}_q)$, the bit length of q itself is already g times shorter than the bit length of r, again resulting in a shorter loop in Miller's algorithm. The possibility of using $T = q$ is already present in [7], but for a very restricted family of curves. This observation leads to the following theorem.

Theorem 2. *Let C be a hyperelliptic curve over $\mathbb{F}_q$ and $r \mid \#J_C(\mathbb{F}_q)$ a large prime. Let $\mathbb{G}_2 = J_C[r] \cap \mathrm{Ker}(\varphi - [q])$ and $\mathbb{G}_1 = J_C[r] \cap \mathrm{Ker}(\varphi - [1])$, then*

$$a(\cdot, \cdot) : \mathbb{G}_2 \times \mathbb{G}_1 \to \mu_r : (\overline{D}_2, \overline{D}_1) \mapsto f_{q,D_2}(D_1)$$

with $D_2 = \rho(\overline{D}_2)$ and $D_1 \in \overline{D}_1$ such that $\mathrm{supp}(D_1) \cap \mathrm{supp}(D_2) = \emptyset$, defines a non-degenerate, bilinear pairing called the hyperelliptic Ate pairing. Furthermore, the relation with the reduced Tate-Lichtenbaum pairing is as follows:

$$e(\overline{D}_2, \overline{D}_1) = a(\overline{D}_2, \overline{D}_1)^{kq^{k-1}} . \tag{2}$$

Note that in Theorem 2, the divisor D_2 is assumed to be reduced and the function f_{q,D_2} is evaluated at the divisor D_1 and not only at $\epsilon(\overline{D}_1)$ (but see Lemma 6). Furthermore, the image of the hyperelliptic Ate pairing already is μ_r so no final exponentiation is required. The proof of Theorem 2 follows from the following four lemmata. The first lemma shows that the Ate pairing indeed maps into μ_r.

Lemma 2. *Let $\overline{D}_2 \in \mathbb{G}_2$, $D_2 = \rho(\overline{D}_2)$ and $\overline{D}_1 \in \mathbb{G}_1$, $D_1 \in \overline{D}_1$ with $\mathrm{supp}(D_1) \cap \mathrm{supp}(D_2) = \emptyset$, then we have $f_{q,D_2}(D_1) \in \mu_r$.*

PROOF: Let $h \in \mathbb{F}_q(C)^*$ with $\mathrm{supp}(\mathrm{div}(h)) \cap \mathrm{supp}(\mathrm{div}(f_{q,D_2})) = \emptyset$, then using Weil reciprocity we obtain

$$\begin{aligned} f_{q,D_2}(\mathrm{div}(h)) &= h(\mathrm{div}(f_{q,D_2})) \\ &= h(qD_2 - [q]D_2) = h(qD_2 - \varphi(D_2)) \\ &= \frac{h(qD_2)}{h(\varphi(D_2))} = \frac{h(D_2)^q}{h(D_2)^q} = 1 , \end{aligned}$$

therefore

$$f_{q,D_2}(D + \mathrm{div}(h)) = f_{q,D_2}(D)f_{q,D_2}(\mathrm{div}(h)) = f_{q,D_2}(D) . \tag{3}$$

As D_1 is defined over $\mathbb{F}_q$ and $\overline{D}_1 \in \mathbb{G}_1$, we obtain

$$f_{q,D_2}(D_1)^r = f_{q,D_2}(rD_1) = f_{q,D_2}(0) = 1$$

since $rD_1 \sim 0$. Using (3) again, we conclude that $f_{q,D_2}(D_1)$ only depends on $\overline{D}_1$ and not on the representative chosen. $\qquad\square$

The following three lemmata show that the Ate pairing can indeed be related to the reduced Tate pairing.

Lemma 3. *Given $\overline{D}_1, \overline{D}_2 \in J_C(\mathbb{F}_{q^k})[r]$, $D_2 = \rho(\overline{D}_2)$ and $D_1 \in \overline{D}_1$ such that $\mathrm{supp}(D_1) \cap \mathrm{supp}(D_2) = \emptyset$, we have*

$$e(\overline{D}_2, \overline{D}_1) = f_{q^k, D_2}(D_1) .$$

PROOF: By definition of the reduced Tate-Lichtenbaum pairing, we have to compute

$$e(\overline{D}_2, \overline{D}_1) = f_{r,D_2}(D_1)^{(q^k-1)/r} = f_{q^k-1,D_2}(D_1) \ ,$$

where the last equality follows from (1) with $N = q^k - 1$. Up to this point the divisor D_2 does not even have to be reduced: indeed, take $D_2' = D_2 + \mathrm{div}(h)$ for some $h \in \mathbb{F}_{q^k}(C)$, then $f_{q^k-1,D_2'} = cf_{q^k-1,D_2}h^{q^k-1}$ for some constant factor $c \in \mathbb{F}_{q^k}$. Since D_1 has degree 0, the constant c is irrelevant and the factor $h^{q^k-1}(D_1)$ clearly equals 1 since $h(D_1) \in \mathbb{F}_{q^k}^*$.

When D_2 is reduced, we have that $\mathrm{div}(f_{q^k,D_2}) = q^k D_2 - [q^k]D_2 = (q^k - 1)D_2$ and $\mathrm{div}(f_{q^k-1,D_2}) = (q^k - 1)D_2 - [q^k - 1]D_2 = (q^k - 1)D_2$ since $D_2 \in J_C[r]$, so without loss of generality we can take $f_{q^k-1,D_2} = f_{q^k,D_2}$, which ends the proof. $\qquad\square$

An easy calculation [4, Lemma 2] proves the following lemma.

Lemma 4. *For any divisor D we can choose $f_{q^k,D}$ such that*

$$f_{q^k,D} = \prod_{i=0}^{k-1} \left(f_{q,[q^i]D}\right)^{q^{k-i-1}} \ . \tag{4}$$

For $D_2 = \rho(\overline{D}_2)$ with $\overline{D}_2 \in G_2$, each of the factors in the right hand side of (4) can be expressed in terms of f_{q,D_2}. To see this, note that $\varphi(D_2) = [q]D_2$ and $\varphi^i(D_2) = [q^i]D_2$, so it suffices to relate $f_{q,\varphi^i(D_2)}$ with f_{q,D_2} as in the following lemma.

Lemma 5. *Let D be a reduced divisor and ψ a purely inseparable map on C with $\psi(P_\infty) = P_\infty$. Then $\psi(D)$ is also reduced and we can take*

$$f_{n,\psi(D)} \circ \psi = f_{n,D}^{\deg(\psi)} \ .$$

PROOF: Let $D = \sum_{i=1}^{m}(P_i) - m(P_\infty)$ be reduced then $\psi(D) = \sum_{i=1}^{m}(\psi(P_i)) - m(P_\infty)$, where we used the fact that $\psi(P_\infty) = P_\infty$. Since ψ is assumed to be purely inseparable we have $\psi(P_i) \neq P_\infty$ and $\psi(P_i) \neq \iota(\psi(P_j))$ for $i \neq j$, i.e. $\psi(D)$ is again reduced. By definition we have $\mathrm{div}(f_{n,\psi(D)}) = n(\psi(D)) - ([n]\psi(D))$. Since ψ is purely inseparable we have

$$\psi^*\left(\mathrm{div}(f_{n,\psi(D)})\right) = n\psi^*(\psi(D)) - \psi^*([n]\psi(D)) = n(\deg\psi)D - \psi^*(\psi([n]D))$$
$$= n(\deg\psi)D - (\deg\psi)([n]D) = \mathrm{div}(f_{n,D}^{\deg(\psi)}) \ .$$

The non-trivial part is the equality $[n]\psi(D) = \psi([n]D)$, which follows from the fact that both sides are reduced divisors (since ψ maps a reduced divisor to a reduced divisor) and that they are linearly equivalent. Indeed,

$$[n]\psi(D) = n\psi(D) + \mathrm{div}(h_n) = \psi(nD) + \mathrm{div}(h_n)$$
$$= \psi([n]D + \mathrm{div}(g_n)) + \mathrm{div}(h_n) = \psi([n]D) + \mathrm{div}(\psi_* g_n) + \mathrm{div}(h_n) \ ,$$

for suitable functions $h_n, g_n \in \overline{\mathbb{F}}_q(C)$. Furthermore,

$$\psi^*\left(\mathrm{div}(f_{n,\psi(D)})\right) = \mathrm{div}\left(\psi^*(f_{n,\psi(D)})\right) = \mathrm{div}(f_{n,\psi(D)} \circ \psi) \ ,$$

so we can take $f_{n,\psi(D)} \circ \psi = f_{n,D}^{\deg(\psi)}$. $\qquad\square$

PROOF OF THEOREM 2:   Since $D_1 \in \mathbb{G}_1$ and fixed under φ, and $D_2 \in \mathbb{G}_2$ is reduced (so $\varphi(D_2) = [q]D_2$), Lemma 5 implies

$$f_{q,[q^i]D_2}(D_1) = f_{q,\varphi^i(D_2)}(D_1) = f_{q,\varphi^i(D_2)}(\varphi^i(D_1)) = (f_{q,D_2}(D_1))^{q^i} ,$$

and using Lemma 4, we obtain

$$f_{q^k,D_2}(D_1) = \prod_{i=0}^{k-1} \left(f_{q,[q^i]D_2}(D_1)\right)^{q^{k-i-1}} = (f_{q,D_2}(D_1))^{kq^{k-1}} . \tag{5}$$

Substituting the above in Lemma 3, we recover Equation (2)

$$e(\overline{D}_2, \overline{D}_1) = (f_{q,D_2}(D_1))^{kq^{k-1}}$$

This equation shows that $f_{q,D_2}(D_1)$ defines a non-degenerate bilinear pairing, since $e(\overline{D}_2, \overline{D}_1)$ is non-degenerate and bilinear. Furthermore, since $f_{q,D_2}(D_1) \in \mu_r$ by Lemma 2, the hyperelliptic Ate pairing is automatically reduced, i.e. no final exponentiation is needed. □

An important remark is that all optimisations that rely on the final powering, such as denominator elimination and ignoring the point at infinity in the evaluation, should be reexamined. It is not hard to see that the first simply no longer holds, whereas the second can be salvaged if the function f_{q,D_2} is properly normalised as in the following lemma.

Lemma 6. Let $\overline{D}_2 \in \mathbb{G}_2$ and $\overline{D}_1 \in \mathbb{G}_1$ with $\operatorname{supp}(\epsilon(\overline{D}_1)) \cap \operatorname{supp}(\epsilon(\overline{D}_2)) = \emptyset$ and let $D_2 = \rho(\overline{D}_2)$, then if $\operatorname{lc}_\infty(f_{q,D_2}) = 1$ with respect to any $\mathbb{F}_q$-rational uniformizer u_∞ then

$$a(\overline{D}_2, \overline{D}_1) = f_{q,D_2}(\epsilon(\overline{D}_1)) . \tag{6}$$

PROOF: By definition we have $\operatorname{div}(f_{q,D_2}) = qD_2 - [q]D_2 = qD_2 - \varphi(D_2)$ since $D_2 \in \mathbb{G}_2$ is reduced and thus $\operatorname{ord}_{P_\infty}(f_{q,D_2}) = -m_2(q-1)$, with $m_2 = \deg(\epsilon(\overline{D}_2))$. This implies that $\operatorname{lc}_\infty(f_{q,D_2}) = 1$ is independent of the choice of $\mathbb{F}_q$-rational uniformizer. Indeed, let u'_∞ be any other $\mathbb{F}_q$-rational uniformizer, then

$$\operatorname{lc}'_\infty(f_{q,D_2}) = \operatorname{lc}_\infty(u'_\infty)^{m_2(q-1)}\operatorname{lc}_\infty(f_{q,D_2}) = \operatorname{lc}_\infty(f_{q,D_2}) .$$

Let $D'_1 \in \overline{D}_1$ such that $\operatorname{supp}(D'_1) \cap (\operatorname{supp}(\operatorname{div}(f_{q,D_2})) \cup \operatorname{supp}(\operatorname{div}(u_\infty)))$ and define $\tilde{f}_{q,D_2} = f_{q,D_2} \cdot u_\infty^{m_2(q-1)}$. The divisor of $\tilde{f}_{q,D_2}$ is

$$\operatorname{div}(\tilde{f}_{q,D_2}) = q\epsilon(\overline{D}_2) - \epsilon(\varphi(\overline{D}_2)) + m_2(q-1) \cdot (\operatorname{div}(u_\infty) - P_\infty)$$

which does not contain P_∞, and it is easy to adapt the proof of Lemma 2 to show that $\tilde{f}_{q,D_2}(\overline{D}_1)$ does not depend on the choice of representative of $\overline{D}_1$.

By construction of D'_1, both $f_{q,D_2}(D'_1)$ and $\tilde{f}_{q,D_2}(D'_1)$ are well defined and

$$\tilde{f}_{q,D_2}(D'_1) = f_{q,D_2}(D'_1) \cdot (u_\infty(D'_1))^{m_2(q-1)} = f_{q,D_2}(D'_1)$$

since $u_\infty(D_1')$ is in $\mathbb{F}_q$. From this, we obtain

$$a(\overline{D}_2, \overline{D}_1) = f_{q,D_2}(D_1') = \tilde{f}_{q,D_2}(D_1') = \tilde{f}_{q,D_2}(\overline{D}_1)$$

$$= \frac{f_{q,D_2}(\epsilon(\overline{D}_1)) \cdot (u_\infty^{m_2}(\epsilon(\overline{D}_1)))^{q-1}}{\mathrm{lc}_\infty(\tilde{f}_{q,D_2})^{\deg(\epsilon(\overline{D}_1))}} = f_{q,D_2}(\epsilon(\overline{D}_1))$$

since $\mathrm{lc}_\infty(\tilde{f}_{q,D_2}) = \mathrm{lc}_\infty(f_{q,D_2}) = 1$ by construction of $\tilde{f}_{q,D_2}$.   □

4 Ate Pairing on Superspecial Curves

In this section, we investigate whether the hyperelliptic Ate pairing can also be defined on $\mathbb{G}_1 \times \mathbb{G}_2$. Recall that a curve C is said to have p-rank zero if $J_C[p] = \{0\}$, i.e. the p-torsion is trivial. An immediate consequence of the absence of p-torsion is that the dual of Frobenius $\hat{\varphi}$ (also called Verschiebung) is purely inseparable. Indeed, $\mathrm{Ker}(\hat{\varphi}) \subset J_C[q]$ since $\hat{\varphi}$ has degree q and thus $\mathrm{Ker}(\hat{\varphi}) = \{0\}$. Since $\hat{\varphi} \circ \varphi = [q]$, we conclude that $\hat{\varphi}$ acts as $\hat{\varphi}(\overline{D}_1) = [q]\overline{D}_1$ for $\overline{D}_1 \in \mathbb{G}_1$ and $\hat{\varphi}(\overline{D}_2) = \overline{D}_2$ for $\overline{D}_2 \in \mathbb{G}_2$.

However, p-rank zero is not restrictive enough for our purposes, since Lemma 5 holds for a purely inseparable map on the curve C, whereas Verschiebung is defined on the Jacobian. A curve C is called superspecial when its Jacobian J_C is isomorphic to E^g with E a supersingular elliptic curve. Note that this is more restrictive than supersingularity, since this only requires J_C to be isogenous to E^g. As an example of superspecial curves we mention the family described by Duursma-Lee [7].

For a superspecial curve, we can write $\hat{\varphi} = \varphi \circ \alpha$ for an automorphism $\alpha \in \mathrm{Aut}(C)$. Note that this automorphism is necessarily defined over $\mathbb{F}_q$, since $\varphi = \hat{\hat{\varphi}} = \hat{\alpha} \circ \hat{\varphi}$ and thus $\alpha \circ \varphi = \varphi \circ \alpha$.

Analysing the various lemmata used in proving Theorem 2, we immediately run into a problem since Lemma 2 is no longer valid. Indeed, let $D_1 = \rho(\overline{D}_1)$ with $D_1 \in \overline{D}_1$ and let $h \in \mathbb{F}_{q^k}(C)^*$, then

$$f_{q,D_1}(\mathrm{div}(h)) = h(qD_1 - [q]D_1) = h(qD_1 - \hat{\varphi}(D_1)) = \frac{h(D_1)^q}{h(\alpha(D_1))}.$$

This shows that even if h would be $\mathbb{F}_q$-rational, $f_{q,D_1}(\mathrm{div}(h))$ still is not 1, so $f_{q,D_1}(D_2)$ with $D_2 \in \overline{D}_2$ is not independent of the representative chosen.

On the other hand, it is easy to verify that Lemma 3 and 4 remain valid when D_1 and D_2 are swapped. Furthermore, since $\hat{\varphi}$ is given by purely inseparable map on C, Lemma 5 still applies. As a result we can prove the following theorem, circumventing the fact that Lemma 2 no longer holds.

Theorem 3. *Let C be a superspecial curve over $\mathbb{F}_q$ and r a large prime with $r \mid \#J_C(\mathbb{F}_q)$. Let $\mathbb{G}_1 = J_C[r] \cap \mathrm{Ker}(\varphi - [1])$ and $\mathbb{G}_2 = J_C[r] \cap \mathrm{Ker}(\varphi - [q])$, then*

$$\hat{a}(\cdot, \cdot) : \mathbb{G}_1 \times \mathbb{G}_2 \to \mu_r : (\overline{D}_1, \overline{D}_2) \mapsto f_{q,D_1}(\epsilon(\overline{D}_2))^d$$

with $D_1 = \rho(\overline{D}_1)$, $d = \gcd(k, q^k - 1)$, $\mathrm{lc}_\infty(f_{q,D_1}) = 1$ *and assuming that* $\mathrm{supp}(D_1) \cap \mathrm{supp}(\epsilon(\overline{D}_2)) = \emptyset$, *defines a non-degenerate, bilinear pairing called the superspecial Ate pairing. Furthermore, the relation with the reduced Tate-Lichtenbaum pairing is as follows:*

$$e(\overline{D}_1, \overline{D}_2) = \hat{a}(\overline{D}_1, \overline{D}_2)^{(k/d)q^{k-1}} . \tag{7}$$

PROOF: Combining Lemma 1, Lemma 3 and Lemma 4 it suffices to compute

$$e(\overline{D}_1, \overline{D}_2) = f_{q^k, D_1}(\epsilon(\overline{D}_2)) = \prod_{i=0}^{k-1} \left(f_{q, [q^i]D_1}(\epsilon(\overline{D}_2)) \right)^{q^{k-i-1}} , \tag{8}$$

where $D_1 = \rho(\overline{D}_1)$. Applying Lemma 5 to $\hat{\varphi}^i$ we conclude that $f_{q, \hat{\varphi}^i(D_1)} \circ \hat{\varphi}^i = f_{q, D_1}^{q^i}$. Since D_1 is reduced and $\overline{D}_1 \in \mathbb{G}_1$, we have $\hat{\varphi}^i(D_1) = [q^i]D_1$. Furthermore, let $D_2 = \rho(\overline{D}_2)$, then since D_2 is reduced and $\overline{D}_2 \in \mathbb{G}_2$, we have $\hat{\varphi}(D_2) = D_2$. Combined with $\hat{\varphi}(P_\infty) = P_\infty$, we conclude that $\hat{\varphi}(\epsilon(\overline{D}_2)) = \epsilon(\overline{D}_2)$. Substituting this in (8) leads to

$$e(\overline{D}_1, \overline{D}_2) = f_{q, D_1}(\epsilon(\overline{D}_2))^{kq^{k-1}} = f_{q, D_1}(\epsilon(\overline{D}_2))^{d \cdot (k/d)q^{k-1}} .$$

Since the left hand side is an r-th root of unity and $\gcd((k/d)q^{k-1}, q^k - 1) = 1$, we conclude that $f_{q, D_1}(\epsilon(\overline{D}_2))^d$ also is an r-th root of unity. Furthermore, $e(\overline{D}_1, \overline{D}_2)$ is non-degenerate and bilinear, so we finally conclude that the superspecial Ate pairing also defines a non-degenerate bilinear pairing. □

The above theorem has been proved by Galbraith et al. [10] in the special case of supersingular elliptic curves in characteristic 2 and 3 using explicit computations.

5   Conclusion

In this paper we have introduced two new pairings on hyperelliptic curves, by generalising the Ate pairing on elliptic curves. The first version applies to all algebraic curves, whereas the second requires the curve to be superspecial, e.g. the Duursma-Lee curves. To prove that both versions are well-defined, we introduced a proper theoretical framework explaining several simpler results in the literature which were proved using ad hoc methods.

The most important property of the Ate pairings is that no final exponentiation is necessary. This raises security questions with respect to pairing inversion and Verheul's results on the computational Diffie-Hellman problem, especially when so-called degenerate divisors are used. The precise security implications of the Ate pairings are currently unknown and much more research is needed.

Acknowledgements

The authors would like to thank Bas Edixhoven and Ben Moonen for their expertise on superspecial curves. Robert Granger would like to thank Alfred

Menezes for his invitation to visit the Centre for Applied Cryptographic Research at the University of Waterloo in May 2006, where this work was initiated.

References

1. R. Avanzi. Aspects of Hyperelliptic Curves over Large Prime Fields in Software Implementation. In M. Joye and J.-J. Quisquater, editor, *CHES*, volume 3156 of *Lecture Notes in Computer Science*, pages 133–147. Springer, 2004.
2. R. Avanzi, H. Cohen, C. Doche, G. Frey, T. Lange, K. Nguyen, and F. Vercauteren. *Handbook of elliptic and hyperelliptic curve cryptography*. Discrete Mathematics and its Applications (Boca Raton). Chapman & Hall/CRC, Boca Raton, FL, 2006.
3. R. Avanzi, N. Thériault, and Z. Wang. Rethinking low genus hyperelliptic jacobian arithmetic over binary fields: Interplay of field arithmetic and explicit formulae. Technical report, CACR, 2006. CACR 2006-07.
4. P. S. L. M. Barreto, S. Galbraith, C. O hEigeartaigh, and M. Scott. Efficient pairing computation on supersingular abelian varieties. *Designs, Codes and Cryptography*, to be published, 2005.
5. P. S. L. M. Barreto, H. Y. Kim, B. Lynn, and M. Scott. Efficient algorithms for pairing-based cryptosystems. In Moti Yung, editor, *CRYPTO*, volume 2442 of *Lecture Notes in Computer Science*, pages 354–368. Springer, 2002.
6. D. G. Cantor. Computing in the Jacobian of a hyperelliptic curve. *Math. Comp.*, 48(177):95–101, 1987.
7. I. M. Duursma and Hyang-Sook Lee. Tate Pairing Implementation for Hyperelliptic Curves $y^2 = x^p - x + d$. In C.-S. Laih, editor, *ASIACRYPT*, volume 2894 of *Lecture Notes in Computer Science*, pages 111–123. Springer, 2003.
8. G. Frey and T. Lange. Fast Bilinear Maps from the Tate-Lichtenbaum Pairing on Hyperelliptic Curves. In F. Hess, S. Pauli, M. Pohst, editors, *ANTS VII*, volume 4076 of *Lecture Notes in Computer Science*, pages 466–479. Springer, 2006.
9. G. Frey and H-G. Rück. A remark concerning m-divisibility and the discrete logarithm in the divisor class group of curves. *Math. Comp.*, 62(206):865–874, 1994.
10. S. Galbraith, C. O hEigeartaigh, and C. Sheedy. Simplified pairing computation and security implications. To appear in *J. Math. Crypt.*, 2007.
11. P. Gaudry, F. Hess and N. P. Smart. Constructive and Destructive Facets of Weil Descent on Elliptic Curves. *J. Cryptology.*, 15(1):19–46, 2002.
12. P. Gaudry, E. Thomé, N. Thériault and C. Diem. A double large prime variation for small genus hyperelliptic index calculus. *Math. Comp.*, 76(257), 475–492, 2007.
13. R. Granger, D. Page, and N. Smart. High security pairing-based cryptography revisited. In F. Hess, S. Pauli, M. Pohst, editors, *ANTS-VII*, volume 4076 of *Lecture Notes in Computer Science*, pages 480–494. Springer, 2006.
14. C. Guyot, K. Kaveh, and V. M. Patankar. Explicit algorithm for the arithmetic on the hyperelliptic Jacobians of genus 3. *J. Ramanujan Math. Soc.*, 19(2):75–115, 2004.
15. F. Hess. Computing Riemann-Roch spaces in algebraic function fields and related topics. *J. Symb. Comp.*, 33(4):425–445, 2002.
16. F. Hess. A Note on the Tate Pairing of Curves over Finite Fields. *Arch. Math.*, 82:28–32, 2004.
17. F. Hess, N. Smart, and F. Vercauteren. The Eta-pairing revisited. *IEEE Transactions on Information Theory*, 52(10):4595–4602, 2006.

18. T. Lange. Formulae for arithmetic on genus 2 hyperelliptic curves. *Appl. Algebra Engrg. Comm. Comput.*, 15(5):295–328, 2005.
19. N. Koblitz and A. Menezes. Pairing-Based Cryptography at High Security Levels. In Nigel Smart, editor, *IMA Int. Conf.*, volume 3796 of *Lecture Notes in Computer Science*, pages 13–36. Springer, 2005.
20. V. S. Miller. Short programs for functions on curves. Unpublished manuscript 1986. Available at `http://crypto.stanford.edu/miller/miller.pdf`.
21. V. S. Miller. The Weil pairing, and its efficient calculation. *J. Cryptology*, 17(4):235–261, 2004.
22. J. H. Silverman. *The arithmetic of elliptic curves*, volume 106 of *Graduate Texts in Mathematics*. Springer-Verlag, New York, 1986.
23. H. Stichtenoth. *Algebraic function fields and codes*. Universitext. Springer-Verlag, Berlin, 1993.
24. E. Verheul. Evidence that XTR is more Secure than Supersingular Elliptic Curve Cryptosystems. In B. Pfitzmann, editor, *EUROCRYPT*, volume of 2045 *Lecture Notes in Computer Science*, pages 195–210. Springer, 2001.
25. N. Yui. On the Jacobian varieties of hyperelliptic curves over fields of characteristic $p > 2$. *J. Algebra*, 52(2):378–410, 1978.

A   Performance Estimates

A.1   Miller's Algorithm

We here give an expanded version of Algorithm 1, tailored for the Ate pairing. From the point of view of computational efficiency, a good choice of uniformizer is $u_\infty = x^g/y$. Let $p(x)$ be a polynomial in x, then with this choice of u_∞ we have:

$$\mathrm{lc}_\infty(p(x)) = \mathrm{lc}(p(x))$$

$$\mathrm{lc}_\infty(y - p(x)) = \begin{cases} 1 & \text{if } \deg(p(x)) \leq g \\ -\mathrm{lc}(p(x)) & \text{if } \deg(p(x)) > g \end{cases}$$

where $\mathrm{lc}(p(x))$ is the leading coefficient (in the variable x) of $p(x)$. It is then easy to obtain $\mathrm{lc}_\infty(f_{q,D_2})$ from the computations in Miller's algorithm.

A more detailed description of Algorithm 1 is given in Algorithm 2. The computations coming from Cantor's algorithm can be replaced by explicit formulae, with some minor changes as $\tilde{v}_1(x)$ must be computed completely, both in the addition and the doubling formulae (in explicit formulae, the computation of $\tilde{v}_1(x)$ is avoided to reduce costs).

Remark 3. For genus 2, computing $res(c_i(x), u_b(x))$ is relatively inexpensive compared with polynomial operations, and it is more efficient to compute the resultant every time we accumulate on c_i rather than working with polynomials (squaring c_i in the doubling step will then become a single field operation). For all other genera, it is more efficient to accumulate c_1 and c_2 as polynomials and to compute resultants only in the final step of Algorithm 2.

Algorithm 2. Miller's algorithm for hyperelliptic curves (detailed)

Inputs: $n \in \mathbb{N}$ and $D_a, D_b \in J_C$ reduced with disjoint affine support, $D_a = [u_a(x), v_a(x)]$, $D_b = [u_b(x), v_b(x)]$

Outputs: $f_{n,D_a}(D_b)$

   Write n as $\sum_{j=0}^{s} n_j 2^j$, with $n_j \in \{0, 1\}$ and $n_s = 1$.

   $D = [u(x), v(x)] \leftarrow D_a$, $c_1(x) \leftarrow 1$, $c_2(x) \leftarrow 1$, $c_3 \leftarrow 1$.

   for $j = s - 1$ down to 0 **do**

      $c_1(x) \leftarrow c_1(x)^2 \bmod u_b(x)$

      $c_2(x) \leftarrow c_2(x)^2 \bmod u_b(x)$

      $c_3 \leftarrow c_3^2$

      $d(x) \leftarrow \gcd(u(x), 2v(x) + h(x))$

      $[\tilde{u}_1(x), \tilde{v}_1(x)] \leftarrow 2D - \mathrm{div}(d(x))$

      $c_1(x) \leftarrow c_1(x) \cdot d(x) \bmod u_b(x)$

      $j \leftarrow 1$

      while $\deg(\tilde{u}_j) > g$ **do**

         $\tilde{u}_{j+1}(x) = Monic\left(\frac{\tilde{v}_j(x)^2 + h(x)\tilde{v}_j(x) - f(x)}{\tilde{u}_j(x)}\right)$.

         $\tilde{v}_{j+1}(x) = -\tilde{v}_j(x) - h(x) \bmod \tilde{u}_{j+1}(x)$.

         $c_1(x) \leftarrow c_1(x) \cdot (v_b(x) - \tilde{v}_j(x)) \bmod u_b(x)$

         $c_2(x) \leftarrow c_2(x) \cdot \tilde{u}_{j+1}(x) \bmod u_b(x)$

         $c_3 \leftarrow c_3 \cdot \mathrm{lc}_\infty(y - \tilde{v}_j)$

         $j \leftarrow j + 1$

      end while

      $D = [u(x), v(x)] \leftarrow [\tilde{u}_j(x), \tilde{v}_j(x)]$

      if $n_j = 1$ **then**

         $d(x) \leftarrow \gcd(u(x), u_a(x), v(x) + v_a(x) + h(x))$

         $[\tilde{u}_1(x), \tilde{v}_1(x)] \leftarrow D + D_a - \mathrm{div}(d(x))$

         $c_1(x) \leftarrow c_1(x) \cdot d(x) \bmod u_b(x)$

         $j \leftarrow 1$

         while $\deg(\tilde{u}_j) > g$ **do**

            $\tilde{u}_{j+1}(x) = Monic\left(\frac{\tilde{v}_j(x)^2 + h(x)\tilde{v}_j(x) - f(x)}{\tilde{u}_j(x)}\right)$.

            $\tilde{v}_{j+1}(x) = -\tilde{v}_j(x) - h(x) \bmod \tilde{u}_{j+1}(x)$.

            $c_1(x) \leftarrow c_1(x) \cdot (v_b(x) - \tilde{v}_j(x)) \bmod u_b(x)$

            $c_2(x) \leftarrow c_2(x) \cdot \tilde{u}_{j+1}(x) \bmod u_b(x)$

            $c_3 \leftarrow c_3 \cdot \mathrm{lc}_\infty(y - \tilde{v}_j)$

            $j \leftarrow j + 1$

         end while

         $D = [u(x), v(x)] \leftarrow [\tilde{u}_j(x), \tilde{v}_j(x)]$

      end if

   end for

   $c \leftarrow \frac{res(c_1(x), u_b(x))}{c_3 \cdot res(c_2(x), u_b(x))}$

   Return c.

A.2   Operation Count

In general, one cannot assume that f_{q,D_2} obtained from the computations of Algorithm 1 is normalised to have $\mathrm{lc}_\infty(f_{q,D_2}) = 1$. The evaluation of $a(\overline{D}_2, \overline{D}_1)$ in Lemma 6 is then computed as

$$a(\overline{D}_2, \overline{D}_1) = \frac{f_{q,D_2}\left(\epsilon(\overline{D}_1)\right)}{\mathrm{lc}_\infty(f_{q,D_2})^{m_1}} .$$ (9)

Tables 1 and 2 give the cost in field operations for the doubling and addition steps of Algorithm 2, for general divisors. The row "first & last" takes into account the cost of the resultants and final multiplications and inversion, as well as the operations saved by having $c_1 = c_2 = c_3 = 1$ in the first doubling step.

Table 1. Costs involved in Miller's algorithm to compute $a(\overline{D}_2, \overline{D}_1)$ using general divisors

		genus 2	genus 3
addition	$\mathbb{F}_q$	$7kM$	$32kM$
	$\mathbb{F}_{q^k}$	$1I + 29M + 5S$	$1I + 91M + 6S$
doubling	$\mathbb{F}_q$	$7kM$	$42kM$
	$\mathbb{F}_{q^k}$	$1I + 29M + 9S$	$1I + 88M + 22S$
first & last	$\mathbb{F}_q$	0	$-8kM$
	$\mathbb{F}_{q^k}$	$1I - 1M - 2S$	$1I + 7M - 13S$

Table 2. Costs involved in Miller's algorithm to compute $\hat{a}(\overline{D}_1, \overline{D}_2)$ for superspecial curves using general divisors

		genus 2	genus 3
addition	$\mathbb{F}_q$	$1I + (25 + 3k)M + 3S$	$1I + (67 + 12k)M + 6S$
	$\mathbb{F}_{q^k}$	$8M + 2S$	$44M$
doubling	$\mathbb{F}_q$	$1I + (25 + 3k)M + 6S$	$1I + (64 + 12k)M + 10S$
	$\mathbb{F}_{q^k}$	$8M + 4S$	$54M + 12S$
first & last	$\mathbb{F}_q$	$(k - 1)M - 1S$	$(k - 1)M - 1S$
	$\mathbb{F}_{q^k}$	$1I - 1M - 2S$	$1I - 1M - 12S$

Table 3. Costs involved in Miller's algorithm to compute $a(\overline{D}_2, \overline{D}_1)$ using degenerate divisors

		genus 2	genus 3
addition	$\mathbb{F}_q$	$4kM$	$13kM$
	$\mathbb{F}_{q^k}$	$1I + 27M + 3S$	$1I + 69M + 6S$
doubling	$\mathbb{F}_q$	$4kM$	$13kM$
	$\mathbb{F}_{q^k}$	$1I + 27M + 7S$	$1I + 66M + 11S$
first & last	$\mathbb{F}_q$	$1M + 1S$	$2M + 2S$
	$\mathbb{F}_{q^k}$	$1I - 1M - 2S$	$1I - 1M - 2S$

Tables 3 and 4 give the cost in field operations for the doubling and addition steps of Algorithm 2, for degenerate divisors, i.e. for divisors whose support is a single point (together with the point at infinity).

Table 4. Costs involved in Miller's algorithm to compute $\hat{a}(\overline{D}_1, \overline{D}_2)$ for superspecial curves using degenerate divisors

		genus 2	genus 3
addition	$\mathbb{F}_q$	$1I + (25 + 4k)M + 3S$	$1I + (67 + 13k)M + 6S$
	$\mathbb{F}_{q^k}$	$2M$	$2M$
doubling	$\mathbb{F}_q$	$1I + (25 + 4k)M + 6S$	$1I + (64 + 13k)M + 6S$
	$\mathbb{F}_{q^k}$	$2M + 2S$	$2M + 2S$
first & last	$\mathbb{F}_q$	$(k - 1)M - 1S$	$(k - 1)M - 1S$
	$\mathbb{F}_{q^k}$	$1I - 1S$	$1I + 1M$

Each addition (respectively doubling) step uses the fastest known explicit formulae in affine coordinates, adapted to include the computation of $\tilde{v}_1(x)$. For the genus three addition, we use the formulae of [14] with the resultant replaced by Cramer's rule (as was done for characteristic 2 in [3]). For the final computations with the resultants, we go back to the resultant computation of [14].

A.3   Performance Comparison

In this section we provide precise operation counts for three security levels: 80, 128 and 192-bit security. The sizes of the finite fields and the security parameters k are chosen such that both the DLP in the Jacobian of the curve $J_C(\mathbb{F}_q)$ and the DLP in the embedding field $\mathbb{F}_{q^k}$ are infeasible.

Following [13] we restrict to the use of so-called pairing friendly finite fields, i.e. $\mathbb{F}_q$ is a prime field with $q \equiv 1 \bmod 12$ and k of the form $2^i 3^j$. For these fields, the cost of the required operations of multiplication, squaring, and inversion can each be expressed simply in terms of base field operations [13], where m, s and i denote the cost of a multiplication, squaring and inversion respectively in $\mathbb{F}_q$.

Bearing in mind that for the same security, the base field will be smaller for higher genus, we must account for this in our cost estimates. We therefore express all costs in terms of the number of $\mathbb{F}_{q_3}$ multiplications we need to perform, where q_i is the base field cardinality of the genus i curve. Using basic Karatsuba, we thus have $M_{q_i} = (q_i/q_3)^{1.585} \cdot M_{q_3}$ for $i = 1, 2$. This estimate is likely to be slightly smaller than what is recorded in practice [1] and so will lead our results to underestimate the genus one operation counts slightly; however we believe they are sufficient for comparison purposes.

For simplicity we assume that a squaring costs the same as a multiplication, and that one inversion is equivalent to ten multiplications. We also assume half as many additions as doublings in Algorithm 2.

Table 6 gives the results of our performance estimates. Of the right-most five columns, the left two are based on the formulae given in Table 1 and 2 and Algorithm 2, while the third and fourth are based on Table 3 and 4. The final column we computed using the estimates in [17], together with the final powering cost estimates from [13], taking the minimum over the choice of Ate or twisted Ate, average or small trace, and quadratic or sextic twist.

Table 5. Cost of $\mathbb{F}_{q^k}$ operations in terms of $\mathbb{F}_q$ operations

k	Mul	Sqr	Inv
6	$15m$	$15s$	$21m+13s+i$
12	$45m$	$45s$	$51m+43s+i$
16	$81m$	$81s$	$90m+90s+i$
24	$135m$	$135s$	$141m+133s+i$
32	$243m$	$243s$	$252m+252s+i$
48	$405m$	$405s$	$411m+403s+i$
54	$375m$	$375s$	$591m+343s+i$

Table 6. Number of $\mathbb{F}_{q^3}$ multiplications to compute the Ate pairing

Security	g	q	k	MOV	number of $\mathbb{F}_{q^3}$ muls				
					ordinary	superspecial	ordinary degenerate	superspecial degenerate	fastest Ate
80	1	172	6	1032					3.12×10^4
	2	86	12	1032	3.79×10^5	1.21×10^5	3.34×10^5	4.92×10^4	
	3	64	16	1024	8.89×10^5	4.82×10^5	6.30×10^5	5.39×10^4	
128	1	256	12	3072					8.63×10^4
	2	128	24	3072	1.64×10^6	4.97×10^5	1.45×10^6	1.78×10^5	
	3	96	32	3072	3.91×10^6	2.12×10^6	2.80×10^6	1.89×10^5	
192	1	384	24	9216					3.87×10^5
	2	192	48	9216	6.68×10^6	1.99×10^6	5.94×10^6	6.60×10^5	
	3	152	54	8208	9.64×10^6	5.18×10^6	6.90×10^6	4.65×10^5	

The table indicates that the Ate pairing for elliptic curves, can be an order of magnitude faster than the basic version of the Ate pairing described in this paper. The reason for the Ate pairing being particularly fast in the elliptic case is the availability of twists, as well as very short traces. Whether high degree twists can be utilised for the hyperelliptic Ate pairing remains open. When using degenerate divisors however, the Ate pairing for superspecial curves with genus two and three is certainly comparable to the genus one case.

Ideal Multipartite Secret Sharing Schemes[*]

Oriol Farràs, Jaume Martí-Farré, and Carles Padró

Dept. of Applied Maths. IV, Technical University of Catalonia, Barcelona
{ofarras,jaumem,cpadro}@ma4.upc.edu

Abstract. Multipartite secret sharing schemes are those having a multipartite access structure, in which the set of participants is divided into several parts and all participants in the same part play an equivalent role. Several particular families of multipartite schemes, such as the weighted threshold schemes, the hierarchical and the compartmented schemes, and the ones with bipartite or tripartite access structure have been considered in the literature. The characterization of the access structures of ideal secret sharing schemes is one of the main open problems in secret sharing. In this work, the characterization of ideal multipartite access structures is studied with all generality. Our results are based on the well-known connections between ideal secret sharing schemes and matroids. One of the main contributions of this paper is the application of discrete polymatroids to secret sharing. They are proved to be a powerful tool to study the properties of multipartite matroids. In this way, we obtain some necessary conditions and some sufficient conditions for a multipartite access structure to be ideal.

Our results can be summarized as follows. First, we present a characterization of matroid-related multipartite access structures in terms of discrete polymatroids. As a consequence of this characterization, a necessary condition for a multipartite access structure to be ideal is obtained. Second, we use linear representations of discrete polymatroids to characterize the linearly representable multipartite matroids. In this way we obtain a sufficient condition for a multipartite access structure to be ideal. Finally, we apply our general results to obtain a complete characterization of ideal tripartite access structures, which was until now an open problem.

Keywords: Secret sharing, Ideal secret sharing schemes, Ideal access structures, Multipartite secret sharing, Multipartite matroids, Discrete polymatroids.

1 Introduction

In a *secret sharing scheme*, every *participant* receives a *share* of a *secret value*. Only the *qualified sets* of participants, which form the *access structure* of the

---

[*] This work was partially supported by the Spanish Ministry of Education and Science under projects TIC 2003-00866 and TSI2006-02731. This work was done partly while the third author was in a sabbatical stay at CWI, Amsterdam. This stay was funded by the *Secretaría de Estado de Educación y Universidades* of the Spanish Ministry of Education.

M. Naor (Ed.): EUROCRYPT 2007, LNCS 4515, pp. 448–465, 2007.
© International Association for Cryptology Research 2007

scheme, can recover the secret value from their shares. This paper deals exclusively with *unconditionally secure perfect* secret sharing schemes, that is, the shares of the participants in an unqualified set do not provide any information about the secret value. The reader will find in [34] an excellent introduction to secret sharing. Observe that the access structure of a secret sharing scheme on a set P of participants is a *monotone increasing* family $\Gamma \subseteq \mathcal{P}(P)$, where $\mathcal{P}(P)$ is the power set of P. That is, every subset of P containing a qualified subset is itself qualified.

Secret sharing was introduced in 1979 by Shamir [31] and Blakley [4], who independently presented two different methods to construct *threshold* secret sharing schemes. Their qualified subsets are those having at least a given number of participants. The threshold schemes proposed in [4,31] are *ideal*, that is, the share of every participant has the same length as the secret, which is the best possible situation in a perfect scheme [16].

Dealing only with threshold access structures can be a serious limitation in some applications of secret sharing. In his seminal paper [31], Shamir made the first attempt to overcome this by proposing a construction of *weighted threshold schemes*. In such a scheme, every participant has a weight (a positive integer) and the sets whose weight sum is greater than a given threshold are qualified. The proposed construction is very simple: take a threshold scheme and give to every participant as many shares as its weight. Nevertheless, the obtained scheme is not ideal anymore. Ito, Saito, and Nishizeki [14] proved, in a constructive way, that there exists a secret sharing scheme for every access structure, but the schemes that are obtained by this method are very far from ideal. Benaloh and Leichter [3] proved that there exist access structures that do not admit any ideal scheme and, as a consequence of the results in [9,11] and other works, in some cases the shares must be much larger than the secret. Actually, very little is known about the construction of efficient secret sharing schemes for general access structures and, in particular, there is a wide gap between the best known lower and upper bounds on the length of the shares.

Due to the difficulty of finding efficient secret sharing schemes for general access structures, it is worthwhile to find families of access structures that admit ideal schemes and have other useful properties for the applications of secret sharing. Brickell [7] proposed a method to construct ideal secret sharing schemes for access structures other than the threshold ones. This method provides ideal schemes for *multilevel* and *compartmented* access structures, two families that were proposed by Simmons [32] because of their interesting applications. These access structures are *multipartite*, that is, the set of participants is divided into several parts and all participants in the same part play an equivalent role. Multipartite access structures are useful in scenarios in which the participants can be divided into different classes, such as hierarchical organizations, or actions that require the agreement of different parties. Other constructions of ideal secret sharing schemes for different classes of multipartite access structures have been presented in [25,35,36].

The natural step beyond the construction of ideal schemes for particular structures is the search of a characterization of the *ideal access structures*, that is, the access structures of ideal secret sharing schemes. This is one of the most important open problems in secret sharing. As a consequence of the results by Brickell [7], and Brickell and Davenport [8], this open problem has important connections with matroid theory. Some basic concepts about matroids and their connection to secret sharing are recalled in Section 4.1.

Brickell and Davenport [8] proved that every ideal secret sharing scheme on a set P of participants determines a matroid $\mathcal{M}$ with ground set $Q = P \cup \{p_0\}$. This matroid determines the access structure of the scheme. Namely, $A \subseteq P$ is a minimal qualified subset if and only if $A \cup \{p_0\}$ is a circuit of $\mathcal{M}$. In this situation, we say that this access structure is *matroid-related* or, more specifically, *related to the matroid* $\mathcal{M}$. Therefore, a necessary condition for an access structure to be ideal is obtained.

Theorem 1. (Brickell and Davenport [8]) *The access structure of every ideal secret sharing scheme is matroid-related.*

The method to construct ideal schemes proposed by Brickell [7], which is based on linear algebra, provides a sufficient condition for an access structure to be ideal.

Theorem 2. (Brickell [7]) *There exists an ideal secret sharing scheme for every access structure that is related to a linearly representable matroid.*

The minimal qualified subsets of a matroid-related access structure form a *matroid port*, a combinatorial object introduced by Lehman [17] in 1964, much before secret sharing was invented. Seymour [29] presented in 1976 a forbidden minor characterization of matroid ports, which has been used recently to obtain new results on the characterization of matroid-related access structures [21]. The *information rate* of a secret sharing scheme is the ratio between the length of the secret and the maximum length of the shares. The main result in [21] is a generalization of Theorem 1.

Theorem 3. (Martí-Farré and Padró [21]) *The access structure of every secret sharing scheme with information rate greater than 2/3 is matroid-related.*

2   Related Work

Due to the difficulty of finding general results, the characterization of ideal access structures has been studied for several particular classes of access structures: the access structures on sets of four [34] and five [15] participants, the ones defined by graphs [6,8,9], the bipartite access structures [28], those with three or four minimal qualified subsets [18], the ones with intersection number equal to one [19], the access structures with rank three [20], and the weighted threshold access structures [2]. In most of these families, all the matroids that are related to access structures in the family are representable, and then the matroid-related access structures coincide with the ideal ones. This, combined with Theorem 3, implies that the optimal information rate of every non-ideal access structure in those families is at most 2/3.

Multipartite access structures were first introduced by Shamir [31] in his introductory work, in which weighted threshold access structures were considered. These structures have been studied also in [23,28] and a characterization of the ideal weighted access structures has been presented in [2]. Brickell [7] constructed ideal secret sharing schemes for several different kinds of multipartite access structures that had been previously considered by Simmons [32]. Other constructions of ideal schemes for these and other multipartite structures have been presented in [12,25,35,36]. A complete characterization of ideal bipartite access structures was given in [28] and, independently, in [24,26]. Partial results on the characterization of tripartite access structures have been presented in [2,10,12]. The first attempt to provide general results on the characterization of ideal multipartite access structures has been made recently by Herranz and Sáez [12]. They present some necessary conditions for a multipartite access structure to be ideal, which generalize the ones given in [10] for the tripartite case. In addition, they present a wide family of ideal tripartite access structures.

3   Our Results

In this paper, we study the characterization of the *ideal multipartite access structures*. Since we can always consider as many parts as participants, every access structure is multipartite, and hence we are not dealing here with a particular family of structures, but with the general problem of the characterization of the ideal access structures. Of course, we do not solve this long-standing open problem. Nevertheless, we present some new results by looking at it under a different point of view. Namely, we investigate the conditions given in Theorems 1 and 2 by taking into account that the set of participants can be divided into several parts formed by participants playing an equivalent role in the structure. We introduce the natural concept of *multipartite matroid*, which applies to the matroids that are defined from ideal multipartite secret sharing schemes. The study of multipartite matroids leads to discrete polymatroids, which appear to be a very powerful tool to characterize the matroid-related multipartite access structures. Even though our results can be applied to the general case, their most meaningful consequences are obtained when applied to some particular families of multipartite access structures. Specifically, in the case that the number of parts is significantly smaller than the number of participants, or in situations in which the parts are distributed in some special way as, for instance, in hierarchical access structures. In particular, we present a complete characterization of the ideal tripartite access structures, which was an open question until now. Our main contributions are described with more detail in the following.

First, we investigate how the necessary condition in Theorem 1 can be applied to multipartite access structures. Consequently, we study the properties of matroid-related multipartite access structures. The partition in the set of participants of a matroid-related access structure extends to the set of points of the corresponding matroid. This leads us to introduce the natural concept of *multipartite matroid*. We point out that every multipartite matroid with m parts defines a *discrete polymatroid* on a set of m points. Discrete polymatroids are a particular class of polymatroids. In the same way as matroids abstract the combinatorial properties of a collection of vectors in a vector space, discrete polymatroids abstract the combinatorial properties of a collection of subspaces in a vector space. Discrete polymatroids have been thoroughly studied by Herzog and Hibi [13], and some of the results in that paper are used here. By using discrete polymatroids, we present in Theorem 8 a characterization of matroid-related multipartite access structures, which implies a necessary condition for a multipartite access structure to be ideal. We present some examples showing that this necessary condition is a useful tool to prove that a given multipartite structure is not ideal.

Second, we study the application of Theorem 2 to multipartite access structures. Therefore, we study the existence of linear representations for multipartite matroids, and we relate them to linear representations of discrete polymatroids. In the same way as in a linear representation of a matroid a vector is assigned to each point in the ground set, a subspace is assigned to each point in a linear representation of a discrete polymatroid. We prove in Theorem 13 that a multipartite matroid is linearly representable if and only if the corresponding discrete polymatroid is linearly representable. This implies a sufficient condition for a multipartite access structure to be ideal. We think that Theorem 13 is interesting not only for its implications in secret sharing, but also as a result about representability of matroids. This result is specially useful if the number of parts is small. For instance, a tripartite matroid can have many points, but, as a consequence of our result, we only have to find three suitable subspaces of a vector space to prove that it is linearly representable.

And third, we apply our general results to the tripartite case, and we present a complete characterization of the ideal tripartite access structures. By using Theorem 8, we characterize the matroid-related tripartite access structures. Theorem 13 is used to prove that all matroids related to these structures are linearly representable, and hence that all matroid-related tripartite access structures are ideal. Moreover, as a consequence of Theorem 3, the optimal information rate of every non-ideal tripartite access structure is at most $2/3$. The application of our general results to the tripartite case requires to solve some non-trivial problems. Therefore, our characterization of the ideal tripartite access structures is not a simple corollary of the main theorems in this paper.

We observe that the last result above cannot be extended to m-partite access structures with $m \geq 4$, because there does not exist any ideal secret sharing scheme defining the Vamos matroid [1,30,33], which is quadripartite. Hence, there exist matroid-related quadripartite access structures that are not ideal. Nevertheless, this does not mean that our general results are not useful for m-partite access structures with $m \geq 4$, as it is demonstrated with some examples.

After the results in this paper, the open problems about the characterization of ideal multipartite access structures are as difficult as the open problems in the general case. That is, closing the gap between the necessary and the sufficient conditions requires to solve very difficult problems about representations of matroids and polymatroids.

4   Multipartite Access Structures, Multipartite Matroids, and Discrete Polymatroids

4.1   Ideal Secret Sharing Schemes and Matroids

As a consequence of the results by Brickell [7], and Brickell and Davenport [8], the characterization of the *ideal* access structures, that is the access structures of ideal schemes, has important connections with matroid theory.

To illustrate these connections, we describe the construction of ideal secret sharing schemes due to Brickell [7]. Given a set P of participants, consider a special participant $p_0 \notin P$, which is usually called *dealer*, and $Q = P \cup \{p_0\}$. Every mapping $\psi \colon Q \to E$, where E is a vector space over some finite field $\mathbb{K}$, determines an ideal secret sharing scheme Σ_ψ on the set P of participants. Given a secret value $s_0 \in \mathbb{K}$, a random vector $\mathbf{x} \in E$ such that the dot product $\mathbf{x} \cdot \psi(p_0)$ is equal to s_0 is chosen uniformly at random. The share of the participant $i \in P$ is the value $s_i = \mathbf{x} \cdot \psi(i) \in \mathbb{K}$. A subset $A \subseteq P$ is in the access structure Γ of the scheme Σ_ψ if and only if the vector $\psi(p_0)$ is a linear combination of the vectors in $\{\psi(i) : i \in A\}$. The ideal schemes of this form are called $\mathbb{K}$-*vector space secret sharing schemes*, and their access structures are called $\mathbb{K}$-*vector space access structures*.

The access structure of Σ_ψ is determined by the *rank function* $r \colon \mathcal{P}(Q) \to \mathbb{Z}$, where $\mathcal{P}(Q)$ is the power set of Q and, for every $X \subseteq Q$, the value $r(X)$ is the dimension of the subspace of E spanned by the set $\{\psi(i) : i \in X\}$. Actually, a subset $A \subseteq P$ is qualified if and only if $r(A \cup \{p_0\}) = r(A)$. It is easy to check that the function r satisfies

1. $0 \le r(X) \le |X|$ for every $X \subseteq Q$, and
2. r is *monotone increasing*: if $X \subseteq Y \subseteq Q$, then $r(X) \le r(Y)$, and
3. r is *submodular*: $r(X \cup Y) + r(X \cap Y) \le r(X) + r(Y)$ for every pair of subsets X, Y of Q.

Matroids are combinatorial objects that abstract and generalize many concepts from linear algebra, including ranks, independent sets, bases, and subspaces. The reader is referred to [27,37] for general references on matroid theory. One of the many possible equivalent definitions for this concept says that a matroid is a pair (Q, r) formed by a finite set Q, the *ground set*, and a *rank function* $r \colon \mathcal{P}(Q) \to \mathbb{Z}$ satisfying the properties above. A matroid $\mathcal{M} = (Q, r)$ is said to be $\mathbb{K}$-*linearly representable* if there exists a $\mathbb{K}$-vector space E and a mapping $\psi \colon Q \to E$ assigning a vector to each element in Q such that the rank function r can be defined from ψ as before.

For a matroid $\mathcal{M} = (Q, r)$ and a point $p_0 \in Q$, we define the access structure $\Gamma_{p_0}(\mathcal{M})$ on the set of participants $P = Q - \{p_0\}$ by $\Gamma_{p_0}(\mathcal{M}) = \{A \subseteq P : r(A \cup \{p_0\}) = r(A)\}$. The access structures of this form are called *matroid-related* (the definition we gave in the Introduction for this concept is equivalent to this one). If the access structure $\Gamma_{p_0}(\mathcal{M})$ is *connected*, that is, if every participant is in a minimal qualified subset, then the matroid $\mathcal{M}$ is univocally determined by $\Gamma_{p_0}(\mathcal{M})$. Observe that Γ is a $\mathbb{K}$-vector space access structure if and only if $\Gamma = \Gamma_{p_0}(\mathcal{M})$ for some $\mathbb{K}$-linearly representable matroid $\mathcal{M}$. Therefore, as a consequence of the construction by Brickell [7], we obtain Theorem 2, a sufficient condition for an access structure to be ideal.

Brickell and Davenport [8] proved that this sufficient condition is not very far from being necessary. Specifically, they proved that every ideal secret sharing scheme on a set P of participants determines a matroid $\mathcal{M}$ with ground set $Q = P \cup \{p_0\}$ such that the access structure of the scheme is $\Gamma_{p_0}(\mathcal{M})$. Therefore, a necessary condition for an access structure to be ideal is obtained (Theorem 1).

Matroids that are obtained from ideal secret sharing schemes are said to be *secret sharing representable* (or *ss-representable* for short). Therefore, an access structure is ideal if and only if it is related to a ss-representable matroid. Since there exist non-ss-representable matroids, the necessary condition in Theorem 1 is not sufficient. The first example, the Vamos matroid, was found by Seymour [30]. Other proofs of this fact were presented in [1,33]. Many other examples non-ss-representable matroids were given by Matúš [22]. In addition, the sufficient condition in Theorem 2 is not necessary because of the non-Pappus matroid, which is not linearly representable but was proved to be ss-representable by Simonis and Ashikhmin [33].

At this point, two open problems arise that are central in the characterization of ideal access structures. First, the characterization of matroid-related access structures and, second, the characterization of ss-representable matroids.

A number of important results and interesting ideas for future research on the characterization of ss-representable matroids can be found in the works by Simonis and Ashikhmin [33] and Matúš [22]. The first one deals with the geometric structure that lies behind ss-representations of matroids. The second one analyzes the algebraic properties that the matroid induces in all its ss-representations. These properties make it possible to find some restrictions on the ss-representations of a given matroid and, in some cases, to exclude the existence of such representations. By using these tools, Matúš [22] presented an infinite family of non-ss-representable matroids with rank three.

4.2   Matroids, Integer Polymatroids, and Discrete Polymatroids

Matroids have been defined in Section 4.1 by using the rank function. There exist many other definitions. We present in the following the ones based on independent sets and on bases. The equivalence between them, which is proved in [27], will be useful to obtain our results.

Let $\mathcal{M} = (Q, r)$ be a matroid. The subsets $X \subseteq Q$ with $r(X) = |X|$ are said to be *independent*. The family $\mathcal{I} \subseteq \mathcal{P}(Q)$ of the independent sets of $\mathcal{M}$ is a nonempty family of subsets characterized by the following two properties.

1. If $I \in \mathcal{I}$ and $I' \subseteq I$, then $I' \in \mathcal{I}$, and
2. if I_1 and I_2 are in $\mathcal{I}$ and $|I_1| < |I_2|$, then there exists $x \in I_2 - I_1$ such that $I_1 \cup \{x\} \in \mathcal{I}$.

The *bases* of the matroid $\mathcal{M}$ are the maximally independent sets. Similarly to the independent sets, the nonempty family $\mathcal{B}$ of the bases determines the matroid. Moreover, a nonempty subset $\mathcal{B} \subseteq \mathcal{P}(Q)$ is the family of bases of a matroid on Q if and only if the following *exchange condition* is satisfied.

– For every $B_1, B_2 \in \mathcal{B}$ and $x \in B_1 - B_2$, there exists $y \in B_2 - B_1$ such that $(B_1 - \{x\}) \cup \{y\}$ is in $\mathcal{B}$.

All bases have the same number of elements, which is the *rank* of $\mathcal{M}$ and is denoted $r(\mathcal{M})$. Actually, $r(\mathcal{M}) = r(Q)$. The *dependent* sets are those that are not independent, and a *circuit* is a minimally dependent set. A matroid is said to be *connected* if, for every two points $x, y \in Q$, there exists a circuit C with $x, y \in C$.

If E is a $\mathbb{K}$-vector space and $\psi \colon Q \to E$ is a $\mathbb{K}$-linear representation of the matroid $\mathcal{M} = (Q, r)$, then a subset $X \subseteq Q$ is independent (respectively, a basis) if and only if the multiset $\{\psi(i) : i \in X\}$, where some values may be repeated, is a linearly independent set of vectors in E (respectively, a basis of the subspace of E spanned by $\psi(Q)$).

A *polymatroid* is a pair $\mathcal{Z} = (J, h)$ formed by a finite set J, the *ground set*, and a *rank function* $h \colon \mathcal{P}(J) \to \mathbb{R}$ satisfying

1. $h(\emptyset) = 0$, and
2. h is *monotone increasing*: if $X \subseteq Y \subseteq J$, then $h(X) \leq h(Y)$, and
3. h is *submodular*: if $X, Y \subseteq J$, then $h(X \cup Y) + h(X \cap Y) \leq h(X) + h(Y)$.

If the rank function h is integer-valued, we say that $\mathcal{Z}$ is an *integral polymatroid*. The reader is referred to [37] for more information about polymatroids.

The following example of an integral polymatroid illustrates the similarity with matroids. In the same way as matroids abstract some properties of collections of vectors, integral polymatroids do the same with collections of subspaces. Let E be a $\mathbb{K}$-vector space, and $V_1, \ldots, V_m$ subspaces of E. It is not difficult to check that the mapping $h \colon \mathcal{P}(\{1, \ldots, m\}) \to \mathbb{Z}$ defined by $h(X) = \dim(\sum_{i \in X} V_i)$ is the rank function of an integral polymatroid $\mathcal{Z} = (\{1, \ldots, m\}, h)$. The integral polymatroids that can be defined in this way are said to be $\mathbb{K}$-*linearly representable*.

Discrete polymatroids were introduced by Herzog and Hibi [13]. They are closely related to integral polymatroids. In addition, we show in the following that discrete polymatroids are extremely useful to study multipartite matroids, and hence they are a very important tool in the characterization of ideal multipartite access structures.

We need to introduce some notation. For every integer $m \geq 1$, we consider the set $J_m = \{1, \ldots, m\}$. Let $\mathbb{Z}_+^m$ denote the set of vectors $u = (u_1, \ldots, u_m) \in \mathbb{Z}^m$ with $u_i \geq 0$ for every $i \in J_m$. If $u, v \in \mathbb{Z}_+^m$, we write $u \leq v$ if $u_i \leq v_i$ for every $i \in J_m$, and we write $u < v$ if $u \leq v$ and $u \neq v$. The vector $w = u \vee v$ is defined by $w_i = \max\{u_i, v_i\}$. The *modulus* of a vector $u \in \mathbb{Z}_+^m$ is $|u| = u_1 + \cdots + u_m$. For every subset $X \subseteq J_m$, we write $u(X) = (u_i)_{i \in X} \in \mathbb{Z}_+^{|X|}$ and $|u(X)| = \sum_{i \in X} u_i$.

A *discrete polymatroid* with *ground set* J_m is a nonempty finite set of vectors $\mathcal{D} \subset \mathbb{Z}_+^m$ satisfying

1. if $u \in \mathcal{D}$ and $v \in \mathbb{Z}_+^m$ is such that $v \leq u$, then $v \in \mathcal{D}$, and
2. for every pair of vectors $u, v \in \mathcal{D}$ with $|u| < |v|$, there exists $w \in \mathcal{D}$ with $u < w \leq u \vee v$.

A *basis* of a discrete polymatroid $\mathcal{D}$ is a maximal element in $\mathcal{D}$, that is, a vector $u \in \mathcal{D}$ such that there does not exist any $v \in \mathcal{D}$ with $u < v$. Similarly to matroids, all bases have the same modulus. In addition, a discrete polymatroid is determined by its bases. Specifically, in [13, Theorem 2.3] it is proved that a nonempty subset $\mathcal{B} \subset \mathbb{Z}_+^m$ is the family of bases of a discrete polymatroid if and only if it satisfies the following *exchange condition*.

– For every $u \in \mathcal{B}$ and $v \in \mathcal{B}$ with $u_i > v_i$, there exists $j \in J_m$ such that $u_j < v_j$ and $u - \mathbf{e}_i + \mathbf{e}_j \in \mathcal{B}$, where $\mathbf{e}_i$ denotes the i-th vector of the canonical basis of $\mathbb{R}^m$.

The mapping $h \colon \mathcal{P}(J_m) \to \mathbb{Z}$ defined by $h(X) = \max\{|u(X)| : u \in \mathcal{D}\}$ is called the *rank function* of the discrete polymatroid $\mathcal{D}$. As a consequence of a result by Herzog and Hibi [13, Theorem 3.4], there is a one-to-one correspondence between discrete polymatroids and integral polymatroids, as it is stated in the following proposition. Because of that, from now on we will deal only with discrete polymatroids.

Proposition 4. *A mapping $h \colon \mathcal{P}(J_m) \to \mathbb{Z}$ is the rank function of a discrete polymatroid $\mathcal{D} \subset \mathbb{Z}_+^m$ with ground set J_m if and only if (J_m, h) is an integral polymatroid. In addition, a discrete polymatroid $\mathcal{D}$ is univocally determined from its rank function h because $\mathcal{D} = \{u \in \mathbb{Z}_+^m : |u(X)| \leq h(X) \text{ for every } X \subseteq J_m\}$.*

4.3   Multipartite Access Structures and Multipartite Matroids

An *m-partition* $\Pi = (X_1, \ldots, X_m)$ of a set X is a disjoint family of m nonempty subsets of X with $X = X_1 \cup \cdots \cup X_m$. Let $\Lambda \subseteq \mathcal{P}(X)$ be a family of subsets of X. For a permutation σ on X, we define $\sigma(\Lambda) = \{\sigma(A) : A \in \Lambda\} \subseteq \mathcal{P}(X)$. A family of subsets $\Lambda \subseteq \mathcal{P}(X)$ is said to be *Π-partite* if $\sigma(\Lambda) = \Lambda$ for every permutation σ such that $\sigma(X_i) = X_i$ for every $X_i \in \Pi$. We say that Λ is *m-partite* if it is Π-partite for some m-partition Π.

These concepts can be applied to access structures Γ, which are actually families of subsets of the set of participants P, and they can be applied as well to the family of independent sets of a matroid. A matroid $\mathcal{M} = (Q, r)$ is *Π-partite* if its family of independent subsets $\mathcal{I} \subseteq \mathcal{P}(Q)$ is Π-partite.

If a multipartite access structure is matroid-related, then the corresponding matroid is multipartite for a similar partition. Specifically, we have the following result.

Lemma 5. *Let $\mathcal{M} = (Q, r)$ be a connected matroid and, for a point $p_0 \in Q$, consider the partitions $\Pi = (P_1, \ldots, P_m)$ and $\Pi_0 = (\{p_0\}, P_1, \ldots, P_m)$ of the sets $P = Q - \{p_0\}$ and Q, respectively. Then the matroid-related connected access structure $\Gamma = \Gamma_{p_0}(\mathcal{M})$ on P is Π-partite if and only if the matroid $\mathcal{M} = (Q, r)$ is Π_0-partite.*

The members of a Π-partite family of subsets are determined by the number of elements they have in each part. We formalize this in the following and we obtain a compact way to represent a multipartite family of subsets. Let $\Pi = (X_1, \ldots, X_m)$ be a partition of a set X. For every $A \subseteq X$ and $i \in J_m$, we define $\Pi_i(A) = |A \cap X_i|$. The partition Π defines a mapping $\Pi \colon \mathcal{P}(X) \to \mathbb{Z}_+^m$ by considering $\Pi(A) = (\Pi_1(A), \ldots, \Pi_m(A))$. If a family $\Lambda \subseteq \mathcal{P}(X)$ of subsets is Π-partite, then $A \in \Lambda$ if and only if $\Pi(A) \in \Pi(\Lambda)$. That is, Λ is completely determined by the set of vectors $\Pi(\Lambda) \subset \mathbb{Z}_+^m$, and hence we can describe an m-partite family of subsets by using vectors in $\mathbb{Z}_+^m$. The following result shows the close connection between multipartite matroids and discrete polymatroids. It can be easily proved by using Proposition 4 and the properties of the independent sets of a matroid.

Proposition 6. *Let $\Pi = (Q_1, \ldots, Q_m)$ be an m-partition of a set Q and let $\mathcal{I} \subseteq \mathcal{P}(Q)$ be a Π-partite family of subsets. Then $\mathcal{I}$ is the family of independent sets of a Π-partite matroid $\mathcal{M} = (Q, r)$ if and only if $\Pi(\mathcal{I}) \subset \mathbb{Z}_+^m$ is a discrete polymatroid. In addition, if $\mathcal{M} = (Q, r)$ is a Π-partite matroid and $h \colon \mathcal{P}(J_m) \to \mathbb{Z}$ is the rank function of the discrete polymatroid $\Pi(\mathcal{I}) \subset \mathbb{Z}_+^m$, then $h(X) = r(\bigcup_{i \in X} Q_i)$ for every $X \subseteq J_m$.*

For a Π-partite matroid $\mathcal{M} = (Q, \mathcal{I})$, we say that $\Pi(\mathcal{I}) \subset \mathbb{Z}_+^m$ is the *discrete polymatroid associated with* $\mathcal{M}$. Clearly, a Π-partite matroid is univocally determined from its associated discrete polymatroid and the partition Π.

5   Matroid-Related Multipartite Access Structures

By using the connection between multipartite matroids and discrete polymatroids we discussed in the previous section, we present a characterization of matroid-related multipartite access structures based on discrete polymatroids. This characterization provides a necessary condition for a multipartite access structure to be ideal.

For every integer $m \geq 1$, we consider the sets $J_m = \{1, \ldots, m\}$ and $J_m' = \{0, 1, \ldots, m\}$. Let $\mathcal{D} \subset \mathbb{Z}_+^m$ be a discrete polymatroid with ground set J_m and rank function $h \colon \mathcal{P}(J_m) \to \mathbb{Z}$. We say that a discrete polymatroid $\mathcal{D}' \subset \mathbb{Z}_+^{m+1}$ with ground set J_m' *completes* $\mathcal{D}$ if its rank function $h' \colon \mathcal{P}(J_m') \to \mathbb{Z}$ is such that $h'(X) = h(X)$ for every $X \subseteq J_m$ while $h'(\{0\}) = 1$ and $h'(J_m') = h(J_m)$.

Since the rank function of $\mathcal{D}'$ is an extension of the one of $\mathcal{D}$, both will be usually denoted by h. For a polymatroid $\mathcal{D}'$ that completes $\mathcal{D}$, consider the family $\Delta(\mathcal{D}') = \{X \subseteq J_m : h(X \cup \{0\}) = h(X)\} \subseteq \mathcal{P}(J_m)$. Given a discrete polymatroid $\mathcal{D}$ with ground set J_m, every completion $\mathcal{D}'$ of $\mathcal{D}$ is determined by $\Delta(\mathcal{D}')$. The next proposition characterizes the families of subsets $\Delta \subseteq \mathcal{P}(J_m)$ for which there exists $\mathcal{D}'$ with $\Delta = \Delta(\mathcal{D}')$. This result will be very useful in the characterization of ideal tripartite access structures.

Proposition 7. *Let $\mathcal{D}$ be a discrete polymatroid with ground set J_m and rank function h. Consider $\Delta \subseteq \mathcal{P}(J_m)$. Then there exists a completion $\mathcal{D}'$ of $\mathcal{D}$ with $\Delta = \Delta(\mathcal{D}')$ if and only if the following conditions are satisfied.*

1. *The family Δ is monotone increasing, $\emptyset \notin \Delta$, and $J_m \in \Delta$.*
2. *If $X \subset Y \subseteq J_m$ and $X \notin \Delta$ while $Y \in \Delta$, then $h(X) < h(Y)$.*
3. *If $X, Y \in \Delta$ and $X \cap Y \notin \Delta$, then $h(X \cup Y) + h(X \cap Y) < h(X) + h(Y)$.*

We say that $\Delta \subseteq \mathcal{P}(J_m)$ is $\mathcal{D}$-*compatible* if it satisfies the conditions in Proposition 7. For every $X \subseteq J_m$, we define the discrete polymatroid $\mathcal{D}(X)$ with ground set X by $\mathcal{D}(X) = \{u(X) : u \in \mathcal{D}\} \subset \mathbb{Z}_+^{|X|}$, and we consider the set of vectors $\mathcal{B}(X) \subset \mathbb{Z}_+^m$ such that $u \in \mathcal{B}(X)$ if and only if $u(X)$ is a basis of $\mathcal{D}(X)$ and $u_i = 0$ for every $i \in J_m - X$. Finally, for a family $\Delta \subseteq \mathcal{P}(J_m)$, we define $\mathcal{G}(\Delta) = \bigcup_{X \in \Delta} \mathcal{B}(X) \subset \mathbb{Z}_+^m$. Our characterization of matroid-related multipartite access structures is given in the following theorem. Since every ideal access structure must be matroid-related, this result provides a necessary condition for a multipartite access structure to be ideal. Moreover, by Theorem 3, this a a a necessary condition for a multipartite access structure to admit a secret sharing scheme with information rate greater than $2/3$.

Theorem 8. *Let Π be an m-partition of P and let Γ be a connected Π-partite access structure on P. Then Γ is matroid-related if and only if there exist a discrete polymatroid $\mathcal{D}$ with ground set J_m and a $\mathcal{D}$-compatible family $\Delta \subseteq \mathcal{P}(J_m)$ such that*

$$\Gamma = \{A \subseteq P : \Pi(A) \geq u \text{ for some vector } u \in \mathcal{G}(\Delta)\},$$

or, equivalently, the family $\min \Gamma$ of the minimal qualified subsets of Γ is determined by

$$\Pi(\min \Gamma) = \bigcup_{X \in \Delta} \{u \in \mathcal{B}(X) : |u(Y)| < h(Y) \text{ for every } Y \in \Delta \text{ with } Y \subsetneq X\},$$

where h is the rank function of the discrete polymatroid $\mathcal{D}$.

Proof. Let $\Pi = (P_1, \ldots, P_m)$ and $\Pi_0 = (\{p_0\}, P_1, \ldots, P_m)$ be partitions of the sets P and $Q = P \cup \{p_0\}$, respectively. Let $\mathcal{M} = (Q, r)$ be a connected Π_0-partite matroid and let $\mathcal{D}' = \Pi_0(\mathcal{I}) \subset \mathbb{Z}_+^{m+1}$ be the discrete polymatroid with ground set J_m' associated with $\mathcal{M}$. Observe that, since $\mathcal{M}$ is connected, $\mathcal{D}'$ completes the discrete polymatroid $\mathcal{D} = \mathcal{D}'(J_m)$. Consider the matroid-related Π-partite access structure $\Gamma_{p_0}(\mathcal{M})$. We only have to prove that a subset $A \subseteq P$ is in $\Gamma_{p_0}(\mathcal{M})$ if and only if $\Pi(A) \geq u$ for some vector $u \in \mathcal{G}(\Delta(\mathcal{D}'))$.

Consider a vector $u = (u_1, \ldots, u_m) \in \mathcal{G}(\Delta(\mathcal{D}'))$ and $A \subseteq P$ with $\Pi(A) \geq u$. Then there exists $X \subseteq J_m$ such that $X \in \Delta(\mathcal{D}')$ and $u(X)$ is a basis of $\mathcal{D}(X)$. We can suppose that $X = \{1, \ldots, r\}$, and hence $u = (u_1, \ldots, u_r, 0, \ldots, 0)$. Consider a subset $B \subseteq A$ with $\Pi(B) = u$. Since $\Pi_0(B) = \tilde{u} = (0, u_1, \ldots, u_r, 0, \ldots, 0) \in \mathcal{D}'$, we deduce that B is an independent set of the matroid $\mathcal{M}$. On the other hand, $\Pi_0(B \cup \{p_0\}) = (1, u_1, \ldots, u_r, 0, \ldots, 0) \notin \mathcal{D}'$ because $\tilde{u}(X)$ is a basis of $\mathcal{D}'(X)$ and $h(X \cup \{0\}) = h(X)$. Therefore, $B \cup \{p_0\}$ is a dependent set of $\mathcal{M}$. This, together with the independence of B, implies that $B \in \Gamma_{p_0}(\mathcal{M})$ and, hence, $A \in \Gamma_{p_0}(\mathcal{M})$.

Let $A \subseteq P$ be a minimal qualified subset of $\Gamma_{p_0}(\mathcal{M})$ and let $X = \{i \in J_m : A \cap P_i \neq \emptyset\}$. We can suppose that $X = \{1, \ldots, r\}$. Consider $u = \Pi_0(A) = (0, u_1, \ldots, u_r, 0, \ldots, 0)$. Observe that $u \in \mathcal{D}'$ because A is an independent set of $\mathcal{M}$. The proof is concluded by checking that $X \in \Delta(\mathcal{D}')$ and that $u(X)$ is a basis of $\mathcal{D}'(X)$. If, on the contrary, $u(X)$ is not a basis of $\mathcal{D}'(X)$, we can suppose without loss of generality that $v = (0, u_1 + 1, u_2, \ldots, u_r, 0, \ldots, 0) \in \mathcal{D}'$. Since A is a minimal qualified subset of $\Gamma_{p_0}(\mathcal{M})$, the set $A \cup \{p_0\}$ is a circuit of $\mathcal{M}$ and, hence, $B = (A \cup \{p_0\}) - \{p_1\}$ is an independent set of $\mathcal{M}$ for every $p_1 \in A \cap P_1$. Therefore, $w = \Pi_0(B) = (1, u_1 - 1, u_2, \ldots, u_r, 0, \ldots, 0) \in \mathcal{D}'$. Since $|v| > |w|$, there exists $x \in \mathcal{D}'$ with $w < x \leq w \vee v$. This implies that $x = (1, u_1, u_2, \ldots, u_r, 0, \ldots, 0) = \Pi_0(A \cup \{p_0\}) \in \mathcal{D}'$, a contradiction. Therefore, $u(X)$ is a basis of $\mathcal{D}'(X)$, and this implies $h(X \cup \{0\}) = h(X)$ because $(1, u_1, u_2, \ldots, u_r, 0, \ldots, 0) \notin \mathcal{D}'$. Hence, $X \in \Delta(\mathcal{D}')$.   □

The condition in Theorem 8 seems very involved and difficult to check. Nevertheless, as we see in the following corollaries and examples, it provides useful tools to check that a given multipartite access structure is not ideal. An important point to be taken into account is that, given a connected matroid-related multipartite access structure Γ, the discrete polymatroid $\mathcal{D}$ and the family of subsets Δ whose existence is proved in Theorem 8 are univocally determined. Effectively, since Γ is connected and matroid-related, there exists a unique matroid $\mathcal{M}$ with $\Gamma = \Gamma_{p_0}(\mathcal{M})$, which determines $\mathcal{D}$ and Δ. Therefore, we can write $\mathcal{D}(\Gamma)$ and $\Delta(\Gamma)$ to represent these objects. For a partition $\Pi = (P_1, \ldots, P_m)$ of a set P, the *support* of a subset $A \subseteq P$ is $\mathrm{supp}(A) = \{i \in J_m : A \cap P_i \neq \emptyset\} \subseteq J_m$. Observe that, if Γ is a matroid-related Π-partite access structure, then $\Delta(\Gamma) = \mathrm{supp}(\Gamma) = \{\mathrm{supp}(A) : A \in \Gamma\}$.

Corollary 9. *Let Γ be a matroid-related m-partite access structure. For every $X \subseteq J_m$, all minimal qualified subsets $A \in \min \Gamma$ with $\mathrm{supp}(A) = X$ have the same cardinality.*

Example 10. Let Γ be a 4-partite access structure with $\Pi(\min \Gamma) = \{(2, 2, 1, 1), (1, 3, 1, 2), (2, 1, 2, 1), (1, 1, 2, 2)\}$. From Corollary 9, Γ is not matroid-related, and hence it is not ideal. Moreover, by Theorem 3, its optimal information rate is at most $2/3$.

Corollary 11. *Let Γ be a connected matroid-related m-partite access structure and consider the discrete polymatroid $\mathcal{D} = \mathcal{D}(\Gamma)$ and the $\mathcal{D}$-compatible family $\Delta = \Delta(\Gamma)$. Let h be the rank function of $\mathcal{D}$. For every $X \in \Delta$ and $A \subseteq \bigcup_{i \in X} P_i$, if $|A| = h(X)$ and $|A \cap (\bigcup_{i \in Y} P_i)| \leq h(Y)$ for all $Y \subseteq X$, then $A \in \Gamma$.*

Example 12. Let Γ be a quadripartite access structure such that

$$\Pi(\min \Gamma) = \{u \in \mathbb{Z}_+^4 : (1,1,1,1) \leq u \leq (3,4,4,4) \text{ and } |u| = 8\} \cup \{(4,0,0,0)\}.$$

We claim that Γ is not matroid-related. Assume the Γ is matroid-related and consider $\mathcal{D} = \mathcal{D}(\Gamma)$ and $\Delta = \Delta(\Gamma)$. Observe that $\min \Delta = \{\{1\}\}$. In addition, from Theorem 8, if $u \in \mathcal{B}(J_4)$, then $u \in \Pi(\min \Gamma)$ or there exist $Y \subsetneq J_4$ and $v \in \mathcal{B}(Y)$ such that $v < u$ and $v \in \Pi(\min \Gamma)$. Therefore, the family of bases of $\mathcal{D}$ is $\mathcal{B} = \mathcal{B}(J_4) = \{u \in \mathbb{Z}_+^4 : (1,1,1,1) \leq u \leq (4,4,4,4) \text{ and } |u| = 8\}$. Moreover, $h(X) = \max\{|u(X)| : u \in \mathcal{D}\} = \max\{|u(X)| : u \in \mathcal{B}\}$ for every $X \subseteq J_4$. Therefore, $h(X) = 4$ if $|X| = 1$, and $h(X) = 6$ if $|X| = 2$, and $h(X) = 7$ if $|X| = 3$, and $h(J_4) = 8$. Since $\{1,2\} \in \Delta$, by Corollary 11, $(3,3,0,0) \in \Pi(\Gamma)$, a contradiction.

6   Representable Multipartite Matroids

Let $\mathbb{K}$ be a field, E a $\mathbb{K}$-vector space, and $V_1, \ldots, V_m$ subspaces of E. It is not difficult to check that the mapping $h: \mathcal{P}(J_m) \to \mathbb{Z}$ defined by $h(X) = \dim(\sum_{i \in X} V_i)$ is the rank function of a discrete polymatroid $\mathcal{D} \subset \mathbb{Z}_+^m$. In this situation, we say that $\mathcal{D}$ is $\mathbb{K}$-*linearly representable* and the subspaces $V_1, \ldots, V_m$ are a $\mathbb{K}$-*linear representation* of $\mathcal{D}$. The main result of this section is the following theorem.

Theorem 13. *Let $\mathcal{M} = (Q, r)$ be a Π-partite matroid such that $|Q| = n$ and $r(\mathcal{M}) = k$. Let $\mathcal{D} = \Pi(\mathcal{I})$ be its associated discrete polymatroid. If $\mathcal{M}$ is $\mathbb{K}$-linearly representable, then so is $\mathcal{D}$. In addition, if $\mathcal{D}$ is $\mathbb{K}$-representable, then $\mathcal{M}$ is $\mathbb{L}$-linearly representable for every field extension $\mathbb{L}$ of $\mathbb{K}$ such that $|\mathbb{L}| > \binom{n}{k} \cdot k$.*

The first claim in the statement is not difficult to prove. Let $\Pi = (Q_1, \ldots, Q_r)$ be a partition of Q and let $\mathcal{M} = (Q, r)$ be a Π-partite matroid with $r(\mathcal{M}) = k$ and $|Q| = n$. Consider the discrete polymatroid $\mathcal{D} = \Pi(\mathcal{I}) \subset \mathbb{Z}_+^m$ and its rank function $h: \mathcal{P}(J_m) \to \mathbb{Z}$. Suppose that $\mathcal{M}$ is represented over the field $\mathbb{K}$ by a matrix M. For every $i \in J_m$, consider the subspace V_i spanned by the columns of M corresponding to the points in Q_i. Then $h(X) = r(\cup_{i \in X} Q_i) = \dim(\sum_{i \in X} V_i)$ for every $X \subseteq J_m$. Therefore, the subspaces $V_1, \ldots, V_m$ are a $\mathbb{K}$-representation of the discrete polymatroid $\mathcal{D}$.

The proof for the second claim in the theorem is much more involved and needs several partial results. Clearly, it is enough to prove that, for every finite field with $|\mathbb{K}| > \binom{n}{k} \cdot k$, the matroid $\mathcal{M}$ is $\mathbb{K}$-linearly representable if the discrete polymatroid $\mathcal{D} = \Pi(\mathcal{I})$ is $\mathbb{K}$-linearly representable.

Assume that $|\mathbb{K}| > \binom{n}{k} \cdot k$ and that $\mathcal{D}$ is $\mathbb{K}$-linearly representable. Then there exists a $\mathbb{K}$-linear representation of $\mathcal{D}$ consisting of subspaces $V_1, \ldots, V_m$ of the

$\mathbb{K}$-vector space $E = \mathbb{K}^k$, where $k = h(J_m) = r(\mathcal{M})$. The proof of the following lemma is not given here due to space limitations. It will be included in the full version of the paper.

Lemma 14. *For every basis u of $\mathcal{D}$, there exists a basis $B = B_1 \cup \cdots \cup B_m$ of the vector space E such that $B_i \subset V_i$ and $|B_i| = u_i$ for every $i \in J_m$, and $B_i \cap B_j = \emptyset$ if $i \neq j$.*

For every $i \in J_m$, take $k_i = \dim V_i$ and $n_i = |Q_i|$. Then $n = n_1 + \cdots + n_m$. Consider the space $\mathbf{M}$ of all $k \times n$ matrices over $\mathbb{K}$ of the form $(M_1|M_2|\cdots|M_m)$, where M_i is a $k \times n_i$ matrix whose columns are vectors in V_i. Observe that the columns of every matrix $M \in \mathbf{M}$ can be indexed by the elements in Q, corresponding the columns of M_i to the points in Q_i. The proof of Theorem 13 is concluded by proving that there exists a matrix $M \in \mathbf{M}$ whose columns are a $\mathbb{K}$-linear representation of the matroid $\mathcal{M}$.

Lemma 15. *If $A \subseteq Q$ is a dependent subset of the matroid $\mathcal{M}$, then, for every $M \in \mathbf{M}$, the columns of M corresponding to the elements in A are linearly dependent.*

Proof. Since $u = \Pi(A) \notin \mathcal{D}$, there exists $X \subseteq J_m$ such that $|u(X)| > h(X) = \dim(\sum_{j \in X} V_j)$. Then the columns of M corresponding to the elements in $A \cap (\cup_{j \in X} Q_j)$ must be linearly dependent. □

Therefore, Lemma 17 concludes the proof of Theorem 13. The following technical lemma is needed to prove it. Recall that, over a finite field $\mathbb{K}$, there exist nonzero polynomials $p \in \mathbb{K}[X_1, \ldots, X_N]$ on N variables such that $p(x_1, \ldots, x_N) = 0$ for every $(x_1, \ldots, x_N) \in \mathbb{K}^N$.

Lemma 16. *Let $p \in \mathbb{K}[X_1, \ldots, X_N]$ be a nonzero polynomial on N variables of degree $d < |\mathbb{K}|$. Then, there exists a point $(x_1, \ldots, x_N)$ in $\mathbb{K}^N$ such that $p(x_1, \ldots, x_N) \neq 0$.*

Lemma 17. *There exists a matrix $M \in \mathbf{M}$ such that, for every basis $B \subseteq Q$ of the matroid $\mathcal{M}$, the corresponding columns of M are linearly independent.*

Proof. By fixing a basis of V_i for every $i \in J_m$, we obtain one-to-one mappings $\phi_i \colon \mathbb{K}^{k_i} \to V_i \subseteq \mathbb{K}^k$. Let $N = \sum_{i=1}^m k_i n_i$. By using the mappings ϕ_i, we can construct a one-to-one mapping $\Psi \colon \mathbb{K}^N = (\mathbb{K}^{k_1})^{n_1} \times \cdots \times (\mathbb{K}^{k_m})^{n_m} \to \mathbf{M}$. That is, by choosing an element in $\mathbb{K}^N$, we obtain n_i vectors in V_i for every $i \in J_m$. For every basis $B \subseteq Q$ of the matroid $\mathcal{M}$, we consider the mapping $f_B \colon \mathbb{K}^N \to \mathbb{K}$ defined by $f_B(\mathbf{x}) = \det(\Psi(\mathbf{x})_B)$, where $\Psi(\mathbf{x})_B$ is the square submatrix of $\Psi(\mathbf{x})$ formed by the k columns corresponding to the elements in B. Clearly, f_B is a polynomial on at most N variables and of degree k, because every entry of the matrix $\Psi(\mathbf{x})_B$ is linear, that is, an homogeneous polynomial of degree 1. Let B be a basis of $\mathcal{M}$ and $u = \Pi(B) \in \mathbb{Z}_+^m$. From Lemma 14, there exists a basis of $\mathbb{K}^k$ of the form $\widetilde{B} = B_1 \cup \cdots \cup B_m$ with $B_i \subset V_i$ and $|B_i| = u_i$ for every $i \in J_m$. By placing the vectors in $\widetilde{B}$ in the suitable positions in a matrix $M \in \mathbf{M}$, we can find a vector $\mathbf{x}_B \in \mathbb{K}^N$ such that $f_B(\mathbf{x}_B) \neq 0$, and hence the polynomial f_B is nonzero for every basis B of $\mathcal{M}$. Therefore, if $\mathcal{B}(\mathcal{M})$ is the family of bases of the matroid $\mathcal{M}$, the polynomial $\mathbf{f} = \prod_{B \in \mathcal{B}(\mathcal{M})} f_B$ is a nonzero polynomial on N variables of degree at most $\binom{n}{k} \cdot k < |\mathbb{K}|$, because $|\mathcal{B}(\mathcal{M})| \leq \binom{n}{k}$. From Lemma 16, there exists a point $\mathbf{x}_0 \in \mathbb{K}^N$ such that $\mathbf{f}(\mathbf{x}_0) \neq 0$, and hence $f_B(\mathbf{x}_0) \neq 0$ for every basis B of $\mathcal{M}$. Clearly, the matrix $\Psi(\mathbf{x}_0)$ is the one we are looking for. $\quad\square$

7 Tripartite Access Structures

In this section, we apply our general results on ideal multipartite access structures to completely characterize the ideal tripartite access structures. The characterization of ideal bipartite access structures was done previously in [28], but only partial results [2,10,12] were known about the tripartite case.

We begin by characterizing the matroid-related tripartite access structures. Afterwards, we prove that all matroids related to those access structures are representable. Therefore, all matroid-related tripartite access structures are vector space access structures, and hence ideal. We obtain in this way a characterization of the ideal tripartite access structures. In addition, as a consequence of Theorem 3, the optimal information rate of every non-ideal tripartite access structure is at most $2/3$.

7.1 Characterizing Matroid-Related Tripartite Access Structures

The values of a rank function $h \colon \mathcal{P}(J_3) \to \mathbb{Z}$ of a discrete polymatroid $\mathcal{D}$ with ground set J_3 will be denoted by $r_i = h(\{i\})$, where $i \in J_3$, and $s_i = h(\{j, k\})$ if $\{i, j, k\} = J_3$, and $s = h(J_3)$. Given integer values r_i, s_i, and s, they univocally determine a discrete polymatroid with ground set J_3 if and only if, for every i, j, k with $\{i, j, k\} = J_3$,

1. $s > 0$, and $0 \leq r_i \leq s_j \leq s$, and
2. $s_i \leq r_j + r_k$, and $s \leq s_i + r_i$, and $s + r_i \leq s_j + s_k$.

Let $\mathcal{D}$ be a discrete polymatroid with ground set J_3. From Proposition 7, a family $\Delta \subseteq \mathcal{P}(J_3)$ is $\mathcal{D}$-compatible if and only if the following conditions are satisfied for every i, j, k with $\{i, j, k\} = J_3$.

1. Δ is monotone increasing, $\emptyset \notin \Delta$, and $J_3 \in \Delta$.
2. $r_i > 0$ if $\{i\} \in \Delta$, and $r_i < s_j$ if $\{i\} \notin \Delta$ and $\{i, k\} \in \Delta$, and $s_i < s$ if $\{j, k\} \notin \Delta$.
3. $s_i < r_j + r_k$ if $\{\{j\}, \{k\}\} \subset \Delta$.
4. $s + r_i < s_j + s_k$ if $\{i\} \notin \Delta$ and $\{\{i, j\}, \{i, k\}\} \subset \Delta$.
5. $s < s_i + r_i$ if $\{\{i\}, \{j, k\}\} \subset \Delta$.

From Theorem 8, a tripartite access structure Γ is matroid-related if and only if there exist integers r_i, s_i, s and a family $\Delta \subseteq J_3$ in the above conditions such that a subset $A \subseteq P$ is in Γ if and only if $\Pi(A) \geq u$ for some $u \in \bigcup_{X \in \Delta} \mathcal{B}(X)$, where

- $\mathcal{B}(J_3) = \{v \in \mathbb{Z}_+^m : (s - s_1, s - s_2, s - s_3) \leq v \leq (r_1, r_2, r_3) \text{ and } |v| = s\}$,
- $\mathcal{B}(\{1,2\}) = \{v \in \mathbb{Z}_+^m : (s_3 - r_2, s_3 - r_1, 0) \leq v \leq (r_1, r_2, 0) \text{ and } |v| = s_3\}$,
- $\mathcal{B}(\{1\}) = \{(r_1, 0, 0)\}$,

and the other sets $\mathcal{B}(X)$ are defined symmetrically.

7.2 All Matroid-Related Tripartite Access Structures Are Ideal

Let $\mathcal{D}$ be a discrete polymatroid with ground set J_3 that is represented over the field $\mathbb{K}$ by three subspaces V_1, V_2, V_3 of a vector space E. If r_i, s_i and s are the integer values of the rank function of $\mathcal{D}$, then $r_i = \dim V_i$ for every $i \in J_3$, and $s_i = \dim(V_j + V_k)$ if $\{i, j, k\} = J_3$, and $s = \dim(V_1 + V_2 + V_3)$. If $\{i, j, k\} = J_3$, consider $t_i = r_j + r_k - s_i = \dim(V_j \cap V_k)$. Observe that $t = \dim(V_1 \cap V_2 \cap V_3)$ is not determined in general by $\mathcal{D}$. That is, there can exist different representations of $\mathcal{D}$ with different values of t. Nevertheless, there exist some restrictions on this value. Of course, $t \leq t_i$ for every $i \in J_3$. In addition, since $(V_1 \cap V_3) + (V_2 \cap V_3) \subseteq (V_1 + V_2) \cap V_3$, we have that $\dim((V_1 + V_2) \cap V_3) - \dim((V_1 \cap V_3) + (V_2 \cap V_3)) = \sum s_i - \sum r_i - (s - t) \geq 0$. Therefore, $\max\{0, s - \sum s_i + \sum r_i\} \leq t \leq \min\{t_1, t_2, t_3\}$. The proof of the following result will appear in the full version of the paper.

Proposition 18. *Let $\mathcal{D}$ be a discrete polymatroid with ground set J_3. Consider an integer t such that $\max\{0, s - \sum s_i + \sum r_i\} \leq t \leq \min\{t_1, t_2, t_3\}$ and take $\ell = \sum s_i - \sum r_i - (s - t)$. Let $\mathbb{K}$ be a field with $|\mathbb{K}| > s_3 + \ell$. Then there exists a $\mathbb{K}$-representation of $\mathcal{D}$ given by subspaces $V_1, V_2, V_3 \subseteq E = \mathbb{K}^s$ with $\dim(V_1 \cap V_2 \cap V_3) = t$.*

As a consequence of Proposition 18, every discrete polymatroid with ground set J_m with $m \leq 3$ is representable over fields of all characteristics. This and Theorem 13 implies that every m-partite matroid with $m \leq 3$ is representable over fields of all characteristics.

Theorem 19 concludes the characterization of ideal tripartite access structures. This result is not a direct consequence of Proposition 18, because the matroids that define tripartite access structures are in general quadripartite, being one of the parts formed by a single point. Therefore, Theorem 19 is proved by showing that every discrete polymatroid $\mathcal{D}'$ with ground set J_3' and $h(\{p_0\}) = 1$ is linearly representable over finite fields of every characteristic. We sketch in the following the proof of this fact. First, a linear representation of the discrete polymatroid $\mathcal{D} = \mathcal{D}'(J_3)$, whose existence is given by Proposition 18, is considered. Afterwards, we have to check that it is possible to find a vector $\mathbf{x}_0$ such that the subspace $V_0 = \langle \mathbf{x}_0 \rangle$, together with the subspaces V_1, V_2, V_3 representing $\mathcal{D} = \mathcal{D}'(J_3)$, form a linear representation of $\mathcal{D}'$. This is done by a case-by-case analysis depending on the family $\Delta(\mathcal{D}')$, and in every case a suitable representation of $\mathcal{D}$ has to be chosen.

Theorem 19. *Every matroid-related tripartite access structure is ideal. More specifically, every matroid-related tripartite access structure is a vector space access structure over finite fields of all positive characteristics.*

Example 20. We prove that the tripartite access structure Γ with

$$\Pi(\min \Gamma) = \{(3,0,0),(2,0,4),(2,4,2),(2,3,3),(1,4,3),(1,3,4)\}.$$

is ideal. Assuming that this is so, we determine $\mathcal{D} = \mathcal{D}(\Gamma)$ and $\Delta = \Delta(\Gamma)$. Observe that $\Delta = \text{supp}(\Gamma) = \{\{1\},\{1,2\},\{1,3\},J_3\}$, and hence $\Pi(\min \Gamma) \subseteq \mathcal{B}(\{1\}) \cup \mathcal{B}(\{1,2\}) \cup \mathcal{B}(\{1,3\}) \cup \mathcal{B}(J_3)$. It is easy to see that $r_1 = 3$, $r_2 = r_3 = 4$, $s_2 = 6$ and $s = 8$. Since there is not any minimal subset in $\mathcal{B}(\{1,2\})$, it follows that $\mathcal{B}(\{1,2\})$ has only one element $(s_3 - r_2, r_2, 0) = (r_1, s_3 - r_1, 0)$, which does not correspond to any minimal qualified subset, and hence $s_3 = 7$. All subsets in $\mathcal{B}(J_3)$ have at least one participant in the first partition, so $s - s_1 = 1$ and $s_1 = 7$. Since the parameters satisfy the above restrictions and Γ coincides with the access structure determined by these parameters, Γ is a matroid-related access structure. Therefore, it is a vector space access structure by Theorem 19.

References

1. A. Beimel, N. Livne. On Matroids and Non-ideal Secret Sharing. *Third Theory of Cryptography Conference, TCC 2006. Lecture Notes in Comput. Sci.* **3876** (2006) 482–501.
2. A. Beimel, T. Tassa, E. Weinreb. Characterizing Ideal Weighted Threshold Secret Sharing. *Second Theory of Cryptography Conference, TCC 2005. Lecture Notes in Comput. Sci.* **3378** (2005) 600–619.
3. J. Benaloh, J. Leichter. Generalized secret sharing and monotone functions. *Advances in Cryptology, CRYPTO'88. Lecture Notes in Comput. Sci.* **403** (1990) 27–35.
4. G.R. Blakley. Safeguarding cryptographic keys. *AFIPS Conference Proceedings*. **48** (1979) 313–317.
5. C. Blundo, A. De Santis, R. De Simone, U. Vaccaro. Tight bounds on the information rate of secret sharing schemes. *Des. Codes Cryptogr.* **11** (1997) 107–122.
6. C. Blundo, A. De Santis, L. Gargano, U. Vaccaro. On the information rate of secret sharing schemes. *Advances in Cryptology, CRYPTO'92. Lecture Notes in Comput. Sci.* **740** (1993) 148–167.
7. E.F. Brickell. Some ideal secret sharing schemes. *J. Combin. Math. and Combin. Comput.* **9** (1989) 105–113.
8. E.F. Brickell, D.M. Davenport. On the classification of ideal secret sharing schemes. *J. Cryptology* **4** (1991) 123–134.
9. R.M. Capocelli, A. De Santis, L. Gargano, U. Vaccaro. On the size of shares of secret sharing schemes. *J. Cryptology* **6** (1993) 157–168.
10. M.J. Collins. A Note on Ideal Tripartite Access Structures. *Cryptology ePrint Archive*, Report **2002/193**, http://eprint.iacr.org/2002/193.
11. L. Csirmaz. The size of a share must be large. *J. Cryptology* **10** (1997) 223–231.
12. J. Herranz, G. Sáez. New Results on Multipartite Access Structures. *IEE Proceedings on Information Security* **153** (2006) 153–162.
13. J. Herzog, T. Hibi. Discrete polymatroids. *J. Algebraic Combin.* **16** (2002) 239–268.
14. M. Ito, A. Saito, T. Nishizeki. Secret sharing scheme realizing any access structure. *Proc. IEEE Globecom'87.* (1987) 99–102.
15. W.-A. Jackson, K.M. Martin. Perfect secret sharing schemes on five participants. *Des. Codes Cryptogr.* **9** (1996) 267–286.

16. E.D. Karnin, J.W. Greene, M.E. Hellman. On secret sharing systems. *IEEE Trans. Inform. Theory* **29** (1983) 35–41.
17. A. Lehman. A solution of the Shannon switching game. *J. Soc. Indust. Appl. Math.* **12** (1964) 687–725.
18. J. Martí-Farré, C. Padró. Secret sharing schemes with three or four minimal qualified subsets. *Des. Codes Cryptogr.* **34** (2005) 17–34.
19. J. Martí-Farré, C. Padró. Secret sharing schemes on access structures with intersection number equal to one. *Discrete Applied Mathematics* **154** (2006) 552–563.
20. J.Martí-Farré, C. Padró. Ideal secret sharing schemes whose minimal qualified subsets have at most three participants. *Fifth Conference on Security and Cryptography for Networks, SCN 2006, Lecture Notes in Comput. Sci.*, **4116** (2006) 201–215.
21. J. Martí-Farré, C. Padró. On Secret Sharing Schemes, Matroids and Polymatroids. *Fourth IACR Theory of Cryptography Conference TCC 2007, Lecture Notes in Comput. Sci.* **4392** (2007) 273–290.
22. F. Matúš. Matroid representations by partitions. *Discrete Math.* **203** (1999) 169–194.
23. P. Morillo, C. Padró, G. Sáez, J. L. Villar. Weighted Threshold Secret Sharing Schemes. *Inf. Process. Lett.* **70** (1999) 211–216.
24. S.-L. Ng. A Representation of a Family of Secret Sharing Matroids. *Des. Codes Cryptogr.* **30** (2003) 5–19.
25. S.-L. Ng. Ideal secret sharing schemes with multipartite access structures. *IEE Proc.-Commun.* **153** (2006) 165–168.
26. S.-L. Ng, M. Walker. On the composition of matroids and ideal secret sharing schemes. *Des. Codes Cryptogr.* **24** (2001) 49–67.
27. J.G. Oxley. *Matroid theory*. Oxford Science Publications. The Clarendon Press, Oxford University Press, New York, 1992.
28. C. Padró, G. Sáez. Secret sharing schemes with bipartite access structure. *IEEE Trans. Inform. Theory* **46** (2000) 2596–2604.
29. P.D. Seymour. A forbidden minor characterization of matroid ports. *Quart. J. Math. Oxford Ser.* **27** (1976) 407–413.
30. P.D. Seymour. On secret-sharing matroids. *J. Combin. Theory Ser. B*, **56** (1992) pp. 69–73.
31. A. Shamir. How to share a secret. *Commun. of the ACM*, **22** (1979) pp. 612–613.
32. G. J. Simmons. How to (Really) Share a Secret. *Advances in Cryptology – CRYPTO'88, Lecture Notes in Comput. Sci.* **403** (1990) 390–448.
33. J. Simonis, A. Ashikhmin. Almost affine codes. *Des. Codes Cryptogr.* **14** (1998) pp. 179–197.
34. D.R. Stinson. An explication of secret sharing schemes. *Des. Codes Cryptogr.* **2** (1992) 357–390.
35. T. Tassa. Hierarchical Threshold Secret Sharing. *Theory of Cryptography, First Theory of Cryptography Conference, TCC 2004. Lecture Notes in Comput. Sci.* **2951** (2004) 473–490.
36. T. Tassa, N. Dyn. Multipartite Secret Sharing by Bivariate Interpolation. *33rd International Colloquium on Automata, Languages and Programming, ICALP 2006 – Lecture Notes in Comput. Sci.* **4052** (2006) 288–299.
37. D.J.A. Welsh. *Matroid Theory*. Academic Press, London, 1976.

Non-wafer-Scale Sieving Hardware for the NFS: Another Attempt to Cope with 1024-Bit

Willi Geiselmann[1] and Rainer Steinwandt[2]

[1] IAKS, Fakultät für Informatik, Universität Karlsruhe (TH), Am Fasanengarten 5,
76128 Karlsruhe, Germany
geiselma@ira.uka.de
[2] Department of Mathematical Sciences, Florida Atlantic University,
777 Glades Road, Boca Raton, FL 33431, USA
rsteinwa@fau.edu

Abstract. Significant progress in the design of special purpose hardware for supporting the Number Field Sieve (NFS) has been made. From a practical cryptanalytic point of view, however, none of the published proposals for coping with the sieving step is satisfying. Even for the best known designs, the technological obstacles faced for the parameters expected for a 1024-bit RSA modulus are significant.

Below we present a new hardware design for implementing the sieving step. The suggested chips are of moderate size and the inter-chip communication does not seem unrealistic. According to our preliminary analysis of the 1024-bit case, we expect the new design to be about 2 to 3.5 times slower than TWIRL (a wafer-scale design). Due to the more moderate technological requirements, however, from a practical cryptanalytic point of view the new design seems to be no less attractive than TWIRL.

Keywords: RSA, cryptanalytic hardware, factoring integers, NFS.

1 Introduction

Even for the best known factoring algorithms, coping with the complexity of a factorization of a 1024-bit RSA modulus looks extraordinary challenging. In an attempt to bring such a record factorization closer to what is currently feasible, various hardware designs to support implementations of the Number Field Sieve (NFS) have been devised. While theoretical advances in the design of factoring algorithms are more desirable, at the moment these special purpose designs for speeding up time-critical computations in the NFS seem to be the most promising approach for practically challenging a 1024-bit RSA modulus. After a series of works on the linear algebra step of the NFS [1,13,8,5,6], one may adopt the position that the linear algebra step expected for a 1024-bit factorization is by now close to or in reach of current technology.

On the other hand, none of the suggested designs implementing the sieving step of the NFS is really satisfying:

- TWINKLE [15,12] builds on an opto-electronic hybrid design where no promising parameter set for the 1024-bit case has been proposed.

M. Naor (Ed.): EUROCRYPT 2007, LNCS 4515, pp. 466–481, 2007.

- For designs building on a mesh architecture, no promising specification for the 1024-bit case is known (cf. [1,7,9,10]).
- SHARK [3] imposes the use of an elaborate butterfly transport system, whose implementation is far from trivial.
- TWIRL [16,14] seems to be the currently best-explored design. Unfortunately, it is a wafer-scale design building on a quite complex layout.

In an attempt to reduce the layout complexity, Geiselmann et al. [4] recently proposed to combine a modified TWIRL with an "ECM engine": For 1024-bit parameters of interest, [4] argues that an optimized implementation of the Elliptic Curve Method (ECM) is capable of efficiently computing all factorizations of (semi-)smooth norms occurring in the sieving step. The idea is that in this way the circuitry for TWIRL's "diary part", which stores large prime factors of norms, can be removed. The design we present below also relies on this idea: We do not store any prime factors encountered during relation collection and assume a postprocessing of the sieving output with an ECM engine as described in [4]. However, unlike TWIRL, the device proposed below is a non-wafer-scale design.

After having recalled some facts on the sieving step in the NFS in Section 2, in Section 3 we describe our design that builds on ideas of several published proposals: Like the mesh-based proposals, we implement a version of line sieving where each sieving line is split into consecutive subintervals. To overcome the need of a wafer-scale design, we distribute (the majority of) the factor bases on moderately sized chips. The circuitry on these chips produces the arithmetic progressions needed for sieving and is inspired by TWIRL. Eventually, to combine the sieving contributions of the different factor base elements, we use a central unit whose structure reminds of the linear algebra design proposed in [6]. In our preliminary analysis of the 1024-bit case, for the ease of comparability we adopt the technological parameters and the NFS parameters from [16]. Summarizing, we expect our device to be about 2 to 3.5 times slower than TWIRL. On the other hand, the maximal chip size involved is 493 mm^2 and also the interconnection circuitry among these chips does not seem utopian. From a practical point of view, this new design appears to be no less attractive than the existing hardware designs for implementing the sieving step.

2 Preliminaries: Sieving in the NFS

For the purposes of this paper, it is sufficient to recall the basic set-up of so-called line sieving in the NFS. For a more thorough discussion of the NFS we refer to the standard reference [11].

2.1 Line Sieving

In a precomputation phase of the NFS two univariate polynomials $f_1(x), f_2(x) \in \mathbb{Z}[x]$ with integer coefficients are determined that have a root m modulo n in common:

$$f_1(m) \equiv f_2(m) \equiv 0 \pmod{n}$$

$b \leftarrow 0$
repeat
   $b \leftarrow b + 1$
   for $i \leftarrow [1, 2]$
      $s_i(a) \leftarrow 0 \quad (\forall a : -A \leq a < A)$
      for $(p, r) \leftarrow P_i$
         $s_i(br + kp) \leftarrow s_i(br + kp) + \log_{\sqrt{2}}(p) \quad (\forall k : -A \leq br + kp < A)$
   for $a \leftarrow \{-A \leq a < A : \gcd(a, b) = 1, s_1(a) > T_1, \text{ and } s_2(a) > T_2\}$
      check if both $F_1(a, b)$ and $F_2(a, b)$ are smooth
until enough pairs (a, b) with both $F_1(a, b)$ and $F_2(a, b)$ smooth are found

Fig. 1. Line sieving

A typical choice is to have $f_1(x)$ of degree $d \geq 5$ and $f_2(x)$ to be monic and linear, i.e., $f_2(x) = x - m$. By setting $F_1(x, y) := y^d \cdot f_1(x/y)$ resp. $F_2(x, y) := y \cdot f_2(x/y)$, two homogeneous polynomials $F_1(x, y), F_2(x, y) \in \mathbb{Z}[x, y]$ are derived. Now everything related to the polynomial $f_1(x)$ resp. $F_1(x, y)$ is said to belong to the *algebraic side*, whereas everything related to the polynomial $f_2(x)$ resp. $F_2(x, y)$ is referred to as belonging to the *rational side*. In particular, for given smoothness bounds $B_1, B_2 \in \mathbb{N}_0$ the sets

$$P_i := \{(p, r) : f_i(r) \equiv 0 \pmod{p}, \ p \text{ prime}, \ p < B_i, \ 0 \leq r < p\} \subseteq \mathbb{N}^2 \quad (i = 1, 2)$$

are known as algebraic and rational *factor base*, respectively.

Throughout the relation collection step, pairs of integers $(a, b) \in \mathbb{Z} \times \mathbb{N}$ with $\gcd(a, b) = 1$ are to be found, so that the values $F_1(a, b)$ and $F_2(a, b)$ are *smooth*. This means that the values $F_1(a, b)$ and $F_2(a, b)$ both factor over the primes $< B_1$ resp. $< B_2$, except for a small number of prime factors. At this, the precise number of 'extra' prime factors on the rational and algebraic side is not necessarily identical. The actual computation of (a, b)-pairs where both $F_1(a, b)$ and $F_2(a, b)$ are smooth can be performed by means of a sieving process, e.g., over a rectangular region $-A \leq a < A, 0 < b \leq B$ with $A, B \in \mathbb{N}$. For organizing this sieving process, different techniques are known, and for our purposes we focus on simple *line sieving* as outlined in Figure 1. At this, the thresholds T_i correspond to the bitlength of the remaining cofactor on the algebraic and rational side, respectively. The T_i-values are to be updated several times throughout the sieving. For the sake of efficiency, in an actual implementation the values $\log_{\sqrt{2}}(p)$ are usually replaced by an integer approximation. Also the use of base $\sqrt{2}$-logarithms is certainly not mandatory. In analogy to [16], in the sequel we will use a 10-bit counter for summing up approximations $\lceil \log_{\sqrt{2}}(p) \rceil$. It is worth noting that testing the norms $F_1(a, b), F_2(a, b)$ for smoothness and in case of smoothness recovering their prime factors is computationally non-trivial. For the device proposed below, we rely on a design as presented in [4], which uses an optimized ECM implementation to perform the required norm factorizations in connection with a TWIRL-based realization of the sieving step.

2.2   Choice of 1024-Bit Parameters

Deducing a reliable estimate for the NFS parameters suitable for a factorization of a 1024-bit RSA modulus is a non-trivial problem in its own and outside the scope of this paper. Already for the sake of comparability, here we adopt parameters from [16]. Summarizing, for the sequel the following parameter choices are of interest:

- On the algebraic side, the smoothness bound $B_1 = 2.6 \cdot 10^{10}$ is used.
- On the rational side, the smoothness bound $B_2 = 3.5 \cdot 10^9$ is used.
- The sieving region $-A < a \leq A, 0 < b \leq B$ uses $A = 5.5 \cdot 10^{14}$ and $B = 2.7 \cdot 10^8$.
- The algebraic and rational polynomials are chosen of degree 5 and 1, respectively, as specified in [16, Appendix B.2].

For further details and a discussion on how to identify suitable NFS parameters, we refer to [14]. With the mentioned parameters, the factor bases are of size $|P_1| \approx 1.134 \cdot 10^9$ and $|P_2| \approx 1.673 \cdot 10^8$, respectively.

3   The Proposed Design: Main Components

For the sake of clarity, in this section we only discuss the basic structure of our design. Parameter choices we made for the case of a 1024-bit factorization are indicated in double brackets $\langle\!\langle \cdot \rangle\!\rangle$, but a discussion of implementation details is postponed to Section 4. The basic organization of the sieving process is analogous to [7,9]. Namely, we divide each sieving line in subintervals of $S \langle\!\langle = 2^{26} \rangle\!\rangle$ consecutive sieve locations. Switching to the next subinterval within one sieving line can be done with local operations only. However, to switch to a different sieving line, i.e., to increase the b-value, new data is to be loaded into the device, and our running time analysis has to take this into account.

   At a high level, the architecture of our design relies on two types of components, which we detail in the sequel: a) a collection unit that is in charge of updating the rational and algebraic sieving counters and b) stations that compute the arithmetic progressions needed for updating the counters.

3.1   Collection Unit

For each value in the current sieving interval, this part of our device hosts an algebraic and a rational DRAM counter for summing up the respective $\log_{\sqrt{2}}(p)$-values. Each of these counters has a size of $b \langle\!\langle = 10 \rangle\!\rangle$ bit, and the counters are distributed onto a number $c \langle\!\langle = 2^{14} \rangle\!\rangle$ of identical processors. We refer to these processors as *counting units*, and each counting unit is in charge of $S/c \langle\!\langle = 2^{12} \rangle\!\rangle$ consecutive sieve locations. It is not necessary to place all counting units on a single chip, and we distribute them onto a small number $\gamma \langle\!\langle = 4 \rangle\!\rangle$ of chips.

   These γ chips are all organized in the same manner: we arrange the respective counting units in two-dimensional arrays of size $\sigma \times \sigma \langle\!\langle = 2^5 \times 2^5 \rangle\!\rangle$, yielding a total number of $c/(\gamma\sigma^2) \langle\!\langle = 4 \rangle\!\rangle$ arrays per chip. Each array is organized as depicted

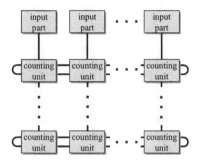

Fig. 2. Organization of one array of counting units

in Figure 2: The counting units in each row are connected through a circular bus, whereas the counting units within a column are connected through a unidirectional bus, originating in an *input part*. This structure is reminiscent of the linear algebra design in [6].

The input parts receive $(\log_{\sqrt{2}}(p), r)$-values from external *stations* (see below) with the r-value indicating to which of the counters in the array the respective $\log_{\sqrt{2}}(p)$-value is to be added. In each clock cycle, a received $(\log_{\sqrt{2}}(p), r)$-values passes (along with an algebraic/rational flag) on to the next row over the vertical bus. Then each counting unit checks whether the pair received on the vertical bus is to be handled in that row. If yes, the packet is removed from the vertical bus and via the circular horizontal bus transported to the correct counting unit. The latter then removes the received packet from the horizontal bus and adds the $\log_{\sqrt{2}}(p)$-value to the appropriate counter.

3.2 Computing the Arithmetic Progressions

Similarly, as in [16], to handle the arithmetic progressions for the $(\log_{\sqrt{2}}(p), r)$-pairs we use different types of circuits, and refer to these as *stations*. In dependence on the size of the prime number p, we distinguish four types of stations, whose structure is reminiscent of the stations in TWIRL.

Largish stations. These are in charge of primes p greater than some bound $B_{\text{largish}} \langle\!\langle = 1.5 \cdot 10^8 \rangle\!\rangle$, where $B_{\text{largish}} > S$. The majority of primes in the factor base is handled in this way. They "hit" no more than once per sieving interval, and the design of largish stations reflects this. Each such station handles a certain number $n_{\text{largish}} \langle\!\langle = 10^5 \rangle\!\rangle$ of factor base elements, which are stored in a sequence of DRAM banks as sketched in Figure 3. Each of the memory banks is operated as a stack, random access is not needed.

First, we initialize the sieving line defined through a specific b-value (starting with $b = 1$): For each factor base element (p, r) we replace r with $br \mod p$ (see Figure 1). This precomputation is performed on an external PC and the modified (p, r)-pairs are then loaded in the mentioned series of DRAM banks. The first DRAM bank holds all (p, r)-pairs that "hit" in the first sieving interval

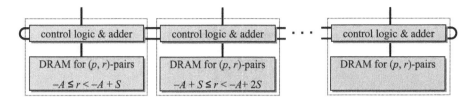

Fig. 3. Largish station

of size S, the second DRAM all those with an r-value indicating a hit in the second subinterval of size S, etc. The number n_banks of DRAM banks we need is $n_\text{banks} = \lceil p_\text{max}/S \rceil + 1 \langle\!\langle \leq 389 \rangle\!\rangle$ with p_max being the maximal prime handled by the station.[1] The number of entries $n_\text{entries} \langle\!\langle \leq 10^5 \rangle\!\rangle$ per DRAM bank has to suffice for holding all "hits" that can occur in a single subinterval of size S.

Now, for sieving the first subinterval, the first DRAM bank is read sequentially (or in small blocks of $\ell_\text{largish} \langle\!\langle = 2 \rangle\!\rangle$ values). The $\log_{\sqrt{2}}(p)$-approximations are constant within one unit, as the prime numbers handled in a unit are of approximately equal size. Along with the r-values (and a rational/algebraic flag), the approximation for $\log_{\sqrt{2}}(p)$ is sent over a unidirectional[2] bus to the appropriate array of the collection unit. Further on, an updated entry is written into the DRAM bank that handles the subinterval where the next "hit" for this progression occurs. More specifically, we proceed as follows: With the adder residing next to each DRAM bank, we compute the new r-value as $r \leftarrow r + p$. Now, choosing $S \langle\!\langle = 2^{26} \rangle\!\rangle$ as power of 2, the most significant bits of the new r-value can serve as counter indicating the number of "hops", that the updated (p, r)-pair has to travel among the cyclically connected DRAM banks. Once the pair has arrived at its destination DRAM, which handles the subinterval for the next "hit", the control logic associated to that DRAM bank removes the packet from the cyclic bus and appends it at the end of the entries currently stored in that DRAM. If there is no space left in this DRAM, the pair is deleted and lost for the entire sieving line. This never happened in our simulations.

Once a complete subinterval (i. e., a DRAM bank) has been processed, the unit proceeds to the (cyclic) successor of that DRAM and processes it in the same manner. In this way, the complete sieving line is processed.

Medium stations. For prime numbers that are smaller than the sieving interval size S, the respective arithmetic progressions may encounter several hits within one subinterval. For some bound $B_\text{medium} \langle\!\langle = 2^{13} \rangle\!\rangle$, we handle the primes $B_\text{medium} < p < B_\text{largish}$ as follows.

In one station $n_\text{medium} \langle\!\langle \approx 10^5 \rangle\!\rangle$ pairs (p, r) are stored in a DRAM bank. As for the largish stations, to start a new sieving line, the r-value is to be initialized according to Figure 1. Unlike for largish primes, now we have only one DRAM bank in the station, and in order to save memory—or rather chip area—we sort

---

[1] Using one more DRAM bank than $\lceil p_\text{max}/S \rceil$ avoids the problem of having to read and write from one DRAM bank at the same time.

[2] Only for initializing a new sieving line this bus is operated in the opposite direction.

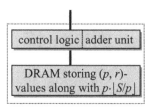

Fig. 4. Medium station

the (p, r)-pairs according to p: In this way, storing the difference between primes is sufficient to recover the next p-value. As sketched in Figure 4, next to the DRAM bank and control logic, we also have an adder unit. The latter consists of an array of $\ell_{\mathrm{medium}}\langle\!\langle= 128\rangle\!\rangle$ adders.

The DRAM will be processed sequentially in blocks of $\ell_{\mathrm{medium}}\langle\!\langle= 128\rangle\!\rangle$ entries. After reading such a block of (p, r)-values, it is forwarded to the addition unit, where the needed primes p are reconstructed from the stored differences. Also, similarly as in the largish stations, the needed $\log_{\sqrt{2}}(p)$-approximations are determined here. The adders now compute all values $r + k \cdot p$ that are relevant for the current subinterval, i. e., p is added as long as the obtained value is still smaller than $S\langle\!\langle= 2^{26}\rangle\!\rangle$, which for S being a power of 2 can be tested by observing a single bit. The respective $r + k \cdot p$-values are transmitted to the appropriate array of the collection unit—together with the $\log_{\sqrt{2}}(p)$-approximation and a rational/algebraic flag. In parallel to the computation of these $\ell_{\mathrm{medium}}\langle\!\langle= 128\rangle\!\rangle$ arithmetic progressions, the r-values stored in the DRAM are updated for the next subinterval. To this aim, along with each (p, r)-entry we also store the (precomputed) value $p \cdot \lfloor S/p \rfloor$ in the DRAM. Knowing this value, updating an r-value for the next sieving interval reduces to computing $r \leftarrow r + \lfloor S/p \rfloor \cdot p$. If this value does not exceed S yet, p has to be added. Eventually, we subtract S from the obtained new r-value.

To keep the number of pins of the chips holding the collection unit within acceptable boundaries, the medium stations will be hosted on the same chips as the collection unit. If the collection unit is distributed over several chips, we have to duplicate the medium stations accordingly. Also, as the medium stations are expected to produce relations at a very high rate, we equip the (unidirectional) buses into the collection unit's arrays with a "panic feedback flag". This allows the collection unit to put a medium station on hold until the buses and buffers can cope with new $(\log_{\sqrt{2}}(p), r)$-pairs again.

Smallish stations. We refer to factor base elements (p, r) with $p \leq B_{\mathrm{medium}}$ as *smallish*, and handle them in basically the same type of stations as just discussed. However, we do without a difference coding here. Progressions computed by smallish stations produce several hits within a subinterval, even within one array of the collection unit. Consequently we duplicate the smallish stations, so that on each chip of the collection unit all smallish primes can be handled locally.

4   Performance and Parameters for the 1024-Bit Case

In this section we discuss more details of our design, when dealing with NFS parameters for a 1024-bit factorization as described in Section 2.2. For choosing and optimizing the design parameters specified below, we relied on simulations by means of a computer algebra system [2] and a heuristic approach. We did not invoke a rigorous mathematical optimization and do not claim that our parameter choices are "the best possible".

In addition to the NFS parameter choices mentioned in Section 2.2, we fix the subinterval size $S := 2^{26}$ that specifies the number of consecutive sieve locations processed by our device at once. As outlined in the previous section, the progressions corresponding to the different types of primes are generated in different types of stations. Below we first describe the structure of the stations and their placement within the device. Section 4.2 details the structure of the collection unit.

4.1   Stations for the 1024-Bit Case

To keep the amount of inter-chip communication at a reasonable level, we subdivide the stations for largish primes into three types. While the first type describes stations that are placed on a chip different from the chips hosting the collection unit, the other two types reside on the same chips as the collection unit. Similarly, we use two different types of medium stations, both residing on the same chips as the collection unit. For the sake of comparability with [16], for estimating the space complexities we assume a 0.13 μm process with a DRAM bit occupying about 0.2 μm^2 and a transistor occupying about 2.8 μm^2 of silicon.

Largish Stations

Type I. We use this type of stations for largish primes $> 1.5 \cdot 10^8$. As described before, the factor base elements are distributed onto different DRAMs, so that all primes of this station relevant for the processed size $S = 2^{26}$ subinterval are stored in one DRAM. For the chosen subinterval size $S = 2^{26}$, we choose the DRAM bank large enough to store up to $100,000$ (p, r)-pairs. For each such pair (p, r) the respective prime $p < 2^{35}$ and r-value (mod 2^{26}) are stored, yielding a total of 34+26 bit per DRAM entry.[3]

The DRAM is read sequentially, on average reading two pairs per clock cycle. Each r-value is sent—together with the 4-bit value $\lceil \log_{\sqrt{2}}(p) \rceil - 55$ that is chosen to be constant for the whole station—to a small *routing network* on the same chip (see below). An adder (with input widths 35 and 26 bit) calculates the next hit for p in the current sieving line. As described in Section 3.2, the pair $(p, r+p)$ is forwarded—through one of the two cyclic buses connecting all DRAM banks of the station—to the DRAM bank in charge of the subinterval where p hits next. More specifically, we send the value $(p, (r + p) \bmod 2^{26})$ to the DRAM

---

[3] The least significant bit of p is known to be 1.

bank that is $(r + p)$ div 2^{26} "hops" away. To implement this routing operation, adjacent to each DRAM two adders (shared among four DRAMs), a decrement and compare unit, and the memory cells for the two buses of width up to 69 each[4] are needed.

All in all, we estimate that this logic can be realized with 4000 transistors per DRAM bank. Together with 6,000,000 bits of storage space, the size of one DRAM bank with update logic is estimated to be 1.2 mm^2. To handle both the algebraic and the rational factor base elements with $p > 1.5 \cdot 10^8$, we use 256 largish stations of Type I (156 algebraic and 99 rational ones). The number of DRAM banks per station varies from 4 up to 389, yielding a total of 13,440 DRAM banks. We distribute the Type I stations on 32 chips, each holding 8 stations with ≈ 420 DRAM banks.

On each chip one *routing network* collects the 16 outputs of the 8 stations and distributes them to the correct array of size 2^{22} on the collection unit. This routing network is realized through a butterfly network with 16 inputs. Each of the four stages of the butterfly network has 16 buffers to store up to 30 pairs $(r, \lceil \log_{\sqrt{2}}(p) \rceil - 55)$. If one of the buffers is full, a panic flag informs the parent nodes to stop sending data. The panic flags of the input nodes of the network stop the corresponding station from producing further pairs. According to our simulations, one station outputs on average 98,000 $(r, \lceil \log_{\sqrt{2}}(p) \rceil - 55)$-pairs per subinterval and all the pairs of the 8 stations on one chip are routed to the correct destination within about 52,000 clock cycles. The butterfly network requires $4 \times 16 \times 30$ buffers for 30-bit values (realized as latches) and the routing and control logic. We estimate that 300,000 transistors with an area of less than 1 mm^2 should be sufficient. The output of the butterfly network—16 pairs comprised of r mod 2^{22} and the 4-bit encoding of the corresponding $\lceil \log_{\sqrt{2}}(p) \rceil$-value—is sent (across chip borders) to the correct array of size 2^{22} of the collection unit.

Summarizing, the largish stations of Type I can produce the needed progressions for the primes $p > 1.5 \cdot 10^8$ in approximately 52,000 clock cycles. For this, we need 32 chips, each having a size of ≈ 472 mm^2 and each outputting $16 \cdot 28 = 448$ bit per clock cycle. Each of the outputs has a fixed destination—an array of size 2^{22} in the collection unit where the hit is processed.

Type II. To keep the amount of inter-chip communication at a reasonable level, we introduce a slightly different type of largish stations, which are in charge of all factor base elements with primes $4 \cdot 10^7 < p < 1.5 \cdot 10^8$. These Type II stations are placed on the chips holding the collection unit: Our collection unit will be distributed onto 4 chips, so we need 4 copies of each of these Type II largish stations. As one chip of the collection unit handles only one quarter of the total subinterval of size $S = 2^{26}$, the Type II largish stations are designed for a sieving interval size of 2^{24}. The overall structure is identical to the Type I case just discussed. However, reflecting the reduced subinterval size 2^{24}, the number of (p, r)-pairs per DRAM is reduced to 50,000. Finally, the calculation of the next hit has to be modified, so that the subsequent three subintervals of size 2^{24} are skipped.

---

[4] We need up to 9 bit for the "hop counter".

With this strategy, the needed arithmetic progressions for the primes $4 \cdot 10^7 < p < 1.5 \cdot 10^8$ can be generated by $4 \cdot 44$ stations with a total of $4 \cdot 290$ DRAMs of size 0.6 mm^2 each. To route the outputs of these 44 stations *on the same chip* to the correct array of the collection unit, four truncated butterfly networks are used. In each of the two stages of the 16 input network, buffers of size two are sufficient to cope with the 11 inputs per clock cycle on average.

Type III. To handle the primes in the range $1.5 \cdot 10^7 < p < 4 \cdot 10^7$, we use a third type of largish stations. The overall structure is the same as for Type I and II. As for the Type II station, the number of factor base elements per DRAM is 50,000 and Type III stations are placed on the same chips as the collection unit. However, the size of the sieving subinterval handled by Type III stations is reduced to 2^{23}. Consequently, we need in total 8 copies (i. e., 2 per chip) of each Type III largish station, and each of these largish stations is in charge of two arrays of the collection unit.

To process all the primes $1.5 \cdot 10^7 < p < 4 \cdot 10^7$, we use $8 \cdot 16$ largish stations of Type III with $8 \cdot 76$ DRAMs of size 0.6 mm^2 each. The outputs of these 16 stations are sent to the correct one of the two related arrays of the collection unit. A switching unit (butterfly network with depth 1) with 16 inputs and $16 \cdot 8$ buffers can handle this.

Medium Stations

Type I. This type of medium stations is in charge of primes in the range $2^{20} < p < 1.5 \cdot 10^7$. In analogy to the largish stations of Type II, each medium station of Type I handles a sieving subinterval of size 2^{24}. Consequently, there are four copies of each medium station of Type I—one on each chip of the collection unit. In total, each chip of the collection unit hosts 20 medium stations of Type I, where each station is equipped with $4.0 \cdot 10^6$ bit of DRAM. The first prime hitting the respective subinterval is stored in full, and for the remaining primes a simple difference coding is used. Storing the difference between successive primes instead of the primes itself allows us to reduce the memory required for a factor base element to 44 bit. On average, from each DRAM, two factor base elements are read per clock cycle.

For each of the respective primes, all relations within the subinterval of the chip (of size 2^{24}) are calculated using several adders, and the hits are reported to the relevant array of the collection unit. Additionally, the corresponding hit in the next subinterval of the device (of size $S = 2^{26}$) is produced and written back into the DRAM. To perform this operation, along with a (p, r)-pair the value $p \cdot \lfloor S/p \rfloor$ is stored in the DRAM. Applying a difference coding as for the p-values, 12 bit suffice for encoding $p \cdot \lfloor S/p \rfloor$—this includes a flag to indicate a new "starting value". To implement the arithmetic for updating the r-values, two adders (with inputs of $(7/24)$ and $(11/24)$ bit) are used that derive the p- and $p \cdot \lfloor S/p \rfloor$-value from the difference encoding, and two 24-bit adders are used to update the r-value for the subsequent sieving interval of size $S = 2^{26}$. As we want to process the factor base elements at a rate of *two* pairs per clock cycle, for each station, we need two quadruples with the mentioned adders. They perform the necessary update within one clock cycle (in a pipeline structure). In total, we estimate the logic for the updating to require no more than 10,000 transistors per DRAM bank. Together with the $4.0 \cdot 10^6$ bit of DRAM, this amounts to a silicon area of $\approx 0.83\,\mathrm{mm}^2$.

The adders mentioned so far are only in charge of updating the DRAM entries. To determine the hits within the subinterval of size 2^{24} handled by a station, the (p, r)-pairs of each station travel, through two cyclic buses, along a chain of 8 adders (of width 24 bit). The first free adder removes the pair from the bus and calculates all the hits of p in the current subinterval of size 2^{24}. The buses of two adjacent chains of 8 adders are connected; if the workload of the two adder chains is not balanced, (p, r)-pairs will change the station. On average, we expect medium stations of Type I to emit 32 pairs ($\lceil\log_{\sqrt{2}}(p)\rceil, r$) per clock cycle (and a maximum of 40). The outputs of two adjacent stations (at a maximum 16 per clock cycle, on average 12) are sent not only to the correct array of the collection unit, but even to the correct quarter of it. This is performed by a butterfly network with 16 inputs; in each node of the network 6 buffers are enough to cope with the inputs.

Type II. The arithmetic progressions for the factor base elements (p, r) with $2^{13} < p < 2^{20}$ are stored in a similar way as in the medium stations of Type I. However, for the medium stations of Type II, *two* DRAM banks of the same size as before are used to store the $\approx 162,000$ pairs representing the first hit *within each subinterval of size* 2^{24}. The update into the next sieving interval (of size S) is realized in the same way as for the Type I stations. Differing from the handling of the primes $> 2^{20}$, however, only the pair (p, r) for the first hit in each array of the collection unit is sent to the collection unit. The other hits are calculated there, i.e., within the collection unit.

Smallish Stations

For each array of the collection unit, the pairs (p, r) with primes $p < 2^{13}$ are stored in a separate DRAM together with the value $\ell \cdot p$, so that $r + \ell \cdot p$ or $r + \ell \cdot p + p$ is the first hit in the next subinterval of size $S = 2^{26}$, and two more numbers for an update to the next row of the array (interval size 2^{17}) and to the next processor (interval size 2^{12}). The update to the next sieving interval is performed with one adder within 3 clock cycles and the first hit for the subinterval is sent to the array of the collection unit for further processing in the same way as the primes $2^{13} < p < 2^{20}$ handled by the medium stations of Type II. We have four smallish stations on each chip, and they easily fit on a silicon area of $0.4\,\mathrm{mm}^2$; there are some 2050 smallish primes. In total, one smallish station requires no more than 1,500 transistors and $2.9 \cdot 10^6$ bit of DRAM. It fits on a silicon area of $\approx 0.06\,\mathrm{mm}^2$.

4.2   Collection Unit for the 1024-Bit Case

The main part of the collection unit consists of 128×128 processors, each in charge of a subinterval of the sieving region of size 2^{12}. This set of processors is split into 16 arrays of 32×32 processors, and distributing the collection unit onto four chips means to place four of these arrays on each chip. The processors within an array are connected through horizontal and vertical buses to transport the $\log_{\sqrt{2}}(p)$-approximations and the index r to the processor in charge. In addition, an algebraic/rational flag is needed, so that we know which of the two counters per sieving location is to be be updated. Each processor stores the algebraic and rational counters in a DRAM holding 2^{12} words of 20 bit each—10 bit for the algebraic and 10 bit for the rational counter.

Array Structure

As in Section 3.1, we refer to the individual processors within an array as *counting units*. The counting units within one array are connected through vertical and cyclic horizontal buses, basically as indicated in Figure 2. More specifically, in each column of the array, we place one vertical bus, that is running top to bottom for columns with an even number and bottom to top otherwise.

Handling data of largish stations. At the top of each of the 32 columns we have an input unit that is connected to one of the 32 chips holding largish stations of Type I. The input unit translates received $(r, \lceil \log_{\sqrt{2}}(p) \rceil - 55)$-values into pairs $(r, \lceil \log_{\sqrt{2}}(p) \rceil)$. Moreover, the two pairs at top of column $2 \cdot i$ and $2 \cdot i + 1$ (for $0 \leq i < 16$) are exchanged if the distances of both pairs to their target row are larger than 15. The resulting values are put onto the vertical buses.

The outputs of the largish stations on the same chip (Type II and III) are put onto the vertical buses after/before row 16. These outputs (on average 20 per clock cycle; 24 as a maximum) are put onto a bus, so that the distance to the target row is at most 16. The pairs are stored in a buffer of size 4 if the appropriate bus is not free. If the buffer is full, a "panic flag" stops the corresponding node of the butterfly network to produce outputs. According to simulations, a panic flag is set in some 2000 cases and delays the output of the largish stations of Type II and III by a few hundred clock cycles.

The target address of the packets on the vertical bus are compared with the actual row number and removed from the vertical bus if they are equal. The $(r, \lceil \log_{\sqrt{2}}(p) \rceil)$-values are then transferred to one of the two cyclic horizontal buses running in opposite directions. Using a buffer of size 4 here seems to be sufficient (in our simulations, less than 0.4 pairs were lost in a sieving interval of size $S = 2^{26}$). The counting unit reads the addresses on both horizontal buses, transfers the pair to its own buffer and removes it from the bus, if a packet has reached its target processor, i. e., the correct counting unit. If there is no space left in the buffers of the processor, it is possible to leave the entry on the bus—it will return to the same position after 32 clock cycles.

Handling data of medium stations. The progressions of the medium stations are input at the left and right side of the array, directly into the correct row. The routing to the correct row is performed by an extra structure, adjacent to the array.

– Progressions output by the medium stations of Type I are sent to the correct quarter of the array by the station. In each quarter of the array (8 rows) at most 10 inputs arrive (8 on average) per clock cycle. On either side of the array, an 8 input/8 output butterfly network distributes 5 inputs to the correct row.

- Progressions that are output by medium stations of Type II are stored in two DRAMs, one on the left and one on the right side of the array. On either side, two 48 bit buses transport $(p, r, \lceil \log_{\sqrt{2}}(p) \rceil)$-values to the correct rows. Along each of theses buses, in each row, the data is forwarded unchanged to the next row if the target of the r-value is not in this row.

   If the data has reached a suitable row, it is checked if $p > 2^{17}$ (then, there is only one hit per row). In this case, $(r, \lceil \log_{\sqrt{2}}(p) \rceil)$ is sent to its destination via one of two horizontal "medium prime buses", and $(p, r + p, \lceil \log_{\sqrt{2}}(p) \rceil)$ is forwarded to the next row. The pairs for primes $p < 2^{17}$ are transferred to the adder unit of this row to produce all hits within this row and feed them into the array. When the adder unit has finished with the prime p, the data is forwarded to the next row through one of the two vertical buses.

Handling data of smallish stations. The hits for smallish primes are counted in a separate array of processors and DRAMs: The slow access time of DRAM (6 clock cycles) does not allow to store all hits of a subinterval of size 2^{12} in one DRAM. Therefore we double all DRAM counters, so that while processing the smallish progressions for the current sieving interval, the medium and largish progressions of the next sieving interval can already be processed in the other DRAM bank. We switch the role of the two DRAM banks for each sieving interval, so that effectively the smallish progressions are always "one sieving interval ahead". More specifically, instead of one array of 32×32 processors, we now have two such arrays, which are merged so that each processor in one array is adjacent to one of the other array. One of these arrays contains the logic needed for handling the smallish primes. In each row of this array, the 16 processors on the left side of the array are connected through one cyclic bus with one input node at the left side. The same connection is established for the right half of the processors of this row.

   The smallish primes are split into two types (Type I: $1024 \leq p < 2^{13}$ and Type II: $p < 1024$). Both types are stored together in one DRAM as described in Section 4.1. For each p, we also store the value $\lceil \log_{\sqrt{2}}(p) \rceil$. The data is distributed to the left and right half of the array and sent on either side of the array through a vertical (56 bit) bus to 32 *progression generators*. All but the first of these progression generators calculate $r_0 := r + \lfloor 2^{17}/p \rfloor \cdot p$. If $r_0 > 2^{17}$, then $(r_0 \pmod{2^{17}}, p)$ is the first pair to be reported in the row of this progression generator, otherwise p is added to r_0 to obtain the first element to be reported in this row. This value is used within the actual row and in addition forwarded to the progression generator of the subsequent row. For smallish primes of Type I, each progression generator calculates the first hit in each processor (using a 12 bit adder and the value $\lfloor 2^{12}/p \rfloor \cdot p$) and sends the triple $(p, r, \lceil \log_{\sqrt{2}}(p) \rceil)$, along with an algebraic/rational flag and a 1-bit flag indicating the type of the smallish prime, through a cyclic 36 bit bus to the processors of its half of its row. All the progressions of primes of Type II within each half of one line are generated by the corresponding progression generator, and the value $(r, \lceil \log_{\sqrt{2}}(p) \rceil)$ along with the algebraic/rational flag is sent to the target processor through the horizontal 36 bit bus.

Each target processor stores the reported $(r, \lceil \log_{\sqrt{2}}(p) \rceil)$-pairs of Type II in one of its 4 buffers and adds the $\lceil \log_{\sqrt{2}}(p) \rceil$-values into its DRAM. For smallish primes of of Type I, on average every 16 clock cycles a hit can be reported, and these $\lceil \log_{\sqrt{2}}(p) \rceil$-values are added to the DRAM with higher priority than for the smallish primes of Type II. Therefore a buffer of size 2 is sufficient for the smallish primes of Type I.

Area Estimate

Each of the counting units requires ≈ 2800 transistors for the largish and medium sized primes and ≈ 1500 transistors for the smallish primes plus two times 82,000 bit of DRAM. The input units for the largish primes require some 1250 transistors per column of one array, and the units for the input of the medium primes 8750 transistors per row. To generate the smallish primes, in addition, some 4400 transistors per row are necessary. Thus, the total area of one array of 32×32 counting units is approximately 26 mm^2 for the medium and largish primes and 22 mm^2 for the smallish primes. Summarizing, each of the four chips holding collection units has a size of 493 mm^2 and consists of:

- 44 largish stations of Type II (180 mm^2),
- $2 \cdot 16$ largish stations of Type III ($2 \cdot 46$ mm^2),
- 20 medium stations of Type I (20 mm^2),
- $4 \cdot 2$ medium stations of Type II ($4 \cdot 2$ mm^2),
- $4 \cdot 1$ smallish stations ($4 \cdot 0.06$ mm^2),
- 4 arrays of collection units ($4 \cdot 48$ mm^2).

One subinterval of size $S = 2^{26}$ is processed within 53,000 clock cycles.

4.3   Combination of the Chips for the 1024-Bit Case

One complete sieving device is comprised of 36 chips: 32 chips (each of size 472 mm^2) holding the largish stations of Type I plus 4 chips (each of size 493 mm^2) hosting the collection unit. Each chip holding largish stations of Type I, per clock cycle sends 16 pairs (with 28 bit each) to one of the 16 arrays of counting units distributed over the four chips holding the collection unit. The collection unit as a whole, i.e., totaling all four chips, receives $4 \cdot 32$ pairs (3584 bit) per clock cycle. The 36 chips can be placed in a regular, grid-like structure, so that the maximum distance any pair has to travel is 5 times the distance between adjacent chips. Implementing this communication across chip borders is non-trivial, but does not appear utopian. The necessary wiring still seems significantly easier to realize than SHARK's transport system [3]. Finally, as we do not store factors found during sieving, the sieving reports output by our device are fed into an ECM engine as described in [4]. In this way, the needed norm factorizations can be obtained without affecting the sieving time in a relevant manner. Including one ECM chip for computing and factoring the norms, the silicon area needed for one complete device is about 172 cm^2.

One sieving line is split into $16.4 \cdot 10^6$ subintervals of size 2^{26} and at a clocking rate of 600 MHz can be processed in less than 25 minutes. The time needed for switching to the next sieving line, i.e., loading new pairs into the DRAMs, requires some 0.035 seconds and is negligible. Similarly, the time needed for outputting the (candidate) relations identified in the completed sieving line is not significant. Using the same 33% saving as in TWIRL [16, Appendix A.5], with 8300 of the above devices, the sieving step for a 1024 bit number can be expected to be completed within one year. Comparing the sieving time/chip area of our design and TWIRL, we see that our device requires by a factor of 3.5 more silicon area than TWIRL. Unlike TWIRL, however, our design is not wafer-scale.

Optimizing parameters. More research is needed for finding optimal parameters for our design: For instance, after a simple modification, the largish units of Type I can output the pairs of two DRAM banks (two consecutive subintervals of size 2^{26}) within 52,000 clock cycles. If the four chips holding the collection units are doubled, the silicon area for this modified device increases by roughly 20 cm^2 and halves the processing time. Using this simple modification, the sieving for a 1024 bit number can be expected to be completed within one year using only 2.0 times the silicon area of TWIRL.

5   Conclusion and Future Work

The hardware design proposed above uses only chips of moderate size (493 mm^2 and 472 mm^2) without paying for this in a drastic loss of performance: Compared to TWIRL, only a factor 2–3.5 in performance is lost. The inter-chip communication required is non-trivial, but still seems doable and easier to realize than the transport system for SHARK. Thus, from a practical cryptanalytic point of view, the new design seems to deserve a more detailed exploration.

So far we did not explore the cost of a prototype for, say, 512 bit or 768 bit numbers, which seems a worthwhile next step. Not only for this, the inclusion of fault detection and fault recovery mechanisms deserves further exploration. Finally, the results achieved so far seem to justify a closer look at our design when allowing more advanced fab technology, say involving a 90 nm process.

References

1. Daniel J. Bernstein. Circuits for Integer Factorization: a Proposal. At the time of writing available electronically at `http://cr.yp.to/papers/nfscircuit.pdf`, 2001.
2. Wieb Bosma, John J. Cannon, and Catherine Playoust. The Magma Algebra System I: The User Language. *Journal of Symbolic Computation*, 24:235–265, 1997.
3. Jens Franke, Thorsten Kleinjung, Christof Paar, Jan Pelzl, Christine Priplata, and Colin Stahlke. SHARK: A Realizable Special Hardware Sieving Device for Factoring 1024-Bit Integers. In Josyula R. Rao and Berk Sunar, editors, *Cryptographic Hardware and Embedded Systems; CHES 2005 Proceedings*, volume 3659 of *Lecture Notes in Computer Science*, pages 119–130. Springer, 2005.

4. Willi Geiselmann, Fabian Januszewski, Hubert Köpfer, Jan Pelzl, and Rainer Steinwandt. A Simpler Sieving Device: Combining ECM and TWIRL. In Min Surp Rhee and Byoungcheon Lee, editors, *Information Security and Cryptology; ICISC 2006 Proceedings*, volume 4296 of *Lecture Notes in Computer Science*, pages 118–135. Springer, 2006.

5. Willi Geiselmann, Hubert Köpfer, Rainer Steinwandt, and Eran Tromer. Improved Routing-Based Linear Algebra for the Number Field Sieve. In *Proceedings of ITCC '05 – Track on Embedded Cryptographic Systems*. IEEE Computer Society, 2005.

6. Willi Geiselmann, Adi Shamir, Rainer Steinwandt, and Eran Tromer. Scalable Hardware for Sparse Systems of Linear Equations, with Applications to Integer Factorization. In Josyula R. Rao and Berk Sunar, editors, *Cryptographic Hardware and Embedded Systems; CHES 2005 Proceedings*, volume 3659 of *Lecture Notes in Computer Science*, pages 131–146. Springer, 2005.

7. Willi Geiselmann and Rainer Steinwandt. A Dedicated Sieving Hardware. In Yvo G. Desmedt, editor, *Public Key Cryptography — PKC 2003*, volume 2567 of *Lecture Notes in Computer Science*, pages 254–266. Springer, 2003.

8. Willi Geiselmann and Rainer Steinwandt. Hardware for Solving Sparse Systems of Linear Equations over GF(2). In Colin D. Walter, Çetin K. Koç, and Christof Paar, editors, *Cryptographic Hardware and Embedded Systems; CHES 2003 Proceedings*, volume 2779 of *Lecture Notes in Computer Science*, pages 51–61. Springer, 2003.

9. Willi Geiselmann and Rainer Steinwandt. Yet Another Sieving Device. In Tatsuaki Okamoto, editor, *Topics in Cryptology — CT-RSA 2004*, volume 2964 of *Lecture Notes in Computer Science*, pages 278–291. Springer, 2004.

10. Tetsuya Izu, Noboru Kunihiro, Kazuo Ohta, and Takeshi Shimoyama. Analysis on the Clockwise Transposition Routing for Dedicated Factoring Devices. In Jooseok Song, Taekyoung Kwon, and Moti Yung, editors, *Information Security Applications: 6th International Workshop, WISA 2005*, volume 3786 of *Lecture Notes in Computer Science*, pages 232–242. Springer, 2006.

11. Arjen K. Lenstra and Jr. Hendrik W. Lenstra, editors. *The development of the number field sieve*, volume 1554 of *Lecture Notes in Mathematics*. Springer, 1993.

12. Arjen K. Lenstra and Adi Shamir. Analysis and Optimization of the TWINKLE Factoring Device. In Bart Preneel, editor, *Advances in Cryptology — EURO-CRYPT 2000*, volume 1807 of *Lecture Notes in Computer Science*, pages 35–52. Springer, 2000.

13. Arjen K. Lenstra, Adi Shamir, Jim Tomlinson, and Eran Tromer. Analysis of Bernstein's Factorization Circuit. In Yuliang Zheng, editor, *Advances in Cryptology — ASIACRYPT 2002*, volume 2501 of *Lecture Notes in Computer Science*, pages 1–26. Springer, 2002.

14. Arjen K. Lenstra, Eran Tromer, Adi Shamir, Wil Kortsmit, Bruce Dodson, James Hughes, and Paul C. Leyland. Factoring Estimates for a 1024-Bit RSA Modulus. In Chi-Sung Laih, editor, *Advances in Cryptology — ASIACRYPT 2003*, volume 2894 of *Lecture Notes in Computer Science*, pages 55–74. Springer, 2003.

15. Adi Shamir. Factoring Large Numbers with the TWINKLE Device. In Çetin K. Koç and Christof Paar, editors, *Cryptographic Hardware and Embedded Systems. First International Workshop, CHES'99*, volume 1717 of *Lecture Notes in Computer Science*, pages 2–12. Springer, 1999.

16. Adi Shamir and Eran Tromer. Factoring Large Numbers with the TWIRL Device. In Dan Boneh, editor, *Advances in Cryptology — CRYPTO 2003*, volume 2729 of *Lecture Notes in Computer Science*, pages 1–26. Springer, 2003.

Divisible E-Cash Systems Can Be Truly Anonymous*

Sébastien Canard[1] and Aline Gouget[2]

[1] France Télécom R&D, 42 rue des Coutures, F-14066 Caen, France
[2] Gemalto, 6, rue de la Verrerie, F-92190 Meudon, France

Abstract. This paper presents an off-line divisible e-cash scheme where a user can withdraw a divisible coin of monetary value 2^L that he can parceled and spend anonymously and unlinkably. We present the construction of a security tag that allows to protect the anonymity of honest users and to revoke anonymity only in case of cheat for protocols based on a binary tree structure without using a trusted third party. This is the first divisible e-cash scheme that provides both full unlinkability and anonymity without requiring a trusted third party.

1 Introduction

Electronic cash systems allow users to withdraw electronic coins from a bank, and then to pay a merchant using electronic coins preferably without communicating with the bank or a trusted party during the payment. Finally, the merchant deposits the spent coins to the bank.

Electronic cash provides user anonymity against both the bank and the merchant during a purchase in order to emulate the perceived anonymity of regular cash transaction. It must be impossible to link two spending protocols and a spending protocol to a withdrawal protocol.

As it is easy to duplicate electronic data, an e-cash system must prevent a user from double-spending. Ideally, the anonymity of honest users must be protected and the identity of cheaters must be recovered without using a trusted third party. An electronic payment system must also prevent a merchant from depositing the same coin twice.

To be practical, an e-cash system must be based on efficient protocols. The most critical protocol is the spending phase between the user and the merchant that must be reasonably efficient. It should also be possible to withdraw or spend several coins more efficiently than repeating several times a single withdrawal or spending protocol.

1.1 Related Works

The compact E-cash scheme [4] allows to withdraw efficiently a wallet containing 2^L coins and provides all the security properties mentioned above. One solution

---

* This work has been partially financially supported by the European Commission through the IST Program under Contract IST-2002-507932 ECRYPT and by the French Ministry of Research RNRT Project "CRYPTO++" .

M. Naor (Ed.): EUROCRYPT 2007, LNCS 4515, pp. 482–497, 2007.

to improve the efficiency of the spending phase is to manage a wallet that contains coins with several monetary values as it was done in [8]; the main drawback is that the user must choose during the withdrawal protocol how many coins he wants for each monetary value.

Divisible e-cash schemes allow a user to withdraw a coin of monetary value 2^L and then to spend this coin in several times by dividing the value of the coin. The aim is to allow a user to efficiently spend a coin of monetary value 2^ℓ, $0 \leq \ell \leq L$, (i.e. more efficiently than repeating 2^ℓ times a spending protocol). Many off-line *divisible e-cash* systems have been proposed in the literature [22,23,13,14,21,9,20,19] providing part of the security properties mentioned above. The first practical divisible e-cash system was proposed by Okamoto [21] and improved by Chan *et al.* in [9]. Both schemes provide anonymity of users but not unlinkability since it is still possible to link several spends from a single divisible coin.

The first *unlinkable divisible* e-cash system that fulfills the usual properties of anonymity and unlinkability was proposed in [20] and improved in [19]. The main drawback of these two systems is that they require a trusted third party to get the identity of the user in case of double-spend detection: this is consequently what we can call a *fair* divisible e-cash system. Moreover, the unlinkability provided by [20,19] is not strong since the merchant and the bank know which part of the withdrawn divisible coin the user is spending which is an information leak on the user.

None of the divisible e-cash schemes of the state of the art provides simultaneously strong unlinkability and truly anonymity of users.

1.2   Our Contribution

We present a strong unlinkable and anonymous divisible off-line e-cash system without trusted third party. We first provide a generic construction and next apply it to the construction of Nakanishi and Sugiyama [20]. Our system is the first that provides the user anonymity such that it is impossible for anybody to make any link between spends and withdraws. Furthermore, our construction does not require a trusted third party to revoke the anonymity of a user that has spent twice the same coin. From a theoretical point of view, the identity of the user can only be revealed when such a case happens. This is the first divisible e-cash system providing this security property.

1.3   Organization of the Paper

This paper is organized as follows. Section 2 describes the security model and requirements for a divisible e-cash system. In Section 3, we present the general principle of the construction. Section 4 is the main one: it contains the new divisible e-cash called $\mathcal{DCS}$. Finally, in Section 5, we give the security proofs of our construction.

2   Security Model

We adopt the model of divisible e-cash system without trusted third party. The three usual players are the user $\mathcal{U}$, the bank $\mathcal{B}$ and the merchant $\mathcal{M}$. The security parameter is denoted by k.

2.1   Algorithms

- ParamKeyGen(k): a probabilistic algorithm outputting the parameters of the system $Params$ ($Params$ contains the parameter k).
- BKeyGen($Params$): a probabilistic algorithm executed by $\mathcal{B}$ outputting the key pair $(sk_\mathcal{B}, pk_\mathcal{B})$.
- KeyGen($Params$): a probabilistic algorithm executed by $\mathcal{U}$ (resp. $\mathcal{M}$) outputting $(sk_\mathcal{U}, pk_\mathcal{U})$ (resp. $(sk_\mathcal{M}, pk_\mathcal{M})$).
- Withdraw($\mathcal{B}(sk_\mathcal{B}, pk_\mathcal{B}, pk_\mathcal{U}, Params)$, $\mathcal{U}(sk_\mathcal{U}, pk_\mathcal{U}, pk_\mathcal{B}, Params)$): an interactive protocol between $\mathcal{B}$ and $\mathcal{U}$. At the end, either $\mathcal{U}$ gets a divisible coin $\mathcal{C}$ of monetary value 2^L (L belongs to $Params$) and outputs OK, or $\mathcal{U}$ outputs $\perp$. The output of $\mathcal{B}$ is either its view $\mathcal{V}_\mathcal{B}^{\mathtt{Withdraw}}$ of the protocol (including $pk_\mathcal{U}$), or $\perp$.
- Spend($\mathcal{U}(2^\ell, pk_\mathcal{M}, \mathcal{C}, Params)$, $\mathcal{M}(sk_\mathcal{M}, pk_\mathcal{B}, Params)$): an interactive protocol between $\mathcal{U}$ and $\mathcal{M}$. At the end, either $\mathcal{M}$ obtains a master serial number S and a proof of validity Π and outputs (S, Π) or $\mathcal{M}$ outputs $\perp$. Either $\mathcal{U}$ updates $\mathcal{C}$ by saving the part of the divisible coin he spent (i.e. the value S) and outputs OK, or $\mathcal{U}$ outputs $\perp$.
- Deposit ($\mathcal{M}((S, \Pi), sk_\mathcal{M}, pk_\mathcal{M}, pk_\mathcal{B}, Params)$, $\mathcal{B}(pk_\mathcal{M}, Params)$): an interactive protocol between $\mathcal{M}$ and $\mathcal{B}$. During the deposit, $\mathcal{B}$ receives (S, Π) from $\mathcal{M}$, checks that it is fresh and that Π is correct. If not, $\mathcal{B}$ outputs $\perp_1$. Else $\mathcal{B}$ computes 2^ℓ serial numbers $\widetilde{S}_1, \ldots, \widetilde{S}_{2^\ell}$ from (S, Π) and $Params$. If one of the serial number $(\widetilde{S}_i, S', \Pi')$ already belongs to $\mathcal{L}$, then the bank outputs $(\perp_2, S, \Pi, S', \Pi')$. Otherwise, $\mathcal{B}$ adds $(\widetilde{S}_i, S, \Pi)$, $1 \leq i \leq 2^\ell$, to its list $\mathcal{L}$ of spent coins, credits $\mathcal{M}$'s account, and returns $\mathcal{L}$. $\mathcal{M}$'s output is OK or $\perp$.
- Identify($(S_1, \Pi_1), (S_2, \Pi_2), Params$): a deterministic algorithm executed by $\mathcal{B}$ that outputs a public key $pk_\mathcal{U}$ and a proof Π_G. If $\mathcal{M}$s who had submitted Π_1 and Π_2 are not malicious, then Π_G is evidence that $pk_\mathcal{U}$ is the registered public key of a user that double-spent a coin.
- VerifyGuilt($pk_\mathcal{U}, \Pi_G, Params$): a deterministic algorithm executed by any actor that outputs 1 if the proof is correct and 0 otherwise. This verification permits anyone to be sure that the user with public key $pk_\mathcal{U}$ is guilty of double-spending a coin.

2.2   Notions of Security

In the following, it is assumed that the overlying experiment has run the algorithm ParamKeyGen on input k to obtain the parameters $Params$.

- **Unforgeability.** Let $\mathcal{A}$ be a p.p.t. Turing Machine. At the start of the game, $\mathcal{A}$ is given the public key $pk_\mathcal{B}$ and $Params$. Suppose that $\mathcal{A}$ interacts K times with an honest bank during withdrawal protocols, then the probability that the number of valid coins that has been spent is at least $2^L K + 1$ is negligible.
- **Unlinkability.** Let $\mathcal{A}$ be a p.p.t. Turing Machine. At the start of the game, $\mathcal{A}$ is given the key pair $(pk_\mathcal{B}, sk_\mathcal{B})$ and $Params$. At the end, $\mathcal{A}$ chooses two honest users 0 and 1. A bit b is secretly and randomly chosen. Then, a spending protocol is played by $\mathcal{A}$ with user b (it is assumed that both honest users still have unspent coins). Finally, $\mathcal{A}$ outputs a bit b'. We require that for every $\mathcal{A}$ playing this game, the probability that $b = b'$ differs from $1/2$ by a fraction that is at most negligible.
- **Identification of double-spenders.** Let $\mathcal{A}$ be a p.p.t. Turing Machine. At the start of the game, $\mathcal{A}$ is given the public key $pk_\mathcal{B}$ and $Params$. The probability that a `Deposit` protocol between an honest merchant and an honest bank outputs $(\perp_2, S, \Pi, S', \Pi')$ such that the output of `Identify` algorithm on inputs (S, Π, S', Π') is not the public key $pk_\mathcal{U}$ of a corrupted user is negligible.
- **Exculpability.** Let $\mathcal{A}$ be a p.p.t. Turing Machine. At the start of the game, $\mathcal{A}$ is given the key pair $(pk_\mathcal{B}, sk_\mathcal{B})$ and $Params$. During the game, $\mathcal{A}$ interacts with honest users to supply them coins. At the end, $\mathcal{A}$ constructs two spent coins (S_1, Π_1) and (S_2, Π_2). The probability that the outputs of the `Identify` algorithm on inputs (S_1, Π_1) and (S_2, Π_2) is the public key $pk_\mathcal{U}$ of an honest user together with a valid proof Π_G is negligible.

Remark 1. Notice that the exculpability property implies that the bank cannot create withdrawals for which the user has not participated. We don't need any extra security property, such as the proposal in [28].

3 General Description

In an anonymous e-cash system without a trusted third party, spending a single coin consists in generating a valid serial number S to allow double-spending detection and a valid security tag T masking the identity of the spender. The spender has to prove that S and T are well-formed without giving any information about his identity. In particular, the identity of the spender must be recovered only in case of double-spending by using the security tag T.

The main motivation of divisible e-cash is to provide a method to withdraw or spend several coins more efficiently than repeating several times a single withdrawal or spending protocol. We provide a general approach to construct divisible e-cash systems strongly unlinkable and truly anonymous (the user identity can be recovered only in case of fraud). This construction can be applied using several basic cryptographic tools.

3.1 Truly Anonymous E-Cash Scheme Based on Binary Trees

The general principle of our construction is derived from the classical binary tree approach [21,9,20] with slight modifications. Each divisible coin of monetary value 2^L is assigned to a binary tree of $L + 2$ levels. The tree root (level 0) with monetary value 2^L is assigned to a serial number denoted by $N_{0,0}$. Any other node has a monetary value corresponding to half of the amount of its parent node, except for the leaves that have no monetary value: they are "dead" leaves. For every level i, $0 \leq i \leq L$, the 2^i nodes are assigned serial numbers denoted by $N_{i,j}$ with $1 \leq j \leq 2^i$, except for the "dead" leaves that are not related to any serial number. Any *divisible* e-cash system should verify the divisibility rule.

Definition 1. *When a node N is used, none of descendant and ancestor nodes of N can be used, and no node can be used more than once.*

This rule is satisfied if, and only if, over-spending is protected. The general principle of our proposal consists in using a single master serial number from which several serial numbers can be derived. Thus, each node of the tree, which includes the leaves, is also related to a particular value called a *tag key*. During the spending protocol, the identity of the spender is encrypted with a tag key in such a way that the decryption key can be derived only in case of a double-spending. Using the binary tree approach, each node of the tree is related to a tag key with the following properties.

- The root tag key and the identity of the user are signed (in a blind manner) by the bank during the withdrawal protocol.
- From the tag key of a node N, it is possible for everyone to compute the tag keys related to the descendant nodes of N. It consequently exists a public deterministic function $\mathcal{F}$ that takes as input a tag key K_{i,b_0} (where i is the level of the targeted node in the tree and $b_0 \in \{0, 1\}$ depends on the position of K in the tree[1]), a bit b (0 for left and 1 for right) and possibly some public parameters $Params$ and that outputs a new tag key $K_{i+1,b}$.

$$\mathcal{F} : (K_{i,b_0}, b, Params) \longrightarrow K_{i+1,b} = \mathcal{F}(K_{i,b_0}, b, Params).$$

- From the tag key of a node, it is impossible (without the knowledge of the root tag key) to compute a tag key which is not related to a descendant of the targeted node.
- The serial number of a particular node is the concatenation of the two children tag keys. Notation is given in Figure 1.

During the spending protocol, the user computes the tag key of the node he wants to spend. This tag key is used to compute the security tag, i.e. the encryption of the spender identity. This encryption should be verifiable and should include randomness. This randomness should be provided by the merchant to ensure the freshness of the spending, i.e., to prevent merchant from sending twice the same coin to the bank. The user also computes the tag keys related to the two direct descendants of the spent node. The concatenation of these two keys is the serial number of the spent coin. This serial number is transmitted during the spend protocol. Later, the bank will compute all the serial numbers of the leaves

---

[1] $b_0 = 0$ if and only if the targeted node belongs to the left subtree of its ancestor.

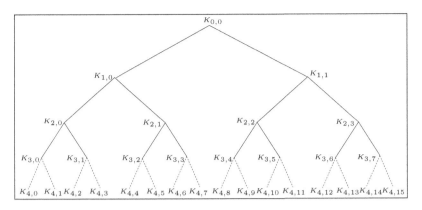

Fig. 1. General principle - Tree of keys

of the tree in order to detect a possible double-spending. If a double-spending is detected, then the bank has access to the encryption of the identity (from one spending) and the corresponding decryption key (from the other spending). Then, the bank can easily find the identity of the cheater.

Example 1. Assume $\mathcal{U}$ wants to spend four coins. Then, $\mathcal{U}$ selects four unitary coins, e.g. those associated to the node $K_{1,0}$. The user $\mathcal{U}$ sends to $\mathcal{M}$ the values $T = E_{K_{1,0}}(Id, R)$, $LK = K_{2,0}$, $RK = K_{2,1}$, and $S = LK \| RK$. The random value R used in the encryption scheme is computed using values sent by the merchant. The user must also prove that the coins are signed by the bank and that it will be possible to identify a double-spender. Consequently, the spending protocol consists also in computing a zero-knowledge proof of knowledge Φ that corresponds to the predicates:

- T is well-formed, i.e. $E_{K_{1,0}}(Id, R)$ has been computed using:
  - the tag key $K_{1,0}$ derived using $\mathcal{F}$ on inputs the root tag key $K_{0,0}$ signed by the bank,
  - the random R that has been chosen by the merchant,
  - the identity Id signed by the bank.
- LK and RK are well-formed, i.e., $K_{2,0}$ and $K_{2,1}$ are both derived from $K_{1,0}$ using $\mathcal{F}$.
- If LK and RK are well-formed, this implies that the serial number S is also well-formed.

To construct a truly anonymous divisible e-cash system, it is then necessary to provide a function $\mathcal{F}$, a verifiable encryption scheme E and a proof Φ. We give an example in Section 4.

3.2   Useful Tools

Proofs of Knowledge. We use zero-knowledge proofs of knowledge constructed over a cyclic group $\mathcal{G}$ either of prime order q or of unknown order: proof of equality of two known representations [10,6], proofs of knowledge of a discrete logarithm [26,17], of a representation, of a double discrete logarithm $PK(\alpha/z = g^{\alpha} \wedge y = g_1^{g_2^{\alpha}})$ [27,20], proof of the "or" statement $PK(\alpha/T_1 = h_1^{\alpha} \vee T_2 = h_2^{\alpha})$ [11,25]. We also need a proof of knowledge of one out of two double discrete logarithm $PK(\alpha/T_1 = g^{h_1^{\alpha}} \vee y = g^{h_2^{\alpha}})$ which is a combination of the two above proofs. These proofs can also be used non interactively by using the Fiat-Shamir heuristic [16].

Camenisch-Lysyanskaya Signature Schemes. These signature schemes are proposed in [5] with in addition some specific protocols:

- an efficient protocol between a user $\mathcal{U}$ and a signer $\mathcal{S}$ that permits $\mathcal{U}$ to obtain from $\mathcal{S}$ a signature σ of some commitment C on values $(x_1, \ldots, x_l)$ unknown from $\mathcal{S}$. $\mathcal{S}$ computes $\texttt{CLSign}(C)$ and $\mathcal{U}$ gets $\sigma = \texttt{Sign}(x_1, \ldots, x_l)$ that can be verified by $\texttt{Verif}(\sigma, (x_1, \ldots, x_l)) = 1$.
- an efficient proof of knowledge of a signature on committed values, denoted by $PK(\alpha_1, \ldots, \alpha_l, \beta/\beta = \texttt{Sign}(\alpha_1, \ldots, \alpha_l))$.

These constructions are quite close to group signature schemes. This is the case of the two following examples, one based on the ACJT signature scheme [1], secure under the Flexible RSA assumption [15], and the other based on the BBS one [2], secure under the q-SDH assumption [2].

4   Divisible E-Cash System $\mathcal{DCS}$

We apply the general construction presented in Section 3.1 to the binary tree used in the system described in [20]. The function $\mathcal{F}$ is chosen to be the modular exponentiation. For each level i, there are three linked generators $g_{i,0}$ for "left", $g_{i,1}$ for "right" and $g_{i,2}$ to compute the security tag. For a node at level $i - 1$ represented by the tag key denoted by K_{i-1,b_0}, the tag key of, e.g. the left children, is $K_{i,0} = g_{i,0}^{K_{i-1,b_0}}$. For the tag key $K_{i,b}$ and a random value R computing using merchant data, the encryption of the user identity $pk_{\mathcal{U}}$ is defined to be $pk_{\mathcal{U}} g_{i+1,2}^{K_{i,b} \cdot R}$. In the following, we assume that $\mathcal{H}$ is a collision-resistant hash function.

4.1   Setup

We consider a group $\mathcal{G}$ of order $o_{\mathcal{G}}$. The elements h_0, h_1, h_2 are random generators of $\mathcal{G}$. $\mathcal{G}_1 = \langle g_1 \rangle$ is a subgroup of $\mathbb{Z}_{o_{\mathcal{G}}}^*$ and each group $\mathcal{G}_i = \langle g_i \rangle$ must be a subgroup of $\mathbb{Z}_{o_{i+1}}^*$ where o_{i+1} is the order of $\mathcal{G}_{i+1}$. For example [20], it is possible to take $\mathcal{G}_i$ as a subgroup of $\mathbb{Z}_{o_{i+1}}^*$ for the prime $o_{i+1} = 2o_i + 1$ with all i. As a consequence, the group $\mathcal{G}_i$ is related to the level i of the tree. The following generators are randomly chosen: g in $\mathcal{G}$, $g_{1,0}, g_{1,1}, g_{1,2}$ in $\mathcal{G}_1$, $g_{2,0}, g_{2,1}, g_{2,2}$ in $\mathcal{G}_2$, $\ldots$, $g_{L+1,0}$, $g_{L+1,1}, g_{L+1,2}$ in $\mathcal{G}_{L+1}$ whose discrete logarithms to the base $g_1, g_2, \ldots, g_{L+1}$ are unknown, respectively. All these data compose the public parameters $Params$

of the system and can be computed by the bank. The bank $\mathcal{B}$ computes the key pair $(sk_{\mathcal{B}}, pk_{\mathcal{B}})$ of a Camenisch-Lysyanskaya signature scheme that will permit it to sign a divisible coin, using the CLSign algorithm.

A user $\mathcal{U}$ (resp. a merchant $\mathcal{M}$) can compute its key pair $(sk_{\mathcal{U}}, pk_{\mathcal{U}})$ (resp. $(sk_{\mathcal{M}}, pk_{\mathcal{M}})$) by choosing randomly $u \in [0, o_{\mathcal{G}}[$ (resp. $m \in [0, o_{\mathcal{G}}[)$ and computing g^u (resp. g^m). The value u (resp m) is the private key $sk_{\mathcal{U}}$ (resp. $sk_{\mathcal{M}}$) and g^u (resp. g^m) is equal to the public key $pk_{\mathcal{U}}$ (resp. $pk_{\mathcal{M}}$).

4.2   Withdrawal Protocol

During a withdrawal protocol, $\mathcal{U}$ interacts with $\mathcal{B}$. $\mathcal{U}$'s inputs are $pk_{\mathcal{B}}$, $sk_{\mathcal{U}}$, $pk_{\mathcal{U}}$ and $Params$, and $\mathcal{B}$'s inputs are $pk_{\mathcal{U}}$, $sk_{\mathcal{B}}$, $pk_{\mathcal{B}}$ and $Params$.

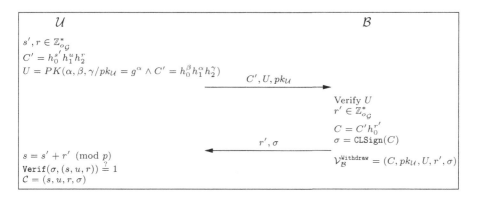

Fig. 2. Withdrawal protocol

The withdrawal protocol permits $\mathcal{U}$ to obtain a new divisible coin by interacting with $\mathcal{B}$ as described in Figure 2. A divisible coin corresponds to a (blind) CL signature done by $\mathcal{B}$ on a secret s and the secret key u of $\mathcal{U}$. Both $\mathcal{U}$ and $\mathcal{B}$ participate to the randomness of the secret s. At the end of the Withdraw protocol, $\mathcal{U}$ gets a divisible coin $\mathcal{C} = (s, u, r, \sigma = \text{Sign}(s, u, r))$.

4.3   Spending Protocol

When $\mathcal{U}$ wants to spend to $\mathcal{M}$ a sub-coin of value 2^{ℓ} ($\ell = L - i$) from his divisible coin $\mathcal{C}$, he chooses an unspent node of the level i, e.g. the node $N_{i,j}$. A spending protocol of the node $N_{i,j}$ consists in the following.

1. $\mathcal{M}$ sends to $\mathcal{U}$ a random value $rand$ and $\mathcal{U}$ computes $R = \mathcal{H}(pk_{\mathcal{M}} \| rand)$.
2. $\mathcal{U}$ randomly chooses $\tilde{g}, \tilde{h} \in \mathcal{G}$, $\tilde{g}_1 \in \mathcal{G}_1$, $\tilde{g}_2 \in \mathcal{G}_2$, ..., $\tilde{g}_{i+1} \in \mathcal{G}_{i+1}$.
3. $\mathcal{U}$ executes the algorithm presented in Figure 3 (in pseudo-code) for the node $N_{i,j}$, outputting the values[2] $(\widetilde{V}_0, \ldots, \widetilde{V}_i, V)$, using the path from the

---

[2] The values $\widetilde{V}_0, \ldots, \widetilde{V}_i$ are computed to prove that the value V is well computed. See proof Φ below and [20].

Input: i, j
Output: $(\widetilde{V}_0, \ldots, \widetilde{V}_i, V)$

  $\tilde{r} \leftarrow \text{Rand}()$, $V \leftarrow g^s$, $\widetilde{V}_0 \leftarrow \tilde{g}^s \tilde{h}^{\tilde{r}}$, $\text{CurrentNode} \leftarrow root$
  If $i = 0$, then return $(\widetilde{V}_0, V)$
  $a \leftarrow 1, b \leftarrow 2^i$
  For $k = 1$ to i
    $\widetilde{V}_k \leftarrow \tilde{g}_k^V$
    If $a \leq j \leq a + (b - a - 1)/2$, then   \\ $N_{i,j}$belongs to leftSubTree(CurrentNode)
      $V \leftarrow (g_{k,0})^V$, $b \leftarrow a + (b - a - 1)/2$ \\ $\text{CurrentNode} \leftarrow$ leftSon(CurrentNode)
    Else \\ $N_{i,j}$ belongs to rightSubTree(CurrentNode)
      $V \leftarrow (g_{k,1})^V$, $a = a + (b - a + 1)/2$ \\ $\text{CurrentNode} \leftarrow$ rightSon(CurrentNode)
  return $(\widetilde{V}_0, \ldots, \widetilde{V}_i, V)$

Fig. 3. Spending protocol - Computation of V

root tree to the node $N_{i,j}$. Next, $\mathcal{U}$ computes the security tag: $LK = g_{i+1,0}^V$, $RK = g_{i+1,1}^V$, $T = pk_{\mathcal{U}} g_{i+1,2}^{V \cdot R}$ and $S = LK \| RK$.

Example 2. Assume $\mathcal{U}$ wants to spend four coins (the same as in Example 1. The user $\mathcal{U}$ sends to the merchant $\mathcal{M}$ the values $LK = g_{2,0}^{g_{1,0}^s}$, $RK = g_{2,1}^{g_{1,0}^s}$, $T = pk_{\mathcal{U}}(g_{2,2}^{R \cdot g_{1,0}^s})$ and $S = LK \| RK$ since $V = g_{1,0}^{g^s}$.

4. $\mathcal{U}$ proves to $\mathcal{M}$ the validity of LK, RK, T (and thus the validity of S) using a non-interactive zero-knowledge proof of knowledge of a signature of $\mathcal{B}$ on the values (s, u, r) and that the value LK, RK, T are correctly computed. This proof of knowledge is constructed from a zero-knowledge proof of knowledge using the Fiat-Shamir heuristic. This proof is as follows:

$$\Phi = PK\Big(\sigma, s, u, r, \tilde{r}, \alpha_1, \ldots, \alpha_{i+1}, \beta \; / $$
$$\sigma = \text{Sign}(s, u, r) \; \wedge \; \widetilde{V}_0 = \tilde{g}^s \tilde{h}^{\tilde{r}} \; \wedge \; \widetilde{V}_1 = \tilde{g}_1^{g^s} \; \wedge \; \widetilde{V}_1 = \tilde{g}_1^{\alpha_1} \; \wedge$$
$$(\widetilde{V}_2 = \tilde{g}_2^{g_{1,0}^{\alpha_1}} \; \vee \; \widetilde{V}_2 = \tilde{g}_2^{g_{1,1}^{\alpha_1}}) \; \wedge \; \widetilde{V}_2 = \tilde{g}_2^{\alpha_2} \; \wedge \; \ldots \; \wedge$$
$$(\widetilde{V}_{i+1} = \tilde{g}_{i+1}^{g_{i,0}^{\alpha_i}} \; \vee \; \widetilde{V}_{i+1} = \tilde{g}_{i+1}^{g_{i,1}^{\alpha_i}}) \; \wedge \; \widetilde{V}_{i+1} = \tilde{g}_{i+1}^{\alpha_{i+1}} \; \wedge$$
$$LK = g_{i+1,0}^{\alpha_{i+1}} \wedge RK = g_{i+1,1}^{\alpha_{i+1}} \wedge T = pk_{\mathcal{U}} g_{i+1,2}^{R \cdot \alpha_{i+1}} \Big)$$

5. $\mathcal{U}$ sends the spent coins (S, Π) to $\mathcal{M}$, with $\Pi = \{2^\ell, T, \Phi, R, \widetilde{V}_0, \ldots, \widetilde{V}_i\}$.

4.4 Deposit Protocol

When $\mathcal{M}$ wants to deposit a coin (S, Π) to $\mathcal{B}$, $\mathcal{M}$ just sends the coin (S, Π) to $\mathcal{B}$. The proof Π should include the monetary value 2^ℓ of the divisible coin, the security tag T, the proof of knowledge Φ and the random data R provided by the merchant. $\mathcal{B}$ checks the validity of Φ and the consistency with S. If (S, Π) is not a valid coin, $\mathcal{B}$ rejects the deposit. Else, $\mathcal{B}$ computes, from S, 2^ℓ serial numbers $\widetilde{S}_{k_1}, \ldots \widetilde{S}_{k_{2^\ell}}$ corresponding to the $2^{\ell+1}$ dead leaves of the sub-tree. This is done by applying several modular exponentiation functions to S, using the right generators. $\mathcal{B}$ has to deal with 2^ℓ unitary coins $(\widetilde{S}_{k_j}, S, \Pi)$, $1 \leq j \leq 2^\ell$.

For every unitary coin $(\widetilde{S}_{k_j}, S, \Pi)$, $\mathcal{B}$ checks if there is already an entry $(\widetilde{S}_{k_j}, S', \Pi')$ in the database. If there is no entry in the database for the serial number $\widetilde{S}_{k_j}$, then $\mathcal{B}$ accepts the deposit of the coin $(\widetilde{S}_{k_j}, S, \Pi)$, credits the $pk_{\mathcal{M}}$'s account and add $(\widetilde{S}_{k_j}, S, \Pi)$ to the database of spent coins. Else, there is an entry $(\widetilde{S}_{k_j}, S', \Pi')$ in the database. Then, $\mathcal{B}$ checks the freshness of merchant randomness R in Π compared to Π'. If it not fresh, $\mathcal{M}$ is a cheat and $\mathcal{B}$ refused the deposit. If R is fresh, $\mathcal{B}$ accepts the deposit of the coin $(\widetilde{S}_{k_j}, S, \Pi)$, credits the $pk_{\mathcal{M}}$'s account and add $(\widetilde{S}_{k_j}, S, \Pi, S', \Pi',)$ to the list of double-spenders. For every entry of the database of double-spenders, $\mathcal{B}$ will executes the Identify algorithm.

4.5   Identify

Assume that a double detection has been done. Then $\mathcal{B}$ knows two accepted spending $(2^{I_1}, S_1 = LK_1 \| RK_1, T_1, R_1, \Phi_1)$ with $I_1 = L - i_1$ and $(2^{I_2}, S_2 = LK_2 \| RK_2, T_2, R_2, \Phi_2)$ with $I_2 = L - i_2$ such that e.g. S_1 is an ancestor of S_2 or $S_1 = S_2$. If $S_1 = S_2$ then the bank can directly get the public key $pk_{\mathcal{U}}$ by computing $\left(T_1^{R_2} / T_2^{R_1} \right)^{1/(R_2 - R_1)} = pk_{\mathcal{U}}$. If S_1 is an ancestor of S_2, then the bank computes the masking value $g_{I_2+1,2}^{V_2}$ (s.t. $T_2 = pk_{\mathcal{U}} g_{I_2+1,2}^{R_2 \cdot V_2}$) from the knowledge of LK_1 and RK_1 and the path[3] from $N_{i_1}^{j_1}$ up to $N_{i_2}^{j_2}$ as described in Figure 4. Then, $\mathcal{B}$ computes the public key $pk_{\mathcal{U}}$ as follows: $(T_2)^{\frac{1}{R_2}} / g_{I_2+1,2}^{V_2} = pk_{\mathcal{U}}$.

```
Input: i₁, j₁, i₂, j₂
Output: V₂
        CurrentNode ← N_{i₁}^{j₁}
        If N_{i₂}^{j₂} belongs to leftSubTree(CurrentNode), then
            V₂ ← LK₁;  CurrentNode ← leftSon(CurrentNode);
        Else
            V₂ ← RK₁;  CurrentNode ← rightSon(CurrentNode);
        For k = i₁ + 2 to i₂ do
            If N_{i₂}^{j₂} belongs to leftSubTree(CurrentNode) , then
                V₂ ← (g_{k,0})^{V₂};  CurrentNode ← leftSon(CurrentNode)
            Else
                V₂ ← (g_{k,1})^{V₂};  CurrentNode ← rightSon(CurrentNode)
            k = k + 1
        return V₂
```

Fig. 4. Identify protocol - Computation of V_2

---

[3] The values $N_{i_1}^{j_1}$ and $N_{i_2}^{j_2}$ are not know by $\mathcal{B}$ but $\mathcal{B}$ knows the path from $N_{i_1}^{j_1}$ up to $N_{i_2}^{j_2}$ since it knows the path used to compute the colliding serial numbers.

4.6   Verify Guilt

The algorithm VerifyGuilt can be executed by any actor from the parameters of the system $Params$ and a proof Π_G. One can parse the proof Π_G as $\left((2^{\ell_1}, S_1, R_1, T_1, \Pi_1), (2^{\ell_2}, S_2, R_2, T_2, \Pi_2)\right)$ and next run Identify on these values. If the algorithm Identify returns a public key $pk_{\mathcal{U}}$, then one can check if Π_1 is consistent with $(2^{\ell_1}, S_1, R_1, T_1)$ and if Π_2 is consistent with $(2^{\ell_2}, S_2, R_2, T_2)$. If both are consistent then accept, else reject.

5   Security Arguments

In this section, we provide the Theorem that stipulates that the $\mathcal{DCS}$ scheme is a secure divisible e-cash system.

Theorem 1. *In the random oracle model, the $\mathcal{DCS}$ scheme is secure:*

- *If the CL signature scheme is unforgeable, then $\mathcal{DCS}$ is unforgeable.*
- *Under the DDH assumption, $\mathcal{DCS}$ is unlinkable.*
- *If the CL signature scheme is unforgeable, then $\mathcal{DCS}$ permits the identification of double-spenders.*
- *Under the DL assumption (and the Flexible RSA assumption if $\mathcal{DCS}$ relies on the ACJT scheme), $\mathcal{DCS}$ has the exculpability property.*

Proof. We have to show that $\mathcal{DCS}$ verifies all security properties.

Unforgeability.   We want to show that if an adversary $\mathcal{A}$ is able to break the unforgeability of our construction, then it is possible to break the unforgeability of the CL signature scheme under adaptive chosen message attack.

We can interact with $\mathcal{A}$ during the withdrawal protocol by playing the role of an honest bank with access to the signature oracle. After each successful spending executed by $\mathcal{A}$, we extract, using standard technique, the values (u, s, r, σ) satisfying the relation embedded into the valid proof of knowledge Π. Since there are more spent coins than $\mathcal{A}$ can legitimately own, and since there is no detection of double-spending (by assumption), then it is necessary that, among all extracted values $(u_j, s_j, r_j, \sigma_j)$, one signature σ on a message $m = (s, u, r)$ is unknown and does not come from the signature oracle. Thus, this one more signature is a signature (forgery) in the CL's scheme on the message $m = (u, s, r)$.

As the CL signature scheme is proven secure against adaptive chosen message attacks under the Flexible RSA assumption (if the scheme relies on the ACJT scheme) or the q-SDH (if the scheme relies on the BBS scheme), it follows that $\mathcal{A}$ cannot succeed with non negligible probability.

Because our proof requires rewinding to extract s' and r from an adversary $\mathcal{A}$, our proof is valid only against sequential attacks. Indeed, in a concurrent setting where the attacker is allowed to interact with the bank in an arbitrarily interleaving manner, our machine may be forced to rewind an exponential number of times. This drawback can be overcome by using for instance well-know techniques [12] which would require from the user to encrypt s' and r in a verifiable manner [7].

Unlinkability. We want to show that if an adversary $\mathcal{A}$ is able to break the unlinkability of our construction, then it is possible to break an instance of the Diffie-Hellman problem. In fact, we use a variant of the Diffie-Hellman problem, called Matching Multi Diffie-Hellman (MMDH) problem, and we prove in Appendix A that if someone is able to solve the MMDH problem, then it is possible to solve a given instance of the DDH problem.

We can interact with $\mathcal{A}$ during the withdraw protocol by playing the role of an honest user except for the two first interactions where we use the MMDH instance. During spending protocols, we can interact with $\mathcal{A}$ by playing the role on an honest user, except when the divisible coin corresponds to one of the two divisible coins associated with the MMDH instance to be solved.

We can win the game when $\mathcal{A}$ chooses the two first users (corresponding to the MMDH instance) and thus use the MMDH instance during the execution of the final spend. If $\mathcal{A}$ does not choose users i_0 and i_1 for the challenge we need to play again the game.

We denote by q_U the average number of users created by $\mathcal{A}$. Our success probability is $\epsilon' = 1 - (1 - (1/2 + \epsilon/2))^{q_U} \equiv 1/2 + q_U \epsilon/2$ within polynomial $\mathcal{T}' = q_U \mathcal{T} + \tau$, where τ is polynomial.

Remark 2. In the simulation, we use the instance of the MMDH problem to interact with $\mathcal{A}$. We also need to choose a value for the bit b. If our choice of b is correct, then there is no problem and we will be able to conclude with the advantage ϵ of $\mathcal{A}$. If this choice is uncorrect, $\mathcal{A}$ has a probability exactly equal to $1/2$ as ours. Repeating the game many times, our success probability of solving the MMDH instance is greater than $1/2$.

Identification of Double-spenders. We want to show that if an adversary $\mathcal{A}$ is able to break the identification of double-spenders property, then it is possible to break the unforgeability of the CL signature scheme.

We have access to a signature oracle taking as input a commitment and outputting a signature on committed values. We interact with $\mathcal{A}$ during withdrawal protocols by playing the role of an honest bank. We also interact with $\mathcal{A}$ during spending protocols playing the role of the merchant. Note that there is no honest users in the game. After each successful spending executed by $\mathcal{A}$, we extract the values (u, s, r, σ) satisfying the relation embedded into the valid proof of knowledge Π. When there is a double-spending, i.e. $(\perp_1, S_1, \Pi_1), (S_2, \Pi_2)$, that means that there exist a valid serial number $\tilde{S}$ which can be computed from both S_1 and S_2. Furthermore, the proof Π_1 is consistent with S_1 and the proof Π_2 is consistent with S_2 and $R_1 \neq R_2$ where R_1 is the random chosen by the merchant in Π_1 and R_2 is the random chosen by the merchant in Π_2. Both Π_1 and Π_2 contains a proof of knowledge of a signature of the bank on the master serial number seed s used to generate S_1, S_2 and $\tilde{S}$. Thus, these two signatures σ_1 and σ_2 are such that at least one of the two is different from the signatures obtained during the execution of the Withdrawal protocols submitted to the signature oracle. This signature (σ_1 or σ_2) is thus a forgery on CL signature scheme. As the CL signature scheme is proven secure against adaptive chosen message attacks, it follows that $\mathcal{A}$ cannot succeed with non-negligible probability.

Exculpability.   The adversary $\mathcal{A}$ wins the game if he can falsely accuse an honest user of a double-spending. This means that the adversary can interact with honest users to obtain spending from them and he wins if he can produce one spend (S', T', Π') related to a valid one (S', T', Π') and such that the output of $\mathtt{Identify}((S, T, \Pi), (S', T', \Pi'))$ is a public key $pk_\mathcal{U}$ of a honest user (with non negligible probability).

The security proof of the exculpability involves forking lemma-like technique for an attacker that exploits both valid spending played by honest users and valid withdrawals played by honest users when the extractability of the RO proofs-of-knowledge relies on the DL assumption in order to falsely accuse an honest user. If the Camenisch-Lysysanskaya scheme of the withdrawal protocol uses a group of unknown order, then the exculpability relies on both the DL assumption for an attacker that exploits valid spendings played by honest users in order to falsely accuse an honest user, and on the factorization assumption to ensure the non-malleability and the soundness of the proof of knowledge Φ (see [3]).

6   Conclusion

In this paper, we present the first off-line divisible e-cash scheme that provides strong unlinkability and truly anonymity. We introduced the idea of using a security tag in a divisible e-cash scheme. The anonymity of users is achieved without impacting the performance of the spending protocol and without using a trusted third party. The spending protocol exploits the binary structure underlying the divisible coin in order to get an efficient spending protocol. However, even if the new scheme permits the spending of multiple coins at a time, it uses double-exponentiation proofs for the spending phase which is still a little expensive. Thus, for a small number of coins at a time, the spending is still expensive. Another possible improvement for the scheme could be to find a method to detect double spending without computing 2^L serial numbers for a divisible coin of monetary value 2^L.

Acknowledgements

We are grateful to Pascal Paillier and Jacques Traoré for their suggestions of improvement, and to Serge Fehr and anonymous referees for their valuable comments. We also wish to mention that a similar work has been independently done by Jan Camenisch, Markulf Kohlweiss, Anna Lysyanskaya and Maria Meyerovich.

References

1. G. Ateniese, J. Camenisch, M. Joye, and G. Tsudik. A Practical and Provably Secure Coalition-resistant Group Signature Scheme. Advances in Cryptology - Crypto'00, volume 1880 of LNCS, pages 255-270, 2000.
2. D. Boneh, X. Boyen and H. Shacham. Short Group Signatures using Strong Diffie Hellman. Advances in Cryptology - Crypto'04, volume 3152 of LNCS, pages 41-55, 2004.
3. F. Boudot and J. Traoré. Efficient Publicly Verifiable Secret Sharing Schemes with Fast or Delayed Recovery. ICISC'99, volume 1726 of LNCS, pages 87-102, 1999.
4. J. Camenisch, S. Hohenberger, and A. Lysyanskaya. Compact E-cash. Advances in Cryptology - Eurocrypt'05, volume 3494 of LNCS, pages 302-321, 2005.
5. J. Camenisch and A. Lysyanskaya. Signature Schemes and Anonymous Credentials from Bilinear Maps. Advances in Cryptology - Crypto'04, volume 3152 of LNCS, pages 56-72, 2004.
6. J. Camenisch and M. Michels. Proving in Zero-knowledge that a Number is the Product of Two Safe Primes. Advances in Cryptology - Eurocrypt'99, volume 1592 of LNCS, pages 107-122, 1999.
7. J. Camenisch and V. Shoup. Practical Verifiable Encryption and Decryption of Discrete Logarithms. In D. Boneh, editor, Advances in Cryptology - Crypto '03, volume 2729 of LNCS, pages 126-144. Springer, 2003.
8. S. Canard, A. Gouget, and E. Hufschmitt. A Handy Multi-coupon System. Applied Cryptography and Network Security - ACNS 2006, volume 3989 of LNCS, pages 66-81, 2006.
9. A.H. Chan, Y. Frankel, and Y. Tsiounis. Easy Come - Easy Go Divisible Cash. Advances in Cryptology - Eurocrypt'98, volume 1403 of LNCS, pages 561-575, 1998.
10. D. Chaum and T. Pedersen. Transferred Cash Grows in Size. Advances in Cryptology - Eurocrypt'92, volume 658 of LNCS, pages 390-407, 1993.
11. R. Cramer, I. Damgard, and B. Schoenmakers. Proofs of Partial Knowledge and Simplified Design of Witness Hiding Protocols. Advances in Cryptology - Crypto'94, volume 839 of LNCS, pages 174-187, 1994.
12. I. Damgard. Efficient Concurrent Zero-knowledge in the Auxiliary String Model. Advances in Cryptology - Eurocrypt '00, volume 1807 of LNCS, pages 418-430, 2000.
13. S. D'Amingo, and G. Di Crescenzo. Methodology for Digital Money based on General Cryptographic Tools. Advances in Cryptology - Eurocrypt'94, volume 950 of LNCS, pages 156-170, 1994.
14. T. Eng, and T. Okamoto. Single-term Divisible Coins. Advances in Cryptology - Eurocrypt'94, volume 950 of LNCS, pages 306-319, 1994.
15. E. Fujisaki and T. Okamoto. Statistical Zero-knowledge Protocols to Prove Modular Polynomial Relations. Advances in Cryptology - Crypto'97, volume 1294 of LNCS, pages 16-30, 1997.
16. A. Fiat and A. Shamir. How to Prove Yourself: Practical Solutions to Identification and Signature Problems. Advances in Cryptology - Crypto'86, volume 263 of LNCS, pages 186-194, 1986.
17. M. Girault, G. Poupard and J. Stern. On the Fly Authentication and Signature Schemes Based on Groups of Unknown Order. Advances in Cryptology - Journal of Cryptology, Volume 19, Number 4. Pages 463-487, Springer-Verlag, 2006.

18. H. Handschuh, Y. Tsiounis, and M. Yung. Decision Oracles are Equivalent to Matching Oracles. Public Key Cryptography PKC '99, volume 1560 of LNCS, pages 276-289. Springer, 1999.
19. T. Nakanishi, M. Shiota, and Y. Sugiyama. An Unlinkable Divisible Electronic Cash with User's Less Computations using Active Trustees. ISITA 2002, 2002.
20. T. Nakanishi and Y. Sugiyama. Unlinkable Divisible Electronic Cash. ISW'00, pages 121-134, 2000.
21. T. Okamoto. An Efficient Divisible Electronic Cash Scheme. Advances in Cryptology - Crypto'95, volume 963 of LNCS, pages 438-451, 1995.
22. T. Okamoto, K. Ohta. Universal Electronic Cash. Advances in Cryptology - Crypto'91, volume 576 of LNCS, pages 324-337, 1992.
23. J.C. Pailles. New Protocols for Electronic Money. Advances in Cryptology - Asiacrypt'92, volume 718 of LNCS, pages 263-274, 1993.
24. D. Pointcheval and J. Stern. Security Arguments for Digital Signatures and Blind Signatures. Journal of Cryptology, Volume 13 - Number 3. Pages 361-396, Springer-Verlag, 2000.
25. A. De Santis, G. Di Crescenzo, G. Persiano, and M. Yung. On Monotone Formula Closure of SZK. FOCS 1994, pages 454-465, 1994.
26. C. P. Schnorr. Efficient Identification and Signatures for Smart Cards. Advances in Cryptology - Crypto'89, volume 435 of LNCS, pages 239-252, 1990.
27. M. Stadler. Publicly Verifiable Secret Sharing. Advances in Cryptology - Crypto'96, volume 1070 of LNCS, pages 190-199, 1996.
28. M. Trolin. A stronger definition for anonymous electronic cash. Cryptology ePrint Archive: Report 2006/241. 2006.

A   Matching Multi Diffie-Hellman problem

The problem underlying the property of unlinkability for $\mathcal{DCS}$ is the *Matching Multi Diffie-Hellman problem (MMDH)*. We show that MMDH can be used to solve the *Decisional Diffie-Hellman problem (DDH)*.

Decisional Diffie-Hellman (DDH) problem: given a random generator $g \in \mathcal{G}$ where $\mathcal{G}$ has prime order and the values h^x, h^y, h^z, the problem consists in deciding if $xy = z$ or not.

Matching Multi Diffie-Hellman (MMDH) problem: let $\mathcal{H}$, $\mathcal{H}_1$ and $\mathcal{H}_2$ be groups of prime order such that $\mathcal{H}_1$ is a subgroup of $\mathbb{Z}_o^*$ where o is the order of $\mathcal{H}_2$. Given three random generators $h \in \mathcal{H}$, $h_1 \in \mathcal{H}_1$ and $h_2 \in \mathcal{H}_2$ and the values $h^{\alpha_0}, h^{\alpha_1}, h_2^{h_1^{\alpha_b}}$ and $h_2^{h_1^{\alpha_{\bar{b}}}}$ where $b \in \{0,1\}$, the problem consists in deciding if $b = 0$ or 1.

Decisional Multi Diffie-Hellman (DMDH) problem: let $\mathcal{H}$, $\mathcal{H}_1$ and $\mathcal{H}_2$ be groups of prime order such that $\mathcal{H}_1$ is a subgroup of $\mathbb{Z}_o^*$ where o is the order of $\mathcal{H}_2$. Given three random generators $h \in \mathcal{H}$, $h_1 \in \mathcal{H}_1$ and $h_2 \in \mathcal{H}_2$ and the values $h^\alpha, h_2^{h_1^\beta}$, the problem consists in deciding if $\alpha = \beta$ or not.

Derived Decisional Diffie-Hellman (DDDH) problem: given random generators $g_1, g_2 \in \mathcal{G}$ where $\mathcal{G}$ has prime order and the values g_1^a, g_2^b, the problem consists in deciding if $a = b$ or not.

The problem MMDH is at least as difficult as DMDH. In fact, the MMDH is the matching problem related to the decisional one DMDH. Therefore, Handschuh, Tsiounis and Yung show [18] that decision oracles are equivalent to matching oracles, which can be applied to our context.

The problem DMDH is at least as difficult as DDDH. Indeed, given an instance (g_1, g_2, g_1^a, g_2^b) of the DDDH problem, we can transform it into an instance $(h = g_1, h_1, h_2 = g_2, h^\alpha = g_1^a, h_1^{h_2^\beta} = h_1^{g_2^b})$ where h_1 is taken at random, of the DMDH problem. Thus, $a = b$ if and only if $\alpha = \beta$.

The problem DDDH is at least as difficult as DDH. Indeed, given an instance (g, g^x, g^y, g^z) of the DDH problem, we can transform is into an instance $(g_1 = g, g_2 = g^x, g_1 = g^x, g_2 = g^z)$ of the DDDH problem. Thus, we have $z = xy$ if and only if $a = b$.

We deduce that MMDH is at least as difficult as DDH.

A Fast and Key-Efficient Reduction of Chosen-Ciphertext to Known-Plaintext Security*

Ueli Maurer and Johan Sjödin

Department of Computer Science, ETH Zurich, CH-8092 Zurich, Switzerland
{maurer,sjoedin}@inf.ethz.ch

Abstract. Motivated by the quest for reducing assumptions in security proofs in cryptography, this paper is concerned with designing efficient symmetric encryption and authentication schemes based on any *weak* pseudorandom function (PRF) which can be much more efficiently implemented than PRFs. Damgård and Nielsen (CRYPTO '02) have shown how to construct an efficient symmetric encryption scheme based on any weak PRF that is provably secure against chosen-*plaintext* attacks. The main ingredient is a range-extension construction for weak PRFs. By using well-known techniques, they also showed how their scheme can be made secure against the stronger chosen-*ciphertext* attacks.

The results of our paper are three-fold. First, we give a range-extension construction for weak PRFs that is optimal within a large and natural class of reductions (especially all known today). Second, we propose a construction of a regular PRF from any weak PRF. Third, these two results imply a (for long messages) much more efficient chosen-ciphertext secure encryption scheme than the one proposed by Damgård and Nielsen. The results also give answers to open questions posed by Naor and Reingold (CRYPTO '98) and by Damgård and Nielsen.

1 Introduction

1.1 Weakening of Cryptographic Assumptions

A general goal in cryptography is to prove the security of cryptographic systems under assumptions that are as weak as possible. Provably secure encryption and authentication schemes based on a *pseudorandom function* (PRF) [11] have been studied extensively [10]. Informally, a PRF is an efficient function with a secret key that cannot be efficiently distinguished from a uniform random function even when it can be queried adaptively (i.e., under a chosen-plaintext attack (CPA)).

The notion of a PRF is very strong and, indeed, it is unclear whether functions such as block ciphers proposed in the literature have this very strong security

---

* This work was partially supported by the Zurich Information Security Center. It represents the views of the authors.

M. Naor (Ed.): EUROCRYPT 2007, LNCS 4515, pp. 498–516, 2007.

property.[1] When designing cryptographic schemes, it is prudent to postulate weaker properties as this makes it more likely that a certain function has such properties or, equivalently, there are potentially more efficient implementations for the weaker requirement compared to the stronger.

A very promising weaker notion of pseudorandomness, proposed by Naor and Reingold [18] (see also [19,1,8,20,22]), is the *weak* PRF (WPRF). Informally, a WPRF is a function with a secret key that cannot be efficiently distinguished from a uniform random function when given a sequence of *random* inputs and the corresponding outputs (i.e., under a *known-plaintext attack* (KPA)). Highly efficient candidates for WPRFs are described in [7] (cf. [19]), although these are not targeted at this particular security notion explicitly. It is an interesting open problem for further research how much block-cipher design can benefit from this weakening of the desired security goal.

While the design of WPRFs has not been studied as extensively as for PRFs, a concrete argument showing that WPRFs are substantially weaker than PRFs is that WPRFs can have rather strong structural properties which are known to be devastating for PRFs. For instance, if $\mathcal{G}$ is a group of prime order p in which the Decisional Diffie-Hellman (DDH) [9] assumption holds, then

$$F : \mathbb{Z}_p \times \mathcal{G} \to \mathcal{G} \quad \text{defined by} \quad F_k(x) \stackrel{\text{def}}{=} F(k, x) = x^k, \tag{1}$$

where k denotes the secret key, is a WPRF that commutes (i.e., $F_k(F_{k'}(x)) = F_{k'}(F_k(x))$) [17]. A WPRF can also be self inverse (i.e., $F_k(F_k(x)) = x$), have a small fraction of bad points (e.g. $F_k(x) = x$ or $F_k(x) = k$), and have related outputs (e.g. $F_k(x\|1) = F_k(x\|0)$ for all x). Due to such structural flaws, most encryption and authentication schemes based on a PRF become insecure if the PRF is simply replaced by a WPRF (for examples see [8]).

In this paper, we propose provably secure encryption and authentication schemes, for the strongest security notion, under the sole assumption of a WPRF. Of course, the security could be based on even weaker assumptions like the one-wayness of certain functions (as PRFs can be obtained from any one-way function [12,11]), but these schemes are not of practical interest due to their inefficiency.

1.2   Contributions and Related Work

The main motivation for this paper is Damgård and Nielsen's elegant work on WPRFs [8]. In their paper, the Pseudorandom Tree (PRT) construction was introduced for transforming any WPRF $F\colon\{0,1\}^n \times \{0,1\}^n \to \{0,1\}^n$ (where the first argument is the key input) into a variable-output-length[2] (VOL) WPRF

$$\mathrm{PRT}^F : \{0,1\}^{3n} \times \{0,1\}^n \times \mathbb{N} \to \{0,1\}^*.$$

---

[1] For example, the design criteria for AES did not include a requirement that a candidate proposal be a PRF, only that it be secure as a block cipher in certain modes of operation, against certain types of attacks.

[2] For a VOL function family $V\colon \mathcal{K} \times \{0,1\}^n \times \mathbb{N} \to \{0,1\}^*$, $|V_k(x, l)| = l$ for all k, x, l.

They also proposed an efficient CPA-secure[3] symmetric encryption scheme based on PRT^F, that is defined by encrypting a message $m \in \{0,1\}^*$ under a key $k \in \{0,1\}^{3n}$ and some auxiliary uniform randomness $r \in \{0,1\}^n$, as

$$(k, r, m) \mapsto \left(r, \mathrm{PRT}_k^F(r, |m|) \oplus m\right). \tag{2}$$

To point out the efficiency of this encryption scheme (and also as a reference for the schemes presented in this work), let us compare it with standard modes of operation such as CBC and CTR. Whereas CBC and CTR invoke the underlying block cipher once per message block to encrypt/decrypt, this scheme invokes the underlying function F once per message block to encrypt/decrypt and roughly $2 \cdot \log_2(b)$ times (where b is the number of message blocks) for generating more key material from the initial key (see below). The key generation can be done offline, so that the throughput is exactly the same as for CBC and CTR. However, whereas CBC and CTR are CPA-secure if the underlying block cipher is a PRF, the Damgård-Nielsen scheme is CPA-secure even when the underlying function is a WPRF, and as WPRFs can be more efficiently implementable than PRFs, their scheme can also be the overall most efficient one. Unfortunately, these modes of operations are not secure against the stronger *chosen-ciphertext attack* (CCA)[4]. In [18, p. 279], Naor and Reingold posed an open problem of how to construct an efficient CCA-secure encryption scheme based on any WPRF. Damgård and Nielsen showed (using well-known techniques) how their CPA-secure scheme can be transformed to a CCA-secure one. Their open question [8, p. 464] whether this can be done more efficiently has been the main motivation for this work.

Before we present our results, let us briefly describe the underlying idea of the PRT-construction (illustrated in Fig. 1(a) on page 506). In a first step, some key material $k_1, \ldots, k_d$ is generated from the initial key k by invoking F in an iterative manner, and then the output blocks are derived by applying F_{k_i}, for $i \in \{1, \ldots, d\}$, iteratively to the input or a previously derived output block. For constructions of this type it is crucial for the security and the efficiency (in terms of the number of applications of F relative to the output length) that this is scheduled in the right way. Recently, two more constructions of this type, the Expanded PRT (ERT) (see Fig. 1(a)) and the Factorial Tree (FCT), were proposed in [16]. However, as we point out in Sect. 3.2, the latter and more efficient construction of the two turns out to be flawed. A natural problem that arises is to find the most efficient VOL-WPRF construction (of this type).

The contributions of this paper are the following:

1. THE ICT-CONSTRUCTION – A VOL-WPRF FROM ANY WPRF: Our Increasing Chain Tree (ICT) construction (see Fig. 1(b)) is more efficient than PRT and ERT (with d generated keys ICT expands the input by a factor of $2^d - 1$, whereas PRT and ERT expand by roughly $1.44^d - 1$ and $1.73^d - 1$, respectively), and ICT also uses a shorter initial key (by a factor of 3).

---

[3] Here, CPA formalizes an adversary's inability, given access to an encryption oracle, to distinguish between two plaintexts given the encryption of one of them.

[4] In a CCA, the adversary has access to an encryption and decryption oracle.

Interestingly, the generated key sequence $k_1, \ldots, k_d$ is not pseudorandom as opposed to the case for PRT and ERT. Indeed, we give strong arguments that ICT is optimal within the large and natural class of constructions described above, and hence also that it is optimal to use ICT instead of PRT in (2).

2. THE IC-CONSTRUCTION – A PRF FROM ANY WPRF: Our Increasing Chain (IC) construction is similar in nature to Goldreich, Goldwasser, and Micali's (GGM) [11] construction of a PRF from any PRG, but it is more than twice as efficient than first transforming the WPRF into a PRG and then applying the GGM-construction. It is also more efficient than the strengthening of a WPRF to a PRF given in [19][5]. This solves their open problem [18, p. 278] whether a more efficient strengthening exists positively. Interestingly, if we instantiate the IC-construction with the DDH-based WPRF F defined in (1), we get Naor and Reingold's [20] highly efficient PRF based on the DDH assumption but with a non-trivial[6] reduction of the key-material by a factor of roughly the input length of the PRF.

3. CCA-SECURE ENCRYPTION BASED ON ANY WPRF: The above results combined with a Wegman-Carter [25] based message authentication code (MAC) and the well-known encrypt-then-MAC method [15,5], yield a CCA-secure encryption scheme from any WPRF that is substantially more efficient than the CCA-secure encryption scheme proposed by Damgård-Nielsen (their number of applications to the WPRF for the MACing is linear in the message length whereas ours is constant). We observe that for our purposes a much weaker primitive than the MAC, namely a *weak* MAC (WMAC)[7], is sufficient (encrypt-then-WMAC actually does the job). This raises the question of constructing possibly efficient WMACs from any WPRF.

4. NON-ADAPTIVE[8] CCA-SECURE ENCRYPTION BASED ON ANY WPRF AND WMAC: Although this type of security may (like CPA-security) be unsatisfactory in practice, the exact requirements for achieving standard security notions are interesting in their own right. It might also motivate further research on basing strong primitives on weak assumptions. Non-adaptive CCA-security has been studied under stronger assumptions in [18].

2   Preliminaries

2.1   Notation and Definitions

Let $s \xleftarrow{\$} S$ denote that s is selected uniformly at random from the set S. If $\mathcal{D}$ is a probability distributions over S then $s \leftarrow \mathcal{D}$ denotes the operation of selecting s at random according to $\mathcal{D}$. If x and y are two bitstrings, $x\|y$ denotes their concatenation, $x[i]$ the i-th bit of x, $x[i,j] \stackrel{\text{def}}{=} x[i]\|x[i+1]\| \cdots \|x[j]$ for $i < j$, and $x[i,i] \stackrel{\text{def}}{=} x[i]$. For two functions f and g, $f \circ g\,(x) \stackrel{\text{def}}{=} f(g(x))$. A function

---

[5] In that work, the PRF is reduced – via a pseudorandom synthesizer – to a WPRF.

[6] The key is not replaced by a pseudorandom sequence based on F.

[7] A WMAC is unforgeable under a *known-plaintext attack* (see [18]).

[8] Here the adversary has no oracle access after the challenge (ciphertext) is presented.

has *variable-input-length* (VIL) if the domain is $\{0,1\}^{\leq N} \stackrel{\text{def}}{=} \cup_{i=1}^{N}\{0,1\}^i$ (for some $N > 1$), and a function $f : \{0,1\}^n \times \mathbb{N} \to \{0,1\}^*$ has *variable-output-length* (VOL) if for all all x and l, $|f(x,l)| = l$ and $f(x, l+1) = f(x,l)\|b$ for some bit b. Let $\mathcal{R}_{N,n}$ and $\mathcal{R}_{\leq N,n}$ denote uniform random functions with range $\{0,1\}^n$, and domain $\{0,1\}^N$ and $\{0,1\}^{\leq N}$, respectively. Let $\mathcal{R}_{n,*}$ denote a VOL-function $\{0,1\}^n \times \mathbb{N} \to \{0,1\}^*$ for which $\mathcal{R}_{n,*}(\cdot, l)$ is a uniform random function $\{0,1\}^n \to \{0,1\}^l$ for all l. Abusing notation, we refer to $\mathcal{R}_{n,*}$ as a uniform random VOL-function. We let $\Pr[\Pi : \mathcal{E}]$ denote the probability of event $\mathcal{E}$ in random experiment Π. $A^{\mathcal{O}}$ denotes an algorithm A with access to an oracle $\mathcal{O}$.

2.2   Cryptographic Functions

CONCRETE SECURITY.   We state our results in the concrete security framework, which was formalized for the following primitives by Bellare, Kilian, and Rogaway [4]. Let $\mathcal{O}^f$ denote the oracle which, if invoked, returns $(r, f(r))$ for a uniform random input r of the function f. The **w**-*advantage* of adversary A for $F : \mathcal{K} \times \{0,1\}^N \to \{0,1\}^n$ with $\mathbf{w} \in \{\mathbf{prf}, \mathbf{wprf}, \mathbf{mac}, \mathbf{wmac}\}$ is defined as:

$$\mathbf{Adv}_{F,A}^{\mathbf{prf}} \stackrel{\text{def}}{=} \left| \Pr\left[k \stackrel{\$}{\leftarrow} \mathcal{K}, b \leftarrow A^{F_k} : b = 1\right] - \Pr\left[\mathbf{R} \leftarrow \mathcal{R}_{N,n}, b \leftarrow A^{\mathbf{R}} : b = 1\right] \right|$$

$$\mathbf{Adv}_{F,A}^{\mathbf{wprf}} \stackrel{\text{def}}{=} \left| \Pr\left[k \stackrel{\$}{\leftarrow} \mathcal{K}, b \leftarrow A^{\mathcal{O}^{F_k}} : b = 1\right] - \Pr\left[\mathbf{R} \leftarrow \mathcal{R}_{N,n}, b \leftarrow A^{\mathcal{O}^{\mathbf{R}}} : b = 1\right] \right|$$

$$\mathbf{Adv}_{F,A}^{\mathbf{mac}} \stackrel{\text{def}}{=} \left| \Pr\left[k \stackrel{\$}{\leftarrow} \mathcal{K}, (m,\tau) \leftarrow A^{F_k}, b = \begin{cases} 1 \text{ if } \tau = F_k(m), m \text{ "new"} \\ 0 \text{ otherwise} \end{cases} : b = 1\right] \right|$$

$$\mathbf{Adv}_{F,A}^{\mathbf{wmac}} \stackrel{\text{def}}{=} \left| \Pr\left[k \stackrel{\$}{\leftarrow} \mathcal{K}, (m,\tau) \leftarrow A^{\mathcal{O}^{F_k}}, b = \begin{cases} 1 \text{ if } \tau = F_k(m), m \text{ "new"} \\ 0 \text{ otherwise} \end{cases} : b = 1\right] \right|$$

where "m *new*" stands for the event that m is distinct from the inputs to F_k. The maximal **w**-advantages are defined as $\mathbf{Adv}_F^{\mathbf{w}}(t,q) \stackrel{\text{def}}{=} \max_A\{\mathbf{Adv}_{F,A}^{\mathbf{w}}\}$, where the maximum is taken over all A restricted to time-complexity[9] t and q (respectively $q - 1$ if $\mathbf{w} \in \{\mathbf{mac}, \mathbf{wmac}\}$) invocations of its oracle.

VIL-FUNCTION FAMILIES.   For a VIL-function family $F : \mathcal{K} \times \{0,1\}^{\leq N} \to \{0,1\}^n$, the **vil-mac**-advantage $\mathbf{Adv}_{F,A}^{\mathbf{vil\text{-}mac}}$ is defined like the **mac**-advantage, except that the adversary A may query inputs of any length ($\leq N$). Let $\mathcal{O}_{\mathbf{vil}}^f$ (for some VIL-function f) denote an oracle that on input $l \leq N$ generates a uniform random input $r \in \{0,1\}^l$ and outputs $(r, f_k(r))$. The **vil-wmac**-advantage $\mathbf{Adv}_{F,A}^{\mathbf{vil\text{-}wmac}}$ is defined like the **wmac**-advantage except that the oracle $\mathcal{O}$ is replaced by $\mathcal{O}_{\mathbf{vil}}$. For $\mathbf{w} \in \{\mathbf{mac}, \mathbf{wmac}\}$, the maximal advantage is defined as $\mathbf{Adv}_F^{\mathbf{vil\text{-}w}}(t, q, \mu) \stackrel{\text{def}}{=} \max_A\{\mathbf{Adv}_{F,A}^{\mathbf{vil\text{-}w}}\}$, where the maximum is taken over all A with time-complexity t, making at most $q - 1$ oracle invocations such that the total length of the inputs to F (including the forgery message) is at most μ bits.

---

[9] I.e., t is the worst-case total running time (including the length of A) of the experiment in which A interacts with its oracle (in some fixed RAM model of computation).

VOL-FUNCTION FAMILIES. Let $\mathcal{O}_{\text{vol}}^f$ denote the oracle that on input $l \in \mathbb{N}$ outputs $(r, f(r, l))$ for a uniform random $r \in \{0,1\}^n$. For a VOL-function family $F : \mathcal{K} \times \{0,1\}^n \times \mathbb{N} \to \{0,1\}^*$, the **vol-wprf**-advantage of A for F is

$$\mathbf{Adv}_{F,A}^{\text{vol-wprf}} \overset{\text{def}}{=} \Pr\left[k \leftarrow \mathcal{K}, b \leftarrow A^{\mathcal{O}_{\text{vol}}^{F_k}} : b = 1\right] - \Pr\left[\mathbf{R} \leftarrow \mathcal{R}_{n,*}, b \leftarrow A^{\mathcal{O}_{\text{vol}}^{\mathbf{R}}} : b = 1\right],$$

and by maximizing over all A, restricted to time-complexity t and at most q oracle queries whose sum totals at most μ, we get the maximal **vol-wprf**-advantage $\mathbf{Adv}_F^{\text{vol-wprf}}(t, q, \mu) \overset{\text{def}}{=} \max_A \{\mathbf{Adv}_{F,A}^{\text{vol-wprf}}\}$.

3   The IC- and ICT-Construction

In this section, we propose the IC-construction, for transforming a WPRF into a PRF, and the ICT-construction, for transforming a WPRF into a VOL-WPRF. Throughout, let $F : \{0,1\}^n \times \{0,1\}^n \to \{0,1\}^n$ denote a function family.[10]

3.1   A PRF from Any WPRF

The IC-construction transforms $F : \{0,1\}^n \times \{0,1\}^n \to \{0,1\}^n$ into

$$\text{IC}^F : (\{0,1\}^n \times \{0,1\}^n \times \{0,1\}^n) \times \{0,1\}^N \to \{0,1\}^n,$$

for some fixed N, where $\text{IC}_{k_1, r, \tau_1}^F(x)$ is defined by the following algorithm:

> **if** $|x| > 1$ **then**
>> **for** $i = 2$ to $|x|$ **do** $k_i = F_{k_{i-1}}(r)$
>> **for** $i = 1$ to $|x|$ **do**
>>> **if** $x[i] = 1$ **then**
>>>> $\tau_{i+1} = F_{k_i}(\tau_i)$
>>> **else**
>>>> $\tau_{i+1} = \tau_i$
>> **return** $\tau_{|x|}$

The following theorem states that IC^F is a PRF if F is a WPRF, even if the r-value of the initial key is not kept secret. Note that F is invoked at most $2N - 1$ times. However, the first $N - 1$ invocations can be pre-processed and cached, and hence at most N invocations are necessary or, to be precise, as many invocations as there are ones in the input.

Theorem 1. *For any t, q, and input length N of* IC^F

$$\mathbf{Adv}_{\text{IC}^F}^{\text{prf}}(t, q) \leq N \cdot \left(\mathbf{Adv}_F^{\text{wprf}}(t, q) + \frac{q(q+1)}{2^{n+1}}\right).$$

---

[10] For simplicity, we choose the key-length to be the same as the input length. We refer to [8] for constructing such a WPRF from any WPRF.

Proof. Let Π_0 denote the following random experiment for an adversary A with time-complexity t which makes at most q queries to its oracle:

$$(k_1, r, \tau_1) \xleftarrow{\$} \{0,1\}^n \times \{0,1\}^n \times \{0,1\}^n, \ b \leftarrow A^{\mathrm{IC}^F_{k_1, r, \tau_1}}.$$

Note that for any query x issued by A and any $s \in \{1, \ldots, N\}$, the sequence $(\tau_1, \ldots, \tau_s)$ (resulting from the second for-loop) does not depend on $x[s, N]$. Hence, $(\tau_1, \ldots, \tau_s)$ can be reused for any other query x' for which $x'[1, s-1] = x[1, s-1]$. We assume that $\mathrm{IC}^F_{k_1, r, \tau_1}$ reuses previously computed τ-values (for saving calls to F) whenever possible, by maintaining a look-up table with all the entries $(x[1, s], \tau_{s+1})$ for which x is a query to $\mathrm{IC}^F_{k_1, r, \tau_1}$, $s \in \{1, \ldots, N\}$, and $x[s] = 1$. We also assume that the calls to F in the first for-loop are pre-processed and cached. For $j = 1, \ldots, N$, let Π_{2j-1} be the same experiment as Π_{2j-2} except that F_{k_j} is replaced by a random function $\mathbf{R}_j$, and let Π_{2j} be the same experiment as Π_{2j-1} except that for each query x issued by A, for which $x[j] = 1$ and $x[1, j]$ is not in the look-up table, the output of $\mathbf{R}_j$ is replaced by a uniform random $R \in \{0,1\}^n$ and $(x[1, j], R)$ is inserted into the table. Let S_i be the event that $b = 1$ in Π_i, for $i = 0, \ldots, 2N$. Now, as Π_{2N} is equivalent to $[\mathbf{R} \leftarrow \mathcal{R}_{N,n}, \ b \leftarrow A^{\mathbf{R}}]$, we get

$$\mathbf{Adv}^{\mathrm{prf}}_{\mathrm{IC}^F, A} \overset{\mathrm{def}}{=} |\Pr[S_0] - \Pr[S_{2N}]|$$

$$\leq \sum_{j=1}^{N} |\Pr[S_{2j-2}] - \Pr[S_{2j-1}]| + \sum_{j=1}^{N} |\Pr[S_{2j-1}] - \Pr[S_{2j}]|$$

$$\leq \sum_{j=1}^{N} \mathbf{Adv}^{\mathrm{wprf}}_{F}(t, \min\{q+1, 2^{j-1}+1\}) + \sum_{j=1}^{N} \min\left\{ \frac{(q+1)q}{2^{n+1}}, \frac{(2^{j-1}+1)2^{j-1}}{2^{n+1}} \right\}$$

$$\leq N \cdot \left(\mathbf{Adv}^{\mathrm{wprf}}_{F}(t, q+1) + \frac{(q+1)q}{2^{n+1}} \right),$$

due to the triangle inequality and the following two facts. First, for $j = 1, \ldots, N$, A can be transformed to a WPRF distinguisher A' for F with time-complexity t, making at most $\min(q+1, 2^{j-1}+1)$ oracle invocations, and having advantage at least $|\Pr[S_{2j-2}] - \Pr[S_{2j-1}]|$. A' with oracle T, simulates the experiment Π_{2j-2} if T is an instance of F and Π_{2j-1} if T is a random function $\mathbf{R}$ (which is possible as all queries to F_{k_j} in Π_{2j-2} and to $\mathbf{R}_j$ in Π_{2j-1} are distributed uniformly at random). Finally, A' decides as A does. Second, Π_{2j-1} and Π_{2j} are equivalent experiments as long as no collision among the inputs on which $\mathbf{R}_j$ is invoked occurs. As $\mathbf{R}_j$ is invoked on at most $\min\{q+1, 2^{j-1}+1\}$ inputs and these are all random, the probability of this event is upper bounded by $\min\{(q+1)q/2^{n+1}, (2^{j-1}+1)2^{j-1}/2^{n+1}\}$. ☐

KEY REDUCTION OF NAOR-REINGOLD'S DDH-BASED PRF. In [20], Naor and Reingold presented an efficient construction of a PRF based on the DDH assumption. It is easy to verify, that IC^F with F as defined in (1) is the same PRF but with a significantly shorter key by a factor of roughly N (recall that N is the input length of IC^F). To be more precise, the first for-loop generates a sequence $k_1, \ldots, k_N$ of keys from the initial key (k_1, r, τ_1) and the second for-loop exactly corresponds to the Naor-Reingold construction with $k_1, \ldots, k_N$ as its key. Note that the reduction is non-trivial in the sense that $k_1, \ldots, k_N$ is not generated from a PRG based on F. For instance $F_{k_1}^{-1}(k_2) = F_{k_2}^{-1}(k_3)$ holds which can easily be verified given k_1, k_2, and k_3.

THE GGM-APPROACH. An alternative (but less efficient) approach to obtain a PRF from any WPRF F is to first transform F into a pseudorandom generator (PRG) and then apply the so-called GGM-construction [11] (which transforms a PRG into a PRF). Informally, a PRG is an efficient deterministic function mapping a truly random string (or seed) to a longer string which is computationally indistinguishable from random. Let us briefly describe the GGM-construction. It transforms a length-doubling PRG G into a PRF (say with N-bits input) as

$$GGM_k(x) \overset{\text{def}}{=} G_{x[1]} \circ \ldots \circ G_{x[N]}(k),$$

where $G_0(k)$ and $G_1(k)$ denote the left and right half of $G(k)$, respectively. The most efficient construction of a length doubling PRG G from F, that we are aware of, uses 3 and 4 invocations to F, respectively, for computing G_0 and G_1:

$$G(k_1\|r\|x) \overset{\text{def}}{=} x\|F_{k_1}(x)\|F_{k_2}(x)\|F_{k_2} \circ F_{k_1}(x)\|F_{k_3}(x)\|r,$$

where $k_2 = F_{k_1}(r)$ and $k_3 = F_{k_2}(r)$. The proof that G is a PRG if F is a WPRF follows directly from Theorem 2 (see next section) and the fact that $G(k_1\|r\|x) = x\| IC_{k,r}^F(x, 4n)\|r$. Hence, to get a PRF with N-bits input and n-bits output, we roughly need $4N$ invocations of F per call in the worst case (cf. the efficiency of IC^F above).

Remark 1. Let us briefly point out a method for improving the computation time of IC^F, at the cost of generating and storing more keys (say N' keys instead of N). On input x (of length N), x is first injectively mapped to a N'-bit string x' of Hamming weight at most some (fixed) c, satisfying

$$\sum_{i=0}^{c} \binom{N'}{i} \geq 2^N.$$

Then IC^F is invoked on x' and the result is output. Here, F is invoked at most c (as opposed to N) times, as there are at most c ones in the input.

3.2   A VOL-WPRF from Any WPRF

The ICT-construction is illustrated in Fig. 1(b) and is defined as

$$ICT^F : (\{0,1\}^n \times \{0,1\}^n) \times \{0,1\}^n \times \mathbb{N} \to \{0,1\}^*$$
$$((k,r),x,l) \mapsto \left(IC_{k,r,x}^F(\langle 1 \rangle) \| IC_{k,r,x}^F(\langle 2 \rangle) \| \cdots \| IC_{k,r,x}^F(\langle \lceil l/n \rceil \rangle) \right)[1,l],$$

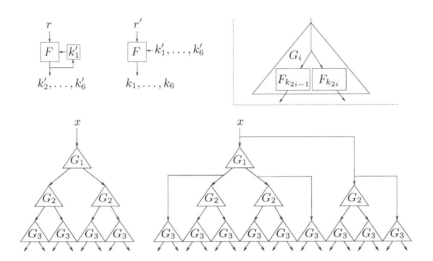

(a) Computation of $\mathrm{PRT}^F_{k'_1,r,r'}(x,14n)$ (bottom left) and $\mathrm{ERT}^F_{k'_1,r,r'}(x,26n)$ (bottom right), i.e., the maximal sized output using 6 generated keys $k_1,\ldots,k_6$ (upper left). Here every output of G_i (defined upper right) for $i=1,2,3$ is part of the global output.

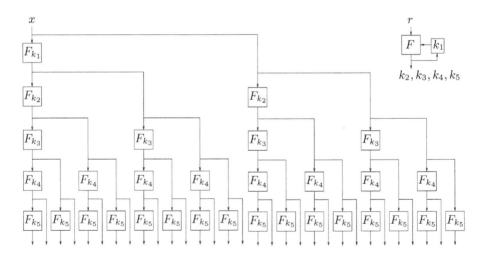

(b) Computation of $\mathrm{ICT}^F_{k_1,r}(x,31n)$, i.e., the output of maximal size using 5 generated keys $k_1,\ldots,k_5$ (upper right). Here every output of F – except for the generated keys (upper right) – is part of the global output. We stress that the order of the output blocks are not the same as presented in the text.

Fig. 1. Illustration of (a) PRT [8], ERT [16], and (b) ICT (of this paper). The generated key sequence $k_1,k_2,\ldots$ is not pseudorandom in (b) as opposed to in (a) (see Sect. 3.1).

where $\langle i \rangle$ denotes the reversed standard bit encoding of the integer i (e.g. $\langle 0 \rangle = 0, \langle 1 \rangle = 1, \langle 2 \rangle = 01, \langle 3 \rangle = 11, \langle 4 \rangle = 001$). Note that $\mathrm{IC}^F_{k,r,x}(\langle 0 \rangle) = x$ can not be part of the output, as x is the input. It is easy to verify, see Fig. 1(b), that $\mathrm{ICT}^F_{k,r}(x,l)$ needs $d - 1 = \lfloor \log_2(\lceil l/n \rceil) \rfloor$ calls to F for computing (or pre-computing) the needed keys $k_1, \ldots, k_d$ and further $\lceil l/n \rceil$ calls for computing the output (i.e., one call per output block). The next theorem states that ICT^F is a VOL-WPRF if F is a WPRF. As for IC, the r-value of the key need not be kept secret.

Theorem 2. *For any t, q, and μ*

$$\mathbf{Adv}^{\text{vol-wprf}}_{\mathrm{ICT}^F}(t, q, \mu) \leq d_{\max} \cdot \mathbf{Adv}^{\text{wprf}}_F(t, q(2^{d_{\max}-1}+1)) + \frac{4^{d_{\max}} \cdot q^2}{2^n},$$

where $d_{\max} = \lfloor \log_2(\lceil l_{\max}/n \rceil) \rfloor + 1$ and $l_{\max} \leq \mu$ is the maximum allowed output length of ICT^F.

Proof. Let Π_0 denote the following random experiment for an adversary A with time-complexity t that make at most q queries whose sum is at most μ:

$$(k, r) \xleftarrow{\$} \{0, 1\}^n \times \{0, 1\}^n, b \leftarrow A^{\mathcal{O}^{\mathrm{ICT}^F_{k,r}}_{\text{vol}}}.$$

Let d denote the maximal number of generated keys (for F), needed for answering the queries to $\mathrm{ICT}^F_{k,r}$ issued by A. Note that the j-th instantiation of F, i.e., F_{k_j}, for $j \in \{1, \ldots, d\}$, is queried at most $q_j = q \cdot (2^{j-1} + 1)$ times. For $j = 1, \ldots, d$, let Π_{2j-1} denote the same random experiment as Π_{2j-2} except that F_{k_j} is replaced by a random function $\mathbf{R}_j$, and let Π_{2j} be the same experiment as Π_{2j-1} except that the outputs of $\mathbf{R}_j$ are replaced by uniform random n-bit strings. Furthermore, let Π_{2d+1} denote the random experiment $[\mathbf{R} \xleftarrow{\$} \mathcal{R}_{n,*}, b \leftarrow A^{\mathcal{O}^{\mathbf{R}}_{\text{vol}}}]$. Now, for $i = 0, \ldots, 2d + 1$, let S_i denote the event that $b = 1$ in Π_i. We get

$$\mathbf{Adv}^{\text{vol-wprf}}_{A, \mathrm{ICT}^F} = |\Pr[S_0] - \Pr[S_{2d+1}]|$$

$$\leq \sum_{j=1}^d |\Pr[S_{2j-2}] - \Pr[S_{2j-1}]| + \sum_{j=1}^d |\Pr[S_{2j-1}] - \Pr[S_{2j}]| + |\Pr[S_{2d}] - \Pr[S_{2d+1}]|$$

$$\leq \sum_{j=1}^d \mathbf{Adv}^{\text{wprf}}_F(t, q_j) + \sum_{j=1}^d \frac{q_j^2}{2^{n+1}} + \frac{q^2}{2^{n+1}} \leq d \cdot \mathbf{Adv}^{\text{wprf}}_F(t, q(2^{d-1}+1)) + \frac{q^2 4^d}{2^n},$$

using the triangle inequality and the following facts. As Π_{2d} and Π_{2d+1} are equivalent experiments as long as the input part of the samples returned by the oracle are distinct, we get $|\Pr[S_{2d}] - \Pr[S_{2d+1}]| \leq q^2/2^{n+1}$. Furthermore, as Π_{2j-1} and Π_{2j} are equivalent as long as the random inputs to $\mathbf{R}_j$ are all distinct, it holds that $|\Pr[S_{2j-1}] - \Pr[S_{2j}]| \leq q_j^2/2^{n+1}$. Finally, A can be transformed into a WPRF distinguisher A' for F with time-complexity t, that makes q_j oracle queries and has advantage $|\Pr[S_{2j-2}] - \Pr[S_{2j-1}]|$. A' with oracle $\mathcal{O}^T$ simply simulates the random experiment that is equivalent to Π_{2j-2} if T is an instance of F and to Π_{2j-1} if T is a random function $\mathbf{R}$ (this is possible as the inputs to F_{k_j} in Π_{2j-2} and to $\mathbf{R}_j$ in Π_{2j-1} are distributed uniformly at random). Finally, A' decides as A does. $\square$

THE FCT-CONSTRUCTION IS FLAWED. Let us point out that the security proof of FCT (in [16]) is flawed. The maximal sized output of FCT^F for two generated keys k_1 and k_2 is defined as

$$x \mapsto F_{k_1}(x)\|F_{k_2}(x)\|F_{k_2} \circ F_{k_1}(x)\|F_{k_1} \circ F_{k_2}(x). \tag{3}$$

Clearly, the construction is insecure for any WPRF F that commutes (i.e., for which $F_{k_2} \circ F_{k_1}(x) = F_{k_1} \circ F_{k_2}(x)$ for all k_1, k_2, and x). As such WPRFs exist under the DDH assumption (see (1)), a fix of the security proof would contradict the assumption and thus be a major breakthrough in number theory.[11]

COMPARING ICT WITH OTHER CONSTRUCTIONS. The idea behind PRT of [8], ERT of [16], and ICT is to first generate keys $k_1, \ldots, k_d$ from the initial key (and F) and then to derive the output blocks sequentially by invoking F_{k_i}, with $i \in \{1, \ldots, d\}$, to the input or a previously computed output block (see Fig. 1). ICT is superior to PRT and ERT for three reasons. First, the initial key of ICT is n bits (plus n bits that may be publicly known) versus $3n$ bits for PRT and ERT. Second, ICT needs $d - 1$ invocations of F to generate the d keys $k_1, \ldots, k_d$ whereas PRT and ERT needs $2d - 1$. Third, the maximal output size using $k_1, \ldots, k_d$ is $(2^d-1)n$ for ICT, roughly $(3^{\frac{d}{2}}-1)n$ for ERT, and roughly $(2^{\frac{d}{2}+1}-2)n$ for PRT.[12] For all constructions, the keys needed for computing outputs of length bounded by some fixed value (say $l_{\max}$) can be pre-processed, such that one call of F is needed per output block. But whereas ICT needs to store say $s \stackrel{\mathrm{def}}{=} \lfloor \log_2(\lceil l_{\max}/n \rceil) \rfloor + 1$ keys, ERT and PRT store about $\lceil 1.26s \rceil$ and $2s$ keys, respectively. The factor in front of the **wprf**-advantage in the security reduction reduces correspondingly, i.e., for s as defined above we roughly have

$$\mathbf{Adv}_{\mathrm{ICT}^F}^{\mathrm{vol\text{-}wprf}}(t,q) \leq \quad s \cdot \mathbf{Adv}_F^{\mathrm{wprf}}(t, 2^{s-1}q) \quad + \; 4^s q^2 / 2^n$$

$$\mathbf{Adv}_{\mathrm{ERT}^F}^{\mathrm{vol\text{-}wprf}}(t,q) \leq 1.26s \cdot \mathbf{Adv}_F^{\mathrm{wprf}}(t, 2^{s-1}q/3) \; + \; 4^s q^2 / (2^n \cdot 9)$$

$$\mathbf{Adv}_{\mathrm{PRT}^F}^{\mathrm{vol\text{-}wprf}}(t,q) \leq \quad 2s \cdot \mathbf{Adv}_F^{\mathrm{wprf}}(t, 2^{s-1}q/4) \; + \; 4^s q^2 / (2^n \cdot 16).$$

OPTIMALITY OF THE ICT-CONSTRUCTION. In [22], it is shown that there is no black-box proof of the security for constructions that expands more than ICT (for any fixed number of generated keys). Here, we show something stronger for the constructions with log-time random access to output blocks, i.e., for the rather balanced constructions where the maximal length of the composition chains are in $O(\log(l))$ for output length l, namely that ICT is optimal within that class of constructions under the *inverse* DDH (IDDH) assumption [2].

---

[11] However, information theoreticly (and even in Minicrypt, i.e., under the assumption that one-way functions exist but no public-key cryptography) (3) is secure [21,22].

[12] The latter two values are exact if d is even. Otherwise $(2\cdot 3^{\frac{d-1}{2}}-1)n$ and $(3\cdot 2^{\frac{d-1}{2}}-2)n$ are exact, respectively.

To be more precise, note that – for $l = 3n$ – the value $\text{ICT}_{k_1,r}^{F}(x, l)$ is derived by first computing $k_2 = F_{k_1}(r)$ and then returning

$$y := F_{k_1}(x) \| F_{k_2}(x) \| F_{k_2} \circ F_{k_1}(x).$$

For $l = 7n$, an extra key $k_3 = F_{k_2}(r)$ is derived and

$$y \| F_{k_3}(x) \| F_{k_3} \circ F_{k_1}(x) \| F_{k_3} \circ F_{k_2}(x) \| F_{k_3} \circ F_{k_2} \circ F_{k_1}(x)$$

is returned. A natural question is whether more can be output before a new key needs to be generated, i.e., for a fixed number of generated keys (say k_1, k_2, and k_3), can we output more than ICT^{F} maximally can (i.e., more than $7n$ bits) by invoking the instantiations (i.e., $F_{k_1}, F_{k_2}, F_{k_3}$) *one* more time than ICT^{F} does (i.e., 8 times instead of 7). The answer turns out to be "no" unless the IDDH assumption is false, since otherwise there is a WPRF F, described in (4), which with high probability both commutes and is self inverse, i.e., for all $k \neq k'$

$$\Pr_{x}[F_k \circ F_{k'}(x) = F_{k'} \circ F_k(x)] \approx 1/4 \quad \text{and} \quad \Pr_{x}[F_k \circ F_k(x) = x] \approx 1/2.$$

If F is used and more is output at least two output blocks will (by the pigeonhole principle) have the same value with high probability (which is unlikely for a uniform random VOL-function). F is defined for a group $\mathcal{G}$ of prime order ρ as

$$F : \mathbb{Z}_\rho \times \mathcal{G} \to \mathcal{G} \quad \text{and} \quad F_k(x) \stackrel{\text{def}}{=} \begin{cases} x^k & \text{if } x \in P_1 \\ x^{k-1} & \text{if } x \in P_2 \end{cases}, \tag{4}$$

where $k \cdot k^{-1} = 1 \pmod{\rho}$ and $\{P_1, P_2\}$ is a partition of $\mathcal{G}$ in roughly equal sized sets (where we assume that it is efficient to decide whether $x \in P_1$ or not). A proof that F is a WPRF if the IDDH assumption holds in $\mathcal{G}$ is given in [14].

4   Applications

In this section, we optimize Damgård and Nielsen's CPA-secure encryption scheme (2) by using ICT instead of PRT. Then, we first make the scheme CCA-secure by applying IC and the well-known encrypt-then-MAC technique (actually we prove that what we call encrypt-then-WMAC does the job here), and, second, we make it non-adaptive CCA-secure by using a (fixed-input-length) WMAC for authenticating the auxiliary uniform randomness.

4.1   Symmetric Encryption

A symmetric encryption scheme $\mathcal{SE} = (E, D)$ consists of two efficient algorithms. The (randomized) encryption algorithm E maps a key k and a message m to a ciphertext $c = E_k(m)$, and the deterministic decryption algorithm D maps a key k and a ciphertext $c = E_k(m)$ to the message $m = D_k(c)$. There are several notions for privacy and integrity of $\mathcal{SE}$ (for an overview, we refer to [5,13,3]). We consider the IND-PX-CY notion (for X,Y $\in \{0, 1, 2\}$), introduced in [13].

Definition 1 (IND-PX-CY). *For an encryption scheme* $\mathcal{SE} = (E, D)$ *(with message space $\mathcal{M}$ and keyspace $\mathcal{K}$), the* **ind-px-cy**-*advantage of an adversary A (with $x, y \in \{0, 1, 2\}$) is defined as follows (where $\perp$ denotes no oracle):*

$$\mathbf{Adv}_{\mathcal{SE}, A}^{\text{ind-p}x\text{-c}y}$$

$$\stackrel{def}{=} 2 \cdot \Pr\left[k \stackrel{\$}{\leftarrow} \mathcal{K}, (m_0, m_1) \leftarrow A^{\mathcal{O}_1, \mathcal{O}_2}, b \stackrel{\$}{\leftarrow} \{0, 1\}, c \leftarrow E_k(m_b), \hat{b} \leftarrow A^{\mathcal{O}_1', \mathcal{O}_2'}(c) : \hat{b} = b\right] - 1,$$

where $(\mathcal{O}_1, \mathcal{O}_1') = \begin{cases} (\perp, \perp) & \text{if } x = 0 \\ (E_k, \perp) & \text{if } x = 1 \\ (E_k, E_k) & \text{if } x = 2 \end{cases}$, $(\mathcal{O}_2, \mathcal{O}_2') = \begin{cases} (\perp, \perp) & \text{if } y = 0 \\ (D_k, \perp) & \text{if } y = 1 \\ (D_k, D_k) & \text{if } y = 2 \end{cases}$,

$m_0, m_1 \in \mathcal{M}$ *with* $|m_0| = |m_1|$, *and A does not query $\mathcal{O}_2'$ with c. By maximizing over all A restricted to time-complexity t, at most $q - 1$ encryption queries of total length at most $(\mu - |m_0|)$ bits, and q' decryption queries of total length at most μ' bits, we let* $\mathbf{Adv}_{\mathcal{SE}}^{\text{ind-p}x\text{-c}y}(t, q, \mu, q', \mu') \stackrel{def}{=} \max_A\{\mathbf{Adv}_{\mathcal{SE}, A}^{\text{ind-p}x\text{-c}y}\}$ *(where one typically drops the parameters (q', μ') if $y = 0$).*

The IND-P2-C0, IND-P2-C2, and IND-P1-C1 notions are often referred to as IND-CPA, (adaptive) IND-CCA, and non-adaptive IND-CCA, respectively.

The *integrity of ciphertext* (INT-CTXT) [5] notion formalizes an adversary's inability – given access to an encryption oracle – to create a new valid ciphertext:

Definition 2 (INT-CTXT). *For $\mathcal{SE} = (E, D)$ (with message space $\mathcal{M}$ and key space $\mathcal{K}$), let D_k^* denote an algorithm that on input c outputs 1 iff c is a valid ciphertext under the key k, i.e., there exists $m \in \mathcal{M}$ such that $D_k(c) = m$.*

$$\mathbf{Adv}_{\mathcal{SE}, A}^{\text{int-ctxt}} \stackrel{def}{=} \Pr\left[k \stackrel{\$}{\leftarrow} \mathcal{K}, A^{E_k, D_k^*}, b \stackrel{def}{=} \begin{cases} 1 & \text{If } \exists i \; \forall j : D_k^*(y_i) = 1 \wedge y_i \neq c_j \\ 0 & \text{otherwise} \end{cases} : b = 1\right],$$

where $c_1, \ldots, c_q$ denote the outputs from E_k and $y_1, \ldots, y_{q'}$ denote A's queries to D_k^*. By maximizing over all A with time-complexity t, that makes at most q queries to E_k of total length at most μ bits, and at most q' queries to D_k^* of total length at most μ' bits, we let $\mathbf{Adv}_{\mathcal{SE}}^{\text{int-ctxt}}(t, q, \mu, q', \mu') \stackrel{def}{=} \max_A\{\mathbf{Adv}_{\mathcal{SE}, A}^{\text{int-ctxt}}\}$.

4.2   A CPA-Secure Encryption Scheme

In [8], Damgård and Nielsen introduced an IND-P2-C0-secure encryption scheme based on any VOL-WPRF $V : \{0, 1\}^\kappa \times \{0, 1\}^n \times \mathbb{N} \to \{0, 1\}^*$. To be precise, their encryption scheme $\mathcal{SE}_1$ is defined by encrypting a message $m \in \{0, 1\}^*$, under the key $k \in \{0, 1\}^\kappa$ and some auxiliary uniform randomness $r \in \{0, 1\}^n$ as

$$(k, r, m) \mapsto \left(r, V_k(r, |m|) \oplus m\right). \qquad (\mathcal{SE}_1) \quad (5)$$

The following proposition originates from [8]. We give the proof for completeness.

Proposition 1. *For any t, q, and μ*

$$\mathbf{Adv}_{\mathcal{SE}_1}^{\text{ind-p2-c0}}(t, q, \mu) \leq 2 \cdot \mathbf{Adv}_V^{\text{vol-wprf}}(t, q, \mu) + \frac{q - 1}{2^{n-1}}.$$

Proof. For $\mathcal{SE}_1 = (E, D)$, let Π_0 denote the IND-P2-C0 random experiment

$$k \xleftarrow{\$} \{0,1\}^\kappa, \ (x_0, x_1) \leftarrow A^{E_k}, \ b \xleftarrow{\$} \{0,1\}, \ y \leftarrow E_k(x_b), \ \hat{b} \leftarrow A^{E_k}(y),$$

where A is any adversary with resources (t, q, μ). Furthermore, let Π_1 be the same experiment as Π_0, except that V_k is replaced by a uniform random VOL-function $\mathbf{R}_{n,*}$. Let Π_2 be the same experiment as Π_1, except that the input y to the adversary is replaced by a truly random string y' (of length $|y|$). For $i = 0, 1, 2$, let S_i denote the event that $\hat{b} = b$ in experiment Π_i. Then

$$\mathbf{Adv}_{\mathcal{SE}_1, A}^{\text{ind-p2-c0}} \stackrel{\text{def}}{=} 2 \cdot \Pr[S_0] - 1 = 2 \cdot \Pr[S_2] - 1 + 2 \cdot \sum_{i=0}^{1} (\Pr[S_i] - \Pr[S_{i+1}])$$

$$\leq 2 \cdot \frac{1}{2} - 1 + 2 \cdot \mathbf{Adv}_V^{\text{vol-wprf}}(t, q, \mu) + 2 \cdot \frac{q-1}{2^n},$$

where the inequality follows from the following three facts. First, A can be transformed into VOL-WPRF distinguisher A' for V with advantage $\Pr[S_0] - \Pr[S_1]$ and resources (t, q, μ). A' with oracle T simply simulates the experiment Π_0 if T is an instance of V and Π_1 if T is a uniform random VOL-function $\mathbf{R}$ (this is possible as the inputs to V_k in Π_0 and to $\mathbf{R}_{n,*}$ in Π_1 are distributed uniformly at random), and then A' returns whatever A does. Second, Π_1 and Π_2 are equivalent experiments as long as the random input to $\mathbf{R}_{n,*}$ in the computation of y is different from the other random inputs to $\mathbf{R}_{n,*}$, an event upper bounded by $(q-1)/2^n$. Third, $\Pr[S_2] = 1/2$ since b is independent of y. □

Remark 2. Given the strong optimality arguments for ICT, it is clear that (2) is optimal when ICT is used (instead of PRT) unless a significantly different approach for range extension of WPRFs is invented.

4.3 A CCA-Secure Encryption Scheme

The well-known encrypt-then-MAC method is a general technique for constructing an INT-CTXT- and IND-P2-C2-secure encryption scheme from any IND-P2-C0-secure encryption scheme $\mathcal{SE} = (Enc, Dec)$ and any VIL-MAC W. The idea is to simply encrypt with Enc and then authenticate the ciphertext using W [15,5]. Here, we note that for the IND-P2-C0-secure scheme $\mathcal{SE}_1$ based on any VOL-WPRF $V : \{0,1\}^{\kappa_1} \times \{0,1\}^n \times \mathbb{N} \to \{0,1\}^*$, it is sufficient if $W : \{0,1\}^{\kappa_2} \times \{0,1\}^* \to \{0,1\}^\ell$ is a VIL-WMAC (as the ciphertexts of $\mathcal{SE}_1$ are pseudorandom). To be precise, the scheme $\mathcal{SE}_2$, defined by encrypting $m \in \{0,1\}^*$ under a key $(k_1, k_2) \in \{0,1\}^{\kappa_1} \times \{0,1\}^{\kappa_2}$ and auxiliary uniform randomness $r \in \{0,1\}^n$ as

$$\left((k_1, k_2), r, m\right) \mapsto \left(r, \underbrace{V_{k_1}(r, |m|) \oplus m}_{c}, W_{k_2}(r\|c)\right), \qquad (\mathcal{SE}_2) \quad (6)$$

is IND-P2-C2 secure if V is a VIL-WPRF and W is a VIL-WMAC:

Theorem 3. *For any t, q, μ, q', and μ'*

$$\mathbf{Adv}_{\mathcal{SE}_2}^{\text{int-ctxt}}(t, q, \mu, q', \mu') \leq \min \left\{ q' \cdot \mathbf{Adv}_W^{\text{vil-mac}}(t, q, \mu + qn + \mu'), \right.$$

$$\left. \mathbf{Adv}_V^{\text{vol-wprf}}(t, q, \mu) + \frac{q^2}{2^{n+1}} + q' \cdot \mathbf{Adv}_W^{\text{vil-wmac}}(t, q, \mu + qn + \mu') \right\}$$

$$\mathbf{Adv}_{\mathcal{SE}_2}^{\text{ind-p2-c2}}(t, q, \mu, q', \mu') \leq 2\, \mathbf{Adv}_{\mathcal{SE}_2}^{\text{int-ctxt}}(t, q, \mu, q', \mu') + \mathbf{Adv}_{\mathcal{SE}_1}^{\text{ind-p2-c0}}(t, q, \mu).$$

Proof. The proof of the first inequality consists of two parts. For the first part, i.e., $\mathbf{Adv}_{\mathcal{SE}_2}^{\text{int-ctxt}}(t, q, \mu, q', \mu') \leq q' \cdot \mathbf{Adv}_W^{\text{vil-mac}}(t, q, \mu + qn + \mu')$, we refer to [5]. For the second part, let Π_0 denote the INT-CTXT random experiment

$$(k_1, k_2) \overset{\$}{\leftarrow} \{0, 1\}^{\kappa_1} \times \{0, 1\}^{\kappa_2}, \ A^{E_{k_1, k_2}, D_{k_1, k_2}^*}$$

for $\mathcal{SE}_2 = (E, D)$ and any adversary A with resources (t, q, μ, q', μ'). Furthermore, let Π_1 be defined as Π_0 except that V_{k_1} has been replaced by a uniform random VOL-function $\mathbf{R}_{n,*}$ and let Π_2 be defined as Π_1 except that the output of $\mathbf{R}_{n,*}$ is replaced by a truly random string (no matter of the input). For $i = 0, 1, 2$, let $\mathcal{E}_i$ denote the event that D_{k_1, k_2}^* outputs 1 in Π_i. Then

$$\mathbf{Adv}_{\mathcal{SE}_2, A}^{\text{int-ctxt}} \overset{\text{def}}{=} \Pr[\mathcal{E}_0] = \Big(\Pr[\mathcal{E}_0] - \Pr[\mathcal{E}_1] \Big) + \Big(\Pr[\mathcal{E}_1] - \Pr[\mathcal{E}_2] \Big) + \Pr[\mathcal{E}_2]$$

$$\leq \mathbf{Adv}_V^{\text{vol-wprf}}(t, q, \mu) + \frac{(q-1)q}{2^{n+1}} + q' \cdot \mathbf{Adv}_W^{\text{vil-wmac}}(t, q, \mu + qn + \mu'),$$

due to the following three facts. First, A implies a VOL-WPRF distinguisher A' for V with advantage $|\Pr[\mathcal{E}_0] - \Pr[\mathcal{E}_1]|$ and resources (t, q, μ). A' with oracle T simply simulates Π_0 if T is an instance of V and Π_1 if T is a uniform random VOL-function $\mathbf{R}$ (this is possible as the inputs to V_{k_1} in Π_0 and to $\mathbf{R}_{n,*}$ in Π_1 are distributed uniformly at random), and then A' outputs 1 if and only if A is successful. Second, Π_1 and Π_2 are equivalent experiments unless the auxiliary random r-values returned by the encryption oracle are not all distinct, an event upper bounded by $q(q-1)/2^{n+1}$. Third, from A we can construct a VIL-WMAC-forger A'' for W with advantage $\Pr[\mathcal{E}_2]/q'$ and resources $(t, q, \mu + q\mu + \mu')$. A'' simply picks a random element $i \in \{1, \dots, q'\}$ and starts simulating Π_2 – except for invoking D_{k_1, k_2}^* on A's queries – by using its own oracle in place of W_{k_2} (this is possible as all inputs to W_{k_2} in Π_2 are distributed uniformly at random). However, once A makes its i-th query to D_{k_1, k_2}^* (if at all), A'' stops the simulation and returns it as its forgery.

For proving the second inequality, let Π_0' denote the IND-P2-C2 experiment

$$(k_1, k_2) \overset{\$}{\leftarrow} \{0, 1\}^{\kappa_1} \times \{0, 1\}^{\kappa_2},$$

$$(x_0, x_1) \leftarrow A^{E_{k_1, k_2}, D_{k_1, k_2}}, \ b \overset{\$}{\leftarrow} \{0, 1\}, y \leftarrow E_{k_1, k_2}(x_b), \ \hat{b} \leftarrow A^{E_{k_1, k_2}, D_{k_1, k_2}}(y),$$

for $\mathcal{SE}_2 = (E, D)$ and any adversary A with resources (t, q, μ, q', μ'). Without loss of generality, we assume that A does not query D_{k_1, k_2} with an output from E_{k_1, k_2}.

Let Π_1' be the same experiment as Π_0', except that all queries to D_{k_1,k_2} are rejected. Moreover, for $i = 0, 1$, let S_i denote the event that $\hat{b} = b$ in Π_i'. Then

$$\mathbf{Adv}_{\mathcal{SE}_2, A}^{\text{ind-p2-c2}} \stackrel{\text{def}}{=} 2 \cdot \Pr[S_0] - 1 = 2 \cdot \left(\Pr[S_0] - \Pr[S_1] \right) + 2 \cdot \Pr[S_1] - 1$$

$$\leq 2 \cdot \Pr[\mathcal{E}] + \mathbf{Adv}_{\mathcal{SE}_2}^{\text{ind-p2-c0}}(t, q, \mu) \leq 2 \cdot \Pr[\mathcal{E}] + \mathbf{Adv}_{\mathcal{SE}_1}^{\text{ind-p2-c0}}(t, q, \mu),$$

where $\mathcal{E}$ denotes the event that a query to D_{k_1,k_2} in Π_1' (or Π_0') is a valid ciphertext. The first inequality follows from the the fact that Π_0' and Π_1' are equivalent experiments unless $\mathcal{E}$ occurs, and that Π_1' is equivalent to the corresponding IND-P2-C0 experiment for $\mathcal{SE}_2$ (in which the VIL-WMAC is superfluous by Proposition 1). It remains to show that

$$\Pr[\mathcal{E}] \leq \mathbf{Adv}_{\mathcal{SE}_2}^{\text{int-ctxt}}(t, q, \mu, q', \mu').$$

This is the case as A can trivially be transformed into a INT-CTXT adversary A''' (for $\mathcal{SE}_2$) using the same resources and having advantage $\Pr[\mathcal{E}]$. A''' simply runs A, by answering its encryption queries with its own encryption oracle and rejecting all decryption queries. In addition, A''' forwards A's decryption queries to its D^* oracle. If A presents its challenge input (m_0, m_1), A''' flips a coin b, queries its encryption oracle with m_b, and returns the result to A. $\square$

Remark 3. The above result leads to an interesting open question for further research, namely, how efficient constructions are there of a VIL-WMAC W based on any WPRF F. One approach – for constructing W – would be to first transform F into the PRF $\text{IC}^F : \{0,1\}^{3n} \times \{0,1\}^N \to \{0,1\}^n$ (see Sect. 3.1) and then apply the following rather standard method [25,23,6] for constructing a VIL-MAC (and thus also a VIL-WMAC) from any PRF. Simply hash the message using an ε-almost universal (AU) hash function $H : \mathcal{K} \times \{0,1\}^* \to \{0,1\}^N$ (i.e., for all distinct $m, m' \in \{0,1\}^*$, $\Pr[k' \leftarrow \mathcal{K} : H_{k'}(m) = H_{k'}(m')] \leq \varepsilon$ [24]) and then apply IC^F to the result: $W_{k,k'}(x) \stackrel{\text{def}}{=} \text{IC}_k^F \circ H_{k'}(x)$.[13] This method is appealing since H exists unconditionally and IC^F is invoked on "short" inputs (of size N). There are 2^{1-N}-AU hash functions, with $5N$-bit key size and maximal input length 2^N, that should do for most practical applications (see [25]).

Remark 4. By combining (6) with $V = \text{ICT}^F$ and a W (as defined above), we get a CCA-secure encryption scheme from any WPRF F. In [8], Damgård and Nielsen also proposed to use the encrypt-then-MAC method for achieving CCA-security of $\mathcal{SE}_1$. However, their approach for constructing the VIL-MAC from any WPRF introduces a too large overhead for the solution to be practical. The number of applications of the WPRF per evaluation is in the order of the message length. The approach we give in Remark 3 is more efficient using at most N applications of the WPRF independently of the message length, where typically $N \ll n$ (recall that n is the block length of F). Whereas this additive overhead is of little concern for "long" messages, it is an open problem whether it can be improved for "short" messages.

---

[13] For any $Q : \mathcal{K}' \times \{0,1\}^N \to \{0,1\}^n$ and ϵ-AU hash function $H : \mathcal{K} \times \{0,1\}^* \to \{0,1\}^N$, $\mathbf{Adv}_{Q \circ H}^{\text{vil-mac}}(t, q, \mu) \leq \mathbf{Adv}_Q^{\text{prf}}(t, q) + q(q-1)\varepsilon/2 + 1/2^n$ (see [6]).

4.4   A Non-adaptive CCA-Secure Encryption Scheme

To achieve IND-P2-C1-security of $\mathcal{SE}_1$, we note that it is sufficient to WMAC the auxiliary randomness r. This has the advantage (over $\mathcal{SE}_2$) that the WMAC does not need to have VIL. To be precise, for $V : \{0,1\}^{\kappa_1} \times \{0,1\}^n \times \mathbb{N} \to \{0,1\}^*$ and $W : \{0,1\}^{\kappa_2} \times \{0,1\}^n \to \{0,1\}^\ell$, let $\mathcal{SE}_3$ denote the encryption scheme defined by encrypting a message $m \in \{0,1\}^*$ under the key $(k_1, k_2) \in \{0,1\}^{\kappa_1} \times \{0,1\}^{\kappa_2}$ and some auxiliary uniform random string $r \in \{0,1\}^n$ as

$$((k_1, k_2), r, m) \mapsto \Big(r, V_{k_1}(r, |m|) \oplus m, W_{k_2}(r) \Big). \qquad (\mathcal{SE}_3) \quad (7)$$

Theorem 4. *For any* t, q, μ, q', *and* μ'

$$\mathbf{Adv}_{\mathcal{SE}_3}^{\text{ind-p2-c1}}(t, q, \mu, q', \mu') \le 2 \cdot q' \cdot \mathbf{Adv}_W^{\text{wmac}}(t, q) + \mathbf{Adv}_{\mathcal{SE}_1}^{\text{ind-p2-c0}}(t, q, \mu+q\mu').$$

Proof. For $\mathcal{SE}_3 = (E, D)$, let Π_0 denote the IND-P2-C1 random experiment for any adversary A with resources (t, q, μ, q', μ'), i.e.,

$$(k_1, k_2) \xleftarrow{\$} \{0,1\}^{\kappa_1} \times \{0,1\}^{\kappa_2},$$

$$(x_0, x_1) \leftarrow A^{E_{k_1,k_2}, D_{k_1,k_2}}, b \xleftarrow{\$} \{0,1\}, y \leftarrow E_{k_1,k_2}(x_b), \ \hat{b} \leftarrow A^{E_{k_1,k_2}}(y).$$

Let Π_1 be the same same random experiment as Π_0 except for replacing A with an adversary B (described next) that has the same advantage as A and does not issue any query to D_{k_1,k_2} for which the auxiliary random part is the same as for a ciphertext returned previously by E_{k_1,k_2}. To be precise, let $\ell_{\max}$ denote the maximal length of the second input part of the decryption queries issued by A (clearly $\ell_{\max} < \mu'$). The adversary B simply runs A and for each encryption query m issued by A, B appends zeroes such that it is of length $l_{\max}$, i.e., $m' := m \| 0^{\ell_{\max} - |m|}$, and then queries the encryption oracle with m'. On output (r, c', w) from the encryption oracle, B returns $(r, c'[1, |m|], w)$ to A (and stores $(m', (r, c', w))$ in a look-up table. If A queries some decryption query, say (r, c, w'), for which r occurs in the look-up table as $(m', (r, c, w))$, B returns $c \oplus c'[1, |c|] \oplus m'[1, |c|]$ if $w = w'$ and otherwise rejects. When A presents its challenge input (m_0, m_1), B flips a coin b, queries its encryption oracle with m_b, and returns the result to A. Finally, B decides as A does. Further, let Π_2 be the same experiment as Π_1 except that all queries to D_{k_1,k_2} are rejected.

Moreover, for $i = 0, 1$, let S_i denote the event that $\hat{b} = b$ in Π_i. Then

$$\mathbf{Adv}_{\mathcal{SE}_3, A}^{\text{ind-p2-c1}} \overset{\text{def}}{=} 2 \cdot \Pr[S_0] - 1 = 2 \cdot \Pr[S_2] - 1 + 2 \cdot \sum_{i=0}^{1} \Big(\Pr[S_i] - \Pr[S_{i+1}] \Big)$$

$$\le \mathbf{Adv}_{\mathcal{SE}_3}^{\text{ind-p2-c0}}(t, q, \mu+q\mu') + 2 \cdot \Pr[\mathcal{E}] \le \mathbf{Adv}_{\mathcal{SE}_1}^{\text{ind-p2-c0}}(t, q, \mu+q\mu') + 2 \cdot \Pr[\mathcal{E}],$$

by the following three facts. First, $\Pr[S_0] = \Pr[S_1]$ as B decides as A does. Second, Π_1 and Π_2 are equivalent experiments unless the event $\mathcal{E}$ occurs that B queries a valid ciphertext to its decryption oracle. It follows that

$$\Pr[S_1] - \Pr[S_2] \le \Pr[\mathcal{E}] \le q' \cdot \mathbf{Adv}_W^{\text{wmac}}(t, q),$$

as B can be transformed to the following forger B' for W with advantage at least $\Pr[\mathcal{E}]/q'$. B' simply picks a random $i \in \{1, \ldots, q'\}$ and starts running B, answering its encryption queries with help of its own oracle and the decryption queries by rejection. When B (if at all) issues its i-th decryption query (r_i, c_i, w_i), B' returns (r_i, w_i) as its forgery (without making any extra calls to its encryption oracle). Third, Π_2 corresponds to the IND-P2-C0 experiment (in which the WMAC W is superfluous by Proposition 1). □

Remark 5. Combining (7) with $V = \mathrm{ICT}^F$ and $W = \mathrm{IC}^F \circ H$ results in an IND-P2-C1-secure scheme based on any WPRF F, but with the advantage that the ε-AU hash function H only is applied on fixed-sized strings (of length n). Alternatively, using $W = \mathrm{IC}^F$ saves the call to H and results in $n/2$ overhead applications on average (as IC^F is then invoked on random inputs).

5 Conclusions

We have proposed two constructions, the Increasing Chain Tree (ICT) and the Increasing Chain (IC). Whereas ICT extends the output length of weak PRFs in an optimal way (within a natural class of extensions) and optimizes Damgård and Nielsen's CPA-secure encryption scheme based on any weak PRF [8], IC is a construction of a regular PRF from any weak PRF that, in particular, reduces the key-material of Naor-Reingold's efficient PRF based on the DDH assumption [20]. By combining IC and ICT, we get a CCA-secure encryption scheme based on any weak PRF that is indeed much more efficient than the CCA-secure scheme proposed by Damgård and Nielsen (especially for "long" messages). It is an open problem to construct efficient schemes for "short" messages. Another interesting question is how to construct efficient weak MACs based on weak PRFs.

Although several highly efficient candidates for weak PRFs exist, none were targeted at this particular security notion explicitly. It is an interesting open problem for further research how much block-cipher design can benefit from this weakening of the desired security goal.

References

1. W. Aiello, S. Rajagopalan, and R. Venkatesan. High-speed pseudorandom number generation with small memory. In *Fast Software Encryption*, volume 1636 of *LNCS*, pages 290–304. Springer, 1999.
2. F. Bao, R. H. Deng, and H. Zhu. Variations of Diffie-Hellman problem. In *ICICS '03*, volume 2836 of *LNCS*, pages 301–312. Springer, 2003.
3. M. Bellare, A. Desai, E. Jokipii, and P. Rogaway. A concrete security treatment of symmetric encryption. In *Proc. of the 38th Symposium on Foundations of Computer Science*, pages 394–403. IEEE, 1997.
4. M. Bellare, J. Kilian, and P. Rogaway. The security of cipher block chaining. In *Advances in Cryptology — CRYPTO '94*, volume 839 of *LNCS*, pages 341–358. Springer, 1994.

5. M. Bellare and C. Namprempre. Authenticated encryption: Relations among notions and analysis of the generic composition paradigm. In *Advances in Cryptology — ASIACRYPT '00*, volume 1976 of *LNCS*, pages 531–545. Springer, 2000.
6. J. Black, S. Halevi, H. Krawczyk, T. Krovetz, and P. Rogaway. Umac: Fast and secure message authentication. In *Advances in Cryptology — CRYPTO '99*, volume 1666 of *LNCS*, pages 313–328. Springer, 1999.
7. A. Blum, M. L. Furst, M. J. Kearns, and R. J. Lipton. Cryptographic primitives based on hard learning problems. In *Advances in Cryptology — CRYPTO '93*, volume 773 of *LNCS*, pages 278–291. Springer, 1993.
8. I. Damgård and J. B. Nielsen. Expanding pseudorandom functions; or: From known-plaintext security to chosen-plaintext security. In *Advances in Cryptology — CRYPTO '02*, volume 2442 of *LNCS*, pages 449–464. Springer, 2002.
9. W. Diffie and M. Hellman. New directions in cryptography. *IEEE Transactions on Information Theory*, IT-22(6):644–654, 1976.
10. O. Goldreich. *Foundations of Cryptography – Volume II – Basic Applications*. Cambridge University Press, 2004.
11. O. Goldreich, S. Goldwasser, and S. Micali. How to construct random functions. *J. ACM*, 33(4):792–807, 1986.
12. J. Håstad, R. Impagliazzo, L. A. Levin, and M. Luby. A pseudorandom generator from any one-way function. *SIAM J. Comput.*, 28(4):1364–1396, 1999.
13. J. Katz and M. Yung. Complete characterization of security notions for probabilistic private-key encryption. In *Proc. of the 32nd Annual Symposium on Theory of Computing*, pages 245–254. ACM, 2000.
14. M. Keller. Constructing weak pseudorandom functions with prescribed structure, 2006. Semester Thesis, ETH Zurich.
15. S. Kent and R. Atkinson. IP encapsulating security payload (ESP), November 1998. Request for Comments 2406.
16. K. Minematsu and Y. Tsunoo. Expanding weak PRF with small key size. In *ICISC '05*, volume 3935 of *LNCS*, pages 284–298. Springer, 2005.
17. M. Naor, B. Pinkas, and O. Reingold. Distributed pseudo-random functions and KDCs. In *Advances in Cryptology — EUROCRYPT '99*, volume 1592 of *LNCS*, pages 327–346. Springer, 1999.
18. M. Naor and O. Reingold. From unpredictability to indistinguishability: A simple construction of pseudo-random functions from MACs. In *Advances in Cryptology — CRYPTO '98*, LNCS, pages 267–282. Springer, 1998.
19. M. Naor and O. Reingold. Synthesizers and their application to the parallel construction of pseudo-random functions. *J. Comp. Sys. Sci.*, 58(2):336–375, 1999.
20. M. Naor and O. Reingold. Number-theoretic constructions of efficient pseudorandom functions. *J. of the ACM*, 51(2):231–262, 2004.
21. K. Pietrzak and J. Sjödin. Weak pseudorandom functions in minicrypt, November 2006. Manuscript.
22. K. Pietrzak and J. Sjödin. Domain extension for weak PRFs; the good, the bad, and the ugly. In *Advances in Cryptology — EUROCRYPT '07*, LNCS. Springer, 2007. This proceedings.
23. V. Shoup. On fast and provably secure message authentication based on universal hashing. In *Advances in Cryptology — CRYPTO '96*, volume 1109 of *LNCS*, pages 313–328. Springer, 1996.
24. D. R. Stinson. Universal hashing and authentication codes. In *Advances in Cryptology — CRYPTO '91*, volume 576 of *LNCS*, pages 74–85. Springer, 1992.
25. M. N. Wegman and J. L. Carter. New hash functions and their use in authentication and set equality. *J. Comp. Sys. Sci.*, 22:265–279, 1981.

Range Extension for Weak PRFs;
The Good, the Bad, and the Ugly

Krzysztof Pietrzak[1,*] and Johan Sjödin[2,**]

[1] CWI Amsterdam
pietrzak@cwi.nl
[2] Department of Computer Science, ETH Zurich, CH-8092 Zurich, Switzerland
sjoedin@inf.ethz.ch

Abstract. We investigate a general class of (black-box) constructions for range extension of weak pseudorandom functions: a construction based on m independent functions $F_1, \ldots, F_m$ is given by a set of strings over $\{1, \ldots, m\}^*$, where for example $\{\langle 2 \rangle, \langle 1, 2 \rangle\}$ corresponds to the function $X \mapsto [F_2(X), F_2(F_1(X))]$. All efficient constructions for range expansion of weak pseudorandom functions that we are aware of are of this form.

We completely classify such constructions as *good*, *bad* or *ugly*, where the good constructions are those whose security can be proven via a black-box reduction, the bad constructions are those whose *in*security can be proven via a black-box reduction, and the ugly constructions are those which are neither good nor bad.

Our classification shows that the range expansion from [10] is optimal, in the sense that it achieves the best possible expansion ($2^m - 1$ when using m keys).

Along the way we show that for weak *quasirandom* functions (i.e. in the information theoretic setting), all constructions which are not bad – in particular all the ugly ones – are secure.

1   Introduction

PSEUDORANDOMNESS, introduced by Blum and Micali, is a crucial concept in theoretical computer science in general, and cryptography in particular. Informally, an object is pseudorandom if no efficient adversary can distinguish it from a truly random one. The most popular pseudorandom objects are pseudorandom generators (PRG), functions (PRF), and permutations (PRP). A PRG is a function $prg : \{0,1\}^n \to \{0,1\}^m$ where $m > n$ and no efficient A can distinguish $prg(U_n)$ from U_m, where U_i denotes the uniform distribution over i bit strings. A PRF is a family of functions $F : \{0,1\}^\ell \times \{0,1\}^n \to \{0,1\}^m$, where no efficient

---

* Supported by DIAMANT, the Dutch national mathematics cluster for discrete interactive and algorithmic algebra and number theory. This work was partially done while the author was a postdoc at the Ecole Normale Supérieure, Paris.
** This work was partially supported by the Zurich Information Security Center. It represents the views of the authors.

M. Naor (Ed.): EUROCRYPT 2007, LNCS 4515, pp. 517–533, 2007.

A can distinguish $F(U_\ell,.)$ from a uniformly random function. *Weak* PRFs, are defined similarly to PRFs, but where the adversary gets only to see the outputs on random inputs (and not on inputs of his choice). PRGs, PRFs, and PRPs are equivalent, i.e. black-box reducible, to one-way functions [4,3,6]. Unfortunately these reductions are quite inefficient, and therefore practical pseudorandom objects are either constructed from scratch (like the AES block-cipher, which is supposed to be a PRP) or from stronger assumptions than OWFs (in particular number theoretic assumptions like Decisional Diffie-Hellman).

RANGE EXTENSION FOR PRGS AND PRFS. From a PRG $prg : \{0,1\}^n \to \{0,1\}^{2n}$ one can efficiently construct a PRG with a larger range: on input $X \in \{0,1\}^n$ compute $Y_L \| Y_R \leftarrow prg(X)$ and output the $4n$-bit string $Z \leftarrow prg(Y_L) \| prg(Y_R)$. One can now recursively apply prg on input Z in order to get a pseudorandom $8n$-bit string and so on. The security of this construction follows by a simple hybrid argument.

From a PRF $prf : \{0,1\}^\ell \times \{0,1\}^n \to \{0,1\}^n$ we can get a PRF $prf' : \{0,1\}^{\ell t} \times \{0,1\}^n \to \{0,1\}^{nt}$ with larger range as

$$prf'(k_1, \ldots, k_t, x) = prf(k_1, x) \| \ldots \| prf(k_t, x)$$

This construction also works for *weak* PRFs, but is not very practical as the number of keys is linear in the expansion factor. Let $bin(i)$ denote the binary representation of i padded with 0's to the length $\lceil \log t \rceil$. The following construction of a $\{0,1\}^\ell \times \{0,1\}^{n - \lceil \log t \rceil} \to \{0,1\}^{nt}$ function

$$prf''(k, x) = prf(k, x \| bin(0)) \| \ldots \| prf(k, x \| bin(t-1))$$

just needs a single key, and prf'' is easily seen to be a PRF if prf is. Unfortunately this construction does not work for weak PRFs (just consider a weak PRF where the output does not depend on the last input bit).

RANGE EXTENSION FOR WEAK PRFS. Efficient range extension for weak PRFs has been investigated in [2,10,11]. All constructions considered in these papers can be defined by an ordered set α of strings over $[m] \stackrel{\text{def}}{=} \{1, \ldots, m\}$. The input to the construction are m keys $k_1, \ldots, k_m$ for the fixed output length PRF F, and a single input x to F. Each string $s \in \alpha$ now defines how to compute a part of the output, for example $s = \langle 2, 1, 3 \rangle$ corresponds to the value $F_{k_3}(F_{k_1}(F_{k_2}(x)))$, thus the expansion factor is the size of α. We give a formal definition for such constructions, which we call expansions, in Section 3.

CLASSIFYING EXPANSIONS. Not all expansions are secure in the sense of being a weak PRF whenever the underlying component F is a weak PRF. Before we continue, the reader might take a look at the three expansions given in the figure below, and try to answer the following question: if F is a weak PRF, which of the three length doubling constructions will also be a weak PRF (here k_1, k_2 are two random independent keys).

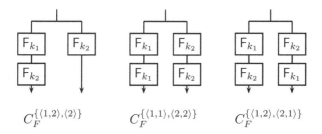

$$C_F^{\{\langle 1,2\rangle,\langle 2\rangle\}} \qquad\qquad C_F^{\{\langle 1,1\rangle,\langle 2,2\rangle\}} \qquad\qquad C_F^{\{\langle 1,2\rangle,\langle 2,1\rangle\}}$$

In this paper we exactly classify which expansions are secure and which are not (cf. Theorem 1). Interestingly there are three (and not just two) natural classes which come up, we will call them good, bad, and ugly (the three constructions in the figure above are simple examples of a good, a bad, and an ugly expansion). We call expansions whose security can be proven by a black-box reduction[1] *good*. We call an expansion *bad*, if its *in*security can be proven by a black-box reduction. There are also expansion which are neither good nor bad, we call them *ugly*.

MORE ON THE NOTION OF WEAK (PSEUDO/QUASI)RANDOM FUNCTIONS. A function is a pseudorandom function if

(i) It cannot be distinguished from a uniformly random function by any efficient distinguisher.
(ii) It can be efficiently computed.

In this paper we also consider the setting where (ii) is not necessarily satisfied, as the function is realized by some oracle, we call such functions simply random functions (RF). If (i) only holds for distinguishers which may query the function on random inputs, we prepend the term "weak" (like weak PRF). Functions which cannot be distinguished from random by any (and not just any efficient) distinguisher making some bounded number of queries are called quasirandom functions (QRF).[2] In particular any function which is a RF relative to a PSPACE oracle is a QRF.[3]

We use the term random*ized* function to denote a function which is not deterministic. This could be an efficient family of functions, where a function is sampled by choosing a random key. It could also be an oracle implementing

---

[1] In such a reduction one constructs an efficient adversary A, such that for every adversary B which breaks the security (as a weak PRF) of the expansion, the adversary A, given black-box access to B, breaks the security of the underlying randomized function (here A and B have only black-box access to the randomized function). Having black-box access to some component means that one only can query it on inputs of ones choice to get some output, but one does not get to see a description (say as a Turing machine) of the component.

[2] In the literature one often refers to such functions a almost k-wise independent functions, where k is a bound on the number of queries.

[3] This is the case as relative to a PSPACE oracle no computational hardness, and thus no pseudorandomness, exists. So if we have a RF relative to a PSPACE oracle, its randomness must be information theoretic, which means it is a QRF.

a function, where the oracle uses randomness. Clearly, any random function is a randomized function as a deterministic function is easily distinguished from random, the converse is not true in general.

1.1 Related Work

OPTIMAL EXPANSIONS. Efficient range expansion for weak PRFs have been investigated by Damgård and Nielsen [2]. They prove that there are good expansions which achieve an exponential expansion factor of roughly $2^{m/2} - 1$ (using m keys). This has been improved to roughly $3^{m/2} - 1$ in [11] and to $2^m - 1$ in [10]. From our classification it follows (Corollary 1) that $2^m - 1$ is indeed the best possible.[4]

EXPANSIONS IN MINICRYPT. In [13], we show that in Minicrypt, i.e. under the assumption that public-key cryptography does not exist[5], *some* ugly constructions[6] are secure. We do not know if relative to this assumption *all* ugly constructions are secure (in this paper we show that relative to a PSPACE oracle all ugly constructions are secure).

1.2 Applications

Weak PRFs are a strictly weaker primitive than PRFs, and thus requiring that some construction (like AES) is only a weak PRF is less of an assumption than assuming it to be a "regular" PRF.[7] Still, for many applications, weak PRFs are enough. An example is symmetric encryption [12,2,10]. The scheme defined by encrypting a message M as $(r, F(k, r) \oplus M)$, where r is sampled uniformly at random, is IND-CPA secure if F is a weak PRF [12]. There is some overhead as the ciphertext is $|r|$ bits longer than the plaintext, but using range extension for weak PRFs, a message of any length can be encrypted [2], and thus the overhead is independent of the message length.

In particular, when using the optimal expansion from [10] in the above scheme one needs $m = \lceil \log_2(|M|/n + 1) \rceil$ shared keys (n being the block-length) for the (fixed output length) weak PRF (those keys can also be computed by expanding a single key, see [10]). This expansion has a "depth" of m, by which we mean that to compute some elements of the output, one will have to invoke the weak

---

[4] In [10], it is shown – under the Inverse Decisional Diffie Hellman (IDDH) assumption – that their expansion α of size $2^m - 1$ is optimal for expansions containing strings of length logarithmic in the expansion factor (this corresponds to log-time random access to the output blocks). However, this still leaves open the possibility that a different expansion of larger size exists. In fact, [11] claim to have found a construction with a better expansion, but their proof is flawed (see [10]).

[5] This means relative to an oracle where one-way functions do exist, but key-agreement does not, such an oracle was constructed by Impagliazzo and Rudich [5].

[6] In particular (using notation introduced in the next section) $\alpha = \{\langle 1, 2 \rangle, \langle 2, 1 \rangle\}$.

[7] Block ciphers like AES are usually not only assumed to be PRFs, but even super pseudorandom-permutations, i.e. indistinguishable from a uniformly random permutation when adaptively queried from both directions.

PRF up to m times sequentially. Let us, however, stress that to compute all $2^m - 1$ outputs, one only needs a total number of $2^m - 1$ invocations. This is no contradiction, as if we compute an element with depth c, all the $c - 1$ values computed on the way will also be part of the output.

Although a depth of m is only logarithmic in the expansion factor, this might already be too much (say, due to hardware restrictions). We show (Corollary 2) that if we require a smaller depth $c < m$, then the best expansion factor we can get is $\sum_{i=0}^{c} \binom{m}{c} - 1$. Note that for $m = c$, this indeed gives the $2^m - 1$ bound.

2   Basic Definitions

By $_L X$ and $_R X$ we denote the left and right half of a bit string X of even length, respectively. We denote with $[m]$ the set $\{1, \dots, m\}$.

An *expansion* α is a set of strings over an alphabet $[m]$ for some $m \in \mathbb{N}$. Consider an expansion $\alpha = \{s_1, \dots, s_t\}$, each $s_i \in [m]^*$. With $s_i[j]$ we denote the j'th letter of s_i. We denote with $\#\alpha \stackrel{\text{def}}{=} m$ the alphabet size, with $|\alpha| \stackrel{\text{def}}{=} t$ the size, with $\|\alpha\| = \sum_{i=1}^{t} |s_i|$ the total length, and for $1 \leq i \leq m$ with α_i the number of occurrences of the letter i in α. Note that $\sum_{i=1}^{\#\alpha} \alpha_i = \|\alpha\|$.

For an expansion $\alpha, \#\alpha = m, |\alpha| = t$, and functions $F_1, \dots, F_m$, each $\mathcal{X} \to \mathcal{X}$, we define the function

$$C^{\alpha}_{F_1, \dots, F_m} = \mathcal{X} \to \mathcal{X}^t$$

as follows. On input $X \in \mathcal{X}$, the i'th component ($i \in [t]$) of the output is computed using s_i as

$$F_{s_i[|s_i|]}(F_{s_i[|s_i|-1]}(\dots F_{s_i[2]}(F_{s_i[1]}(X)) \dots)).$$

We will refer to the above computation as the evaluation of the i'th *chain*. For a randomized function F, we denote with C^{α}_F the function $C^{\alpha}_{F_1, \dots, F_m}$ where $m = \#\alpha$ and each F_i as an independent instantiation of F.

3   The Good, the Bad and the Ugly

We classify the expansions into three classes depending on the security they guarantee for C^{α}_F.

THE GOOD: α is good if the security of C^{α}_F as a weak random function can be efficiently black-box reduced to the security of F as a weak random function.[8] So whenever F is a weak random function, also C^{α}_F is, and moreover this holds relative to any oracle.

THE BAD: α is bad if there is an efficient construction F' which uses some function F as a black-box, such that the security of F' as a weak random function

---

[8] The reduction being efficient means that from any adversary A which breaks the security of C^{α}_F, we construct an adversary B where $B^{A,F}$ breaks the security of F, and the size of B (as an oracle circuit) is polynomial in the size of α and the range of F.

can be efficiently black-box reduced to the security of F as a weak random function, but $C_{F'}^{\alpha}$ is not a weak random function.

THE UGLY: α is ugly if it is neither good nor bad.

We now give a simple classification of all expansions into three classes $\mathfrak{G}, \mathfrak{B}$ and $\mathfrak{U}$, which by Theorem 1 below are exactly the good, the bad, and the ugly expansions.

Definition 1. *An expansion* $\alpha = \{s_1, \ldots, s_t\}$ *is*

- *of type* $\mathfrak{B}$ *if it does contain a string with two consecutive identical letters or two identical strings, i.e.*

$$\exists i, k \text{ where } s_i[k] = s_i[k+1] \quad or \quad \exists i, j, 1 \leq i < j \leq m : s_i = s_j.$$

- *of type* $\mathfrak{G}$ *if it is not of type* $\mathfrak{B}$ *and whenever a letter c appears before a letter d in some $s \in \alpha$, then d does not appear before c in any string $s' \in \alpha$, i.e.*[9]

$$\forall s, s' \in \alpha, i, j, i', j' : s[i] = s'[i'] \wedge s[j] = s'[j'] \wedge i < j \Rightarrow i' < j'.$$

- *of type* $\mathfrak{U}$ *if it is not of type* $\mathfrak{G}$ *or* $\mathfrak{B}$.

Theorem 1 (main)

(i) An expansion is **good** *if and only if it is of type* $\mathfrak{G}$.
(ii) An expansion is **bad** *if and only if it is of type* $\mathfrak{B}$.
(iii) An expansion is **ugly** *if and only if it is of type* $\mathfrak{U}$.

That $\mathfrak{G}$ expansions are good and $\mathfrak{B}$ expansions are bad follows by rather simple black-box reductions (Lemmata 1 and 2), the "only if" part is much harder. In order to show that the $\mathfrak{U}$ expansions are ugly, one has to come up with an oracle implementing a random function, such that relative to this oracle the expansion is not secure (thus it is not good), and another oracle relative to which it is secure (thus it is not bad). For the latter oracle we use a PSPACE oracle, as we show (Theorem 2) that for QRFs (recall that any RF is a QRF relative to a PSPACE oracle) any expansion which is not of type $\mathfrak{B}$, is secure. The following table summarizes the proof of the theorem.

$\mathfrak{G}$	$\mathfrak{U}$	$\mathfrak{B}$
good by Lemma 2 (and [10])	not good by Lemma 3	
not bad by Theorem 2		bad by Lemma 1

So Theorem 1.(i) follows from Lemma 2 and 3, Theorem 1.(ii) follows from Theorem 2 and Lemma 1, and Theorem 1.(iii) follows from (i) and (ii).

COROLLARIES. For every m, [10] construct a good expansion of size $2^m - 1$ using m keys: let α contain all $2^m - 1$ distinct s (of length at least 1) over $[m]$ where $s[i-1] < s[i]$ for all $2 \leq i \leq |s|$. From our classification it follows that this is best possible, and moreover, this expansion is the unique good expansion of size $2^m - 1$ (up to relabellings of the keys).

---

[9] Note that we do not require $c \neq d$, so this condition implies that no letter appears more than once in any string.

Corollary 1. *For any m and α with alphabet size $\#\alpha = m$, if α is good then*

$$|\alpha| \leq 2^m - 1,$$

and this is tight for $\alpha = \{s \in [m]^* \; ; \; s[1] < s[2] < \ldots < s[|s|]\}$.

For some $c < m$, consider the expansion we get by removing all $s \in \alpha$ of length more than c from the optimal expansion just described. This expansion is still good, and it is not hard to show that it is the best good expansion of depth c using m keys.

Corollary 2. *For any $m, c \leq m$, and α with alphabet size $\#\alpha = m$, if α is good then*

$$|\alpha| \leq \sum_{i=0}^{c} \binom{m}{i} - 1,$$

and this is tight for $\alpha = \{s \in [m]^* \; ; \; s[1] < s[2] < \ldots < s[|s|], |s| \leq c\}$.

Note that Corollary 1 is just a special case of Corollary 2 for the case $c = m$.

4   The Bad Expansions Are Exactly $\mathfrak{B}$

To prove that expansions outside of $\mathfrak{B}$ are not bad, we use the random systems framework of Maurer [7]. Here we only give a rather informal and restricted exposition of the framework, in particular we only consider known-plaintext attacks (KPA), as this is the only attack relevant for this paper.

NOTATION. We use capital calligraphic letters like $\mathcal{X}$ to denote sets, capital letters like X to denote random variables, and small letters like x denote concrete values. To save on notation we write X^i for $X_1, X_2, \ldots, X_i$.

RANDOM SYSTEMS. Informally, a *random system* is a system which takes inputs $X_1, X_2, \ldots$ and generates, for each new input X_i, an output Y_i which depends probabilistically on the inputs and outputs seen so far. We define random systems in terms of the distribution of the outputs Y_i conditioned on $X^i Y^{i-1}$, more formally: An $(\mathcal{X}, \mathcal{Y})$-*random system* $\mathbf{F}$ is a sequence of conditional probability distributions $\mathsf{P}^{\mathbf{F}}_{Y_i|X^i Y^{i-1}}$ for $i \geq 1$. Here we denote by $\mathsf{P}^{\mathbf{F}}_{Y_i|X^i Y^{i-1}}(y_i, x^i, y^{i-1})$ the probability that $\mathbf{F}$ will output $y_i \in \mathcal{Y}$ on input $x_i \in \mathcal{X}$ conditioned on the fact that $\mathbf{F}$ did output $y_j \in \mathcal{Y}$ on input $x_j \in \mathcal{X}$ for $j = 1, \ldots, i-1$.

Uniformly random functions (URFs) are random systems which will be important in this paper, throughout $\mathbf{R}_{n,m}$ will denote a URF $\{0,1\}^n \to \{0,1\}^m$.

CONDITIONS FOR RANDOM SYSTEMS. With $\mathbf{F}^{\mathcal{A}}$ we denote the random system $\mathbf{F}$, but which additionally defines an internal binary random variable after each query (called a condition). Let $A_i \in \{0,1\}$ denote the condition after the i'th query. We set $A_0 = 0$ and require the condition to be monotone which means that $A_i = 1 \Rightarrow A_{i+1} = 1$ (i.e. when the condition failed, it will never hold again). Let $\bar{a}_i$ denote the event $A_i = 1$, then with $\nu^{\mathsf{KPA}}(\mathbf{F}^{\mathcal{A}}, \bar{a}_k)$ we denote the probability of the event $\bar{a}_k$ occurring when $\mathbf{F}^{\mathcal{A}}$ is queried on random inputs, i.e.

$$\nu^{\mathsf{KPA}}(\mathbf{F}^{\mathcal{A}}, \bar{a}_k) \stackrel{\text{def}}{=} \sum_{x^k \in \mathcal{X}^k} \Pr[X^k = x^k] \cdot \Pr[\bar{a}_k \text{ holds in } \mathbf{F}^{\mathcal{A}}(x^k)]$$

$$= \frac{1}{|\mathcal{X}|^k} \sum_{x^k \in \mathcal{X}^k} \Pr[\bar{a}_k \text{ holds in } \mathbf{F}^{\mathcal{A}}(x^k)].$$

INDISTINGUISHABILITY. For $(\mathcal{X}, \mathcal{Y})$-random systems $\mathbf{F}$ and $\mathbf{G}$, we denote with $\Delta_k^{\mathsf{KPA}}(\mathbf{F}, \mathbf{G})$ the distinguishing advantage of any unbounded distinguisher in a k query known-plaintext attack. This advantage is simply the statistical distance, i.e. with X^k being uniformly random over $\mathcal{X}^k$

$$\Delta_k^{\mathsf{KPA}}(\mathbf{F}, \mathbf{G}) \stackrel{\text{def}}{=} \frac{1}{2} \sum_{x^k \in \mathcal{X}^k, y^k \in \mathcal{Y}^k} \Pr[X^k = x^k] \cdot \left| \Pr[\mathbf{F}(x^k) = y^k] - \Pr[\mathbf{G}(x^k) = y^k] \right|$$

$$= \frac{1}{2 \cdot |\mathcal{X}|^k} \sum_{x^k \in \mathcal{X}^k, y^k \in \mathcal{Y}^k} \left| \Pr[\mathbf{F}(x^k) = y^k] - \Pr[\mathbf{G}(x^k) = y^k] \right|.$$

$\mathbf{F}^{\mathcal{A}} \stackrel{\circ}{=} \mathbf{G}^{\mathcal{B}}$ denotes that $\mathbf{F}^{\mathcal{A}}$ is equivalent to $\mathbf{G}^{\mathcal{B}}$ while the respective condition holds:

$$\mathbf{F}^{\mathcal{A}} \stackrel{\circ}{=} \mathbf{G}^{\mathcal{B}} \iff \forall x^i, y^i : \Pr^{\mathbf{F}^{\mathcal{A}}}_{a_i \wedge Y^i | X^i}(y^i, x^i) = \Pr^{\mathbf{G}^{\mathcal{B}}}_{b_i \wedge Y^i | X^i}(y^i, x^i).$$

We say that $\mathbf{F}^{\mathcal{A}}$ is dominated by $\mathbf{G}$, which is denoted by $\mathbf{F}^{\mathcal{A}} \preceq \mathbf{G}$, if on any input x^i and for any possible output y^i the probability that $\mathbf{F}^{\mathcal{A}}(x^i)$ output y^i and the condition $\mathcal{A}$ holds, is at most the probability that $\mathbf{G}(x^i) = y^i$.

$$\mathbf{F}^{\mathcal{A}} \preceq \mathbf{G} \iff \forall x^i, y^i : \Pr^{\mathbf{F}^{\mathcal{A}}}_{a_i \wedge Y^i | X^i}(y^i, x^i) \leq \Pr^{\mathbf{G}}_{Y^i | X^i}(y^i, x^i)$$

or equivalently $\forall x^i, y^i : \Pr^{\mathbf{F}^{\mathcal{A}}}_{a_i \wedge Y_i | X^i Y^{i-1} a_{i-1}}(y_i, x^i, y^{i-1}) \leq \Pr^{\mathbf{G}}_{Y_i | X^i Y^{i-1}}(y_i, x^i, y^{i-1})$

Note that $\mathbf{F}^{\mathcal{A}} \stackrel{\circ}{=} \mathbf{G}^{\mathcal{B}}$ implies $\mathbf{F}^{\mathcal{A}} \preceq \mathbf{G}$ and $\mathbf{G}^{\mathcal{B}} \preceq \mathbf{F}$. The following are the two main propositions of the framework (restricted to the case of KPA attacks).

Proposition 1. *If* $\mathbf{F}^{\mathcal{A}} \preceq \mathbf{G}$ *then* $\Delta_q^{\mathsf{KPA}}(\mathbf{F}, \mathbf{G}) \leq \nu^{\mathsf{KPA}}(\mathbf{F}^{\mathcal{A}}, \bar{a}_q)$.

Proposition 2. *For any random systems* $\mathbf{F}$ *and* $\mathbf{G}$*, there exist conditions* $\mathcal{A}$ *and* $\mathcal{B}$ *such that*

$$\mathbf{F}^{\mathcal{A}} \stackrel{\circ}{=} \mathbf{G}^{\mathcal{B}} \quad and \quad \Delta_q^{\mathsf{KPA}}(\mathbf{F}, \mathbf{G}) = \nu^{\mathsf{KPA}}(\mathbf{F}^{\mathcal{A}}, \bar{a}_q) = \nu^{\mathsf{KPA}}(\mathbf{G}^{\mathcal{B}}, \bar{b}_q).$$

Proposition 1 is quite easy to prove and appeared in the original paper [7]. Proposition 2 is from (the yet unpublished) [9], a weaker version of this proposition appeared in [8].

4.1 Expansions Not in $\mathfrak{B}$ Are Not Bad

By the following theorem, $C^{\alpha}_{\mathbf{F}_1, \dots, \mathbf{F}_m}$ is a weak quasirandom function whenever the $\mathbf{F}_i$'s are weak QRFs and α is not in $\mathfrak{B}$. The distance of the output of $C^{\alpha}_{\mathbf{F}_1, \dots, \mathbf{F}_m}$

on q random queries can be upper bounded by the sum of the distances of the $\mathbf{F}_i$'s on $q\alpha_i$ random queries (recall that α_i is the number of invocations of $\mathbf{F}_i$ on an invocation of $C^\alpha_{\mathbf{F}_1,\ldots,\mathbf{F}_m}$), plus some term which is small unless $q \cdot \|\alpha\|$ is in the order of $2^{n/2}$.

Theorem 2. *For any expansion $\alpha = \{s_1,\ldots,s_t\}$ which is not of type $\mathfrak{B}$, any randomized functions $\mathbf{F}_i : \{0,1\}^n \to \{0,1\}^n, 1 \le i \le \#\alpha := m$, and every $q \in \mathbb{N}$:*

$$\Delta^{\mathsf{KPA}}_q(C^\alpha_{\mathbf{F}_1,\ldots,\mathbf{F}_m}, \mathbf{R}_{n,n\cdot t}) \le \sum_{i=1}^m \Delta^{\mathsf{KPA}}_{q\cdot\alpha_i}(\mathbf{F}_i, \mathbf{R}_{n,n}) + \frac{q^2\|\alpha\|^2}{2^n}.$$

Proof. To save on notation let

$$\mathbf{I} \stackrel{\text{def}}{=} C^\alpha_{\mathbf{R}_1,\ldots,\mathbf{R}_m} \quad \text{and} \quad \mathbf{C} \stackrel{\text{def}}{=} C^\alpha_{\mathbf{F}_1,\ldots,\mathbf{F}_m},$$

where the $\mathbf{R}_i$'s are independent instantiations of $\mathbf{R}_{n,n}$. By the triangle inequality

$$\Delta^{\mathsf{KPA}}_q(\mathbf{C}, \mathbf{R}_{n,n\cdot t}) \le \Delta^{\mathsf{KPA}}_q(\mathbf{C}, \mathbf{I}) + \Delta^{\mathsf{KPA}}_q(\mathbf{I}, \mathbf{R}_{n,n\cdot t}). \tag{1}$$

The theorem follows from the two claims below, which bound the two terms on the right hand side of (1) respectively.

Claim 1

$$\Delta^{\mathsf{KPA}}_q(\mathbf{I}, \mathbf{R}_{n,n\cdot t}) \le \frac{q^2\|\alpha\|^2}{2^{n+1}}$$

Proof (of Claim 1) We define a condition $\mathcal{D}$ on $\mathbf{I}$ as follows: the condition is satisfied as long as for all $i, 1 \le i \le m$, there was no nontrivial collision on the inputs to the component $\mathbf{R}_i$. Here the trivial collisions are the "unavoidable" collisions which occur when two chains have the same prefix. For example in $C^{\{\langle 1,2,3,4\rangle, \langle 1,2,4,3\rangle\}}_{\mathbf{R}_1,\ldots,\mathbf{R}_4}$ the inputs to $\mathbf{R}_1$ in the two different chains are always identical, the same holds for the inputs to $\mathbf{R}_2$ (but not for $\mathbf{R}_3$ or $\mathbf{R}_4$). We now show (using that α is not of type $\mathfrak{B}$) that this condition satisfies $\mathbf{I}^\mathcal{D} \preceq \mathbf{R}_{n,n\cdot t}$, i.e.

$$\forall x^i, y^i : \mathsf{Pr}^{\mathbf{I}}_{Y_i \wedge d_i | X^i Y^{i-1} \wedge d_{i-1}}(y_i, x^i, y^{i-1}) \le \mathsf{Pr}^{\mathbf{R}_{n,n\cdot t}}_{Y_i | X^i Y^{i-1}}(y_i, x^i, y^{i-1}) = 2^{-n\cdot t}. \tag{2}$$

Assume we invoke $\mathbf{I}$ on the i'th query $x_i \in \{0,1\}^n$, and that d_{i-1}, i.e. the condition was satisfied after the $(i-1)$'th query. We evaluate the $t = |\alpha|$ chains of $\mathbf{I} = C^\alpha_{\mathbf{R}_1,\ldots,\mathbf{R}_m}$ one by one and assume that the s_i's are ordered by increasing length.[10] For any j, when computing the j'th chain we stop just before we invoke the last component $\mathbf{R}_{s_j[|s_j|]}$. Now, if the input to this component is fresh (i.e. $\mathbf{R}_{s_j[|s_j|]}$ was never invoked on that input before), then every output has probability exactly 2^{-n}. The probability that we get fresh inputs (to the last components) on all t chains and the outputs will be consistent with y_i in all chains is thus at most $2^{-n\cdot t}$. On the other hand, if at some point we have an

---

[10] This will only be important if one chain is the prefix on another.

input which is not fresh, then there has been a collision. Now, as no two chains are equivalent (as α is not in $\mathfrak{B}$) and we process them by increasing length, it follows that this collision was a nontrivial one, and thus $\bar{d}_i$. This concludes the proof of (2). The first step of

$$\Delta_q^{\mathsf{KPA}}(\mathbf{I}, \mathbf{R}_{n,n \cdot t}) \leq \nu^{\mathsf{KPA}}(\mathbf{I}^{\mathcal{D}}, \bar{d}_q) \leq \frac{\sum_{i=1}^{m}(q \cdot \alpha_i)^2}{2^{n+1}} \leq \frac{q^2 \|\alpha\|^2}{2^{n+1}} \tag{3}$$

follows by Proposition 1 using $\mathbf{I}^{\mathcal{D}} \preceq \mathbf{R}_{n,n \cdot t}$. The second step follows by the birthday bound: the fact that $\bar{d}_q$ means that at some point for some $i \in [m]$ the uniformly random output of $\mathbf{R}_i$ did collide with some "old" input to $\mathbf{R}_i$. As $\mathbf{R}_i$ is invoked $q \cdot \alpha_i$ times, the probability that there will be a collision is at most $(q \cdot \alpha_i)^2 / 2^{n+1}$. To get the probability that there will be a collision for any $\mathbf{R}_i, i \in [m]$, we take the union bound. $\triangle$

Claim 2

$$\Delta_q^{\mathsf{KPA}}(\mathbf{C}, \mathbf{I}) \leq \frac{q^2 \|\alpha\|^2}{2^{n+1}} + \sum_{i=1}^{m} \Delta_{q \cdot \alpha_i}^{\mathsf{KPA}}(\mathbf{F}_i, \mathbf{R}_{n,n})$$

Proof (of Claim 2) For every $i, 1 \leq i \leq m$, let $\mathcal{A}^i$ and $\mathcal{B}^i$ be conditions such that (the existence of such conditions follows by Proposition 2)

$$\mathbf{F}_i^{\mathcal{A}^i} \stackrel{\circ}{=} \mathbf{R}_{n,n}^{\mathcal{B}^i} \quad \text{and} \quad \Delta_q^{\mathsf{KPA}}(\mathbf{F}_i, \mathbf{R}_{n,n}) = \nu^{\mathsf{KPA}}(\mathbf{R}_{n,n}^{\mathcal{B}^i}, \bar{b}_q^i) = \nu^{\mathsf{KPA}}(\mathbf{F}_i^{\mathcal{A}^i}, \bar{a}_q^i). \tag{4}$$

To save on notation let $\mathcal{B} \stackrel{\text{def}}{=} \mathcal{B}^1 \wedge \cdots \wedge \mathcal{B}^m$, $\mathcal{A} \stackrel{\text{def}}{=} \mathcal{A}^1 \wedge \cdots \wedge \mathcal{A}^m$ and $q_i = q \cdot \alpha_i$. . As for all $\mathbf{F}_i^{\mathcal{A}^i} \equiv \mathbf{R}_{n,n}^{\mathcal{B}^i}$ for $1 \leq i \leq m$, it follows that

$$\mathbf{C}^{\mathcal{A}} \stackrel{\circ}{=} \mathbf{I}^{\mathcal{B}}. \tag{5}$$

Let $b \Rightarrow_q d$ denote the event defined on $\mathbf{I}^{\mathcal{B} \wedge \mathcal{D}}$ which holds if at any timepoint up to after the q'th query, either $\mathcal{D}$ holds or $\mathcal{B}$ does not hold (or equivalently, either $\mathcal{D}$ does not fail, or it only fails after $\mathcal{B}$ fails). The first step below follows by Proposition 1 using (5). The last step follows by the union bound and observing that $\bar{b}_q \vee \bar{d}_q$ holds iff $\bar{d}_q \vee [\bar{b}_q \wedge [b \Rightarrow_q d]]$.

$$\begin{aligned}\Delta_q^{\mathsf{KPA}}(\mathbf{C}, \mathbf{I}) &\leq \nu^{\mathsf{KPA}}(\mathbf{I}^{\mathcal{B}}, \bar{b}_q) \\ &\leq \nu^{\mathsf{KPA}}(\mathbf{I}^{\mathcal{B} \wedge \mathcal{D}}, \bar{b}_q \vee \bar{d}_q) \\ &\leq \nu^{\mathsf{KPA}}(\mathbf{I}^{\mathcal{D}}, \bar{d}_q) + \nu^{\mathsf{KPA}}(\mathbf{I}^{\mathcal{B} \wedge \mathcal{D}}, \bar{b}_q \wedge [b \Rightarrow_q d]). \end{aligned} \tag{6}$$

We can bound the first term of (6) using (3) as $\nu^{\mathsf{KPA}}(\mathbf{I}^{\mathcal{D}}, \bar{d}_q) \leq q^2 \|\alpha\|^2 / 2^{n+1}$. We now bound the second term of (6), using $\bar{b}_q \iff \bar{b}_{q_1}^1 \vee \ldots \vee \bar{b}_{q_m}^m$ in the first inequality, and the union bound in the second step:

$$\begin{aligned}\nu^{\mathsf{KPA}}(\mathbf{I}^{\mathcal{B} \wedge \mathcal{D}}, \bar{b}_q \wedge [b \Rightarrow_q d]) &= \nu^{\mathsf{KPA}}(\mathbf{I}^{\mathcal{B} \wedge \mathcal{D}}, [\bar{b}_{q_1}^1 \vee \ldots \vee \bar{b}_{q_m}^m] \wedge [b \Rightarrow_q d]) \\ &\leq \sum_{i=1}^{m} \nu^{\mathsf{KPA}}(\mathbf{I}^{\mathcal{B} \wedge \mathcal{D}}, \bar{b}_{q_i}^i \wedge [b \Rightarrow_q d]). \end{aligned}$$

The term $\nu^{\mathsf{KPA}}(\mathbf{I}^{\mathcal{B}\wedge\mathcal{D}}, \bar{b}^i_{q_i} \wedge [b \Rightarrow_q d])$ is the probability that when querying $\mathbf{I}$ on q random inputs, the condition $\mathcal{B}^i$ defined on $\mathbf{R}^i$ will fail, and it will do so before $\mathcal{D}$ fails. Now, as long as $\mathcal{D}$ holds, $\mathbf{R}^i$ is invoked on uniformly random inputs: the inputs are either part of the global input (which is random in a KPA attack), or it is the output of some URF $\mathbf{R}^j$. It is important to note that in this case always $j \neq i$,[11] so $\mathbf{R}_i$ is never invoked on its own output, which guarantees that (while $\mathcal{D}$ holds) the inputs to $\mathbf{R}_i$ are not only random, but also independent of $\mathbf{R}_i$. So the probability that $\bar{b}^i_{q_i} \wedge [b \Rightarrow_q d]$ in $\mathbf{I}$ can be upper bounded by the probability that $\bar{b}_{q_i}$ in $\mathbf{R}^{\mathcal{B}^i}_i$ in a normal KPA attack, i.e.

$$\nu^{\mathsf{KPA}}(\mathbf{I}^{\mathcal{B}\wedge\mathcal{D}}, \bar{b}^i_{q_i} \wedge [b \Rightarrow_q d]) \leq \nu^{\mathsf{KPA}}(\mathbf{R}^{\mathcal{B}^i}_i, \bar{b}^i_{q_i}) = \Delta^{\mathsf{KPA}}_{q_i}(\mathbf{F}_i, \mathbf{R}_{n,n}),$$

where the second step follows by (4).   △

□

4.2   Type $\mathfrak{B}$ Expansions Are Bad

Lemma 1. *Expansions of type $\mathfrak{B}$ are bad.*

Proof. To prove the lemma we show a black-box construction of a random function G^P based on a permutation P such that:

(i) The security of G^P as a weak random function can be black-box reduced to the security of P as a random permutation.
(ii) For every bad expansion α, $C^\alpha_{G^P}$ is not a weak random function.

Note that we assume that G has access to an oracle which implements a random permutation,[12] and not just a weak RF as required by the lemma. We can do this as random permutations and weak random functions are equivalent, in the sense that both can be constructed from (and imply the existence of) functions which are hard to invert on random inputs[13] via a black-box reduction [3,4,6].

To simplify the argument, in the proof we assume that the random permutation $P : \{0,1\}^{2n} \to \{0,1\}^{2n}$ is a uniformly random permutation (URP). As by definition a random permutation is indistinguishable from a URP, this does not change the statement. $G^P(X) : \{0,1\}^{2n} \to \{0,1\}^{2n}$ is defined as follows, first let $Y = {}_LY\|{}_RY \leftarrow P^{-1}(X)$, now

$$G^P(X) = \begin{cases} 0^{2n} & \text{if } {}_LY = 0^n \text{ or } X = 0^{2n} \\ P(0^n\|{}_RX) & \text{otherwise.} \end{cases}$$

We first prove statement (i), namely that G^P is a weak random function (in fact, as we assume that P is a URP, we can even show that G^P is a weak quasirandom function).

---

[11] This is because α is not of type $\mathfrak{B}$ and thus no $s \in \alpha$ has two identical consecutive letters.

[12] A random permutation is a random bijective function (with same range and domain).

[13] Such functions are called one-way functions in the special (and most interesting) case where the function can be efficiently computed in forward direction.

Claim 3

$$\Delta_q^{\mathsf{KPA}}(G^P, \mathbf{R}_{2n,2n}) \le \frac{3q^2}{2^n}$$

Proof (of Claim 3). By the triangle inequality

$$\Delta_q^{\mathsf{KPA}}(G^P, \mathbf{R}_{2n,2n}) \le \Delta_q^{\mathsf{KPA}}(G^P, P) + \Delta_q^{\mathsf{KPA}}(P, \mathbf{R}_{2n,2n}) \tag{7}$$

G^P is equivalent to P unless we happen to query G^P on input 0^{2n} or an input X where the first n bits of $P^{-1}(X)$ are 0^n. For a random X, this happens with probability $\le 2^{-2n} + 2^{-n}$. By the union bound

$$\Delta_q^{\mathsf{KPA}}(G^P, P) \le \frac{2q}{2^n}. \tag{8}$$

By the so called PRF/PRP switching lemma (see e.g. [1]) we have

$$\Delta_q^{\mathsf{KPA}}(P, \mathbf{R}_{2n,2n}) \le \frac{q^2}{2^{2n+1}}. \tag{9}$$

The claim follows from (7), (8), and (9). $\qquad\qquad\qquad\qquad\qquad\qquad\triangle$

Now we prove statement (ii), i.e. that for every bad expansion α, $C_{G^P}^\alpha$ is not a weak random function. Recall that $\alpha = \{s_1, \ldots, s_t\}$ is bad if either $s_i = s_j$ for some $i \ne j$ or there is a s_i with two consecutive identical letters, i.e. for some $j : s_i[j] = s_i[j+1]$. When $s_i = s_j$ then also the i'th and j'th tuple in the output of $C_{G^P}^\alpha(X)$ are identical for any X, and thus easy to distinguish from random.

We now consider the other case. Let α be any expansion where for some element $s \in \alpha$ we have for some j that $s[j] = s[j+1]$. As we prove a negative statement, we can without loss of generality assume that s is the only element in α. We claim that $C_{G^P}^\alpha$ is not random as for any $m = \#\alpha$ instantiations $G_1^P, \ldots, G_m^P$ of G^P and any X we have $C_{G_1^P, \ldots, G_m^P}^\alpha(X) = 0^{2n}$. To see this let $X_0 = X$ and for $i = 1, \ldots, |s| : X_i = G_{s[i]}^P(X_{i-1})$, then $C_{G_1^P, \ldots, G_m^P}^\alpha(X) = X_{|s|}$. Now by the definition of G^P, for any Z and $i \in [m]$, $G_i^P(G_i^P(Z)) = 0^{2n}$, in particular $X_{j+1} = G_{s[j+1]}^P(G_{s[j]}^P(X_{j-1})) = 0^{2n}$, and as $G_i^P(0^{2n}) = 0^{2n}$ for any i we get $X_\ell = 0^{2n}$ for all $\ell \ge j$. For concreteness let us illustrate this computation on the example $\alpha = \{\langle 1, 2, 2, 3 \rangle\}$. Here P_2 is the P used by G_2^P, and $X_3 = 0^{2n}$ holds as ${}_L P_2^{-1}(X_2) = 0^n$.

$$X = X_0 \overset{G_1^P}{\to} X_1 \overset{G_2^P}{\to} X_2 = P_2(0^n \| {}_R X_1) \overset{G_2^P}{\to} X_3 = 0^{2n} \overset{G_3^P}{\to} X_4 = 0^{2n}. \qquad \square$$

5   The Good Expansions Are Exactly ©

5.1   Type © Expansions Are Good

The following lemma is from [10], for completeness we give a proof in the appendix.

Lemma 2. *Expansions of type © are good.*

5.2   Expansions Not in $\mathfrak{G}$ Are Not Good

Lemma 3. *Expansions not in $\mathfrak{G}$ are not good.*

By Lemma 1 expansions of type $\mathfrak{B}$ are not good. It remains to show that there exists an oracle $\mathcal{O}$ relative to which a weak random function $F^{\mathcal{O}}$ exists, but where for any expansion $\alpha = \{s_1, \ldots, s_t\}$ of type $\mathfrak{U}$ the function $C^{\alpha}_{F^{\mathcal{O}}}$ is not weakly random. The oracle $\mathcal{O}$ we construct will consist of two parts, which can be accessed by setting the first part of the input to either "eval" or "break".

Let n be our security parameter (think of $\mathcal{O}$ as a family of oracles, one for each $n \in \mathbb{N}$). Let $m = \max_i |s_i|$ and $\ell = n^3 m$. Let $F^{\mathcal{O}} : \{0,1\}^n \times \{0,1\}^\ell \to \{0,1\}^\ell$ ($\mathcal{O}$ still to be defined)

$$F^{\mathcal{O}}(k, x) = \mathcal{O}(eval, k, x).$$

We will often write the key as a subscript $F^{\mathcal{O}}_k(.) = F^{\mathcal{O}}(k, .)$. The all zero string 0^n is excluded from the valid keys as later 0^n will have the special meaning of "no key".

We now define the "eval" part of the oracle. Initially, $2^{mn} - 1$ disjoint subsets of $\{0,1\}^\ell$, each of size 2^n, are sampled. Each such set corresponds to an ordered sequence of at most m (and at least one) keys, the set corresponding to the keys $k_1, \ldots, k_{m'}, m' \le m$ is denoted $S_{0^{(m-m')n} \| k_1 \| k_2 \| \ldots \| k_{m'}}$. With S_0 we denote the elements from $\{0,1\}^\ell$ which are in no set, i.e. $S_0 = \{0,1\}^\ell \setminus \bigcup_{x \in \{0,1\}^{mn} \setminus 0^{mn}} S_x$ (we have $|S_0| = 2^{mn^3} - 2^{mn^2} + 2^n$, i.e. all but a 2^{-n} fraction of elements from $\{0,1\}^\ell$ are in S_0).

Now for any key k, $\mathcal{O}(eval, k, .)$ maps the elements from S_0 at random to $S_{0^{(m-1)n} \| k}$. As for the inputs not in S_0, for any key k and keys $k_1, \ldots, k_{m'}$, $\mathcal{O}(eval, k, .)$ is defined as a random bijective function from $S_{0^{(m-m')n} \| k_1 \| k_2 \| \ldots \| k_{m'}}$ to $S_{0^{(m-m'-1)n} \| k_1 \| k_2 \| \ldots \| k_{m'} \| k}$ (where for $m = m'$, we shift the leftmost key out, i.e. we map $S_{k_1 \| k_2 \| \ldots \| k_m}$ to $S_{k_2 \| \ldots \| k_m \| k}$). Note that this means that for any $t \le m$ and $x \in S_0$ a value computed as $y = F^{\mathcal{O}}_{k_t}(F^{\mathcal{O}}_{k_{t-1}} \ldots F^{\mathcal{O}}_{k_1}(x))$ is in $S_{0^{(m-t)n} \| k_1 \| \ldots \| k_t}$. For a computationally bounded distinguisher, this y will look random, but the computationally unbounded "break" part of the oracle (defined below) can learn the keys $k_1, \ldots, k_t$ used.

We now define the "break" part of the oracle. $\mathcal{O}(break, .)$ is a $(\{0,1\}^\ell)^2 \to \{0,1\}$ function and defined as follows. For any $Y_1 \in S_{0^{m-m'n} \| a_1 \| \ldots \| a_{m'}}$ and $Y_2 \in S_{0^{m-m''n} \| b_1 \| \ldots \| b_{m''}}$, we define $\mathcal{O}(break, \{Y_1, Y_2\}) = 1$ if there are i, i', j, j' where

$$a_i = b_{i'} \qquad a_j = b_{j'} \qquad i < j \qquad i' > j', .$$

and $\mathcal{O}(break, \{Y_1, Y_2\}) = 0$ otherwise. In particular, $\mathcal{O}(break, \{Y_1, Y_2\})$ outputs 0 if either $Y_1 \in S_0$ or $Y_2 \in S_0$.

Claim 4. *For any α of type $\mathfrak{U}$, $C^{\alpha}_{F^{\mathcal{O}}}$ is not a weak random function (relative to the oracle $\mathcal{O}$).*

Proof (of Claim 4). Let $X \in \{0,1\}^\ell$ be a random input, and $Y = C^{\alpha}_{F^{\mathcal{O}}}(X)$. Let Y_i denote the i'th ℓ-bit block of $Y \overset{\text{def}}{=} Y_1 \| \ldots \| Y_t$. As $\alpha = \{s_1, \ldots, s_t\}$ is of type

$\mathfrak{U}$, there are i, j and letters c, d such that $s_i = *c*d*$, and $s_j = *d*c*$, where each $*$ is a wildcard, i.e. stands for "any" string. As $Y_i \in S_{*s_i} = S_{*c*d*}$ and $Y_j \in S_{*s_j} = S_{*d*c*}$, it follows that $\mathcal{O}(break, \{Y_i, Y_j\}) = 1$. On the other hand, for a random $Y' = Y'_1 \| \ldots \| Y'_t$ the probability that $\mathcal{O}(break, \{Y'_i, Y'_j\}) = 1$ is very small: we get a rough (but already exponentially small) upper bound on this probability by using that the oracle will output 0 whenever Y'_i is in S_0, i.e.

$$\Pr[\mathcal{O}(break, \{Y'_i, Y'_j\}) = 1] \leq \Pr[Y'_i \notin S_0] < 1/2^n.$$

Thus we can distinguish the output Y of $C^\alpha_{F\mathcal{O}}$ from random Y' with advantage almost 1.   △

Claim 5. $F^\mathcal{O}$ *is a weak random function relative to* $\mathcal{O}$.

Proof (sketch of Claim 5). Clearly the function $F^\mathcal{O}$ is a random function relative to the oracle $\mathcal{O}(eval, ., .)$ alone (i.e. where there is no $\mathcal{O}(break, .)$).

Now we will show that adding the oracle $\mathcal{O}(break, .)$ will not break the security of $F^\mathcal{O}$ as a weak random function (but note that it trivially does break the security of $F^\mathcal{O}$ as a (non weak) random function[14]), as if an adversary $A^\mathcal{O}$ can distinguish $F^\mathcal{O}(k, .)$ from random on random inputs and access to the oracle $\mathcal{O}(break, .)$, then there is an adversary $B^{\mathcal{O}, A}$ which uses A as "black-box" and which can distinguish $F^\mathcal{O}$ without querying the oracle $\mathcal{O}(break, .)$ at all (this is a contradiction as $F^\mathcal{O}$ is a random function relative to $\mathcal{O}(eval, .)$ alone). The adversary $B^{\mathcal{O}, A}$ on input $Q = \{(X_1, Y_1), \ldots, (X_q, Y_q)\}$ (where the X_i's are random and the Y_i's are either random or $Y_i = F^\mathcal{O}(k, X_i)$ for a random k) runs A on input Q. Here A has no access to the oracle $\mathcal{O}$, but B controls A's oracle gates. B initializes an empty set T, this T will be used to remember the queries made by A. Whenever A requests the output of $\mathcal{O}(eval, .)$ on some input k, x, $B^{\mathcal{O}, A}$ correctly answers with $y = \mathcal{O}(eval, k, x)$ and adds (k, x, y) to T. When A requests the output of $\mathcal{O}(break, .)$ on an input $\{Y, Y'\}$, B guesses the answer itself, and we will show that $B^{\mathcal{O}, A}$ can indeed guess $\mathcal{O}(break, \{Y, Y'\})$ correctly with high probability. We now describe how $B^{\mathcal{O}, A}$ guesses $\mathcal{O}(break, \{Y, Y'\})$.

$B^{\mathcal{O}, A}$ first looks up the sequence $(k_1, x_1, y_1), \ldots, (k_t, x_t, y_t) \in T$ where $Y = y_t$ and for $i = 2, \ldots, t : x_i = y_{i-1}$ (where t is maximal, i.e. $(k, x, x_1) \notin T$ for any k, x). Similarly it looks up the sequence $(k'_1, x'_1, y'_1), \ldots, (k'_{t'}, x'_{t'}, y'_{t'})$ where $y'_{t'} = Y'$. Note that this means that Y and Y' were computed as

$$Y = F^\mathcal{O}_{k_t}(F^\mathcal{O}_{k_{t-1}}(\ldots F^\mathcal{O}_1(x_1)\ldots)) \qquad Y' = F^\mathcal{O}_{k'_{t'}}(F^\mathcal{O}_{k'_{t'-1}}(\ldots F^\mathcal{O}_1(x'_1)\ldots)). \quad (10)$$

Now, if there are i, j, i', j' where $i < j \leq m, j' < i' \leq m$ and $k_i = k'_{i'}, k_j = k'_{j'}$, then $B^{\mathcal{O}, A}$ guesses that $\mathcal{O}(break, \{Y, Y'\})$ is 1 and guesses that it is 0 otherwise.

---

[14] Having chosen plaintext access to a function $T(.)$, we pick some key k and evaluate $C^{\langle 1,2 \rangle, \langle 2,1 \rangle}_{T(.), F^\mathcal{O}(k, .)}$ on some input X to get an output $Y = Y_1 \| Y_2$. As $\{\langle 1, 2 \rangle, \langle 2, 1 \rangle\}$ is ugly, if $T(.)$ is of the form $F^\mathcal{O}(k', .)$, then $\mathcal{O}(break, \{Y_1, Y_2\})$ will be 1, and if $T(.)$ is a URF, then $\mathcal{O}(break, \{Y_1, Y_2\})$ will almost certainly be 0. Thus we can distinguish $F^\mathcal{O}(k, .)$ with random k from a URF.

When the guess is 1, it is always correct by the definition of $\mathcal{O}(break, .)$. So we must show that when the guess is 0 then $\mathcal{O}(break, \{Y, Y'\}) = 1$ is very unlikely.

First we assume that the random $X_1, \ldots, X_q \in Q$ are all in S_0, this will hold but with probability $q/2^n$. Next, we assume that for the case where the Y_i are computed as $F^{\mathcal{O}}(k, X_i)$, A never makes a query $\mathcal{O}(eval, k, X)$ for any X. As k is random this will be true with probability at least $q/2^n$. Now if $B^{\mathcal{O},A}$ wrongly guesses that $\mathcal{O}(break, \{Y, Y'\})$ is 0, then the initial input x_1 or x'_1 from (10) was not in S_0. As x_1 and x'_1 were not received as an output from $\mathcal{O}$ (otherwise we could extend one of the sequences of (10)), A has guessed a value outside of S_0. As S_0 is a random subset which covers all but a $1/2^n$ fraction of possible inputs, the probability that A could have guessed an x_1 outside of S_0 is at most $1/2^n$ (same for x'_1). $\triangle$

The lemma follows from the two claims above.

Acknowledgments

We would like to the thank the Eurocrypt committee for their suggestions.

References

1. Mihir Bellare and Phillip Rogaway. The security of triple encryption and a framework for code-based game-playing proofs. In *Advances in Cryptology — EUROCRYPT '06*, volume 4004 of *LNCS*, pages 409–426. Springer, 2006.
2. Ivan Damgård and Jesper B. Nielsen. Expanding pseudorandom functions; or: From known-plaintext security to chosen-plaintext security. In *Advances in Cryptology — CRYPTO '02*, volume 2442 of *LNCS*, pages 449–464. Springer, 2002.
3. Oded Goldreich, Shafi Goldwasser, and Silvio Micali. How to construct random functions. *J. ACM*, 33(4):792–807, 1986.
4. Johan Håstad, Russell Impagliazzo, Leonid A. Levin, and Michael Luby. A pseudorandom generator from any one-way function. *SIAM J. Comput.*, 28(4):1364–1396, 1999.
5. Russell Impagliazzo and Steven Rudich. Limits on the provable consequences of one-way permutations. In *Proc, 21th ACM Symposium on the Theory of Computing (STOC)*, pages 44–61, 1989.
6. Michael Luby and Charles Rackoff. Pseudo-random permutation generators and cryptographic composition. In *Proc, 18th ACM Symposium on the Theory of Computing (STOC)*, pages 356–363, 1986.
7. Ueli Maurer. Indistinguishability of random systems. In *Advances in Cryptology — EUROCRYPT '02*, volume 2332 of *LNCS*, pages 110–132. Springer, 2002.
8. Ueli Maurer and Krzysztof Pietrzak. Composition of random systems: When two weak make one strong. In *Theory of Cryptograpy — TCC '04*, volume 2951 of *LNCS*, pages 410–427. Springer, 2004.
9. Ueli Maurer, Krzysztof Pietrzak, and Renato Renner. Indistinguishability amplification, 2006. Cryptology ePrint Archive: Report 2006/456, 2006.
10. Ueli Maurer and Johan Sjödin. A fast and key-efficient reduction of chosen-ciphertext to known-plaintext security. In *Advances in Cryptology — EUROCRYPT '07*, LNCS. Springer, 2007. This proceedings.

11. Kazuhiko Minematsu and Yukiyasu Tsunoo. Expanding weak PRF with small key size. In *ICISC '05*, volume 3935 of *LNCS*, pages 284–298. Springer, 2005.
12. Moni Naor and Omer Reingold. From unpredictability to indistinguishability: A simple construction of pseudo-random functions from MACs. In *Advances in Cryptology — CRYPTO '98*, LNCS, pages 267–282. Springer, 1998.
13. Krzysztof Pietrzak and Johan Sjödin. Weak pseudorandom functions in minicrypt, November 2006. Manuscript.

A   Proof of Lemma 2

Proof (of Lemma 2). To show that expansions of type $\mathfrak{G}$ are good, we must show that for any expansion α of type $\mathfrak{G}$, the security of C_F^α as a weak random function can be black-box reduced to the security of F as a weak random function.

Let $\mathbf{Adv}_q^A(F, G)$ denote the advantage of the distinguisher A to distinguish the randomized function F from G in a q query known-plaintext attack. More formally, consider the random variable $Q^F = (X_1, \ldots, X_q, Y_1, \ldots, Y_q)$ where the X_i's are uniformly random and $Y_i = F'(X_i)$ for an instantiation F' of F, then

$$\mathbf{Adv}_q^A(F, G) = \Pr[A(Q^F) \to 1] - \Pr[A(Q^G) \to 1].$$

We prove the following statement:

For any expansion α of type $\mathfrak{G}$, any randomized function F with range and domain $\{0,1\}^n$, there exists an adversary B such that for any adversary A

$$\mathbf{Adv}_{q \cdot \alpha_{max}}^{B^{A,F}}(F, \mathbf{R}_{n,n}) \geq \frac{\mathbf{Adv}_q^A(C_F^\alpha, \mathbf{R}_{n,n \cdot t})}{\#\alpha} - \frac{q^2 \cdot \alpha_{max}^2}{2^n}. \tag{11}$$

Where $\alpha_{max} = \max(\alpha_1, \ldots, \alpha_{\#\alpha})$. Moreover B only uses A and F as a black-box and it is efficient (basically, all that B has to do is to simulate C_F^α on q inputs and it invokes A only once).

So if A breaks the security of C_F^α as a weak RF, then B breaks the security of the underlying F as a weak RF. For the special case of pseudorandom functions, this statement implies that if F is a weak PRF, then so is C_F^α. We now prove (11).

Consider an expansion $\alpha = \{s_1, \ldots, s_t\}$ of type $\mathfrak{G}$. We can assume without loss of generality that for all $s \in \alpha$ and $0 < i < j \leq t$ it holds that $s[i] < s[j]$ (as we can always permute the letters of an α of type $\mathfrak{G}$ so that this holds).

For the proof it will be convenient to introduce a new random system. With $\mathbf{B}_{a,b}$ we denote a random beacon $\{0,1\}^a \to \{0,1\}^b$, this system is simply a random source which outputs a new uniformly random value in $\{0,1\}^b$ on each input. As $\mathbf{B}_{a,b}$ and $\mathbf{R}_{a,b}$ have exactly the same output distribution unless queried twice on the same input, it is easy to show that for any A (e.g. using the framework from section 4 for the second step)

$$\mathbf{Adv}_q^A(\mathbf{B}_{a,b}, \mathbf{R}_{a,b}) \leq \Delta_q^{\mathsf{KPA}}(\mathbf{B}_{a,b}, \mathbf{R}_{a,b}) \leq \frac{q^2}{2^{a+1}}. \tag{12}$$

Let $m := \#\alpha$ and consider the hybrid systems $C_i \stackrel{\text{def}}{=} C^\alpha_{\mathbf{B}_1,\ldots,\mathbf{B}_i,F_{i+1},\ldots,F_m}$, where each $\mathbf{B}_i$ denotes an instantiation of $\mathbf{B}_{n,n}$. As $C_0 \equiv C^\alpha_{F_1,\ldots,F_m}$, $C_m \equiv C^\alpha_{\mathbf{B}_1,\ldots,\mathbf{B}_m} \equiv \mathbf{B}_{n,t\cdot n}$ we have

$$\mathbf{Adv}_q^A(C^\alpha_{F_1,\ldots,F_m}, \mathbf{B}_{n,t\cdot n}) = \sum_{i=1}^m \mathbf{Adv}_q^A(C_{i-1}, C_i). \tag{13}$$

For $i \in [m]$ let $B_i^{A,F}$ be an adversary which on input $(X_1,\ldots,X_{q\alpha_i},Y_1,\ldots,Y_{q\alpha_i})$ simulates the computation of $C^\alpha_{\mathbf{B}_1,\ldots,\mathbf{B}_{i-1},T,F_{i+1},\ldots,F_m}$ (T to be defined) on q random inputs $X_1',\ldots,X_q'$ to get outputs $Y_1',\ldots,Y_q'$, and then outputs the output of $A(X_1',\ldots,X_q',Y_1',\ldots,Y_q')$. In this simulation the component T is only queried on uniformly random inputs[15]. Instead of choosing those inputs at random, we require that $B_i^{A,F}$ uses the values $X_1,X_2,\ldots$ if it has to define the random values which are used as inputs to T. Now, if T is a beacon $\mathbf{B}_{n,n}$, then $C^\alpha_{\mathbf{B}_1,\ldots,\mathbf{B}_{i-1},T,F_{i+1},\ldots,F_m}$ is C_i, and if T is an instance of F then it is C_{i-1}, so

$$\mathbf{Adv}_{q\alpha_i}^{B_i^{A,F}}(F, \mathbf{B}_{n,n}) = \mathbf{Adv}_q^A(C_{i-1}, C_i). \tag{14}$$

Now consider an adversary $B^{A,F}$ which first chooses a random $i \in [m]$ and then runs $B_i^{A,F}$. Using (14) in the second and (13) in the third step, we get:

$$\begin{aligned}
\mathbf{Adv}_{q\cdot\alpha_{max}}^{B^{A,F}}(F, \mathbf{B}_{n,n}) &= \frac{1}{m}\sum_{i=1}^m \mathbf{Adv}_{q\alpha_i}^{B_i^{A,F}}(F, \mathbf{B}_{n,n}) \\
&= \frac{1}{m}\sum_{i=1}^m \mathbf{Adv}_q^A(C_{i-1}, C_i) \\
&= \frac{\mathbf{Adv}_q^A(C^\alpha_{F_1,\ldots,F_m}, \mathbf{B}_{n,t\cdot n})}{m}.
\end{aligned} \tag{15}$$

To conclude the proof of (11) we must "replace" the beacons $\mathbf{B}$ in (15) by URFs $\mathbf{R}$. Below we use the triangle inequality in the first and third, and (15) in the second step. In the last step we use (12) twice.

$$\begin{aligned}
&\mathbf{Adv}_{q\cdot\alpha_{max}}^{B^{A,F}}(F, \mathbf{R}_{n,n}) \\
&\geq \mathbf{Adv}_{q\cdot\alpha_{max}}^{B^{A,F}}(F, \mathbf{B}_{n,n}) - \mathbf{Adv}_{q\cdot\alpha_{max}}^{B^{A,F}}(\mathbf{B}_{n,n}, \mathbf{R}_{n,n}) \\
&= \frac{\mathbf{Adv}_q^A(C^\alpha_{F_1,\ldots,F_m}, \mathbf{B}_{n,t\cdot n})}{m} - \mathbf{Adv}_{q\cdot\alpha_{max}}^{B^{A,F}}(\mathbf{B}_{n,n}, \mathbf{R}_{n,n}) \\
&\geq \frac{\mathbf{Adv}_q^A(C^\alpha_{F_1,\ldots,F_m}, \mathbf{R}_{n,t\cdot n})}{m} - \frac{\mathbf{Adv}_q^A(\mathbf{B}_{n,t\cdot n}, \mathbf{R}_{n,t\cdot n})}{m} - \mathbf{Adv}_{q\cdot\alpha_{max}}^{B^{A,F}}(\mathbf{B}_{n,n}, \mathbf{R}_{n,n}) \\
&\geq \frac{\mathbf{Adv}_q^A(C^\alpha_{F_1,\ldots,F_m}, \mathbf{R}_{n,t\cdot n})}{m} - \underbrace{\frac{q^2}{2^{n+1}\cdot m} - \frac{q^2\cdot\alpha_{max}^2}{2^{n+1}}}_{q^2\cdot\alpha_{max}^2/2^n} \qquad\qquad \square
\end{aligned}$$

---

[15] As $s[i] < s[j]$ if $i < j$, so T is invoked on either the global input or on the output of some $\mathbf{B}_j, j < i$.

Feistel Networks Made Public, and Applications

Yevgeniy Dodis and Prashant Puniya

Department of Computer Science,
Courant Institute of Mathematical Sciences,
New-York University
{dodis,puniya}@cs.nyu.edu

Abstract. Feistel Network, consisting of a repeated application of the Feistel Transform, gives a very convenient and popular method for designing "cryptographically strong" permutations from corresponding "cryptographically strong" functions. Up to now, all usages of the Feistel Network, including the celebrated Luby-Rackoff's result, critically rely on (a) *the (pseudo)randomness of round functions*; and (b) *the secrecy of (at least some of) the intermediate round values* appearing during the Feistel computation. Moreover, a small constant number of Feistel rounds was typically sufficient to guarantee security under assumptions (a) and (b). In this work we consider several natural scenarios where at least one of the above assumptions does not hold, and show that a constant, or even logarithmic number of rounds is *provably insufficient* to handle such applications, implying that a new method of analysis is needed.

On a positive side, we develop a new combinatorial understanding of Feistel networks, which makes them applicable to situations when the round functions are merely *unpredictable* rather than (pseudo)random and/or when the intermediate round values may be leaked to the adversary (either through an attack or because the application *requires* it). In essence, our results show that in any such scenario a super-logarithmic number of Feistel rounds is *necessary and sufficient* to guarantee security.

Of independent interest, our technique yields a novel domain extension method for messages authentication codes and other related primitives, settling a question studied by An and Bellare in CRYPTO 1999.

Keywords: Feistel Network, Verifiable Random Functions/Permutations, PRFs, PRPs, MACs, Domain Extension.

1   Introduction

Feistel Networks are extremely popular tools in designing "cryptographically strong" permutations from corresponding "cryptographically strong" functions. Such networks consist of several iterative applications of a simple Feistel permutation $\Psi_f(x_L \parallel x_R) = x_R \parallel x_L \oplus f(x_R)$, with different (pseudo)independent round functions f used at each round. Among their applications, they are commonly used in the design of popular block ciphers, such as DES, as well as other constructs, such as popular padding schemes OAEP [2] or PSS-R [3]. In particular, the celebrated result of Luby and Rackoff [12] shows that three (resp.

M. Naor (Ed.): EUROCRYPT 2007, LNCS 4515, pp. 534–554, 2007.

four) rounds of the Feistel transform are sufficient to turn a pseudorandom function (PRF) family into a pseudorandom permutation (PRP) family (resp. strong PRP family). There has been a lot of subsequent work (e.g., [20,23,15,22]) on improving various aspects of the Luby-Rackoff's result (referred to as "LR" from now on). However, all these results crucially relied on:

(a) *the (pseudo)randomness of round functions*; and
(b) *the secrecy of (at least some of) the intermediate round values appearing during the Feistel computation*

In this work we consider several natural scenarios where at least one of the above assumptions does not hold, and show that a fundamentally new analysis technique is needed for such applications. But first let us motivate our study.

IS UNPREDICTABILITY ENOUGH?   We start with the assumption regarding pseudorandomness of round functions. This assumption is quite strong, since practical block ciphers certainly do not use PRFs as their round functions. Instead, they heuristically use considerably more than the three-six rounds predicted by the LR and all the subsequent "theoretical justifications". Thus, a large disconnect still remains to be bridged. Clearly, though, we need to assume some security property of the round function, but can a weaker property be enough to guarantee security?

In the context of domain extension of message authentication codes, An and Bellare [1] studied a natural question whether *unpredictability* — a much weaker property than pseudorandomness — can at least guarantee the unpredictability of the resulting Feistel permutation. Although not as strong as pseudorandomness, this will at least guarantee some minimal security of block ciphers (see [7]), is enough for basic message authentication, and anyway doubles the domain of the unpredictable function, which is useful (and non-trivial!) by itself. [1] gave a negative answer for the case of three rounds, and suggested that "even more rounds do not appear to help". This result indicates that previous "LR-type techniques" are insufficient to handle unpredictability (since in the case of PRFs three rounds are enough), and also leaves open the question whether more Feistel rounds will eventually be enough to preserve unpredictability. Our work will completely resolve this question. Along the way, it will prove that Feistel Networks could serve as domain extenders for message authentication codes.

IS IT SAFE TO LEAK INTERMEDIATE RESULTS?   Another crucial reason for the validity of the LR result is the fact that all the intermediate round values are never leaked to the attacker. In fact, the *key* to the argument, and most of the subsequent results, is that the attacker effectively gets no information about most of these values in case a PRF is used for the round function, and simple attacks (which we later generalize to many more rounds) are possible to invalidate the LR result in case the intermediate values are leaked. Unfortunately, for many natural applications this assumption (or conclusion!) can not be enforced, and a totally new argument is needed. We give several examples.

Starting with the simplest (but also least interesting) example, intermediate values might be inadvertently leaked through an attack. For example, one might imagine a smartcard implementing a block cipher via the Feistel network using a secure chip implementing a PRF. In this case the attacker might be able to observe the communication between the smartcard and the chip, although it is unable to break the security of the chip. More realistically, when the round functions are not PRFs, the attacker might get a lot of information about the intermediate values anyway, even without extra attack capabilities. For example, in the case of unpredictable functions (UFs) mentioned above, we will construct provably secure UFs such that the output of the Feistel Network completely leaks *all* the intermediate round values. Although artificial, this example illustrates that weaker assumptions on the round functions can no longer guarantee the secrecy of intermediate values.

For yet another example, the round function might simply be public to begin with. This happens when one considers the question of implementing an ideal cipher from a random oracle, considered by the authors in TCC'06 [6]. In this case the round function is a publicly accessible random oracle, and is certainly freely available to the attacker. To see the difference with the usual block cipher setting where four round are enough, [6] showed that even five Feistel rounds are not sufficient to built an ideal cipher, although conjectured that a larger constant number of rounds is sufficient. The authors also showed a weaker positive implication in the so called "honest-but-curious model", although only for a super-logarithmic number of rounds (as they also showed, reducing the number of rounds in this model would imply the security in the usual, "malicious" model). As a final example (not considered in prior work), the attacker might get hold of the intermediate values because the *application requires to reveal such values*. This happens when one tries to add *verifiability* to PRFs and PRPs (or their unpredictable analogs), which we now describe in more detail.

VERIFIABLE RANDOM FUNCTIONS AND PERMUTATIONS. We consider the problem of constructing *verifiable random permutations* (VRPs) from *verifiable random functions* (VRFs). VRFs and VRPs are verifiable analogs of PRFs and PRPs, respectively. Let us concentrate on VRFs first. Intuitively, regular PRFs have a limitation that one must trust the owner of the secret key that a given PRF value is correctly computed. And even when done so, a party receiving a correct PRF value cannot later convince some other party that the value is indeed correct (i.e., PRF values are "non-transferable"). In fact, since the function values are supposed to be (pseudo)random, it seems that such verifiability of outputs of a PRP would contradict its pseudorandomness. The way out of this contradiction was provided by Micali, Rabin and Vadhan [17], who introduced the notion of a VRF. Unlike PRFs, a VRF owner must be able to provide a short proof that any given VRF output is computed correctly. This implies that the VRF owner must publish a public key allowing others to verify the validity of such proofs. However, every "unopened" VRF value (i.e., one for which no proof was given yet) should still look indistinguishable from random, even if many other values were "opened" (by giving their proofs). Additionally, the public key should commit the owner of the VRF to all its function values in a unique way, even if the owner tries to select an "improper" public key. Micali et al. [17]

also gave a secure construction of a VRF based on the RSA assumption. Since then, several more efficient constructions of VRFs have been proposed based on various cryptographic assumptions; see [13,5,8].

The notion of a VRP, which we introduce in this paper, naturally adds verifiability to PRPs, in exactly the same natural way as VRFs do to PRFs. We will describe some applications of VRPs later (and more in [7]), but here let us concentrate on the relation between VRFs and VRPs. On the one hand, it is easy to see that a VRP (on a "non-trivial domain") is also a VRF, just like in the PRF/PRP case. On a first look, we might hope that the converse implication holds as well, by simply applying the Luby-Rackoff result to VRFs in place of PRFs. However, a moment of reflection shows that this is not the case. Indeed, the proof for the iterated Feistel construction *must include all the VRF values for the intermediate rounds*, together with their proofs. Thus, the attacker can legally obtain all the intermediate round values for every input/output that he queries, except for the one on which he is being "challenged". This rules out the LR-type proof for this application. More critically, even the recent proof of [6] (implementing the ideal cipher from a random oracle in the "honest-but-curious" model) appears to be "fundamentally inapplicable" as well. Indeed, that proof crucially used the fact that truly random functions (in fact, random oracles) are used in all the intermediate rounds: for example, to derive various birthday bounds used to argue that certain "undesirable" events are unlikely to happen. One might then hope that a similar argument might be carried out by replacing all the VRFs by truly random function as well. However, such "wishful replacement" is prevented by the fact that we are required to prove the correctness of each intermediate round value, and we (provably) *cannot provide such proofs when we use a totally random function in place of a VRF* (which is "committed" to by its public key). To put it differently, with a random function we have no hope of simulating the VRF proofs that are "legally expected" by an adversary attacking the VRP construction. Thus, again, a new technique is needed.

VERIFIABLE UNPREDICTABLE FUNCTIONS AND PERMUTATIONS. We also consider the natural combination of the scenarios we considered so far, exemplified by the task of constructing *verifiable unpredictable permutations* (VUPs) from *verifiable unpredictable functions* (VUFs) [17] (also called *unique signature schemes* [11,13]). A VUF is defined in essentially the same way as VRFs, except that the pseudorandomness requirement for VRFs is replaced by a weaker unpredictability requirement. Similarly, VUPs, introduced in this paper, are either the permutation analogs of VUFs, or, alternatively, unpredictable analogs of VRPs. Of course, as a VRP is also a VUP, we could attempt to build a VUP by actually building a VRP via the Feistel construction applied to a VRF, as suggested in the previous paragraph. However, this seems quite wasteful since VUFs appear to be much easier to construct than VRFs. Indeed, although in theory VUFs are equivalent to VRFs [17], the "Goldreich-Levin-type" reduction from VUFs to VRFs in [17] is *extremely* inefficient (it loses exponential security and forces the authors to combine it with another inefficient tree construction). Moreover, several previous papers [17,13] constructed *efficient* VUFs based on relatively standard *computational* assumptions, while all the *efficient* VRF constructions [5,8] are based on very ad hoc *decisional* assumptions. Thus, it is natural to study the security of the Feistel network when applied to VUFs. In this case, not only the round functions cannot be assumed pseudorandom, but also all the intermediate values must be leaked together with their proofs of correctness, making this setting the most challenging to analyze.

OTHER RELATED WORK. Several prior works tried to relax the security of some of the round functions. For example, Naor and Reingold showed that the first and the fourth round could use pairwise independent hash functions instead of PRFs. In a different vein, Maurer et al. [14] studied the case when the PRFs used are only non-adaptively secure. Already in this setting, the authors showed that it is unlikely that four Feistel rounds would yield a PRP (although this is true in the so called "information-theoretic" setting). However, in these results at least some of the round functions are still assumed random. In terms of leaking intermediate results, Reyzin and Ramzan showed that in a four-round construction it is safe to give the attacker *oracle access* to the second and third (but not first and fourth) round functions. This is incomparable to our setting: we leak intermediate results actually happening during the Feistel computation, and for *all* the rounds. Finally, we already mentioned the paper by the authors [6], which showed how to deal with public intermediate results when *truly random* round functions are used. As we argued, however, this technique is insufficient to deal with unpredictability, and cannot even be applied to the case of VRFs (because one cannot simulate the proofs of correctness for a truly random function).

1.1   Our Results

In this work we develop a new understanding of the Feistel Network which allows us to analyze the situations when the intermediate round values may be leaked to the adversary, and also handle cases when the round values are merely unpredictable rather than pseudorandom. In our modeling, a k-round Feistel Network is applied to k members $f_1 \ldots f_k$ independently selected from some (not necessarily pseudorandom) function family C, resulting in a Feistel permutation π. Whenever an attacker makes a forward (resp. backward) query to π (resp. π^{-1}), we assume that it learns all the intermediate values (as we mentioned, this is either required by the application, or may anyway happen with unpredictable functions).

NEGATIVE RESULT. As our first result, we show a simple attack allowing an adversary to compute any value $\pi^{-1}(y)$ by making at most exponential in k number of *forward* queries to π. Since such an inversion should be unlikely (with polynomially many queries) even for an unpredictable permutation, this immediately means that at least a superlogarithmic number of Feistel rounds (in the security parameter λ) is *necessary* to guarantee security for *any* of the applications we consider. Aside from showing the *tightness of all our positive results* described below, this result partially explains *why practical block ciphers use significantly more than 3-6 rounds* predicted by all the previous "theoretical justifications" of the Feistel Network. Indeed, since all such ciphers heuristically use round functions which are not PRFs, and we just showed that even unpredictable round functions might leak a lot (or even *all*) of the intermediate results, the simple attack we present might have been quite applicable if a small constant number of rounds was used!

MATCHING POSITIVE RESULT. On a positive side, we show a general combinatorial property of the Feistel Network which makes essentially no assumptions (such as pseudorandomness) about the round functions used in the Feistel construction, and allows us to apply it to a wide variety of situations described above, where the previous techniques (including that of [6]) failed. In essence, for any $s \leq k/2$, we show that if an attacker, making a sub-exponential in s number of (forward or backward) queries to the construction and always learning all the intermediate round values, can cause a non-trivial collision somewhere between rounds s and $k - s$, then the attacker can also find a simple (and nontrivial) XOR condition on a constant (up to six) number of the round values of the queries he has made. This means that if a function family C is such that it is provably hard for an efficient attacker to find such a non-trivial XOR condition, — and we call such families *5-XOR resistant* (see Section 4), — then it is very unlikely that the attacker can cause any collisions between rounds s and $k - s$ (as long as s, and thus k, are super-logarithmic in the security parameter λ). And once no such collisions are possible, we show that is possible to directly argue the security of the Feistel Network for our applications. In particular, as even mere unpredictability is enough to establish 5-XOR resistance, we conclude that super-logarithmic number of Feistel rounds is *necessary and sufficient* to yield

- a (strong) unpredictable permutation (UP) from any unpredictable function (UF).
- a strong PRP from any PRF, which remains secure even if all the round values are made public.
- a strong VUP from any VUF.
- a strong VRP from any VRF.

These results are in sharp contrast with the "LR-type" results where a constant number of rounds was sufficient, but also give the first theoretical justification regarding the usage of Feistel Networks not satisfying assumptions (a) or (b) mentioned earlier. For the case of block ciphers, our justification seems to match more closely the number of rounds heuristically used in practical constructions.

IMPLICATIONS TO DOMAIN EXTENSION. Since the Feistel Network doubles the length of its input, our results could also be viewed in relation to the question of domain extension of UFs, VUFs and VRFs. In practice, the question of domain extension is typically handled by a collision-resistant hash function (CRHF): it uses only one call the the underlying n-bit primitive f and does not require the secret key to grow. However, the existence of a CRHF is a theoretically strong assumption, which does not seem to follow from the mere existence of UFs, VRFs or VUFs. This is especially true for UFs, whose existence follows from the existence of mere one-way functions and, hence, can even be "black-box separated" from CRHFs [24]. Thus, it makes sense to consider the question of domain extension *without introducing new assumptions.*

For PRFs, this question is easily solved by using (almost) universal hash functions (instead of CRHFs) to hash the message to n bits before applying the n-bit PRF. However, this technique fails for UFs, VUFs and VRFs: in the case of unpredictability because the output reveals information about the hash key, and for VRFs because it is unclear how to provide proofs of correctness without revealing the hash key. Another attempt (which works for digital signatures) is to use target collision-resistant hash functions [21] in place of CRHFs, but such functions have to be freshly chosen for each new input, which will break the unique provability of UFs, VUFs and VRFs. (Additionally, the hash key should also be authenticated, which further decreases the bandwidth.) In case the underlying n-bit primitive f is shrinking (say, to $n - a$ bits), one can use some variant of the cascade (or Merkle-Damgård) construction. Indeed, this was formally analyzed for MACs by [1,16]. However, the cost of this method is one evaluation of f per a input bits. In particular, in case the output of f is also equal to n, which is natural if one wants to extend the domain of a UF given by a block cipher, this method is either inapplicable or very inefficient.[1]

In contrast, our method builds a UF/VUF/VRF from $2n$ to $2n$ bits from the one from n to n bits, by using $k = \omega(\log \lambda)$ evaluations of f, albeit also at the price of increasing the secret key by the same amount. This answers the question left open by An and Bellare [1] (who only showed that three rounds are insufficient): *Feistel Network is a good domain extender for MACs if and only if it uses super-logarithmic number of rounds!*

Moreover, in the context of UFs (and VUFs), where one wants to minimize the output length as well, we notice that the output length can be easily reduced from $2n$ to n. This is done by simply dropping the "left half" of the k-round Feistel network output! The justification for this optimization follows by noticing that in this case the attacker will only make forward queries to the Feistel construction. For such attackers, we can extend our main combinatorial lemma as follows. For any $s \leq k$, if a 5-XOR resistant family is used to implement the round functions and the attacker made less than exponential in s number of queries, then the attacker has a negligible chance to cause any collisions between rounds s and k (as opposed to $k - s$ we had when backward queries were allowed). From this, one can derive that $k = \omega(\log \lambda)$ Feistel rounds is enough to turn a UF (or VUF) from n to n bits into one from $2n$ to n bits. Moreover, in the case of UFs we expect that one would use a (possibly heuristic) pseudorandom generator to derive the k round keys (much like in the case of block ciphers), meaning that the only effective cost is k computations of the basic UF. Once the domain is doubled, however, one can use the cascade methods [1,16] to increase it further without increasing the key or the output length.

---

[1] In principle, such length-preserving f can be "truncated" by a bits, but this loses an exponential factor in a in terms of exact security. Thus, to double the input length, one would have to evaluate f at least $\Omega(n/\log \lambda)$ times.

OTHER APPLICATIONS. In the full version [7], we illustrate several applications of our results. We describe only a couple here due to the space constraints.

As a simple, but illustrative application, we notice that VRPs immediately yield *non-interactive, setup-free, perfectly-binding commitments schemes*. The sender chooses a random key pair (SK, PK) for a VRP π. To commit to m (in the domain of the VRP), the sender sends PK and the value $c = \pi_{SK}(m)$ to the receiver. To open m, the sender sends m and the proof that $c = \pi_{SK}(m)$, which the receiver can check using the public key PK. The hiding property of this construction trivially follows for the security of VRPs. As for binding, it follows from the fact that π is a permutation even for an *adversarial choice of PK*. As we can see, it is not clear how to achieve binding *directly* using plain VRFs. However, given our (non-trivial) equivalence between VRFs and VRPs, we get that VRFs are also sufficient for building non-interactive, perfectly binding commitment schemes without setup. Alternatively, to commit to a single bit b, one can use VUPs augmented with the Goldreich-Levin bit [10]. Here the sender would pick a random r and x, and send PK, r, $\pi_{SK}(x)$, and $(x \cdot r) \oplus b$, where $x \cdot r$ denotes the inner product modulo 2. Using our equivalence between VUPs and VUFs, we see that VUFs are sufficient as well.

We remark that the best general constructions of such commitments schemes was previously based on one-way permutations (using the hardcore bit) [4], since Naor's construction from one-way functions [19] is either interactive, or non-setup-free. Since the assumption of one-way permutations is incompatible with VUFs or VRFs, our new construction is not implied by prior work.

Micali and Rivest [18] suggested the following elegant way to perform *non-interactive lottery* (with the main application in micropayments). The merchant publishes a public key PK for a VRF f, the user chooses a ticket x, and wins if some predicate about $f(x)$ is true (for example, if $f(x)$ is less than some threshold t). Since f looks random to the user, the user cannot significantly bias his odds no matter what x he chooses. Similarly, since the merchant is committed to f by the public key PK, they merchant cannot lie about the value $f(x)$. However even in this case, nothing stops the merchant from publishing a "non-balanced" VRF (meaning choosing a specific f such that $f(x)$ is "far from random" even for *random* x). In the extreme case, a constant function $f(x) = c$, where c is selected so that the predicate does not hold. We need "balancedness" to ensure that the merchant not only cannot change the value of f after the commitment, but also that the user has a fair chance of winning when he chooses a *random* x, no matter which f the merchant selects. VRPs perfectly solve this problem.

Moreover, VRPs have an extra advantage that one can *precisely* know the number of possible winners: it is exactly equal to the number of strings y satisfying the given predicate. Thus, one can always allocate a given number of prizes and never worry that with some small probability there will be more winners than prizes.

We briefly mention some other applications described in [7]. For example, UPs are enough to argue weaker "fall-back" security properties for some applications of block ciphers, which is nice in case the PRP assumption on the block cipher turns out incorrect. VRPs, or sometimes even VUPs, can be useful in several applications where plain VRFs are insufficient. For example, to implement so called "invariant signatures" needed by Goldwasser and Ostrovsky [11] in constructing non-interactive zero-knowledge proofs. Additionally, VRPs could be useful for adding verifiability to some application of PRPs (where, again, PRFs are not sufficient). For example, to construct verifiable CBC encryption or decryption, or to "truthfully", yet efficiently, sample certain verifiable huge (pseudo)random objects [9], such as random constant-degree expanders. Finally, our construction of VRPs from VRFs could lead to a "proof-transferable" implementation of the Ideal Cipher Model using a semi-trusted third party. We refer to [7] for more details, and hope that more applications of our constructs and techniques will be found.

2   Definitions and Preliminaries

Let λ denote the security parameter. We use $negl(\lambda)$ to denote a negligible function of λ. Fibonacci(k) denotes the kth Fibonacci number, and thus Fibonacci(k) $= \mathcal{O}(1.618^k)$.

Now we give informal definitions of the various primitives that we use in this paper. For formal definitions, see full version [7]. We start by defining the notion of *pseudorandom functions* (PRFs). We use a slightly non-standard definition of PRFs that is convenient to prove our results. However, this definition is equivalent to the usual definition.

In the new PRF attack game, the attacker A_f runs in three stages: (1) In the experimentation phase, it is allowed to query a PRF sampled from the PRF family. (2) In the challenge phase, it sends an unqueried PRF query and in response the challenger sends either the PRF output or a random output with equal probability. (3) In the analysis phase, the attacker again gets oracle access to the PRF, but cannot query it on the challenge query. At the end of the attack, A_f has to guess if the challenge response was random or pseudorandom. The attacker A_f wins if it guesses correctly. Similar to the notion of PRFs, we can define the notion of *(strong) pseudorandom permutations* (PRPs). Here the attacker has oracle access to both the forward as well as inverse PRP, but the attack game is otherwise similar to that for PRFs.

A slightly weaker notion than PRFs is that of *Unpredictable Functions* (UFs). Unpredictable functions are also popularly known as (deterministic) *Message Authentication Codes* (MACs). In this case, the UF attacker is allowed to query an unpredictable function from the UF family, and it needs to predict the output of the UF on an unqueried input at the end of the interaction. The advantage of the UF adversary is the maximum probability with which it predicts correctly. In an analogous fashion, we can also define the notion of *Unpredictable Permutations* (UPs), where the attacker has oracle access to both the forward and inverse permutation and has to predict an unqueried input/output pair.

We can define verifiable analogs of each of the above primitives. Thus, we get *verifiable random functions*, *verifiable random permutations*, *verifiable unpredictable functions* and *verifiable unpredictable permutations*. In each case, the primitive takes a public/private key pair, and consists of three algorithms (Gen, Prove, Verify). The Gen algorithm outputs a public/private key pair. The Prove algorithm allows the private key owner to compute the function/permutation output as well as give a proof of correctness. Finally, the Verify algorithm allows anyone who knows the public key to verify the correctness of an input/output pair by observing the corresponding proof.

Each of these primitives satisfies two properties: (1) *Correctness*, i.e. one can verify correct input/output pairs, and (2) *Soundness*, i.e. one cannot prove two distinct outputs for the same input, even for an *adversarially chosen public key*. Additionally, these primitives satisfy the natural analogs of the pseudorandomness/unpredictability definition of the corresponding non-verifiable primitive (except the attacker also gets the proofs for all the values except for the challenge).

The *Feistel transformation* using $f : \{0,1\}^n \to \{0,1\}^n$ is a permutation Ψ_f on $2n$ bits defined as, $\Psi_f(x) \stackrel{def}{=} x_R \parallel x_L \oplus f(x_R)$. The symbols x_L and x_R denote the left and right halves of $2n$ bit string x. We will often call the construction based on k iterated applications of the Feistel transformation, a k-round LR construction, and denote it by $\Psi_{f_1 \ldots f_k}$ (or Ψ_k when $f_1 \ldots f_k$ are clear from context) where $f_1 \ldots f_k$ are the *round functions* used. If the input to Ψ_k is $x = R_0 \| R_1$, for $R_0, R_1 \in \{0,1\}^n$, then the k-round LR construction Ψ_k generates k more n-bit values $R_2 \ldots R_{k+1}$ (one after each application of a round function, i.e. $R_i = f_{i-1}(R_{i-1}) \oplus R_{i-2}$ for $i = 2 \ldots (k+1)$). We will refer to the n-bit values $R_0, R_1 \ldots R_k, R_{k+1}$ as the *round values* of the LR construction.

3   Insecurity of $\mathcal{O}(\log \lambda)$-Round Feistel

We will demonstrate here that upto a logarithmic number of Feistel rounds do not suffice for any of our results. In order to make our proof precise, we show a simple adversary that is able to find the input corresponding to any permutation output $y \in \{0,1\}^{2n}$ by making polynomially many *forward* queries and observing the intermediate round values.

Theorem 1. *For the k round Feistel construction Ψ_k that uses $k = \mathcal{O}(\log \lambda)$ round functions, there exists a probabilistic polynomial time adversary A_π that takes oracle access to Ψ_k (while also gets access to the intermediate round values of Ψ_k). The adversary A_π makes $\mathcal{O}(\mathsf{Fibonacci}(k)) = poly(\lambda)$ forward queries to Ψ_k and with high probability finds the input corresponding to an output y without actually making that query.*

Proof: The adversary A_π starts by choosing a permutation output y, that it will try to invert Ψ_k on. For concreteness, we assume that $y = 0^{2n}$ (anything else works just as well). We will describe the recursive subroutine that the attacker A_π is based on. Say the round functions of Ψ_k are $f_1 \ldots f_k$. The recursive function that we describe is $E(j, Y)$, where j is the number of rounds in the Feistel construction and Y is a $2n$ bit value, and the task of $E(j, Y)$ is to find the input such that the jth and $(j+1)^{th}$ round values are Y_L and Y_R (the left and right halves of Y), respectively.

- $\mathbf{E(1, Y)}$: Choose a random $R'_0 \leftarrow \{0,1\}^n$. Make the forward query $R'_0 \parallel Y_L$ to Ψ_1, where the 2nd round value is R'_2. Now the 1st and 2nd round values for the input $R'_2 \oplus R'_0 \oplus Y_R \parallel Y_L$ are Y_L and Y_R.
- $\mathbf{E(j, Y)}$, $\mathbf{j > 1}$: Perform the following steps,
    • Make a random query $R_0 \parallel R_1 \leftarrow \{0,1\}^{2n}$, and say the $2n$ bit value at the jth round is is $R_j \parallel R_{j+1}$. Then, $f_j(R_j) = (R_{j-1} \oplus R_{j+1})$.
    • Run $E(j-2, (f_{j-1}(R_{j-1}) \oplus Y_L) \parallel R_{j-1})$ and the $2n$ bit value at the $(j-1)^{th}$ round is $R_{j-1} \parallel Y_L$. Hence $f_j(Y_L) = R_{j-1} \oplus R_{j+1}$.
    • Run $E((j-1), (f_j(Y_L) \oplus Y_R) \parallel Y_L)$, and the jth and $(j+1)^{th}$ round values are Y_L and Y_R, respectively.

The adversary A_π essentially runs the algorithm $E(k, 0^{2n})$. Now we need to make sure that the adversary A_π does not query on the input corresponding to the output 0^{2n}. But since all the queries made in the recursive algorithm are essentially chosen at random, we know that the probability of this happening is $\frac{q}{2^{2n}}$. Hence, the probability that A_π succeeds is at least $\left(1 - \frac{q}{2^{2n}}\right)$.   ▢

We note that the above attacker works in a scenario where it can only make forward queries to the Feistel construction Ψ_k. In case it can make inverse queries as well, it is possible to design a similar attacker that succeeds in $\mathcal{O}(\mathsf{Fibonacci}(k/2))$ queries. If the number of rounds $k = \mathcal{O}(\log \lambda)$, then the number of queries needed by either of these attackers is polynomial in the security parameter λ.

It is easy to see how such an attacker can be utilized in three of the four scenarios, if we use the Feistel construction for each of these cases.

- *PRP construction with public round values*: By definition, for a PRP we should not be able to invert an output without actually querying the construction on it.
- *VRP (VUP) construction using VRFs (VUFs)*: In order to provide the proofs for the VRP (VUP), the VRP (VUP) construction will need to reveal all intermediate VRF (VUF) inputs/outputs and the corresponding proofs.

On the first look, it seems that when we use a Feistel construction with *unpredictable functions* in each round to construct an *unpredictable permutation* (UP), the UP adversary cannot make use of the above attacker since it does not have access to all the intermediate round values. However, we will show that if certain pathological (but secure) unpredictable functions are used as round functions, then the UP adversary can infer *all* the round values simply by observing the output of the Feistel construction!

Lemma 1. *For any $k \leq \frac{n}{\omega(\log \lambda)}$ (in particular, if $k = \mathcal{O}(\log \lambda)$), there exist k secure unpredictable functions $f_1 \ldots f_k$, such that by querying the k-round Feistel construction $\Psi_{f_1 \ldots f_k}$ on any input, an attacker can always efficiently learn all the intermediate round values (even when it does not have access to the intermediate round values).*

Proof: Let $\{g_i : \{0,1\}^n \to \{0,1\}^{n/k}\}_{i \in \{1...k\}}$ be k secure unpredictable functions. For $i \in \{1, k\}$, we will define the functions $f_i : \{0,1\}^n \to \{0,1\}^n$ as $f_i(x) = 0^{(i-2)\cdot(n/k)} \| x_{i-1} \| g_i(x) \| 0^{(k-i)\cdot(n/k)}$, where x_{i-1} denotes the $(i-1)^{th}$ (n/k) bit block in the input x. Each of the functions f_i is a secure unpredictable function if the corresponding function g_i is a secure UF.

Consider a query $(R_0 \| R_1) \in \{0,1\}^{2n}$ made to the Feistel construction $\Psi_{f_1...f_k}$. Now we will consider both R_0 and R_1 as consisting of k blocks of length (n/k) each, which we will denote by $R_0 = R_0^1 \| \ldots \| R_0^k$ and $R_1 = R_1^1 \| \ldots \| R_1^k$. Denote the round values generated in computing the output of this construction as $(R_0, R_1) \ldots (R_k, R_{k+1})$, where $R_k \| R_{k+1}$ is the output of this construction. If the number of rounds k in the Feistel construction is even, then we note that the output of the construction is:

$$R_k = (g_1(R_1) \oplus R_0^1 \oplus R_1^1) \| \ldots \| (g_{k-1}(R_{k-1}) \oplus R_0^{k-1}) \| R_0^k$$
$$R_{k+1} = (g_1(R_1) \oplus R_0^1 \oplus R_1^1) \| \ldots \| (g_k(R_k) \oplus R_1^k)$$

If number of rounds k is odd, then the output of the Feistel construction is,

$$R_k = (g_1(R_1) \oplus R_0^1 \oplus R_1^1) \| \ldots \| (g_{k-1}(R_{k-1}) \oplus R_1^{k-1}) \| R_1^k$$
$$R_{k+1} = (g_1(R_1) \oplus R_0^1 \oplus R_1^1) \| \ldots \| (g_k(R_k) \oplus R_0^k)$$

Now it is easy to find each of the round function outputs (and hence the intermediate round values) by simply observing the right half of the output of the Feistel construction. $\qquad\qquad\boxed{}$

Thus, we see that if the number of rounds in the Feistel construction (using UFs) used to construct *unpredictable permutations* is $k = \mathcal{O}(\log \lambda)$, then the resulting construction is insecure (since all the intermediate round values may be visible and we can apply theorem 1). Even if we attempt to shrink the output length of this MAC construction by chopping the left half of the output, it would be possible to retrieve all intermediate round values by simply observing the MAC output. In fact, even for $k = \omega(\log \lambda)$ (but less than $n/\omega(\log \lambda)$) rounds it might be possible to retrieve all intermediate round values, and hence a new proof technique is needed.

4 A Combinatorial Property of the Feistel Construction

In this section, we will prove a general combinatorial lemma about the k round LR-construction Ψ_k, that uses arbitrary round functions $f_1 \ldots f_k$. We will see in the following section that this lemma is crucial in deriving each of our results using the Feistel construction.

Consider an arbitrary ordered sequence of q forward/inverse permutation queries made to the construction Ψ_k, each of which is a $2n$ bit string. Denote the $(k+2)$ n-bit round values associated with the ith query as $R_0^i, R_1^i \ldots R_k^i, R_{k+1}^i$, where $R_0^i \| R_1^i$ (resp. $R_k^i \| R_{k+1}^i$) is the input if this is a forward (resp. inverse) query. We say that such a sequence of queries produces an sth round value collision, if the sth round value collides for two different permutation queries from this query sequence. That is, we have that $R_s^i = R_s^j$ for $i, j \in \{1 \ldots q\}$ and $R_0^i \parallel R_1^i \neq R_0^j \parallel R_1^j$.

We essentially show that if any such sequence of q queries produces a rth round value collision for any $r \in \{s \ldots (k-s)\}$ (where $s \leq (k/2)$), then one of the following must hold:

1. The number of queries q is exponential in s.
2. For this sequence of queries, there is at least one new round function evaluation such that the new round value generated can be represented as a bit-by-bit XOR of upto 5 previously existing round values.

We refer to the second condition above as the 5-*XOR condition*. We label the queries in the order they are made, i.e. query i is made before query $i + 1$ for $i = 1 \ldots q - 1$. By a "new round function evaluation", we mean when a round function is evaluated on an input (i.e. the corresponding round value) to which it was not applied in an earlier query. If the ith query is a forward (inverse) query and the round function evaluation $f_j(R_j^i)$ is a new one, then the new round value generated as a result is R_{j+1}^i (resp. R_{j-1}^i). The 5-*XOR condition* essentially states that for at least one such new round function evaluation, the new round value generated can be represented as the bit-by-bit XOR of upto 5 previously existing round values. Here previously existing round values include round values from previous queries and round values in the same query that were generated earlier (depending on whether this is a forward/inverse query). Our combinatorial result is formalized in the main lemma below (where, for future convenience, we denote R_j^i by $R[i, j]$).

Lemma 2. *Let Ψ_k be a k round LR construction that uses fixed and arbitrary round functions $f_1 \ldots f_k$. For any $s \leq \frac{k}{2}$, and any ordered sequence of $q = o(1.3803^{\frac{s}{2}})$ forward/inverse queries, with associated round values $R[i, 0], \ldots, R[i, k + 1]$ for $i = 1 \ldots q$, if the 5-XOR condition does not hold for this sequence of queries then there is no r^{th} round value collision for these queries, for all $r \in \{s \ldots (k - s)\}$.*

Note that lemma 2 simply states a structural property of the k-round LR construction that holds irrespective of the round functions used in the construction. The proof of this lemma is quite technical and we omit it here due to space constraints (see [7]).

Next, we state a more restricted version of the combinatorial lemma, when the adversary only makes forward queries to the Feistel construction. This lemma (whose proof can also be found in [7]) will be useful when we attempt *domain extension of MACs* in the next section. We give an intuition of the proof of this lemma, which is quite similar to the proof of Lemma 2 (though slightly simpler).

Lemma 3. *Let Ψ_k be a k-round LR construction that uses fixed and arbitrary round functions $f_1 \ldots f_k$. For any round number s, and any ordered sequence of $q = o(1.3803^{\frac{s}{2}})$ forward queries, with associated round values $R[i, 0], \ldots, R[i, k + 1]$ for $i = 1 \ldots q$, if the 5-XOR condition does not hold for this sequence of forward queries then there is no r^{th} round value collision for these queries, for all $r \geq s$.*

Proof Intuition: Consider a sequence of q queries for which the rth round values of two queries collide, while the 5-XOR condition does not hold. Without loss of generality, we can assume that one of queries involved in the rth round value collision is the last one (i.e. the qth query) [2]. We will label the queries $1 \cdots q$, in the order in which they were made. Thus for the round value $R[i, j]$, all the round values $R[i', j']$, with $(i' < i)$ or $(i' = i) \wedge (j' < j)$, were generated before $R[i, j]$. We denote by $\mathsf{p}(i, j)$, the least query number such that $R[\mathsf{p}(i, j), j] = R[i, j]$.

Our main argument consists of four steps which all rely on the fact that the 5-XOR condition does not hold. We start by showing that if the round value $R[q, r]$ collides with the rth round value in an earlier query, then all the round values $R[q, 1] \ldots R[q, (r - 1)]$ collide with corresponding round values in earlier queries as well. That is,

$$(\mathsf{p}(q, r) < q) \Rightarrow (\mathsf{p}(q, 1) < q) \wedge \ldots \wedge (\mathsf{p}(q, (r - 1)), (r - 1))$$

In order to see this, consider the round value $R[q, (r - 1)]$. We know that $(f_{r-1}(R[q, (r - 1)]) = R[q, (r - 2)] \oplus R[q, r])$. Now since both the round values $R[q, (r - 2)]$ and $R[q, r]$ were generated before $R[q, (r - 1)]$ (the former because this is a forward query and the latter because $\mathsf{p}(q, r) < q$), it must be the case that $\mathsf{p}(q, (r - 1)) < q$ since otherwise the 5-XOR condition will be satisfied. Now we can apply the same argument to the round values $R[q, (r - 2)]$ down to $R[q, 1]$ to get the desired result. Moreover, we can also show that these queries $\mathsf{p}(q, 1) \ldots \mathsf{p}(q, r)$ could only have been made in certain restricted orders. In particular, we show that there is a $j \in \{1 \ldots r\}$ such that

$$\mathsf{p}(q, 1) > \ldots > \mathsf{p}(q, j) < \ldots < \mathsf{p}(q, r)$$

In order to see this consider any three consecutive round values $R[q, (i-1)], R[q, i]$ and $R[q, (i + 1)]$, corresponding to queries $\mathsf{p}(q, (i - 1)), \mathsf{p}(q, i)$ and $\mathsf{p}(q, (i + 1))$. We know that $f_i(R[\mathsf{p}(q, i), i]) = R[\mathsf{p}(q, (i - 1)), (i - 1)] \oplus R[\mathsf{p}(q, (i + 1)), (i + 1)]$. If it were the case that $\mathsf{p}(q, (i - 1)) < \mathsf{p}(q, i)$ and $\mathsf{p}(q, (i + 1)) < \mathsf{p}(q, i)$, then this would imply a 5-XOR condition. The only orders that do not have such a query triple are the ones specified above.

Now we can deduce that at least one of these two strictly descending/ascending query sequence, i.e. $\mathsf{p}(q, 1) > \ldots > \mathsf{p}(q, j)$ or $\mathsf{p}(q, j) < \ldots < \mathsf{p}(q, r)$, consists of at least $(r/2)$ queries. Without loss of generality, as the longer sequence of queries is $\mathsf{p}(q, 1) > \ldots > \mathsf{p}(q, j)$. We consider any of the queries $\mathsf{p}(q, \ell)$ for $\ell \in \{1 \ldots (j-2)\}$, and show that each of the round values $R[\mathsf{p}(q, \ell), 1] \ldots R[\mathsf{p}(q, \ell), (\ell-1)]$ collide with the corresponding round value in an earlier query. The first step of this argument, i.e. showing that $R[\mathsf{p}(q, \ell), (\ell - 1)]$ collides with the corresponding round value in an earlier query, is the tricky step in this part, beyond which the argument is similar to the first step. Thus, we show that

$$(\mathsf{p}(\mathsf{p}(q, \ell), 1) < \mathsf{p}(q, \ell)) \wedge \ldots \wedge (\mathsf{p}(\mathsf{p}(q, \ell), (\ell - 1)) < \mathsf{p}(q, \ell))$$

---

[2] In addition, we assume that the query sequence does not consist of any duplicate queries.

Next, we show that the queries $p(p(q, \ell), 1) \ldots p(p(q, \ell), (\ell - 1))$ occur only in a *strictly descending* order. Additionally, we also show that the first $(\ell - 2)$ of these queries were made strictly in between the queries $p(q, (\ell + 1))$ and $p(q, \ell)$. That is, we show that

$$p(q, (\ell + 1)) < p(p(q, \ell), (\ell - 2)) < \ldots < p(p(q, \ell), 1) < p(q, \ell)$$

Note that this is the really crucial step of the argument since we have essentially shown that each of the queries $p(p(q, \ell), 1) \ldots p(p(q, \ell), (\ell - 2))$ is distinct from any of the queries $p(q, 1) \ldots p(q, j)$ (since they occur strictly in between two consecutive queries in the latter sequence). In addition, we are also able to prove that these queries are in strict descending order (unlike the queries $p(q, 1) \ldots p(q, r)$).

We notice that the above technique can again be applied to the strictly descending sequence of queries, $p(p(q, \ell), 1) \ldots p(p(q, \ell), (\ell - 2))$. In this manner, we can continue this argument recursively and derive a recurrence equation to count the number of queries whose existence we prove (which can all shown to be different using the technique above) as follows:

$$q \geq \mathcal{Q}(r/2), \text{ where } \mathcal{Q}(i) = i + \sum_{\ell=2}^{i-2} \mathcal{Q}(\ell - 2)$$

Upon solving this recurrence, we get that $q = \omega(1.3803^{r/2})$.   □

In our applications, we will be interested in using the LR construction with round functions that resist the 5-XOR condition, when any adaptive adversary makes a polynomial number of queries to the construction while having access to all the intermediate round values. We will specify this as a property of families of functions from which the round functions are independently derived. Hence, let us begin by describing a *function family*. A *function family* C is a set of functions along with a distribution defined on this set. For such a family, $f \leftarrow C$ denotes sampling a function according to the distribution specified by C. A function family is called a *5-XOR resistant function family* if the LR construction using independently sampled functions from this family resists the 5-XOR condition when queried a polynomial number of times by any adaptive adversary.

Definition 1 (5-XOR resistant function family). *A function family $C_{(k,n)}$, that consists of length preserving functions on n bits, is a 5-XOR resistant function family if for any adversary A,*

$$\Pr \left[\begin{array}{c} A \text{ 5-XOR condition} \\ \text{holds in } (A \longleftrightarrow \Psi_{f_1 \ldots f_k}) \end{array} \middle| \ f_1 \ldots f_k \leftarrow C_{(k,n)} \right] \leq \epsilon_{xor} = negl(\lambda)$$

Here the advantage ϵ_{xor} of the adversary A depends on the running time of A and the security parameter λ. The running time of A, the input length n and number of Feistel rounds k are all polynomial functions of λ.

By applying Lemma 2 to a LR construction using round functions independently sampled from a 5-XOR resistant function family, we can derive the following corollary.

Corollary 1. *Let Ψ_k be a k-round LR construction that uses round functions that are independently sampled from a 5-XOR resistant function family consisting of functions on n bits. For any adversary A that adaptively makes permutation queries to Ψ_k, while observing the intermediate round values, it holds that*

- *if A makes both forward/inverse queries, then for any round number $s \leq (k/2)$ with $s = \omega(\log \lambda)$,*

$$\Pr\left[\begin{array}{c} \exists \; r^{th} \text{ round value collision during } A \leftrightarrow \Psi_k \\ \text{for some } r \in \{s \ldots (k-s)\} \end{array}\right] \leq \epsilon_{xor}$$

- *if A makes only forward queries, then for any round number $s = \omega(\log \lambda)$,*

$$\Pr\left[\begin{array}{c} \exists \; r^{th} \text{ round value collision during } A \leftrightarrow \Psi_k \\ \text{for some } r \in \{s \ldots k\} \end{array}\right] \leq \epsilon_{xor}$$

Here the bound ϵ_{xor} denotes the maximum advantage of the XOR finding adversary that runs in time $\mathcal{O}(t_A + (q_A k)^5)$, where t_A is the running time of the adversary A and q_A denotes the number of queries made by it. Also, t_A, q_A and the input length n are all polynomial in λ.

This corollary is easily proved since the 5-XOR finding adversary simply runs the collision finding adversary, and performs a brute-force search for a 5-XOR condition when it finds a round value collision. From Lemma 2, such a 5-XOR condition is guaranteed to exist. In fact, we will make use of this corollary in each of the results that we present in the next section, since each of these function families will turn out to be 5-XOR resistant (the proof of this result can also be found in [7]; here we just state the result, although briefly sketching the case of UFs inside the proof of Theorem 3).

Theorem 2. *For each of the primitives: (1) unpredictable functions, (2) pseudo-random functions, (2) verifiable unpredictable functions, and (4) verifiable random functions; a function family that yields an independent random sample of the appropriate primitive is a 5-XOR resistant function family.*

5   Implications

All the cryptographic applications of the Feistel construction until recently have relied on all or some of the round functions not being visible to the adversary. In the previous section, we proved a combinatorial property of the Feistel construction where the internal round function values were visible to the adversary. Now we will describe how this property can be applied to a variety of scenarios to yield new or improved cryptographic constructions than before.

We get the following constructions using this new technique: (1) secure construction of *unpredictable permutations* from *unpredictable functions*, (2) more resilient construction of *pseudorandom permutations* from *pseudorandom functions*, (3) construction of *verifiable unpredictable permutations* from *verifiable unpredictable functions*, and (4) construction of *verifiable random permutations* from *verifiable random functions*.

In each case, the proof consists of three parts: (1) showing that the function family under consideration is a 5-XOR function family (see Theorem 2); (2) using Corollary 1 to show that the corresponding permutation construction is unlikely to have collisions at "advanced" rounds; and (3) show that the lack of such collisions implies that the construction is secure. All the proofs are given in [7].

5.1   Unpredictable Permutations and More Resilient PRPs

As a first implication of our combinatorial result, we can see that an $\omega(\log(\lambda))$-round LR construction with independent PRFs in each round gives a more resilient construction of PRPs that remain secure even if the intermediate round values are visible to the attacker. We defer further details of this application to the full version [7].

We saw in Section 3 that observing the output of a $k = n/\omega(\log \lambda)$ round Feistel construction with unpredictable round functions may leak all the intermediate round values. Even for realistic UFs, some partial information about the intermediate round values may be leaked through the output. As we discussed earlier, in such a case none of the previous proof techniques are applicable. We will prove a much stronger result here, by showing that if we use a super-logarithmic number of rounds in the Feistel construction, then the resulting UP construction is secure even if the adversary gets all the intermediate round values along with the permutation output.

The UP construction $\Psi_{U,k}$ that we propose consists of $k = \omega(\log \lambda)$ rounds of the Feistel construction using independent *unpredictable functions* $f_1 \ldots f_k \leftarrow F$. The following theorem essentially states that this construction is a secure UP construction. Due to space constraints, we omit the formal proof of this theorem (see [7]) and give a short proof intuition here.

Theorem 3. *Given a UP adversary A_π (with advantage ϵ_π) in the unpredictability game against the UP construction $\Psi_{U,k}$ (using round functions from UF family $\mathcal{F}$), one can build a UF adversary A_f that has comparable advantage (to ϵ_π) in the UF attack game against a UF sampled from $\mathcal{F}$. Quantitatively, we show $\epsilon_\pi = \mathcal{O}\left(\epsilon_f \cdot (qk)^6\right)$, where ϵ_f denotes the maximum advantage of a UF adversary running in time $\mathcal{O}(t + (qk)^5)$ against a UF sampled from $\mathcal{F}$. Here t, q are the running time and the number of queries made by A_π, respectively.*

Proof Intuition: Consider the UP adversary A_π that has advantage ϵ_π against the construction $\Psi_{U,k}$, based on the k-round LR construction with independently sampled UFs from the family F in each round. We can consider two cases.

Case 1: the sequence of queries made by A_π satisfies the 5-XOR condition with probability at least $\epsilon_\pi/2$. However, this means that our UF family in not 5-XOR resistant, contradicting Theorem 2. To get the exact security bound (alternatively, to sketch the proof Theorem 2 for the case of UFs), we can construct an attacker A_f for the UF as follows. A_f proceeds by plugging in its challenge UF randomly as any one of the round functions $f_1 \ldots f_k$, say f_i. It then chooses at random a query made by A_π as the query in which it will try to predict the output of f_i. It honestly computes the LR construction for this query until the round function f_i, getting round value R_i. At this point, it chooses a random XOR representation from all round values that already exist and outputs this as its prediction for $f_i(R_i)$. Since the UP attacker A_π forces a 5-XOR condition with non-negligible probability, then A_f also succeeds with non-negligible probability (precisely, the advantage ϵ_f of A_f is $\Omega(\epsilon_\pi/(qk)^6)$).

Case 2: alternatively, A_π wins the UP attack game with advantage at least $\epsilon_\pi/2$ without its queries satisfying the 5-XOR condition. In this case we construct A_f as follows. It will attempt to predict a fresh UF value for the middle round function $f_{k/2}$ (and will choose the remaining functions by itself). It will simulate A_π in the obvious manner, using its oracle to find out the middle values $f_{k/2}$. When the adversary A_π outputs a prediction (X, Y) for some fresh UP input/output, A_f computes the LR construction "forward" honestly to get $R_{k/2}$ from X, and "backward" honestly to get $R_{k/2+1}$ from Y. It then outputs $R_{k/2-1} \oplus R_{k/2+1}$ as its prediction for $f_{k/2}(R_{k/2})$, winning if A_π won *and the value $R_{k/2}$ is "fresh"*. Thus, A_f could only fail if it already made the query $f_{k/2}(R_{k/2})$ in order to respond to one of the queries of A_π. However, this would imply a $(k/2)^{th}$ round collision, and we can use the combinatorial Lemma 2 to show that the 5-XOR condition must have been true, contradicting our assumption on A_π that the 5-XOR condition was false.   ⊡

DOMAIN EXTENSION OF MACs.   The above result can also be viewed as a construction of MACs from $2n$ to $2n$ bits using MACs from n to n bits. We observe that it is possible to reduce the output length in the above construction to n by simply dropping the left half of the output. Using this technique, we get a MAC construction from $2n$ to n bits. To briefly justify it, in the usual MAC attack game the attacker can only make forward queries. From corollary 1, we get that for any $s = \omega(\log \lambda)$ no efficient attacker can cause a collision on any round value $r \in \{s \ldots k\}$ with non-negligible probability. Thus, a proof of security for this MAC will proceed by plugging in the target n- to n-bit MAC in the last round function of the Feistel construction, and arguing that the attacker predicting the $2n$- to n-bit constructed MAC must also forge this last-round n-to n-bit MAC. This is done using a similar proof technique to that for Theorem 3 (albeit using second part of Corollary 1 to argue that no collision occurs at the last round).

MORE RESILIENT PRPs.   Similarly to the above, we show that $\omega(\log \lambda)$ Feistel rounds yeilds a construction of PRPs from PRFs that remains secure even if the PRF input/output pairs used in the intermediate rounds are visible to an attacker. We denote the corresponding k-round construction by $\Psi_{R,k}$, and show the following quantitative result in [7].

Theorem 4. *Given a PRP adversary A_π with advantage ϵ_π in the "extended PRP" attack game against $\Psi_{R,k}$ (using round functions from PRF family $\mathcal{F}$), one can build a PRF adversary A_f having comparable advantage (to ϵ_π in the PRF attack game against a PRF sampled from the PRF family $\mathcal{F}$. Quantitatively, we show that $\epsilon_\pi = \mathcal{O}\left(qk\epsilon_f + \frac{(qk)^6}{2^n}\right)$, where ϵ_f denotes the maximum advantage of a PRF adversary running in time $\mathcal{O}(t + (qk)^5)$ against a PRF sampled from $\mathcal{F}$. Here t, q are the running time and number of queries made by A_π, respectively.*

5.2   Verifiable Unpredictable/Pseudorandom Permutations

The VRP and VUP constructions that we propose are essentially the same, except that we use VRFs as round function in one case and VUFs in the other. Our VRP (resp. VUP) construction $\Psi_{VR,k}$ (resp. $\Psi_{VU,k}$) uses a k-round Feistel construction using independent VRFs (resp. VUFs) $f_1 \ldots f_k \leftarrow \mathcal{F}$ as round functions. The public/private keys of $\Psi_{VR,k}$ (resp. $\Psi_{VU,k}$) are simply the concatenation of the public/private keys of the k VRFs (resp. VUFs). The Prove functionality for $\Psi_{VR,k}$ (resp. $\Psi_{VU,k}$) simply gives the permutation output, and gives all intermediate round values along with the VRF (resp. VUF) proofs as its proof. The Verify functionality simply checks if all intermediate VRF (resp. VUF) proofs verify correctly.

We then prove the three properties of the VRP (resp. VUP) construction: *Completeness, Soundness* (or unique proofs) and *Pseudorandomness* (resp. *Unpredictability*). The Completeness and Soundness properties in each case are a direct consequence of the corresponding VRF (resp. VUF) properties. Here the *Pseudorandomness* (resp. *Unpredictability*) property are proven in much the same way as Theorem 4 (resp. Theorem 3); see [7] for formal proofs.

Theorem 5. *Given a VRP (resp. VUP) adversary A_π with advantage ϵ_π in the pseudorandomness (resp. unpredictability) game against the VRP (resp. VUP) construction $\Psi_{VR,k}$ (resp. $\Psi_{VU,k}$) using VRFs (resp. VUFs) sampled from the VRF (resp. VUF) family $\mathcal{F}$ as round functions, one can build a VRF (resp. VUF) adversary A_f that has comparable advantage (to ϵ_π) in the pseudorandomness (resp. unpredictability) game against a VRF (resp. VUF) sampled from $\mathcal{F}$. Quantitatively, we show $\epsilon_\pi = \mathcal{O}\left(qk\epsilon_f + \frac{(qk)^6}{2^n}\right)$ (resp. $\mathcal{O}(q^6 k^7 \cdot \epsilon_f)$), where ϵ_f denotes the maximum advantage of a VRF (resp. VUF) adversary running in time $\mathcal{O}(t + (qk)^5)$ (resp. $\mathcal{O}(t + (qk)^5)$) against a VRF (resp. VUF) sampled from $\mathcal{F}$. Here t, q are the running time and number of queries made by A_π, respectively.*

Acknowledgments. We would like to thank Rafail Ostrovsky and Shabsi Walfish for several helpful discussions.

References

1. Jee Hea An and Mihir Bellare, *Constructing VIL-MACs from FIL-MACs: Message Authentication under Weakened Assumptions*, CRYPTO 1999: 252-269.
2. M. Bellare and P. Rogaway, *Optimal Asymmetric Encryption*, Proceedings of Eurocrypt'94, LNCS vol. 950, Springer-Verlag, 1994, pp. 92–111.
3. M. Bellare and P. Rogaway, *The exact security of digital signatures - How to sign with RSA and Rabin*. Proceedings of Eurocrypt'96, LNCS vol. 1070, Springer-Verlag, 1996, pp. 399-416.
4. Manuel Blum, *Coin Flipping by Telephone - A Protocol for Solving Impossible Problems*, COMPCON 1982: 133-137.
5. Y. Dodis, *Efficient construction of (distributed) verifiable random functions*, In *Proceedings of 6th International Workshop on Theory and Practice in Public Key Cryptography*, pp 1 -17, 2003.
6. Y. Dodis and P. Puniya, *On the relation between Ideal Cipher and Random Oracle Models*, In *Theory of Cryptography Conference 2006*.
7. Y. Dodis and P. Puniya, *Feistel Networks made Public, and Applications*, Full Version, available from IACR EPrint Archive.
8. Y. Dodis and A. Yampolskiy, *A Verifiable Random Function With Short Proofs and Keys*, In *Workshop on Public Key Cryptography (PKC)*, January 2005.
9. Oded Goldreich, Shafi Goldwasser and Asaf Nussboim, *On the Implementation of Huge Random Objects*, FOCS 2003: 68-79.
10. Oded Goldreich and Leonid A. Levin, *A Hard-Core Predicate for all One-Way Functions*, STOC 1989: 25-32.
11. Shafi Goldwasser and Rafail Ostrovsky, *Invariant Signatures and Non-Interactive Zero-Knowledge Proofs are Equivalent (Extended Abstract)*, in *CRYPTO 1992*: 228-245.
12. M. Luby and C. Rackoff, *How to construct pseudo-random permutations from pseudo-random functions*, in *SIAM Journal on Computing*, Vol. 17, No. 2, April 1988.
13. A. Lysyanskaya, *Unique Signatures and verifiable random functions from DH-DDH assumption*, in *Proceedings of the 22nd Annual International Conference on Advances in Cryptography (CRYPTO)*, pp. 597 612, 2002.
14. Ueli M. Maurer, Yvonne Anne Oswald, Krzysztof Pietrzak and Johan Sj?din, *Luby-Rackoff Ciphers from Weak Round Functions?*, EUROCRYPT 2006: 391-408.
15. Ueli M. Maurer and Krzysztof Pietrzak, *The Security of Many-Round Luby-Rackoff Pseudo-Random Permutations*, in *EUROCRYPT 2003*, 544-561.
16. Ueli M. Maurer and Johan Sj?din, *Single-Key AIL-MACs from Any FIL-MAC*, ICALP 2005: 472-484.
17. S. Micali, M. Rabin and S. Vadhan, *Verifiable Random functions*, In *Proceedings of the 40th IEEE Symposium on Foundations of Computer Science*, pp. 120 -130, 1999.
18. Silvio Micali and Ronald L. Rivest, *Micropayments Revisited*, CT-RSA 2002, 149-163.
19. Moni Naor, *Bit Commitment Using Pseudo-Randomness*, CRYPTO 1989: 128-136.
20. Moni Naor and Omer Reingold, *On the construction of pseudo-random permutations: Luby-Rackoff revisited*, in *Journal of Cryptology*, vol 12, 1999, pp. 29-66.
21. Moni Naor and Moti Yung, *Universal One-Way Hash Functions and their Cryptographic Applications*, STOC 1989: 33-43.

22. Jacques Patarin, *Security of Random Feistel Schemes with 5 or More Rounds*, in *CRYPTO 2004*, 106-122.
23. Z. Ramzan and L. Reyzin, *On the Round Security of Symmetric-Key Cryptographic Primitives*, in Advances in Cryptography - Crypto, LNCS vol. 1880, Springer-Verlag, 2000.
24. Daniel R. Simon, *Finding Collisions on a One-Way Street: Can Secure Hash Functions Be Based on General Assumptions?*, EUROCRYPT 1998: 334-345.

Oblivious-Transfer Amplification

Jürg Wullschleger

ETH Zürich, Switzerland
wjuerg@inf.ethz.ch

Abstract. Oblivious transfer (OT) is a primitive of paramount importance in cryptography or, more precisely, two- and multi-party computation due to its universality. Unfortunately, OT cannot be achieved in an unconditionally secure way for both parties from scratch. Therefore, it is a natural question what information-theoretic primitives or computational assumptions OT *can* be based on.

The results in our paper are threefold. First, we give an optimal proof for the standard protocol to realize unconditionally secure OT from a weak variant of OT called *universal OT*, for which a malicious receiver can virtually obtain any possible information he wants, as long as he does not get all the information. This result is based on a novel distributed leftover hash lemma which is of independent interest.

Second, we give conditions for when OT can be obtained from a faulty variant of OT called *weak OT*, for which it can occur that any of the parties obtains too much information, or the result is incorrect. These bounds and protocols, which correct on previous results by Damgård *et. al.*, are of central interest since in most known realizations of OT from weak primitives, such as noisy channels, a weak OT is constructed first.

Finally, we carry over our results to the computational setting and show how a weak OT that is sometimes incorrect and is only mildly secure against computationally bounded adversaries can be strengthened.

Keywords: oblivious-transfer amplification, universal oblivious transfer, weak oblivious transfer, computational weak oblivious transfer, distributed leftover hash lemma, hard-core lemma.

1 Introduction

The goal of *multi-party computation*, introduced in [42], is to allow two parties to carry out a computation in such a way that no party has to reveal unnecessary information about her input. A primitive of particular importance in this context is *oblivious transfer* (OT) [39,36,18]. *Chosen one-out-of-two string oblivious transfer*, $\binom{2}{1}$-OT^n for short, is a primitive where the sender sends two strings x_0 and x_1 of length n and the receiver's input is a choice bit c; the latter then learns x_c but gets no information about the other string x_{1-c}. One reason for the importance of OT is its *universality*, i.e., it allows for carrying out *any* two-party computation [32]. Unfortunately, OT is impossible to achieve in an unconditionally secure way from scratch, i.e., between parties connected by a noiseless

M. Naor (Ed.): EUROCRYPT 2007, LNCS 4515, pp. 555–572, 2007.

channel. However, if some additional weak primitives are available such as noisy channels or noisy correlations, then unconditional security can often be achieved [12,11,17,15,13,40,16,35]. Most of these protocols first implement a weak version of OT, and then strengthen it to achieve OT. In [20,23] it was shown that such a strengthening is sometimes also needed in the computational setting.

In this paper we study how weak versions of OT can be *amplified* to OT.

1.1 Previous Work

Various weak versions of OT have been proposed. In most of them, only the receiver's side is weak, such as α-1-2 *slightly OT* from [12], or only the sender's side is weak, such as *XOT*, *GOT* or *UOT with repetitions* from [6,7]. All of these primitives were shown to be strong enough to imply OT. In [8], a more general primitive called *Universal OT*, (α)-$\binom{2}{1}$-UOT^n for short, has been proposed, where α specifies a lower bound on the amount of uncertainty a (possibly malicious) receiver has over *both* inputs, measured in collision- or min-entropy. Unfortunately, the security proof contained an error that was corrected in [16]. It was shown that $\binom{2}{1}$-OT^ℓ can be implemented from one instance of (α)-$\binom{2}{1}$-UOT^n with an error of at most ε if $\ell \leq \frac{1}{4}\alpha - \frac{3}{4}\log(1/\varepsilon) - 1$.

Weak OT, introduced in [17], is a weak version of $\binom{2}{1}$-OT^1 where *both* players may obtain additional information about the other player's input, and where the output may have some errors. It is used as a tool to construct OT out of *unfair primitives*, i.e., primitives where the adversary is more powerful than the honest participant, such as the *unfair noisy channel*. Weak OT is denoted as (p, q, ε)-WOT, where p is the maximal probability that the sender gets side information about the receiver's input, q the maximal probability that the receiver gets side information about the sender's input, and ε is the maximal probability that an error occurs. Using a simple simulation argument, it was shown in [17] that there cannot exist a protocol that implements $\binom{2}{1}$-OT^1 from (p, q, ε)-WOT if $p+q+2\varepsilon \geq 1$. For $\varepsilon = 0$, they give a protocol secure against active adversaries that implements $\binom{2}{1}$-OT^1 from $(p, q, 0)$-WOT for $p + q < 1$, which is optimal. Furthermore, for the case where p, q, and ε are bigger than 0, a protocol is presented that is secure against passive adversaries for $p + q + 2\varepsilon < 0.45$. Weak OT was later generalized in [15] to *(special) generalized weak OT*, in order to improve the reduction of $\binom{2}{1}$-OT^1 to unfair noisy channels.

In [20], a reduction of $\binom{2}{1}$-OT^1 to (p, q, ε)-WOT in the computational setting was presented. These results were used to show that OT can be based on *collections of dense trapdoor permutations*.

1.2 Problems with the Definition of Weak OT in [17]

While [17] does not give a formal definition of (p, q, ε)-WOT, [15] formally defines (p, q, ε)-WOT by giving an ideal functionality. Their definition implicitly makes two assumptions. It requires that, firstly, the players do not get information about whether an error occurred, and secondly, that the event that an error occurs is independent from the events that the players get side information. These assumptions are rather unnatural and in most of the cases where (p, q, ε)-WOT is used, they cannot be satisfied. For example, neither the simulation of (p, q, ε)-WOT for $p + q + 2\varepsilon = 1$, nor the application to the unfair noisy channel satisfy these assumptions.

Unfortunately, if we remove these two assumptions from the definition of (p, q, ε)-WOT, the E-Reduce protocol from [17] gets insecure, because it depends on the fact that the two events are independent. The following example illustrates the problem: Even though $(0, 1/2, 1/4)$-WOT can be simulated, by applying R-Reduce(3000, E-Reduce(10, (0, 1/2, 1/4)-WOT)) (using the reductions R-Reduce and E-Reduce as defined in [17]) we get a $(0, 0.06, 0.06)$-WOT, which implies $\binom{2}{1}$-OT^1. We would get an information-theoretic secure $\binom{2}{1}$-OT^1 from scratch, which is impossible.

Directly affected by this problem are Lemma 5 and Theorem 2 in [17] and Lemma 6 in [15]. Indirectly affected are Lemma 11 and Theorem 3 in [17], and Lemma 1, 4, 5 and 7 in [15], as they rely on Lemma 5 in [17].

1.3   Contribution

In the first part, we show how to implement $\binom{2}{1}$-OT$^\ell$ from one instance of (α)-$\binom{2}{1}$-UOT^n for $\ell \leq \frac{\alpha}{2} - 3 \log \frac{1}{\varepsilon}$ with an error of at most 2ε. This improves the bound of [16] by a factor of two, at the cost of a slightly bigger error term, and is asymptotically optimal for the standard protocol using 2-universal hashing. The proof makes use of a new *distributed leftover hash lemma*, which is a generalization of the leftover hash lemma and of independent interest.

In the second part, we will look at reductions of $\binom{2}{1}$-OT^1 to (p, q, ε)-WOT, for new, weaker definitions of (p, q, ε)-WOT. Using a different E-Reduce protocol that also works for our definitions, we show for the special case where $p = 0$ ($q = 0$), that $\binom{2}{1}$-OT^1 can efficiently be implemented from (p, q, ε)-WOT if $\sqrt{q} + 2\varepsilon < 1$ ($\sqrt{p} + 2\varepsilon < 1$), secure against passive adversaries. For the general case, we show that if $p + q + 2\varepsilon \leq 0.24$ or $\max(p + 22q + 44\varepsilon, 22p + q + 44\varepsilon, 7\sqrt{p+q} + 2\varepsilon) < 1$, $\binom{2}{1}$-OT^1 can efficiently be implemented from (p, q, ε)-WOT secure against passive adversaries. This fixes Lemma 5 and Theorem 2 in [17] and gives some new bounds, but does not reach the bound of $p + q + 2\varepsilon < 0.45$ from [17].

Finally, we apply these results to the computational case, and show, using the uniform hard-core lemma from [26], how an OT which may contain errors and which is only mildly *computationally* secure against the two players can be amplified to a computationally-secure OT. In particular, we show that if (p, q, ε)-WOT can be amplified to $\binom{2}{1}$-OT^1 in the information-theoretic setting, then also the computational version of (p, q, ε)-WOT which we call (p, q, ε)-compWOT can be amplified to a computationally-secure $\binom{2}{1}$-OT^1, *using the same protocol*. In combination with our information-theoretic results, we get a way to amplify (p, q, ε)-compWOT. Our results generalize the results presented in [20], as we cover a much bigger region for the values p, q and ε, and in our case the security for both players may only be computational.

A more detailed analysis of all these results can be found in [41].

2   Preliminaries

Let X and X' be two random variables distributed over the same domain $\mathcal{X}$. The *advantage* of an algorithm $A : \mathcal{X} \to \{0,1\}$ to distinguish X from X' is defined as $\mathrm{Adv}^A(X, X') := \big| \Pr[A(X) = 1] - \Pr[A(X') = 1] \big|$. The *statistical distance* between X and X' is defined as $\Delta(X, X') = \frac{1}{2} \sum_{x \in \mathcal{X}} \big| \Pr[X = x] - \Pr[X = x] \big|$. It is easy to see that $\Delta(X, X') = \max_A \mathrm{Adv}^A(X, X')$. We say that a random variable X *over* $\mathcal{X}$ *is ε-close to uniform with respect to* Y, if $\Delta(P_{XY}, P_U P_Y) \leq \varepsilon$, where P_U is the uniform distribution over $\mathcal{X}$.

Definition 1. *Let P_{XY} be a distribution over $\{0,1\} \times \mathcal{Y}$. The* maximal bit-prediction advantage *of X from Y is* $\mathrm{PredAdv}(X \mid Y) := 2 \cdot \max_f \Pr[f(Y) = X] - 1$.

In other words, if $\mathrm{PredAdv}(X \mid Y) = \delta$, then we have for all functions $f : \mathcal{Y} \to \{0,1\}$ that $\Pr[f(Y) = X] \leq (1+\delta)/2$. It is easy to see that there exists an event $\mathcal{E}$ with $\Pr[\mathcal{E}] = \mathrm{PredAdv}(X \mid Y)$, such that if $\mathcal{E}$ occurs, then X is a function of Y and if $\mathcal{E}$ does not occur, then X is uniform conditioned on Y. Furthermore, we have $\mathrm{PredAdv}(X \mid Y) = 2 \cdot \Delta(P_{XY}, P_U P_Y)$, where P_U is the uniform distribution over $\{0,1\}$. Let $H_\infty(X \mid Y) = \min_{xy:P_{XY}(x,y)>0} - \log P_{X|Y}(x \mid y)$ be the *conditional min-entropy* of X given Y. A function $h : \mathcal{R} \times \mathcal{X} \to \{0,1\}^m$ is called a 2-*universal hash function* [10], if for all $x_0 \neq x_1 \in \mathcal{X}$, we have $\Pr[h(R, x_0) = h(R, x_1)] \leq 2^{-m}$, if R is uniform over $\mathcal{R}$.

We say that a function $f : \mathbb{N} \to \mathbb{N}$ is *polynomial in k*, denoted by $\mathrm{poly}(k)$, if there exists a constant $c > 0$ such that $f(k) \in O(k^c)$. A function $f : \mathbb{N} \to [0,1]$ is *negligible in k*, denoted by $\mathrm{negl}(k)$, if for all $c > 0$, $f(k) \in o(k^{-c})$.

2.1   Definition of Security

A $\mathcal{W}$-*hybrid protocol* is a sequence of interactions between two players. In each step, the players may apply a randomized function on their data, and send the result to the other player. They may also use the functionality $\mathcal{W}$ by sending input to $\mathcal{W}$ which gives them an output back according to the specification of $\mathcal{W}$. In the last stage the players output a randomized function of their data. A protocol is *efficient* if it can be executed using two polynomial time turing machines.

In the *semi-honest model*, the adversary is *passive*, which means that she follows the protocol, but outputs her entire view, i.e., all the information she has obtained during the execution of the protocol. In the *malicious model* the adversary is *active*, which means that he may change his behavior in an arbitrary way. Our definitions for the security of a protocol are based on the standard *real vs. ideal* paradigm of [34] and [1] (see also [9]). The idea behind the definition is that anything an adversary can achieve in the *real life protocol*, he could also achieve by another attack in an *ideal world*, i.e., where the players only have black-box access to the functionality they try to achieve. If the executions in the real and the ideal settings are *statistically indistinguishable* (the statistical distance is smaller than ε), we call the protocol *secure with an error of at most ε*, if they are only *computationally indistinguishable* (any efficient algorithm has negligible advantage in distinguishing them), we call the protocol *computationally secure*.

We will only look at a *fully randomized* version of $\binom{2}{1}$-OT^n denoted by $\binom{2}{1}$-ROT^n. $\binom{2}{1}$-ROT^n is equivalent to $\binom{2}{1}$-OT^n, which was shown in [4] and formally proved in [2]. Our definition of $\binom{2}{1}$-ROT^n is similar to the definitions in [16] and [14].

Definition 2 (Randomized oblivious transfer, malicious model). *A protocol Π between a sender and a receiver where the sender outputs $(X_0, X_1) \in \{0,1\}^n \times \{0,1\}^n$ and the receiver outputs $(C, Y) \in \{0,1\} \times \{0,1\}^n$ securely implements $\binom{2}{1}$-ROT^n in the malicious model with an error of at most ε, if the following conditions are satisfied:*

- *(Correctness) If both players are honest, then $\Pr[Y \neq X_C] \leq \varepsilon$.*
- *(Security for the sender) For an honest sender and any (malicious) receiver with output $\overline{V}$, there exists a random variable $\overline{C} \in \{0,1\}$, such that $X_{1-\overline{C}}$ is ε-close to uniform with respect to $(\overline{C}, X_{\overline{C}}, \overline{V})$.*
- *(Security for the receiver) For an honest receiver and any (malicious) sender with output $\overline{U}$, C is ε-close to uniform with respect to $\overline{U}$.*

In the semi-honest model, we additionally require that $\overline{C} = C$, because we also require the adversary in the ideal world to be semi-honest.

3   Distributed Randomness Extraction

In order to get an optimal bound for the reduction from $\binom{2}{1}$-OT^1 to (α)-$\binom{2}{1}$-UOT^n, we will need a generalization of Lemma 1, the *leftover hash lemma*. Since this is of independent interest, we present it in a separate section.

Lemma 1 tells us how many almost-random bits can be extracted from an imperfect source of randomness X, if some additional uniform randomness is present. It is also known as *privacy amplification*. See also [3,25].

Lemma 1 (Leftover hash lemma [5,30]). *Let X be a random variable over $\mathcal{X}$ and let $m > 0$. Let $h : \mathcal{S} \times \mathcal{X} \to \{0,1\}^m$ be a 2-universal hash function. If $m \leq H_\infty(X) - 2\log(1/\varepsilon) + 2$, then for S uniform over $\mathcal{S}$, $h(S, X)$ is ε-close to uniform with respect to S.*

We now generalize the setting and let two players independently extract randomness from two *dependent* random variables. Lemma 1 tells us that if the length of the extracted strings are smaller than the min-entropy of these random variables, then each of the extracted strings is close to uniform. However, the two strings might still be dependent on each other. Lemma 2 now says that if the total length of the extracted strings is smaller than the overall min-entropy, then the two strings are also almost independent. The obtained bound is optimal. The proof is very similar to a standard proof of the leftover hash lemma.

Lemma 2 (Distributed leftover hash lemma). *Let X and Y be random variables over $\mathcal{X}$ and $\mathcal{Y}$, and let $m, n > 0$. Let $g : \mathcal{S} \times \mathcal{X} \to \{0,1\}^m$ and $h : \mathcal{R} \times \mathcal{Y} \to \{0,1\}^n$ be 2-universal hash functions. If*

$$\min\left(H_\infty(X) - m, H_\infty(Y) - n, H_\infty(XY) - m - n\right) \geq 2\log(1/\varepsilon),$$

then, for (S, R) uniform over $\mathcal{S} \times \mathcal{R}$, $(g(S, X), h(R, Y))$ is ε-close to uniform with respect to (S, R).

Proof. For any W having distribution P_W over $\mathcal{W}$, and W' being uniformly distributed over $\mathcal{W}$, we have

$$\Delta(W, W') = \frac{1}{2}\sum_w \left|P_W(w) - \frac{1}{|\mathcal{W}|}\right| = \frac{1}{2}\sqrt{\left(\sum_w \left|P_W(w) - \frac{1}{|\mathcal{W}|}\right|\right)^2}$$

$$\leq \frac{1}{2}\sqrt{|\mathcal{W}|}\sqrt{\sum_w \left(P_W(w) - \frac{1}{|\mathcal{W}|}\right)^2} = \frac{1}{2}\sqrt{|\mathcal{W}|}\sqrt{\sum_w P_W^2(w) - \frac{1}{|\mathcal{W}|}}.$$

Here we used that $\left(\sum_{i=1}^n a_i\right)^2 \leq n\sum_{i=1}^n a_i^2$, which follows from Cauchy-Schwarz. Let $V = g(S, X)$, $V' = h(R, Y)$ and U, U' be two uniform random variables over $\{0,1\}^m$ and $\{0,1\}^n$. Choosing $W := (V, V', S, R)$ and $W' := (U, U', S, R)$ in the above inequality, we get

$$\Delta((V, V', S, R), (U, U', S, R))$$
$$\leq \frac{1}{2}\sqrt{|\mathcal{S}||\mathcal{R}|2^{m+n}}\sqrt{\sum_{vv'sr} P_{VV'SR}^2(v, v', s, r) - \frac{1}{|\mathcal{S}||\mathcal{R}|2^{m+n}}}.$$

Since $\sum_x P_X^2(x)$ is the *collision probability*[1] of a random variable X, we have for (X_0, Y_0) and (X_1, Y_1) independently distributed according to P_{XY} and for uniformly random $S_0, S_1, R_0,$ and R_1 that

$$\sum_{vv'sr} P_{VV'SR}^2(v, v', s, r)$$
$$= \Pr[g(X_0, S_0) = g(X_1, S_1) \wedge h(Y_0, R_0) = h(Y_1, R_1) \wedge S_0 = S_1 \wedge R_0 = R_1]$$
$$= \Pr[S_0 = S_1 \wedge R_0 = R_1]\Pr[g(X_0, S_0) = g(X_1, S_0) \wedge h(Y_0, R_0) = h(Y_1, R_0)].$$

Because g and h are 2-universal hash functions, we have

$$\Pr[g(X_0, S_0) = g(X_1, S_0) \wedge h(Y_0, R_0) = h(Y_1, R_0)]$$
$$\leq \Pr[X_0 = X_1 \wedge Y_0 = Y_1] + 2^{-m}\Pr[X_0 \neq X_1 \wedge Y_0 = Y_1]$$
$$+ 2^{-n}\Pr[X_0 = X_1 \wedge Y_0 \neq Y_1] + 2^{-m-n}$$
$$= (1 + 3\varepsilon^2)2^{-m-n},$$

from which follows that $\Delta((V, V', S, R), (U, U', S, R)) \leq \frac{\sqrt{3}}{2}\varepsilon.$ □

---

[1] Let X_0 and X_1 be distributed according to P_X. The collision probability is $\Pr[X_0 = X_1] = \sum P_X(x)^2$.

4   Universal Oblivious Transfer

In this section, we give an implementation of $\binom{2}{1}$-ROT$^\ell$ that uses one instance of *universal oblivious transfer (UOT)*, that allows ℓ to be roughly twice as large as in [16], at the cost of a slightly larger error term.

UOT is a weak version of ROT that allows a malicious receiver to obtain more information than what he would be allowed in ROT. For simplicity, we only define a perfect version of UOT. The definition (and also the proof of Theorem 1) can easily be adapted to the statistical case.

Definition 3 (Universal oblivious transfer, malicious model). *A protocol Π between a sender and a receiver where the sender outputs $(X_0, X_1) \in \{0,1\}^n \times \{0,1\}^n$ and the receiver outputs $(C, Y) \in \{0,1\} \times \{0,1\}^n$ securely implements (α)-$\binom{2}{1}$-UOT^n in the malicious model, if the following conditions are satisfied:*

- *(Correctness) If both players are honest, then $Y = X_C$.*
- *(Security for the sender) For an honest sender and any (malicious) receiver with output $\overline{V}$, we have $H_\infty(X_0, X_1 \mid \overline{V}) \geq \alpha$.*
- *(Security for the receiver) For an honest receiver and any (malicious) sender with output $\overline{U}$, C is uniform with respect to $\overline{U}$.*

We will use the same protocol as [6,8,7,16]. Note that this protocol is only secure in the the malicious, but not to the semi-honest model.

Protocol ROTfromUOT(α, n, ℓ)
Let $(U_0, U_1) \in \{0,1\}^\ell \times \{0,1\}^\ell$ be the senders output and $(C, Y) \in \{0,1\} \times \{0,1\}^\ell$ the receivers output. Let $h : \mathcal{R} \times \{0,1\}^n \to \{0,1\}^\ell$ be a 2-universal hash function.

1. Both players execute (α)-$\binom{2}{1}$-UOT^n. The sender receives (X_0, X_1), and the receiver receives (C, W).
2. The sender chooses $R_0, R_1 \in \mathcal{R}$ at random and sends (R_0, R_1) to the receiver.
3. The sender outputs $(U_0, U_1) := (h(R_0, X_0), h(R_1, X_1))$, and the receiver outputs $(C, Y) := (C, h(R_C, W))$.

To prove that the protocol is secure for the sender, we will define an additional random variable $A \in \{0, 1, 2\}$ that distinguishes between three cases. (We assume that the receiver gets to know A, which may only help him.) We will show that the protocol is secure in all three cases. It is easy to see that for this protocol the bound we obtain in Theorem 1 is asymptotically optimal.

Theorem 1. *Let $\varepsilon > 0$. Protocol ROTfromUOT(α, n, ℓ) securely implements $\binom{2}{1}$-ROT$^\ell$ with an error of at most 2ε out of one instance of (α)-$\binom{2}{1}$-UOT^n in the malicious model, if $\ell \leq \alpha/2 - 3\log(1/\varepsilon)$.*

Proof. Obviously the protocol satisfies correctness. Let the sender be honest. Let $\overline{V}'$ be the output of (α)-$\binom{2}{1}$-UOT^n to the (malicious) receiver. We will implicitly condition on $\overline{V}' = v'$. After the execution of (α)-$\binom{2}{1}$-UOT^n, we have $H_\infty(X_0 X_1) \geq \alpha$. Let $S_i := \{x_i \in \mathcal{X}_i : \Pr[X_i = x_i] \leq 2^{-\alpha/2}\}$, for $i \in \{0, 1\}$. We define the random variable A as follows. Let $A = 2$ if $(X_0 \in S_0) \wedge (X_1 \in S_1)$, let $A = 0$ if $(X_0 \notin S_0) \wedge (X_1 \in S_1)$, let $A = 1$ if $(X_1 \notin S_1) \wedge (X_0 \in S_0)$, and let A be chosen uniformly at random in $\{0, 1\}$ if $(X_0 \notin S_0) \wedge (X_1 \notin S_1)$. If $\Pr[A = 2] \leq \varepsilon$, we will ignore the event $A = 2$. Therefore, we redefine A for this event to take on the value 3. We end up with a random variable A that takes on the value 2 with probability 0 or at least ε, and which takes on the value 3 with probability at most ε. Let $\overline{C} = \min(A, 1)$.

- If the event $A = i$ occurs for $i \in \{0, 1\}$, we have $\overline{C} = i$. All $x_i \in S_i$ have $\Pr[X_i = x_i \wedge A = i] = 0$. For all $x_i \notin S_i$ we have $\Pr[X_i = x_i \wedge A = i] \geq \Pr[X_i = x_i]/2 \geq 2^{-\alpha/2-1}$. It follows that

$$\Pr[X_{1-i} = x_{1-i} \mid X_i = x_i \wedge A = i] = \frac{\Pr[X_{1-i} = x_{1-i} \wedge X_i = x_i \wedge A = i]}{\Pr[X_i = x_i \wedge A = i]}$$
$$\leq 2^{-\alpha}/2^{-\alpha/2-1} = 2^{-\alpha/2+1} ,$$

  and hence, $H_\infty(X_{1-\overline{C}} \mid X_{\overline{C}}, A = i) \geq \alpha/2 - 1$. Since $R_{1-\overline{C}}$ is chosen independently of the rest, it follows from Lemma 1 that, given $A = i$, the distribution of $U_{1-\overline{C}}$ is ε-close to uniform with respect to $(R_0, R_1, U_{\overline{C}})$.
- If the event $A = 2$ occurs, we have $\overline{C} = 1$, $\Pr[A = 2] \geq \varepsilon$, $\Pr[X_0 = x_0 \wedge X_1 = x_1 \mid A = 2] \leq 2^{-\alpha}/\varepsilon$, and $\Pr[X_i = x_i \mid A = 2] \leq 2^{-\alpha/2}/\varepsilon$, for $i \in \{0, 1\}$. It follows that $H_\infty(X_0 \mid A = 2) \geq \alpha/2 - \log(1/\varepsilon)$, $H_\infty(X_1 \mid A = 2) \geq \alpha/2 - \log(1/\varepsilon)$, and $H_\infty(X_0 X_1 \mid A = 2) \geq \alpha - \log(1/\varepsilon)$. Since R_0 and R_1 are chosen independently of the rest, it follows from Lemma 2 that given $A = 2$, (U_0, U_1) is ε-close to uniform with respect to (R_0, R_1), from which follows that $U_{1-\overline{C}}$ is ε-close to uniform with respect to $(R_0, R_1, U_{\overline{C}})$.

Therefore, for all $a \in \{0, 1, 2\}$, given $A = a$, there exists a $\overline{C}$ such that the distribution of $U_{1-\overline{C}}$ is ε-close to uniform with respect to $(R_0, R_1, \overline{C}, U_{\overline{C}})$. It follows that $U_{1-\overline{C}}$ is also ε-close to uniform with respect to $(R_0, R_1, \overline{C}, U_{\overline{C}})$ given $A < 3$, and since $\Pr[A = 3] \leq \varepsilon$, $U_{1-\overline{C}}$ is 2ε-close to uniform with respect to $(R_0, R_1, \overline{C}, U_{\overline{C}})$. Because this holds for all $v' \in \mathcal{V}'$, and because $\overline{V}$ is a randomized function of $(R_0, R_1, \overline{V}')$, $U_{1-\overline{C}}$ is also 2ε-close to uniform with respect to $(\overline{C}, U_{\overline{C}}, \overline{V})$.

Let the receiver be honest, and let $\overline{U}'$ be the output of (α)-$\binom{2}{1}$-UOT^n to a (malicious) sender. From the security of (α)-$\binom{2}{1}$-UOT^n follows that C is uniform with respect to $\overline{U}'$. Since the receiver does not send any messages to the sender, C is also uniform with respect to $\overline{U}$. $\qquad\square$

5 Weak Oblivious Transfer

In this section we show how ROT can be implemented using many instances of *weak oblivious transfer (WOT)*, which is a weak version of ROT where *both* players may get additional information, and where the output may be incorrect. We start by giving two new, weaker definitions of WOT for both models.

Definition 4 (Weak oblivious transfer, semi-honest model). *Let Π be a protocol between a sender and a receiver that outputs $(X_0, X_1) \in \{0,1\} \times \{0,1\}$ to the sender and $(C, Y) \in \{0,1\} \times \{0,1\}$ to the receiver. Let U be the view of the semi-honest sender, and let V be the view of the semi-honest receiver. Let $E := X_C \oplus Y$. Π implements (p, q, ε)-WOT in the semi-honest model, if*

- *(Correctness)* $\Pr[Y \neq X_C] \leq \varepsilon$.
- *(Security for the sender)* $\mathrm{PredAdv}(X_{1-C} \mid V, E) \leq q$.
- *(Security for the receiver)* $\mathrm{PredAdv}(C \mid U, E) \leq p$.

Since C and Y are part of V, (C, X_C, V) is a function of (V, E). Note that for the protocols we present here, it would be sufficient to require $\mathrm{PredAdv}(C \mid U) \leq p$ for the security for the receiver. We do not use this definition in order to get a stronger Theorem 6 that is easier to proof.

Definition 5 (Weak oblivious transfer, malicious model). *Let Π be a protocol between a sender and a receiver that outputs (X_0, X_1) to the sender and (C, Y) to the receiver. Π implements (p, q, ε)-WOT in the malicious model, if*

- *(Correctness)* $\Pr[Y \neq X_C] \leq \varepsilon$.
- (Security for the sender) *For an honest sender and any (malicious) receiver with output $\overline{V}$, there exists a $\overline{C}$, such that $\mathrm{PredAdv}(X_{1-C} \mid \overline{C}, X_{\overline{C}}, \overline{V}) \leq q$.*
- (Security for the receiver) *For an honest receiver and any (malicious) sender with output $\overline{U}$, we have $\mathrm{PredAdv}(C \mid \overline{U}) \leq p$.*

It is easy to see that in both models $(\varepsilon, \varepsilon, \varepsilon)$-WOT implies $\binom{2}{1}$-ROT[1] with an error of at most ε.

Besides the fact that we only consider a randomized version of WOT, our definitions of (p, q, ε)-WOT differ from the definitions used in [17] and [15] in the fact that we do not specify exactly what a malicious player may receive, but we only require that his output should not give too much information about X_{1-C} and C. This means that a malicious player may, for example, always receive whether an error occurred in the transmission or not, if that information is independent of the inputs. The most important difference is, however, that our definitions do not require that the error must occur independently of the event that a player gets side information, which is very important when we want to apply it. Note that our definitions still are quite close to the definitions from [17,15], because there exist events with probability $1 - p$ and $1 - q$, such that if they occur, then the adversary does not get any side information.

In order to improve the achievable range of the reductions, *Generalized WOT* (GWOT) was introduced in [15]. Our weaker definitions of WOT imply that, at least for the moment, the usage of GWOT does not give any advantage over WOT.

Notice that the impossibility result, Lemma 1 in [17], only works for our weaker definitions of WOT.

5.1   Basic Protocols for WOT Amplification

To achieve $\binom{2}{1}$-OT[1] from (p, q, ε)-WOT, we will use the reductions R-Reduce, S-Reduce and E-Reduce. Protocol R-Reduce is used to reduce the parameter p, and Protocol S-Reduce is used to reduce the parameter q. Both protocols where already used in [12,17,15,20], as well as in [22,33] to build OT-combiners. It is easy to verify that these protocols are also secure when our definitions of (p, q, ε)-WOT is used. (Notice that R-Reduce and S-Reduce, as well as E-Reduce below, use a non-randomized WOT as input. Therefore, we have to apply first the protocol presented in [4,2] that converts ROT into OT.)

Lemma 3 ([17]). *Protocol R-Reduce$(n, \mathcal{W})$ implements a (p', q', ε')-WOT in the semi-honest and the malicious model out of n instances of (p, q, ε)-WOT, where $p' = 1 - (1 - p)^n \leq np$, $q' = q^n \leq e^{-n(1-q)}$, and $\varepsilon' = (1 - (1 - 2\varepsilon)^n)/2 \leq n\varepsilon$.*

Protocol S-Reduce$(n, \mathcal{W})$ implements a (p', q', ε')-WOT in the semi-honest and the malicious model out of n instances of (p, q, ε)-WOT, where $p' = p^n \leq e^{-n(1-p)}$, $q' = 1 - (1 - q)^n \leq np$, and $\varepsilon' = (1 - (1 - 2\varepsilon)^n)/2 \leq n\varepsilon$.

Protocol E-Reduce was also used in [20] and is an one-way variant of Protocol E-Reduce presented in [17]. It is only secure in the semi-honest model.

Protocol E-Reduce$(n, \mathcal{W})$
The sender has input $(x_0, x_1) \in \{0, 1\} \times \{0, 1\}$, and the receiver $c \in \{0, 1\}$.

1. They execute $\mathcal{W}$ n times, using x_0, x_1 and c as input in the ith execution. The receiver receives y_i.
2. The receiver outputs $y := \text{majority}(y_1, \ldots, y_n)$.

Lemma 4. *Protocol E-Reduce$(n, \mathcal{W})$ implements (p', q', ε')-WOT in the semi-honest model out of n instances of (p, q, ε)-WOT, where $p' = 1 - (1 - p)^n \leq np$, $q' = 1 - (1 - q)^n \leq nq$ and $\varepsilon' = \sum_{i=\lceil n/2 \rceil}^{n} \binom{n}{i} \varepsilon^i (1 - \varepsilon)^{n-i} \leq e^{-2n(1/2 - \varepsilon)^2}$.*

The proof of Lemma 4 is straightforward. The last inequality follows from the Chernoff-Hoeffding bound.

5.2   WOT Amplification for $\varepsilon = 0$

If $p, q > 0$, but $\varepsilon = 0$, we only need Protocols R-Reduce and S-Reduce. As they are the same as in [17], their result for this case also holds for our definitions. The bound is optimal. For a more detailed analysis, see [41].

Theorem 2 ([17]). *If $p + q \leq 1 - 1/\text{poly}(k)$, then $(2^{-k}, 2^{-k}, 0)$-WOT can be efficiently implemented using $(p, q, 0)$-WOT secure in the semi-honest and the malicious model.*

5.3   WOT Amplification for $p = 0$ or $q = 0$

The special case where $\varepsilon > 0$, but either $p = 0$ or $q = 0$ has not been considered in [17]. There is a strong connection of this problem to the *one-way key-agreement problem* studied in [28], as well as to the *statistical distance polarization problem* studied in [37,38]. We use the same protocol as Lemma 3.1.12 in [38].

Theorem 3. *For constant p, q, and ε with $p = 0 \;\wedge\; \sqrt{q} + 2\varepsilon < 1$ or $q = 0 \;\wedge\; \sqrt{p} + 2\varepsilon < 1$, $(2^{-k}, 2^{-k}, 2^{-k})$-WOT can efficiently be implemented using (p, q, ε)-WOT secure in the semi-honest model.*

Proof. We will only show the theorem for $q = 0$. For $p = 0$ it is symmetric.

Let $\beta = p$, and $\alpha = 1 - 2\varepsilon$. Let $\lambda = \min(\alpha^2/\beta, 2)$, $\ell = \lceil \log_\lambda 4k \rceil$ and $m = \lambda^\ell/(2\alpha^{2\ell}) \le (\alpha^{2\ell}/\beta^\ell)/(2\alpha^{2\ell}) = 1/(2\beta^\ell)$. From $\sqrt{p} + 2\varepsilon < 1$ follows that $\beta < \alpha^2$ and hence, $1 < \lambda \le 2$. Notice that m is polynomial in k, since $\ell = O(\log k)$. We use the reductions $\mathcal{W}' = $ S-Reduce$(\ell, \mathcal{W})$, $\mathcal{W}'' = $ E-Reduce$(m, \mathcal{W}')$, and $\mathcal{W}''' = $ S-Reduce$(k, \mathcal{W}'')$. Since $\mathcal{W}$ is a $(\beta, 0, (1 - \alpha)/2)$-WOT, $\mathcal{W}'$ is a $(\beta', 0, (1 - \alpha')/2)$-WOT, where $\beta' = \beta^\ell$ and $\alpha' = \alpha^\ell$. $\mathcal{W}''$ is a $(\beta'', 0, (1 - \alpha'')/2)$-WOT with $\beta'' \le m\beta' \le 1/2$ and

$$\alpha'' \ge 1 - 2\exp\left(-\frac{\lambda^\ell}{2\alpha^{2\ell}} \cdot \frac{(\alpha^\ell)^2}{2}\right) = 1 - 2\exp\left(-\frac{\lambda^\ell}{4}\right) \ge 1 - 2e^{-k}\,.$$

Finally, $\mathcal{W}'''$ is a $(\beta''', 0, (1 - \alpha''')/2)$-WOT with $\alpha''' \ge (1 - 2e^{-k})^k \ge 1 - 2ke^{-k} \ge 1 - 2^{-k}$ and $\beta''' \le 2^{-k}$, as long as k is sufficiently large, which can be achieved by artificially increasing k at the start. □

5.4   WOT Amplification for $p, q, \varepsilon > 0$

To find a good protocol for the general case is much harder. We start with the case where all values are smaller than $1/50$.

Lemma 5. *In the semi-honest model, $(2^{-k}, 2^{-k}, 2^{-k})$-WOT can efficiently and securely be implemented using $O(k^{2+\log(3)})$ instances of $(1/50, 1/50, 1/50)$-WOT.*

Proof. We iterate the reduction $\mathcal{W}' := $ S-Reduce$(2, $ R-Reduce$(2, $ E-Reduce$(3, \mathcal{W})))$ t times. In every iteration, we have $p' \le (2 \cdot (3p))^2 = 36p^2$, $q' \le 2 \cdot ((3q)^2) = 18q^2$, and $\varepsilon' \le 2 \cdot 2 \cdot (3\varepsilon^2 - 2\varepsilon^3) \le 12\varepsilon^2$, from which follows that after t iterations, we have $\max(p', q', \varepsilon') \le (36/50)^{2^t}$. To achieve $\max(p', q', \varepsilon') \le 2^{-k}$, we choose $t := \lceil \log k / \log(50/36) \rceil \le \log(3 \cdot k) + 1 = \log(6 \cdot k)$. We need at most $12^t \le (6 \cdot k)^{\log(12)} = O(k^{2+\log(3)})$ instances of $\mathcal{W}$. □

The following Lemma 6 is a corrected version of Lemma 5 in [17]. Since our Protocol E-Reduce is different, we are only able to achieve a smaller bound. As in [17], we obtain our bound using a simulation.

Let $l_i(p, q)$ be a function such that for all p, q and $\varepsilon < l_i(p, q)$, $\binom{2}{1}$-ROT^1 can be implemented using (p, q, ε)-WOT. Using $l_i(p, q)$, we define $l_{i+1}(p, q) := \max(S_\varepsilon^{-1}(l_i(S_p(p), S_q(q))), R_\varepsilon^{-1}(l_i(R_p(p), R_q(q))), E_\varepsilon^{-1}(l_i(E_p(p), E_q(q))))$, where $S_p(p) := p^2$, $S_q(q) := 1 - (1-q)^2$, $S_\varepsilon^{-1}(\varepsilon) := (1 - \sqrt{1 - 2\varepsilon})/2$, $R_p(p) := 1 - (1-p)^2$, $R_q(q) := q^2$, $R_\varepsilon^{-1}(\varepsilon) := (1 - \sqrt{1 - 2\varepsilon})/2$, $E_p(p) := 1 - (1 - p)^3$, $E_q(q) := 1 - (1 - q)^3$, and $E_\varepsilon^{-1}(\varepsilon)$ is the inverse of $E_\varepsilon(\varepsilon) := 3\varepsilon - 2\varepsilon^3$.

Now for all p, q and $\varepsilon < l_{i+1}(p, q)$, $\binom{2}{1}$-ROT^1 can be implemented using (p, q, ε)-WOT, since one of the protocols S-Reduce$(2, \mathcal{W})$, R-Reduce$(2, \mathcal{W})$, or E-Reduce$(3, \mathcal{W})$ achieves $\varepsilon' < l_i(p', q')$, from which we can achieve $\binom{2}{1}$-ROT^1.

From Lemma 5 follows that $l_0(p, q) := 0.02 - p - q$ satisfies our condition. Iterating 8 times, we get that for all p, q, $l_8(p, q) \geq (0.15 - p - q)/2$. Using $l'_0(p, q) := (0.15 - p - q)/2$ and iterating 11 times, we get $l'_{11}(p, q)$. Since for all p, q we have $l'_{11}(p, q) \geq (0.24 - p - q)/2$, we get

Lemma 6. *If $p + q + 2\varepsilon \leq 0.24$, then $(2^{-k}, 2^{-k}, 2^{-k})$-WOT can efficiently be implemented using (p, q, ε)-WOT secure in the semi-honest model.*

Often (p, q, ε)-WOT will be applied when one of the three values is big, while the others are small. We will now give bounds for these three cases.

Lemma 7. *If $p + 22q + 44\varepsilon < 1 - 1/\operatorname{poly}(k)$, then $(2^{-k}, 2^{-k}, 2^{-k})$-WOT can efficiently be implemented using (p, q, ε)-WOT secure in the semi-honest model.*

Proof. We apply $\mathcal{W}' = \mathsf{S\text{-}Reduce}(n, \mathcal{W})$ for $n = \lceil \ln(20)/(1 - p) \rceil$. From Lemmas 3 follows directly that we obtain a (p', q', ε')-WOT with $p' + q' + 2\varepsilon' \leq 0.24$. The lemma follows now from Lemma 6. □

Lemma 8. *If $22p + q + 44\varepsilon < 1 - 1/\operatorname{poly}(k)$, then $(2^{-k}, 2^{-k}, 2^{-k})$-WOT can efficiently be implemented using (p, q, ε)-WOT secure in the semi-honest model.*

Lemma 9. *If $7\sqrt{p + q} + 2\varepsilon < 1 - 1/\operatorname{poly}(k)$, then $(2^{-k}, 2^{-k}, 2^{-k})$-WOT can efficiently be implemented using (p, q, ε)-WOT secure in the semi-honest model.*

Proof. We apply $\mathcal{W}' = \mathsf{E\text{-}Reduce}(n, \mathcal{W})$ for $n = \lceil \ln(50)/(2(\frac{1}{2} - \varepsilon)^2) \rceil$. From Lemma 4 follows directly that we obtain a (p', q', ε')-WOT with $p' + q' + 2\varepsilon' \leq 0.24$. The lemma follows now from Lemma 6. □

Theorem 4. *If $p + q + 2\varepsilon \leq 0.24$, or $\min(p + 22q + 44\varepsilon, 22p + q + 44\varepsilon, 7\sqrt{p + q} + 2\varepsilon) \leq 1 - 1/\operatorname{poly}(k)$, then $(2^{-k}, 2^{-k}, 2^{-k})$-WOT can efficiently be implemented using (p, q, ε)-WOT secure in the semi-honest model.*

6   Computationally Secure Weak Oblivious Transfer

Even though the protocols from the last section are purely information-theoretic, we can also use them in the computational semi-honest model, as we will see in this section. The main tool to show this will be a *pseudo-randomness extraction theorem* (Theorem 5), that is a modified version of Theorem 7.3 from [27]. It is based on the *uniform hard-core lemma* from [26], which is a uniform variant of the hard-core lemma from [29].

6.1   Pseudo-randomness Extraction

The main difference of Theorem 5 compared to the (implicit) extraction lemma in [24,25] and the extraction lemma in [21] is that it allows the adversary to gain some additional knowledge during the extraction (expressed by the function Leak), which is needed for our application.

Besides a simplification, the main difference of our Theorem 5 to Theorem 7.3 from [27] is that we allow the functions Ext and Leak also to depend on the values Z_i. Intuitively, Theorem 5 says the following: if we have an information-theoretic protocol (modeled by the two functions Ext and Leak), that converts many instances of X over which an adversary having Z has only partial knowledge, into an X' over which the adversary has almost no knowledge, and if we have a computational protocol (modeled by the function $f(W)$ and the predicate $P(W)$), where an adversary having $f(W)$ has only partial *computational* knowledge about $P(W)$, then the modified information-theoretic protocol, where every instance of X is replaced with $P(W)$ and every instance of Z with $f(W)$, will produce a value over which the adversary has almost no computation knowledge.

Theorem 5 (Pseudo-randomness Extraction Theorem, Modified Theorem 7.3 in [27]). *Let the functions* $f : \{0,1\}^k \to \{0,1\}^\ell$, $P : \{0,1\}^k \to \{0,1\}$, *and* $\beta : \mathbb{N} \to [0,1]$ *computable in time* poly(k) *be given. Assume that every polynomial time algorithm B satisfies*

$$\Pr[B(f(W)) = P(W)] \leq (1 + \beta(k))/2$$

for all but finitely many k, for a uniform random $W \in \{0,1\}^k$. Further, let also functions $n(k)$, $s(k)$,

$$\text{Ext} : \{0,1\}^{\ell \cdot n} \times \{0,1\}^n \times \{0,1\}^s \to \{0,1\}^t ,$$

$$\text{Leak} : \{0,1\}^{\ell \cdot n} \times \{0,1\}^n \times \{0,1\}^s \to \{0,1\}^{t'} ,$$

be given which are computable in time poly(k), *and satisfy the following: for any distribution P_{XZ} over $\{0,1\} \times \{0,1\}^\ell$ where* PredAdv$(X \mid Z) \leq \beta(k)$, *the output of* Ext(Z^n, X^n, R) *is $\varepsilon(k)$-close to uniform with respect to* Leak(Z^n, X^n, R) *(where $R \in \{0,1\}^s$ is chosen uniformly at random). Then, no polynomial time algorithm A, which gets as input*

$$\text{Leak}((f(w_1), \dots, f(w_n)), (P(w_1), \dots, P(w_n)), R) ,$$

(where $(w_1, \dots, w_n)$ are chosen uniformly at random) distinguishes

$$\text{Ext}((f(w_1), \dots, f(w_n)), (P(w_1), \dots, P(w_n)), R)$$

from a uniform random string of length t with advantage $\varepsilon(k) + \gamma(k)$, for any non-negligible function $\gamma(k)$.

The proof of Theorem 5 is very similar to the proof of Theorem 7.3 in [27] and can be found in [41]. Note that our proof makes an additional step that has been missing in the proof of Theorem 7.3 in [27].

6.2 Computational-WOT Amplification

We will denote the computational version of (p, q, ε)-WOT by (p, q, ε)-compWOT. The difference to the information-theoretic definition is that now we require the algorithms that guess X_{1-C} or C to be efficient, i.e., to run in polynomial time.

Definition 6 (Computationally secure weak oblivious transfer, semi-honest model). *Let functions $\varepsilon : \mathbb{N} \to [0, 1/2]$, $p : \mathbb{N} \to [0, 1]$, and $q : \mathbb{N} \to [0, 1]$ computable in time $\mathrm{poly}(k)$ be given. Let Π be a protocol between a sender and a receiver. On input 1^k, Π outputs $(X_0, X_1) \in \{0, 1\} \times \{0, 1\}$ to the sender and $(C, Y) \in \{0, 1\} \times \{0, 1\}$ to the receiver. Let U be the view of a semi-honest sender, and let V be the view of a semi-honest receiver. Let $E := X_C \oplus Y$. Π implements $(p(k), q(k), \varepsilon(k))$-compWOT in the semi-honest model, if*

- *(Efficiency) Π can be executed in time $\mathrm{poly}(k)$.*
- *(Correctness) $\Pr[Y \neq X_C] \leq \varepsilon(k)$ for all k.*
- *(Security for the sender) All polynomial time algorithms A satisfy*

$$\Pr[A(V, E) = X_{1-C}] \leq (1 + q(k))/2$$

for all but finitely many k.
- *(Security for the receiver) All polynomial time algorithms A satisfy*

$$\Pr[A(U, E) = C] \leq (1 + p(k))/2$$

for all but finitely many k.

We apply Theorem 5 twice to get Theorem 6, which says that if we have a protocol that implements (p, q, ε)-compWOT, and an efficient information-theoretic protocol that implements $\binom{2}{1}$-ROT^1 from (p, q, ε)-WOT secure in the semi-honest model, then we can construct a protocol that implements $\binom{2}{1}$-ROT^1 computationally secure in the semi-honest model.

Theorem 6. *Let the functions $\varepsilon(k)$, $p(k)$, and $q(k)$ computable in time $\mathrm{poly}(k)$ be given. Let a protocol Π achieve (p, q, ε)-compWOT and let an efficient information-theoretic protocol Π' be given which takes 1^k as input and implements $(2^{-k}, 2^{-k}, 2^{-k})$-WOT from (p, q, ε)-WOT secure in the semi-honest model. Then, protocol Π', where every instance of (p, q, ε)-WOT is replaced by an independent outcome of Π, implements $\binom{2}{1}$-ROT^1 computationally secure in the semi-honest model.*

Proof. Let $W = (W_S, W_R)$ be the randomness used in Π by the sender and the receiver, and let Z be the communication. (X_0, X_1) and (C, Y) are the output to the honest sender and receiver, respectively. $U = (X_0, X_1, Z, W_S)$ and $V = (C, Y, Z, W_R)$ are the views of the semi-honest sender and receiver, respectively. Let $E := Y \oplus X_C$. Note that all these values are functions of W.

In the protocol Π', the sender receives $(X_0, X_1)^n$, which are her output from the n independent instances of Π, and the receiver receives $(C, Y)^n$. The sender outputs (X_0^*, X_1^*) and the receiver (C^*, Y^*). Let $R = (R_S, R_R)$ be the randomness used in Π' by both players, and let Z' be the communication produced by Π'. $V^* = (E^*, C^*, Y^*, V^n, Z', R_R)$ is the view of the semi-honest receiver after the execution of Π', and $U^* = (E^*, X_0^*, X_1^*, U^n, Z', R_S)$ the view of the semi-honest sender. Let $E^* := Y^* \oplus X_{C^*}^*$. Note that the values E^*, X_0^*, X_1^*, C^*, Y^*, V^*, U^* and Z' are functions of $((X_0, X_1, C, Y)^n, R)$.

First of all, the resulting protocol will be correct and efficient, as every outcome of Π satisfies $\Pr[Y \neq X_C] \leq \varepsilon$.

For the security of the sender, we define the following functions: let $f(W) := (V, E)$ and $P(W) := X_{1-C}$. Since $X_C = E \oplus Y$, it is possible to simulate the protocol Π' using the values $(V, E)^n$, $(X_{1-C})^n$, and R. Therefore, we can define $\mathrm{Ext}((V, E)^n, (X_{1-C})^n, R) := X_{1-C*}^*$ and $\mathrm{Leak}((V, E)^n, (X_{1-C})^n, R) := V^*$. Since Π' implements $(\mathrm{negl}(k), \mathrm{negl}(k), \mathrm{negl}(k))$-WOT, the functions Ext and Leak satisfy the extraction requirements from Theorem 5 with $\varepsilon(k) = \mathrm{negl}(k)$. Furthermore, Ext and Leak can be computed efficiently, since the protocol Π' is efficient. From the security condition of compWOT follows that every polynomial-time algorithm B satisfies $\Pr[B(f(W)) = P(W)] \leq (1 + q(k))/2$ for all but finitely many k, for W chosen uniformly at random. Theorem 5 tells us that no polynomial time algorithm A, which gets as input $\mathrm{Leak}((V, E)^n, (X_{1-C})^n, R)$, distinguishes $\mathrm{Ext}((V, E)^n, (X_{1-C})^n, R)$ from a uniform random bit with advantage $\mathrm{negl}(k) + \gamma(k)$, for any non-negligible function $\gamma(k)$, from which follows that the protocol is computationally secure for the sender.

For the security of the receiver, we define the following functions: let $f(W) := (U, E)$ and $P(W) := C$. Since $X_C = E \oplus Y$, it is possible to simulate the protocol Π' using the values $(U, E)^n$, C^n, and R. Therefore, we can define $\mathrm{Ext}((U, E)^n, C^n, R) := C^*$, and $\mathrm{Leak}((U, E)^n, C^n, R) := U^*$. Since Π' implements $(\mathrm{negl}(k), \mathrm{negl}(k), \mathrm{negl}(k))$-WOT, the functions Ext and Leak satisfy the extraction requirements from Theorem 5 with $\varepsilon(k) = \mathrm{negl}(k)$. Furthermore, Ext and Leak can be computed efficiently, since the protocol Π' is efficient. From the security condition of compWOT follows that every polynomial time algorithm A satisfies $\Pr[A(f(W)) = P(W)] \leq (1 + p(k))/2$ for all but finitely many k, for W chosen uniformly at random. Theorem 5 tells us that no polynomial time algorithm B, which gets as input $\mathrm{Leak}((U, E)^n, C^n, R)$, distinguishes $\mathrm{Ext}((U, E)^n, C^n, R)$ from a uniform random bit with advantage $\mathrm{negl}(k) + \gamma(k)$, for any non-negligible function $\gamma(k)$, from which follows that the protocol is computationally secure for the receiver.   □

Together with the information-theoretic protocols presented in Section 5, (Theorems 2, 3 and 4) we get a way to implement ROT based on compWOT, computationally secure in the semi-honest model. From [31] follows that such a protocol implies one-way functions. Using the compiler from [19], we get an implementation of OT computationally secure in the malicious model. The following corollary follows.

Corollary 1. *Let the functions $\varepsilon(k)$, $p(k)$, and $q(k)$ computable in time $\mathrm{poly}(k)$ be given, such that either for all k $\varepsilon = 0 \wedge p + q < 1 - 1/\mathrm{poly}(k)$ or $p + q + 2\varepsilon \leq 0.24$ or $\min(p + 22q + 44\varepsilon, 22p + q + 44\varepsilon, 7\sqrt{p+q} + 2\varepsilon) < 1 - 1/\mathrm{poly}(k)$, or, for constant functions $p(k)$, $q(k)$ and $\varepsilon(k)$, $(p = 0) \wedge (\sqrt{q} + 2\varepsilon < 1)$ or $(q = 0) \wedge (\sqrt{p} + 2\varepsilon < 1)$. If there exists a protocol Π that achieve (p, q, ε)-compWOT computationally secure in the semi-honest model, then there exists a protocol that implements $\binom{2}{1}$-OT^1 computationally secure in the malicious model.*

Corollary 1 generalizes results from [20], because it covers a much wider range of values for p, q, and ε, and it allows the security for both players to be only computational.

Acknowledgments

I would like to thank Thomas Holenstein and Stefan Wolf for helpful discussions, and Ivan Damgård and Louis Salvail for answering my questions about their work. I also thank Serge Fehr, Iftach Haitner, Melanie Raemy, Christian Schaffner and anonymous referees for giving helpful comments on this work.

I was supported by the Swiss National Science Foundation (SNF).

References

1. D. Beaver. Foundations of secure interactive computing. In *Advances in Cryptology — CRYPTO '91*, volume 1233 of *LNCS*, pages 377–391. Springer-Verlag, 1992.
2. D. Beaver. Precomputing oblivious transfer. In *Advances in Cryptology — EUROCRYPT '95*, volume 963 of *LNCS*, pages 97–109. Springer-Verlag, 1995.
3. C. H. Bennett, G. Brassard, C. Crépeau, and U. Maurer. Generalized privacy amplification. *IEEE Transactions on Information Theory*, 41, 1995.
4. C. H. Bennett, G. Brassard, C. Crépeau, and H. Skubiszewska. Practical quantum oblivious transfer. In *Advances in Cryptology — CRYPTO '91*, volume 576 of *LNCS*, pages 351–366. Springer, 1992.
5. C. H. Bennett, G. Brassard, and J.-M. Robert. Privacy amplification by public discussion. *SIAM Journal on Computing*, 17(2):210–229, 1988.
6. G. Brassard and C. Crépeau. Oblivious transfers and privacy amplification. In *Advances in Cryptology — EUROCRYPT '97*, volume 1233 of *LNCS*, pages 334–347. Springer-Verlag, 1997.
7. G. Brassard, C. Crépeau, and S. Wolf. Oblivious transfers and privacy amplification. *Journal of Cryptology*, 16(4):219–237, 2003.
8. C. Cachin. On the foundations of oblivious transfer. In *Advances in Cryptology — EUROCRYPT '98*, volume 1403 of *LNCS*, pages 361–374. Springer-Verlag, 1998.
9. R. Canetti. Security and composition of multiparty cryptographic protocols. *Journal of Cryptology*, 13(1):143–202, 2000.
10. J. L. Carter and M. N. Wegman. Universal classes of hash functions. *Journal of Computer and System Sciences*, 18:143–154, 1979.
11. C. Crépeau. Efficient cryptographic protocols based on noisy channels. In *Advances in Cryptology — CRYPTO '97*, volume 1233 of *LNCS*, pages 306–317. Springer-Verlag, 1997.
12. C. Crépeau and J. Kilian. Achieving oblivious transfer using weakened security assumptions (extended abstract). In *Proceedings of the 29th Annual IEEE Symposium on Foundations of Computer Science (FOCS '88)*, pages 42–52, 1988.
13. C. Crépeau, K. Morozov, and S. Wolf. Efficient unconditional oblivious transfer from almost any noisy channel. In *Proceedings of Fourth Conference on Security in Communication Networks (SCN)*, volume 3352 of *LNCS*, pages 47–59. Springer-Verlag, 2004.

14. C. Crépeau, G. Savvides, C. Schaffner, and J. Wullschleger. Information-theoretic conditions for two-party secure function evaluation. In *Advances in Cryptology — EUROCRYPT '06*, volume 4004 of *LNCS*, pages 538–554. Springer-Verlag, 2006. Full version available at http://eprint.iacr.org/2006/183.

15. I. Damgård, S. Fehr, K. Morozov, and L. Salvail. Unfair noisy channels and oblivious transfer. In *Theory of Cryptography Conference — TCC '04*, volume 2951 of *LNCS*, pages 355–373. Springer-Verlag, 2004.

16. I. Damgard, S. Fehr, L. Salvail, and C. Schaffner. Cryptography in the bounded quantum-storage model. In *Proceedings of the 46th Annual IEEE Symposium on Foundations of Computer Science (FOCS '05)*, pages 449–458. IEEE Computer Society, 2005.

17. I. Damgård, J. Kilian, and L. Salvail. On the (im)possibility of basing oblivious transfer and bit commitment on weakened security assumptions. In *Advances in Cryptology — EUROCRYPT '99*, volume 1592 of *LNCS*, pages 56–73. Springer-Verlag, 1999.

18. S. Even, O. Goldreich, and A. Lempel. A randomized protocol for signing contracts. *Commun. ACM*, 28(6):637–647, 1985.

19. O. Goldreich, S. Micali, and A. Wigderson. How to play any mental game. In *Proceedings of the 21st Annual ACM Symposium on Theory of Computing (STOC '87)*, pages 218–229. ACM Press, 1987.

20. I. Haitner. Implementing oblivious transfer using collection of dense trapdoor permutations. In *Theory of Cryptography Conference — TCC '04*, volume 2951 of *LNCS*, pages 394–409. Springer-Verlag, 2004.

21. I. Haitner, D. Harnik, and O. Reingold. On the power of the randomized iterate. In *Advances in Cryptology — CRYPTO '06*, volume 4117 of *LNCS*, pages 21–40. Springer-Verlag, 2006.

22. D. Harnik, J. Kilian, M. Naor, O. Reingold, and A. Rosen. On robust combiners for oblivious transfer and other primitives. In *Advances in Cryptology — EURO-CRYPT '05*, volume 3494 of *LNCS*, pages 96–113, 2005.

23. D. Harnik, M. Naor, O. Reingold, and A. Rosen. Completeness in two-party secure computation: a computational view. In *Proceedings of the 36th Annual ACM Symposium on Theory of Computing (STOC '04)*, pages 252–261. ACM Press, 2004.

24. J. Håstad. Pseudo-random generators under uniform assumptions. In *Proceedings of the 22st Annual ACM Symposium on Theory of Computing (STOC '90)*, pages 395–404. ACM Press, 1990.

25. J. Håstad, R. Impagliazzo, L. A. Levin, and M. Luby. A pseudorandom generator from any one-way function. *SIAM J. Comput.*, 28(4):1364–1396, 1999.

26. T. Holenstein. Key agreement from weak bit agreement. In *Proceedings of the 37th ACM Symposium on Theory of Computing (STOC '05)*, pages 664–673. ACM Press, 2005.

27. T. Holenstein. *Strengthening key agreement using hard-core sets*. PhD thesis, ETH Zurich, Switzerland, 2006. Reprint as vol. 7 of *ETH Series in Information Security and Cryptography*, Hartung-Gorre Verlag.

28. T. Holenstein and R. Renner. One-way secret-key agreement and applications to circuit polarization and immunization of public-key encryption. In *Advances in Cryptology — CRYPTO '05*, volume 3621 of *LNCS*, pages 478–493. Springer-Verlag, 2005.

29. R. Impagliazzo. Hard-core distributions for somewhat hard problems. In *Proceedings of the 36th Annual IEEE Symposium on Foundations of Computer Science (FOCS '95)*, pages 538–545. IEEE Computer Society, 1995.

30. R. Impagliazzo, L. A. Levin, and M. Luby. Pseudo-random generation from one-way functions. In *Proceedings of the 21st Annual ACM Symposium on Theory of Computing (STOC '89)*, pages 12–24. ACM Press, 1989.

31. R. Impagliazzo and M. Luby. One-way functions are essential for complexity based cryptography. In *Proceedings of the 30th Annual IEEE Symposium on Foundations of Computer Science (FOCS '89)*, pages 230–235, 1989.

32. J. Kilian. Founding cryptography on oblivious transfer. In *Proceedings of the 20th Annual ACM Symposium on Theory of Computing (STOC '88)*, pages 20–31. ACM Press, 1988.

33. R. Meier, B. Przydatek, and J. Wullschleger. Robuster combiners for oblivious transfer. In *Theory of Cryptography Conference — TCC '07*, LNCS. Springer-Verlag, 2007.

34. S. Micali and P. Rogaway. Secure computation (abstract). In *Advances in Cryptology — CRYPTO '91*, volume 576 of *LNCS*, pages 392–404. Springer-Verlag, 1992.

35. A. Nascimento and A. Winter. On the oblivious transfer capacity of noisy correlations. In *Proceedings of the IEEE International Symposium on Information Theory (ISIT '06)*, 2006.

36. M. O. Rabin. How to exchange secrets by oblivious transfer. Technical Report TR-81, Harvard Aiken Computation Laboratory, 1981.

37. A. Sahai and S. Vadhan. Manipulating statistical difference. In *Randomization Methods in Algorithm Design (DIMACS Workshop '97)*, volume 43 of *DIMACS Series in Discrete Mathematics and Theoretical Computer Science*, pages 251–270. American Mathematical Society, 1999.

38. S. Vadhan. *A study of statistical zero-knowledge proofs*. PhD thesis, Massachusets Institute of Technology, USA, 1999.

39. S. Wiesner. Conjugate coding. *SIGACT News*, 15(1):78–88, 1983.

40. S. Wolf and J. Wullschleger. Zero-error information and applications in cryptography. In *Proceedings of 2004 IEEE Information Theory Workshop (ITW '04)*, 2004.

41. J. Wullschleger. *Oblivious-Transfer Amplification*. PhD thesis, ETH Zurich, Switzerland, 2007.

42. A. C. Yao. Protocols for secure computations. In *Proceedings of the 23rd Annual IEEE Symposium on Foundations of Computer Science (FOCS '82)*, pages 160–164, 1982.

Simulatable Adaptive Oblivious Transfer

Jan Camenisch[1], Gregory Neven[2,3], and abhi shelat[1]

[1] IBM Research, Zurich Research Laboratory, CH-8803 Rüschlikon
[2] Katholieke Universiteit Leuven, Dept. of Electrical Engineering, B-3001 Heverlee
[3] Ecole Normale Supérieure, Département d'Informatique, 75230 Paris Cedex 05

Abstract. We study an *adaptive* variant of oblivious transfer in which a
sender has N messages, of which a receiver can adaptively choose to receive k one-after-the-other, in such a way that (a) the sender learns nothing about the receiver's selections, and (b) the receiver only learns about
the k requested messages. We propose two practical protocols for this
primitive that achieve a stronger security notion than previous schemes
with comparable efficiency. In particular, by requiring full simulatabil-
ity for both sender and receiver security, our notion prohibits a subtle
selective-failure attack not addressed by the security notions achieved by
previous practical schemes.

Our first protocol is a very efficient generic construction from unique
blind signatures in the random oracle model. The second construction
does not assume random oracles, but achieves remarkable efficiency with
only a constant number of group elements sent during each transfer. This
second construction uses novel techniques for building efficient simulatable protocols.

1 Introduction

The *oblivious transfer* (OT) primitive, introduced by Rabin [Rab81], and extended by Even, Goldreich, and Lempel [EGL85] and Brassard, Crépeau and
Robert [BCR87] is deceptively simple: there is a sender S with messages $M_1, \ldots,$
M_N and a receiver R with a selection value $\sigma \in \{1, \ldots, N\}$. The receiver wishes
to retrieve M_σ from S in such a way that (1) the sender does not "learn" anything about the receiver's choice σ and (2) the receiver "learns" only M_σ and
nothing about any other message M_i for $i \neq \sigma$. Part of the allure of OT is
that it is *complete*, i.e., if OT can be realized, virtually any secure multiparty
computation can be [GMW87, CK90].

In this paper, we consider an *adaptive* version of oblivious transfer in which
the sender and receiver first run an initialization phase during which the sender
commits to a "database" containing her messages. Later on, the sender and
receiver interact as before so that the receiver can retrieve some message M_σ. In
addition, we allow the receiver to interact with the sender $k-1$ additional times,
one interaction after the other, in order to retrieve additional values from the
sender's database. Notice here that we specifically model the situation in which
the receiver's selection in the ith phase can *depend* on the messages retrieved in
the prior $i-1$ phases. This type of adaptive OT problem is central to a variety

M. Naor (Ed.): EUROCRYPT 2007, LNCS 4515, pp. 573–590, 2007.

of practical problems such as patent searches, treasure hunting, location-based services, oblivious search, and medical databases [NP99b].

The practicality of this adaptive OT problem also drives the need for efficient solutions to it. Ideally, a protocol should only require communication linear in N and the security parameter κ during the initialization phase (so that the sender commits to the N messages), and an amount of communication of $O(\max(\kappa, \log N))$ during each transfer phase (so that the receiver can use cryptography and encode the index of his choice).[1] In the race to achieve these efficiency parameters, however, we must also not overlook—or worse, *settle* for less-than-ideal security properties.

1.1   Security Definitions of Oblivious Transfer

An important contribution of this work is that it achieves a stronger simulation-based security notion at very little cost with respect to existing schemes that achieve weaker notions. We briefly summarize the various security notions for OT presented in the literature, and how our notion extends them.

Honest-but-curious model. In this model, all parties are assumed to follow the protocol honestly. Security guarantees that after the protocol completes, a curious participant cannot analyze the transcript of the protocol to learn anything else. Any protocol in the honest-but-curious model can be transformed into fully-simulatable protocols, albeit at the cost of adding complexity assumptions and requiring costly general zero-knowledge proofs for each protocol step.

Half-simulation. This notion, introduced by Naor and Pinkas [NP05], considers malicious senders and receivers, but handles their security separately. Receiver security is defined by requiring that the sender's view of the protocol when the receiver chooses index σ is indistinguishable from a view of the protocol when the receiver chooses σ'. Sender security, on the other hand, involves a stronger notion. The requirement follows the real-world/ideal-world paradigm and guarantees that any malicious receiver in the real world can be mapped to a receiver in an idealized game in which the OT is implemented by a trusted party. Usually, this requires that receivers are efficiently "simulatable," thus we refer to this notion as *half-simulation*.

The Problem of Selective Failure. We argue that the definition of half-simulation described above does not imply all properties that one may expect from a $OT^N_{k\times 1}$ scheme. Notice that a cheating sender can always make the current transfer fail by sending bogus messages. However, we would not expect him to be able to cause failure based on some property of the receiver's selection. Of course, the sener can also prevent the receiver from retrieving M_σ by replacing it

---

[1] In practice, we assume that $\kappa > \log(N)$—so that the protocol can encode the receiver's selection—but otherwise that κ is chosen purely for the sake of security. In this sense, $O(\kappa)$ is both conceptually and practically different than $O(\mathrm{polylog}(N))$.

with a random value during the initialization phase. But again, the sender should not be able to make this decision anew at each transfer phase. For example, the sender should not be able to make the first transfer fail for $\sigma = 1$ but succeed for $\sigma \in \{2, \ldots, N\}$, and to make the second transfer fail for $\sigma = 2$ but succeed for $\sigma \in \{1, 3, \ldots, N\}$. The receiver could publicly complain whenever a transfer fails, but by doing so it gives up the privacy of its query. Causing transfers to fail may on the long term harm the sender's business, but relying on such arguments to dismiss the problem is terribly naive. A desperate patent search database may *choose* to make faster money by selling a company's recent queries to competitors than by continuing to run its service.

We refer to this issue as the *selective-failure* problem. To see why it is not covered by the half-simulation notion described above, it suffices to observe that the notion of receiver security only *hides* the message received by the receiver from the cheating sender's view. A scheme that is vulnerable to selective-failure attacks does not give the cheating sender any additional advantage in breaking the receiver's privacy, and may therefore be secure under such a notion. (This illustrates the classic argument from work in secure multiparty computation that achieving just privacy is not enough; both privacy and correctness must be achieved simultaneously.) In fact, the schemes of [NP05] are secure under half-simulation, yet vulnerable to selective-failure attacks. In an earlier version [NP99b], the same authors recognize this problem and remark that it can be fixed, but do not give formal support of their claim. A main contribution of this work is to show that it can be done without major sacrifices in efficiency.

Simulatable OT. The security notion that we consider employs the real-world/ideal-world paradigm for both receiver and sender security. We extend the functionality of the trusted party such that at each transfer, the sender inputs a bit b indicating whether it wants the transfer to succeed or fail. This models the capability of a sender in the real world to make the transfer fail by sending bogus messages, but does not enable it to do so based on the receiver's input σ. Moreover, for security we require indistinguishability of the combined outputs of the sender and the receiver, rather than only of the output of the dishonest party. The output of the honest receiver is assumed to consist of all the messages $M_{\sigma_1}, \ldots, M_{\sigma_k}$ that it received. This security notion excludes selective-failure attacks in the real world, because the ideal-world sender is unable to perform such attacks, which will lead to noticeable differences in the receiver's output in the real and ideal world.

Finally, we observe that simulatable oblivious transfer is used as a primitive to build many other cryptographic protocols [Gol04]. By building an efficient OT protocol with such simulation, we take the first steps at realizing many other interesting cryptographic protocols.

1.2   Construction Overview

Our random-oracle protocol. Our first construction is a black-box construction using any unique blind signature scheme. By *unique*, we mean that for all public keys and messages there exists at most one valid signature. First, the sender generates a key pair (pk, sk) for the blind signature scheme, and "commits" to each message in its database by XOR-ing the message M_i with $H(i, s_i)$, where s_i is the unique signature of the message i under pk. Intuitively, we're using s_i as a key to unlock the message M_i. To retrieve the "key" to a message M_i, the sender and receiver engage in the blind signature protocol for message i. By the unforgeability of the signature scheme, a malicious receiver will be unable to unlock more than k such messages. By the blindness of the scheme, the sender learns nothing about which messages have been requested.

The random oracle serves four purposes. First, it serves as a one-time pad to perfectly hide the messages. Second, it allows a simulator to extract the sender's original messages from the commitments so that we can prove receiver-security. Third, in the proof of sender-security, it allows the simulator to both extract the receiver's choice and, via programming the random oracle, to make the receiver open the commitment to an arbitrary message. Finally, it allows us to extract forgeries of the blind signature scheme from a malicious receiver who is able to break sender-security.

Our standard-model protocol. There are three main ideas behind the standard protocol in §4. At a very high level, just as in the random oracle protocol, the sender uses a unique signature of i as a key to encrypt M_i in the initialization phase. However, unlike the random-oracle protocol, we observe here that we only need a blind signature scheme which allows signatures on a small, *a-priori fixed* message space $\{1, \ldots, N\}$.

The second idea concerns the fact that after engaging in the blind-signing protocol, a receiver can easily check whether the sender has sent the correct response during the transfer phase by verifying the signature it received. While seemingly a feature, this property becomes a problem during the simulation of a malicious receiver. Namely, the simulator must commit to N random values during the initialize phase, and later during the transfer phase, open any one of these values to an arbitrary value (the correct message M_i received from the trusted party during simulation). In the random oracle model, this is possible via programming the random oracle. In the standard model, a typical solution would be to use a trapdoor commitment. However, a standard trapdoor commitment is unlikely to work here because most of these require the opener to send the actual committed value when it opens the commitment. This is not possible in our OT setting since the sender does not know which commitment is being opened.

Our solution is to modify the "blind-signing" protocol so that, instead of returning a signature to the user, a one-way function (a bilinear pairing in our case) of the signature is returned. To protect against a malicious sender, the sender then proves in zero-knowledge that the value returned is computed correctly. In the security proof, we will return a random value to the receiver and fake the zero-knowledge proof.

The final idea behind our construction concerns a malicious receiver who may use an invalid input to the "blind-signature protocol" in order to, say, retrieve a signature on a value outside of $\{1, \ldots, N\}$. This is a real concern, since such an attack potentially allows a malicious receiver to learn the product $M_i \cdot M_j$ which violates the security notion. In order to prevent such cheating, we require the receiver to prove in zero-knowledge that (a) it knows the input it is requesting a signature for, and (b) that the input is valid for the protocol. While this is conceptually simple, the problem is that the size of such a theorem statement, and therefore the time and communication complexity of such a zero-knowledge proof, could potentially be linear in N. For our stated efficiency goals, we need a proof of constant size. To solve this final problem, we observe that the input to the blind signature process is a small set—i.e., only has N possible values. Thus, the sender can sign all N possible input messages (using a different signing key x) to the blind signature protocol and publish them in the initialization phase. During the transfer phase, the receiver blinds one of these inputs and then gives a zero-knowledge proof of knowledge that it knows a signature of this blinded input value. Following the work of Camenisch and Lysyanskaya [CL04], there are very efficient proofs for such statements which are constant size.

Finally, in order to support receiver security, the sender provides a proof of knowledge of the "commitment key" used to commit to its input message. This key can thus be extracted from the proof of knowledge and use it to compute messages to send to the trusted party.

1.3   Related Work

The concept of oblivious transfer was proposed by Rabin [Rab81] (but considered earlier by Wiesner [Wie83]) and further generalized to one-out-of-two OT (OT_1^2) by Even, Goldreich and Lempel [EGL85] and one-out-of-N OT (OT_1^N) by Brassard, Crépeau and Robert [BCR87]. A complete history of the work on OT is beyond our scope. In particular, here we do not mention constructions of OT which are based on generic zero-knowlege techniques or setup assumptions. See Goldreich [Gol04] for more details.

Bellare and Micali [BM90] presented practical implementations of OT_1^2 under the honest-but-curious notion and later Naor and Pinkas [NP01] did the same under the half-simulation definition. Brassard et al. [BCR87] showed how to implement OT_1^N using N applications of a OT_1^2 protocol. Under half-simulation, Naor and Pinkas [NP99a] gave a more efficient construction requiring only $\log N$ OT_1^2 executions. Several direct 2-message OT_1^N protocols (also under half-simulation) have been proposed in various works [NP01, AIR01, Kal05].

The first adaptive k-out-of-N oblivious transfer ($OT_{k \times 1}^N$) protocol is due to Naor and Pinkas [NP99b]. Their scheme is secure under half-simulation and involves $O(\log N)$ invocations of a OT_1^2 protocol during the transfer stage. Using optimistic parameters, this translates into a protocol with $O(\log N)$ rounds and at least $O(k \log N)$ communication complexity during the transfer phase. The same authors also propose a protocol requiring 2 invocations of a $OT_1^{\sqrt{N}}$ protocol. Laur and Lipmaa [LL06] build an $OT_{k \times 1}^N$ in which k must be a constant. Their security notion specifically *tolerates* selective-failure, and the efficiency of their construction depends on the efficiency of the fully-simulatable OT_1^N and the equivocable (i.e., trapdoor) list commitment scheme which are used as primitives.

In the random oracle model, Ogata and Kurosawa [OK04] and Chu and Tzeng [CT05] propose two efficient $OT_{k\times 1}^N$ schemes satisfying half-simulation which require $O(k)$ computation and communication during the transfer stage. Our first generic $OT_{k\times 1}^N$ construction based on unique blind signatures covers both schemes as special cases, offers full simulation-security, and fixes minor technical problems to prevent certain attacks. Prior to our work, Malkhi and Sella [MS03] observed a relation between OT and blind signatures, but did not give a generic transformation between the two. They present a direct OT_1^N protocol (also in the random oracle model) based on Chaum's blind signatures [Cha88]. Their scheme could be seen as a $OT_{k\times 1}^N$ protocol as well, but it has communication complexity $O(\kappa N)$ in the transfer phase. Their scheme is not an instantiation of our generic construction.

$OT_{k\times 1}^N$ can always be achieved by publishing commitments to the N data items, and executing k OT_1^N protocols on the N pieces of opening information. This solution incurs costs of $O(\kappa N)$ in each transfer phase.

Naor and Pinkas [NP05] demonstrate a way to transform a singe-server private-information retrieval scheme (PIR) into an oblivious transfer scheme with sublinear-in-N communication complexity. This transformation is in the half-simulation model and the dozen or so constructions of OT from PIR seem to also be in this model. Moreover, there are no adaptive PIR schemes known.

2   Definitions

If $k \in \mathbb{N}$, then 1^k is the string consisting of k ones. The empty string is denoted ε. If A is a randomized algorithm, then $y \xleftarrow{\$} \mathsf{A}(x)$ denotes the assignment to y of the output of A on input x when run with fresh random coins. Unless noted, all algorithms are probabilistic polynomial-time (PPT) and we implicitly assume they take an extra parameter 1^κ. A function $\nu : \mathbb{N} \to [0,1]$ is *negligible* if for all $c \in \mathbb{N}$ there exists a $\kappa_c \in \mathbb{N}$ such that $\nu(\kappa) < \kappa^{-c}$ for all $\kappa > \kappa_c$.

2.1   Blind Signatures

A blind signature scheme $\mathcal{BS}$ is a tuple of PPT algorithms $(\mathsf{Kg}, \mathsf{Sign}, \mathsf{User}, \mathsf{Vf})$. The signer generates a key pair via the key generation algorithm $(pk, sk) \xleftarrow{\$} \mathsf{Kg}(1^\kappa)$. To obtain a signature on a message M, the user and signer engage in an interactive signing protocol dictated by the $\mathsf{User}(pk, M)$ and $\mathsf{Sign}(sk)$ algorithms. At the end of the protocol, the User algorithm returns a signature s or $\perp$ to indicate rejection. The verification algorithm $\mathsf{Vf}(pk, M, s)$ returns 1 if the signature is deemed valid and 0 otherwise. Correctness requires that $\mathsf{Vf}(pk, M, s) = 1$ for all (pk, sk) output by the Kg algorithm, for all $M \in \{0,1\}^*$ and for all signatures output by $\mathsf{User}(pk, M)$ after interacting with $\mathsf{Sign}(sk)$. We say that $\mathcal{BS}$ is *unique* [GO92] if for each public key $pk \in \{0,1\}^*$ and each message $M \in \{0,1\}^*$ there exists at most one signature $s \in \{0,1\}^*$ such that $\mathsf{Vf}(pk, M, s) = 1$.

The security of blind signatures is twofold. On the one hand, *one-more un-forgeability* [PS96] requires that no adversary can output $n + 1$ valid message-signature pairs after being given the public key as input and after at most n interactions with a signing oracle. We say that $\mathcal{BS}$ is unforgeable if no PPT adversary has non-negligible probability of winning this game.

Blindness, on the other hand, requires that the signer cannot tell apart the message it is signing. The notion was first formalized by Juels et al. [JLO97], and was later strengthened to *dishonest-key blindness* [ANN06, Oka06]. In this work, we further strengthen the definition to *selective-failure blindness*. It is defined through the following game. The adversary first outputs a public key pk and two messages M_0, M_1. It is then given black-box access to two instances of the user algorithm, the first implementing $\mathsf{User}(pk, M_b)$ and the second implementing $\mathsf{User}(pk, M_{1-b})$ for a random bit $b \xleftarrow{\$} \{0, 1\}$. Eventually, these algorithms produce local output s_b and s_{1-b}, respectively. If $s_b \neq \bot$ and $s_{1-b} \neq \bot$, then the adversary is given the pair (s_0, s_1); if $s_b = \bot$ and $s_{1-b} \neq \bot$, then it is given $(\bot, \varepsilon)$; if $s_b \neq \bot$ and $s_{1-b} = \bot$, then it is given $(\varepsilon, \bot)$; and if $s_b = s_{1-b} = \bot$ it is given $(\bot, \bot)$. The adversary then guesses the bit b. The scheme $\mathcal{BS}$ is said to be selective-failure blind if no PPT adversary has a non-negligible advantage in winning the above game.

2.2   Simulatable Adaptive Oblivious Transfer

An adaptive k-out-of-N oblivious transfer scheme $\mathcal{OT}^N_{k \times 1}$ is a tuple of four PPT algorithms $(\mathsf{S_I}, \mathsf{R_I}, \mathsf{S_T}, \mathsf{R_T})$. During the initialization phase, the sender and receiver perform an interactive protocol where the sender runs the $\mathsf{S_I}$ algorithm on input messages $M_1, \ldots, M_N$, while the receiver runs the $\mathsf{R_I}$ algorithm without input. At the end of the initialization protocol, the $\mathsf{S_I}$ and $\mathsf{R_I}$ algorithm produce as local outputs state information S_0 and R_0, respectively. During the i-th transfer, $1 \leq i \leq k$, the sender and receiver engage in a selection protocol dictated by the $\mathsf{S_T}$ and $\mathsf{R_T}$ algorithms. The sender runs $\mathsf{S_T}(S_{i-1})$ to obtain updated state information S_i, while the receiver runs the $\mathsf{R_T}$ algorithm on input state information R_{i-1} and the index σ_i of the message it wishes to receive, to obtain updated state information R_i and the retrieved message M'_{σ_i}. Correctness requires that $M'_{\sigma_i} = M_{\sigma_i}$ for all messages $M_1, \ldots, M_N$, for all selections $\sigma_1, \ldots, \sigma_k \in \{1, \ldots, N\}$ and for all coin tosses of the algorithms.

To capture security of an $\mathcal{OT}^N_{k \times 1}$ scheme, we employ the real-world/ideal-world paradigm. Below, we describe a real experiment in which the parties run the protocol, while in the ideal experiment the functionality is implemented through a trusted third party. For the sake of simplicity, we do not explicitly include auxiliary inputs to the parties. This can be done, and indeed must be done for sequential composition of the primitive, and our protocols achieve this notion as well.

Real experiment. We first explain the experiment for arbitrary sender and receiver algorithms $\widehat{\mathsf{S}}$ and $\widehat{\mathsf{R}}$. The experiment $\mathbf{Real}_{\widehat{\mathsf{S}}, \widehat{\mathsf{R}}}(N, k, M_1, \ldots, M_N, \Sigma)$ proceeds as follows. $\widehat{\mathsf{S}}$ is given messages $(M_1, \ldots, M_N)$ as input and interacts with $\widehat{\mathsf{R}}(\Sigma)$, where Σ is an adaptive selection algorithm that, on input messages $M_{\sigma_1}, \ldots, M_{\sigma_{i-1}}$, outputs the index σ_i of the next message to be queried. In their first run,

$\widehat{S}$ and $\widehat{R}$ produce initial states S_0 and R_0 respectively. Next, the sender and receiver engage in k interactions. In the i-th interaction for $1 \leq i \leq k$, the sender and receiver interact by running $S_i \overset{\$}{\leftarrow} \widehat{S}(S_{i-1})$ and $(R_i, M_i^*) \overset{\$}{\leftarrow} \widehat{R}(R_{i-1})$, and update their states to S_i and R_i, respectively. Note that M_i^* may be different from M_{σ_i} when either participant cheats. At the end of the k-th interaction, sender and receiver output strings S_k and R_k respectively. The output of the **Real**$_{\widehat{S},\widehat{R}}$ experiment is the tuple (S_k, R_k).

For an $OT_{k \times 1}^N$ scheme (S_I, S_T, R_I, R_T), define the honest sender S algorithm as the one which runs $S_I(M_1, \ldots, M_N)$ in the initialization phase, runs S_T in all following interactions, and always outputs $S_k = \varepsilon$ as its final output. Define the honest receiver R as the algorithm which runs R_I in the initialization phase, runs $R_T(R_{i-1}, \sigma_i)$ and in the i-th interaction, where Σ is used to generate the index σ_i, and returns the list of received messages $R_k = (M_{\sigma_1}, \ldots, M_{\sigma_k})$ as its final output.

Ideal experiment. In experiment **Ideal**$_{\widehat{S}',\widehat{R}'}(N, k, M_1, \ldots, M_N, \Sigma)$, the (possibly cheating) sender algorithm $\widehat{S}'(M_1, \ldots, M_N)$ generates messages $M_1^*, \ldots, M_N^*$ and hands these to the trusted party T. In each of the k transfer phases, T receives a bit b_i from the sender $\widehat{S}'$ and an index σ_i^* from the (possibly cheating) receiver $\widehat{R}'(\Sigma)$. If $b_i = 1$ and $\sigma_i^* \in \{1, \ldots, N\}$, then T hands $M_{\sigma_i^*}^*$ to the receiver; otherwise, it hands $\perp$ to the receiver. At the end of the k-th transfer, $\widehat{S}'$ and $\widehat{R}'$ output a string S_k and R_k; the output of the experiment is the pair (S_k, R_k).

As above, define the ideal sender $S'(M_1, \ldots, M_N)$ as one who sends messages $M_1, \ldots, M_N$ to the trusted party in the initialization phase, sends $b_i = 1$ in all transfer phases, and uses $S_k = \varepsilon$ as its final output. Define the honest ideal receiver R' as the algorithm which generates its selection indices σ_i through Σ and submits these to the trusted party. Its final output consists of all the messages it received $R_k = (M_{\sigma_i}, \ldots, M_{\sigma_N})$.

Sender security. We say that $OT_{k \times 1}^N$ is sender-secure if for any PPT real-world cheating receiver $\widehat{R}$ there exists a PPT ideal-world receiver $\widehat{R}'$ such that for any polynomial $N_m(\kappa)$, any $N \in [1, N_m(\kappa)]$, any $k \in \{1, \ldots, N\}$, any messages $M_1, \ldots, M_N$, and any selection strategy Σ, the advantage of any PPT distinguisher in distinguishing the distributions

$$\mathbf{Real}_{S,\widehat{R}}(N, k, M_1, \ldots, M_N, \Sigma) \quad \text{and} \quad \mathbf{Ideal}_{S',\widehat{R}'}(N, k, M_1, \ldots, M_N, \Sigma)$$

is negligible in κ.

Receiver security. We say that $OT_{k \times 1}^N$ is receiver-secure if for any PPT real-world cheating sender $\widehat{S}$ there exists a PPT ideal-world sender $\widehat{S}'$ such that for any polynomial $N_m(\kappa)$, any $N \in [1, N_m(\kappa)]$, any $k \in \{1, \ldots, N\}$, any messages $M_1, \ldots, M_N$, and any selection strategy Σ, the advantage of any PPT distinguisher in distinguishing the distributions

$$\mathbf{Real}_{\widehat{S},R}(N, k, M_1, \ldots, M_N, \Sigma) \quad \text{and} \quad \mathbf{Ideal}_{\widehat{S}',R'}(N, k, M_1, \ldots, M_N, \Sigma)$$

is negligible in κ.

3   A Generic Construction in the Random Oracle Model

In this section, we describe a generic yet very efficient way of constructing adaptive k-out-of-N OT schemes from unique blind signature schemes, and prove its security in the random oracle model.

3.1   The Construction

To any unique blind signature scheme $\mathcal{BS} = (\mathsf{Kg}, \mathsf{Sign}, \mathsf{User}, \mathsf{Vf})$, we associate the $OT^N_{k \times 1}$ scheme as depicted in Fig. 1. The security of the oblivious transfer scheme

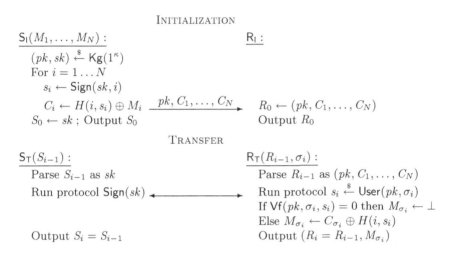

INITIALIZATION

$\mathsf{S_I}(M_1, \ldots, M_N):$

$\quad (pk, sk) \xleftarrow{\$} \mathsf{Kg}(1^\kappa)$

$\quad$ For $i = 1 \ldots N$

$\quad\quad s_i \leftarrow \mathsf{Sign}(sk, i)$

$\quad\quad C_i \leftarrow H(i, s_i) \oplus M_i$ $\quad\xrightarrow{\ pk, C_1, \ldots, C_N\ }$

$\quad S_0 \leftarrow sk$; Output S_0

$\mathsf{R_I}:$

$\quad R_0 \leftarrow (pk, C_1, \ldots, C_N)$

$\quad$ Output R_0

TRANSFER

$\mathsf{S_T}(S_{i-1}):$

$\quad$ Parse S_{i-1} as sk

$\quad$ Run protocol $\mathsf{Sign}(sk)$ $\xleftarrow{\hspace{3cm}}$

$\quad$ Output $S_i = S_{i-1}$

$\mathsf{R_T}(R_{i-1}, \sigma_i):$

$\quad$ Parse R_{i-1} as $(pk, C_1, \ldots, C_N)$

$\quad$ Run protocol $s_i \xleftarrow{\$} \mathsf{User}(pk, \sigma_i)$

$\quad$ If $\mathsf{Vf}(pk, \sigma_i, s_i) = 0$ then $M_{\sigma_i} \leftarrow \bot$

$\quad$ Else $M_{\sigma_i} \leftarrow C_{\sigma_i} \oplus H(i, s_i)$

$\quad$ Output $(R_i = R_{i-1}, M_{\sigma_i})$

Fig. 1. A construction of $OT^N_{k \times 1}$ using a random oracle H and any unique blind signature scheme $\mathcal{BS} = (\mathsf{Kg}, \mathsf{Sign}, \mathsf{User}, \mathsf{Vf})$

follows from that of the blind signature scheme. In particular, Theorem 1 states that the sender's security is implied by the one-more unforgeability of $\mathcal{BS}$, while Theorem 2 states that the receiver's security follows from the selective-failure blindness of $\mathcal{BS}$. We provide brief proof sketches below; detailed proofs can be found in the full version [CNS07].

Theorem 1.   *If the blind signature scheme $\mathcal{BS}$ is unforgeable, then the $OT^N_{k \times 1}$ depicted in Fig. 1 is sender-secure in the random oracle model.*

Proof (Sketch). The idea of this proof is that the ideal-world receiver $\widehat{\mathsf{R}}'$ runs the real-world receiver $\widehat{\mathsf{R}}$ on random ciphertexts $C_1, \ldots, C_N$. $\widehat{\mathsf{R}}'$ observes the random oracle queries made by $\widehat{\mathsf{R}}$, trying to parse them as $H(\sigma_i, s_i)$ such that $\mathsf{Vf}(pk, \sigma_i, s_i)$ $=1$. If this succeeds, then it requests M_{σ_i} from the trusted party and returns $C_{\sigma_i} \oplus M_{\sigma_i}$. If there are more than k such queries, then $\widehat{\mathsf{R}}$ has forged the blind signature scheme.

Theorem 2. *If the blind signature scheme $\mathcal{BS}$ is selective-failure blind, then the $\mathcal{OT}^N_{k \times 1}$ scheme depicted in Fig. 1 is receiver-secure in the random oracle model.*

Proof (Sketch). For any real-world cheating sender $\widehat{\mathsf{S}}$, consider the ideal-world cheating sender $\widehat{\mathsf{S}}'$ that, when $\widehat{\mathsf{S}}$ outputs $(pk, C_1, \ldots, C_N)$, goes over all random oracle queries made by $\widehat{\mathsf{S}}$ and tries to parse them as (i, s_i) such that $\mathsf{Vf}(pk, i, s_i) = 1$. If this succeeds it sets $M_i \leftarrow C_i \oplus H(i, s_i)$, all other messages are chosen at random. It then submits $M_1, \ldots, M_N$ to the trusted party. At the i-th transfer, $\widehat{\mathsf{S}}'$ runs $\widehat{\mathsf{S}}$ against $(R_i, M_i^*) \overset{\$}{\leftarrow} \mathsf{R}_\mathsf{T}(R_{i-1}, 1)$. If $M_i^* = \perp$ then $\widehat{\mathsf{S}}'$ submits a zero bit to the trusted party, indicating that the present transfer should fail, otherwise it sends a one. The selective-failure blindness of $\mathcal{BS}$ ensures that $\widehat{\mathsf{S}}$ cannot distinguish a query for index 1 from any other query, and that it cannot make R_T fail depending on the value of its selection.

Instantiations. Many blind signature schemes exist, but only the schemes of Chaum [Cha88, BNPS03] and Boldyreva [Bol03] seem to be unique. Both are efficient two-round schemes which result in round-optimal adaptive oblivious transfer protocols.

The instantiation of our generic construction with Chaum's blind signature scheme coincides with the direct OT scheme of Ogata-Kurosawa [OK04]. However, special precautions must be taken to ensure that Chaum's scheme is selective-failure blind. For example, the sender must use a prime exponent e greater than the modulus n [ANN06], or must provide a non-interactive proof that $\gcd(e, n) = 1$ [CPP07]. Anna Lysyanskaya suggests having the receiver send e to the sender. This solution is much more efficient than the previous two, but would require re-proving the security of the $\mathcal{OT}^N_{k \times 1}$ scheme since it is no longer an instance of our generic construction. In any case, the authors of [OK04] overlooked this need, which creates the possibility for attacks on the receiver's security of their protocol. For example, a cheating sender could choose $e = 2$ and distinguish between transfers for σ_i and σ_i' for which $H(\sigma_i)$ is a square modulo n and $H(\sigma_i')$ is not.

When instantiated with Boldyreva's blind signature scheme [Bol03] based on pairings, our generic construction coincides with the direct OT scheme of Chu-Tzeng [CT05]. A similar issue concerning the dishonest-key blindness of the scheme arises here, but was also overlooked. The sender could for example choose the group to be of non-prime order and break the receiver's security in a similar way as demonstrated above for the scheme of [OK04]. One can strengthen Boldyreva's blind signature scheme to provide selective-failure blindness by letting the user algorithm check that the group is of prime order p and that the generator is of full order p.

4   Simulatable Adaptive OT in the Standard Model

Computational assumptions. Our protocol presented in this section requires bilinear groups and associated hardness assumptions. Let Pg be a pairing group generator that on input 1^κ outputs descriptions of multiplicative groups $\mathbb{G}_1, \mathbb{G}_\mathsf{T}$

of prime order p where $|p| = \kappa$. Let $\mathbb{G}_1^* = \mathbb{G}_1 \setminus \{1\}$ and let $g \in \mathbb{G}_1^*$. The generated groups are such that there exists an admissible bilinear map $e : \mathbb{G}_1 \times \mathbb{G}_1 \to \mathbb{G}_T$, meaning that (1) for all $a, b \in \mathbb{Z}_p$ it holds that $e(g^a, g^b) = e(g, g)^{ab}$; (2) $e(g, g) \neq 1$; and (3) the bilinear map is efficiently computable.

Definition 1 (Strong Diffie-Hellman Assumption [BB04]). *We say that the ℓ-SDH assumption associated to a pairing generator* Pg *holds if for all PPT adversaries A, the probability that $A(g, g^x, \ldots, g^{x^\ell})$ where $(\mathbb{G}_1, \mathbb{G}_T) \xleftarrow{\$} \mathsf{Pg}(1^\kappa)$, $g \xleftarrow{\$} \mathbb{G}_1^*$ and $x \xleftarrow{\$} \mathbb{Z}_p$, outputs a pair $(c, g^{1/(x+c)})$ where $c \in \mathbb{Z}_p$ in negligible in κ.*

Definition 2 (Power Decisional Diffie-Hellman Assumption). *We say that the ℓ-PPDH assumption associated to* Pg *holds if for all PPT adversaries A, the probability that A on input $(g, g^x, g^{x^2}, \ldots, g^{x^\ell}, H)$ where $(\mathbb{G}_1, \mathbb{G}_T) \xleftarrow{\$} \mathsf{Pg}(1^\kappa)$, $g \xleftarrow{\$} \mathbb{G}_1^*$, $x \xleftarrow{\$} \mathbb{Z}_p$, $H \xleftarrow{\$} \mathbb{G}_T$, distinguishes the vector $T = (H^x, H^{x^2}, \ldots, H^{x^\ell})$ from a random vector $T \xleftarrow{\$} \mathbb{G}_T^\ell$ is negligible in κ.*

Boneh-Boyen signatures. We modify the weakly-secure signature scheme of Boneh and Boyen [BB04] as follows. The scheme uses a pairing generator Pg as defined above. The signer's secret key is $x \xleftarrow{\$} \mathbb{Z}_p$, the corresponding public key is $y = g^x$. The signature on a message M is $s \leftarrow g^{1/(x+M)}$; verification is done by checking that $e(s, y \cdot g^M) = e(g, g)$. This scheme is similar to the Dodis and Yampolskiy verifiable random function [DY05].

Security under weak chosen-message attack is defined through the following game. The adversary begins by outputting ℓ messages $M_1, \ldots, M_\ell$. The challenger generates a fresh key pair and gives the public key to the adversary, together with signatures $s_1, \ldots, s_\ell$ on $M_1, \ldots, M_\ell$. The adversary wins if it succeeds in outputing a valid signature s on a message $M \notin \{M_1, \ldots, M_\ell\}$. The scheme is said to be unforgeable under weak chosen-message attack if no PPT adversary A has non-negligible probability of winning this game. An easy adaptation of the proof of [BB04] can be used to show that this scheme is unforgeable under weak chosen-message attack if the $(\ell + 1)$-SDH assumption holds.

Zero-knowledge proofs. We use definitions from [BG92, CDM00]. A pair of interacting algorithms (P, V) is a proof of knowledge (PoK) for a relation $R = \{(\alpha, \beta)\} \subseteq \{0,1\}^* \times \{0,1\}^*$ with knowledge error $\kappa \in [0, 1]$ if (1) for all $(\alpha, \beta) \in R$, $\mathsf{V}(\alpha)$ accepts a conversation with $\mathsf{P}(\beta)$ with probability 1; and (2) there exists an expected polynomial-time algorithm E, called the *knowledge extractor*, such that if a cheating prover $\widehat{\mathsf{P}}$ has probability ϵ of convincing V to accept α, then E, when given rewindable black-box access to $\widehat{\mathsf{P}}$, outputs a witness β for α with probability $\epsilon - \kappa$.

A proof system (P, V) is *perfect zero-knowledge* if there exists a PPT algorithm Sim, called the *simulator*, such that for any polynomial-time cheating verifier $\widehat{\mathsf{V}}$ and for any $(\alpha, \beta) \in R$, the outputs of $\widehat{\mathsf{V}}(\alpha)$ after interacting with $\mathsf{P}(\beta)$ and that of $\mathsf{Sim}^{\widehat{\mathsf{V}}(\alpha)}(\alpha)$ are identically distributed.

INITIALIZATION

$\underline{S_I(1^\kappa, M_1, \ldots, M_N):}$ $\underline{R_I(1^\kappa):}$

$(\mathbb{G}_1, \mathbb{G}_T) \xleftarrow{\$} Pg(1^\kappa)$

$g, h \xleftarrow{\$} \mathbb{G}_1^*$; $H \leftarrow e(g, h)$

$x \xleftarrow{\$} \mathbb{Z}_p$; $y \leftarrow g^x$; $pk \leftarrow (g, H, y)$

For $i = 1, \ldots, N$ do

  $A_i \leftarrow g^{1/(x+i)}$

  $B_i \leftarrow e(h, A_i) \cdot M_i$

  $C_i \leftarrow (A_i, B_i)$ $\xrightarrow{\quad pk, C_1, \ldots, C_N \quad}$

$S_0 \leftarrow (h, pk)$ $\xrightarrow{\quad PoK\{(h): H = e(g,h)\} \quad}$ $R_0 \leftarrow (pk, C_1, \ldots, C_N)$

TRANSFER

$\underline{S_T(S_{i-1}):}$ $\underline{R_T(R_{i-1}, \sigma_i):}$

                                                             $v \xleftarrow{\$} \mathbb{Z}_p$; $V \leftarrow (A_{\sigma_i})^v$

                       $\xleftarrow{\qquad\qquad V \qquad\qquad}$

$\xleftarrow{\quad PoK\{(\sigma_i, v): e(V, y) = e(V, g)^{-\sigma_i} e(g, g)^v\} \quad}$

$W \leftarrow e(h, V)$ $\xrightarrow{\qquad\qquad W \qquad\qquad}$

      $\xrightarrow{\quad PoM\{(h): H = e(g,h) \wedge W = e(h,V)\} \quad}$

$S_i = S_{i-1}$ $M \leftarrow B_{\sigma_i}/(W^{1/v})$

                                                      $R_i = R_{i-1}$

Fig. 2. Our $OT_{k\times1}^N$ protocol in the standard model associated to pairing generator Pg. We use notation by Camenisch and Stadler [CS97] for the zero-knowledge protocols. They can all be done efficiently (in four rounds and $O(\kappa)$ communication) by using the transformation of [CDM00]. The protocols are given in detail in the full version [CNS07].

Our protocol in the standard model is depicted in Fig. 2. All zero-knowledge proofs can be performed efficiently in four rounds and with $O(\kappa)$ communication using the transformation of [CDM00]. The detailed protocols are provided in the full version [CNS07]. We assume that the messages M_i are elements of the target group $\mathbb{G}_T$.[2] The protocol is easily seen to be correct by observing that $W = e(h, A_{\sigma_i})^v$, so therefore $B_{\sigma_i}/W^{1/v} = M_{\sigma_i}$.

We now provide some intuition into the protocol. Each pair (A_i, B_i) can be seen as an ElGamal encryption [ElG85] in $\mathbb{G}_T$ of M_i under public key H. But instead of using random elements from $\mathbb{G}_T$ as the first component, our protocol uses verifiably random [DY05] values $A_i = g^{1/(x+i)}$. It is this verifiability that during the transfer phase allows the sender to check that the receiver is indeed asking for the decryption key for one particular ciphertext, and not for some combination of ciphertexts.

---

[2] This is a standard assumption we borrow from the literature on Identity-Based Encryption. The target group is usually a subgroup of a larger prime field. Thus, depending on implementation, it may be necessary to "hash" the data messages into this subgroup. Alternatively, one can extract a random pad from the element in the target group and use $\oplus$ to encrypt the message.

Receiver security. We demonstrate the receiver security of our scheme by proving the stronger property of unconditional statistical indistinguishability. Briefly, the ideal-world sender can extract h from the proof of knowledge in the initialization phase, allowing it to decrypt the messages to send to the trusted party. During the transfer phase, it plays the role of an honest receiver and asks for a randomly selected index. If the real-world sender succeeds in the final proof of membership (PoM) of the well-formedness of W, then the ideal sender sends $b = 1$ to its trusted-party T to indicate continue.

Notice how the sender's response W is simultaneously determined by the initialization phase, unpredictable by the receiver during the transfer phase, but yet *verifiable* once it has been received (albeit, via a zero-knowledge proof). Intuitively, these three properties prevent the selective-failure attack.

Theorem 3. *The $OT^N_{k \times 1}$ protocol in Fig. 2 is unconditionally receiver-secure.*

Proof. We show that for every real-world cheating sender $\widehat{\mathsf{S}}$ there exists an ideal-world cheating sender $\widehat{\mathsf{S}}'$ such that no distinguisher D, regardless of its running time, has non-negligible probability to distinguish the distributions $\mathbf{Real}_{\widehat{\mathsf{S}},\mathsf{R}}(N, k, M_1, \ldots, M_N, \Sigma)$ and $\mathbf{Ideal}_{\widehat{\mathsf{S}}',\mathsf{R}'}(N, k, M_1, \ldots, M_N, \Sigma)$. We do so by considering a sequence of distributions $\mathbf{Game\text{-}0}, \ldots, \mathbf{Game\text{-}3}$ such that for some $\widehat{\mathsf{S}}'$ that we construct, $\mathbf{Game\text{-}0} = \mathbf{Real}_{\widehat{\mathsf{S}},\mathsf{R}}$ and $\mathbf{Game\text{-}3} = \mathbf{Ideal}_{\widehat{\mathsf{S}}',\mathsf{R}'}$, and by demonstrating the statistical difference in the distribution for each game transition. Below, we use the shorthand notation

$$\Pr[\mathbf{Game\text{-}i}] = \Pr\left[\mathsf{D}(X) = 1 : X \xleftarrow{\$} \mathbf{Game\text{-}i} \right] .$$

Game-0: This is the distribution corresponding to $\mathbf{Real}_{\widehat{\mathsf{S}},\mathsf{R}}$, i.e., the game where the cheating sender $\widehat{\mathsf{S}}$ is run against an honest receiver R with selection strategy Σ. Obviously, $\Pr[\mathbf{Game\text{-}0}] = \Pr\left[\mathsf{D}(X) = 1 : X \xleftarrow{\$} \mathbf{Real}_{\widehat{\mathsf{S}},\mathsf{R}} \right]$.

Game-1: In this game the extractor E_1 for the first proof of knowledge is used to extract from $\widehat{\mathsf{S}}$ the element h such that $e(g, h) = H$. If the extractor fails, then the output of **Game-1** is $\perp$; otherwise, the execution of $\widehat{\mathsf{S}}$ continues as in the previous game, interacting with $\mathsf{R}(\Sigma)$. The difference between the two output distributions is given by the knowledge error of the PoK, i.e.,

$$\Pr[\mathbf{Game\text{-}1}] - \Pr[\mathbf{Game\text{-}0}] \leq \frac{1}{p} .$$

Game-2: This game is identical to the previous one, except that during the transfer phase the value V sent by the receiver is replaced by picking a random v' and sending $V' \leftarrow A_1^v$. The witness $(v', 1)$ is used during the second PoK. Since V and V' are both uniformly distributed over $\mathbb{G}_1$, and by the perfect witness-indistinguishability of the PoK (implied by the perfect zero-knowledge property), we have that $\Pr[\mathbf{Game\text{-}2}] = \Pr[\mathbf{Game\text{-}1}]$.

Game-3: In this game, we introduce an ideal-world sender $\widehat{\mathsf{S}}'$ which incorporates the steps from the previous game. Algorithm $\widehat{\mathsf{S}}'$ uses E_1 to extract h from $\widehat{\mathsf{S}}$, decrypts M_i^* as $B_i/e(h, A_i)$ for $i = 1, \ldots, N$ and submits $M_1^*, \ldots, M_N^*$ to the trusted party T. As in **Game-2**, during the transfer phase, $\widehat{S}'$ feeds $V' \xleftarrow{\$} A_1^{v'}$ to $\widehat{S}$ and uses $(v', 1)$ as a witness in the PoK. It plays the role of the verifier in the final PoM of W. If $\widehat{S}$ convinces $\widehat{S}'$ that W is correctly formed, then $\widehat{S}'$ sends 1 to the trusted party, otherwise it sends 0. When $\widehat{S}$ outputs its final state S_k, $\widehat{S}'$ outputs S_k as well.

One can syntactically see that

$$\Pr[\textbf{Game-3}] \;=\; \Pr[\textbf{Game-2}] \;=\; \Pr\left[\mathsf{D}(X) = 1 \;:\; X \xleftarrow{\$} \textbf{Ideal}_{\widehat{S}', R'} \right].$$

Summing up, we have that the advantage of the distinguisher D is given by

$$\Pr\left[\mathsf{D}(X) = 1 \;:\; X \xleftarrow{\$} \textbf{Ideal}_{\widehat{S}', R'} \right] - \Pr\left[\mathsf{D}(X) = 1 \;:\; X \xleftarrow{\$} \textbf{Real}_{\widehat{S}, R} \right] \;\leq\; \frac{1}{p}.$$

Sender security. The following theorem states the sender-security of our second construction.

Theorem 4. *If the $(N+1)$-SDH assumption and the $(N+1)$-PDDH assumptions associated to Pg hold, then the $OT_{k \times 1}^N$ protocol depicted in Fig. 2 is sender-secure.*

Proof. Given a real cheating receiver $\widehat{R}$, we construct an ideal-world cheating receiver $\widehat{R}'$ such that no algorithm D can distinguish between the distributions $\textbf{Real}_{S, \widehat{R}}(N, k, M_1, \ldots, M_N, \Sigma)$ and $\textbf{Ideal}_{S', \widehat{R}'}(N, k, M_1, \ldots, M_N, \Sigma)$. We again do so by considering a sequence of hybrid distributions and investigate the differences between successive ones.

Game-0: This is the distribution corresponding to $\widehat{R}$ being run against the honest sender $S(M_1, \ldots, M_N)$. Obviously, we have that $\Pr[\textbf{Game-0}] = \Pr\left[\mathsf{D}(X) = 1 \;:\; X \xleftarrow{\$} \textbf{Real}_{S, \widehat{R}} \right].$

Game-1: This game differs from the previous one in that at each transfer the extractor E_2 of the second PoK is used to extract from $\widehat{R}$ the witness (σ_i, v). If the extraction fails, **Game-1** outputs $\perp$. Because the PoK is perfect zero-knowledge, the difference on the distribution with the previous game is statistical (i.e., independent of the distinguisher's running time) and given by k times the knowledge error, or $\Pr[\textbf{Game-1}] - \Pr[\textbf{Game-0}] \leq k/p$. Note that the time required to execute these k extractions is k times the time of doing a single extraction, because the transfer protocols can only run sequentially, rather than concurrently. One would have to resort to concurrent zero-knowledge protocols [DNS04] to remove this restriction.

Game-2: This game is identical to the previous one, except that **Game-2** returns $\perp$ if the extracted value $\sigma_i \notin \{1, \ldots, N\}$ during any of the transfers. One can see that in this case $s = V^{1/v}$ is a forged Boneh-Boyen signature on message σ_i. The difference between **Game-1** and **Game-2** is bounded by the following claim, which we prove below:

Claim (1). If the $(N + 1)$-SDH assumption associated to Pg holds, then $\Pr[\mathbf{Game\text{-}2}] - \Pr[\mathbf{Game\text{-}1}]$ is negligible.

Game-3: In this game the PoK of h in the initialization phase is replaced with a simulated proof using Sim_1, the value W returned in each transfer phase is computed as $W \leftarrow (B_{\sigma_i}/M_{\sigma_i})^v$, and the final PoM in the transfer phase is replaced by a simulated proof using Sim_3. Note that now the simulation of the transfer phase no longer requires knowledge of h. However, all of the simulated proofs are proofs of true statements and the change in the computation of W is purely conceptional. Thus by the perfect zero-knowledge property, we have that $\Pr[\mathbf{Game\text{-}3}] = \Pr[\mathbf{Game\text{-}2}]$.

Game-4: Now the values $B_1, \ldots, B_N$ sent to $\widehat{\mathsf{R}}$ in the initialization phase are replaced with random elements from $\mathbb{G}_T$. Now at this point, the second proof in the previous game is a simulated proof of a false statement. Intuitively, if these changes enable a distinguisher D to separate the experiments, then one can solve an instance of the SBDHI problem. This is caputed in the following claim:

Claim (2). If the $(N + 1)$-PDDH assumption associated to Pg holds, then $\Pr[\mathbf{Game\text{-}4}] - \Pr[\mathbf{Game\text{-}3}]$ is negligible.

The ideal-world receiver $\widehat{\mathsf{R}}'$ can be defined as follows. It performs all of the changes to the experiments described in **Game-4** except that at the time of transfer, after having extracted the value of σ_i from $\widehat{\mathsf{R}}$, it queries the trusted party T on index σ_i to obtain message M_{σ_i}. It then uses this message to compute W. Syntactically, we have that

$$\Pr\left[\, D(X) = 1 \;:\; X \xleftarrow{\$} \mathbf{Ideal}_{\mathsf{S}',\widehat{\mathsf{R}}'} \,\right] = \Pr[\mathbf{Game\text{-}4}].$$

Summing up the above equations and inequalities yields that

$$\Pr\left[\, D(X) = 1 \;:\; X \xleftarrow{\$} \mathbf{Ideal}_{\mathsf{S}',\widehat{\mathsf{R}}'} \,\right] - \Pr\left[\, D(X) = 1 \;:\; X \xleftarrow{\$} \mathbf{Real}_{\mathsf{S},\widehat{\mathsf{R}}} \,\right]$$

is negligible. The running time of $\widehat{\mathsf{R}}'$ is that of $\widehat{\mathsf{R}}$ plus that of $O(N^2)$ exponentiations, k extractions and k proof simulations, so is polynomial in κ.

It remains to prove the claims used in the proof above.

Proof (Claim (1)). We prove the claim by constructing an adversary A that breaks the unforgeability under weak chosen-message attack of the modified Boneh-Boyen signature scheme . By the security proof of [BB04], this directly gives rise to an expected polynomial-time adversary with non-negligible advantage in solving the $(N + 1)$-SDH problem.

Given a cheating receiver $\widehat{\mathsf{R}}$ for that distinguishes between **Game-1** and **Game-2** with advantage ϵ_{St}, consider the forger A that outputs messages $M_1 = 1, \ldots, M_N = N$, and on input a public key y and signatures $A_1, \ldots, A_N$ runs the honest sender algorithm using these values for h and $A_1, \ldots, A_N$. At each transfer it uses E_2 to extract from $\widehat{R}$ values (σ_i, v) such that $e(V, y) = e(V, g)^{-\sigma_i} e(g, g)^v$. (This extraction is guaranteed to succeed since we already eliminated failed extractions in the transition from **Game-0** to **Game-1**.) When $\sigma_i \notin \{1, \dots, N\}$ then A outputs $s \leftarrow V^{1/v}$ as its forgery on message $M = \sigma_i$. The forger A wins whenever it extracts a value $\sigma_i \notin \{1, \dots, N\}$ from $\widehat{S}$. Its running time is that of $\widehat{R}$ plus k times the running time of a single extraction, so polynomial in κ.

Proof (Claim (2)). Given an algorithm D with non-negligible probability in distinguishing **Game-2** and **Game-3**, consider the following algorithm A for the PDDH problem for $\ell = N+1$. On input $(u, u^x, \dots, u^{x^{N+1}}, V)$ and a vector $(T_1, \dots, T_{N+1})$, A proceeds as follows. For ease of notation, let $T_0 = V$. Let f be the polynomial defined as $f(X) = \prod_{i=1}^{N}(X + i) = \sum_{i=0}^{N} c_i X^i$. Then A sets $g \leftarrow u^{f(x)} = \prod_{i=0}^{N}(u^{x^i})^{c_i}$ and $y \leftarrow g^x = \prod_{i=0}^{N}(u^{x^{i+1}})^{c_i}$. If f_i is the polynomial defined by $f_i(X) = f(X)/(X + i) = \sum_{j=0}^{N-1} c_{i,j} X^j$, then A can also compute the values $A_i = g^{1/(x+i)}$ as $A_i \leftarrow \prod_{j=0}^{N-1}(u^{x^j})^{c_{i,j}}$. It then sets $H \leftarrow V^{f(x)} = \prod_{i=0}^{N} T_i^{c_i}$, and computes $B_i = H^{1/(x+i)}$ as $B_i \leftarrow \prod_{j=0}^{N-1} T_i^{c_{i,j}}$, and continues the simulation of $\widehat{R}$'s environment as in **Game-3** and **Game-4**, i.e., at each transfer extracting (σ_i, v), computing $W \leftarrow (B_{\sigma_i}/M_{\sigma_i})$ and simulating the PoM. When $\widehat{R}$ outputs its final state R_k, algorithm A runs $b \xleftarrow{\$} D(\varepsilon, R_k)$ and outputs b.

In the case that $T_i = V^{x^i}$ one can see that the environment that A created for $\widehat{S}$ is exactly that of **Game-3**. In the case that $T_1, \dots, T_N$ are random elements of $\mathbb{G}_T$, then one can easily see that this environment is exactly that of **Game-4**. Therefore, if D has non-negligible advantage in distinguishing the outputs of **Game-3** and **Game-4**, then A has non-negligible advantage in solving the $(N + 1)$-PDDH problem. The running time of A is at most that of the distinguisher D plus that of $O(N^2)$ exponentiations, of $k + 1$ simulated proofs, and of k extractions.

Acknowledgements

The authors would like to thank Xavier Boyen, Christian Cachin, Anna Lysyanskaya, Benny Pinkas, Alon Rosen and the anonymous referees for their useful comments and discussions. Gregory Neven is a Postdoctoral Fellow of the Research Foundation Flanders (FWO-Vlaanderen). This work was supported in part by the European Commission through the IST Program under Contract IST-2002-507932 ECRYPT and Contract IST-2002-507591 PRIME.

References

[AIR01]   W. Aiello, Y. Ishai, and O. Reingold. : Priced oblivious transfer: How to sell digital goods. In *EUROCRYPT 2001*, p. 119–135.

[ANN06]   M. Abdalla, C. Namprempre, and G. Neven. On the (im)possibility of blind message authentication codes. In *CT-RSA 2006*, p. 262–279.

[BB04]   D. Boneh and X. Boyen. Short signatures without random oracles. In *EUROCRYPT 2004*, p. 56–73.

[BCR87]   G. Brassard, C. Crépeau, and J.M. Robert. All-or-nothing disclosure of secrets. In *CRYPTO'86*, p. 234–238.

[BG92]   M. Bellare and O. Goldreich. On defining proofs of knowledge. In *CRYPTO'92*, p. 390–420.

[BM90]   M. Bellare and S. Micali. Non-interactive oblivious transfer and applications. In *CRYPTO'89*, p. 547–557.

[BNPS03]   M. Bellare, C. Namprempre, D. Pointcheval, and M. Semanko. The one-more-RSA-inversion problems and the security of Chaum's blind signature scheme. *J. Cryptology*, 16(3):185–215, 2003.

[Bol03]   A. Boldyreva. Threshold signatures, multisignatures and blind signatures based on the gap-Diffie-Hellman-group signature scheme. In *PKC 2003*, p. 31–46.

[BR93]   M. Bellare and P. Rogaway. Random oracles are practical: A paradigm for designing efficient protocols. In *ACM CCS 93*, p. 62–73.

[Can00]   R. Canetti. Security and composition of multi-party cryptographic protocols. *J. Cryptology*, 13(1):143–202, 2000.

[CDM00]   R. Cramer, I. Damgård, and P. MacKenzie. Efficient zero-knowledge proofs of knowledge without intractability assumptions. In *PKC 2000*, p. 354–372.

[Cha88]   D. Chaum. Blind signature systems. U.S. Patent #4,759,063, 1988.

[CK90]   C. Crépeau and J. Kilian. Weakening security assumptions and oblivious transfer. In *CRYPTO'88*, p. 2–7.

[CL04]   J. Camenisch and A. Lysyanskaya. Signature schemes and anonymous credentials from bilinear maps. In *CRYPTO 2004*, p. 56–72.

[CNS07]   J. Camenisch, G. Neven, and a. shelat. Simulatable Adaptive Oblivious Transfer. Cryptology ePrint Archive, 2007.

[CPP07]   D. Catalano, D. Pointcheval, and T. Pornin. Trapdoor hard-to-invert group isomorphisms and their application to password-based authentication. To appear in *J. Cryptology*, 2007.

[CS97]   J. Camenisch and M. Stadler. Efficient group signature schemes for large groups. In *CRYPTO'97*, p. 410–424.

[CT05]   C.-K. Chu and W.-G. Tzeng. Efficient k-out-of-n oblivious transfer schemes with adaptive and non-adaptive queries. In *PKC 2005*, p. 172–183.

[DNS04]   C. Dwork, M. Naor, and A. Sahai. Concurrent zero-knowledge. *J. ACM*, 51(6):851–898, 2004.

[DY05]   Y. Dodis and A. Yampolskiy. A verifiable random function with short proofs and keys. In *PKC 2005*, p. 416–431.

[EGL85]   S. Even, O. Goldreich, and A. Lempel. A randomized protocol for signing contracts. *Communications of the ACM*, 28(6):637–647, 1985.

[ElG85]   T. ElGamal. A public key cryptosystem and signature scheme based on discrete logarithms. *IEEE Transactions on Information Theory*, 31:469–472, 1985.

[Gol04]   O. Goldreich. Foundations of Cryptography, Volume 2. Cambridge University Press, 2004.

[GMW87]   O. Goldreich, S. Micali, and A. Wigderson. How to play any mental game, or a completeness theorem for protocols with honest majority. In *19th ACM STOC*, p. 218–229.

[GO92]   S. Goldwasser and R. Ostrovsky. Invariant signatures and non-interactive zero-knowledge proofs are equivalent. In *CRYPTO'92*, p. 228–245.

[JLO97]   A. Juels, M. Luby, and R. Ostrovsky. Security of blind digital signatures (Extended abstract). In *CRYPTO'97*, p. 150–164.

[Kal05]   Y. Kalai. Smooth projective hashing and two-message oblivious transfer. In *EUROCRYPT 2005*, p. 78–95.

[LL06]   S. Laur and H. Lipmaa. On security of sublinear oblivious transfer. Cryptology ePrint Archive, 2006.

[MS03]   D. Malkhi and Y. Sella. Oblivious transfer based on blind signatures. Technical Report 2003-31, Leibniz Center, Hebrew University, 2003.

[MSK02]   S. Mitsunari, R. Sakai, and M. Kasahara. A new traitor tracing. *IEICE Transactions Fundamentals*, E85-A(2):481–84, 2002.

[NP99a]   M. Naor and B. Pinkas. Oblivious transfer and polynomial evaluation. In *31st ACM STOC*, p. 245–254, 1999.

[NP99b]   M. Naor and B. Pinkas. Oblivious transfer with adaptive queries. In *CRYPTO'99*, p. 573–590.

[NP01]   M. Naor and B. Pinkas. Efficient oblivious transfer protocols. In *12th SODA*, p. 448–457, 2001.

[NP05]   M. Naor and B. Pinkas. Computationally secure oblivious transfer. *J. Cryptology*, 18, 2005.

[OK04]   W. Ogata and K. Kurosawa. Oblivious keyword search. *J. Complexity*, 20(2-3):356–371, 2004.

[Oka06]   T. Okamoto. Efficient blind and partially blind signatures without random oracles. In *TCC 2006*, LNCS, p. 80–99.

[OS04]   W. Ogata and R. Sasahara. k out of n oblivious transfer without random oracles. *IEICE Transactions*, 87-A(1):147–151, 2004.

[PS96]   D. Pointcheval and J. Stern. Security proofs for signature schemes. In *EUROCRYPT'96*, p. 387–398.

[Rab81]   M. Rabin. How to exchange secrets by oblivious transfer. Technical Report TR-81, Harvard Aiken Computation Laboratory, 1981.

[Wie83]   S. Wiesner. Conjugate Coding. *SIGACT News*, 15, 1983. . 78–88.

Author Index

Bauer, Aurélie   361
Boyen, Xavier   210, 394

Camenisch, Jan   246, 573
Canard, Sébastien   482
Chen, Hao   291
Cramer, Ronald   291, 329

Damgård, Ivan   329, 412
Deng, Yi   148
Dodis, Yevgeniy   534
Dubois, Vivien   264

Enge, Andreas   379

Farràs, Oriol   448
Fouque, Pierre-Alain   264

Gaudry, Pierrick   379
Geiselmann, Willi   466
Goldwasser, Shafi   291
Gouget, Aline   482
Granger, Robert   430

Haan, Robbert de   291, 329
Hess, Florian   430
Hohenberger, Susan   246
Hsiao, Chun-Yuan   169

Jarecki, Stanisław   97
Joux, Antoine   361

Katz, Jonathan   115, 311
Koo, Chiu-Yuen   311

Lenstra, Arjen   1
Lin, Dongdai   148
Lindell, Yehuda   52
Lu, Chi-Jen   169

Martí-Farré, Jaume   448
Maurer, Ueli   498
Minder, Lorenz   347

Neven, Gregory   573

Ong, Shien Jin   187
Oyono, Roger   430

Padró, Carles   448
Pedersen, Michael Østergaard   246
Pietrzak, Krzysztof   23, 517
Pinkas, Benny   52
Preneel, Bart   276
Puniya, Prashant   534

Reyzin, Leonid   169
Ristenpart, Thomas   228

shelat, abhi   573
Shmatikov, Vitaly   97
Shokrollahi, Amin   347
Sjödin, Johan   498, 517
Steinberger, John P.   34
Steinwandt, Rainer   466
Stern, Jacques   264
Stevens, Marc   1

Thériault, Nicolas   430
Thorbek, Rune   412

Vadhan, Salil   187
Vaikuntanathan, Vinod   291
Vercauteren, Frederik   430

Weger, Benne de   1
Woodruff, David P.   79
Wu, Hongjun   276
Wullschleger, Jürg   555

Yilek, Scott   228
Yung, Moti   129

Zhao, Yunlei   129

Lecture Notes in Computer Science

For information about Vols. 1–4381

please contact your bookseller or Springer

Vol. 4515: M. Naor (Ed.), Advances in Cryptology - EU-ROCRYPT 2007. XIII, 591 pages. 2007.

Vol. 4510: P. Van Hentenryck, L. Wolsey (Eds.), Integration of AI and OR Techniques in Constraint Programming for Combinatorial Optimization Problems. X, 391 pages. 2007.

Vol. 4506: D. Zeng, I. Gotham, K. Komatsu, C. Lynch, M. Thurmond, D. Madigan, B. Lober, J. Kvach, H. Chen (Eds.), Intelligence and Security Informatics: Biosurveillance. XI, 234 pages. 2007.

Vol. 4504: J. Huang, R. Kowalczyk, Z. Maamar, D. Martin, I. Müller, S. Stoutenburg, K.P. Sycara (Eds.), Service-Oriented Computing: Agents, Semantics, and Engineering. X, 175 pages. 2007.

Vol. 4493: D. Liu, S. Fei, Z. Hou, H. Zhang, C. Sun (Eds.), Advances in Neural Networks – ISNN 2007, Part III. XXVI, 1215 pages. 2007.

Vol. 4492: D. Liu, S. Fei, Z. Hou, H. Zhang, C. Sun (Eds.), Advances in Neural Networks – ISNN 2007, Part II. XXVII, 1321 pages. 2007.

Vol. 4491: D. Liu, S. Fei, Z.-G. Hou, H. Zhang, C. Sun (Eds.), Advances in Neural Networks – ISNN 2007, Part I. LIV, 1365 pages. 2007.

Vol. 4486: M. Bernardo, J. Hillston (Eds.), Formal Methods for Performance Evaluation. VII, 469 pages. 2007.

Vol. 4484: J.-Y. Cai, S.B. Cooper, H. Zhu (Eds.), Theory and Applications of Models of Computation. XIII, 772 pages. 2007.

Vol. 4483: C. Baral, G. Brewka, J. Schlipf (Eds.), Logic Programming and Nonmonotonic Reasoning. IX, 327 pages. 2007. (Sublibrary LNAI).

Vol. 4482: A. An, J. Stefanowski, S. Ramanna, C.J. Butz, W. Pedrycz, G. Wang (Eds.), Rough Sets, Fuzzy Sets, Data Mining and Granular Computing. XIV, 585 pages. 2007. (Sublibrary LNAI).

Vol. 4481: J.T. Yao, P. Lingras, W.-Z. Wu, M. Szczuka, N.J. Cercone, D. Ślęzak (Eds.), Rough Sets and Knowledge Technology. XIV, 576 pages. 2007. (Sublibrary LNAI).

Vol. 4480: A. LaMarca, M. Langheinrich, K.N. Truong (Eds.), Pervasive Computing. XIII, 369 pages. 2007.

Vol. 4479: I.F. Akyildiz, R. Sivakumar, E. Ekici, J.C.d. Oliveira, J. McNair (Eds.), NETWORKING 2007. Ad Hoc and Sensor Networks, Wireless Networks, Next Generation Internet. XXVII, 1252 pages. 2007.

Vol. 4472: M. Haindl, J. Kittler, F. Roli (Eds.), Multiple Classifier Systems. XI, 524 pages. 2007.

Vol. 4471: P. Cesar, K. Chorianopoulos, J.F. Jensen (Eds.), Interactive TV: a Shared Experience. XIII, 236 pages. 2007.

Vol. 4470: Q. Wang, D. Pfahl, D.M. Raffo (Eds.), Software Process Change – Meeting the Challenge. XI, 346 pages. 2007.

Vol. 4464: E. Dawson, D.S. Wong (Eds.), Information Security Practice and Experience. XIII, 361 pages. 2007.

Vol. 4463: I. Măndoiu, A. Zelikovsky (Eds.), Bioinformatics Research and Applications. XV, 653 pages. 2007. (Sublibrary LNBI).

Vol. 4462: D. Sauveron, K. Markantonakis, A. Bilas, J.-J. Quisquater (Eds.), Information Security Theory and Practices. XII, 255 pages. 2007.

Vol. 4459: C. Cérin, K.-C. Li (Eds.), Advances in Grid and Pervasive Computing. XVI, 759 pages. 2007.

Vol. 4453: T. Speed, H. Huang (Eds.), Research in Computational Molecular Biology. XVI, 550 pages. 2007. (Sublibrary LNBI).

Vol. 4452: M. Fasli, O. Shehory (Eds.), Agent-Mediated Electronic Commerce. VIII, 249 pages. 2007. (Sublibrary LNAI).

Vol. 4450: T. Okamoto, X. Wang (Eds.), Public Key Cryptography – PKC 2007. XIII, 491 pages. 2007.

Vol. 4448: M. Giacobini et al. (Ed.), Applications of Evolutionary Computing. XXIII, 755 pages. 2007.

Vol. 4447: E. Marchiori, J.H. Moore, J.C. Rajapakse (Eds.), Evolutionary Computation, Machine Learning and Data Mining in Bioinformatics. XI, 302 pages. 2007.

Vol. 4446: C. Cotta, J. van Hemert (Eds.), Evolutionary Computation in Combinatorial Optimization. XII, 241 pages. 2007.

Vol. 4445: M. Ebner, M. O'Neill, A. Ekárt, L. Vanneschi, A.I. Esparcia-Alcázar (Eds.), Genetic Programming. XI, 382 pages. 2007.

Vol. 4444: T. Reps, M. Sagiv, J. Bauer (Eds.), Program Analysis and Compilation, Theory and Practice. X, 361 pages. 2007.

Vol. 4443: R. Kotagiri, P.R. Krishna, M. Mohania, E. Nantajeewarawat (Eds.), Advances in Databases: Concepts, Systems and Applications. XXI, 1126 pages. 2007.

Vol. 4440: B. Liblit, Cooperative Bug Isolation. XV, 101 pages. 2007.

Vol. 4439: W. Abramowicz (Ed.), Business Information Systems. XV, 654 pages. 2007.

Vol. 4438: L. Maicher, A. Sigel, L.M. Garshol (Eds.), Leveraging the Semantics of Topic Maps. X, 257 pages. 2007. (Sublibrary LNAI).

Vol. 4433: E. Şahin, W.M. Spears, A.F.T. Winfield (Eds.), Swarm Robotics. XII, 221 pages. 2007.

Vol. 4432: B. Beliczynski, A. Dzielinski, M. Iwanowski, B. Ribeiro (Eds.), Adaptive and Natural Computing Algorithms, Part II. XXVI, 761 pages. 2007.

Vol. 4431: B. Beliczynski, A. Dzielinski, M. Iwanowski, B. Ribeiro (Eds.), Adaptive and Natural Computing Algorithms, Part I. XXV, 851 pages. 2007.

Vol. 4430: C.C. Yang, D. Zeng, M. Chau, K. Chang, Q. Yang, X. Cheng, J. Wang, F.-Y. Wang, H. Chen (Eds.), Intelligence and Security Informatics. XII, 330 pages. 2007.

Vol. 4429: R. Lu, J.H. Siekmann, C. Ullrich (Eds.), Cognitive Systems. X, 161 pages. 2007. (Sublibrary LNAI).

Vol. 4427: S. Uhlig, K. Papagiannaki, O. Bonaventure (Eds.), Passive and Active Network Measurement. XI, 274 pages. 2007.

Vol. 4426: Z.-H. Zhou, H. Li, Q. Yang (Eds.), Advances in Knowledge Discovery and Data Mining. XXV, 1161 pages. 2007. (Sublibrary LNAI).

Vol. 4425: G. Amati, C. Carpineto, G. Romano (Eds.), Advances in Information Retrieval. XIX, 759 pages. 2007.

Vol. 4424: O. Grumberg, M. Huth (Eds.), Tools and Algorithms for the Construction and Analysis of Systems. XX, 738 pages. 2007.

Vol. 4423: H. Seidl (Ed.), Foundations of Software Science and Computational Structures. XVI, 379 pages. 2007.

Vol. 4422: M.B. Dwyer, A. Lopes (Eds.), Fundamental Approaches to Software Engineering. XV, 440 pages. 2007.

Vol. 4421: R. De Nicola (Ed.), Programming Languages and Systems. XVII, 538 pages. 2007.

Vol. 4420: S. Krishnamurthi, M. Odersky (Eds.), Compiler Construction. XIV, 233 pages. 2007.

Vol. 4419: P.C. Diniz, E. Marques, K. Bertels, M.M. Fernandes, J.M.P. Cardoso (Eds.), Reconfigurable Computing: Architectures, Tools and Applications. XIV, 391 pages. 2007.

Vol. 4418: A. Gagalowicz, W. Philips (Eds.), Computer Vision/Computer Graphics Collaboration Techniques. XV, 620 pages. 2007.

Vol. 4416: A. Bemporad, A. Bicchi, G. Buttazzo (Eds.), Hybrid Systems: Computation and Control. XVII, 797 pages. 2007.

Vol. 4415: P. Lukowicz, L. Thiele, G. Tröster (Eds.), Architecture of Computing Systems - ARCS 2007. X, 297 pages. 2007.

Vol. 4414: S. Hochreiter, R. Wagner (Eds.), Bioinformatics Research and Development. XVI, 482 pages. 2007. (Sublibrary LNBI).

Vol. 4412: F. Stajano, H.J. Kim, J.-S. Chae, S.-D. Kim (Eds.), Ubiquitous Convergence Technology. XI, 302 pages. 2007.

Vol. 4411: R.H. Bordini, M. Dastani, J. Dix, A.E.F. Seghrouchni (Eds.), Programming Multi-Agent Systems. XIV, 249 pages. 2007. (Sublibrary LNAI).

Vol. 4410: A. Branco (Ed.), Anaphora: Analysis, Algorithms and Applications. X, 191 pages. 2007. (Sublibrary LNAI).

Vol. 4409: J.L. Fiadeiro, P.-Y. Schobbens (Eds.), Recent Trends in Algebraic Development Techniques. VII, 171 pages. 2007.

Vol. 4407: G. Puebla (Ed.), Logic-Based Program Synthesis and Transformation. VIII, 237 pages. 2007.

Vol. 4406: W. De Meuter (Ed.), Advances in Smalltalk. VII, 157 pages. 2007.

Vol. 4405: L. Padgham, F. Zambonelli (Eds.), Agent-Oriented Software Engineering VII. XII, 225 pages. 2007.

Vol. 4403: S. Obayashi, K. Deb, C. Poloni, T. Hiroyasu, T. Murata (Eds.), Evolutionary Multi-Criterion Optimization. XIX, 954 pages. 2007.

Vol. 4401: N. Guelfi, D. Buchs (Eds.), Rapid Integration of Software Engineering Techniques. IX, 177 pages. 2007.

Vol. 4400: J.F. Peters, A. Skowron, V.W. Marek, E. Orłowska, R. Słowiński, W. Ziarko (Eds.), Transactions on Rough Sets VII, Part II. X, 381 pages. 2007.

Vol. 4399: T. Kovacs, X. Llorà, K. Takadama, P.L. Lanzi, W. Stolzmann, S.W. Wilson (Eds.), Learning Classifier Systems. XII, 345 pages. 2007. (Sublibrary LNAI).

Vol. 4398: S. Marchand-Maillet, E. Bruno, A. Nürnberger, M. Detyniecki (Eds.), Adaptive Multimedia Retrieval: User, Context, and Feedback. XI, 269 pages. 2007.

Vol. 4397: C. Stephanidis, M. Pieper (Eds.), Universal Access in Ambient Intelligence Environments. XV, 467 pages. 2007.

Vol. 4396: J. García-Vidal, L. Cerdà-Alabern (Eds.), Wireless Systems and Mobility in Next Generation Internet. IX, 271 pages. 2007.

Vol. 4395: M. Daydé, J.M.L.M. Palma, Á.L.G.A. Coutinho, E. Pacitti, J.C. Lopes (Eds.), High Performance Computing for Computational Science - VECPAR 2006. XXIV, 721 pages. 2007.

Vol. 4394: A. Gelbukh (Ed.), Computational Linguistics and Intelligent Text Processing. XVI, 648 pages. 2007.

Vol. 4393: W. Thomas, P. Weil (Eds.), STACS 2007. XVIII, 708 pages. 2007.

Vol. 4392: S.P. Vadhan (Ed.), Theory of Cryptography. XI, 595 pages. 2007.

Vol. 4391: Y. Stylianou, M. Faundez-Zanuy, A. Esposito (Eds.), Progress in Nonlinear Speech Processing. XII, 269 pages. 2007.

Vol. 4390: S.O. Kuznetsov, S. Schmidt (Eds.), Formal Concept Analysis. X, 329 pages. 2007. (Sublibrary LNAI).

Vol. 4389: D. Weyns, H.V.D. Parunak, F. Michel (Eds.), Environments for Multi-Agent Systems III. X, 273 pages. 2007. (Sublibrary LNAI).

Vol. 4385: K. Coninx, K. Luyten, K.A. Schneider (Eds.), Task Models and Diagrams for Users Interface Design. XI, 355 pages. 2007.

Vol. 4384: T. Washio, K. Satoh, H. Takeda, A. Inokuchi (Eds.), New Frontiers in Artificial Intelligence. IX, 401 pages. 2007. (Sublibrary LNAI).

Vol. 4383: E. Bin, A. Ziv, S. Ur (Eds.), Hardware and Software, Verification and Testing. XII, 235 pages. 2007.